Plant Physiology

Plant
Physiology

IRWIN P. TING
University of California, Riverside

ADDISON-WESLEY PUBLISHING COMPANY
Reading, Massachusetts · Menlo Park, California
London · Amsterdam · Don Mills, Ontario · Sydney

SPONSORING EDITOR: James Funston
PRODUCTION EDITOR: Doris Machado
DESIGNER: Marshall Henrichs
ILLUSTRATIONS: ANCO/Boston, Bruce E. Holloway, Kristin Kramer
ART COORDINATOR: Kristin Kramer
COVER DESIGN: Richard Hannus
COVER PHOTOGRAPH: Marshall Henrichs
CHAPTER OPENING PHOTOGRAPHS: Marshall Henrichs

This book is in the

ADDISON-WESLEY SERIES IN THE LIFE SCIENCES

Library of Congress Cataloging in Publication Data

Ting, Irwin P.
  Plant physiology.

  Bibliography: p.
  Includes index.
  1. Plant physiology.   I. Title.
QK711.2.T56   581.1   80-16448
ISBN 0-201-07406-0

ISBN 0-201-07406-0
ABCDEFGHIJK-DO-8987654321

PREFACE

This textbook is a comprehensive treatment of modern plant physiology and is intended for an undergraduate course. It can also be used as a review for beginning graduate courses in plant sciences, and portions of it may be selected for courses in plant biochemistry, crop physiology, and plant physiological ecology.

The book starts with a chapter (Part I) that reviews plant cell structure and function from the viewpoint of physiology. It can be used as an introduction to the material covered in subsequent chapters or as a review of organelle, cell, and plant structure.

Part II includes basic principles of plant physiology that can be discussed and understood in physical terms. Topics included are: properties of water, solutions, colloids, the concept of acidity and pH, and an introduction to cellular energetics, diffusion, centrifugation, and molecular size. Additional chapters on photobiology, plant water relations, stomatal physiology and gas exchange, transport within the plant, the interaction of plants with radiation, and the concept of an energy budget complete this part. These six chapters are written so that they can be used as a single unit to present a great deal of basic plant physiology. These chapters also serve as a foundation for the subsequent discussion of biochemistry (Part III), metabolism (Part IV), and growth and development (Part V).

Part III is a brief but comprehensive treatment of the biochemical constituents of plants. The chemistry and metabolism of nucleic acids, amino acids and proteins, carbohydrates and organic acids, lipids, and natural products are discussed in detail sufficient to allow this part to be used as a short course in plant biochemistry; it can serve as a reference and review as desired.

Part IV has four chapters covering unique aspects of the metabolic activity of plants: mineral nutrition and ion transport, respiration, the photochemical processes of photosynthesis, and photosynthetic metabolism.

Part V covers growth and development and begins with a general discussion of growth, growth kinetics, the growth movements, and nongrowth turgor movement. A complete chapter on the structure, assay, metabolism, and function of the plant hormones leads into a chapter on vegetative growth and development, which ends with a discussion of environmental influences on vegetative growth. This part concludes with a discussion of phytochrome-controlled growth processes and reproductive growth.

For a one-term or one-semester course in plant physiology, Part I on structure and function may be used as a brief introduction, followed by Part II, with Chapters Two and Seven deleted or used as reference only. Part III can be skipped and emphasis placed on Part IV, focusing on the processes of photosynthesis.

Portions of this book may be useful in other plant science courses. For a course in plant biochemistry or plant metabolism, Part III, which covers biochemical constituents and their metabolism, and chapters on respiration and photosynthesis in Part IV are certainly appropriate. For a course in plant physiological ecology, Part II can be used along with Chapter Sixteen and the environmental portions of Part V.

There are review exercises at the end of each chapter. Some of these questions ask the students to summarize and evaluate data in the text or to relate this data to other topics. Other exercises require the student to make numerical calculations. These are provided to help the student develop a quantitative attitude toward plant physiology.

Three appendices have been included. Appendix I is a brief summary of the International System of unit nomenclature. Appendix II is a list that includes (1) the most commonly consulted journals containing original research in plant physiology and (2) the most readily available review periodicals. Appendix III has answers and solutions to the review exercises at the end of each chapter.

Even though my philosophy throughout the text is that an understanding of plant physiology is best obtained by a thorough understanding of the underlying principles of physics and chemistry, the text does not assume more than limited familiarity with the physical sciences. An important objective was to develop the physical and chemical principles of plant physiology briefly but in sufficient depth that comprehension of their application to plant phenomena would be easy. For those chapters that have quantitative sections, the review questions at the end include problems and answers that can be used to test understanding. Also included in the text are tables and graphs that can be used to obtain constants and parameters for problem solving.

Riverside, California   I.P.T.
July 1981

CONTENTS

ACKNOWLEDGMENTS

I want to thank my many students and colleagues of the University of California, Riverside, without whom the preparation of this book would have been impossible. I am especially grateful to R. L. Heath and W. W. Thomson who were always available. Also, I am grateful to W. L. Belser, E. M. Lord, J. B. Mudd, C. P. Pauling, L. M. Shannon, and J. J. Sims, who read chapters or sections, and to students D. P. Brisken, J. C. Carson, K. A. Shackel, and M. A. Stidham, who read and commented on many sections. Katherine Kroll warrants my special thanks for assisting throughout the entire preparation. I also want to thank Joseph Cowles, William T. Jackson, Atila O. Klein, Lee H. Pratt, Albert Ruesink, and David C. Whitenberg for penetrating and constructive reviews of the manuscript. I owe a debt to my first mentors, H. N. Mozingo and W. E. Loomis, and to W. M. Dugger with whom I shared our plant physiology course for almost a decade, and finally to Donna Mead for being there.

Structure and Function of Plant Cells and Tissues

I

PROLOGUE

The idea that structure and function of plant components are related is found in the work of George Haberlandt, who in 1884 published his book *Physiological Plant Anatomy*. His was the first clear statement of the relationship between structure and function, but it was perhaps not so evident that the relationship was a product of evolution and not because of specific design. The structure−function concept is somewhat dangerous because it tends to lead to teleological interpretation, thwarting further scientific investigation. Nevertheless, if it is clearly kept in mind that teleology has no place in science, the ultimate goal of structure−function studies to interpret function on the basis of structure can be met. Such interpretation, however, is just a beginning and final analysis must be based on more dynamic studies, using all the disciplines of science that are available.

The information for all the functions of the plant is housed in the complex DNA polymers. DNA is mostly stored in the nucleus, but some occurs in the double-membrane-bound organelles, chloroplasts, and mitochondria as well. DNA acts as the template for RNA synthesis, which in turn determines the primary structure of the proteins. The proteins regulate virtually all physiological processes. At the cellular level, much of the physiology of the plant is biochemical. The biochemical reactions and metabolic pathways are an integral part of physiology. These are regulated to some extent by strict compartmentation in different subcellular organelles. Thus the CO_2-fixation reactions of photosynthesis occur in plastids called chloroplasts, and the oxidative metabolism of respiration occurs in mitochondria. Other metabolic events are specifically associated with membranes, microbodies, dictyosomes, vacuoles, microtubules, and the other organelles. Even though the organelles are associated with specific and usually unique metabolic events, they frequently do not act alone. As an example, the photorespiration pathway occurs sequentially in chloroplasts, peroxisomes, and mitochondria.

Most plant cells have all the major cellular organelles. Despite this, there is much differentiation among the various plant cells; each functions in a different manner. There are the vessel elements that conduct water during transpiration, sieve elements that function in the long-distance transport of solutes, epidermal cells that give protection to the plant body, sclerenchyma and collenchyma cells that give support, and highly specialized cells with very specific functions, such as the guard cells of stomata, the transfer cells of phloem, and secretory cells. Many of these cell types are highly modified when compared with the simple parenchyma cell and appear to the human observer to be structurally adapted for the functions that we have assigned to them. Such adaptation is evidently the product of a long period of evolution.

Even more complex than the differentiation of cells into differing functional types is the organization of cell types into tissues. The tissues may be quite simple, such as epidermis, or complex, such as xylem and phloem. Both xylem and phloem are composed of at least four different cell types and appear to be highly organized, functional systems. The final and highest organization is into organs, roots, shoots, and leaves.

In the part to follow, there is an overview of the plant-cell organelles, the plant-cell types, the tissue types, and the organs, all presented from a functional perspective. The remaining four parts of the text, which concern the biophysics, the biochemistry, the metabolism, and the development of plants, will contain more detailed discussions of the physiology correlated with the structure of the plant-cell organelles, cell types, tissue types, and organs that make up the functional plant.

CHAPTER ONE

Structure and Function of Plant Cells and Tissues

Cells are the basis of all plant life, and an understanding of plant physiology is only possible with knowledge of cell structure and function. Microscopes were generally available by the sixteenth century, and in 1665 Robert Hooke of England described the cell as a walled structure containing juices. Up until the nineteenth century, cells were considered as entities within a matrix; in 1824 Rene Dutrochet of France generalized that cells were the basic structural and functional units of life. After Robert Brown and others recognized the importance of the nucleus in 1833, Rudolf Virchow of Russia in 1855 stated his now-famous dictum, *Omnis cellula e cellula*—"All cells from cells!"

The cell, as implied in Virchow's dictum, houses the information for the continuity of life. Within the nucleus containing the complex DNA the code for protein synthesis is found. The proteins are largely synthesized outside the nucleus from the information transcribed from the DNA to RNA. Proteins, of course, regulate most cellular activity. Since all of this takes place in the cell, it is the cell that is the basic unit of the living organism. Most processes for growth, development, and maintenance occur in the cell, although the importance of the cell wall should not be underestimated.

Furthermore, it is in the green cells of plants that the complex process of food production from CO_2 and light energy takes place. The primary distinguishing feature of autotrophic plants is that they are green. It is this green chlorophyll that makes the plant a self-sufficient living organism and the subsistence supplier for all other living organisms as well. Thus a thorough understanding of cell structure and function is a necessary prerequisite for understanding plant physiology.

However, it is not the cell alone which makes the plant. Cells do not occur alone but as aggregations that form tissues. Tissues are frequently aggregated to form organs, such as roots, stems, and leaves. And it is this aggregation of functional cells into tissues and organs that makes the green plant. Cells should not be studied as entities unto themselves. The entire organism must be viewed as a functional living thing. The transport and communication between and among cells is an important key to a thorough understanding of plant physiology.

In this first chapter, cells, tissues, and organs will be discussed from the viewpoint of structure related to function. Subsequent chapters will build on these structural relationships.

1.1 Plant Cells

1.1.1 Basic cell structure

A representative plant cell is depicted in the electron micrograph of Fig. 1.1.

Essential features of the plant cell are a rigid cell wall and a large central vacuole. The cell wall may be primary, or, in the case of many functional and sometimes nonliving cells, there may be a secondary wall with lignin deposits. Secondary walls are deposited to the inside of the primary walls. The cytoplasm, which is the living portion of the cell, contains the various organelles and components basic to the life of the cell. A limiting membrane, the cell or plasma membrane, is appressed to the inside of the cell wall, and the interface between the vacuole and cytoplasm is marked by the vacuole membrane (the tonoplast).

Within the cytoplasm, numerous organelles and cytoplasmic structures are present. These include the nucleus with a nucleolus, plastids (such as chloroplasts, leucoplasts, chromoplasts, and amyloplasts), mitochondria, microtubules, microbodies, spherosomes, ribosomes, dictyosomes (Golgi bodies), and a complex membranous network called the endoplasmic reticulum.

Other kinds of structures may be present, even viruses, but the above list encompasses most of the main cellular inclusions. Of course, there is no such entity as a "typical" plant cell. Each plant-cell type will have its own unique features; the typical cell exists only as an illustrative form.

Plant cells differ from animal cells in that they are bounded by a rigid cell wall and have a characteristic large central vacuole. Many plant cells have plastids, organelles unique to the plant kingdom. Vacuoles of animal tissues are smaller and tend to be more numerous. Furthermore, plant-cell walls are in direct contact with each other, giving the plant more rigidity than is found in animal tissues.

Plant cells vary in size, depending on type and function. Some plant cells may be only a few μm in diameter and others, such as certain phloem fibers, may be several decimeters long. A typical parenchymatous cell found in potato tubers will be about 100 μm on a side. Assuming a boxlike structure, the volume would be 10^6 cubic μm, or one-thousandth of a cubic millimeter. It would take about one million cells to fill a cubic centimeter (1 ml) or a billion to fill a liter.

Plant cells, of course, are not boxlike but assume various shapes. It has been shown that spheres, when packed, assume a fourteen-sided structure, a

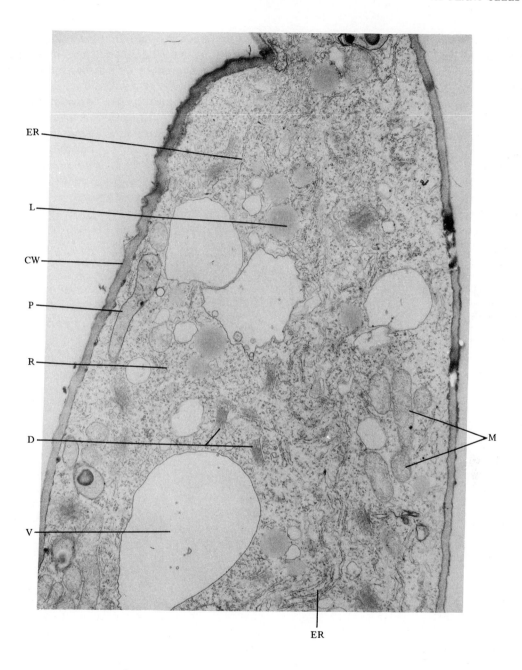

FIGURE 1.1 Electron micrograph of a young cotton-fiber cell, showing many of the cellular inclusions found in plant cells ($\times$ 20,000). ER = endoplasmic reticulum with associated ribosomes. D = dictyosomes. R = ribosomes. ER = endoplasmic reticulum. P = plastid. M = mitochondrion. L = lipid body. CW = cell wall. V = vacuole. Photograph courtesy of Dr. J. D. Berlin, Texas Tech University, Lubbock, Texas.

tetrakaidecahedron. In parenchymatous tissues that have been studied, there is a tendency toward the tetrakaidecahedron form.

1.1.2 Cellular organelles

Although the cell is considered the basic unit of living organisms, most actual cellular functions occur within or on subcellular particles called organelles. In fact, the different metabolic functions of the cell are frequently separated in different organelles. These organelles may be discrete structures with a high degree of organization, such as the nucleus, mitochondria, and chloroplasts, or they may be simply organized membranes. An understanding of the structure of the various subcellular organelles is necessary for a thorough appreciation of cellular function. In this part, the major organelles of the plant cell are described. The details of the metabolic functions of the various organelles are described throughout the remaining chapters.

There are few ways to study organelles other than by isolation, including *in vitro* analysis. Some information can be obtained from microscopy alone, but the ideal studies combine visual analysis with physiological or biochemical studies. In some instances, it is possible to stain organelles or their contents with specific dyes. If colored or electron-dense substances are formed, visualization is possible with the aid of either a light or electron microscope.

Cells are usually broken gently in the presence of protective agents such as buffers to maintain proper pH, osmotic agents to keep the proper tonicity, and antioxidants to prevent oxidative destruction. Relatively strong forces are required to break through plant-cell walls to free the intact organelles. To circumvent this problem, it is possible to form protoplasts first by enzymatically digesting the cell wall. Protoplasts can be broken osmotically or with gentle shear forces, frequently without harming the enclosed organelles.

Organelles are most commonly prepared by differential centrifugation of filtered cell homogenates. Sedimentation rates for nuclei and chloro-plasts are much greater than for microbodies and mitochondria. Thus nuclei and chloroplasts can be readily separated from the smaller mitochondria and microbodies by centrifugation.

Subsequent purification can be accomplished by density-gradient centrifugation of a suspension of organelles. If a suspension of organelles is centrifuged at high speed in sucrose or other gradients, the organelles will reach equilibrium at their own density. With proper choices of centrifugal forces and density gradients, it is possible to isolate and purify most cellular inclusions.

The cell wall

Plants are rigid because of the presence of an inelastic cell wall and high turgor pressure. It is this firm cell wall that gives the plant cell unique properties in comparison with the animal cell. Furthermore, the entire plant body is quite rigid because of the cementing material of the middle lamella holding adjacent cells together.

The primary cell wall is produced from the cytoplasm of the actively growing plant cell and is composed largely of cellulose, hemicelluloses, pectins, and structural protein. Little lignin is present. In many plant cells, once the cell has enlarged and begins to differentiate, a secondary wall is produced and laid down upon the primary wall by apposition. This latter wall gives hardness and much rigidity to the plant body because of the deposition of lignin. The secondary wall has cellulose and hemicellulose in addition to lignin but has little pectin material.

A most important biological property of plant cell walls is that they are very permeable to aqueous solutions. Cell walls, of course, do have charged groups and will bind ionic material, but by and large they are permeable to most substances of biological interest and present little in the way of a transport barrier. It is the limiting plasma membrane and not the cell wall that imparts transport selectivity to the cell.

Pectins can be easily extracted from plant-cell walls with hot water or dilute acidic and basic solutions because they are relatively soluble substances.

Hemicelluloses (e.g., arabans, galactans, etc.), which are polysaccharides with some properties similar to cellulose, are less soluble than pectins but can be extracted with harsh alkaline treatment. Cellulose itself is rather resistant to hydrolysis and extraction. Lignin is extracted only with great difficulty.

The cellulosic structure of the plant-cell wall is reasonably well known. Long chains of 1,4-linked β-glucose form the cellulose strands that occur in bundles, forming micelles. Groups of these micellar strands are further grouped in a higher-order structure to form microfibrils. The microfibrillar groupings make up the plant-cell wall. The cellulose-chain, micellar, and microfibrillar arrangement is shown diagramatically in Fig. 1.2. The microfibrillar arrangement within the plant-cell wall can be observed with the aid of the electron microscope.

In addition to the characteristic polysaccharides, cell walls have a variety of proteins. Extensin, a structural glycoprotein, is a component of cell walls and appears to vary somewhat among species. The structural protein and the various cell-wall polysaccharides are covalently bonded. Enzymatic proteins make up the nonstructural proteins associated with cell walls and include hydrolytic enzymes such as β-glucosidase and galactosidases. Cell walls also have a protein that inhibits the activity of polygalacturonases. The latter are secreted by many plant pathogens during cell-wall decay. Thus it seems likely that the polygalacturonase inhibitor present in cell walls imparts resistance to microbes by plants. Undoubtedly, there are many other kinds of proteins associated

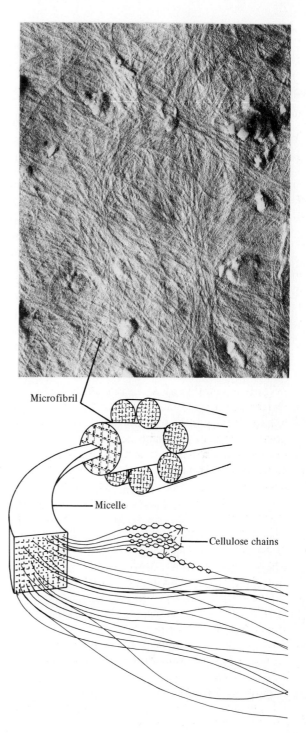

Microfibril

Micelle

Cellulose chains

FIGURE 1.2 Electron micrograph of a cell wall prepared by the freeze-fracture technique showing the microfibril ultrastructure. The view is looking down on the plasma membrane and the microfibrils are below pushed against the membrane ($\times$40,000). The drawing illustrates that each microfibril is composed of cellulose chain micelles. Adapted from J. Bonner and R. W. Galston. 1958. *Principles of Plant Physiology*. W. H. Freeman and Co., San Francisco. Photograph courtesy of Dr. W. W. Thomson, University of California, Riverside.

with cell walls, but unequivocal assignment to cell walls is rather difficult unless there is firm association. The chemical makeup of cell walls is further discussed in Chapter 10.

Plant-cell membranes

Plant-cell membranes were first recognized in the middle of the nineteenth century. During this period, it was observed that plant cells would exclude certain pigments and hence showed selectivity of transport. It is this selectivity of uptake and transport by the limiting cell membrane that separates the contents of the plant cell from the surrounding environment. Other membranes of the cell, including those of the vacuoles (the tonoplasts), chloroplasts, mitochondria, and other cellular organelles, show selectivity of transport as well. The nature and mechanism of such transport selectivity are discussed in Chapter 6. In addition to transport functions, many membranes of the cell present surfaces for important biochemical reactions.

When viewed with the electron microscope, a single membrane appears as two electron-dense contours separated by a light space (Fig. 1.3). It is about 7.5 nanometers (nm) wide, with the electron-dense contours each about 2 nm wide separated by the 3.5-nm inner space.

Some of the plant-cell organelles, such as plastids, mitochondria, and nuclei, are bounded by two membranes. Certain others, such as microbodies and vacuoles, are bounded by only one membrane. Care must be taken to keep in mind that a single membrane is most frequently composed of a lipid-bilayer structure with a protein sheath. The lipid bilayer with protein makes up a single membrane, although some organelles such as spherosomes appear to be bounded by a half-membrane. The characteristic two-membrane organization mentioned above is composed of two of the lipid-bilayer–protein-sheath units lying side by side.

Cellular membranes are composed largely of amphiphatic phospholipids and glycolipids in addition to lipoproteins and nonconjugated proteins.

Usually, they are about 50 percent protein, 40 percent lipid, and a few percent carbohydrate.

Several models have been proposed for membrane structure, most involving a lipid bilayer with associated protein. Models must account for the known chemical composition of the membranes as well as their biochemical, biophysical, and transport properties. Two models that will be discussed are the lipid-bilayer model of Danielli and Davson, first proposed in 1935, and the fluid-mosaic model subsequently proposed by Singer.

The lipid-bilayer model assumes that the nonionic portions of lipid molecules are held together by hydrophobic bonds. Their ionic heads are exposed toward the outside. Associated with the exposed ionic surfaces is protein, bonded either with ionic or covalent bonds. A diagrammatic interpretation of the lipid-bilayer model of Danielli and Davson is illustrated in Fig. 1.3 and seems to be consistent with the electron microscopic view of cellular membranes. To complete such a model, it can be assumed that there are pores of polar and nonpolar regions that facilitate transport of polar and nonpolar substances.

Robertson in 1959 proposed a modification of the Danielli and Davson model in which the lipid bilayer was covered with a sheet of nonlipid material, and he further proposed the concept of a unit membrane in which all of the membranes of the cell were virtually the same. There is, however, much evidence that membranes do differ, and the Robertson unit-membrane concept is not accepted in its entirety.

From thermodynamic considerations of membrane structure, Singer and others proposed in the 1960s that the functional cell membranes are a viscous fluid mosaic rather than a rigid lipid bilayer as is implied in the Danielli and Davson model. The fluid lipid bilayer has within it a mosaic of integral proteins with their polar ends exposed to the aqueous environment and their nonpolar regions buried within the nonpolar lipid portion. Weakly attached to the ionic surface of the lipid portion are peripheral proteins largely bonded by means of ionic bonds. The latter are readily re-

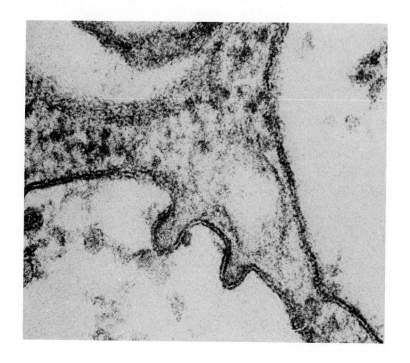

FIGURE 1.3 Electron micrograph showing the typical bilayer membrane of plant cells ($\times$132,000). The membrane is the vacuole membrane, or tonoplast. The double-track, bilayer contour is composed of two outer electron-dense bands presumed to be composed of protein and the heads of the lipid molecules. The inner, light-colored region is composed of the remainder of the lipid molecules. Usually only the tonoplast and plasma membrane show such distinct double tracks. The drawings depict two models for membrane structure, accounting for both the bilayer appearance when viewed with the electron microscope and the chemical properties of membranes. The Danielli and Davson (1935) model is the simplest assuming that the membrane is made of a lipid bilayer and is sheathed in protein. The more complex fluid-mosaic model of Singer (Singer and Nicolson, 1972) incorporates integral and peripheral proteins within and upon the fluid lipid bilayer as explained in the text. Electron micrograph courtesy of Dr. W. W. Thomson, University of California, Riverside.

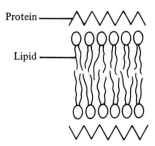

Protein ——

Lipid ——

The Danielli and Davson
lipid bilayer model (1935)

The Singer
fluid-mosaic model (1971)

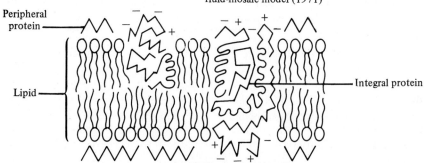

Peripheral
protein ——

Lipid ——

—— Integral protein

TWO MODELS FOR MEMBRANE STRUCTURE

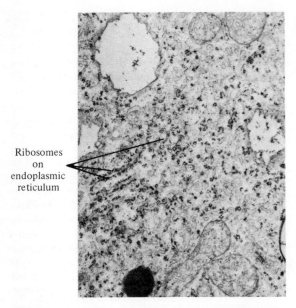

Ribosomes
on
endoplasmic
reticulum

FIGURE 1.4 Electron micrograph showing ribosomes attached to endoplasmic reticulum (×42,120). These function in protein synthesis. Micrograph courtesy of Dr. W. W. Thomson, University of California, Riverside.

moved by washing with dilute salt solutions. Diagrams of the fluid-mosaic model are shown in Fig. 1.3.

Throughout the cell there is a continuous membrane system known as the endoplasmic reticulum. The endoplasmic reticulum probably functions at least in part as a reactive surface for cellular chemistry and may tend to form compartments for metabolic regulation. It is visible in Fig. 1.1.

In the portions of the cell that are actively synthesizing protein, ribosomes are found attached to the endoplasmic reticulum. The endoplasmic reticulum with attached ribosomes is termed rough endoplasmic reticulum, as opposed to smooth endoplasmic reticulum, which is without ribosomes (refer to Fig. 1.4). When endoplasmic reticulum membranes are isolated by centrifugation techniques, they are called microsomes. The role of ribosomes in protein synthesis is discussed in Chapter 8.

The plant-cell nucleus

The nucleus, present in all living functional plant cells except sieve-tube cells, is limited by two membranes. These membranes have relatively large pores on the order of 50 to 100 nm in diameter. Plant-cell nuclei are from 5 to 20 μm in diameter. Figure 1.5 shows plant-cell nuclei plus a diagram of the nuclear membrane. Usually there is only one nucleus per cell (Table 1.1).

The nuclei contain protein, lipid, and, perhaps most important, the nucleic acids, ribonucleic acid (RNA) and deoxyribonucleic acid (DNA). The latter, which is a structural component of the nuclear chromosomes, has the genetic information for the continuity of life and the information for protein assembly. Thus the nucleus is the information center of the cell.

By and large, the nucleus appears to be rather amorphic with little definite structure internally. The matrix is composed of a network of DNA and protein called chromatin. The prominent dark bodies observed in the nucleus are nucleoli. The nucleolus functions in the storage of RNA and plays a role in the synthesis of ribosomal RNA (rRNA). It is on the ribosomes where protein synthesis ultimately takes place.

The internal organization of the nucleus has been studied in much detail. For the most part, the chromosomes make up this organization. The basic building block is the DNA polymer. The amount in the nucleus varies considerably but

TABLE 1.1   Size and number of cellular organelles in a representative plant cell

Organelle	Average diameter range	Expected number per cell
Nucleus	5–20 μm	1
Chloroplasts	5–20 μm	50–200
Mitochondria	1–5 μm	500–2000
Microbodies	1–5 μm	500–2000
Ribosomes	0.025 μm	500,000–5,000,000

Data from Bonner in Bonner and Varner (1976).

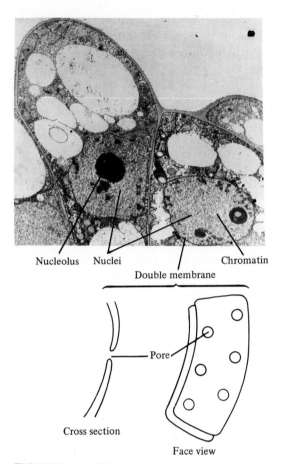

Nucleolus   Nuclei   Chromatin

Double membrane

Pore

Cross section

Face view

FIGURE 1.5 Epidermal cells of a cotton ovule showing large, prominent nuclei with well-defined nucleoli (×10,000). The enlarged cell in the center will develop into a cotton fiber. The accompanying drawing illustrates the double nuclear membrane and pores. Micrograph courtesy of Dr. C. A. Beasley, University of California, Riverside.

mostly represent the protein-coding genes, but some information may exist in the repetitive sequences as well. Some of the latter are known to code for the RNA polymers. In addition, the genes for the histones (basic proteins found in the nucleus—see below) form repetitive sequences.

The double-helical DNA is associated with the histones in the cell nucleus, which together make up the chromatin or nucleoprotein complexes. Although there are several different histones, their binding to DNA is not gene-specific. The nucleoprotein complexes of DNA and histones result in supercoiling and the formation of distinct "beads," called chromomeres, along the DNA strand. Whereas the DNA coil has a diameter of 2.5 nm, the supercoil has a diameter of 10 nm. The supercoil itself will coil, forming a yet larger fibril with a diameter of 20 nm. It can be noted here that the histones represent general repressors of transcription. In part, activation of a DNA sequence comes about by interaction of the basic histones with acidic proteins called hertones. Such interaction frees the histones and exposes the DNA. DNA is, therefore, a much better template for RNA synthesis than is chromatin.

Most of the chromatin is organized into subunits called chromomeres. These are the large, beadlike structures (visible with the light microscope) that make up the chromosomes. The chromosome is the largest organized structure in the nucleus. During RNA synthesis, the chromomeres are transcribed completely. Thus it appears that the chromomeres are the genes.

Additional organization of the chromosomes is not entirely clear. There is some evidence that all the chromosomes of the haploid genome are connected, but this may be an artifact of preparation for microscopy. In addition, the chromosomes may be attached to the nuclear envelope through the centromeres. Thus the nuclear envelope may regulate chromosome metabolism and function (such as replication).

Plant-cell nuclei have been isolated, although with much difficulty. Both weak shear forces that

ranges from 1 to 100 picograms per haploid genome in angiosperms. The DNA of the haploid genome occurs either as unique sequences or in repetitive sequences. The latter may be in excess of one million copies per genome. The unique sequences gently break open the cells and quick, high-speed blending have been used with some success. After being released from the broken cells, the nuclei can be collected on net filters or purified by sucrose-density-gradient centrifugation. Subsequent study of the nuclear contents can be accomplished by lysis of the nuclei with detergents.

Chloroplasts

Chloroplasts are characteristic plant organelles bounded by two bilayer membranes. They are relatively large with dimensions comparable to nuclei, about 5 to 20 μm in diameter. Photosynthetic cells may have 15 to 25 or more chloroplasts.

The structure of a chloroplast is illustrated in Fig. 1.6. The double-membrane system makes up the limiting envelope of the chloroplast and does not appear to be directly connected to the inner-lamellar membrane system. In the terminology of Weier and Stocking, the enclosed membranes of the basic unit structure, the thylakoids, may be stacked into arrays collectively called grana, or they may extend from stack to stack to form intergranal lamellae. Weier and Stocking call these intergranal lamellae frets. As illustrated in the figure, a granal stack may be composed of small, enclosed thylakoids and portions of other larger thylakoids running from granum to granum.

The material surrounding the thylakoids and within which the grana and fretwork are imbedded is called stroma. It contains much soluble protein and is highly hydrophilic.

Studies of chloroplast function have indicated that the CO_2-assimilation reactions of photosynthesis occur within the stroma and the photochemical reactions of photosynthesis, including electron transport to form ATP and NADPH, occur on the lamellar membranes of the thylakoids. As will be discussed in Chapter 15, there appear to be functional differences between the granal membranes and the intergranal fretwork membranes with respect to photochemical reactions.

Freeze-etch and freeze-fracture studies of chloroplast structure have revealed the presence of small particles located on the faces of the thylakoids (Fig. 1.6). These structures are undoubtedly involved in the photochemical reactions of photosynthesis.

The chloroplast stroma contains protein, lipid droplets, starch grains, a variety of soluble metabolites including amino acids and sugars, RNA, and DNA. Chloroplasts have the capacity for protein synthesis.

In higher plants, there is not much evidence for division of chloroplasts, although photomicrographic movies have shown that they are highly pleomorphic. Most chloroplasts probably come from proplastids that are inherited through the cytoplasm. Electron micrographs have shown proplastids, but not mature plastids, in egg cells.

Evidence for cytoplasmic inheritance of chloroplasts comes largely from breeding experiments with albino- and variegated-leafed plants. Flowers from green-leafed plants produce green-leafed plants, whereas flowers from variegated and albino plants produce seed resulting only in variegated and albino plants. The inheritance is independent of the pollen source and seems to be completely maternal.

In addition to chloroplasts, plant cells are likely to have other kinds of plastids. There are starch-storing amyloplasts, pigment-containing chromoplasts, lipid-storing elaioplasts, and colorless plastids called leucoplasts. They are all similar in structure, being bounded by two membranes, but except for chloroplasts there may not be much internal membrane organization. All of the plastids appear to arise from a common proplastic precursor.

Chloroplasts and plastids can be isolated and purified by either aqueous or nonaqueous techniques.

Mitochondria

Mitochondria, like chloroplasts, are double-membrane-bounded organelles found in virtually all plant cells. The structure of plant mitochondria is illustrated in Fig. 1.7. There may be tens or even

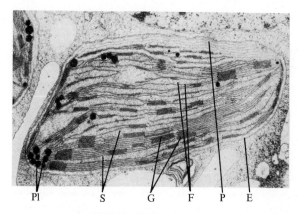

WHOLE CHLOROPLAST

Pl   S   G   F   P   E

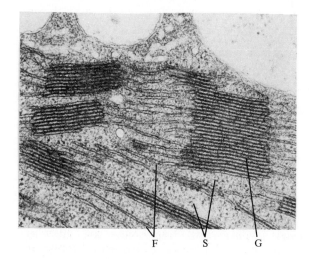

ENLARGED VIEW OF GRANAL REGION

F   S   G

Grana

Frets

FIGURE 1.6 Electron micrograph of a whole chloroplast (×34,160) with enlarged view of a granal region (×23,120). E = chloroplast double membrane or envelope. G = grana. S = stroma. Pl = plastoglobuli. F = fret. P = peripheral reticulum. The drawings show the Weier (Weier et al., 1963) concept for the arrangement of the thylakoids making up the chloroplast membrane system. Drawing adapted from R. G. S. Bidwell. 1979. *Plant Physiology*. Macmillan Co., New York. Electron micrographs courtesy of Dr. W. W. Thomson, University of California, Riverside.

hundreds in each cell. They are mostly elongate and are 1 to 5 μm long, but occasionally plant mitochondria may be nearly spherical. The outer membrane, which is slightly thicker than the inner membrane, has apparent pores. The inner membrane is frequently highly folded to form sheetlike or tubular extensions called cristae. Tubular cristae are most common in plant mitochondria. Located on the cristae are structures that apparently have many of the enzymatic proteins associated with electron transport and oxidative phosphorylation. Within the mitochondria proper

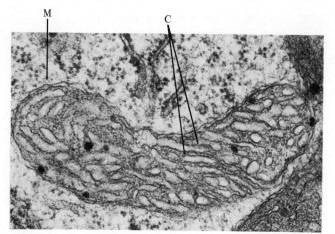

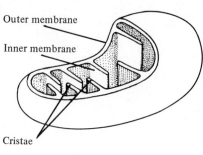

FIGURE 1.7 Electron micrograph of a plant-cell mitochondrion showing double membrane and internal cristae ($\times 250,000$). M = outer membrane. C = cristae. The drawing is a conceptual model of a mitochondrion showing the arrangement of the double membrane to form the outer-membrane and the inner-membrane cristae. Electron micrograph courtesy of Dr. W. W. Thomson, University of California, Riverside.

the matrix is made up of the soluble enzymes that catalyze the reactions of the citric acid cycle of aerobic respiration. Thus the mitochondrion is the subcellular organelle housing the cellular machinery for the aerobic respiratory process that oxidizes acetate, generating the energy for ATP production.

Interestingly, the mitochondrion contains DNA and RNA, and like the chloroplast plays a role in the biosynthesis of certain of its own enzymes.

Microbodies

During the early use of the electron microscope, subcellular organelles of the microbody type were discovered. These spherical organelles, 0.5 to 1.5 μm in diameter, are bounded by a single membrane. The matrix or stroma is rather amorphous, although crystalline bodies are not infrequent. All are characterized by flavoproteins that produce hydrogen peroxide and by the enzyme catalase. Catalase oxidizes peroxide to water and oxygen in a detoxification reaction. Crystalline inclusions within the microbodies may be catalase (Fig. 1.8).

There is evidence that microbodies are formed from smooth endoplasmic reticulum. They contain no DNA but perhaps some RNA. RNA is most likely a residual of the endoplasmic reticulum.

"Microbody" is a structural term. Physiologically and biochemically, microbodies can be differentiated into several types, most notably peroxisomes and glyoxysomes. Peroxisomes are green-leaf microbodies usually associated with chloroplasts. They contain glycolate oxidase and function to a large extent in glycolate oxidation during photorespiration.

Glyoxysomes are microbodies present in seeds that store oil, such as the seeds of cucurbits and castor bean. They contain the enzymes for fatty acid oxidation and the enzymes of the glyoxylate cycle, which is the pathway linking fatty acid degradation to sugar synthesis. Thus glyoxysomes are intimately involved in sugar synthesis from fats during seed germination. Glyoxysomes are characterized biochemically by the presence of malate synthetase and isocitrase, the two enzymes unique to the glyoxylate cycle.

Other types of microbodies have been described,

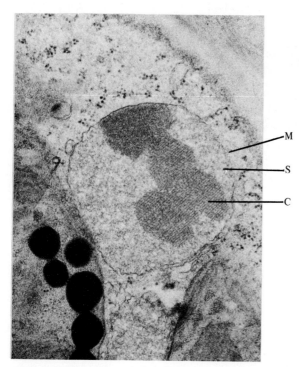

M

S

C

FIGURE 1.8 Electron micrograph of a single-membrane-bounded plant microbody from mesophyll cell of *Citrus* (×81,700). It is quite common to find microbodies closely associated with chloroplasts (the structure below left with the dense plastoglobuli is a chloroplast). M = membrane. S = stroma. C = crystalline inclusion of protein. The crystalline inclusion is probably mostly the enzyme catalase. Micrograph courtesy of Dr. W. W. Thomson, University of California, Riverside.

but they are not as well characterized biochemically and will not be discussed here.

Dictyosomes

Dictyosomes are characteristic subcellular organellelike inclusions that can be viewed in most cells with the electron microscope (refer to Fig. 1.9). They are frequently called Golgi bodies because they were first viewed and described by Camillo Golgi in 1891.

They appear as stacks (up to 20 or more) of en-

closed single-membrane units. Within the stacks (called cisternae) are proteins. Dictyosomes are 1 to 3 μm in diameter and about 0.5 μm thick. There may be from a few to one hundred or more per cell. An electron micrograph and diagram illustrating a dictyosome is shown in Fig. 1.9.

Some electron microscope studies have indicated that dictyosomes may not be free organelles but rather may be connected to the endoplasmic reticulum of the cell. Dictyosomes appear to be most abundant in cells with a secretory function. It is believed, therefore, that their primary cellular function is secretion. Perhaps the most important secretory function is cell-wall deposition. In cells that are rapidly elongating and producing new cell-wall material, dictyosomes are abundant along the cell periphery. There is much evidence from isolated dictyosomes that they contain enzymes that synthesize polysaccharides.

Microtubules

Microtubules and the somewhat smaller microfilaments are common cytoplasmic inclusions of plant cells (see Fig. 1.10). The microtubules are about 24 nm in diameter and can be several μm in length. They appear as hollow tubes but do have material within. Each is composed of from 11 to 13 proteinaceous strands. Microtubules frequently occur in the characteristic 9 + 2 arrangement common in cilia and flagella. Quite often the groups of microtubules have thin threadlike bridges between them.

Microtubules appear to be involved in several complex phenomena within the plant cell. There is good correlation between the internal arrangement of microtubules and cell shape, suggesting that the microtubules may make up a cytoskeleton of the cell. In addition to this role in cell symmetry, the arrangement of the cellulose microfibrils in the cell wall is highly correlated with the arrangement of microtubules adjacent to the cell wall. There is some evidence that microtubules are involved in transport processes in the cell and that they may aid cyclosis.

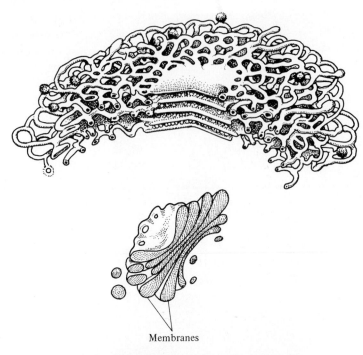

Membranes

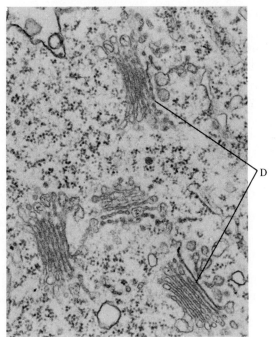

FIGURE 1.9 Electron micrograph of a young cotton fiber showing four dictyosomes (D) (×81,000). Also shown in this micrograph are many ribosomes (refer to Fig. 1.4). The inset is a drawing of a dictyosome based on the appearance in electron micrographs adapted from H. H. Mollenhauer and D. J. Morre. 1966. Golgi apparatus and secretion. *Ann. Rev. Plant Physiol.* 17:27-46. Micrograph courtesy of Dr. J. D. Berlin, Texas Tech University, Lubbock, Texas.

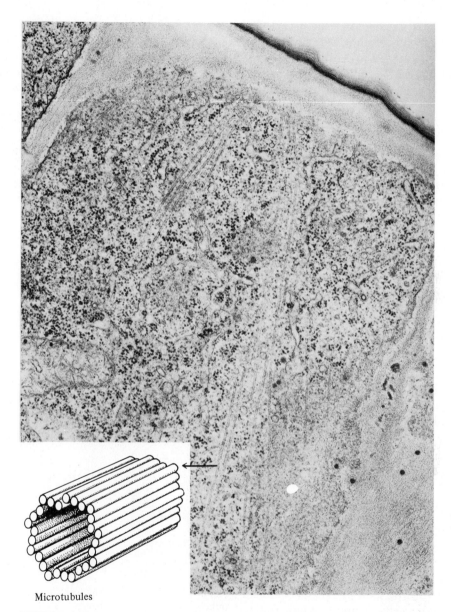

Microtubules

FIGURE 1.10 Electron micrograph showing microtubules running through a young plant cell. There are many ribosomes in the micrograph and a portion of a large mitochondrion in the lower left corner. The drawing is of a single microtubule illustrating the arrangement of the fibrils making up the tubular nature of the microtubule. Electron micrograph courtesy of Dr. W. W. Thomson, University of California, Riverside.

Because the major structure of flagella and cilia is microtubular, the microtubules are involved in cell movement. In addition to this kind of whole-cell movement by flagella and cilia, microtubules aid in movement within the cell. Cyclosis was mentioned above. The spindle fibers, which aid chromosome movement during cell division, are composed largely of microtubules. Thus the microtubules function in chromosome migration during anaphase.

Much of the functioning of the microtubules involves assembly and disassembly of the tubule by subunit organization or disorganization. The subunits are globular and are made of protein and other chemicals. The protein of the microtubule, called tubulin, has been studied in some detail. It is a dimer with a mass of about 110,000 daltons. The subunits are different, with masses of about 53,000 and 56,000 daltons. Tubulin binds two moles of guanidine nucleotides; one is GDP and one is GTP.

The cell-division poison, colchicine, which prevents migration of chromosomes during anaphase of cell division, evidently functions by binding to tubulin. The binding either causes disassembly or prevents assembly of the subunits.

Vacuoles

Large vacuoles are characteristic features of plant cells, where they are much more conspicuous than in animal cells. Vacuoles are bounded by a single limiting membrane that imparts differential-permeability properties. Vacuoles are classically identified as sites of staining with vital dyes such as neutral red. The limiting membrane of the vacuole is called the tonoplast. Matile (1978) considers all the vacuoles of a cell as a unified compartment called the vacuome.

The contents of the plant vacuole are varied and include inorganic ions, sugars, amino acids, organic acids, gums, mucilages, tannins, flavonoids, phenolics, alkaloids, and other nitrogenous compounds. Collectively the contents are referred to as the cell sap.

Vacuoles have been isolated from plant-cell protoplasts (cells without the wall) by osmotic lysis (see Fig. 1.11). Studies with isolated vacuoles have revealed that many contain hydrolytic enzymes such as proteases, lipases, phosphatases, and nucleases. Thus many are comparable to the lysosomes of animal cells, which are vacuoles containing hydrolytic enzymes.

A type of vacuole known as a spherosome is present in some plant cells and is characterized by a limiting membrane different from the basic bilayer membrane in that it appears as a single electron-dense contour when viewed with the aid of the

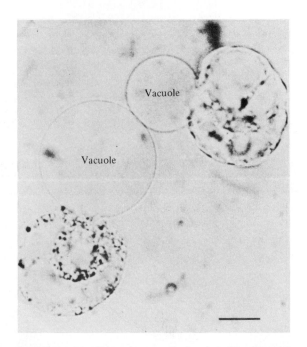

FIGURE 1.11 Light micrograph showing the release of intact vacuoles from tobacco-cell protoplasts. Protoplasts are cells without the cell walls, prepared by digestion of the cell wall with enzymes. Lysis of the protoplasts was done by adding water to the medium containing the protoplasts, causing osmotic shock. The experiment shows that the vacuole is an intact organelle that can be isolated by gently breaking cells. The bar in the lower right of the photograph is 10 μm. Photograph courtesy of Dr. I. J. Mettler, University of California, Riverside.

electron microscope. Spherosomes tend to accumulate lipid and do not stain with vital dyes.

The function of vacuoles is rather complex. Plants have a limited capacity for waste disposal and vacuoles serve as waste-storage containers. In addition, the turgor of the cell is a result of excess water accumulation in the vacuole, which pushes the cytoplasmic contents against the plasma membrane and cell wall.

Autolysis is another function of those vacuoles which have hydrolytic enzymes. Senescence of cells and tissues is frequently accompanied by vacuole bursting, which releases hydrolytic enzymes. One such instance exists when potatoes or apples are injured (e.g., bruised). Phenolics and polyphenoloxidases are released from broken vacuoles and produce polymerization and tanning. Tanning is evident from the browning of potatoes and apples when they are injured.

The cytosol

The cytosol, which is the soluble phase of the cell, is that portion of the cell that cannot be identified as a specific organelle. Enzymes that do not associate with specific organelles when isolated are said to be cytosol enzymes. The concept of a cytosol is operational in that all metabolic functions are most probably associated with membranes in one way or another. Nevertheless, in this text the term cytosol will refer to the soluble phase of the cell.

1.1.3 Types of plant cells

The organelles of the plant cell discussed above make up and are a part of the higher-order organization, the cells. The cells, of course, are not generalized but highly specialized, with much division of labor. Thus there are epidermal cells, storage cells, transport cells, and others. Such organization and specialization is an integral part of plant physiology and much is to be gained by a thorough knowledge of cell structure. In this section, a brief review of cell types is given from the viewpoint of function. The cell types are placed into four func-

tional categories: (1) ground or base cells that function largely for storage; (2) protective cells that function for physical protection; (3) supporting cells that give physical support to the plant; and (4) conducting cells that function for transport. Some cells that fall within a category may be highly modified. As an example, the guard cells making up the stomata are epidermal cells that fall into the protective-cell category. Such anomalies make structure−function studies an important aspect of physiology.

Ground (base) cells

Ground, or base, cells are the simple, relatively undifferentiated cells of plants and may be considered as the characteristic plant cell in form and function. They are known as parenchyma cells. Parenchyma cells are illustrated in Fig. 1.12. All have thin primary cell walls and are alive at the mature stage. Because of the thin primary walls these cells are elastic, allowing expansion and contraction with changes in turgor. Large central vacuoles are typical, and frequently they have prominent starch grains. They are mostly isodiametric in shape, and many assume the tetrakaidecahedron form. They function primarily in storage and, when containing chlorophyll, in photosynthesis. The latter photosynthetic parenchymatous cells are called chlorenchyma cells.

Protective cells

Protective cells include epidermal cells of young primary plant bodies and cork or periderm cells of secondary tissues. Epidermal cells (see Fig. 1.12) are alive at maturity and may contain water-soluble pigments of the anthocyanin or β-cyanin type. Epidermal cells do not contain chloroplasts except in some aquatic species. Cork cells, unlike epidermal cells of primary tissues, are nonliving at maturity.

These protective cells have waxlike deposits on and within the interstices of their cell walls that effectively waterproofs them. The waxlike deposit

CELL TYPES

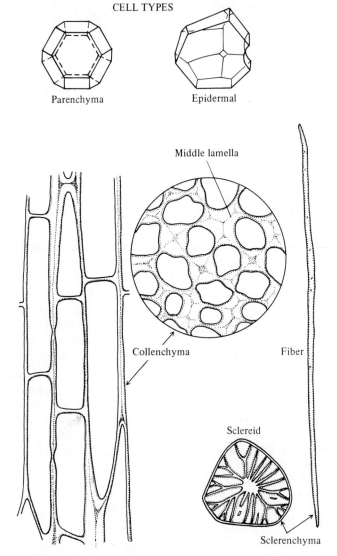

FIGURE 1.12 Examples of protective, storage, and supporting cells. The protective epidermal cells are ordinarily flattened, whereas the storage parenchyma cells are mostly isodiametric. The supporting cells of the collenchyma and sclerenchyma type are frequently elongated, although they may be compact, such as the sclerenchymatous sclereids that give hardness to nuts. The collenchyma cells have unequally thickened primary cell walls. the fibers usually have secondary walls. All adapted from K. Esau. 1977. *Anatomy of Seed Plants*. 2nd ed. John Wiley and Sons, New York.

of the epidermal cell is called cutin and the waxlike deposit of the cork cell is called suberin. The cells which Robert Hooke first observed and described in 1665 were cork cells.

Supporting cells

SCLERENCHYMA CELLS

Sclerenchyma cells function in the support of the plant body. At maturity, they are nonliving and have thick secondary walls. Sclerenchyma cells are of two types, fibers and sclereids. Fibers are elongated cells with thick secondary walls of cellulose and lignin and completely lack a protoplast at maturity. The secondary walls of sclerenchyma cells may have thinly developed areas called pits. The pits frequently serve as passageways from cell to cell.

Sclereids or stone cells are thickened, small cells found in hard plant tissues such as nuts and shells. They may occur singly or in groups. The stone cells found in pear pulp are groups of sclereids. Sclereids, like fibers, may have simple pits. Fibers and sclereids are illustrated in Fig. 1.12.

COLLENCHYMA CELLS

Collenchyma cells, like fibers, are usually elongated support cells but differ from fibers in having primary cell walls and being alive at maturity. They characteristically have unequally thickened primary cell walls. The strands of cells in celery stalks (a leaf petiole) are composed of groups of collenchyma cells and vascular tissue. Collenchyma cells are illustrated in Fig. 1.12.

Conducting cells

The conducting cells of plant tissues are perhaps the most complex. They fall into two groups: the water-conducting tracheids and vessel elements, and the food-conducting sieve cells and sieve-tube cells. Water-conducting cells occur in the xylem tissue, and food-conducting cells are found in the phloem tissue.

TRACHEIDS AND VESSEL ELEMENTS

Tracheids are the basic water-conducting cell type in plants. They are elongated, nonliving cells at maturity and have thickened secondary cell walls of various architecture. Their end walls are tapered and not perforated. They have pits on the side walls, which may be either simple or complex (of the bordered type). Typical tracheids are illustrated in Fig. 1.13. When functional, tracheids are positioned such that adjacent pits are in proper juxtaposition, allowing water and solute flow from cell to cell. Because of great tensile strength, tracheids give strength to the plant body. Thus tracheids function in both water and solute transport and in structural support.

Vessel elements (Fig. 1.13) are highly modified water-conducting cells. Considered by plant evolutionists to be derived from tracheids, they are similar to tracheids but have perforated end walls. Thus vessel elements line up end to end to form elongated vessels up to several meters or more long. Vessel elements occur in most angiosperms but not in gymnosperms. Tracheids are the basic water-conducting cells of gymnosperms.

The pits of vessel elements are frequently more complex than those of tracheids but function similarly. A diagram of a complex bordered pit, such as may be found in the walls of vessel elements or tracheids, is illustrated in Fig. 1.13. A raised architecture of the secondary wall at the pit results in a raised or bordered appearance. The apparent membrane is formed from the primary wall. A torus, which frequently occurs in gymnosperms, may act as a water valve.

SIEVE ELEMENTS—SIEVE CELLS AND SIEVE-TUBE CELLS

Sieve cells are the basic food-conducting cells found in the phloem tissue. They are elongated cells similar to tracheids but are living at maturity. However, they do not have a nucleus when functioning in conduction. Their end walls and frequently their lateral walls have perforated regions called sieve areas. The more highly evolved sieve-

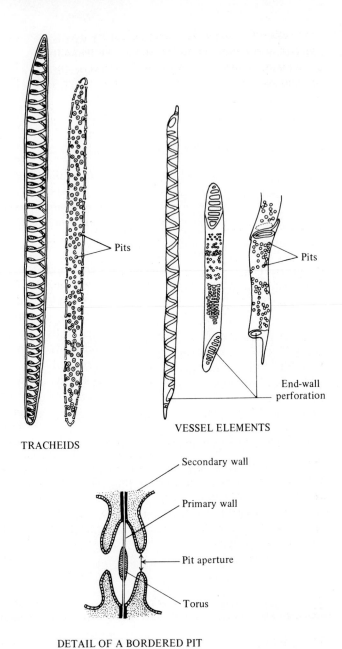

Pits

Pits

End-wall
perforation

VESSEL ELEMENTS

TRACHEIDS

Secondary wall

Primary wall

Pit aperture

Torus

DETAIL OF A BORDERED PIT

FIGURE 1.13 The water-conducting cells of the xylem. The
tracheids with tapered end walls
are more primitive than the vessel
elements that have perforated
end walls. The secondary-wall
depositions may be annular,
spiral, or very extensive with pits.
Whereas the vessel elements line
up end to end to form vessels, the
tracheids do not because their
end walls are not perforated.
Water transport from tracheid to
tracheid is exclusively through
pits of adjacent tracheids. The
tracheids frequently have extensive bordered pits (inset drawing). The border is formed from
the secondary wall. The primary
wall of the tracheid with bordered pits is thin and evidently
presents little resistance to water
transport.

tube cells differ from sieve cells primarily because the end walls have complex sieve regions called sieve plates. Tubes formed by sieve-tube cells lined up end to end are called sieve tubes (refer to Fig. 1.14).

Sieve elements are always associated with at least one living companion cell. The nucleus of the companion cell probably functions as the nucleus of the sieve element.

Sieve elements are about 50 μm long in angiosperms and over 1 mm long in gymnosperms. Crafts (1961) gives an average of 450 μm for 15 angiosperm species and 1450 μm for 58 gymnosperms. They are variable in width, ranging from 0.1 to about 5 μm. As well as being anucleate at maturity, the sieve-tube cell lacks a definite vacuole. There is a plasma membrane appressed to the cell wall and cytoplasm with the usual inclusions.

The cell lumen appears open except for the presence of transcellular strands going through the end-wall sieve plates from sieve-tube cell to sieve-tube cell. The transcellular strands are composed of material originally called slime and now known to be largely protein. This protein is termed P-protein for phloem-protein. When sieve elements are injured, the P-protein tends to precipitate to form slime bodies and slime plugs. This slime plugs the sieve plates and prevents translocation. The plasma membrane is apparently continuous between sieve-tube cells.

A complex carbohydrate called callose appears to line each pore of the sieve plate. Furthermore, when the sieve elements become nonfunctional the pores become plugged with callose.

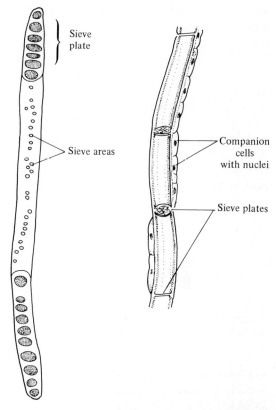

FIGURE 1.14 The food-conducting cells of the phloem. At left is an elongated sieve-tube element with sieve plates on the end walls formed from groups of sieve areas. There are also sieve areas on the side walls through which transport takes place. At the right there are three sieve-tube cells aligned end to end. Food transport occurs by mass flow though the cell lumens and across the sieve plates. Adapted from C. L. Wilson and W. E. Loomis. 1967. *Botany*. 4th ed. Holt, Rinehart, and Winston, New York.

1.2 The Plant Body Plan

The plant body plan is basically a vertical axis that has differentiated into an upper, usually above-ground shoot and a lower below-ground root. Lateral appendages of the vegetative shoot are the branches and leaves. The production of leaves is orderly, and definite arrangements can be ascertained. Leaves may be alternate, opposite, or whorled, and the number produced per each full turn of the stem circumference is quite uniform for each species. Lateral appendages of the shoot may be highly modified to form tendrils, thorns, or other kinds of structures. Once the plant shifts from vegetative growth to reproductive growth, the lateral appendages produced are floral.

The lateral appendages of primary roots are the lateral roots. The root lateral appendages develop from deep tissues, whereas the shoot appendages arise from superficial tissues.

The basic plant body plan is shown in Fig. 1.15, with a vertical axis consisting of an upper shoot with broad-leafed appendages and a lower root with lateral roots. The plant body can be highly modified into a variety of growth forms. Thus plants may be rosette-like, succulent and globular, or reduced to mere shoots with virtually no roots.

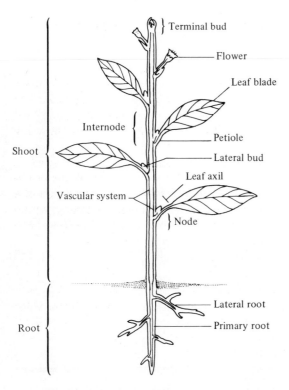

FIGURE 1.15 The basic body plan of a representative angiosperm. The vertical axis is composed of root and shoot. The shoot can be divided into a series of internodes between the nodes. Lateral buds occur just above the leaf bases and nodes. Adapted from C. L. Wilson and W. E. Loomis. 1967. *Botany*. 4th ed. Holt, Rinehart and Winston, New York.

In general the plant is a tissue organism, although the roots, stems, leaves, and floral parts can be considered as organs. The functional subparts of the plant are the tissues; therefore, the student should become thoroughly familiar with the structure of plant tissues before embarking on the study of plant physiology. In addition, the organization of tissues in the various organs, leaves, stems, and roots should be studied.

1.2.1 Tissue types

An assemblage of cells with an apparent common function is referred to as a tissue. The cells may be of a single type, as in a simple tissue, or several types that form complex tissues.

Tissues composed of dividing cells are called meristematic tissues, whereas those with nondividing, nonmeristematic cells are called permanent tissues.

Permanent tissues

The permanent tissues of plants can be separated into two basic types. Simple permanent tissues are those composed mostly of a single cell type. Complex permanent tissues are composed of several cell types.

SIMPLE PERMANENT TISSUES

Protective tissues

Epidermis is the protective tissue of primary plant parts. The epidermis consists of epidermal cells, some of which may be highly modified (see below). An epidermis may be composed of a single layer or it may be multilayered, such as found in many succulent plants. In these plants, the epidermis consists of a single layer of epidermis proper and a multilayered hypodermis.

Epidermal cells may be modified to form kidney or dumbbell-shaped guard cells of stomata, or they may have appendages in the form of hairs, glands, or other sorts of trichomes. These trichomes function largely in protection, excretion, or extending to some extent the boundary layer above the plant surface, retarding water loss and altering heat exchange.

All exposed surfaces of the epidermis have a cutin deposit, which makes up the cuticle. The cuticle has varying degrees of waxiness that makes it relatively impermeable to water.

The epidermis of primary roots is of special interest because root hairs at the tip greatly increase the total surface area (refer ahead to Fig. 6.4). Root hairs are extensions of single epidermal cells and may enlarge the root surface 10- to 20-fold, greatly increasing the capacity for absorption of water and nutrients because of the increased volume of soil occupied.

Parenchyma (storage) tissues

Parenchyma tissues are found throughout the plant body and function primarily in storage. In leaf tissue and some stem tissues, chlorophyll-containing parenchyma (chlorenchyma) functions in photosynthesis.

In stems and roots, specialized parenchyma functions in storage. Toward the center of stems, the parenchyma tissue is called pith and toward the outside it is called cortex. Pith and cortex are similar in appearance and function. Pith and cortex cells may have starch grains, crystals, or other inclusions. In woody plants, the pith and cortex become obliterated because of the woody tissues, and thus pith and cortex are mostly tissues of primary plant bodies.

The mesophyll tissue of leaves is composed of parenchyma tissue. In the usual dicotyledonous leaf, the upper mesophyll is parenchyma tissue having cells arranged in definite rows resembling a palisade. This tissue is referred to as palisade parenchyma. Toward the abaxial surface of the leaf, the parenchyma is more loosely arranged and is referred to as spongy parenchyma. In monocotyledonous plants, the mesophyll parenchyma is usually not as differentiated.

Support tissues

The simple support tissues are of two types, collenchyma and sclerenchyma. Collenchyma tissues are composed of collenchyma cells arranged in bundles, and they give support to primary plant tissues, including leaf blades, petioles, and young stems. Sclerenchyma tissue, made of either fibers or sclereids, gives support to primary or secondary plant tissues.

Sclereids are the hard cells found in nuts, shells, and seeds and also occur as groups of cells in soft parenchyma tissues in fruits. Fibers are found associated with vascular bundles and give support.

COMPLEX PERMANENT TISSUES

Protective (cork) tissue

Cork tissue is the protective tissue of secondary plant tissues and makes up part of the bark. The exposed surfaces have a waxlike deposit called suberin that is similar to cutin, the deposit covering the epidermis.

Cork cells are rather loosely arranged, and at frequent intervals there are definite groups of unsuberized cells that differentiate in the region of a stoma to form a lenticel. Lenticels, openings within the bark, function in gas exchange. In species such as birch, lenticels are prominent.

Cork is composed of three different cell types. The cork cells proper arise from a meristematic tissue called the phellogen, or cork cambium. The cork cambium produces cork cells (called phellem) toward the outside and similar cork cells (called phelloderm) toward the inside. The structure and more details are given later in Fig. 19.13.

Conducting tissue (xylem and phloem)

Xylem is a complex tissue that may be composed of at least four different cell types: fibers, parenchyma, tracheids, and vessel elements. Its function is primarily transport of water and, to a lesser extent, transport of minerals and simple organic compounds. Storage and support are also important roles of xylem tissue. In the primary plant body the xylem arises from the procambium, whereas the secondary xylem arises from the vascular cambium.

Fibers, of course, give support to the xylem tissue, as do the conducting tracheids and vessel elements. The parenchymatous cells function in storage and perhaps some transport. In many xylem tissues, most xylary elements are in contact with living parenchyma cells, even though it is known that living cells are not a requirement for the transport function of the xylem. The radially arranged parenchyma cells are termed rays.

Phloem, like xylem, is composed of four different cell types: parenchyma, fibers, sieve-tube cells, and companion cells. The fibers give support, and the parenchymatous cells (including parenchyma rays) probably function in storage.

Meristematic tissues

Meristematic tissues are groups of cells that remain embryonic and retain the capability of cell division throughout their life. This continuous capability of division gives the plant body an open, indeterminate system of growth as opposed to the typical closed, determinate system of animals that reach a definite size limit.

Meristematic cells are small and thin-walled with dense cytoplasm and few vacuoles. Basically, there are three types of meristematic tissues in plants: apical, intercalary, and lateral. The apical meristems of the root and shoot apex give rise to primary growth. Intercalary meristems are specialized apical meristems that occur within growing stems of monocots. Grasses, for example, have intercalary meristems such that growth occurs within the stem and not at the apex.

Lateral meristems, the vascular and cork cambia, give rise to secondary growth and thus only occur in those plants which have secondary growth, such as the woody dicots.

Certain monocots have specialized thickening meristems that give rise to lateral growth. The lateral growth of the yuccas and palms is the result of secondary thickening meristems.

1.2.2 Plant organs

The primary plant organs of note are the leaves, stems, roots, and floral organs. Organs are defined as an assemblage of tissues that have a common physiological function. The student may note that plant organs are somewhat different in nature from those of animals. Thus animals such as mammals have hearts, lungs, livers, and so on, but plants have no such comparable structures. Superficially, it may seem that the plant organs mentioned above are conceptually more comparable to the arms and legs of animals, and this may be the case. But if for no other reason than convenience, the leaves, stems, and roots will be considered as organs and discussed as such.

Leaves

The typical plant leaf is flattened dorsoventrally, forming a thin blade. It is composed of a stalk, or petiole, and the flattened blade, or lamina. Leaves may have at the base of their petioles leaflike appendages called stipules. Stipules may be modified into a variety of shapes. It appears that the leaf lamina is adapted and structured as a light-gathering organ. Its flattened shape intercepts maximum radiant energy. In addition, the flattened shape forms a good energy radiator, being largely air-cooled. The thinness of the leaf blade allows for maximum carbon-dioxide diffusion throughout the mesophyll.

Typical leaves are arranged on stems either alternately, oppositely, or in whorls. Leaf blades may be either simple (consisting of a single lamina) or compound (in which the leaf blade is dissected into more than one leaflet). The common leaf blades, which are compound, are either palmate or pinnate. In grasses and other monocots, the lamina is usually elongate with parallel venation, unlike the palmate or netted venation of the typical dicot leaf.

Many leaves are highly modified. Some may be in the form of tendrils or have tendril appendages similar to the common pea plant. Others may be modified into pitchers to trap insects or may have insect-attracting glands, such as the sundews. Still others are fleshy and modified for water storage, such as the succulents. In some cases, leaves may be reduced to spines or may even be totally absent. Such modifications have important consequences for plant physiology and will be discussed in subsequent sections.

A typical dicotyledonous leaf is composed of several tissue types. The upper and lower surfaces of the flattened blade are bounded by the epidermis, which may have cells modified into guard cells, forming the stomata. Stomata frequently occur on the lower surface only, particularly in woody dicots, but herbaceous plants will usually have stomata on both surfaces. The entire exposed surface of the leaf is covered by a waxy cuticle that is even found within the substomatal chamber.

Most of the leaf is composed of parenchymatous tissue called mesophyll. In the usual dicotyledonous plant, the mesophyll is differentiated into an upper palisade parenchyma and a lower spongy parenchyma. The mesophyll cells have chloroplasts and function in photosynthesis.

Venation of the leaf is formed from water-conducting xylem tissue and food-conducting phloem tissue. The xylem is toward the upper, or adaxial, surface of the leaf and the phloem is toward the lower, or abaxial, surface. Most leaf veins are surrounded by large parenchymatous cells called bundle sheath cells. The bundle sheath cells with the enlarged xylem and phloem vascular tissue make up the vascular bundle. There may be sclerenchyma fibers either within or on the outside of the vascular bundle. A typical dicotyledonous leaf is illustrated in cross section in Fig. 1.16.

Monocotyledonous leaves differ from dicotyledonous leaves in that the mesophyll is not differentiated into an upper palisade parenchyma and a lower spongy parenchyma. The mesophyll is relatively undifferentiated, similar to the spongy mesophyll, but is tightly packed, similar to the palisade parenchyma. Cross sections of a monocotyledonous leaf and a dicotyledonous leaf are compared in Fig. 1.16.

There are many other modified anatomical leaf plans that have important physiological consequences. For example, submerged aquatic plants frequently have chloroplasts in epidermal cells, unlike those of most land plants (with the exception of the guard cells), and internally there are present very large air spaces within the parenchyma. Such tissue is called aerenchyma and apparently gives buoyancy to the plant and also allows for more efficient gas exchange and/or gas storage.

The leaves of succulent plants usually have large water- and acid-storing mesophyll cells exhibiting little or no differentiation into palisade and spongy parenchyma. In addition, below the epidermis there is usually a well-developed, multilayered hypodermis. These succulents have the modified photosynthetic metabolism known as crassulacean acid metabolism, discussed in Chapter 16.

Stems

The morphology of a young stem is best studied with a dormant, woody twig, as shown in Fig. 1.17. The terminus is the terminal bud; after breaking dormancy it will develop into the next year's growth. Along the twig there are enlarged regions with axillary buds in the axils of old leaf scars. The distance between buds or leaf scars is called an internode and the enlarged regions are the nodes. By observing terminal bud-scale scars it is possible to age a stem, provided that growth is on an annual basis.

Stems may be of several types. In the simplest, there is a single dominant "leader" with side branches as is the case for most trees. More shrubby plants will have many leaders, only a few of which show any appreciable dominance. Stems may be highly modified into climbing and twining structures such as occurs in vines, they may be reduced to almost nothing, or they may even be greatly enlarged with much storage parenchyma tissue as is found in many succulents such as cacti.

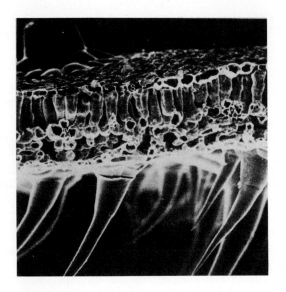

FIGURE 1.16 Scanning electron micrograph of a dicotyledonous leaf showing the upper palisade parenchyma and lower spongy parenchyma of the mesophyll. Note the extensive air spaces between the cells, which are so important for gas diffusion. The drawing also illustrates a typical dicotyledonous leaf. Adapted from C. L. Wilson and W. E. Loomis. 1967. *Botany*. 4th ed. Holt, Rinehart and Winston, New York. The inset compares monocotyledonous and dicotyledonous leaf anatomy. Adapted from D. C. Braungart and R. H. Arnett, Jr. 1962. *An Introduction to Plant Biology*. The C. V. Mosby Co., St. Louis. In the dicot, the mesophyll is differentiated into an upper palisade and a lower spongy mesophyll, whereas there is no such differentiation in the typical monocot. Frequently, the bundle sheath cells of monocots are more prominent. Micrograph courtesy of the New Zealand Department of Scientific and Industrial Research. The micrograph appears in J. Troughton and L. A. Donaldson. 1973. *Probing Plant Structures*. McGraw-Hill, New York.

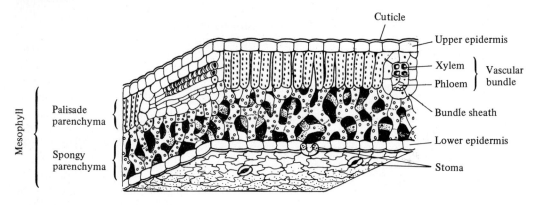

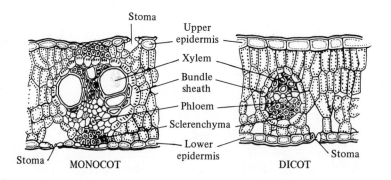

The internal anatomy of the herbaceous plant differs significantly from the woody, dicotyledonous plant. In the monocotyledonous plant, vascular bundles of xylem and phloem are distributed somewhat randomly throughout the stem cortex (ground parenchyma). Figure 1.18 shows an enlarged diagram of a vascular bundle and the distribution of bundles within the stem of a grass plant. Grasses, being herbaceous monocots, do not have any secondary growth and no cambium or cork tissue.

A cross section (Fig. 1.18) of a herbaceous dicotyledonous stem shows a more ordered arrangement of vascular tissue within the ground tissue, or cortex. Vascular bundles within the cortex of the herbaceous dicotyledonous stem are usually not bounded by bundle sheath cells or sclerenchyma tissue and appear more open than do the vascular bundles of the monocot. A cambium may be present within the vascular bundle (the fascicular cambium), and an interfascicular cambium between bundles may form a ring of tissue within the stem. Thus some secondary growth may occur within the herbaceous dicotyledonous stem. If the cambium is present, the ground parenchyma tissue can be differentiated into outer cortex and inner pith.

The young stem of the woody dicot begins development similar to the herbaceous dicot in that vascular bundles are present within the ground-parenchyma tissue and are arranged in an orderly ring. Interfascicular cambium forms and begins to produce xylem toward the inside and phloem toward the outside. Eventually there will be formed concentric rings of secondary xylem and phloem, and the primary xylem and phloem will be obliterated. In the older woody stem a definite bark forms that is composed of phloem and cork tissue.

Roots

The general morphology of a root tip is illustrated in Fig. 1.19. At the very tip is a root cap of loosely bound cells that secrete mucilaginous material. Root caps probably function in lubricating soil

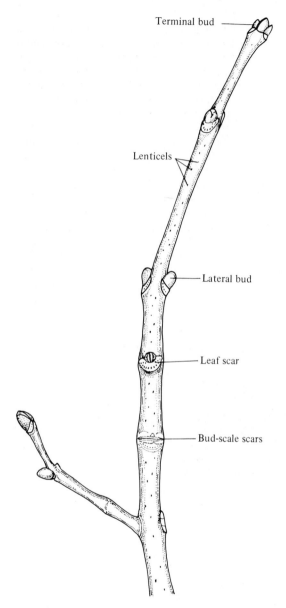

Terminal bud

Lenticels

Lateral bud

Leaf scar

Bud-scale scars

FIGURE 1.17 A typical dormant twig of a woody dicotyledonous plant showing the terminal bud, a lateral bud, leaf scar remaining after leaf abscission, and bud-scale scars. The lenticels are morphological openings of the stem through which gas exchange takes place. Adapted from C. L. Wilson and W. E. Loomis. 1967. *Botany*. 4th ed. Holt, Rinehart and Winston, New York.

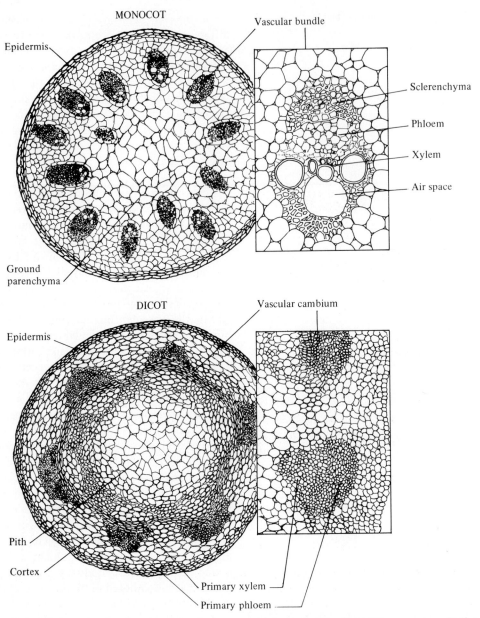

FIGURE 1.18 Diagrams showing the anatomy of a monocotyledonous and a dicotyledonous stem. In the monocot stem, the vascular bundles are scattered throughout the parenchyma, whereas in the dicot stem the vascular bundles are more orderly and form a definite concentric ring. In some dicots, vascular cambium will form between the bundles eventually connecting throughout and forming a solid ring of conducting tissue.

(cross or transverse section)

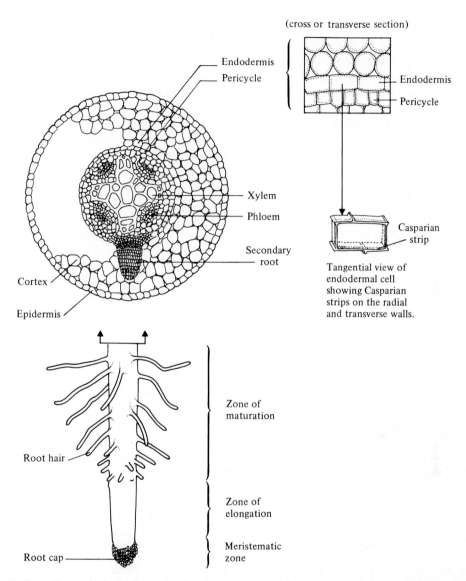

Endodermis

Pericycle

Endodermis

Pericycle

Xylem

Phloem

Secondary root

Casparian strip

Cortex

Epidermis

Tangential view of endodermal cell showing Casparian strips on the radial and transverse walls.

Zone of maturation

Root hair

Zone of elongation

Meristematic zone

Root cap

FIGURE 1.19 General morphology of root tip showing root cap, meristematic zone, zone of elongation, and zone of maturation with root hairs. The cross section of the root, taken just above the root-hair zone in the region of maturation, shows a secondary root forming from the pericycle. The region to the inside of the endodermis is called the stele and includes the pericycle and the conducting tissues, xylem and phloem. The endodermis is made of a single ring of concentric cells that have suberized radial and transverse walls with no intercellular spaces between the cells. Transport from the cortex to the stele must be through the endodermal cells and not around them because of the presence of the suberized Casparian strips of the endodermal cells. Adapted from C. L. Wilson and W. E. Loomis. 1967. *Botany*. 4th ed. Holt, Rinehart and Winston, New York.

during root growth and in protection of the root tip.

Covered by the root cap is the meristematic apex (the region of cell division). Adjacent to the region of cell division, cells elongate in the region of cell elongation. The most proximal portion of the developing root is the region of cell maturation, where the young, elongated root cells differentiate into mature, functional root cells. Epidermal cells in the region of maturation almost always have root hairs that greatly increase the absorptive area of the growing root. Root hairs occur just behind the region of elongation and hence are not broken as the root grows through the soil.

Highly modified aerial roots are common in epiphytic plants such as orchids and vines. Grasses such as maize (corn) have a primary fibrous root system and a secondary adventitious root system arising from the stem tissue. Such adventitious roots arising from stem tissue and not from roots are called prop roots in maize. Of course, it is only the vascular plants which have true roots.

Internally, the root is somewhat different from the stem. Vascular tissue is more centrally located and discrete vascular bundles are not present. A typical root cross section is depicted in Fig. 1.19. The stele, or central core, has the xylem and phloem tissue. Xylem is present in the very center and the phloem is located around the xylem in definite regions. In plants with secondary growth, a vascular cambium may form between the primary xylem and phloem, producing secondary xylem and phloem.

The first differentiated concentric ring of cells is the pericycle, which is the meristematic tissue that gives rise to lateral roots. Lateral appendages of the root arise from deep tissues, unlike the stem, in which lateral appendages develop from superficial meristematic tissue.

The next concentric ring of tissue, also a single sheet of cells, is the endodermis. The lateral walls of the endodermis are suberized such that aqueous solutions (including all mineral ions) must pass through the endodermal cells in order to enter the stele. The suberized endodermis also prevents back-leakage of water from the stele into the cortex.

In older tissue the endodermis is largely obliterated, but cell walls become very thickened, preventing transport from or to the stele other than through living cells. It is generally believed that the presence of the endodermis makes a waterproof barrier between the stele and outer cortex such that hydrostatic pressure can build up in the stele.

Toward the tip of the root the endodermis is open. For this reason, any water or salt uptake in the meristematic zone can enter the stele directly and not be limited to transport through the endodermal cells. However, most evidence indicates that little or no water is taken up at the root apex.

1.3 Plant Classification

The plant kingdom includes a large variety of plants—some 350,000 different species. They are characterized by the presence of a rigid cell wall and the capability of autotrophy. However, if bacteria and fungi are included in the plant kingdom, it means that some are not autotrophic.

A simple scheme for classification is shown in Table 1.2 and divides plants into either vascular or nonvascular. The latter includes all those plants that lack definite supporting and conducting tissue, although the large kelps of the brown algae do have some food-conducting cells. Nonvascular plants include the bacteria, fungi, algae, and the liverworts and mosses. The vascular plants, which are all the rest, can be subdivided into four groups: the Psilopsida, the Lycopsida, the Sphenopsida, and the Pteropsida. The Psilopsida has two living genera, *Psilotum* and *Tmesipteris,* the Lycopsida includes the living club mosses, and the Sphenopsida (horsetails) has a single living genus, *Equisetum.* These vascular plants that do not bear seeds are interesting from an evolutionary viewpoint,

TABLE 1.2 A simple plant classification

NONVASCULAR PLANTS
  Bacteria
  Fungi
  Algae
  Liverworts and Mosses (Bryophytes)
VASCULAR PLANTS
  Psilopsida (Psilophytes)
  Lycopsida (Club Mosses)
  Sphenopsida (Horsetails)
  Pteropsida
    Filicineae (Ferns)
    Gymnospermeae (Conifers and Cycads)
    Angiospermeae (Flowering Plants)
      Dicotyledoneae
      Monocotyledoneae

but represent little more than about 1000 species in all. Virtually all of the extant species are club mosses.

The Pteropsida includes the ferns, the gymnosperms, and the angiosperms. Living ferns do not have seeds but reproduce vegetatively and with spores; however, many fossil ferns apparently were seed-bearing.

The Gymnospermeae (meaning naked seed), with less than 1000 species, includes the cycads and the cone-bearing plants, or conifers. Seeds are borne on scales and not enclosed within carpels as they are in the angiosperms. The Angiospermeae (angios = sac) are those vascular plants that have their seeds enclosed in carpels and that make up by far most of the living plants. Angiosperms are divided into two large groups: the Dicotyledoneae, with nearly 250,000 species, and the Monocotyledoneae, with about 50,000 species. The dicots are named because of the presence of two seed leaves, or cotyledons, and the monocots are named for a single cotyledon on the embryo. Gymnosperms have a variable number of cotyledons.

The scope of the physiology of plants logically includes all those species listed in Table 1.2, from the simplest bacteria to the complex flowering plants. There is relatively little known about the physiology of nonflowering plants except for the bacteria and, to some extent, the fungi. This textbook will, therefore, be a book about flowering plants. Examples and discussion of the physiology of the nonvascular and the nonflowering vascular plants will be limited to a few examples that clarify flowering plant physiology.

Review Exercises

1.1 List some major differences between typical plant and animal cells. Indicate how you think such differences affect the physiology of the cells.

1.2 How do primary cell walls differ from secondary cell walls? Answer from the standpoint of structure and function. Clearly contrast the structural and functional properties of cell walls and cell membranes.

1.3 Many times it is extremely difficult to identify cellular organelles and inclusions in electron micrographs. List those features that you would look for to identify the following organelles: nucleus, plastid, mitochondrion, microbody, dictyosome, and vacuole.

1.4 Measurements taken from an electron micrograph indicated that a boxlike parenchymatous cell is about 100 micrometers (μm) on a side, the average diameter of a chloroplast is 15 μm, a mitochondrion in face view is 1 by 5 μm, and the average diameter of a microbody is 1.5 μm. Compute the volume of the parenchyma cell and also the percentage of the total volume occupied by 10 chloroplasts, 30 mitochondria, and 25 microbodies. If a single large central vacuole had an average diameter of 80 μm, what percentage of the cell would it occupy?

1.5 Tracheids and sieve-tube cells are the main long-distance transport cells of plants. Compare and contrast their structure. From your knowledge of their structure, compare and contrast transport by them.

References

ANDERSON, J. M. 1975. The molecular organization of chloroplast thylakoids. *Biochim. Biophys. Acta* 416: 191–235.

BEEVERS, H. 1977. Microbodies in higher plants. *Ann. Rev. Plant Physiol.* 30:159–193.

BONNER, J., and J. E. VARNER. 1965. *Plant Biochemistry.* 1st ed. Academic Press, New York.

BONNER, J., and J. E. VARNER. 1976. *Plant Biochemistry.* 3d ed. Academic Press, New York.

COPALDI, R. A. 1974. A dynamic model of cell membranes. *Sci. Amer.* 230:26–33.

CRAFTS, A. S. 1961. *Translocation in Plants.* Holt, Rinehart, & Winston, New York.

CUTTER, E. G. 1969. *Plant Anatomy: Experiment and Interpretation,* Part 1. *Cells and Tissues.* Addison-Wesley, Reading, Mass.

DANIELLI, J. F., and H. DAVSON. 1935. A contribution to the theory of permeability of thin membranes. *J. Cell. Comp. Physiol.* 5:495–508.

ESAU, K. 1977. *Anatomy of Seed Plants.* 2d ed. Wiley, New York.

HALL, J. L., T. J. FLOWERS, and R. M. ROBERTS. 1976. *Plant Cell Structure and Metabolism.* Longman, London.

HEPLER, P. K., and B. A. PALEVITZ. 1974. Microtubules and microfilaments. *Ann. Rev. Plant Physiol.* 25: 309–362.

MATILE, P. 1978. Biochemistry and function of the vacuole. *Ann. Rev. Plant Physiol.* 29:193–213.

MOLLENHAUER, H. H., and D. J. MORRE. 1966. Golgi apparatus and plant secretion. *Ann. Rev. Plant Physiol.* 17:27–46.

MORRE, D. J. 1975. Membrane biogenesis. *Ann. Rev. Plant Physiol.* 26:441–481.

NAGL, W. 1976. Nuclear organization. *Ann. Rev. Plant Physiol.* 27:39–69.

PRIDHAM, J. B. (ed.). 1968. *Plant Cell Organelles.* Academic Press, New York.

QUAIL, P. H. 1979. Plant cell fractionation. *Ann. Rev. Plant Physiol.* 30:425–484.

ROBERTSON, J. D. 1959. The ultrastructure of cell membranes and their derivatives. *Biochem. Soc. Symp.* 16:3–43.

SINGER, S. J. 1974. The molecular organization of membranes. *Ann. Rev. Biochem.* 43:805–833.

SINGER, S. J., and G. L. NICOLSON. 1972. The fluid mosaic model of the structure of cell membranes. *Science* 175:720–731.

SNYDER, J. A., and J. R. McINTOSH. 1976. Biochemistry and physiology of microtubules. *Ann. Rev. Biochem.* 45:699–720.

TOLBERT, N. E. 1971. Microbodies—peroxisomes and glyoxysomes. *Ann. Rev. Plant Physiol.* 22:23–44.

TORREY, J. G., and D. T. CLARKSON (eds.). 1975. *The Development and Function of Roots.* Academic Press, New York.

WEIER, T. E., C. R. STOCKING, W. W. THOMSON, and H. DREVER. 1963. The grana as structural units in chloroplasts of mesophyll in *Nicotiana rustica* and *Phaseolus vulgaris. J. Ultrastruct. Res.* 8:122–143.

ZIMMERMANN, M. H., and C. L. BROWN. 1971. *Trees: Structure and Function.* Springer-Verlag, New York.

i

Structure and Function of Plant Cells and Tissues

PROSPECTUS It is the stated goal of structure–function studies to interpret function on the basis of structure. For the most part, such studies with the use of light and electron microscopes have been quite successful. The discovery of a new organelle such as the peroxisome or a new cell type such as the transfer cell stimulates much research that ultimately increases our knowledge. But despite such powerful tools as the transmission and the scanning electron microscopes, structural studies for the most part are static, making the interpretation of dynamic function more difficult. New microscopic techniques that allow for three-dimensional viewing and cinematic observation should bring present interpretation much closer to reality.

Other than the tendency to interpret structure–function in teleological terms, perhaps the greatest concern is the creation of artifacts during tissue preparation. Frequently, rather drastic treatment is required to prepare samples for microscopic observation. Tissues are desiccated, evacuated, and stained with toxic materials. Such treatments can cause alteration of structure because of crystal formation, material deposition, and membrane distortion. This alteration, coupled with the fact that the final image is a two-dimensional, static picture of a three-dimensional, dynamic structure, requires that much care be taken in the interpretation of TEM and SEM pictures.

Finally, it is necessary to keep in mind that drawings are usually idealized from many different preparations. Such "typical" organelles, cells, and tissues frequently do not occur in nature. Any interpretations based on typical components should be qualified properly.

There are many interesting structure–function questions still unresolved. As an example, the origin of mitochondria and plastids that have their own complement of DNA raises questions of primitive symbioses and coevolution. Although we have a fairly clear understanding of the function of nuclei, chloroplasts, and mitochondria, organelles such as microtubules, vacuoles, and even membranes still have not been too well defined functionally. The cell wall is poorly understood. Is it inert or is it dynamic, with living protoplasm? At a higher level of organization, even though the function may be reasonably well understood, the mechanism may not be. For example, we know the xylem functions in water transport and the phloem functions in long-distance solute transport, but the mechanisms are poorly understood. We know the stomata are the portals of gas exchange, but we are just now beginning to appreciate the mechanism of stomatal opening. Transfer cells are presumed to function in transport, but there is virtually no direct evidence for such a role. These and many other questions will be addressed in the next part on biophysics.

Biophysical Processes: Exchanges with the Physical Environment

II

PROLOGUE Biophysics plays an important role in the physiology of plants. Almost without exception plant physiologists adhere to the philosophy that all plant function can ultimately be explained in terms of physics and chemistry. Physics is one of the most appealing areas of plant physiology because of its relative simplicity. It is simple not because the concepts or mathematical formulas are necessarily easy to grasp, but simple because the physics can be explained in mathematical terms that can be tested readily by the experimental method. As an example, a reasonably simple and well-known concept such as diffusion plays an immensely important role in the physiology of plants. Carbon-dioxide uptake by leaves during photosynthesis, water loss by transpiration, and the between- and within-cell transport of solutes are mostly diffusional processes. Even the interaction of enzymes with their substrates is largely dependent on diffusion. Because diffusion can be described precisely by mathematical formulas, it is readily adaptable to the scientific method during physiological studies.

The plant cell is largely an aqueous system, and thus the properties of solutions and colloids become important to the understanding of plant physiology. Similarly, much of the physiology of plants is a function of the interaction of the plant with radiant energy.

Visible light of all qualities is important. Both red and blue are important in photosynthesis. Blue is important for phototropic growth responses and red is important for photomorphogenesis. Both infrared and visible light regulate the energy budget of plants. A thorough knowledge of the physical principles of radiant energy is a prerequisite for plant physiology.

Some of the first investigations of plant processes were physical studies. In 1897 Askenasy conducted experiments giving evidence that water was pulled up stems by the forces of evaporation at the leaf surface. These studies were followed by those of Dixon in 1914 that led directly to our present concept that water movement in the xylem is purely a physical process. The studies of Brown and Escombe in 1900 showed that the extremely high efficiency of gas exchange by plant leaves, which was comparable to free surfaces of equal area, could be explained on physical grounds alone.

Many of the most important advances in plant physiology began when physicists became interested in plant function. Physicists were responsible for our present understanding of the photoreactions of photosynthesis, our understanding of the energy budget of plant leaves, our present concepts of the involvement of phytochrome in the regulation of plant growth and development, and our current concepts of plant water relations, based on both free energy and water potential. In addition, many of our theories about morphogenesis are biophysical theories.

This part, composed of six chapters, covers most of the important physical principles of plant physiology. The first chapter (Chapter 2) is an introduction to some of the physical concepts important to the understanding of plant physiology, such as diffusion, solutions and colloids, acidity, cellular energetics, and the concept of a gradient. Chapter 3 is an introduction to photobiology, outlining the principles necessary to understand energy budgets, photosynthesis, and photomorphogenesis. The next three chapters cover the main biophysical phenomena of plant physiology exclusive of photobiology. These are plant water relations (Chapter 4), gas exchange (Chapter 5), and transport processes (Chapter 6). The final chapter of this part (Chapter 7) discusses the energy budget of plants, including heat load and the dissipation of energy by radiation, convection, and transpiration. In addition, the consequences of heat and cold are discussed.

CHAPTER TWO

Some Basic Biophysical Principles

In this chapter there is a review of some basic biophysical principles important for a thorough understanding of the physiology of plants. All scientists accept the premise that plants obey the established laws of physics and chemistry, and such knowledge is necessary to interpret and understand plant physiology. Throughout this text, the principles discussed here will be used to illustrate the physiology of plants. Thus this chapter can be used as a reference or as an integral part of the course.

2.1 Water

Water (H_2O) must be considered to be one of the most important substances on earth. Covering over 70% of the earth's surface and making up as much as 95% of the matter of living organisms, it is virtually unique among liquids. Water's properties, because of molecular structure, are unlike those of other Group VI A hydrides such as H_2S,

41

TABLE 2.1   Periodic properties of Group VI A hydride compounds in comparison with water

Compound	MW	°C	
		P_F	P_B
H_2Te	129.63	−51	−4
H_2Se	80.98	−64	−42
H_2S	34.08	−83	−62
H_2O	18.02	0.0	+100

Data from *CRC Handbook* (1979).

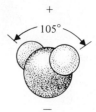

FIGURE 2.1 Diagrammatic representation of water molecule to illustrate the asymmetric structure and dipolar nature. The bond angle between the two hydrogens is 105° rather than a symmetrical 180°.

H_2Se, and H_2Te (Table 2.1). Because of its unique properties, it is a universal solvent. There are few substances that are not soluble in water.

The role of water in living plants is complex. It functions as a solvent for a variety of biological compounds and substances, it helps by hydration to maintain the structure of polymeric compounds such as proteins and nucleic acids, it acts as a substrate or reagent for biochemical reactions, and, because of its presence, it maintains turgidity of cells and tissues.

2.1.1 Molecular structure

Water, consisting of two hydrogen atoms and one oxygen, is a covalently bonded molecule. As a result of the strong attraction of the oxygen nucleus for the hydrogen, electrons from the hydrogen atoms are distorted from their more usual position. The bonding angle between hydrogens is 105°. Because of this asymmetry of hydrogen (see Fig. 2.1), water has a strong dipole moment, i.e., it is highly polarized with a strong separation of positive and negative charge. Because of this polarization, it readily shares its hydrogen with oxygen of other molecules. Thus water is very cohesive, binding strongly with itself, and adhesive, binding strongly with other molecules containing oxygen.

Cohesion

The cohesiveness of water molecules binding with themselves results in a very ordered structure

(Fig. 2.2). When they are frozen as ice, a definite lattice exists (Fig. 2.3). In liquid form, the lattice is altered. As ice melts, hydrogen bonds are broken and water increases in density up to a temperature of 4° C* as the lattice structure collapses and becomes more dense. When the temperature increases from 4° more bonds are broken because of thermal agitation, and water becomes less dense (Fig. 2.4). At 100° hydrogen bonds may break completely, resulting in escape of water molecules as vapor.

There is debate over the exact structure of water, but it seems clear that it is a lattice similar to that shown in Fig. 2.2 and that the structure brought about by hydrogen bonding accounts for its unique properties.

The fact that the maximum density occurs at 4° rather than at a minimum temperature like most substances has a profound effect on aquatic life. Ice at 0° has a density of 0.917 g cm^{-3}† whereas water at 0° has a density of about 0.999 g cm^{-3} and at 4° is 1.0 g cm^{-3}. Hence water freezes from the top down rather than from the bottom up. The frozen water at the surface tends to insulate the water below, preventing large bodies of water from freezing solid and killing aquatic life.

*All temperatures in this book are Celsius (°C) unless otherwise specified.
†The notation g cm^{-3} should be read as grams per cubic centimeter. It can also be expressed as g/cm^3.

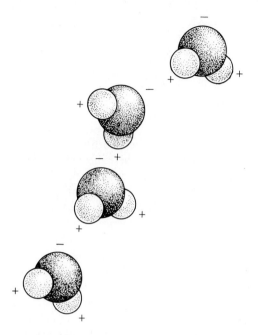

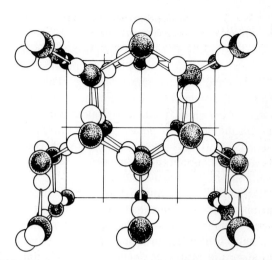

FIGURE 2.3 Diagrammatic representation of ordered, lattice structure of water caused by hydrogen bonding. Drawn after Buswell, A. M. and W. H. Rodebush. 1956. Water, *Sci. Am.,* Vol. 194, p 80.

FIGURE 2.2 Diagrammatic representation of hydrogen bonding of water molecules. There is ionic attraction between the positive pole of the water molecule and the negative pole created by the oxygen. Water molecules will hydrogen-bond to most oxygen-containing molecules such as sugars as well as to other electronegative sites.

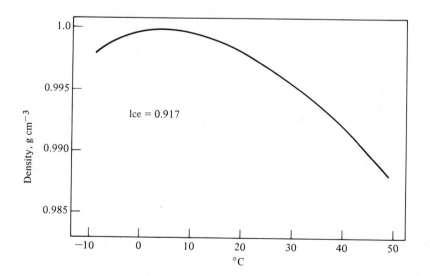

FIGURE 2.4 Graph to show the relationship between the density of water and the temperature of water. The maximum density occurs when the temperature is 4°. Note that ice is much less dense than liquid water.

Adhesion

The adhesiveness of water is explained by hydrogen bonding of water molecules to other polar surfaces such as oxygen-containing molecules. The strong dipole of water exerts electrostatic and gravitational forces on charged electrovalent compounds and on the dipoles of polar, covalent compounds. Thus water will adsorb to substances such as cellulose (cotton) but not to polyesters with few oxygens available for hydrogen bonding.

Virtually all of the interesting, important physical properties of water that are so divergent from other comparable substances can be explained on the basis of its dipolar structure. This asymmetrical structure results in electrostatic attraction for other dipoles and ions and in hydrogen bonding with oxygen-containing molecules. However, the hydrogen bond is relatively easy to break when compared with a covalent bond since it has a bond energy in the range of only 10 to 30 kilojoules (kJ) per mole.

2.1.2 Physical properties

Many of the physical properties of water are important for an understanding of plant water relations. Some of those properties are discussed below.

Heat of vaporization

The latent heat of vaporization, the amount of heat required to convert one unit of liquid to vapor, is excessively high for water as compared with other liquids. The high heat of vaporization, 2435 J g^{-1} at 25° (582 cal g^{-1}), results because of the cohesiveness of water. Compared with water, methane (CH_4) has a heat of vaporization of 577 J g^{-1} at −159° and ethanol (C_2H_5OH) has a heat of vaporization of 854 J g^{-1} at 78.3°. Perhaps the most important consequence of water's high heat of vaporization is the cooling effect as water evaporates from living surfaces. Each gram of water lost by a leaf, for example, requires about 2258 joules and hence this much heat will be lost for each gram of water transpired.

Heat of fusion

The heat of fusion, which is the heat required to change a solid to a liquid, is also high for water (ice) in comparison with other substances. For example, the heat of fusion of water is 333.5 J g^{-1}, whereas for methane it is only 60.7 J g^{-1} at −114.4°.

An interesting application of this high heat of fusion for water is in frost protection. Citrus groves are frequently protected from frost injury by flooding. When the water freezes, the heat liberated during fusion (that is, during freezing) adds heat to the groves and protects the trees.

Viscosity

The viscosity of water, or resistance to flow, is higher in water than in most liquids, again because of the hydrogen bonding. The viscosity of water at 20° (actually at 20.20°), taken as the reference, is 1.0 centipoise. Ethanol has a viscosity at 20° of 1.2 centipoise and hence is slightly more viscous, but ethyl ether has a viscosity of 0.2 centipoise at 20°.

Viscosity decreases with increasing temperature (Fig. 2.5). The increase in tendency to flow (decrease in viscosity) as temperature increases comes about by thermal disruption of the hydrogen bonds. Temperature effects on the transport of liquid within plants can be partially accounted for by viscosity changes.

Volume and density

In the previous section, it was mentioned that water was most dense at about 4° and less dense above and below 4°. The volume of water, expressed as the reciprocal of density, that is, cubic centimeters per gram, is least at 4°. As temperature increases or decreases from 4°, the volume occupied by one gram of water increases. Thus at 0° water will have a volume of 1.00012 cm^3 g^{-1} and ice will be 1.09 cm^3 g^{-1}. Water at 20° will be 1.00177 cm^3 g^{-1}.

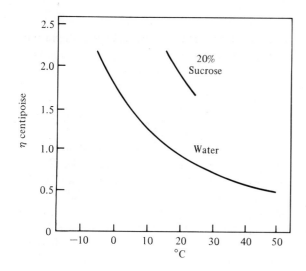

FIGURE 2.5 Graph to show the relationship between the viscosity of water and the temperature of water. As temperature increases, the viscosity of water decreases. A sugar solution is more viscous than pure water.

Surface tension

The surface tension of water—the tendency of the surface to contract and resemble an elastic membrane—is high for water. Water droplets tend to form spheres, a condition resulting in less energy per unit of surface relative to volume. The energy associated with the surface is surface tension and can be expressed in units of dynes per centimeter ($dyn\ cm^{-1}$).

One result of this surface tension is the rise of a liquid within a capillary tube. The height of rise is a direct result of the surface tension. The greater the surface tension the greater the rise in the capillary. Rise by capillarity is very important for consideration of the rise of water in trees. It is further discussed in Chapter 6.

The height of capillary rise (h) within a capillary is given by

$$h = \frac{2\ \gamma \cos \theta}{\rho \cdot g \cdot r},$$

where γ is the surface tension energy (water at 20° = 72 dyn cm^{-1}), $\cos\theta$ is the cosine of the angle of con-tact between the water and the capillary, ρ is the density (water = 1 g cm^{-3}), g is the acceleration due to gravity (981 dyn g^{-1}), and r is the radius of the capillary tube.

Specific heat

Water has the highest specific heat of the common liquids. Water is used as the standard for a calorie, the calorie being the amount of heat required to raise 1 cm^3 of water 1° C between 15.5 and 16.5°. Hence water has a specific heat or heat capacity of 1 cal g^{-1} degree^{-1} (note that 1 calorie = 4.184 joules). By contrast, ethanol has a specific heat of 0.58 cal g^{-1} degree^{-1} at 25°.

This high specific heat has importance in terms of rates of heating and cooling of aqueous bodies. Large bodies of water, such as lakes and oceans, tend to ameliorate adjacent temperatures. Thus climates are usually mild around oceans. For this same reason, dry climates tend to have greater temperature extremes than do humid climates. The rates of heating and cooling of plant tissue are governed partly by the heat capacity of water.

Living organisms containing large quantities of water tend to be protected somewhat from significant temperature changes because of small changes in heat load.

Optical properties

Pure water is a colorless (as well as a tasteless and odorless) liquid; however, because of molecular motion water will tend to scatter short-wavelength light (blue) and transmit the visible long wavelengths (red) with the result that water frequently appears blue. Long-wave heat radiation (infrared) is absorbed by water, but ultraviolet light will penetrate substantially.

2.1.3 Chemical properties

The chemical properties of water are as equally important biologically as are the physical properties. Water will react spontaneously with many inorganic compounds, but frequently energy input is required. For example, water will react with substances commonly occurring in plants, such as Mg and Zn, to form their oxides, but high heat is required. In addition, water will form the hydroxides of CaO and K_2O. Perhaps more important is that oxides of nonmetals will form acids in water.

$$CO_2 + H_2O \rightarrow H_2CO_3$$
$$SO_2 + H_2O \rightarrow H_2SO_3$$

The first equation is important because of the central role of CO_2 in plant metabolism and perhaps as a protoplasmic buffer. The second may be important in atmospheres that are polluted with SO_2 after fossil-fuel combustion.

Hydrolyses involving water are among the most important of biochemical reactions. The basic reaction can be depicted as follows.

$$R-O-R' + H_2O \rightarrow R-OH + R'-OH$$

Hydrolyses are involved in the breaking of peptide, ester, and other glycosidic bonds during intermediary metabolism.

2.2 Solutions

Much of the chemistry taking place in the plant cell is solution chemistry, and many of the biological compounds of importance are soluble in water. In this section, many of the important properties of aqueous solutions are reviewed. The colligative properties of solutions are discussed in some detail since they form the basis for many plant-water relations.

2.2.1 Colligative properties

Solutions are homogeneous mixtures of substances with at least two phases. The solvent, or continuous, phase largely retains its properties. In biological systems the solvent is almost always water. The discontinuous phase (the solute particles) is frequently altered. Of course, solutions can be composed of gases, liquids, or solids dissolved in gases, liquids, or solids. However, we usually think in terms of simple solutions in which gases or solids are dissolved in a liquid. The most common solutions of plants are those of salts, sugars, organic and amino acids, and small-molecule proteins dissolved in water.

Colligative properties of solutions are those properties that depend solely on the number of particles rather than on their kind. This is true, of course, only if there are no interactions among the solutes and if each and every particle acts independently. If interactions occur, correction for effective particles or activity must be made.

In very dilute or ideal solutions, interactions are frequently not important. In some solutions (e.g., sucrose) the solute molecules apparently bind H_2O

molecules, resulting in greater effects than can be predicted on the basis of the number of molecules. Sucrose solutions have a greater vapor-pressure lowering than theory predicts, causing all the colligative properties to be more pronounced.

The main colligative properties of interest to plant physiologists are vapor pressure, boiling point, freezing point, and osmotic pressure.

Vapor pressure

Vapor pressure is the tendency for a liquid to evaporate (the tendency for molecules to escape solution) and is an exponential function of temperature (Fig. 2.6). More precisely, vapor pressure can be defined as the pressure exerted by vapor over a liquid where it is in equilibrium with itself. Boiling occurs when the vapor pressure of the liquid equals the pressure over the liquid. Hence at standard conditions of temperature and pressure water will boil when the vapor pressure reaches 1 atmosphere (atm), or 760 mm Hg pressure.

More conveniently, vapor pressure can be ex-

pressed in units of bars or millibars, where 1 bar equals 0.987 atmospheres. Bars are used rather than atmospheres because the bar is an energy unit (atmospheres and mm of Hg are pressure units) equal to 10^6 dyn cm^{-2}. Modern considerations of vapor pressure tend to use the more understandable energy units consistent with thermodynamic theory.

The expression for the relationship between vapor pressure (p) and temperature is

$$\log p = -\frac{k}{T} + C,$$

where k and C are constants and T is the absolute temperature.

When a solute is added to a pure liquid, it tends to be diluted. This dilution will decrease the vapor pressure of water because of less water per volume. Actually, the reduction in vapor pressure brought about by solutes in solution is more complicated than simple dilution and involves solute–solvent interactions that decrease the free energy of the

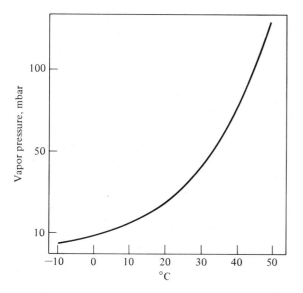

FIGURE 2.6 Graph to illustrate the exponential increase in vapor pressure of water with increasing temperature.

solvent, with a resulting decrease in vapor pressure. The decrease in vapor pressure resulting from adding a solute to a pure solvent can be estimated from Raoult's law:

$$p = \left(\frac{m}{m_s + m}\right) p^\circ,$$

where p° is the vapor pressure of pure solvent, m_s is the moles of solute, and m is the moles of solvent.* Thus a 1-molal sucrose solution will lower the vapor pressure of water at 20° from 23.38 mbar to 22.97 mbar. The calculation is

$$p = \left(\frac{55.6}{1 + 55.6}\right) \ 23.38 = 22.97 \text{ mbar},$$

where $m = 55.6$ or the moles of water in 1000 g of water (1000 g $18g^{-1}$ mol^{-1}) and $m_s = 1$ mole of sucrose.

The boiling point

Solutes tend to raise the boiling point of liquids since solutes reduce the vapor pressure. Boiling, as stated above, comes about when the vapor pressure of a liquid equals the total pressure over the liquid. Under standard conditions of temperature and pressure (STP), water will boil when its vapor pressure equals one bar. Hence a solution with a reduced vapor pressure must be heated to a higher temperature to reach the boiling point.

It can be shown that a 1-molal aqueous solution will have a boiling point 0.52° higher than pure water. Thus the boiling point elevation ($\triangle P_B$) can be estimated from the following expression:

$$\triangle P_B = 0.52m,$$

where m is the molality (moles of solute per 1000 g of solvent). The above expression emphasizes that the boiling point, which is dependent on the vapor pressure, is a function of the number of solute molecules and not of their kind.

*$m/(m_s + m)$ = mole fraction of solvent; $m_s/(m_s + m)$ = mole fraction of solute.

The freezing point

The freezing point is lowered by the presence of solutes because of the reduction in vapor pressure, similar to the boiling point. In fact, a 1-molal solution reduces the freezing point by 1.86°. The freezing point reduction ($\triangle P_F$) because of the presence of a solute can be estimated from

$$\triangle P_F = -1.86m.$$

Because of high solute contents, plant extracts tend to freeze at a few degrees below zero. The cytoplasm in intact plants frequently will not freeze even at 40° below zero. This is because ice nucleation does not occur and the tissue supercools. Ice crystals will, however, form in the intercellular spaces, frequently causing physical damage to the adjacent cells. In addition, ice formation between cells acts as a sink for water, resulting in cellular desiccation.

Osmotic pressure

Osmotic pressure is a colligative property of solutions that can be shown to be directly dependent upon vapor pressure. Osmotic pressure can be simply defined as the pressure necessary to prevent flow of a solvent from pure solvent to a solution through a membrane that is permeable only to the solvent. The defining equation is

$$\pi = C \cdot R \cdot T,$$

where

π = osmotic pressure in bars,

C = mol L^{-1},

R = gas constant = $(0.08) \cdot \dfrac{(L \cdot bar)}{(mol \cdot deg)}$, and

T = absolute temperature in °K (°K = °C + 273°).

A 1-molar solution at 0° C (273° K) will have an osmotic pressure of 22 bars:

$$\pi = (1 \text{ mol } L^{-1}) \cdot (0.08) \cdot \frac{(L \cdot bar)}{(mol \cdot deg)} \cdot (273°) = 22 \text{ bars}.$$

Interactions of solute with solvent will tend to reduce the effective amount of solvent and result in greater osmotic pressures. Interactions of solute with other solute molecules will result in less-than-expected solute activities and lower osmotic pressures than predicted.

Almost all of the solutions of plant cells contain electrolytes that result in properties much different than those of solutions of simple, nonionic chemicals.

2.2.2 Electrolytes

Electrolytes are those substances that upon going into solution will conduct an electric current. Those substances that will ionize are electrolytes. Some common strong electrolytes (complete ionization) in plant tissues are the inorganic ions K^+, Na^+, Ca^{+2}, Mg^{+2}, Mn^{+2}, Zn^{+2}, Cl^-, SO_4^{-2}, and PO_4^{-3}. Common weak electrolytes (incomplete ionization) are the organic cations and anions, such as amino acids and organic acids, and the complex charged polymers, such as proteins and nucleic acids. These electrolytes alter the colligative properties of water in the same way as discussed above.

The behavior of electrolytes was first discussed by Arrhenius in 1880. He assumed ionization in solution of materials such as NaCl, KCl, and $CaSO_4$. Since electrolyte molecules result in more than one particle (ion) per molecule, colligative properties cannot be predicted on the basis of molality alone. For complete ionization and no

interaction, $1m$ NaCl would result in twice the vapor-pressure lowering in water that would be predicted on the basis of molality. Similarly, the freezing point depression would be 3.72°, the boiling point elevation would be 1.04°, and the osmotic pressure would be 43.7 bars. However, only the strong electrolytes will ionize completely and then only when in very dilute solutions such that there will be no interaction of the solute molecules. Thus a $1m$ solution of NaCl actually has an osmotic pressure of 43.2 bars rather than the expected 43.7 bars.

Electron (solute) interactions tend to decrease the effective number of particles in solution. These interactions, known as Debye-Huckel effects, can be described in dilute solutions:

$$\log \gamma = -0.51 \ Z^2 \sqrt{\mu} \ ,$$

where

γ = activity coefficient, or estimate of the effective concentration of the ion,

Z = net charge of the ion, and

μ = ionic strength, defined as $\mu = \frac{1}{2} \Sigma \ C \cdot Z^2$, that is, half the sum of the concentrations of ions times the square of their charge.

Interactions of solute with solvent cause a deviation, such as if there were more particles. Sucrose, which binds water through hydrogen bonding, will have an osmotic pressure greater than predicted. A $1m$ sucrose solution will have an osmotic pressure of 25.1 bars rather than the expected 21.8 bars.

2.3 Colloids

Many of the particles in the aqueous phase of the cell are larger than those that form true solutions, yet they still stay dispersed. Much of protoplasm is composed of these stable suspensions, called colloids. Colloidal chemistry is extremely complex but highly important to the metabolism and biochemistry of the functional cell.

2.3.1 Definition

It is frequently desirable to clearly differentiate between true solutions and colloidal suspensions, even though they may form a smooth gradation. It can be recalled that solutions are defined as homogeneous mixtures of substances. The solute molecules are dispersed to form a discontinuous phase within the continuous (solvent) phase. A solution is quite stable and the particles resist precipitation, even at high centrifugal forces. If the particles do not become dispersed but remain in clumps and easily settle out, either freely or by centrifugal force, the mixture is defined as a suspension.

Under certain circumstances, however, mixtures may form stable suspensions by virtue of their electric properties or water(solvent)-binding capacities. These stable suspensions with discontinuous-phase particles in the size range of 10^{-4} to $1\ \mu m$ are termed colloidal suspensions. It is more than likely that there are multiple ions or molecules making up each colloidal particle, resulting in large surfaces. It is these surfaces of the colloidal particles that make them so interesting and reactive in the biological sense. The stability of the colloid results in large part because of the tendency of the particles to clump and reduce their surface area.

2.3.2 Terminology

Colloids of solid particles in liquid are called sols. A solid colloidal system is called a gel. Reversible sol–gel transformations are common in biological systems. Colloids of solids or liquids in gas are termed aerosols. Liquids in liquids are called emulsions.

Depending on their properties, colloids can be lyophilic (show an affinity for solvent) or lyophobic (no affinity) or, if specifically aqueous, hydrophilic or hydrophobic. Hydrophilic colloids are those in which the particles have an affinity for the water; hydrophobic colloids lack such affinity. It is believed that hydrophilic sols base their stability on water hydration of the particles and hydrophobic sols are stable because of water exclusion and repulsion of particles due to charged surfaces.

Plant gums and mucilages, starch, and proteins tend to form hydrophilic sols or colloids, whereas metallic sols are frequently hydrophobic.

2.3.3 Properties

Perhaps the most important properties of colloids are based on their large surface area. Surfaces of such proportions function as interfaces for chemical and physical reactions. In addition, sol–gel transformations are implicated in the functioning of cytoplasm, particularly during movement. These and other properties of colloids are discussed below.

Surface area

From a biological viewpoint it is the immense surface area of colloids that imparts the important physiological properties of protoplasm. On these surfaces the many and varied biochemical reactions of metabolism occur.

By cutting a 1-cm^3 cube into cubes of $10^{-6}\ \mu m$ (colloidal dimensions), it is possible to calculate the effect on surface area. The number of cubes will increase from one to 10^{15} and the surface area will increase from 6 cm^2 to 600 m^2, an increase of 1,000,000 times. Thus the surface area per unit of mass of protoplasm is immense.

Filterability

Colloids are considered to be filterable since they will pass through the usual laboratory filters. Filter paper with pores 1–$5\ \mu m$ in diameter easily allows colloids to pass, and it is only the newer ultrafilters with pores the size of colloids that retain colloidal particles. Ultrafilters of the proper dimensions allow for the separation of colloidal particles from those in true solution.

Colloidal particles can be removed from suspension by high centrifugation. Forces up to 100,000 times that of gravity are sufficient to remove protein colloids from water. Substances in true solution, of course, will remain dissolved in the water.

The Tyndall effect

Colloids illuminated with a beam of light at right angles to the observer are seen to scatter or diffract the light. This effect, known to anyone who has observed dust in room air, is indicative of a colloid.

The greater the index of diffraction (the capacity to alter the direction of light), the greater will be the visual effect. Because of the more pronounced scattering of short, blue wavelengths of light, colloids frequently appear to be blue when viewed from the side.

Light entering a leaf will be scattered according to the principles outlined above. Such scattering is undoubtedly important for light gathering during photosynthesis and for other cellular photochemical reactions in which light is absorbed by pigments.

Viscosity

Colloids decrease in viscosity with increases in temperature, similar to solutions. Hydrophilic sols increase exponentially in viscosity when the colloidal particles are increased. Thus plant gums, mucilages, and protein sols become highly viscous when concentrated. Hydrophobic sols do not increase in viscosity as much as hydrophilic sols apparently because they lack interaction with the water.

Electrical properties

The micelles of colloidal particles are almost always charged. The positive or negative charges result from the chemical groups associated with the particles. Inorganic colloids may be positively or negatively charged, whereas the organic colloids occurring naturally in plants are usually negatively charged. The micelles of proteins, for example, will have a net negative charge. Attracted to the negative charge will be cations, resulting in an ionic double layer. This ionic double layer is one force giving stability to the colloid.

As well as having the ionic double layer, hydrophilic colloids will also have a water shell or hydration shell associated with them, resulting in even more stability. Thus we can visualize a negatively charged proteinaceous sol with a cation double layer and a hydration shell. Hydrophobic sols will not have a hydration layer; their stability depends solely on the ionic double layer.

Coagulation

Since the stability of colloids depends on the ionic double layer and hydration shell, flocculation or coagulation can be brought about by removing the water layer and neutralizing the charge (the inner layer). Hydrophobic sols will flocculate with charge neutralization only. Proteins will thus tend to clump at their isoelectric points when their net charge is zero or upon treatment with solvents that remove their water shells. These properties are frequently manipulated in protein purification in which unlike proteins are separated.

Salting out is a protein-purification procedure in which the ionic strength of a protein solution is increased by adding ammonium sulfate. When the isoelectric point of each protein in the solution is reached, the proteins will coagulate and can be precipitated by centrifugation. Similarly, proteins can be coagulated by removing the water of hydration with such solvents as alcohol and acetone.

Amphoterism

Polymeric, biological gels frequently may be amphoteric—able to act as either an acid or a base—existing as either positive or negative particles, depending on pH. This is particularly true of the protein sols because of the functional groups of the amino acids. Amino acids have carboxyls and amino groups that may be charged.

At high pH (low proton concentrations) protein sols will be negatively charged, and at low pH (high proton concentrations) protein sols will have positive charge. This is largely a function of the amphoteric amino acids.

$$\underset{NH_3^+}{R-CH-CO_2H} \underset{+H^+}{\overset{-H^+}{\rightleftharpoons}} \underset{NH_2}{R-CH-CO_2H} \underset{+H^+}{\overset{-H^+}{\rightleftharpoons}} \underset{NH_2}{R-CH-CO_2^-}$$

Thixotrophy

Many gels including protoplasm demonstrate thixotropic properties. Agitation will cause a gel to revert to a sol, which upon standing will gel again, a property called thixotropy. An additional prop-erty of interest is hysteresis. Gels retain a memory of sorts. When dried or altered and then returned to the gel form again, properties of the original gel are approached or retained fully.

2.4 pH—Acidity

Many of the chemical and physical reactions occur-ring in plant cells are regulated by acidity, or pH. Virtually all enzymic reactions have rather narrow pH optima, and many of the chemical reactions of cells either give up or take up protons. pH, or more specifically, proton gradients, drive many of the transport and energetic reactions of the cell. Thus pH and the regulation of pH in cells is intimately tied to metabolism.

Solutions that resist change in pH are called buf-fers. Buffers are common in plant cells and for the most part plant cells resist changes in pH, although there are some notable exceptions, such as the acid metabolism of succulent plants. Some of the most common buffers in plants are made of proteins and organic acids. CO_2 undoubtedly is also an impor-tant buffer in plant tissues and fluids.

The usual range of pH in plant fluids is from 5.2 to 6.2, although many plants are known for their high acidity. Plants in the Polygonaceae, Crassulaceae, Cactaceae, Saxifragaceae, Rosaceae, and Gerani-aceae are especially notable for high acid contents. Some plants are distinctly alkaline, but usually the fluids of plants are acidic.

It is well known that the soluble pigments of the anthocyanin group change color with pH. Thus the floral colors of many plants are pH-dependent. The flavones are yellow at high pH but colorless at low pH; many of the other anthocyanins are red at low pH and turn blue when the pH is raised. Antho-cyanins are further discussed in Chapter 12.

pH and acidity are reviewed in the section below. The section can be used for study or review.

2.4.1 Normality

The usual methods of expressing concentration are moles per liter (molarity, M) or moles per 1000 grams of solvent (molality, m). In the case of ioniz-able acids and bases, however, solution results in multiple particles produced from each molecule, and a more informative designation of concentra-tion is normality. Normality (N) is the molarity multiplied by the number of exchangeable hydro-gens or equivalents. Thus a 1-molar solution of HCl would also be 1 normal, but a 1-molar H_2SO_4 solution would be 2 normal by virtue of the two hydrogens. A 1-molar acetic acid solution would be 1 normal, but a 1-molar malic acid solution is 2 normal (acetic acid = CH_3COOH; malic acid = $COOHCHOHCH_2COOH$). It should be clear that a 1-normal solution of malic acid would be 0.5 mole per liter ($0.5M$). This nomenclature is also spoken of as equivalency. One equivalent of malic acid would be 0.5 mole.

2.4.2 pH

Acids are ionizable compounds that give up pro-tons. Strong acids are the mineral acids such as HCl, H_2SO_4, and HNO_3. These tend to ionize com-pletely when in solution:

$$HCl \longrightarrow H^+ + Cl^-.$$

Actually, the proton reacts with H_2O to form a hydronium ion, H_3O^+, but it is conventional to speak simply of protons, H^+.

Weak acids are the organic acids and tend to ionize only partially when in solution. Thus acetic acid ionizes to form acetate ions and protons, but some acetic acid will remain un-ionized:

$$CH_3COOH \rightleftharpoons CH_3COO^- + H^+.$$

Because of the partial ionization, an equilibrium constant (K) can be established. The equilibrium between the free acid, acetic acid, and the acetate anion is given by

$$K = \frac{[CH_3COO^-][H^+]}{[CH_3COOH]}.$$

The equilibrium constant is specific for each acid at a specified temperature. Ionization (or equilibrium) constants for the common plant acids are given in Table 2.2.

Bases are ionizable compounds that tend to take up protons. Thus acetate (CH_3COO^-) is the conjugate base to acetic acid (CH_3COOH). Occasionally, substances that yield hydroxyl ions (OH^-) are

considered bases. This is true if the proton accepted by the base comes from water during the ionization of water:

$$H_2O \rightleftharpoons H^+ + OH^-.$$

The measure of acidity (or basicity) comes from the above equation for water ionization. The ionization constant for the ionization of water is

$$K = \frac{[H^+][OH^-]}{[H_2O]}.$$

Ionization of pure water results in $10^{-7}M$ protons and $10^{-7}M$ hydroxyl ions. Because the amount of water that ionizes to form H^+ and OH^- is very small, there will be essentially no reduction in the total amount of $[H_2O]$ present. For this reason, $[H_2O]$ is set equal to $1M$, and the expression reduces to

$$[H^+][OH^-] = [10^{-7}][10^{-7}] = 10^{-14}.$$

The concentration of protons is normally expressed with the familiar pH scale, defined as the negative logarithm of the hydrogen-ion concentration:

$$pH = -\log[H^+].$$

Hence pure water with equal protons and hydroxyl ions will have a pH of 7.0 because

$$[H^+][OH^-] = 10^{-14}$$
$$[10^{-7}][10^{-7}] = 10^{-14}$$
$$pH = -\log 10^{-7} = 7.0.$$

As an additional example of pH calculation, a solution with a $10^{-3}M$ hydrogen-ion concentration would have a pH equal to 3.0:

$$pH = -\log[H^+] = -\log 10^{-3} = 3.0.$$

Similarly, a solution with a $10^{-5}M$ hydrogen-ion concentration would have a pH of 5.0. It should be apparent that a solution with a $10^{-9}M$ hydroxyl-ion concentration would have a pH of 5.0. This is because the ionization constant for water is 10^{-14}, requiring that

$$[H^+][OH^-] = 10^{-14}.$$

TABLE 2.2 Ionization constants (pK) of some common plant acids

Acid	pK
Acetic	4.75
Ascorbic	4.10
Aspartic	3.86
Caproic	9.82
	4.83
Citric	3.08
	4.74
	5.40
Formic	3.75
Fumaric	3.03
	4.44
Glycolic	3.83
Malic	3.40
	5.11
Succinic	4.16
	5.61

Data from *CRC Handbook* (1979).

Thus if

$$[H^+][10^{-9}] = 10^{-14},$$

then

$$[H^+] = 10^{-5}.$$

Salts are compounds formed from the conjugate base of an acid and the conjugate acid of a base, i.e., from the anion of an acid and the cation of a base. The malate anion from malic acid ionization will form a sodium salt with the sodium from NaOH (or any other sodium):

$$COOHCHOHCH_2COOH + 2NaOH \rightarrow$$

Malic acid

$$COO^-Na^+CHOHCH_2COO^-Na^+ + 2H_2O.$$

Sodium malate

In this example, malate is the anion from the acid, and sodium is the cation from the base (NaOH). Or in other terms, malate is the conjugate base of the acid (malic acid); and sodium is the conjugate acid of the base (NaOH).

In the opposite manner, alanine (an amino acid) can act as the cation to form a salt with the chloride of HCl:

$$CH_3\underset{NH_3^+}{CHCOOH} + HCl \rightarrow CH_3\underset{NH_3^+Cl^-}{CHCOOH} + H^+.$$

2.4.3 Buffers

Solutions that resist changes in pH by absorbing or reacting with either protons or hydroxyl ions are called buffers. Buffers are frequently mixtures of either weak acids and their salts or weak bases and their salts, although strong acids and bases will act as good buffers at low or high pH, respectively.

A solution of acetic acid and sodium acetate will form an effective buffer. These compounds ionize in solution as follows.

$$CH_3COOH \rightleftharpoons CH_3COO^- + H^+ \qquad (1)$$

$$CH_3COONa \rightleftharpoons CH_3COO^- + Na^+ \qquad (2)$$

The buffering action occurs because protons added to the mixture will react with the acetate anions and push reaction (1) to the left, maintaining a nearly constant proton concentration in solution. Undissociated acid will increase but this does not affect the pH. Hydroxyl ions will react with the protons and more protons will be produced by further ionization of acetic acid. The acetate anions and sodium ions from the salt will balance the counter ion of the added protons or hydroxyls to maintain electrical neutrality.

The buffering range and capacity can be estimated from the ionization constant and concentrations of acid and salt in solution. For acetic acid, the ionization constant is

$$K = \frac{[H^+][CH_3COO^-]}{[CH_3COOH]} = 1.75 \times 10^{-5}.$$

The negative logarithm of K, termed the pK, will give the pH of maximum buffering capacity. For the above, the pK is 4.75.

As a general rule, effective buffering will be between one pH unit above and below the pK. Hence acetic acid and acetate buffers will buffer well over the range of 3.75 to 5.75 (Fig. 2.7).

Strong acids and bases are good buffers when at low or high pH simply because small amounts of protons or hydroxyl ions are insignificant compared to that already present from the fully ionized strong acid or base.

The pH of a buffer can be predicted from the pK of the acid and the concentration of the acid and the base in solution from the Henderson–Hasselbalch equation*

$$pH = pK + \log \frac{[base]}{[acid]}.$$

Thus the Henderson–Hasselbalch equation can be used to compute the ratio of base to acid necessary to prepare a buffer solution of any desired pH.

---

*This equation is derived by taking the logarithm of both sides of the equilibrium expression and noting that pH = –log [H$^+$] and pK = –log K.

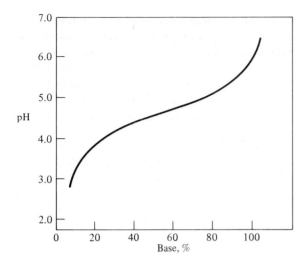

FIGURE 2.7 Graph to show the change in pH of a sodium acetate buffer as the percentage of base (here NaOH) increases. There is effective buffering from about one pH unit above and below the pK. The pK is 4.8 and the buffering range is 3.8 to 5.8.

2.5 Cellular Energetics

The biochemical reactions of cells are driven by free-energy changes, and thus cellular reactions can be evaluated in terms of their equilibrium positions and free energies of reaction. In the reaction

$$y \longrightarrow x,$$

if there is a decrease in free energy (defined as the energy available for work), the reaction will proceed spontaneously. Such a reaction is said to be exergonic. If there is an increase in free energy, that is, of x over y in the above reaction, then for the reaction to proceed an energy input is necessary. The reaction cannot occur spontaneously because it is endergonic.

The free-energy content, or available energy, is based on the well-known, simple thermodynamic expression

$$G = H - T \cdot S,$$

where G is the free energy (the energy available for work), H is the heat content or enthalpy, S is the entropy, and T is the absolute temperature. Because of the difficulty of determining the absolute values of these thermodynamic parameters, it is more common to determine changes ($\triangle$) that occur during a reaction according to the following expression:

$$\triangle G = \triangle H - T \cdot \triangle S.$$

In this expression, $\triangle G$ signifies the change in free energy occurring during the reaction, $\triangle H$ represents the heat exchange that takes place, and $\triangle S$ is the entropy change that occurs. Reactions that have a negative $\triangle G$ are exergonic and occur spontaneously. Those reactions that are endergonic with a positive $\triangle G$ will not occur spontaneously.

For each reaction, a standard free-energy change ($\triangle G^{\circ}$) can be determined. The standard free energy of a reaction is defined as

$$\triangle G^{\circ} = - R \cdot T \cdot \ln K,$$

where K is the equilibrium constant of the reaction ($K = x/y$) when the reactants and products are at unit activity and at pH $= 0$.

The free-energy change ($\triangle G$) of any reaction can be related to the standard free-energy change ($\triangle G^\circ$) through the following expression:

$$\triangle G = \triangle G^\circ + R \cdot T \cdot \ln \frac{x}{y}.$$

In addition, $\triangle G^{\circ\prime}$ is used to indicate the standard free-energy change of the reaction when at pH $= 7.0$. $\triangle G^{\circ\prime}$ is more useful to plant physiologists because most biological reactions take place close to pH 7.0 and not at very low pH.

Using a gas constant of 8.31 joules per mole-degree, a temperature of 25°C, and converting to base-10 logarithms, the following simplified expression is useful to estimate the standard free-energy change ($\triangle G^{\circ\prime}$) at pH 7.0:

$$\triangle G^{\circ\prime} = -5.7 \times 10^3 \text{ J mol}^{-1} \cdot \log K.$$

Oxidation–reduction reactions are frequently evaluated according to their oxidation–reduction (redox) potentials. Oxidation is defined as the loss of electrons and reduction is defined as the gain of electrons.* Those components with more negative potentials are better electron donors, and those with more positive potentials are better acceptors. The component of the redox reaction with the most negative potential will reduce the component with the more positive potential.

The ratio of oxidized component to reduced component in any redox couple will influence the reaction along with pH and temperature. It is common to represent redox potentials as standard potentials at molar concentrations and pH 0. The standard redox potential is indicated as E° and at pH 7.0 as $E^{\circ\prime}$. E can be estimated from $E^{\circ\prime}$ with the following expression:

$$E = E^{\circ\prime} + \frac{R \cdot T}{n \cdot F} \cdot \ln \frac{\text{oxidized component}}{\text{reduced component}},$$

where

$R =$ gas constant (8.314 J degree^{-1} equivalent^{-1});

$T =$ absolute temperature (°K);

$n =$ number of equivalents; and

$F =$ Faraday constant (96,400 J equivalent^{-1} volt^{-1}).

At room temperature (20°), molal concentrations, and converting to base-10 logs, the expression reduces to (in units of millivolts)

$$E = E^{\circ\prime} + 60 \text{ mV} \cdot \log \frac{\text{oxidized component}}{\text{reduced component}}.$$

If the change in standard redox potential ($\triangle E^{\circ\prime}$) that occurs during a balanced oxidation–reduction reaction is known, it can be related to the standard free-energy change ($\triangle G^{\circ\prime}$) by the following expression:

$$\triangle G^{\circ\prime} = -n \cdot F \cdot \triangle E^{\circ\prime}.$$

2.6 Diffusion

Diffusion is an exceedingly important process for living organisms. Photosynthesis depends to a large extent on diffusion of CO_2. Water loss from plants is largely a diffusion process. Uptake of minerals by roots from the soil solution depends in

---

*The basic oxidation–reduction equation is: $A_{oxid} + e \rightleftharpoons A_{red}$, where A is a substance that can accept or give up electrons.

part on diffusion, and virtually all chemical processes including those catalyzed by enzymes depend on molecular collisions brought about by diffusing molecules. Below, the principles of diffusion are discussed from the standpoint of their importance to plant physiological processes.

Diffusion can be defined as the net movement of molecules from a region of high free energy to a region of low free energy. In practice, the free energy (energy available to do work) is governed by the concentration of the molecules that are diffusing, and diffusion can be defined as the movement of molecules from a region of high concentration to a region of low concentration. Gases, liquids, or solids can diffuse within gases, liquids, or solids.

All molecules above absolute zero ($-273°C$ or $0°K$) are in continuous motion because of their kinetic energy. Any one molecule will move in a straight line until it strikes an object—another molecule of the same or a different kind—and recoils in another direction with no loss in energy. The net result of this continuous motion is that molecules of one kind tend to assume spacing such that the average distance between any two molecules is the same. If there are like molecules concentrated in one portion of an enclosed vessel, they will diffuse because of their high concentration until the average distance between them is equal. At this point there will be no more diffusion. Molecular motion and movement still occur because of kinetic energy, but there is no more *net* diffusion.

2.6.1 The driving force for diffusion

The driving force for diffusion can be understood through an understanding of the simple thermodynamics discussed above. If there is a change in free energy ($\triangle G$) when the process goes from G_1 to G_2, that is, if

$$G_2 - G_1 = -\triangle G,$$

the process will tend to proceed spontaneously. If the change in free energy is positive (that is, if $+\triangle G$), the process would require an energy input. In the specific case of diffusion, a high concentration of molecules with a certain amount of order or nonrandomness (that is, at a given level of entropy, S) will tend to become less ordered and more random after diffusing, with the result that entropy increases. The decrease in free energy is accountable for by the increase in entropy. It is this tendency for an increase in randomness during diffusion that is the driving force for diffusion.

2.6.2 Fick's law for diffusion

Fick's law for diffusion is the basic statement describing diffusion, relating the rate to the gradient. In words, Fick's law states that the rate of diffusion or movement of molecules with time is proportional to a diffusion coefficient and the gradient. In terms of the calculus, diffusion is

$$\frac{dQ}{dt} = -D \cdot \frac{dc}{dx} ,$$

where

Q = quantity diffusing,

t = time,

D = diffusion coefficient, and

$\frac{dc}{dx}$ = gradient.

The term dQ/dt of Fick's law is the rate of diffusion or flux. A rate is defined as a process per unit of time. Hence the rate of diffusion, dQ/dt, can be expressed in units of grams (or moles) per second. A flux is defined as the diffusion crossing a unit surface area per unit time. Hence flux (also dQ/dt) is in units of grams (or moles) $cm^{-2}s^{-1}$. Flux is the more usual means of expressing diffusion.

2.6.3 The concept of a gradient

The concept of a gradient is important for the understanding of diffusion processes. By definition, a gradient is a concentration difference between two specified points. It need not, of course, be a concentration difference, but it could be a pressure difference (in the case of a gas) or any other parameter indicative of free energy.

As an example, if we start at some source, say $c_1 = 20$ g water L^{-1} and a sink $c_2 = 6$ g L^{-1}, the concentration difference between the source and sink would be the difference, $\triangle c$.

$$\triangle c = c_1 - c_2 = 20 - 6 = 14$$

To define the gradient, we must know the distance between c_1 and c_2 in some convenient units. If, for example, c_1 and c_2 were two points within a tube 10 cm apart, the gradient would be the concentration at c_1 minus the concentration at c_2 divided by the distance (x), 10 cm:

$$\frac{c_1 - c_2}{x} = \frac{20 - 6}{10} = 1.4 \text{ g L}^{-1} \text{ cm}^{-1}.$$

In other terms, as defined in the expression for Fick's law, the gradient is dc/dx, where c is the concentration and x is the distance. Here it is not necessary to assume that the gradient is necessarily linear (that is, that there is a straight line drop from c_1 to c_2).

The gradient can be illustrated graphically to show the notion of steepness. The two curves in Fig. 2.8 illustrate two gradients, one steeper than the other. It should be apparent that the steeper the gradient, the greater the diffusion rate.

2.6.4 The diffusion coefficient

A thorough comprehension of the factors making up the diffusion coefficient is useful in understanding the diffusion process. In simple terms, D can be defined as

$$D = \frac{K \cdot T}{\sqrt{m} \cdot \eta} \,,$$

where

$K =$ a constant;

$T =$ absolute temperature;

$m =$ molecular weight of the diffusing substance; and

$\eta =$ viscosity of the medium through which diffusion occurs.

The term K is a combination of physical constants that are independent of the diffusing substance.

According to the equation, D is a function of temperature. As the temperature increases, D increases and so does the diffusion rate. Temperature directly influences molecular motion and therefore kinetic energy.

D is inversely proportional to the viscosity of the medium. As the viscosity of the medium through

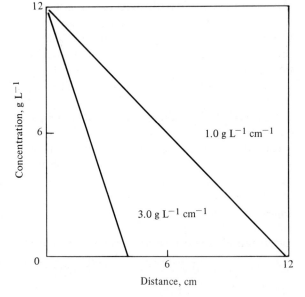

FIGURE 2.8 Diagram to illustrate the concept of steepness of gradient. A gradient is the change in concentration (or comparable parameter) over some distance (dc/dx). The lower curve, depicting a gradient of 3.0 g L^{-1} cm^{-1}, is steeper than the upper curve, with a gradient of 1.0 g L^{-1} cm^{-1}.

which the diffusion is occurring increases, the diffusion rate decreases. The increasing viscosity simply interferes physically with the diffusing molecules. As explained earlier in this chapter, viscosity decreases with increasing temperature.

It is important to keep in mind that the viscosity terms refer to the medium and not to the substance diffusing. Viscosity cannot be a factor for the diffusing substance since viscosity is resistance to flow and is not a factor in diffusion. Flow is movement independent of diffusion.

Water is about 55 times more viscous than air, meaning that substances will diffuse about 55 times faster through air than through water. Such a consideration is important for studies of gas transport in leaves. Without the many large air spaces, gas transport from the atmosphere and to the atmosphere would be slow indeed. Most of the leaf mesophyll cells are in direct contact with air spaces.

D will vary inversely with the density of the diffusing material. The more dense a substance is, the slower it will move. In the expression, relative density is estimated by the square root of the molecular weight of the diffusing substance. Water vapor ($m = 18$ g mol^{-1}) will diffuse faster than CO_2 ($m = 44$ g mol^{-1}). The relative rate of diffusion for water in comparison with CO_2 is 1.56:

$$\frac{\sqrt{44}}{\sqrt{18}} = 1.56.$$

2.7 Centrifugation

Techniques to separate (purify) proteins, particles, and cellular organelles involving centrifugal methods are common in biology. In sedimentation-rate procedures, substances are sedimented differentially according to their relative sizes and shapes. The actual rate of sedimentation in a fluid will be a function of the difference in density between the particle and the centrifugation medium, the viscosity of the medium, the shape of the particle, and the centrifugal force. Thus large proteins will sediment faster than small proteins, and large chloroplasts can be sedimented quite readily in the presence of smaller mitochondria and other cellular particles.

The rate of sedimentation (dx/dt) is given by

$$\frac{dx}{dt} = s \cdot \omega^2 \cdot x,$$

where

x = distance from center of centrifugal rotation;

ω = angular velocity in units of radians per unit time; and

s = sedimentation coefficient, which is a function of the size of the particle.

Because most proteins have s-values between 1 and 100×10^{-13}, s is redefined as the Svedberg unit, $S = 10^{13}s$. Svedberg units are convenient to describe proteins and other subcellular particles (for example, see Chapter 9).

Figure 2.9 is a graph of the relative rates of sedimentation of chloroplasts, microbodies, and mitochondria isolated from spinach leaves and centrifuged in sucrose. Calculated Svedberg units (S) from these data are: chloroplasts = 30,500; microbodies = 9000; and mitochondria = 3000. Table 2.3 compares some representative S-values for some common plant components.

Alternatively, particles can be separated or purified by density-gradient centrifugation in which centrifugation is through a medium of increasing density. If the medium density brackets that of the density of the particles being centrifuged, the particles will come to their own densities within the gradient. As an example, in a sucrose density gradient, spinach chloroplasts will band at a den-

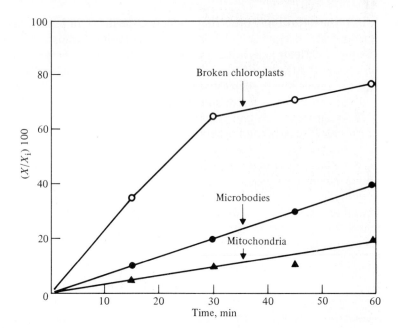

FIGURE 2.9 Graph illustrating the rate of sedimentation of chloroplasts, microbodies, and mitochondria. The ordinate is the relative distance that the organelle has penetrated into the centrifuge tube $[(X/X_i) \ 100]$, plotted against the time of centrifugation. The sedimentation rate is greater for chloroplasts than for microbodies, and the rate for microbodies is greater than for mitochondria. The sedimentation rate is largely a function of the size of the organelles. Modified from an experiment by V. Rocha and I. P. Ting. 1970. Preparation of plant organelles from spinach leaves. *Arch. Biochem. Biophys.* 140:398–407.

sity of $1.17 \ \mathrm{g \ cm^{-3}}$, mitochondria will band at $1.21 \ \mathrm{g \ cm^{-3}}$, and microbodies will band at a density of $1.24 \ \mathrm{g \ cm^{-3}}$. The photograph shown in Fig. 2.10 illustrates the separation and purification of chloroplasts, mitochondria, and microbodies from spinach leaves by sucrose-density-gradient centrifugation.

An effective centrifugation technique to separate plant cell organelles and subcellular particles is through a combination of sedimentation-rate centrifugation and equilibrium-density centrifugation. A sedimentation (S)–density gradient (ρ) graph (called an S–ρ plot) can be used as an aid in designing separation techniques. Here the equilibrium density of a particle is graphed against its sedimentation rate (see Fig. 2.11). As can be seen from the S–ρ graph in Fig. 2.11, it would be difficult to separate intact chloroplasts from mitochondria by equilibrium density alone when in sucrose since both band at about $1.21 \ \mathrm{g \ cm^{-3}}$. However, a combination of sedimentation rate and equilibrium density is an effective means of separation. The sedimentation rates of mitochondria and intact chloro-

TABLE 2.3 Some representative Svedberg units (S) for a few substances

Substance	S	Mass (daltons)
Trypsin inhibitor (soybean protein)	2.3	24,000
Peroxidase (enzyme from horseradish)	3.5	40,000
Lipoxidase (enzyme from soybean)	5.6	98,000
Ascorbate oxidase (enzyme from squash)	6.9	146,000
γ-globulin (protein of barley)	8.7	210,000
Legumin (protein of pea)	12.6	330,000
RNA (*E. coli*)	16.7	500,000
RNA (*E. coli*)	23.0	1,120,000
DNA (bacteria)	30	19,000,000
Tobacco mosaic virus	185	31,340,000

Data from Sober (1970).

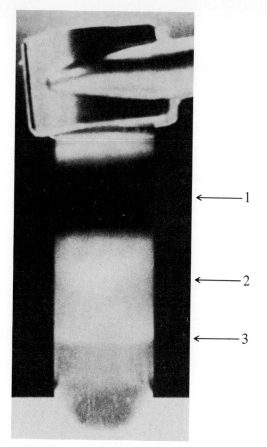

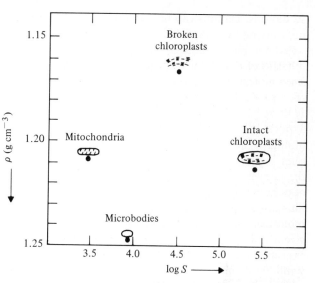

FIGURE 2.11 A sedimentation rate (s)-equilibrium density (ρ) graph or $S - \rho$ graph to visualize the separation of organelles and subcellular particles by a combination of sedimentation-rate and equilibrium-density centrifugation. First, a preliminary separation is conducted by sedimentation-rate centrifugation, and a final separation is then conducted by equilibrium-density centrifugation. See text for details.

FIGURE 2.10 Density-gradient centrifugation of chloroplasts, mitochondria, and microbodies in a linear sucrose gradient prepared from 40% to 80% sucrose. The broken chloroplasts (1) band at a sucrose density of 1.17 g cm⁻³, the mitochondria (2) band at a density of 1.21 g cm⁻³, and the microbodies (3) band at a density of 1.24 g cm⁻³. A combination of sedimentation-rate centrifugation and equilibrium-density centrifugation is an effective means of separating a variety of subcellular organelles and particles.

plasts are vastly different. The intact chloroplasts can first be separated from the mitochondria by sedimentation-rate centrifugation. The mitochondrial fraction obtained by sedimentation rate will contain many intact microbodies (see Fig. 2.11). Mitochondria can then be separated from microbodies by equilibrium-density centrifugation.

2.8 Molecular Size

Molecules and even larger, more complex biological substances such as viruses, chromosomes, ribosomes, and nucleic acids and proteins can be described on the basis of their size. To be precise, they can be described on the basis of their actual mass (Edsall, 1970). By agreement of the International Union of Pure and Applied Chemists (IUPAC), the fundamental unit of mass is the dalton. The dalton is defined relative to the isotope of carbon, ^{12}C. A dalton is a unit of mass equivalent to 1/12 of the mass of ^{12}C. Thus the mass of ^{12}C is

12 daltons. Daltons can be converted to grams by multiplying by 1.66×10^{-24} grams.

Perhaps more frequently we describe molecules in terms of molecular weights. But IUAPC has defined molecular weight very precisely to be the mass of a particular substance relative to the mass of $1/12$ of ^{12}C. Since molecular weight is a ratio, it is a pure, dimensionless number. More frequently, biologists actually mean molar mass (M) when referring to molecular weight where molar mass is the numerical equivalent of molecular weight (and of daltons) but in units of grams per mole. By way of illustration, glucose has a mass of 180 daltons, a molecular weight of 180, and a molar mass of 180 g mol^{-1}. Throughout this text, actual mass will be expressed in units of daltons, and molecular weight, symbolized by M, will mean molar mass and be expressed in units of g mol^{-1}.

In biology, when is dalton used and when is molecular weight used? Many substances of biological interest can be described in terms of mass but not in terms of molecular weight since they are not true molecules. Viruses (composed of protein and nucleic acids), complex proteins, and other polymeric substances such as nucleic acids have water and ions associated with them. Even more complex substances such as chromosomes and ribosomes can be isolated intact and studied. If their size or mass is determined, it should be expressed in units of daltons because they are not true molecules. Molecular weight should be reserved for actual molecules. It should be apparent that the phrase "substance X has a molecular weight of Y daltons" is not a valid statement. As defined here, substance X has a mass of Y daltons and a molecular weight of Y g mol^{-1}. Molecular weight should only be used to describe substances for which there are actual molecular formulas. Daltons can be used for any substance since they represent an expression of mass.

There are many methods available for the determination of mass and molecular weight of biological substances. The most commonly used are estimation from the measurement of colligative properties, estimation from x-ray diffraction studies, calculation from sedimentation rates in an ultracentrifuge, electrophoretic gel techniques that separate partially on the basis of size as well as on charge, chromatographic techniques using sieving gels that separate on the basis of size, light-scattering methods, and direct chemical analysis. Different methods will give quite different results. A first consideration might be the difference between the molecular weight and the formula weight. The formula weight can be determined by adding the atomic weights of the atoms making up the formula of the compound. If the formula is also the true molecular formula, the formula weight is also the molecular weight. Those molecules in which the bonding is covalent ordinarily maintain the same molecular weight regardless of whether they exist as gases, liquids, or solids. However, such substances as salts do not have a constant molecular formula but vary depending on the physical state. Similarly, many of the important biological polymers such as proteins, lipids, and nucleic acids may each have a simple chemical formula, but they exist as salts, polymers, or aggregates in the natural state. Thus molecular weights are not true formula weights.

A second consideration is the difference between a number-average and a weight-average molecular weight, depending on the method of determination. As an example, molecular-weight determinations on the basis of colligative properties give a number-average estimate. This is true because colligative properties depend on the number of molecules, which is the definition of a colligative property. If the molecular weight of a substance is determined first by a colligative property and then by light scattering, a different estimate may be obtained. Light scattering depends on the weight of substances, and the estimate is a weight average.

From the standpoint of the estimation of the size of most polymers and complex substances of biological interest, the difference between molecular weight and mass becomes very important. It is possible, for example, to estimate the size of a polypeptide by the rate at which it passes through a sieving gel. The sieving-gel beads are formed by polymerization of a carbohydrate material. The greater the extent of polymerization, the smaller the pores and interstices of the gel. Small proteins will enter the gels with small pores and large proteins will flow around them. Thus a mixture of small proteins and large proteins can be separated by passing the mixture through a column contain-ing a sieving gel. Large proteins will elute faster than will small proteins. Such columns can be calibrated with proteins or other substances of known size. Since the proteins coming through the column will be hydrated and will have bound ions, the mass can be estimated but the molecular weight cannot. Here the determination would ordinarily be expressed in units of daltons. However, if the same protein was degraded into its component

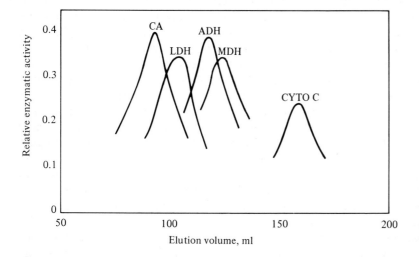

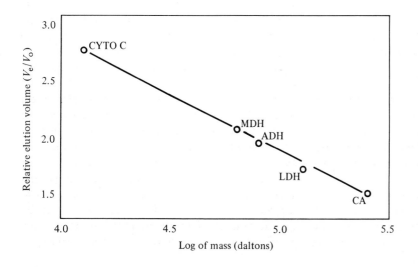

FIGURE 2.12 Separation of proteins with the use of a sieving gel. In the upper panel the elution profile of five proteins of different size is shown. The sequence of elution shown in the graph (upper panel) is CA (catalase, 250,000 daltons), LDH (lactate dehydrogenase, 136,000 daltons), ADH (alcohol dehydrogenase, 84,000 daltons), MDH (malate dehydrogenase, 66,000 daltons), and CYTO C (cytochrome c, 12,500 daltons). When the relative elution volume (V_e/V_o), determined from the volume required to wash a substance from the column (V_e) divided by the void volume (V_o) of the column (see text), is graphed against the logarithm of the mass in daltons, a straight line is obtained (see lower panel). This curve can be used to estimate the size of an unknown protein or other substance once its V_e/V_o is determined experimentally. From an experiment by Victor Rocha, 1970, University of California, Riverside.

amino acid residues, the molecular weight could be estimated either from a knowledge of the exact chemical formula or by assuming that the average molecular weight of an amino acid is about 100 and multiplying by the number of amino acid residues. Such a method will give an estimate of the molecular weight. Similarly, by knowing the number of base residues of a nucleic acid an estimate of the molecular weight can be made. If the determination is made on the basis of gel filtration or centrifugation, mass rather than molecular weight is estimated. Here the units of expression should be daltons, not grams per mole.

The very large biological entities such as ribosomes, membrane components, and organelles can be sized as well, but here we do not ordinarily even use mass in units of daltons. When the determination is by centrifugation the comparative unit is the Svedberg unit, as discussed above under centrifugation.

Figure 2.12 illustrates the elution of proteins from a sieving-gel column. The upper panel shows the separation of catalase, lactate dehydrogenase, alcohol dehydrogenase, malate dehydrogenase, and cytochrome c on the basis of their size. Graphed is the activity of each enzyme in the elution volume as the proteins are washed from the column with buffer. Catalase, with a mass of about 250,000 daltons, elutes from the column faster than lactate dehydrogenase, with a mass of about 136,000 daltons. Similarly, alcohol dehydrogenase, which is 84,000 daltons, elutes after lactate dehydrogenase but before malate dehydrogenase, with a mass of about 66,000 daltons. Cytochrome c, the smallest of the set of 12,500 daltons, elutes last.

The elution profile of a set of proteins with known masses or molecular weights can be used to calibrate a column to be used to estimate the size of a previously unknown protein. Usually a curve is prepared in which the relative elution volume is graphed against the logarithm of the mass in daltons (see lower panel of Fig. 2.12). The relative elution volume is expressed as the ratio of the actual elution volume (taken as the midpoint of the peak) (V_e) to the void volume of the column (V_o). The void volume is the volume of buffer or other fluid required to wash a substance through the column that is completely excluded from the gel beads. It is equal to the volume in the column outside of the beads. Provided that the proteins are approximately spherical, the graph of V_e/V_o plotted against log (daltons) will be a straight line.

Once the sieving-gel column is calibrated, the unknown protein is passed through the column and the relative elution volume is determined. From the V_e/V_o of the unknown protein, its size in daltons can be read from the graph. Sieving gels are commercially available with a wide range of pores that will allow separation of substances with molecular weights from less than 100 to over 500,000.

2.9 Conclusion

The biophysical principles discussed in this chapter are basic to an understanding of the physiology of plants. Because the plant is an aqueous body and lives in an environment that contains water, the understanding of the properties of water and solutions is paramount. Water and its properties can be studied *in vitro*, but the student should keep in mind that the solutions of the plant obey the same principles that were outlined above. Salts and small soluble metabolites affect the colligative properties of the aqueous cytoplasm. Many of these metabolites act as buffers and influence most processes, including transport and the myriad of metabolic reactions that are so dependent on pH.

As will be seen in subsequent chapters of this text, much of plant physiology is biophysical. The coupling of the plant to the environment is best understood using biophysical principles. Many of the internal processes, such as transport, the photochemical events of photosynthesis, and growth and development, are primarily biophysical phenomena. In addition, the student may be surprised to learn that much of growth and development is interpreted with biophysical principles, including such basic processes as cellular expansion. Thus biophysics along with biochemistry forms the very foundation of plant physiology.

Review Exercises

2.1 What would be the vapor pressure over a $2m$ sugar solution at $30°C$? What would be the freezing point of a $0.2m$ sucrose solution? Estimate the boiling point of $0.05m$ solution of sodium chloride. Compute the approximate osmotic pressure of a $0.5M$ solution of glucose. State any assumptions that you have made for the above calculations.

2.2 What would be the effective concentration (activity) of sodium in a 1-millimolar (mM) trisodium citrate solution? As the solution becomes more concentrated, how well will calculation of the activity approximate the actual activity?

2.3 Compute the pH of a $0.01M$ solution of glycolic acid, a $0.01M$ solution of hydrochloric acid, and a $0.01M$ solution of sodium hydroxide. Calculate the pH of a solution composed of $0.01M$ sodium acetate and $0.02M$ acetic acid.

2.4 If the equilibrium constant (K) for the hydrolysis of glucose-6-phosphate is 170, what is the standard free energy of oxidation ($\triangle G°'$)? The $\triangle E°'$ for the oxidation of NADH is 1.14 volts:

$$NADH + H^+ \rightleftharpoons NAD^+ + 2H^+ + 2e^-$$

$$\tfrac{1}{2}O_2 + 2H^+ + 2e^- \rightleftharpoons H_2O$$

$$NADH + \tfrac{1}{2}O_2 + H^+ \rightleftharpoons NAD^+ + H_2O$$

Compute the standard free energy of hydrolysis ($\triangle G°'$) from the $\triangle E°'$

2.5 Given that the diffusion coefficient for water vapor in air is $0.24\ cm^2\ s^{-1}$, predict the diffusion coefficients for carbon dioxide (molar mass = 44 g mol^{-1}), oxygen (molar mass = 32 g mol^{-1}), and hydrogen (molar mass = 2 g mol^{-1}) under the same conditions.

References

ANDERSON, N. G. 1966. The development of zonal centrifuges. *National Cancer Institute Monograph* 21. National Cancer Institute, Bethesda, Md.

CRAFTS, A. S. 1968. Water structure and water in the plant body. In T. T. Kozlowski (ed.), *Water Deficits and Plant Growth,* Vol. 1. Academic Press, New York.

DETERMANN, H. 1968. *Gel chromatography.* Springer-Verlag, New York.

EDSALL, J. T. 1970. Definition of molecular weight. *Nature* 228:888–889.

EISENBERG, D., and W. KAUZMANN. 1969. *The Structure and Properties of Water.* Oxford University Press, New York.

FRANKS, F. (ed.). 1972. *Water: A Comprehensive Treatise.* Plenum Press, New York.

MOROWITZ, H. J. 1970. *Entropy for Biologists.* Academic Press, New York.

NOBEL, P. 1974. *Introduction to Biophysical Plant Physiology.* W. H. Freeman, San Francisco.

RICHARDSON, J. A. 1964. *Physics in Botany.* Sir Isaac Pitman and Sons, Ltd., London.

SOBER, H. A. 1970. *Handbook of Biochemistry.* 2d ed. Chemical Rubber Company, Cleveland.

SLATYER, R. O. 1967. *Plant-Water Relationships.* Academic Press, New York.

SMITH, F. A., and J. A. RAVEN. 1979. Intracellular pH and its regulation. *Ann. Rev. Plant Physiol.* 30:289–311.

SPANNER, D. C. 1964. *Introduction to Thermodynamics.* Academic Press, New York.

VIRGIN, H. I. 1953. Physical properties of protoplasm. *Ann. Rev. Plant Physiol.* 4:363–382.

WEAST, R. C., and M. J. ASTLE (eds.). 1979. *CRC Handbook of Chemistry and Physics.* Vol. 60 (60th ed.). Chemical Rubber Company Press, West Palm Beach, Florida.

ZIMMERMANN, U. 1978. Physics of turgor and osmoregulation. *Ann. Rev. Plant Physiol.* 29:121–148.

CHAPTER THREE

Basics of Photobiology

The natural environment of the plant is full sunlight with radiation of many wavelengths. Much of the physiology of the plant is a direct result of this radiation that strikes the absorbing surfaces. The terminology of radiation is complicated and confusing because of two independent systems that developed. In one, radiation is evaluated from a psychological viewpoint based on the sensitivity of the human eye. This is the photometric system. The other system is based on the energy of radiation independent of visual evaluation. This is the radiometric system. Both have value, but the radiometric system is the preferred system in plant photobiology. In addition, an understanding of radiation becomes difficult because radiation appears to have both wave and particulate properties. But the most important aspect of radiation for the physiology of plants is its interaction with matter, more specifically with the pigments and other chemicals of the plant. This radiation is what drives the many important reactions of the cell, such as those of photosynthesis.

Much of the radiation striking the leaf and other plant surfaces simply heats the leaf (Chapter 7). In this chapter, the terminology of radiation is discussed along with its wavelike and particulate properties. In addition, interactions of radiation with pigments of the plant are discussed along with the use of action and absorption spectra to study photochemical reactions.

The importance of radiation to plants and life in general is readily understood when it is realized that it alone is the source of all energy on earth.

The sun, of course, is the source of radiation that is useful for biological phenomena. The radiation that is discussed throughout this chapter and in subsequent chapters is the radiation created by nuclear reactions in the sun. But even the energy of radioactivity that can be used by us is from the radiation created by nuclear events caused by unstable atoms.

3.1 The Electromagnetic Spectrum

Radiation is composed of a continuous series of waves ranging from very short cosmic rays less than 10^{-14} meters (m) in length to very long radio waves over 10^6 m in length, an effective range of 20 orders of magnitude. This range of radiation waves is called the electromagnetic spectrum. The spectrum is illustrated in Fig. 3.1. It is quite striking to recognize what a small portion of the spectrum visible radiation actually occupies, from 3.9×10^{-7} to 7.7×10^{-7} m, or 390 nm to 770 nm. Neither ultraviolet short waves nor infrared long waves are visible, although people who have had lenses removed from their eyes because of cataracts can see ultraviolet below 390 nm.

The International System of Units (SI = Système International d'Unités) adopted in 1960 dictates that meters (m) are used to measure wavelengths.* Because of the length of radiation waves over the range of biological importance, we normally use 10^{-9} m or nanometers (nm). The archaic millimicrons (mμ) or angstroms (Å) should not be used. Therefore, rather than describing blue light as 5×10^{-7} m, 500 mμ, or 5000 Å, 500 nm will be used.

An alternative expression for wavelength is wave number ($\bar{\nu}$). Wave number is the reciprocal of wavelength in units of m^{-1}. Wave number is frequently used in place of wavelength because it is directly proportional to the energy of the wave. As the wave number increases, the energy associated

with the wave increases. Wavelength is inversely proportional to energy, and energy decreases when wavelength increases.

The energy of a quantum† is given by the following expression:

$$\epsilon = h\nu,$$

where

ϵ = energy associated with a quantum,

h = Planck's constant in units of joule-seconds (6.6254×10^{-34} J-s), and

ν = frequency of the quantum in units of s^{-1}.

Frequency (ν) is related to wavelength (λ) through the fundamental expression

$$c = \lambda\nu \quad \text{or} \quad \nu = \frac{c}{\lambda},$$

where

c = speed of propagation in units of m s^{-1},

λ = wavelength in units of m, and

ν = frequency in units of m^{-1}.

It can be seen from these expressions that energy is directly related to frequency ($\epsilon = h\nu$) and to wave number ($\epsilon = hc\bar{\nu}$) but is inversely related to wavelength ($\epsilon = hc/\lambda$). Table 3.1 relates color, approximate wavelength, and energy of the mean wavelength of each color.

---

*The SI system of unit nomenclature should be used at all times unless there are compelling reasons to deviate. The SI system is described in more detail in Appendix I.

†A quantum (plural = quanta) is defined as the fundamental unit quantity of energy in quantum theory. A photon is the quantum associated with the electromagnetic spectrum. A meson is the comparable unit associated with nuclear fields.

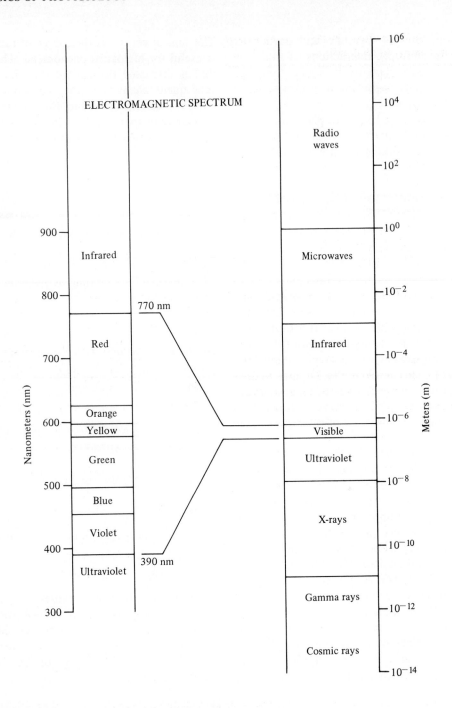

FIGURE 3.1 The electromagnetic spectrum, with an enlarged portion illustrating the wavelength ranges of the various colors.

TABLE 3.1   Visible spectrum with the wavelength range and associated energy in kilojoules

Color	Wavelength range, nm	Energy, kilojoules per einstein
UV	<400	297
Violet	400–425	289
Blue	425–490	259
Green	490–560	222
Yellow	560–580	209
Orange	580–640	197
Red	640–740	172
IR	>740	163

3.2 Radiation Terminology

In general terms, radiation may be defined as energy propagated through space in the form of waves. The terminology is somewhat confusing because of two systems that have developed independently. First, there is a set of terms to describe visible qualities of radiation, such as footcandles, illumination, and lumens. Second, there is a set of terms referring to the energy characteristics of radiation (including the visible), such as joules, watts, ergs, and calories. Unfortunately, these systems are not interchangeable except under defined conditions.

3.2.1 Photometry

Photometric analysis of radiation is based on a visual evaluation. The actual units are based on energy, but they are modified according to the visual sensitivity of the human eye. A "standard eye" as defined by Preston in 1961 is shown in Fig. 3.2 (see Seliger and

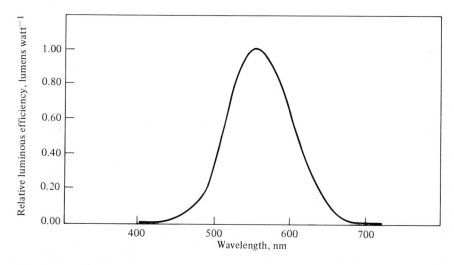

FIGURE 3.2 Graph of the "standard eye" expressed in relative luminous efficiency (lumens per watt). The maximum efficiency is in the green at 556 nm. At 556 nm there are 680 lumens per watt on an absolute scale. The graph illustrates that more watts are required in the blue and red to produce the same luminous power (lumens) than are required in the green. See text for additional information. Adapted from Seliger and McElroy, 1965.

McElroy, 1965, for a detailed development of this concept). The curve of Fig. 3.2 shows a peak in the yellow-green at 556 nm, corresponding to the greatest sensitivity of the light-adapted human eye. This curve tells us that much more red or blue light energy is required to give the same amount of light as green since the eye is much more sensitive in the green than in the red or blue. We know this intuitively since we cannot see beyond violet short waves or red long waves.

Radiant energy is described in three terms:

1. in basic energy units that indicate the amount of energy;
2. as a power (or flux) in units of energy received per unit time; and
3. as the power received per unit area.

In the photometric system, the basic unit of luminous energy is the candela, often referred to as a talbot. A candela has been rigorously defined as 1/60 of the intensity of 1 cm^2 of a blackbody radiator at the temperature of solidification of platinum (2042°K). Thus the candela is defined in terms of the light given off by platinum during solidification.

The power, or luminous flux, is given in units of lumens, a lumen being the flux radiating from a standard candle falling on a unit surface. A lumen is measured in units of talbots per second.

Finally, the power per area (or intensity) is expressed as footcandles; one footcandle is defined as one lumen per square foot. It is the illumination on a surface of one square foot in area with all points receiving a flux of one lumen. In metric units, illuminance or intensity is given as lux; a lux is a lumen per square meter.

An additional term that is frequently encountered is brightness. Brightness is a property of the light emitter and is defined as the flux emitted per area. The common unit is the lambert, defined as the flux emitted by one cm^2 of the source.

In photometry, light energy is defined in a physiological sense. It is important to remember that the light intensity expressed as footcandles or lumens per area cannot be translated in terms of actual energy. It takes more energy in the blue and in the red than it does in the green to produce the same number of lumens, a consequence of being defined in reference to the sensitivity of the human eye.

3.2.2 Radiometry

Except for studies of visual and psychological events, physiologists are rarely interested in the photometric system of defining radiant energy, even when the energy is in the spectral range of visible light. This is because the interest is in how much energy is required for processes as expressed in equal terms

TABLE 3.2 Comparison of radiometric and photometric radiant energy

	RADIOMETRIC UNITS	PHOTOMETRIC UNITS
Radiant (luminous) energy	joule (erg, calorie)	candela (talbot, candle)
Radiant (luminous) power	watt	lumen
Irradiance (illuminance)	watt per square centimeter	footcandle

Definitions—
(1) erg = work for the operation of a force of one dyne through one centimeter of distance (dyne = force required to act on a mass of one gram to cause an acceleration of one cm s^{-1} s^{-1}; (2) joule (J) = 10^7 erg; (3) calorie (cal) = 4.18 J; (4) watt (W) = 10^7 erg s^{-1} = J s^{-1}; (5) candela (cd) = fundamental unit of light energy, 1/60 of the intensity of one cm^2 of the radiation of a blackbody at the temperature of the solidification of platinum (2042°K); (6) lumen (lm) = luminous flux through a unit solid angle (steradian) from one candela (in units of cd s^{-1}): (7) footcandle (fc) = illuminance of an area one square foot all points of which receive one lumen (lm ft^{-2}; 1 lm cm^{-2} = 929 fc); (8) lux = lm m^{-2} (1 lux = 0.093 fc).
  Frequently, radiation is expressed as photon flux, which is a measure of the number of photons per unit area per unit time. Most commonly, the measurement is expressed in einsteins, E (6.023 × 10^{23} photons) or microeinsteins (6.023 × 10^{17}).

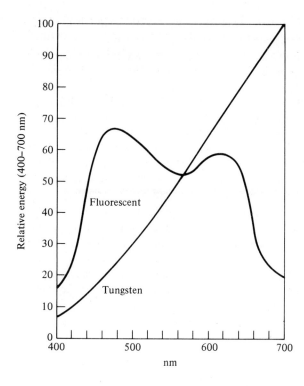

FIGURE 3.3 A comparison of the relative energy output from a "daylight" fluorescent lamp and a tungsten lamp. Whereas the fluorescent lamp has definite peak outputs in the blue and red regions of the spectrum and decreases in the far-red and infrared regions, the tungsten continues to increase in energy output as the wavelength increases. The tungsten lamp puts out more heat energy. The graph illustrates that it is not possible to compare the energy output of the two common types of lamps using a light meter. Adapted from Seliger and McElroy, 1965.

and not in terms relative to visual acuity.

The basic unit of energy in the radiometric system is the joule (J). The joule, therefore, is comparable to the candela. A joule is 10^7 ergs. In addition, we frequently encounter the unit calorie, which is equivalent to 4.18 joules. Thus radiant energy can be expressed as joules, ergs, or calories. Joules is preferred to be consistent with the SI system.

The radiometric unit of power or radiant flux is expressed similarly to luminous flux, as energy per time. Here, a joule per second is defined as a watt. Hence the watt is the unit of power in the radiometric system and is equivalent to the photometric unit of a lumen.

Finally, the intensity or irradiance is expressed as flux or power per area, such as watts cm^{-2} or J cm^{-2} s^{-1}. It is also common to encounter cal cm^{-2} min^{-1} as an expression of irradiance.

In some plant physiological processes such as photosynthesis, the actual number of photons involved in the process is a better index than the total energy involved. As will be explained in a subsequent section, for a photochemical reaction to occur at least one photon must be involved. There is a one-to-one equivalent—one photochemical reaction, one photon. Of course, the efficiency of the reaction can be less than one, but it cannot be greater than one. The unit to express photons is the einstein; one einstein (E) is 6.023×10^{23} photons. More frequently, μE are used ($\mu E = 6.023 \times 10^{17}$ photons). Thus the photon flux is expressed in μE received per unit area per unit time. Most frequently, in photosynthesis studies, the photon flux is expressed as μE m^{-2} s^{-1} over the wavelength range of 400 to 700 nm, which is the photosynthetically active range (PAR).

The units of the radiometric and photometric systems are compared in Table 3.2. The reason for using the radiometric system rather than the photometric system in biological studies can now readily be understood from an inspection of Fig. 3.2. The curve expresses relative luminous efficiency in units of lumens per watt as a function of wavelength of light. At the peak maximum efficiency of 556 nm there are 680 lumens per watt. In the blue at about 450 nm it would require about 20 times more power to produce the same number of lumens since a watt would only result in about 24 lumens. Similarly, in the red at about 650 nm it would require close to 10 times more power to give the same

luminous flux since one watt would only result in about 68 lumens. Much confusion results, therefore, when energy from different light sources is expressed in lumens or footcandles rather than in watts or watts cm^{-2}. Clearly, an equal number of footcandles from a tungsten lamp and from a fluorescent lamp will not result in the same number of watts cm^{-2} (Fig. 3.3).

3.3 The Nature of Radiation

Early research by such great scientists as Newton, Maxwell, and Faraday led to the development of our modern concepts of the dual nature of radiant energy. It was discovered early that radiant energy had both magnetic and electrical properties, and in fact we frequently speak of electromagnetic radiation.

3.3.1 The wave nature of radiation

It was Maxwell who in 1865 clearly defined light (or radiation) as transverse electromagnetic waves (Fig. 3.4). The electrical and magnetic fields are both perpendicular to the direction of the ray and perpendicular to each other.

Radiation in this sense can be described as a wave function whose length and frequency are related to the speed of propagation as described previously:

$$c = \lambda \nu.$$

The constant c (speed) is 3.3×10^8 m s^{-1} in a vacuum. In any medium other than a vacuum, the speed of propagation will be reduced. When c is expressed in units of meters per second, λ is in meters and ν is in reciprocal seconds. The phenomena of refraction, interference, and diffraction are explained in terms of the wave nature of radiation.

Refraction

When radiation strikes a flat surface, it will reflect at an angle equal to the angle of the incident beam. However, when radiation passes from one medium to another of different density, because of a change in speed it will deflect, or in proper terms refract. If the radiation passes from a less dense to a more dense medium (air to water, for example), it will refract *toward* the normal (Fig. 3.5). When passing from a more dense to a less dense medium, it will refract *away from* the normal. The deflection or refraction is greater for short-wave radiation than for long-wave radiation. These features of the

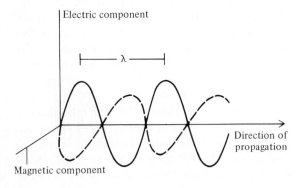

FIGURE 3.4 Diagram illustrating the wave nature of radiation by indicating the magnetic and electric components, each perpendicular to the direction of wave propagation and perpendicular to each other. λ = the wavelength in meters. ν = the number of λ passing a given point per second. The speed of propagation (c) is the product of λ times ν, expressed in units of meters per second.

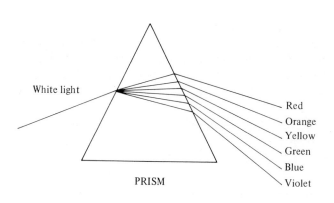

White light

PRISM

Red
Orange
Yellow
Green
Blue
Violet

FIGURE 3.5 Diagram to illustrate how a prism separates white light into component colors of varying wavelengths. When radiation (here the white light) passes from a less dense medium (air) into a more dense medium (the prism), it is refracted toward the normal (perpendicular to the tangent). The shorter the wavelength, the greater the degree of refraction. When the radiation passes from a more dense medium (when leaving the prism) to a less dense medium, refraction is away from the normal. Thus with proper design and correct orientation of a prism, it is possible to separate white light into component colors. The separation is based on the property of refraction.

wavelike nature of light are used to explain the action of a prism in separating wavelengths (Fig. 3.5).

When white light enters a prism (see Fig. 3.5), because of the prism's greater density the light will refract toward the normal. When it leaves the prism the diffraction is away from the normal. Since short-wavelength radiation refracts more than long-wavelength radiation, the prism will tend to separate the light into its component colors.

Diffraction

When two or more radiation waves are at the same place in space at the same time, they interfere. The result is the algebraic sum of the two. If the two are exactly in phase, that is, if the peaks and troughs of the waves coincide, there is a new wave produced with exactly twice the amplitude, a phenomenon termed constructive interference. If the waves are 180° out of phase such that wave peaks correspond with wave troughs, there is destructive interference and the waves cancel each other. This phenomenon of interference is best described with the wave nature of radiation.

Alternate constructive and destructive interference is termed diffraction and can result if a beam strikes a reflecting surface that has a regular array of scattering or reflecting centers. A crystal with a uniform lattice structure is such an object, and with the use of x-rays, crystalline structure can be ascertained by the nature of the diffraction patterns. X-rays are used since their wavelengths are in the range of atomic dimensions.

The x-rays are scattered by the electrons, the amount of scattering being proportional to the density of electrons. In order to determine the structure of the crystal by computation, it is necessary to know the direction of scattering plus the wavelength, amplitude, and the phase. The direction and wavelength can be determined from the position of the spots on the x-ray picture. The amplitude can be estimated from the darkness of the spot. The phase, however, must be determined by more complicated analytical methods. Once the pattern of direction, wavelength, amplitude, and phase is determined, the three-dimensional structure of the molecules of the crystal can be determined by computation.

Figure 3.6 shows a photograph of an x-ray diffraction pattern of a naturally occurring sesquiterpene isolated from a marine red alga (*Laurencia* sp.). The three-dimensional structure of the molecule was determined from the diffraction pattern. The drawing below the diffraction pattern shows a stereoscopic pair, which can be used to view the proposed structure in three dimensions.

Similar x-ray diffraction patterns and subsequent computer analysis have allowed the determination of structure of a variety of biological chemicals, from the complex bromine and chlorine compound shown in Fig. 3.6 to proteins and nucleic acids.

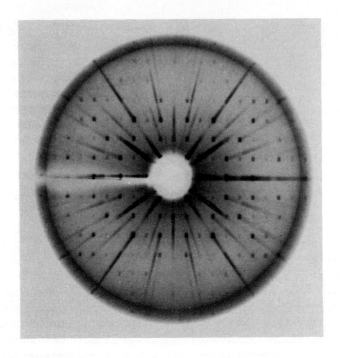

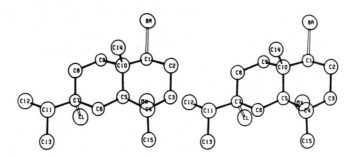

FIGURE 3.6 X-ray diffraction pattern obtained from crystals of selinane (a sesquiterpene) isolated from the red alga, *Laurencia* sp. From the pattern, the three-dimensional structure of the molecule was determined. Note that this natural product contains both bromine and chlorine. The diagram below the x-ray diffraction pattern is a stereoscopic pair drawing of the molecule. If a 3 x 5 file card is placed on edge between the stereoscopic pair and the eyes are focused to bring in a single image, the molecule can be seen in three dimensions. You may also see the 3-D image by staring at each drawing and then crossing your eyes. From an experiment by A. F. Rose, J. J. Sims, R. M. Wing, and G. M. Wiger. 1978. Marine natural products. *Tetrahedron Letters* 29:2533–2536. Photograph of the x-ray diffraction pattern courtesy of Dr. Richard Wing.

Monochromatic light

In many physiological studies, it is necessary to obtain light of narrow bandwidths, i.e., single colors. There may be interest in studying growth phenomena such as germination, flowering, or leaf expansion as a function of light quality, or monochromatic light may be desired for analytical purposes. Many compounds absorb radiation of specific wavelengths (see Section 3.3.6). Such compounds can frequently be analyzed quantitatively by light absorption.

A simple method to obtain narrow-wavelength radiation is with absorption filters. Such filters absorb certain wavelengths and transmit others. They are useful but not very quantitative. Interference filters are somewhat better and work by reflecting all wavelengths except those that are transmitted.

Monochromators isolate narrow wavelengths by using either prisms or diffraction gratings. A glass prism for visible light or a quartz prism for ultraviolet light separates wavelengths by refraction as described above. A diffraction grating is more complex and consists of a grating made of thin, parallel, evenly spaced reflecting strips. The parallel array of reflecting strips causes constructive and destructive interference as described above. From such diffraction patterns the desired wavelengths can be isolated.

Lasers can be used to generate high-intensity monochromatic light. A ruby laser emits light with a wavelength of 694.3 nm, and others are available that even give off ultraviolet light.

Polarized light

Light propagated through space in which the electric fields are all parallel is said to be plane-polarized. Ordinary light has electrical components that are oriented in all directions from the axis. Polaroid lenses and certain types of prisms have the properties necessary for polarizing ordinary light. Plane-polarized light becomes important in biology because asymmetric molecules (molecules with asymmetric centers) have the capacity to rotate plane-polarized light to a new direction.

It is well known that optically active substances can exist in two forms that are mirror images of each other. Those components that rotate plane-polarized light to the right (clockwise) are called dextrorotatory and those compounds that rotate light to the left are called levorotatory.

3.3.2 The particulate nature of radiation

Whereas the wave theory for the nature of radiation best describes electromagnetic energy propagation through space and accounts for the phenomena of refraction, interference, and diffraction, it is best to consider radiation as discrete particles once it interacts with matter. In 1901 Planck concluded that radiation is discontinuous and not wavelike and that it exists as discrete values with energies of $h\nu$. Subsequently, Einstein in 1905 proposed that radiant energy was both absorbed and emitted in discrete steps. As explained in a previous section, each particle of radiation is called a quantum; for the electromagnetic spectrum, the term photon is used.

It should be clear from the previous section that the energy associated with a quantum or photon is equal to $h\nu$. A quantity of energy equal to a mole of photons is referred to as an einstein (E). An einstein is thus Avogadro's number (N_A) times either $h\nu$ or ϵ.

$$E = N_A h\nu = N_A \epsilon$$

The above considerations led to an important principle that applies to the photochemical reactions of plants, the Stark–Einstein Law of the Photochemical Equivalent. This law states that there can be no more than one molecular event or reaction for each quantum absorbed. Such a generalization will become very important for the study of photosynthesis in Chapter 15.

The quantum efficiency (Q) of a photochemical reaction is defined as

$$Q = \frac{\text{molecules reacting}}{\text{quanta absorbed}}.$$

Consistent with the Stark–Einstein law above, the limits of Q are from 0 to 1.0. As Q approaches 1.0, the light-driven reaction increases in efficiency.

3.3.3 Interaction with matter

Excitation

When electromagnetic radiation interacts with matter, several things may occur. Prior to discussing these and their implications for physiology, it is constructive to state the Grotthuss–Draper law, which states that for a photochemical change to occur, a photon must be absorbed. As already mentioned, the Stark–Einstein law limits these events to a one-to-one relationship, that is, one absorption cannot result in more than one event.

Very-high-energy radiation of short wavelengths in the range of cosmic, gamma, and x-rays can bring about actual ionization or destruction of chemical bonds. Fragments or free radicals produced by such ionizing radiation may then participate in further chemical reactions because of the high reactivity of such radicals. It is this electromagnetic radiation of high energy that can cause mutation by means of interaction with nucleic acids.

Long-wave, low-energy radiation such as radio waves can result in rotational and vibrational molecular changes, but they are not of much importance to plants. Molecular changes brought about by infrared radiation with wavelengths greater than about 800 nm result in heat.

Important to our considerations of photobiological reactions of plants is the electromagnetic radiation falling within the wavelength range of 300 to 800 nm. Photons of the proper frequency can be absorbed and result in electronic changes of great biological importance. It is the valence electrons within the outer orbital that are most affected since they can be excited by energy associated with visible photons.

After excitation, several events may occur. First, the electron may return to its original "ground state" with the liberation of energy (that is, reemission of radiation) by heat or light. The latter can be rapid, in the form of fluorescence, or delayed, as in phosphorescence. These events are illustrated in Fig. 3.7.

More important for physiology, however, is the increased reactivity of the excited molecule. The elevation of an electron to a higher-energy orbital after absorption of a photon makes the molecule quite reactive because of the higher-energy state. In fact, we can think of it as a reductant that can donate electrons, a concept that becomes very important in photosynthesis. Since the light-excited molecule is at a higher energy level, it is more reactive and participates in thermal reactions that would ordinarily require more heat.

Energy loss

HEAT

Figure 3.7 illustrates energy changes associated with photon absorption. When a photon with the proper energy level interacts with a pigment (a molecule that absorbs photons), one of the valence electrons is raised to a higher-energy orbital. The excitation is quantized such that exactly the proper amount of energy must be associated with the photon to raise it to the first excited state, termed the first singlet. A greater increment of energy would have to be present to raise the electron to the second excited state, or second singlet.

After excitation the higher-energy electron can be donated in a reaction, leaving an "electron hole," or the energy may be transferred by vibrational and resonance transfer to another molecule, in which case the electron will drop back to a lower-energy orbital in a radiationless transfer. Similarly, it may drop back to a lower-energy orbital (ground state), with energy dissipation in the form of infrared radiation or heat. In this case nothing of photochemical consequence happens.

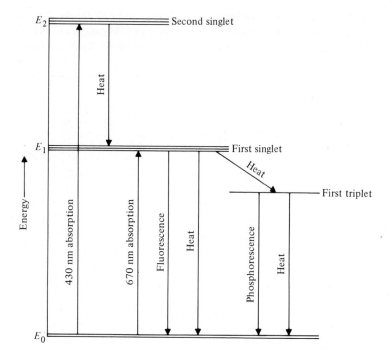

E_2 ══════ Second singlet

Heat

E_1 ══════ First singlet

Heat

First triplet

Energy →

430 nm absorption

670 nm absorption

Fluorescence

Heat

Phosphorescence

Heat

E_0 ══════

FIGURE 3.7 Simple diagram to illustrate some of the possible energy transitions after a pigment molecule absorbs radiation. The molecule used as an example is chlorophyll with blue (430 nm) and red (670 nm) absorption maxima. After absorption of the radiation, the molecule is raised from the ground-state energy level (E_0) to a higher energy level (E_1 or E_2) depending on the energy of the absorbed radiation. The energy can be dissipated in the form of heat radiation, visible light radiation, fluorescence, or delayed phosphorescence. The singlet and triplet levels are energy levels representing higher-energy orbitals. The triplet differs from the singlet in that the electron spin has changed. See text for details.

LIGHT

After excitation, many pigments return to ground state from the excited singlet with the emission of light energy, a process termed fluorescence. In the simplest case, the fluorescence emission would be of the same wavelength as that absorbed. However, in complex pigment molecules there are other losses of energy such that the emission is at a longer wavelength of less energy. Hence the absorption spectrum (that is, a graph of light absorption versus wavelength) will be at a shorter wavelength (higher energy) than is the emission spectrum.

The electrons within an orbital of a stable molecule are paired and assumed to be identical except for the fact that they are spinning in opposite directions. Excitation of an electron by photon absorption results in elevation of the electron to a higher-energy orbital. These higher-energy orbitals are called singlet states. The electron in the singlet state differs from the electron in the remaining ground state by simple virtue of being farther away from the nucleus and in a higher energy state. The two paired electrons now separated by energy state still retain their respective spins.

It is possible, however, that the spin of the excited electron may change such that it is identical to the other member of the pair. This is a somewhat unlikely event accomplished by some loss of energy (see Fig. 3.7). The state attained is called a triplet state. Because it is an unlikely or an improbable event, the probability of return to ground state is diminished. Hence it is somewhat more stable. The triplet-state electron thus remains for a longer period than does the singlet state. When the triplet-state electron returns to ground state it may be accompanied by light emission. This delayed light emission is termed phosphorescence and is of a longer wavelength than fluorescence.

3.3.4 Photochemistry

Obviously, if the pigment molecules lose energy by light emission (either fluorescence or phosphorescence), little can be accomplished in terms of useful work. Yet many pigment molecules of biological systems are fluorescent. The prevention of light emissions by interaction of excited pigment with either solvent or other solute molecules is termed quenching. Therefore, quenching becomes extremely important for the useful work of excited biological pigments since it is in this way that ground state can be reestablished without energy dissipation in the form of useless radiation.

Photochemical reactions occur when the energy of the excited pigment (such as chlorophyll) is used to drive biochemical reactions. As will be discussed in Chapter 15, such energy is used to make ATP and reduce NADP to NADPH. Both ATP and NADPH are high-energy compounds used in energy-requiring biochemical reactions. When these photochemical reactions take place in the cell, fluorescence is quenched.

3.3.5 Absorption Spectra

A graph of light absorption as a function of wavelength is an absorption spectrum. Because of the Grotthus−Draper law requiring absorption of photons before reactions, absorption spectra of biological pigments give useful experimental information about photochemical reactions.

The Beer−Lambert law quantitatively describes light absorption by pigments. The amount of light (I) passing through a pigment solution is related to the amount of light striking the solution (I_o), the thickness of the solution (ℓ), the concentration of the pigment in solution (C), and a factor related to the degree that the pigment absorbs the light, the extinction coefficient (ϵ).

In terms of the calculus, light attenuation as it passes through a solution is described as follows:

$$\frac{dI}{d\ell} = -\epsilon C I.$$

Integration gives $2.302 \log I_o/I = \epsilon \ell C$, where $2.303 \log I_o/I$ is defined as the absorbance, A. Hence the absorbance or absorption of light by a solution is

$$A = \epsilon \ell C.$$

A is related to the absorption coefficient (ϵ), the thickness of the solution (ℓ), and the concentration (C). It can be seen that the concentration of a pigment in solution can be ascertained if A, ℓ, and ϵ are known.

A graph of absorbance against wavelength is an absorption spectrum. An absorption spectrum for the green pigments of plants, the chlorophylls, is shown in Fig. 3.8. It can be seen that chlorophyll absorbs light throughout the visible spectrum but has definite absorption peaks at 670 and 430 nm, in the red and blue. The "shoulders" and smaller peaks are important to the processes of photosynthesis, but we can guess from the absorption spectrum that red and blue light are probably more important than other kinds.

Absorption spectra are frequently determined at the temperatures of liquid nitrogen or liquid helium. The low temperatures reduce vibrational effects at the molecular level that tend to make absorption peaks broad. At very low temperatures, the peaks of the spectrum are quite sharp.

Absorption spectra are very useful in analytical research. Since absorption spectra are characteristic for molecules that absorb light, they can be used for precise identification. The cytochromes, for example, are differentiated on the basis of their absorption spectra. In addition, absorption spectra can be used to advantage in measuring rates of reactions if the spectra change during the course of the reaction. The pyridine nucleotides, NAD and NADP, have an absorption peak at 340 nm when reduced (refer to Fig. 8.12). When oxidized, the 340-nm absorption band disappears. Reactions involving either the oxidation or reduction of these two pyridine nucleotides can be easily followed with an ultraviolet spectrophotometer capable of measuring radiation at 340 nm. Many other reac-

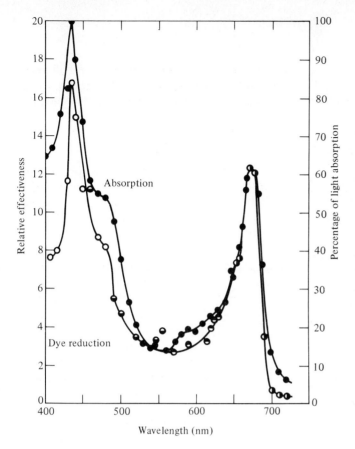

FIGURE 3.8 Graph illustrating an absorption spectrum and an action spectrum. The spectra are from experiments performed on the chloroplasts isolated from Swiss chard (*Beta vulgaris*). The absorption spectrum was determined by measuring the amount of light absorbed by a chloroplast suspension at varying wavelengths of light. The action spectrum for photosynthesis was determined by measuring the reduction of an oxidation–reduction dye (dichlorophenol indophenol) during the oxygen-evolving step of photosynthesis. The greater the reduction of the dye, the greater the photosynthesis. It is apparent from the graph that both absorption of light and photosynthesis are at a maximum in the blue (430 nm) and in the red (670 nm). Such an experiment is evidence for chlorophyll being the pigment that absorbs the light energy for photosynthesis since both the light absorption and the process (photosynthesis) have maxima at the same wavelengths of light. From S. L. Chen. 1952. The action spectrum for the photochemical evolution of oxygen by isolated chloroplasts. *Plant Physiol.* 27:35–48.

tions of biological importance are also measured in this manner.

Molecules that give off radiation have an emission spectrum. An emission spectrum is similar to an absorption spectrum except that it is an indication of the wavelength of light given off rather than that which is absorbed. Many molecules will emit light after absorbing light (fluorescence or phosphorescence). The emission spectrum is shifted to a longer wavelength (less energy) in comparison with the absorption spectrum.

3.3.6 Action spectra

Action spectra relate physiological or other processes to the wavelength of radiation. They are of interest because they relate light quality to the process. In addition, if action spectra and absorption spectra correspond it is possible to assign a pigment to a physiological process or function.

The construction of a proper action spectrum is a difficult task. It is necessary to conduct a series of experiments, each at a different wavelength of light, to obtain an index of the reaction expressed relative to the amount of light energy given. As an example, an action spectrum for photosynthesis could be obtained by measuring photosynthesis by a convenient method such as CO_2 uptake, O_2 evolution, or some other index as a function of energy. Thus it is necessary to obtain photosynthesis per energy unit at different wavelengths. The final spectrum will be a graph of photosynthesis per energy unit plotted against wavelength (an expression of relative effectiveness). Such a graph is illustrated in Fig. 3.8. The peaks on the graph indicate the greatest efficiency of photosynthesis and, according to Fig. 3.8, the greatest efficiency of photosynthesis occurs in the red at 670 nm and in the blue at 430 nm.

An alternative way to express the action spectrum for photosynthesis would be to graph the quantum efficiency (Q) against wavelength. The greatest quantum efficiency, that is, the greatest reaction with the least amount of light energy, would occur at 670 nm and 430 nm.

Comparison of the absorption spectrum for chlorophyll with the action spectrum for photosynthesis shown in Fig. 3.8 suggests that chlorophyll is the pigment that absorbs the light during photosynthesis. Both chlorophyll absorption and photosynthesis have major peaks in the blue and red regions of the spectrum.

Review Exercises

3.1 Explain why electromagnetic energy is sometimes considered to be particulate in nature and is sometimes considered to be a wave function.

3.2 Convert the following into centimeters: 50 nm, 640 μm, 800 mμ, 1000 Å.

3.3 Compute the energy per quantum for electromagnetic radiation of the following wavelengths: 200, 400, 600, and 1000 nm. Express your answer in ergs, joules, and calories. What is the energy in joules per einstein?

3.4 Chlorophyll has an extinction coefficient (ϵ) of 36 ml cm^{-1} mg^{-1} at 652 nm when in 80% acetone. If the absorbance (A) of a solution of chlorophyll measured in a 1-cm cell is 0.32 A units, what is the concentration of chlorophyll in the solution? If the original extract was 50 ml obtained from 1 g of leaf tissue, what is the chlorophyll content in mg chlorophyll per g of tissue? If the leaf has a specific area of 0.02 g cm^{-2}, what is the chlorophyll expressed on an area basis?

3.5 If the potential difference between hydrogen and oxygen is 1.2 electron volts, is there enough energy in red light to oxidize water during photosynthesis? There are 92 kilojoules per electron volt equivalent.

References

BAINBRIDGE, R., G. C. EVANS, and O. RACKMAN. 1966. *Light as an Ecological Factor.* Blackwell Scientific Publ., Oxford.

BICKFORD, E. D., and S. DUNN. 1972. *Light for Plant Growth.* Kent State University Press, Kent, Ohio.

CLAYTON, R. K. 1970. *Light and Living Matter,* Vol. 1, *The Physical Part.* McGraw-Hill, New York.

CLAYTON, R. K. 1971. *Light and Living Matter,* Vol. 2, *The Biological Part.* McGraw-Hill, New York.

DOWNS, R. J., and H. HELLMERS. 1975. *Environment and the Experimental Control of Plant Growth.* Academic Press, New York.

GRUDINSKII, D. M. 1976. *Plant Biophysics.* Israel Program for Scientific Translations, Jerusalem.

NORRIS, K. H. 1968. Evaluation of visible radiation for plant growth. *Ann. Rev. Plant Physiol.* 19:490–499.

OSTER, G. 1968. The chemical effects of light. *Sci. Amer.* 219:158–170.

SELIGER, H. H., and W. D. McELROY. 1965. *Light: Physical and Biological Action.* Academic Press, New York.

SMITH, K. C., and P. C. HANAWALT. 1969. *Molecular Photobiology.* Academic Press, New York.

WEISSKOPF, V. F. 1968. How light interacts with matter. *Sci. Amer.* 219:60–71.

WOLKEN, J. J. 1968. *Photobiology.* Reinhold, New York.

CHAPTER FOUR

Plant Water Relations

4.1 Plant Water

Water is the single most important factor on earth for the life of all living organisms. It is water that limits plant growth in virtually all environments. Slight reductions in water availability result in reductions in such important physiological functions as photosynthesis and respiration. Furthermore, the hydration of tissue is a necessary requirement for actual cell expansion during growth. The importance of water for plant growth can be discussed in three categories: hydration, as a chemical, and as a solvent.

As discussed in Chapter 2, because of hydrogen bonding water will bond to other molecules that have oxygen or other groups that are electronegative. The tertiary structure of many polymeric molecules such as proteins and nucleic acids is a partial function of their water sheath. The reactivity of smaller molecules is modified by hydration.

Protoplasm itself is 80% to 90% water. Drying of protoplasm significantly alters its properties, eventually to the point of irreversible coagulation.

81

Free water of hydration is equally important since it is the hydrostatic pressure of this water that maintains turgidity and partially maintains the form of tissues. Hydrostatic pressure is absolutely necessary for growth since it is this pressure that is the force for cell enlargement. Water-deficient cells tend to be small and compact rather than large and succulent.

4.1.2 Water as a chemical

Water actually enters into biochemical reactions of metabolism as a substrate. The all-important reaction of photosynthesis, which evolves oxygen, results because of the photochemical splitting of water to yield electrons for reducing CO_2 to carbo-

hydrate. Water acts as a hydroxyl (OH) donor for some hydroxylation reactions.

The most common biochemical reaction of water is hydrolysis. Hydrolysis reactions are important in catabolic processes such as fat degradation, protein metabolism, and polysaccharide breakdown.

4.1.3 Water as a solvent

Finally, water functions as a solvent. All life is aqueous and most of the chemical reactions of life take place in aqueous solutions. Protoplasm is an aqueous system. Those few events occurring in hydrophobic regions, such as in membranes, are important from a metabolic standpoint by virtue of water exclusion.

4.2 Free Energy of Water

From the standpoint of plant water relations, one of the most important questions concerns the transport of water, as illustrated by transport from soil to roots, from roots to stems, leaves and the entire plant body, and ultimately from the evaporation surfaces to the atmosphere.

There is essentially only one principle involved. Simply stated, it is that water will move from a region of high free energy to a region of lower free energy. The free energy spoken of is that energy available for work and is defined more rigorously as the partial molal Gibbs free energy.

The term partial molal free energy requires some explanation. In the free-energy concept, each molecule has a total internal energy equal to its kinetic and potential energy. The free energy is the useful energy that, under appropriate conditions, is available for work. The molal free energy is equal to the mean free energy of a molecule times the number of molecules per mole. The term "partial" comes from the fact that thermodynamic quantities are composed of many variables. Here we are referring to the free energy per mole when all other factors such

as temperature are held constant. Thus we use the term "partial" to designate this energy.

In the following discussion, the principles of plant water status will be developed in thermodynamic terms. Nevertheless, keep in mind that, in the simplest terms, "water runs downhill."

The partial molal free energy as defined here can also be expressed as the difference between water affected by chemical, electrical, gravitational, pressure, or other forces and free, pure water.

$$\mu_w - \mu_w° = RT \ln e - RT \ln e°$$

or

$$\Delta \mu_w = RT \ln \frac{e}{e°},$$

where

μ_w = chemical potential of water in question (in J mol^{-1}),

$\mu_w°$ = chemical potential of pure water,

R = gas constant,

T = absolute temperature, and

e and $e°$ = vapor pressures of the water in question and pure water, respectively.

Note that $e/e° \times 100$ is the definition for relative humidity.

Two features of the expression worthy of thought are:

1. If e and $e°$ are equal, that is, if e is also pure water, then $\ln e/e°$ is zero and $\Delta\mu$ is zero. Thus by definition we set the chemical potential of pure water (relative to pure water) equal to zero.

2. If e is less than $e°$ as will be the case if it is not equal to $e°$, then $\ln e/e°$ will be a negative number. Therefore, the $\Delta\mu_w$ will be less than zero, expressed as a negative number.

4.2.1 Water potential

By convention, plant physiologists have redefined the free energy of water by converting the energy units of joules per mole to pressure units of bars. This is done by dividing the chemical-potential equation by the partial molal volume of water, $\overline{V}$,

$$\frac{(18 \text{ g mol}^{-1})}{(1 \text{ g cm}^{-3})} = 18 \text{ cm}^3 \text{ mol}^{-1},$$

and redefining as the water potential, ψ:

$$\psi = \frac{\mu - \mu°}{\overline{V}} = \frac{RT \ln \dfrac{e}{e°}}{\overline{V}} .$$

The units are now joules per cm^3 rather than joules per mole. Joules per cm^3 are equivalent to dynes per cm^2 and 10^6 dynes per cm^2 is one bar. We use bars as the basic unit for water potential, ψ. It is helpful to know that there are 0.987 bars per atmosphere. The pascal is also used as a pressure unit in plant water relations.*

The primary principle to remember is that water always moves from a region of high water potential to a region of low water potential. If we knew the values of the water potential, ψ, in any two regions, we could quickly predict the direction of water transfer. As is standard in any thermodynamic consideration, the process will proceed spontaneously, that is, without energy input, if there is a loss in free energy upon completion of the process. Hence the difference in ψ between source (the region supplying the water) and sink (the receiving region) will be the indicator for water transfer. The free-energy content of the water in the *sink* must be less than the free-energy content of the water in the *source* for spontaneous water transfer to occur. The concept is illustrated in Fig. 4.1.

To ascertain the change in water potential, $\Delta\psi$,

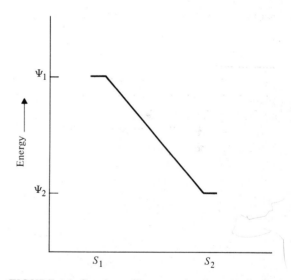

FIGURE 4.1 Graph to illustrate the decrease in free energy as the water potential decreases from ψ_1 to ψ_2. S_1 is the source with ψ_1 and S_2 is the sink with ψ_2. $\Delta\psi$ is the driving force for water transfer from S_1 to S_2. If the difference between S_1 and S_2 is expressed as a distance or equivalent, $(\psi_1 - \psi_2) / (S_1 - S_2)$, or $\Delta\psi/\Delta S$, is a gradient. The greater the slope of the line on the graph, the greater the gradient.

*The pascal (Pa) is a pressure unit equal to the force of one newton acting uniformly over one square meter. A bar is equivalent to 10^5 pascals.

simply subtract the ψ of the sink from the ψ of the source:

$$\triangle\psi = \psi_{sink} - \psi_{source}.$$

The value of $\triangle\psi$ *must* be negative for spontaneous net water transfer.

It needs to be made clear that this thermodynamic approach to water relations only gives information on the potential for water transfer. It says nothing about rates of transfer or whether barriers may be present to prevent transfer. In equilibrium thermodynamics such as this, only the beginning and the end are considered.

A simple direct analogy illustrates the point. Visualize a river flowing downstream. The ψ of the water upstream is greater than the ψ of the water downstream by virtue of its position in a gravitational field, one of the factors affecting ψ. We know there is substantial free energy present because the flow of water can be used to turn a paddle wheel, which in turn can be used to drive a turbine for the production of hydroelectric power. Furthermore, despite the potential difference between upstream and downstream, flow can be prevented by a barrier—a dam. There would be no change in the potential difference, that is, in the $\triangle\psi$, but the dam prevents flow. Finally, water could be pumped upstream against the potential gradient, but this would require energy. Sufficient energy would be required to reverse the ψ such that water would not go against a potential gradient.

The ψ is composed of several component forces (potentials), and it is necessary to ascertain these in order to estimate ψ. Forces or components such as osmotic pressure, hydrostatic pressure, gravitational forces, electrical forces, and absorptive forces come into play.

ψ can be computed as the algebraic sum of its components,

$$\psi = \psi_\pi + \psi_p + \psi_m + \psi\dots,$$

where

ψ_π = osmotic potential,

ψ_p = pressure potential,

ψ_m = absorptive or matric potential, and

$\psi\dots$ = any other force which may influence ψ.

The component potentials of ψ are discussed below.

4.2.2 Component potentials

Osmotic potential

Osmotic pressure was defined in Chapter 2 as that pressure necessary to prevent flow of a solvent, here being water, from a region of high concentration to a region of low concentration through a differentially permeable membrane. This flow of a solvent through a differentially permeable membrane is called osmosis.

Several terms and ideas associated with the concept of osmosis require elaboration. It is explicitly stated in the definition of osmosis that water "flows." However, this notion is somewhat debatable, the main evidence being that the process takes place faster than what can be accounted for by simple diffusion. For our purposes, we will assume that the process of osmosis takes place by pressure flow; the exact process is not important for an understanding.

A differentially permeable membrane is one that is permeable to the solvent and not the solute. Differentially permeable membranes are permeable to certain solutes (or solvents) but not to others. For example, a membrane may allow electrolytes through but not nonelectrolytes. Of course, a membrane is not apt to be completely permeable. Note here that permeability is a property of the membrane and not of solutes or solvent. In this context, semipermeable means the same as differentially permeable.

Osmotic pressure can best be understood by means of a description of an osmometer such as the one shown in Fig. 4.2. An osmometer is a chamber with a differentially permeable membrane attached to a manometer, or capillary tube. The chamber is filled with a solution, e.g., a sugar solution. Any solution will do, provided that the membrane is truly differentially permeable. Recall here that osmotic pressure is a colligative property dependent on the number of solute molecules in solution and not their kind (see Chapter 2). If the chamber is filled with just enough solution so that there is no excess hydrostatic pressure in the bag and is then placed in a beaker of pure water (with $\psi = 0$), the pure water will tend to flow into the chamber from the beaker. The actual hydrostatic pressure created by the excess water in the chamber can be measured by the water-column height within the manometer. At equilibrium the water will have reached a height

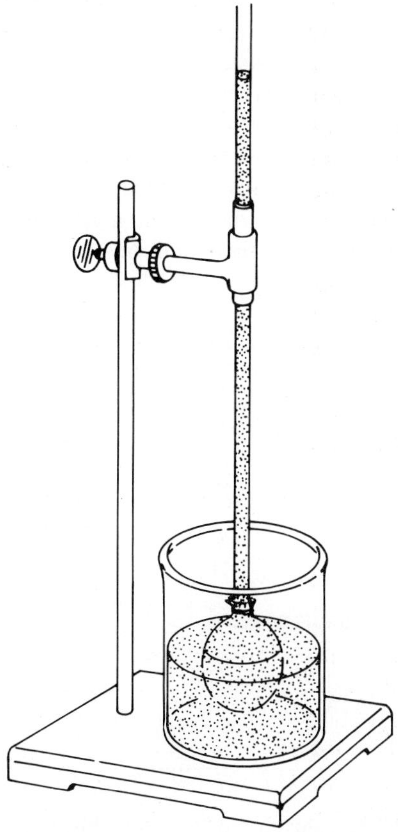

FIGURE 4.2 Diagram of an osmometer, which shows equilibrium hydrostatic pressure in a manometer marking the value of the osmotic pressure. If there is a 1-molal sucrose solution in the chamber and pure water in the beaker, the osmotic pressure will be approximately 23.4 bars at room temperature (20°C).

that depends on the number of solute molecules in solution in the chamber.

The hydrostatic pressure at equilibrium is the osmotic pressure. The actual hydrostatic pressure caused by the presence of the water in the chamber is the turgor pressure. Hence the definition of osmotic pressure is the value of the hydrostatic pressure established in an osmometer when the solution pressure is in equilibrium with pure water. This definition is the same as the previous definition of that pressure necessary to prevent flow from pure water to a solution through a differentially permeable membrane.

It is very important to realize that the real pressure present at any time is the hydrostatic (turgor) pressure and that osmotic pressure is the numerical value of the hydrostatic pressure at equilibrium.

The addition of solutes to water will reduce the free energy of the water and hence reduce the ψ. This is absolutely clear because in an osmotic system water flows from the pure water to the solution, and since water only flows from a high ψ to a low ψ, the free energy of the solution must be less than that of the pure water.

The important value is the osmotic potential, ψ_π, which can be estimated numerically from the osmotic pressure by changing the sign (from positive to negative).

In Chapter 2 we saw that osmotic pressure (π) could be defined as

$$\pi = CRT,$$

where

C = concentration in moles per liter,

R = gas constant, and

T = absolute temperature.

Thus osmotic potential is simply

$$\psi_\pi = -CRT.$$

A $1M$ solution will have an ψ_π of -21.8 bars:

$$\psi_\pi = -(1 \text{ mol L}^{-1}) \cdot (0.08) \cdot \frac{(\text{L} \cdot \text{bar})}{(\text{mol} \cdot \text{deg})} \cdot (273°);$$

$$= -21.8 \text{ bars}.$$

We can thus compute the ψ_π of any solution with a known concentration by

$$\psi_\pi = (-21.8) \cdot (M) \cdot \left(\frac{T}{273}\right).$$

Of course, with the above equation it must be assumed that the solution is ideal.

By way of example, the ψ of a $0.1\,M$ solution of sucrose at $15°$ would be estimated at -2.3 bars. The ψ_π is estimated from

$$\psi_\pi = (-21.8)\,(0.1)\left(\frac{288}{273}\right) = -2.3.$$

Then, since ψ is the algebraic sum of its component potentials (here only ψ_π),

$$\psi = \psi_\pi = -2.3 \text{ bars.}$$

Thus the ψ of a $0.1\,M$ sugar solution would be -2.3 bars if it was open to the atmosphere.

In summary, we have defined osmotic pressure in terms of an equilibrium hydrostatic pressure developed by excess water flow into an osmometer system. Since solutes decrease the free energy of water, we have defined osmotic potential, ψ_π, as the negative value of the osmotic pressure. ψ_π is one of the components of the water potential, ψ.

Since solutes decrease the free energy of water, the value of the osmotic potential will always be negative (or zero in the case of pure water).

Pressure potential

In the osmotic system shown in Fig. 4.2, water will flow from the beaker into the osmometer on a water-potential gradient. The ψ of the water in the beaker is by definition zero:

$$\psi_{\text{beaker}} = 0.$$

The water potential within the osmometer with a $0.1\,M$ sugar solution was computed to be -2.3 bars. Thus the free-energy difference or water-potential difference, $\Delta\psi$, between source and sink is

$$\Delta\psi = \psi_{\text{sink}} - \psi_{\text{source}} = (-2.3) - 0 = -2.3 \text{ bars.}$$

Since the $\Delta\psi$ is negative, water will flow spon-

taneously from source (the beaker) to the sink (within the osmometer chamber). Water will continue to flow into the osmometer until the $\Delta\psi$ is zero or until the two ψ are equal. The water leaving the beaker does not alter the water, and the ψ will remain unchanged at $\psi = 0$. Within the osmometer chamber, however, the hydrostatic pressure that builds up will tend to balance the osmotic pressure.

The actual hydrostatic pressure (turgor) is termed the pressure potential and designated ψ_p. When a positive hydrostatic pressure develops, the ψ_p will be positive. Under certain conditions, ψ_p may be negative. For example, water may be under tension; then the ψ_p will be negative.

In the osmotic system of Fig. 4.2, water will flow into the chamber from the beaker until the ψ of the chamber is zero, reaching equilibrium with the ψ of the beaker water. This will occur when the hydrostatic pressure or pressure potential reaches $+2.3$ bars. At this point, the ψ will be

$$\psi = \psi_\pi + \psi_p;$$
$$\psi = -2.3 + (+2.3) = 0.$$

Therefore, the positive ψ_p develops because the hydrostatic pressure tends to balance the ψ_π.

Another way to visualize ψ_p is illustrated in Fig. 4.3. Here two vessels of pure water are separated

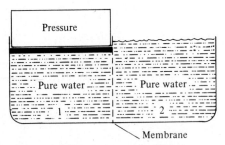

FIGURE 4.3 Illustration to show how pressure increases the water potential, causing a flow of pure water from chamber 1 into the pure water of chamber 2 across a membrane. The force of the piston pushing on the pure water increases the free energy of the water (that is, increases the ψ) such that ψ_1 is greater than ψ_2. Water thus flows from chamber 1 to chamber 2 on a water-potential gradient.

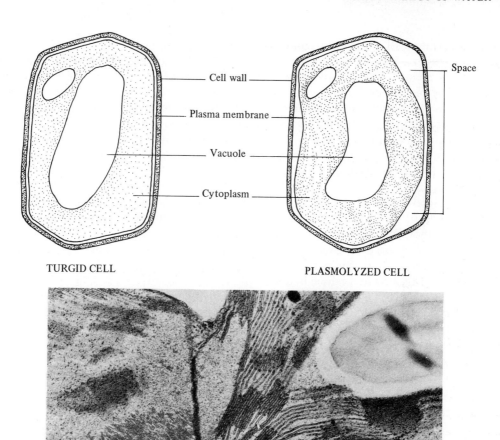

Cell wall

Plasma membrane

Vacuole

Cytoplasm

Space

TURGID CELL

PLASMOLYZED CELL

FIGURE 4.4 Diagrams of turgid and plasmolyzed cells. In the plasmolyzed cell, because of lack of turgor the vacuole is contracted and the plasma membrane has pulled away from the cell wall, leaving a space. The electron micrograph is of mesophyll cells from pinto beans exposed to ozone ($\times 50{,}000$). The ozone causes water leakage and cell plasmolysis. Note that the plasma membrane (pm) has pulled away from the cell wall, leaving a space that is apparent across the entire photo. The structures labelled cf are crystalloid bodies that have appeared in response to the ozone treatment. The three small arrows point to the plasma membrane. From an experiment by Susan Nagahashi, 1975, University of California, Riverside.

by a membrane. One vessel is fitted with a piston in order to apply pressure. Initially, the ψ in each is 0 and hence they are equal. There would be no net water flow. If the piston applies pressure to one vessel, the increase in free energy, ψ, would cause water to flow from left to right. Once again the water flows from a high ψ to a low ψ.

Points to remember about ψ_p, one of the components of ψ, are that it is the result of a real hydrostatic pressure, its numerical value is usually positive but can be negative, and it may balance ψ_π.

Positive ψ_p tends to maintain the turgidity of cells. We assume that, for the most part, solutes contributing to ψ_π and the excess water causing the hydrostatic pressure, or turgor potential, are largely in vacuoles. This turgidity, caused by a positive ψ_p, pushes the cytoplasm against the cell membrane and wall and gives form to the cell. When the ψ_p goes to zero because of water loss, such a cell will plasmolyze (refer to Fig. 4.4). A cell in air, however, may not show plasmolysis if the cell membrane is held to the cell wall by hydration.

Plasmolysis results when $\psi_p = 0$; the cytoplasm pulls away from the cell wall because as water leaves the vacuole becomes flaccid. Incipient plasmolysis is defined as the point at which 50% of the cells are plasmolyzed.

Water transport to and from animal cells and tissues can be described adequately on the basis of osmotic gradients. Because plants have rigid cell walls, water transport in plants is better described on the basis of water-potential gradients, of which ψ_π is only one component. ψ_p is equally important.

Figure 4.5 shows how ψ, ψ_π, and ψ_p change as the cell volume increases when water enters. ψ increases when water enters, and the excess water causes the ψ_p to increase. The ψ_π also increases because the water that enters dilutes the solute concentration. If the cell is equilibrated against pure water with $\psi = 0$, at equilibrium the cell ψ will be 0 also.

Matric potential, ψ_m

The matric potential is that component of the plant water potential that comes about by water adsorption to cellular constituents. Water adsorption to surfaces reduces the free energy of water, and thus the value of ψ_m is negative.

Imbibition, the process of adsorption of water to nearly dry surfaces such as wood, takes place along with the liberation of substantial amounts of heat. When dry seeds take up water, this same process of imbibition is accompanied by heat loss. Heat loss from water during the adsorption process is a direct

FIGURE 4.5 Diagram to illustrate how water potential (ψ), osmotic potential (ψ_π), and pressure potential (ψ_p) change in a cell as water flows in from a pure-water vessel and the cell volume increases. The point 1.0 on the abscissa (relative cell volume) is for the flaccid cell. As water flows in, the cell volume increases and the cell water potential approaches that of the water in the vessel. The pressure potential increases because of the entry of water and there is a slight dilution of the cell contents, causing the osmotic potential to become less negative. When the cell water potential reaches zero, the same as the pure-water vessel, no more water enters.

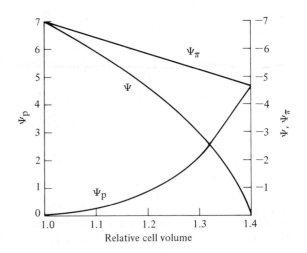

measure of the decrease in free energy of the water.

The swelling associated with imbibition is sufficiently energetic to fracture solid rock; the Egyptians split granite by wetting dry wood wedges.

By and large, matric potential is not as important a component of ψ as are ψ_π and ψ_p. In exceptional cases, such as dry seeds or perhaps succulents with high mucilage content, ψ_m could be important.

Most estimates of ψ_m yield numbers of about 0.1 bar. Many of the laboratory methods to estimate ψ_π also include the matric component.

In soils, ψ_m is the most important component, and except in saline soils ψ_π is insignificant.

Gravitational and electrical components

For the most part, gravitational (ψ_g) and electrical (ψ_e) components of the ψ are not considered. Perhaps in tall trees ψ_g may become a factor; with height, a correction for the acceleration due to gravity may be applied. Assuming a mean acceleration due to gravity (g) of 980 cm s^{-2}, a correction of -0.00031 cm s^{-2} m^{-1}, and a 100-meter-tall tree, the correction is only 0.0032%. Gravitational forces are, therefore, largely ignored in plant water relations.

Electroosmosis, a process by which water moves along electrical gradients, could be of some significance. Since water is a dipole it can readily move in an electric field. Potentials across living membranes, however, are only on the order of 100 millivolts. This voltage is probably insignificant as a force for water transport in plants, and, as a consequence, ψ_e is usually not considered.

Other factors that influence the free energy of water could be considered when obtaining ψ, but for the most part the defining equation is

$$\psi = \psi_p + \psi_\pi.$$

4.3 The Rate of Water Transport

4.3.1 The driving force

The water potential (ψ) developed through thermodynamic considerations gives valuable information about the potential direction of water transport. Provided that there is a water-potential drop from source to sink, water will tend to flow from source to sink spontaneously. The important value is the $\triangle\psi$, estimated from the ψ of the source and sink. The $\triangle\psi$ is considered to be the driving force for all water transport.

The rate at which water is transported must be evaluated from considerations of nonequilibrium thermodynamics. Here we are interested in how fast water is transported, expressed as dw/dt. Units can be grams of water per second or any other convenient units of volume per time.

The flux of water flow, dw/dt, is usually expressed as J_w, where J_w is the volume of water flow through some area of surface per time, for example, g water $\cdot$ (cm^2 of tissue surface)$^{-1}$ s^{-1}. The flux of water flow from cell to cell will be a function of the water permeability of the limiting membranes, L_p, and the driving force for water transport, $\triangle\psi$. Thus water flow can be expressed as

$$J_w = L_p \,\triangle\psi.$$

Since

$$\psi = \psi_p + \psi_\pi,$$

then

$$\triangle\psi = \triangle\psi_p + \triangle\psi_\pi,$$

where the $\triangle$ indicates the difference in each potential between source and sink. Hence water flow, J_w, is

$$J_w = L_p \,(\triangle\psi_p + \triangle\psi_\pi).$$

This expression will suffice to describe water transport between cells or tissues or from the outside to inside the plant, provided that the limiting membranes are truly differentially permeable.

However, this is almost never the case for real-life plant systems. In other words, most membranes will be somewhat leaky with respect to solutes. Very few solutes are completely excluded or completely retained by plant cells. Sucrose, glucose, and many other sugars are readily transported, as are many inorganic ions such as Cl^-, $SO_4^=$, Na^+, and K^+. Certain substances, such as the sugar-alcohols, are transported only to a limited extent.

The osmotic pressure will decrease according to the degree to which the membrane lacks differential permeability with respect to a solute. Any leakiness will alter the theoretical value.

4.3.2 The reflection coefficient

The permeability of a membrane to a substance can be expressed as the ratio of the actual osmotic potential (ψ_π^a) to the theoretical osmotic potential (ψ_π^t):

$$\sigma = \frac{\psi_\pi^a}{\psi_\pi^t} \ .$$

Note that if σ is one, the membrane will not allow the solute to pass, and the actual osmotic potential will equal the theoretical. Leakiness of the membrane will result in σ values less than one. If the σ value is known for a solute, the actual ψ_π^a can be

TABLE 4.1   Examples of reflection coefficients (σ) from pea chloroplasts

Solute	σ
Sucrose	1.0
Erythritol	0.90
Glycerol	0.63
Isoleucine	0.33
Alanine	0.01

Data from Nobel (1970).

obtained from the theoretical ψ_π^t by

$$\psi_\pi^a = \sigma \cdot \psi_\pi^t.$$

σ is called the reflection coefficient and is known for many substances. Table 4.1 illustrates some known σ values for the chloroplast envelope.

The proper expression to describe the flux of water flow from one system to another is, therefore,

$$J_w = L_p \left(\triangle\psi_p + \sigma\triangle\psi_\pi\right).$$

This expression provides for the water permeability (L_p) of the tissue. The greater the permeability, the greater the flux of water. It also takes into account the driving force, $\triangle\psi$, expressed as $\triangle\psi_p + \triangle\psi_\pi$, and corrects for solute leakage or deviations from differential permeability.

4.4 Atmospheric Water

The atmosphere of our earth is composed of a mixture of some eleven gases plus water vapor (Table 4.2). Nitrogen is the most abundant gas, making up 78% of the atmosphere. The next most abundant gas is oxygen at 21%. Argon accounts for nearly 1% and CO_2 for about 0.03%. Other gases, such as neon, helium, krypton, hydrogen, xenon, ozone, and radon, occur in trace quantities. Polluted air may contain varying quantities of hydrogen fluoride, sulfur dioxide, oxides of nitrogen, increased quantities of ozone, complex organic compounds, and particulate matter. The density of air at standard temperature and pressure (STP) is 1.28 mg cm^{-3}.

The actual amount of water in the air varies considerably, but on the average it is 0.02% to 0.4% by weight. If we assume an average of about 0.2%, the average atmospheric vapor pressure at STP calculated from Raoult's law is 2 mbar. The actual vapor pressure of the air becomes important for plant water relations (see Chapter 6) because evaporation from plant surfaces is a function of the vapor pressure of the water in the plant and the vapor pressure of the water in the atmosphere.

The amount of water in the air is conveniently expressed on a relative basis, known as the relative humidity. Relative humidity expresses the amount of water in the air as compared to what the air can actually hold. Since the amount of water the air can hold increases with temperature, relative humidity has meaning only if the temperature is also known.

Relative humidity (RH) is expressed as

$$\% \ RH = \frac{e}{e^\circ} \cdot 100,$$

where

e = vapor pressure of air at some specified temperature, and

e° = vapor pressure of pure water at the same temperature.

The vapor pressure of pure water at STP is 6.1 mbar; thus the average RH the world over is 32.8%.

$$\frac{2.0 \ \text{mbar}}{6.1 \ \text{mbar}} (100) = 32.8\%$$

Vapor pressure increases exponentially with temperature, as shown previously in Fig. 2.6. The vapor pressure of the water in the air, however, does not tend to increase exponentially with temperature because the air also expands with temperature according to the gas laws. The water dilution compensates for the increased pressure. Thus barring any changes in the air mass of a region, the actual vapor pressure of the water in the air stays fairly constant day and night.

Air becomes quite dry as altitude increases. On top of Mt. Whitney, the highest mountain in the continental United States (4420 meters), the vapor pressure decreases from 2 mbar average to about 1.2 mbar, and on top of the highest mountain, Mt. Everest (8850 meters), the vapor pressure would only be 0.62 mbar. This decrease may be important for plants growing at high altitudes.

TABLE 4.2 Percent by weight of atmospheric gases in dry air at sea level

Gas	%
Nitrogen (N_2)	78.09
Oxygen (O_2)	20.95
Argon (Ar)	0.93
Carbon dioxide (CO_2)	0.03
Neon (Ne)	1.8×10^{-3}
Helium (He)	5.24×10^{-4}
Krypton (Kr)	1.0×10^{-4}
Hydrogen (H_2)	5.0×10^{-5}
Xenon (Xe)	8.0×10^{-6}
Ozone (O_3)	1.0×10^{-6}
Radon (Rn)	6.0×10^{-18}

Data from *CRC Handbook* (1979) (see References, Chapter 2).

4.5 Measurement of ψ, ψ_p, ψ_π

4.5.1 Water potential

In large part, methods to estimate water potential are equilibrium methods. Tissue to be estimated is brought into equilibrium with a solution of known water potential.

Weight or volume change

In weight or volume change methods, tissue is placed in a series of graded solutions of known water potential. The latter can be prepared from sucrose, mannitol, or salt solutions (KCl, $CaCl_2$) provided that the water potential is known. Recall that in a free, open solution, the water potential will equal the osmotic potential since no excess pressure can build up. Table 4.3 lists osmotic potentials of sucrose solutions.

The tissue water is allowed to come to equilibrium with the solution. Those tissues that have a greater water potential than the solution will lose weight as water is lost. Conversely, those tissues with a lower water potential will gain weight by water

TABLE 4.3 **The water potential of sucrose solutions at 2°C in bars**

M	ψ	M	ψ
0.0	0.00	1.0	−34.6
0.1	−2.64	1.1	−39.8
0.2	−5.29	1.2	−45.4
0.3	−8.13	1.3	−51.6
0.4	−11.11	1.4	−58.4
0.5	−14.31	1.5	−65.8
0.6	−17.77	1.6	−73.9
0.7	−21.49	1.7	−83.0
0.8	−25.54	1.8	−93.2
0.9	−29.70	1.9	−104.5

From Walter (1931).

absorption. A graph of weight change versus water potential of the standard solutions can then be used to estimate the water potential of the test tissue. The tissue that neither gained nor lost water is in a solution with the same water potential; that is,

$$\triangle\psi = \psi_{tissue} - \psi_{solution} = 0$$

or

$$\psi_{tissue} = \psi_{solution}.$$

In practice, it is not necessary to wait for complete equilibrium since water exchange will begin immediately; it is only weight or volume changes that are of interest. Time required for meaningful measurements will depend on tissue permeability. Clearly, leathery leaves with a thick cuticle will require more time than will thin, mesic leaves (leaves of a moderate-water habitat). Figure 4.6 illustrates the relationship between weight change and solution water potential.

The dye method

A method similar to the one above was introduced by the Russian physiologist V. S. Chardakov and is referred to as the Russian dye method. Tissue is placed in test tubes with a graded series of known osmotic potential and then allowed to come into equilibrium. The method is based on the fact that those solutions that gain water from the tissue will decrease in density and those that lose water to the

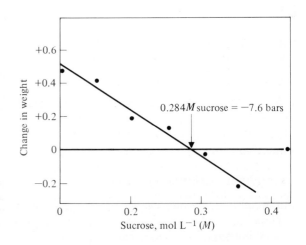

FIGURE 4.6 Diagram of the change in weight of potato-tuber tissue when bathed in different osmotic concentrations of sucrose solutions. The osmotic concentration of sucrose expressed in bars at the point of no weight change (zero line) is equivalent to the water potential of the potato tissue. In this particular experiment by Donald Briskin in 1979 at the University of California, Riverside, there was no weight change in the potato section when placed in 0.284M sucrose. When placed in more dilute sucrose solutions the tuber tissue took up water and increased in weight, and when placed in a more concentrated solution the tuber tissue lost weight. The 0.284M sucrose solution is equivalent to an osmotic potential of –7.6 bars, and it is thereby judged that the water potential of the potato tuber was –7.6 bars.

tissue will become more dense as the solute in the solution concentrates. After equilibrium, the density change in each test tube can be estimated by placing a drop of original test solution stained with a water-soluble dye such as methylene blue in the center of each solution using a pipette. If the drop sinks, the density has decreased because of water gain. Thus the tissue in this solution has a greater water potential and loses water. Similarly, if the drop rises because the standard solution has lost water, thereby becoming more dense, the tissue in it has a lower water potential. The standard solution, which has no change in density, is the one with a water potential comparable to the test tissue.

This method is suited for field use, requiring only two series of graded standard solutions, one stained with a dye and the other used to test the tissue. In the laboratory, the method can be made more quantitative by estimating the density of the solutions with a refractometer.

Pressure equilibrium

Several instruments are available that will estimate tissue water potential by developing a pressure just sufficient to balance the water potential of the tissue.

In its simplest sense, increasing pressure can be applied to a tissue, e.g., a leaf, until water is forced out. The pressure required to force water out of the tissue is equal to the water potential of the tissue. Problems arise when trying to adjust the applied pressure to just balance the water potential but not measure resistance or water from broken cells.

One of the better methods of this type is the Scholander bomb, illustrated in Fig. 4.7. A fresh

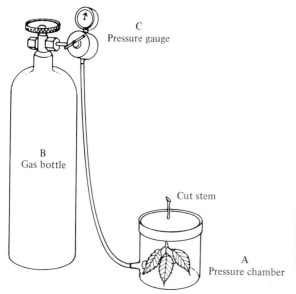

FIGURE 4.7 Diagram of the Scholander pressure-bomb apparatus to measure water potential. A cut stem or leaf is placed in the pressure chamber (A) with the cut surface just exposed. The chamber is pressurized with a nitrogen gas bottle (B) until fluid just appears at the cut surface. The amount of pressure (read from the gas gauge, C) required to just force the tissue fluid from the tissue is assumed to be equal to the water potential of the tissue. The photo is a pressure bomb built by Alan Eckard at University of California, Riverside. From P. F. Scholander, H. T. Hammel, E. D. Bradstreet, and E. A. Hemmingsen. 1965. Sap pressure in vascular plants. *Science* 148:339 346.

leaf or stem is placed in a pressure chamber with petiole or cut stem exposed. The water will be held in the leaf with a force in bars equal to the water potential. Pressure in the chamber is increased until water starts to exude from the cut petiole or stem. The pressure required to barely push the water out of the tissue will just balance the water potential.

Overall, the method has problems of tissue collapse and the complication of pushing water through cells and tissues, but it is a convenient method with a high degree of reliability. Furthermore, the instrument can be readily adapted for field studies.

Vapor equilibrium

Perhaps the most sophisticated method to measure ψ is based on vapor exchange between tissues and the atmosphere. The most reliable method uses a thermocouple psychrometer to estimate vapor pressure. The psychrometer has a temperature-controlled chamber to house the tissue and a wet- and dry-bulb thermocouple to estimate the relative humidity of the chamber atmosphere around the tissue. The measurement is taken when the chamber atmosphere is in equilibrium with the tissue. At this point, the water potential of the chamber atmosphere is equal to the water potential of the tissue since they are in equilibrium.

A drop of water is placed on the wet-bulb thermocouple, and the relative humidity in the closed chamber housing the tissue is estimated by the cooling of the wet bulb. This is the usual psychrometric method of measuring relative humidity. The water potential is then estimated from the basic relationship

$$\psi = RT \ln \frac{e}{e^\circ},$$

where e/e° is the relative humidity.

Disadvantages of the psychrometric method include the fact that salt deposits or dirt on the tissue will act as water sinks causing erroneous readings; precise temperature control is also required (to within 0.001°). In addition, very long equilibration times are required before humidity can be estimated in the chambers. There are, however, instruments available that require less temperature control and that can be calibrated by measuring the rate of cooling of the wet thermocouple as evaporation into the chamber takes place. These instruments are available for field use because temperature control and long equilibrium times are not necessary.

4.5.2 Osmotic potential

The plasmolytic method

One of the oldest methods to estimate osmotic potential of tissues is to find a standard solution of known osmotic potential that will just cause the cells of the test tissue to plasmolyze. In practice, one looks for the point at which 50% of the cells are plasmolyzed, i.e., incipient plasmolysis. It is necessary to observe cells directly and to use a tissue in which plasmolysis can be recognized. The principle of the method is based on the expression $\psi = \psi_p + \psi_\pi$. At plasmolysis the pressure potential equals 0, and $\psi = \psi_\pi$. Here the osmotic potential of the standard solution will equal the osmotic potential of the tissue.

The cryoscopic method

Since osmotic potential is a colligative property, it can be estimated by measuring any of the colligative properties, from vapor pressure to freezing point. The simplest is usually freezing point. As explained in Chapter 2, a 1-molal solution will depress the freezing point 1.86°.

First, cell sap must be expressed from the tissue. This is frequently done by forcing the fluid out under pressure or by preparing an extract from ground tissue. The latter is less satisfactory. Freezing point is then determined on the expressed sap with care taken to account for undercooling, i.e., cooling below the freezing point. The osmotic potential can be calculated from the following relationship:

$$\psi_\pi = \triangle P_F \cdot \left(\frac{-22.7}{1.86}\right).$$

In practice, one usually calibrates freezing point against standard solutions of known osmotic potential.

The vapor-pressure method

Vapor pressure, being a colligative property, can readily be used to estimate osmotic potential. The vapor-equilibrium (psychrometric) method used to estimate water potential described in Section 4.5.1 can be used after the tissue is quick-frozen to break cellular membranes. Freezing disrupts the cells and eliminates the pressure potential. Thus $\psi = \psi_\pi$ and osmotic potential can be estimated directly.

4.5.3 Matric potential

Estimation of the value of matric potential is a rather difficult task. A mixture such as cytoplasm or expressed cell sap will have many colloidal surfaces that adsorb water, and estimation of osmotic potential will include the matric component because water adsorption will reduce the vapor pressure. Osmotic potential and matric potential are indistinguishable by the usual methods.

If only a matric potential is present, those methods designed to measure osmotic or water potential will suffice. The usual method for measuring matric potential in soils is to force water from the soil colloids with an applied pressure. The pressure required to produce free water will be equal to the matric potential of the soil water. Similar methods can be used to estimate matric potential of plant colloids such as gums or mucilages.

4.5.4 Pressure potential

Pressure potential of cells and tissues is the most difficult component to estimate experimentally. The usual method is to obtain values for ψ and ψ_π, and calculate ψ_p:

$$\psi_p = \psi - \psi_\pi.$$

This is a satisfactory method provided that accurate estimates of ψ and ψ_π can be obtained.

Alternative methods are limited by technology. Attempts have been made to attach micromanometers to large cells, and pressure transducers have been employed. By and large, such instruments are not generally available.

4.6 Soil Properties

For the most part, forces holding water to soil particles are matric potentials, although saline or otherwise salty soils may have a measurable osmotic potential.

Soils are composed of minerals and organic matter. The latter, termed humus, is formed primarily from the decomposition products of cellulose and lignin. Soil particles are composed of minerals and organic material of varying sizes. Physical properties of soils govern the water-holding capacity and, more importantly, the water availability to plants. Soil also contains living material, bacteria, algae, invertebrates, and plant roots, all of which contribute to soil properties and to the soil atmosphere. The water availability of soils is a direct function of these properties.

4.6.1 Soil texture

The solid, inorganic portion of the soil is composed of particles of various sizes. Depending on size, soils can be classified according to texture in the following categories: coarse sand, fine sand, silt, and clay. Table 4.4 gives the approximate range of

TABLE 4.4 Soil classification on basis of particle size

Texture	Size (mm)
Coarse sand	2.0 − 0.2
Fine sand	0.2 − 0.02
Silt	0.02 − 0.002
Clay	< 0.002

particle size in each category. As the percentage of clay increases, the soil texture becomes more fine and the water-holding capacity increases. Organic matter also increases water-holding capacity.

4.6.2 Soil structure

Texture is a function of soil particle size whereas structure depends on aggregation and porosity. In the usual soil, the individual soil particles are aggregated to form larger micelles. The ultimate result is a structure with varying amounts of pore space, or porosity. Pore spaces are considered to fall into two major classes, microspaces and macrospaces. The microspaces hold capillary water whereas the macrospaces can hold capillary water as well as free water. The macrospaces contribute mainly to the soil atmosphere. Because of cementing, pores can be large but not continuous.

The heavier (i.e., finer) the soil (e.g., clays), the fewer macrospaces and the more microspaces. Thus in terms of structure as well as texture, clays with small particles have a greater water-holding capacity.

4.6.3 Soil water

The water content of the soil is frequently expressed on a dry-weight basis:

$$\% \text{ soil water} = \frac{\text{FW} - \text{DW}}{\text{DW}} \, 100,$$

where

FW = field weight of wet soil, and

DW = oven dry weight.

A dry-weight basis is logical because the informa-tion desired is frequently how much water the soil can hold as compared to its dry state.

The soil water can be classified as (1) gravitational water, (2) capillary water, or (3) hygroscopic water. Most of the irrigation or precipitation that adds water to the soil is lost by percolation, i.e., passage through macrospaces caused by the forces of gravity. This water, which fills the macrospaces and is lost by gravitational forces, is termed gravitational water. It adds little to available water of the soil.

The smaller spaces of the soil, the microspaces, will retain water against the forces of gravity by capillarity. This water, held by capillary forces within the microspaces, is available for plant growth and is termed capillary water. The forces with which the soil holds capillary water are the matric forces. Tightly bound capillary water, which is largely unavailable for plant growth, is called hygroscopic water. Hygroscopic water can be visualized as thin molecular films of water on soil particles held there by strong forces.

An important feature of soils related to texture and structure is water availability. The amount of capillary water available for plant growth falls between two points, called the field capacity (FC) and the permanent wilting percentage (PWP).

Field capacity

The field capacity is defined as that amount of water held against gravitational forces, expressed as a percentage of dry weight. It can be estimated in soils after irrigation, percolation, and runoff simply by determining the moisture content. Field capacity represents the upper limit for water avail-ability. Obviously the field capacity percentage will vary with soil texture. On the average, clay soils will have a field capacity of about 40–45%, silts about 20%, and sands 5–10%.

Permanent wilting percentage

The permanent wilting percentage (PWP) is the lower limit of water availability in a soil. It is marked by that soil moisture percent at which

TABLE 4.5 Typical soil percentages for field capacity and permanent wilting point

Texture	FC (0.3 bar)	PWP (15 bar)	Available water, %*
Sand	4.5	2.2	2.3
Silt (loam)	18.4	12.6	5.8
Clay	45.1	26.2	18.9

*% water is on a dry-weight basis.
Data from Kramer (1969).

plants wilt and will not recover unless water is added to the soil. Care must be taken here to note that plants frequently wilt during midday when water loss exceeds water uptake. This is called incipient wilting if the plants recover toward the end of the day when water uptake balances water loss. In addition, many plants have leaves that do not show signs of wilting; determining PWP requires the estimation of ψ_p. The concept of the permanent wilting percentage was developed with sunflowers for agricultural purposes and is best used only in such a context. Nevertheless, the idea of a PWP is useful in understanding plant water relations.

Permanent wilting percentages of soils vary with texture. On the average, clays will show a PWP of about 20%, silts about 10%, and sands 3–5%. Despite the fact that plants will show a range of wilting percentages the concept is developed as a soil property and not as a plant property. Table 4.5 summarizes these two cardinal points for soil water and gives the percentage range of availability.

Water availability

It is immediately clear from the preceding discussion that knowledge of the percentage of water in a soil gives little information about availability unless the soil properties are also known. A soil with 15% water would be below the PWP if it were clay and above the FC if it were a sand. Thus for studies of plant water relations, it is more desirable to express water content in terms of water potential. Not only is this more logical but also it is easier because soils become uniform with respect to water. Field capacity for virtually all soils is in the range of −0.1 to −0.3 bars. The permanent wilting point is about −15 bars.

Figure 4.8 shows how the water potential of sand and loamy soils changes with water percentages.

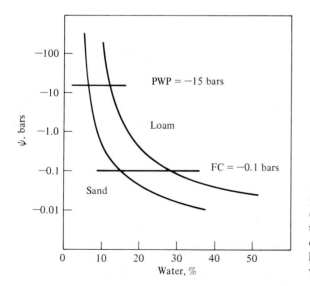

FIGURE 4.8 Diagram to illustrate how soil water potential changes as the percentage of water in the soil changes. As the percentage of water in the soil increases the water also increases. Fine soils such as loams, because of greater surface area of the smaller particles, have higher water potentials as a function of percentage of water than do coarse soils such as sands.

The value of −0.1 bar for field capacity reflects gravitational forces, whereas the PWP of −15 bars reflects the usual lower limit of water potential in plants.

4.6.4 Measurement of soil water

Percentage of water is frequently estimated by collecting soil from the field and storing it in a water-tight container. The soil is weighed in the tared sample container and is then dried for about 48 hours at 105° C. After reweighing the now dry soil, the percentage of moisture can be computed.

Water content can also be estimated by neutron-scattering devices. This method requires a source for neutrons and a detector. It is based on the fact that neutrons are slowed to a greater extent by striking hydrogen atoms than by striking other atoms. Since the hydrogen that varies in soils is mainly the hydrogen associated with water, neutron scattering can be used to estimate water content. Similarly, gamma-ray absorption can be used to estimate soil water. Here, gamma attenuation is used to measure soil density, which will vary with water content.

Water content of soils will affect electrical resistance and capacitance as well as thermal conductivity. All of these properties will vary with water content and can be used to estimate soil water if the instruments are properly calibrated. The most common method measures resistance with a Bouyoucos block. Electrodes are embedded in plaster-of-paris blocks and buried in soil. Water in the soil equilibrates with the block, and changes in electrical resistance are an indication of soil water. These blocks are useful over the range of −0.5 bar to −15 bars. The above methods can be calibrated either in moisture percentages or in potentials.

A tensiometer is a device that can be used to estimate soil moisture potential directly. It consists of a ceramic-capped, water-filled tube connected to a vacuum gauge. The tip is inserted into the soil. Water in the tube equilibrates with the soil water and the pressure drop is measured directly. The pressure drop in bars is equal to the soil water po-

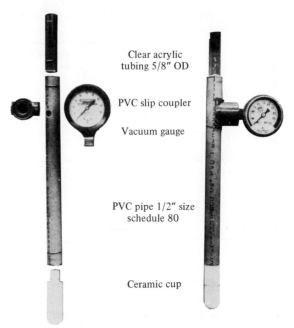

Clear acrylic tubing 5/8″ OD

PVC slip coupler

Vacuum gauge

PVC pipe 1/2″ size schedule 80

Ceramic cup

FIGURE 4.9 Diagram of a soil tensiometer, a device to measure soil water potential. The tube is filled with water and the ceramic tip is inserted into the soil. The water in the tube and cup will come to equilibrium, with the soil water causing a pressure drop, or vacuum, in the tube that can be read with the vacuum gauge. The drop in pressure, measured in bars, is equal to the soil water potential. The photo shows an actual tensiometer constructed at the University of California, Riverside.

tential. Figure 4.9 illustrates a typical tensiometer.

Estimates of soil water potential and matric potential can be made by the vapor-pressure and direct-pressure methods described in Section 4.5 for plant tissue.

4.6.5 Soil water movement

Gravitational water readily moves through soils. By and large, capillary water does not move appreciably in soils except in the very finest clay soils. Thus water from deep water tables is not a source of water for plant growth and development unless, of course, roots penetrate to the water table.

Figure 4.10 illustrates a typical soil water profile.

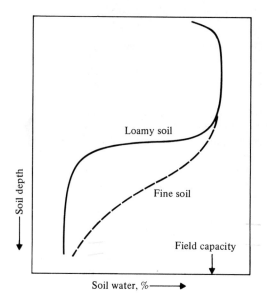

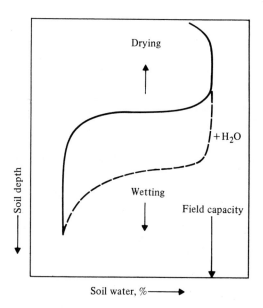

FIGURE 4.10 Diagram to illustrate how the percentage of soil water is apt to change with the depth of the soil (upper panel). At the soil surface, there will probably be less water than just below the surface. In a typical agricultural soil, the percentage of water remains rather constant through the soil until a depth is reached that is relatively dry. A fine soil will not have as sharp a water profile because of some water creep by capillarity. When water is added to the soil, it will tend to increase water at greater depths to a percentage equivalent to the field capacity. Below this depth the usual type of soil water profile will be evident.

After irrigation, water percolates through the soil and brings it to field capacity throughout. Below the region of field capacity there is a sharp drop in soil water. "Creep" in fine soils would show as a slow change from field capacity toward hygroscopic water. If more water is added, it simply tends to lower the profile but does not change its shape.

Review Exercises

4.1 A cell with a pressure potential of 8 bars and an osmotic potential of −16 bars is placed in a beaker of pure water. What is the water potential in the beaker and in the cell initially? In which direction will water flow? After equilibrium, what will be the component potentials and water potentials in the beaker and cell? What assumptions have you made in your calculations?

4.2 If the cell in Review Exercise 4.1 is placed in a beaker that contains a solution with an osmotic potential of −7 bars, in which direction will water flow? Compute your answer using the cell in Review Exercise 4.1 both before and after equilibrium. What will be the values of the water potential, pressure potential, and osmotic potential in the cell at equilibrium?

4.3 Cell A has an osmotic potential of −20 bars and a pressure potential of +6 bars. Cell A is placed adjacent to Cell B, which has an osmotic potential of −16 bars and a pressure potential of +12 bars. In which direction will water flow? What are the final equilibrium potentials?

4.4 Assume a situation in which roots are in a soil that has a matric plus osmotic potential of −1 bar. The roots have an osmotic potential of −10 bars and an initial pressure potential of +7 bars. If the roots come to equilibrium with the soil, what will be the equilibrium potentials of the root? After equilibrium is established, if the soil is irrigated with a salt solution that brings the total water potential of the soil to −5 bars, describe what happens to the root (in terms of potentials) immediately and then after a new equilibrium is established.

4.5 If cell A has a pressure potential of 5 bars and contains 5 ml of $0.5M$ glucose and cell B has a pressure potential of 5 bars and contains 5 ml of $0.5M$ sucrose, in which direction will water flow? Is it possible for water to move from a cell that has a high solute concentration to a cell with a lower solute concentration? Explain and give some examples.

References

BOYER, J. S. 1969. Measurement of the water status of plants. *Ann. Rev. Plant Physiol.* 20:351–364.

GARDNER, W. R. 1965. Dynamic aspects of soil-water availability to plants. *Ann. Rev. Plant Physiol.* 16: 323–342.

HELLEBUST, J. A. 1976. Osmoregulation. *Ann. Rev. Plant Physiol.* 27:485–505.

KOZLOWSKI, T. T. 1964. *Water Metabolism in Plants.* Harper and Row, New York.

KOZLOWSKI, T. T. (ed.). 1968. *Water Deficits and Plant Growth,* Vols. 1, 2, 3. Academic Press, New York.

KRAMER, P. J. 1969. *Plant and Soil Water Relationships: A Modern Synthesis.* McGraw-Hill, New York.

KRAMER, P. J., E. B. KNIPLING, and L. N. MILLER. 1966. Terminology of cell-water relations. *Science* 153: 889–890.

NOBEL, P. 1970. *Introduction to Biophysical Plant Physiology.* Freeman, San Francisco.

PARKER, J. 1969. Further studies of drought resistance in woody plants. *Bot. Rev.* 29:124–201.

PHILIP, J. R. 1966. Plant water relations: Some physical aspects. *Ann. Rev. Plant Physiol.* 17:245–268.

RAY, P. M. 1960. On the theory of osmotic water movement. *Plant Physiol.* 35:783–795.

SLAVIK, B. 1974. *Methods of Studying Plant Water Relations.* Springer-Verlag, New York.

SLAYTER, R. O. 1967. *Plant-water Relationships.* Academic Press, New York.

SPANNER, D. C. 1973. The components of the water potential in plants and soils. *J. Exp. Bot.* 24:816–819.

WALTER, H. 1931. *Die Hydratur der Pflanze.* Gustav Fischer, Stuttgart.

CHAPTER FIVE

Gas Exchange and Stomatal Physiology

One of the most important physiological functions of plants is the exchange of gases between the atmosphere and the leaf surface. The study of gas exchange encompasses not only physiology but anatomy and ecology as well. The most important gases are CO_2, given off during respiration and consumed during photosynthesis, and O_2, consumed during respiration and given off during photosynthesis. Equally as important to the plant, especially from the standpoint of water balance, is the loss of water vapor from the leaf to the atmosphere during transpiration.

Most of the gas exchange taking place occurs through the small pores on the exposed surfaces. These pores, or stomata, are the primary portals of CO_2 entry during photosynthesis and of water loss during transpiration. Because gas exchange plays such a central role in the physiology of plants, a thorough knowledge of the physiology of stomata and the principles of gas exchange are important for a complete understanding of plant physiology.

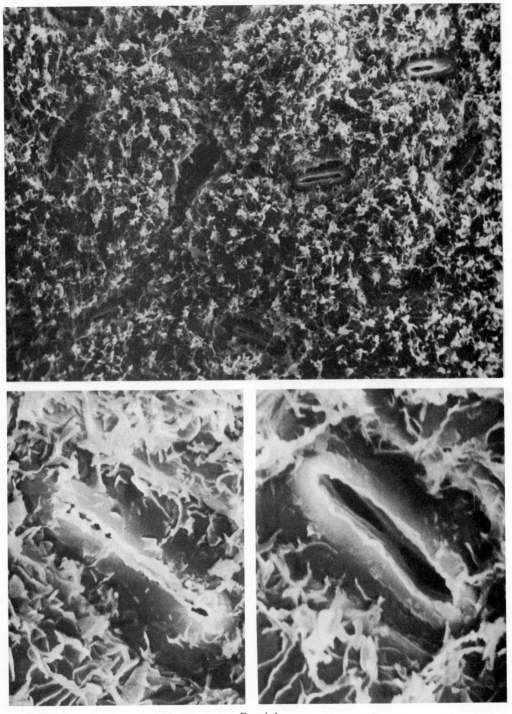

Panel A

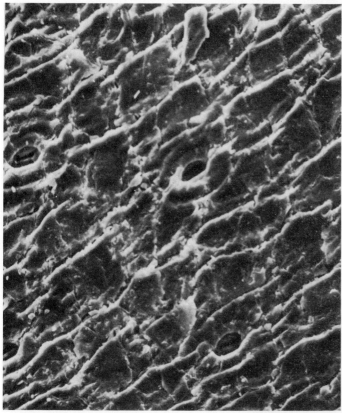

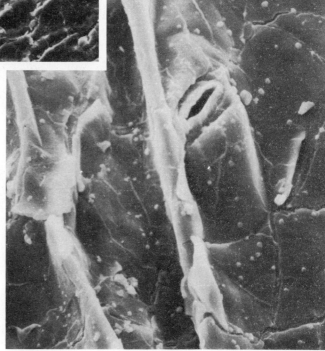

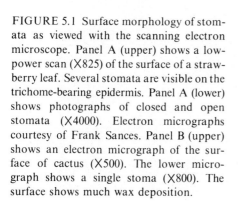

FIGURE 5.1 Surface morphology of stomata as viewed with the scanning electron microscope. Panel A (upper) shows a low-power scan ($\times 825$) of the surface of a strawberry leaf. Several stomata are visible on the trichome-bearing epidermis. Panel A (lower) shows photographs of closed and open stomata ($\times 4000$). Electron micrographs courtesy of Frank Sances. Panel B (upper) shows an electron micrograph of the surface of cactus ($\times 500$). The lower micrograph shows a single stoma ($\times 800$). The surface shows much wax deposition.

Panel B

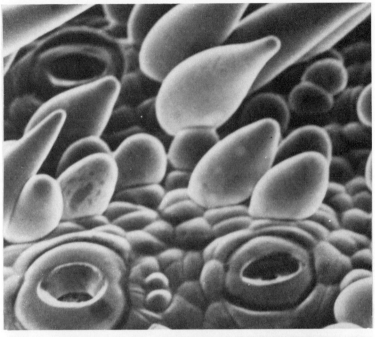

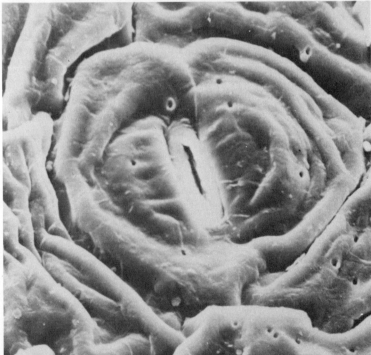

FIGURE 5.1 (cont.)

Panel C (upper) shows stomata and trichome hairs on the surface of a cotton-plant ovule. Photograph courtesy of C. A. Beasley.

Panel C (lower) shows a stoma on the surface of a barrel cactus. The subsidiary cells are very prominent.

Panel C

5.1 Stomatal Physiology

The stomata (singular: stoma = mouth) occurring on plant surfaces are among the most interesting of plant structures. They are present on exposed parts of virtually all plants except algae and fungi and thus occur in liverworts and mosses and all vascular plants. Although generally thought of as being regulatory structures of leaves, they are also found on stems, floral parts, and fruits. Figure 5.1 shows the surface morphology of stomata as seen with the aid of a scanning electron microscope.

On the average there are about 10,000 stomata per cm^2 of leaf surface, although many xeric plants, such as the succulents (cacti, ice plant, *Sedum*, etc.), will average 1000 and some deciduous trees may have as many as 100,000 or more per cm^2. Usually there are more stomata on the lower surface of leaves than there are on the upper surface. Stomata play a major role in the physiology of plants by regulating gas exchange. In general, they are open in the light and closed in the dark.

5.1.1 Morphology

The stomatal apparatus is composed of two guard cells surrounding the pore, two or more adjacent subsidiary cells, and a substomatal cavity. The apparatus is discussed below.

5.1.2 Guard cells

Different kinds of guard cells are known, but basically there are two types: (1) elliptical (kidney-shaped) cells, and (2) the dumbbell-shaped guard cells of grasses. Kidney-shaped guard cells are shown in Fig. 5.2A and the dumbbell-shaped guard cells are illustrated in Fig. 5.2B. Guard cells have a large, well-defined nucleus and small chloroplasts. Therefore, they are unusual in that they have plastids while other epidermal cells of land plants ordinarily do not. They have mitochondria and other common cellular inclusions. Figure 5.3 shows two electron micrograph views of the internal structure of guard cells.

There are few or perhaps no plasmodesmata (see Chapter 6) or cellular connections between guard cells and subsidiary cells, although plasmodesmata do occur between subsidiary cells and adjacent epidermal cells. Because the connecting walls between guard cells and subsidiary cells are thin, transport does not seem to require the presence of cellular connections.

As can be seen in Fig. 5.3B, the elliptical guard cells have thickened ventral walls adjacent to the pore and thin dorsal walls distal to the pore. The thickened ventral walls of the guard cells aid in the opening process. The entire stomatal apparatus is

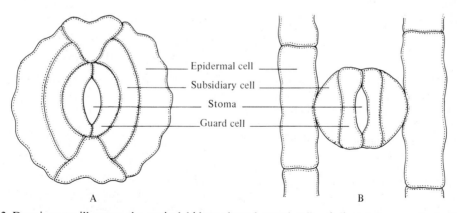

Epidermal cell
Subsidiary cell
Stoma
Guard cell

A   B

FIGURE 5.2 Drawings to illustrate the typical kidney-shaped guard cells of dicotyledonous plants (A) and the dumbbell-shaped guard cells of monocotyledonous plants (B).

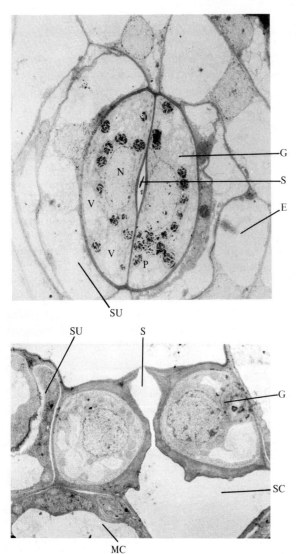

FIGURE 5.3 Electron micrographs of dicotyledonous-plant guard cells making up a stoma. Panel A shows a face view and Panel B shows a cross section through two guard cells (×11,040). G = guard cell. S = stoma. SU = subsidiary cell. E = epidermal cell. MC = mesophyll cell. SC = substomatal cavity. N = nucleus. M = mitochondrion. P = plastid. V = vacuole. Courtesy of Dr. W. W. Thomas, University of California, Riverside.

covered with cutin, which prevents water loss from the guard cells.

The grass-type guard cell shown in face view in Fig. 5.2B is composed of thin-walled, enlarged end portions and a bridge with thickened walls. There are no plasmodesmata connecting the adjacent subsidiary cells. Swelling of the end portions because of water uptake causes opening of the pore.

The guard cells differentiate from protodermal cells early in the expansion of the leaf. After differentiation, slight separation of the ventral walls creates the pore, or stoma. Both types of guard cells have a micellar cell-wall structure with the micelles radiating from the pore to the dorsal guard-cell walls (refer ahead to Fig. 5.19).

5.1.3 Subsidiary cells

All guard cells have adjacent subsidiary cells that apparently aid in the opening process, although in plants such as *Vicia faba* there is little differentiation of the subsidiary cells, which appear comparable to other epidermal cells. Cell walls of subsidiary cells are thin, and there are no chloroplasts present. Plasmodesmata may connect subsidiary cells with adjacent epidermal cells. The guard cells are suspended over the substomatal cavity by the subsidiary cells. A typical arrangement is shown in Fig. 5.4.

Cutin deposits line the exposed surfaces of all cells within the substomatal cavity; it evidently protects against injury and excessive drying.

5.1.4 The size of stomata

Guard cells are generally smaller than the adjacent subsidiary cells and other epidermal cells. An individual elliptical guard cell has surface dimensions on the order of 40×15 μm. Table 5.1 lists some common dimensions. When open, length changes little or not at all and width is slightly smaller because of the increase in height.

The pore aperture is somewhat variable when open but averages about 10 μm in diameter. The

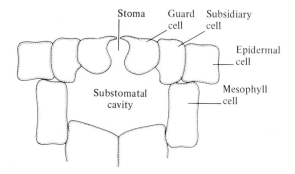

FIGURE 5.4 Drawing of stomatal apparatus showing the attachment of guard cells to subsidiary cells. Note that the guard cells are suspended over the substomatal cavity by the subsidiary cells.

long width of the pore changes little when open. Meidner and Mansfield (1968) reported stomatal aperture widths when open of 10, 12, and 9 μm for *Allium cepa*, *Vicia faba*, and *Ranunculus bulborus*, with no change in length.

5.1.5 The distribution of stomata

As previously stated, stomata occur on most exposed plant surfaces covered by an epidermis. They occur on both surfaces of the leaves of many plants, but on woody plants they may be present just on the lower surfaces. When stomata are on both surfaces, leaves are said to be amphistomatous. When they are just on the lower surface, the leaf is said to be hypostomatous. The epistomatous (upper surface) condition occurs on certain floating aquatic leaves. Submerged leaves normally do not have stomata. Based on an average of 10,000 stomata per cm^2 and a maximum aperture of 10 μm, there

would be 100 μm between stomata, which when open would represent 1% of the total surface area. The remarkable feature of the data shown in Tables 5.2 and 5.3 is the constancy of open area and relative spacing.

TABLE 5.1 Some guard cell dimensions; measurements are in μm

Species	Open	Closed
Allium cepa	38×14	38×17
Vicia faba	40×9	40×11
Ranunculus bulborus	45×13	45×14

Data from Meidner and Mansfield (1968).

TABLE 5.2   Dimensions of stomata from some common plants

Plant	Stomata per cm^2	Aperture* (diameter in μm)	Spacing† (stomatal diameters)
Bean	28,100	5.4	12.6
Begonia	4,000	15.6	11.5
Castor bean	17,600	7.6	11.2
Coleus	14,100	7.9	12.0
English ivy	15,800	8.3	10.9
Geranium	5,900	15.9	9.2
Maize	6,800	13.9	9.9
Oat	2,300	27.5	8.6
Sunflower	15,600	16.5	5.5
Tomato	13,000	10.4	9.5
Wheat	1,400	27.4	11.0
Average	11,327	14.2	10.2

*Aperture is calculated from length $\times$ width and assumes a perfect circle.

†Spacing, expressed in relative stomatal diameters, is calculated from the ratio of the absolute center-to-center distance to the maximum pore diameter. If stomata are 100 μm apart and the aperture is 10 μm, the relative spacing is 10 (100 μm/10 μm). Data from Verduin (1949).

TABLE 5.3 Frequency of stomata and assumed open area of leaf when stomata are open

Species	Stomata per cm²		Open Space, %
	Upper	Lower	
Pinus sylvestris	12,000	12,000	1.2
Larix decidua	1,400	1,600	0.15
Allium cepa	17,500	17,500	2.0
Zea mays	9,800	10,800	0.7
Tilia europea	—	37,000	0.9
Helianthus annuus	12,000	17,500	1.1
Vicia faba	6,500	7,500	1.0
Sedum spectabilis	2,800	3,500	0.32

Open pore area calculated by assuming 6 μm maximum aperture.

Data from Meidner and Mansfield (1968).

A useful measurement of stomatal spacing is determined on the basis of relative diameter, i.e., the ratio of the distance between stomata to the maximum aperture. Verduin (1949) calculated (his Table 4.2) that, on the average, the relative spacing was 10.2 for 14 flowering plant species. This remarkable constancy is important for diffusion through multi-pored surfaces (see Section 5.2).

Since stomata differentiate early in the development of leaves, the number per area and the spacing of stomata will change as the leaf ages. Furthermore, it is known that leaves higher on the plant have more but smaller stomata. For example, Meidner and Mansfield (1968) noted that on wheat, the first leaf below the top flag leaf had 5000 stomata per cm² on the upper surface whereas the third leaf below the flag leaf had 3900 on the upper surface, or 22% fewer. The distribution on a single leaf may vary significantly as well. They also noted that on the upper surface of a corn leaf there were 7700 stomata per cm² proximally, 9800 in the midregion, and 10,800 distally. At each position, there were 8200, 10,800, and 11,800 per cm² on the lower surface.

5.1.6 Diurnal opening and closing cycle

The most important physiological feature of stomata is that they open in response to light and close in darkness (Fig. 5.5). With the exception of certain succulent plants, deviations from the light opening response are not of prime importance in plant biology.

Figure 5.6 depicts the diurnal course of stomatal

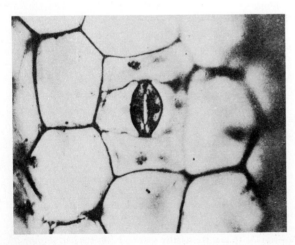 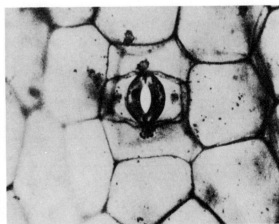

FIGURE 5.5 Photograph of stomata of *Zebrina* leaf samples taken in the dark while closed (left) and in the light while open (right). The epidermis was stripped from the leaf and plunged into absolute alcohol. After fixing the tissue was strained with Congo red. From F. E. Lloyd. 1908. *The Physiology of Stomata.* Carnegie Institution of Washington Publ. No. 82.

opening in a broadleafed, dicotyledonous plant. In idealized form, stomata will open within 30 minutes after the onset of light, usually to their maximum aperture. Throughout the daylight period, they usually remain open. Sometimes there will be slight closing toward the end of the day, and in some plants there may be complete afternoon closing. It is not infrequent to find midday closure, either partially or completely, if plants are water-stressed. The midday closing response is due to guard-cell collapse when leaves wilt. Late-afternoon recovery is common when water uptake once again balances losses. Stomatal closing in the dark may be somewhat slower than opening. It is not uncommon to observe some night opening, particularly toward the end of the dark period.

Some investigators have observed a stomatal-opening rhythm during the daylight period. Such a rhythm results because of periodic daily water stress in leaf cells plus interaction with the regulatory mechanism.

Stomatal opening in succulent plants deviates substantially from the norm. In the common type of succulent, stomata will be closed during the day and open at night. In some, such as the cacti, stomata may be open throughout the dark period and closed all day. In others—for example, *Kalanchoe*—stomata are open at night and toward the end of the day period but are closed in the morning. In others, the typical pattern of day opening and night closing may exist. All variations of day and night opening are known in succulents, and some species change their patterns, depending on the environmental conditions.

Figure 5.7 illustrates stomatal opening in a succulent plant during the course of a 24-hour period. The nocturnal stomatal opening in succulent plants is related to both their nighttime uptake of CO_2 and their organic-acid synthesis. The night stomatal opening, organic-acid synthesis, and night CO_2 uptake is called crassulacean acid metabolism (CAM) and is discussed in Chapter 15.

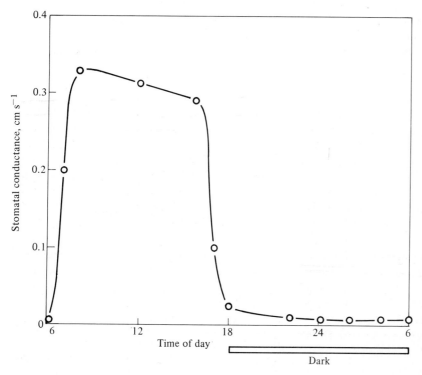

FIGURE 5.6 The diurnal curve of stomatal opening. The data are expressed as stomatal conductance (cm s^{-1}), an indication of the capacity for diffusion through stomata and an indirect measure of stomatal opening. The stomata open rapidly in the light and close at the end of the daylight period. Stomata remain closed throughout the dark period. The data are from *Peperomia*, and were obtained by the author in 1978 at the University of California, Riverside.

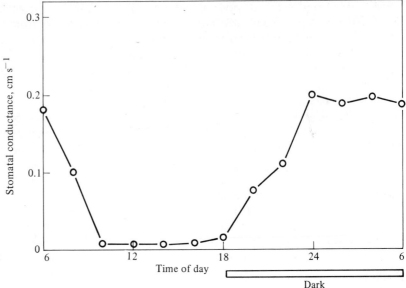

FIGURE 5.7 The diurnal curve of stomatal opening for a Crassulacean acid metabolism (CAM) succulent. The data are expressed as stomatal conductance (cm s^{-1}). Refer to Fig. 5.6. The data are from a cactus obtained by Dr. Zac Hanscom in 1977.

5.2 The Basic Transport Equation

The coupling of the plant to the atmosphere via gas exchange is one of the more important biophysical aspects of plant physiology. The exchange of CO_2 and O_2 during photosynthesis and respiration and the loss of water vapor during transpiration have been studied by plant physiologists for many decades. Recently, through the development of some simple gas-transfer equations, a thorough understanding has been gained. Since most of the gas exchange takes place through the stomata, a consideration of the physics of small-pore diffusion is important. In addition to small-pore diffusion, the principles of evaporation are important for an understanding of water loss by vapor exchange.

In this section the principles of evaporation, discussed from the standpoint of plant water loss and the principles of small-pore diffusion as they relate to stomata, will be covered. After the discussion of evaporation, the gas-transfer equation describing gas diffusion between the leaf and the atmosphere will be developed, based on an analogy with Ohm's law. After the latter discussion there is a section detailing the entire development of the concepts of small-pore diffusion. This detailed section is not necessary for an understanding of gas transfer in plants but is included because of the importance of the process to plant physiology. It can be used as reference or additional study.

5.2.1 Evaporation

A pan of water or any surface saturated with water will lose water according to the laws of evaporation. The only requirement is that there be a free-energy difference between the evaporating surface and the atmosphere (see Chapter 4). If the free energy or water-potential gradient is reversed, condensation will occur. The physical properties of water that should be understood before reading the following discussion are outlined in Chapter 2. Chapters 2 and 4 should be reviewed before proceeding with this section.

Evaporation from pure water is a function of the temperature of the water, the temperature of the atmosphere, and the relative humidity. In still air, only the vapor pressure of the wet surface and the vapor pressure of the gaseous water in the atmosphere need to be known to obtain an indication of the potential for evaporation. If we assume the evaporating surface is free water with a water potential of zero ($\psi = 0$), the following will be the best indication of the potential for evaporation:

$$\Delta e = e^\circ - e,$$

where

e° = vapor pressure of the evaporating surface, and

e = vapor pressure of the water in the air.

The Δe, the difference between the vapor pressure of the surface and the vapor pressure of the air, is the driving force for evaporation. The greater the Δe, the greater the potential for evaporation.

Rather than expressing the e terms in pressure units (for example, mbar), for the calculation of transpiration it is more convenient to express e in terms of actual vapor densities [in units of $g(H_2O) \times 10^{-6} \text{ cm}^{-3}$ (air)]. The saturation vapor density (e°) as a function of temperature can be read from the graph in Fig. 5.8. Alternatively, the saturation vapor pressure in mbar can be read from the graph in Fig. 2.6 (Chapter 2).

As an example, reading from Fig. 5.8 at 25°, the saturation vapor density (e°) is $23 \times 10^{-6} \text{ g cm}^{-3}$. To determine the actual vapor density of the air (e), the relative humidity and air temperature must be known. The relative humidity is the actual amount of water in the air, expressed as a percentage of the saturation value. If the relative humidity is 70% and the air temperature 25°, e is $16 \times 10^{-6} \text{ g cm}^{-3}$, or 70% of $23 \times 10^{-6} \text{ g cm}^{-3}$. Any value for e can readily be computed from e° and the relative humidity:

$$e = e^\circ \times \frac{\% \text{ RH}}{100}.$$

To estimate Δe, care must be taken to determine

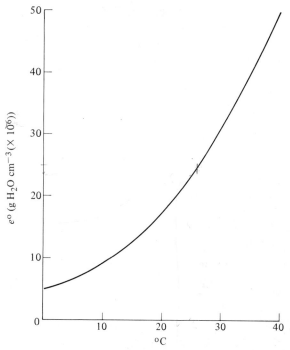

FIGURE 5.8 Graph showing the relationship between the saturation vapor density (e°) and temperature. The units of e° are $g \times 10^{-6} \text{ cm}^{-3}$.

e° for the temperature of the water surface that is evaporating, and e must be estimated from the relative humidity and e° at the air temperature. Only if the evaporating surface and the air are at the same temperature will the e° values be the same.

The index Δe is sometimes called the vapor-pressure deficit, because it represents the amount of water the air can hold. Of course, the latter is only valid if e° is determined for air temperature. It is perhaps better to consider Δe as the driving force for evaporation.

5.2.2 The form of the gas-transfer equation

A simple way to understand the gas-transfer equation for describing CO_2 uptake during photosynthesis and water loss during transpiration is with an analogy to Ohm's law for the flow of electric current. Ohm's law is:

$$I = \frac{E}{R},$$

where

I = current flow,

E = voltage, and

R = resistance to current flow.

E, the voltage, is the electrical potential difference between the source and the sink, or, in other words, the potential difference for electrical flow between one point and another. The greater the voltage, the greater the potential for current flow. The voltage can be considered as the driving force for current flow. The characteristics of R, the resistance, will govern the actual rate of current flow. R can be visualized as the characteristics of the wire through which flow takes place. Long, thin wires have a greater R than short, fat wires. Conducting properties of the wire will also change the resistance.

The basic gas-transfer equation describing the exchange of gases between the plant and the environment takes the same form as Ohm's law:

$$f = \Delta/R,$$

where

f = gas uptake or loss expressed in convenient units such as g (of gas) cm^{-2} (of plant surface) s^{-1},

Δ = gas potential difference between the plant surface and the atmosphere (analogous to voltage) (The Δ will be followed by a symbol to indicate the kind of gas: Δe for water and Δc for CO_2), and

R = resistance to gas transfer.

For transpiration, f would be expressed in units of g (of H_2O loss) cm^{-2} (of leaf surface) s^{-1}. Usually, transpiration is symbolized as T. Thus the expression for transpiration would be: $T = \Delta e / R$.

The driving force for transpiration, Δe, analogous to the voltage of Ohm's law, is the vapor-pressure difference between the leaf and the atmosphere. Thus Δe is identical to the Δe explained above under evaporation. Δe is determined in exactly the same way as is the Δe for evaporation. Thus Δe is calculated from the equation

$$\Delta e = e^\circ - e.$$

It is ordinarily assumed that the water in the leaf evaporating into the substomatal cavity is at 100% relative humidity. This assumption means that the water within the leaf is actually pure water. In fact it is not, but the slight reduction in vapor pressure because of solutes and other forces is not a significant factor and can be ignored for most purposes. Therefore, to ascertain e° for the water in the leaf only the leaf temperature need be measured. The e° value is the saturation vapor density at the leaf temperature.

The vapor density of the air (or vapor pressure) is determined in the same manner as described for e in the section above for evaporation. The relative humidity of the air and the air temperature must be known, and e is then determined from the saturation vapor density (e°) at air temperature, determined from Fig. 5.8 and the relative humidity.

The resistance term (R) of the gas-transfer equation is a function of the stomata and a boundary layer over the leaf that interferes with gas transfer. Therefore, the resistance term can be factored into two component resistances. The most important of the component resistances is that resistance accounted for by the stomata. This resistance is symbolized as R_s and is called the stomatal resistance. R_s is a function of the size, number, and spacing of stomata, and most important a function of whether or not they are open. When they are open, R_s is low; when closed, R_s is high. R_s is a variable resistance because of stomatal opening and closing. It is a function of both the physiological factors (opening, closing) and the physical dimensions of the stomata. The units of R are seconds per centimeter (s cm^{-1}).

In series with the stomatal resistance (R_s) is a boundary-layer resistance (R_a) above the surface.

The boundary layer is considered to be that region above the leaf where the gradient changes linearly, or the region between the leaf and air that is unstirred. It is a property of the leaf surface and air movement. If air is moving over the surface, the boundary-layer resistance is reduced. Figure 5.9 illustrates the relationship between the stomatal resistance and the boundary-layer resistance.

An additional resistance component that can be considered in transpiration is the resistance of the surface that does not have stomata. A certain portion of the water loss from plants takes place directly through the leaf surface, but it is mostly prevented by the presence of cuticle. This resistance

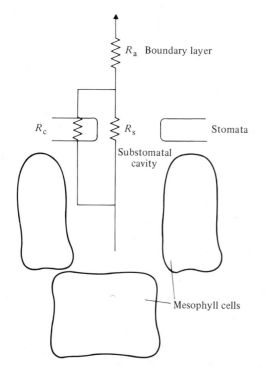

FIGURE 5.9 Electrical analogue model for the diffusion of water vapor from the leaf to the atmosphere during transpiration. The model illustrates the boundary-layer resistance (R_s), the stomatal resistance (R_s), and the cuticular resistance (R_c). R_s and R_c are parallel resistances. R_s and R_a are resistances in series, as are R_c and R_a.

associated with the cuticle is called the cuticular resistance and is symbolized by R_c. As can be seen from the diagram in Fig. 5.9, R_c is a resistance parallel with R_s but in series with R_a. Parallel resistances are not additive directly, but the reciprocals can be added to give the reciprocal of the total resistance: *cuticular resistance*

$$1/R = 1/R_c + 1/R_s.$$

The resistance determined in this way can be added to R_a to give the total leaf resistance.

For the most part, R_c can be ignored because it is very large and does not contribute significantly to the total resistance.* The usual range of R_s for agricultural plants with open stomata is 1 to 2 s cm^{-1}. R_a is on the order of 0.2 s cm^{-1}. R_c can be as high as 500 to 1000 s cm^{-1}.

As shown in Fig. 5.9, the stomatal resistance (R_s) and the boundary-layer resistance (R_a) are in series. Thus they can be added directly to obtain the total resistance, R, provided that the cuticular resistance is ignored. The total resistance (R) of a leaf is

$$R = R_a + R_s.$$

The overall gas-transfer equation for transpiration is

$$T = \frac{\Delta e}{R_a + R_s}.$$

It should be apparent from the above development of the gas-transfer equation that water loss from the plant surface begins by evaporation at the mesophyll cell walls into the substomatal cavity. We have assumed that the relative humidity in the substomatal cavity is 100%. Thus the saturation vapor density (e°) is determined at the leaf temperature using Fig. 5.8. Diffusion of the water vapor from the leaf first meets the variable resistance attributable to the stomata (R_s). R_s will be low if

*As an exercise, prove to yourself that a large R_c is unimportant in the total resistance. Use an R_a of 0.2 s cm^{-1}, an R_s of 1.0 s cm^{-1}, and an R_c of 500 s cm^{-1}. First determine the sum of R_c and R_s using reciprocals and then add to R_a.

the stomata are open and high if the stomata are closed. Second, the water vapor will encounter the boundary-layer resistance (R_a) that further resists water loss from the leaf. In moving air R_a will be less than in still air.

It is possible to extend the gas-transfer equation for transpiration to photosynthesis with only a few modifications. It should be noted that the resistances, R_s and R_a, are physical resistances and can be converted from water to CO_2. The factor 1.56 can be used.* To convert R_s for water vapor to R_s for CO_2, multiply by 1.56. The R_s is greater for CO_2 than for water because CO_2 is a larger molecule and diffuses more slowly.

CO_2 diffuses into the leaf from the air. The driving force (Δc) is the difference between the CO_2 concentration in the atmosphere and the CO_2 concentration in the leaf at the site of CO_2 fixation. On the average, the CO_2 concentration in unpolluted air is 6.47×10^{-7} g cm^{-3} (of air) at STP. It is usually assumed that the concentration of CO_2 at the site of fixation is zero. As a first approximation, therefore, the driving force (Δc) for CO_2 uptake during photosynthesis is taken to be 6.47×10^{-7} g cm^{-3}. If the actual concentration of CO_2 in the leaf is known, this can be taken into account in the computation of Δc.

The first resistance that CO_2 encounters when diffusing into the leaf is the boundary-layer resistance (R_a). As noted above, this is the same resistance encountered by water vapor leaving the leaf, except that it must be corrected for the greater mass of CO_2. Second, the CO_2 will encounter the variable resistance caused by the stomata (R_s). Once again this is the same as that encountered by water but corrected for the mass of CO_2. After entering the substomatal cavity, the CO_2 will go into solution in the fluids of the mesophyll. Within the mesophyll cells, the CO_2 diffuses or is transported to the chloroplast where biochemical reduction takes place. The transport and biochemistry associated with CO_2 fixation are considered as an additional

*The correction factor is the ratio of the square roots of their molecular weights: $\dfrac{\sqrt{CO_2}}{\sqrt{H_2O}} = \dfrac{\sqrt{44}}{\sqrt{18}} = 1.56$.

resistance, known as the mesophyll resistance (R_m). The three resistances to CO_2 uptake (R_a, R_s, and R_m) are illustrated in the diagram in Fig. 5.10. As can be observed from the figure, the resistances are in series and can be added directly. The usual minimum range of R_m is 1.5 to 3.0 s cm^{-1}.

The mesophyll resistance (R_m), sometimes referred to as the residual resistance because it is every component not considered in R_a and R_s, can be factored into two components. There is a physical component attributable to transport and diffu-

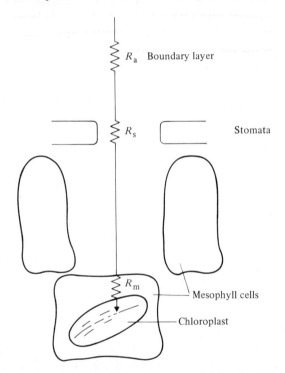

FIGURE 5.10 Electrical analogue model for the diffusion of CO_2 from the atmosphere to the chloroplast during photosynthesis. The model illustrates the boundary-layer resistance (R_a), the stomatal resistance (R_s), and the mesophyll resistance (R_m). The R_a and R_s are the same overall as for water vapor but corrected for the slower-diffusing, larger CO_2 molecule. R_m takes into account both diffusion and transport of CO_2 within the mesophyll cells to the chloroplasts and the biochemistry associated with photosynthesis. R_a, R_s, and R_m are in series and are directly additive. As in water-vapor diffusion, there is a cuticular resistance, R_c.

sion and a biochemical component associated with the chemistry of CO_2 fixation. These two components are poorly understood and will not be discussed further.

The gas-transfer equation for CO_2 uptake during photosynthesis (P) is

$$P = \frac{\triangle c}{R_a + R_s + R_m}.$$

The expression is an extension of the gas-transfer equation for transpiration. Both are based on an analogy with Ohm's law and adequately describe transpiration and photosynthesis when considered at the gas-exchange level.

An alternative method to express the gas-transfer equation is in terms of conductances (G) rather than in terms of resistances (R). Conductance is the reciprocal of resistance and has units of cm s^{-1}:

$$G = \frac{1}{R}.$$

Thus in terms of conductances, the basic gas-transfer equation is:

$$f = G \triangle e$$

As with resistance, the conductance can be factored into boundary-layer conductance (G_a), stomatal conductance (G_s), and a mesophyll conductance (G_m). Conductances are sometimes more useful because they are directly proportional to gas transfer rather than inversely proportional as in the case of the resistances. The data in Fig. 5.6 and 5.7 showing the diurnal stomatal opening curves are expressed in terms of conductances and give an immediate impression of stomatal opening. The actual data are G_s in units of cm s^{-1}. Further discussions of the gas-transfer equations are in Chapters 6 and 16.

Physics of small-pore diffusion

The gas-transfer equations discussed above can also be developed through a consideration of the physics of small-pore diffusion and further analogy with Ohm's law. A model illustrating a small pore

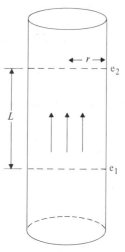

FIGURE 5.11 Model for Fick's Law of Diffusion. Diffusion takes place within a tube with length L and radius r. Diffusion is from e_1 to e_2. As explained in the text, the diffusion will be proportional to the driving force, i.e., the difference between the gas concentration at e_1 and e_2 (that is, the $\triangle e$), and to the cross-sectional area (πr^2) and will be inversely proportional to the length between e_1 and e_2.

and simulating a stoma is depicted in Fig. 5.11. In terms of diffusion through small pores (see Fig. 5.11), the vapor-pressure difference between e_1 and e_2 is analogous to the voltage. The greater the $\triangle e$ (that is, $e^1 - e^2$), the greater the potential for diffusion from 1 to 2. The resistance term, R, is composed of the length (L) between 1 and 2, the cross-sectional area of the tube (πr^2), and a factor D (diffusion coefficient). The diffusion coefficient (D) varies with the temperature, the properties of the gas that is diffusing, and the properties of the medium through which the diffusion takes place.

Thus an expression for diffusion through small pores is

$$Q = \frac{\triangle e}{L/(\pi r^2 D)}.$$

Q is analogous to I or current flow, $\triangle e$ is analogous to the voltage and is the driving force, and $L/(\pi r^2 D)$ is the resistance, R. R can be evaluated as follows: when L increases, there is a greater diffusion path length from 1 to 2. R increases and diffusion is less. As the cross-sectional area of the tube increases, diffusion increases and the term R gets smaller. As D increases, diffusion increases. The diffusion coefficient can be computed precisely. It is discussed in Chapter 2 in detail.

Although R can be computed from a knowledge of characteristics of the system, it is perhaps more convenient to estimate it by experiment. This is done by predicting Q (diffusion) on the basis of Fick's law, $Q = D\Delta e$ (refer to Section 2.6.2). The ratio of the actual diffusion to the predicted diffusion will be R, the resistance.

In summary, the equation $Q = D\Delta e / R$ will be adequate to describe diffusion between two points within a tube. The expression as defined here is known as Fick's Law for Diffusion.

The model for stomatal diffusion is somewhat different from Fick's law and different from that described in Fig. 5.11. A cross section of a typical stoma is shown in Fig. 5.12. The diffusion streamlines shown in the figure depict diffusion through the stoma and illustrate that for water the source is below the stomatal tube at the mesophyll surface and that the atmosphere sink is above the stomatal pore tube. Thus diffusion through a stoma deviates substantially from Fick's law, which describes dif-

fusion within a tube. Because source and sink are greater than for the within-tube model and because the diffusion path length is longer, a substantial correction is necessary. This correction must be added to R or $1/r^2$ in order to increase the diffusion path length. From acoustical theory, the term $1/r$ best corrects for the increase in path length since the increase is proportional to the reciprocal of the tube radius. Two ends require a $1/2r$ correction. Thus the resistance term becomes

$$R = \left(\frac{L}{\pi r^2} + \frac{1}{2r}\right)\frac{1}{D}.$$

The term adequately describes the resistance to diffusion through a single stoma.

The Diameter law

It was discovered by Brown and Escombe in 1900 that diffusion through a single pore was more nearly proportional to the radius or diameter than to the cross-sectional area of the pore. An experimental relationship for diffusion through single pores of stomatal dimensions is shown in Fig. 5.13.

This Diameter law relationship will hold provided that the path length to cross-sectional area of the tube is small, or, in terms of stomata, if the guard

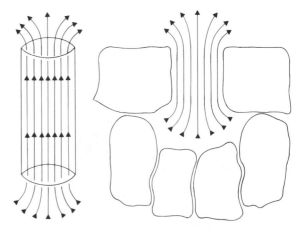

FIGURE 5.12 Model for diffusion through a small tube comparable to a stoma and a model for analogous diffusion through a stoma. Note that the source for the gas is below the opening of the tube (below the stoma in the substomatal cavity) and that the sink is above the opening of the tube (above the stomatal pore). Thus both the source for the gas and the sink for the gas are larger than the tube dimension. This gives the stomata greater capacity for diffusion than would be predicted from the within-tube model (Fick's law) illustrated in Fig. 5.11.

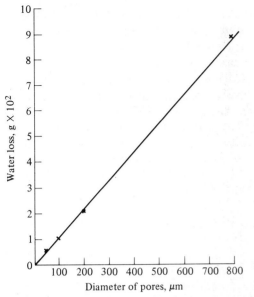

FIGURE 5.13 Brown and Escombe relationship between diffusion through a small pore of stomatal dimensions and the diameter of the pore. Water loss increases linearly with an increase in pore diameter. This relationship of proportionality to pore diameter rather than proportionality to pore area is a consequence of the large source region above the pore. See Fig. 5.12. From I. P. Ting and W. E. Loomis. 1963. Diffusion through stomates. *Amer. J. Bot.* 50:866–872.

cells are not thick. The Diameter law can be shown algebraically as follows:

$$Q = \frac{\Delta e}{(L/\pi r^2 + 1/2r)\,\dfrac{1}{D}} .$$

Rearrangement gives

$$Q = \frac{D\Delta e\,\pi r^2}{L + \pi r/2} .$$

If $r \gg L$, then L plus $\pi r/2$ is approximately equal to $\pi r/2$. Thus

$$Q = \frac{D\Delta e\,\pi r^2}{\pi r/2}$$

and

$$Q = D\Delta e(2r);$$

the diffusion is proportional to $2r$ (the diameter). For elliptical pores, as is the case for most stomata, the product of width times length should be used rather than πr^2.

It should be noted that the Diameter law will only hold under special conditions.

Multipore diffusion

As described in Section 5.1.5, there are on the average 10,000 stomata per square centimeter. Assuming an average maximum aperture of 10 μm, 1% of the area would be open. On first thought, it would seem reasonable to simply multiply the single-pore gas-diffusion equation by the number of stomata per cm^2 (n) to obtain a new expression describing multipore diffusion. Such an expression would be:

$$Q = \frac{(n)\,\Delta e}{R} .$$

When this is calculated, the predicted water loss is estimated to be 50 times greater than an open surface of comparable area, an absurd prediction. The upper limit for water loss by any surface would be the free-water evaporation rate.

To understand this limitation, it is necessary to visualize the vapor layer that forms over the evaporating surface. Figure 5.14 depicts diffusion through single pores and shows the diffusion streamlines plus vapor- or diffusion-shell buildup over the pore. In the multipored situation, the vapor shells that form over each pore eventually coalesce and form a single vapor shell (Fig. 5.14). The coalescence of vapor shells to form a single vapor shell, which is comparable to the free surface, sets the upper limit for diffusion away from the surface at the maximum evaporation of the free surface.

This vapor shell that forms over the multipored surface represents an additional resistance to diffusion and is identical to the boundary-layer resis-

FIGURE 5.14 Drawing to illustrate diffusion stream-lines and vapor shells that build up over pores. The top diagram depicts three pores that are far enough apart to act independently (there is no overlap of diffusion shells). The lower diagram showing five pores illustrates what happens when the pores are so close that the diffusion shells overlap. Coalescence forms a single vapor shell over the entire surface that represents the boundary layer. Thus a multipored surface such as this will act similarly to a free surface of equal area.

tance. If we use R_s as the resistance attributable to the pores, then R_a can be used to indicate the boundary-layer resistance. Thus the resistance to diffusion, R_s and R_a, attributable to the pores and the boundary layer are in series and can be added to obtain the total resistance, R:

$$R = R_a + R_s.$$

The diffusion equation adequate to describe multi-pore diffusion similar to that of a leaf is

$$Q = \frac{\Delta e}{R_a + R_s}.$$

The curve for diffusion as a function of pore diameter of multipore systems such as a leaf epidermis is hyperbolic, as shown in Fig. 5.15. This latter curve is for pores ranging from 2.5 to 20 μm in diameter and 2500 pores per cm². The shape of this curve can be explained based on the principles of small-pore diffusion.

When the pore aperture is small, about 2.5 to 5 μm, the pores are far apart. They will act nearly independently, and the diffusion is nearly linear with pore diameter. The relationship is similar to that shown in Fig. 5.13. However, when the pores become close on a relative basis they interfere, and the total diffusion per pore decreases. Finally, at maximum aperture the diffusion is constant or at a maximum. This is because the diffusion shells have coalesced and the multipored surface is acting as a unit (refer to Fig. 5.14).

An alternative way to visualize the shape of the curve in Fig. 5.15 is in terms of the relative magnitude of the resistances, R_a and R_s. When the stomatal apertures are small, in the range of 2.5 to 5 μm, the R_s is large. R_s is large in relation to R_a and

FIGURE 5.15 Diffusion of water vapor through a model multipore surface analogous to leaf epidermis. The diameter of the pores ranges from 2.5 μm to 20 μm, which is over the range of stomatal dimensions. There are 2500 pores per cm². The graph shows the expected linear increase in diffusion at small apertures and the nearly constant diffusion rate at wide pore apertures. The shape of the curve is predictable from the principles of small-pore diffusion outlined in the text. From I. P. Ting and W. E. Loomis. 1965. Further studies concerning stomatal diffusion. *Plant Physiol.* 40:220–228.

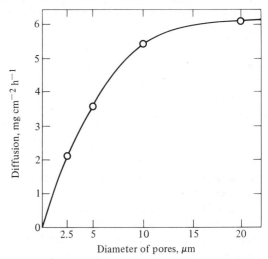

the expression is essentially

$$Q = \frac{\Delta e}{R_s} \, ,$$

and the diffusion is in proportion to the diameter. A linear response of diffusion to diameter of pore is expected. When stomata are wide open, the R_s is low and unimportant compared to R_a. Since R_a is nearly constant and a function of the size of the surface, changing pore diameter does not alter the diffusion. The expression is

$$Q = \frac{\Delta e}{R_a} .$$

That the boundary layer, R_a, must be limiting diffusion can be seen by observation in moving air. Moving air will minimize the boundary-layer resistance, R_a, and once again the expression becomes $Q = \Delta e / R_s$. There will be a linear increase in diffusion with increases in pore diameter. The effect of wind on the boundary layer resistance is shown in Fig. 5.16.

It should be apparent that the gas-transfer equation developed here is identical to that developed in Section 5.2.2.

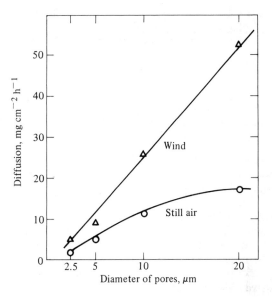

FIGURE 5.16 The effect of wind on the diffusion through a multipore surface comparable to a leaf epidermis. In still air the expected hyperbolic diffusion function is present (see Fig. 5.15). In moving air the diffusion is almost a linear function of the diameter of the pores of the multipore surface. From I. P. Ting and W. E. Loomis. 1965. Further studies concerning stomatal diffusion. *Plant Physiol.* 40:220–228.

5.3 Environmental Factors Affecting Stomatal Opening

5.3.1 Light

As a rule, stomata open in light and close in darkness provided that the leaf is in good water balance. Care must be taken to differentiate between the rate of stomatal opening and the final equilibrium aperture obtained in the light. Both the rate and the final aperture increase with increasing light intensity. The rate of stomatal opening is quite fast, with a maximum aperture obtained within 15 to 60 minutes after exposure to threshold light levels. Threshold light levels are in the range of 1 to 2% of full sunlight. The actual rate of opening and the maximum aperture reached is dependent on the species. Some plants have stomata that require sub-stantially high light levels, whereas others may even open in the dark.

Attempts to measure an action spectrum for stomatal opening have revealed maximum response to blue and red light with a spectrum comparable to photosynthesis. The light opening response is more than likely a photosynthetic response. Guard cells have chloroplasts and perform some of the reactions of photosynthesis. The role of photosynthesis may be in production of energy (that is, ATP) for the energy-dependent cation pumping or for organic solute synthesis and in the reduction of the partial pressure of CO_2. It is known that reduced CO_2 levels are accompanied by stomatal opening (see Section 5.3.2).

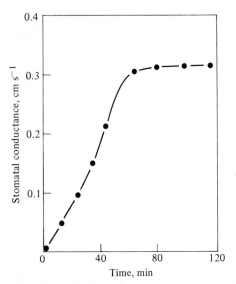

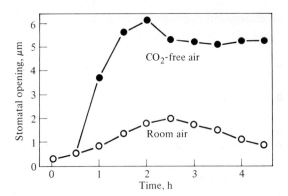

FIGURE 5.17 Early-morning stomatal opening curve for an almond tree, expressed in terms of stomatal conductance (cm s⁻¹). At zero time, the sun hit the leaves. After 60 min the stomata were at their maximum conductance, indicating that maximum stomatal aperture had been reached. At this time (0830), the sun was 20% of the maximum radiation reached during the day (maximum = 1600 microeinsteins m^{-2} s^{-1}).

FIGURE 5.18 Graph to illustrate stomatal opening of *Xanthium* when exposed to CO_2-free air while in the dark. In the experiment, the temperature was increased from 27° to 36° after 14 hours in the dark at 27°. The increased temperature caused some stomatal opening, but the primary response was to the CO_2-free air. Redrawn after graph in T. A. Mansfield. 1965. Studies in stomatal behavior XII. Opening in high temperature in darkness. *J. Exp. Bot.* 16:721–731.

The closing response in darkness may be rapid or slow. In some cases darkness is accompanied by an immediate stomatal closure even faster than the opening response. But in other cases complete closing may take hours.

Figure 5.17 shows the early-morning stomatal opening curve for the stomata on the leaves of an almond tree. Stomatal opening, expressed in conductance, reached a maximum within 60 minutes after direct sun hit the leaves.

5.3.2 Carbon dioxide

As early as 1916, Linsbauer observed that reduced CO_2 levels around a leaf caused stomatal opening regardless of whether the plant was in the light or the dark. It was subsequently shown that for most plants, maximum aperture is achieved at about 0.1 mbar CO_2 pressure, or close to 33% of ambient. This CO_2 concentration is about equal to the CO_2

compensation point, the point where photosynthetic CO_2 uptake just balances CO_2 losses by respiration. A few species such as *Zea mays* have compensation points near zero. In *Zea mays*, the maximum stomatal aperture occurs when the CO_2 concentration is close to zero. It seems that the maximum stomatal aperture occurring in response to reduced CO_2 is when the level is near the CO_2 compensation point for the species.

Whereas low CO_2 levels tend to cause stomatal opening, high CO_2 (above ambient) will tend to close stomata. The stomatal response to CO_2 concentration has led to the conclusion that the role of photosynthesis in stomatal opening is to bring about a decreased CO_2 concentration in the vicinity of the guard cells. Figure 5.18 shows the effect of CO_2 on stomatal aperture.

5.3.3 Water and humidity

Because stomatal opening is a physical process dependent upon guard-cell turgor, any changes in water availability may alter stomatal opening. The midday closure of stomata so frequently observed probably comes about by leaf cells becoming flaccid when water loss exceeds water uptake.

It is expected that water-stressed or wilted plants will have closed stomata. But, because stomatal opening depends to a large extent on relative turgor between guard cells and adjacent epidermal cells, a slight change in leaf water status may result in stomatal opening. Indeed, during the first steps of wilting it is not uncommon to observe slight stomatal opening.

There is evidence suggesting that relative humidity of the air affects stomatal opening independently of its effect on guard cells. Stomata tend to close if exposed to low humidity.

5.3.4 Temperature

It is difficult to separate temperature (and light) effects from water status effects since high temperatures affect water status by drying. There is evidence, however, that stomatal opening is greater at high temperatures than at low temperatures and that some plants have a stomatal-opening threshold temperature of $5-10°C$.

Since stomatal opening and the maintenance of stomatal opening are active processes dependent on plant metabolism, it is expected that temperature

plays a direct role. Temperature effects on the above processes will directly affect rates of stomatal opening and the degree of stomatal opening.

5.3.5 Environmental interaction

It is difficult to separate the effects of individual environmental parameters since all are related to some extent. Strong light is often accompanied by high temperature, which affects water status. Photosynthetic activity takes place in light; thus CO_2 will be low near guard cells in the light. Both temperature and water status affect photosynthesis.

Stomatal opening is the result of a balance of different processes, depending on the environment. Opposing environmental effects such as light, CO_2, and water account in part for rhythmic stomatal behavior. Light causes opening by a reduction in partial pressure of CO_2. Open stomata result in water losses by transpiration and some wilting. Stomata then close, causing internal photosynthetic depletion of CO_2 with the result that stomata open again. During the closing phase some water recovery is expected. The overall pattern is a periodic rhythm of stomatal aperture variation. Such control by CO_2 and water is a good example of control by feedback regulation.

5.4 The Mechanism of Stomatal Opening

Despite over a century of active research and investigation by numerous plant scientists, we still do not know precisely how stomata open and close. It is clear, however, that opening is a response to increased turgor of the guard cells relative to adjacent subsidiary and other epidermal cells. Similarly, closing is accompanied by a decrease in turgor.

The physical mechanism for opening is explainable in terms of the known turgor changes and the microanatomy of the guard cells. Typical elliptical guard cells are illustrated in Fig. 5.19, showing the cellulose micellar arrangement. These micelles tend to radiate out from the pore aperture from the thickened inner ventral wall adjacent to the pore

to the thin dorsal wall adjacent to the subsidiary cells. The greatest extension (when turgid) takes place parallel to the length of the guard cell and is restricted along the width. As turgidity increases the thin dorsal wall will bulge, forcing the ventral wall into a concave position. The net result of increased turgor is the bowing inward of the ventral walls and the opening of the aperture. Consideration of the microanatomy also explains why the width of the aperture changes during opening but the aperture length remains virtually unchanged. The remaining question is what brings about the guard-cell turgor changes that account for the opening. It seems reasonable to conclude that the

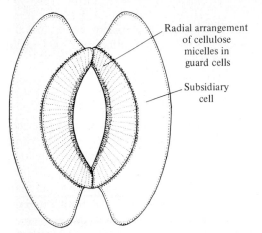

Radial arrangement
of cellulose
micelles in
guard cells

Subsidiary
cell

FIGURE 5.19 Diagram of a stomatal apparatus illustrating the radial arrangement of cellulose micelles in the guard cells. Because of the radial arrangement extending from the thin dorsal wall to the thickened ventral wall, increased turgor of the guard cells causes bowing inward of the ventral walls and opening of the aperture. Adapted from H. Meidner and T. A. Mansfield. 1968. *Physiology of Stomata.* McGraw-Hill, New York.

increase in turgor of the guard cells results from a decrease in water potential. The only reasonable mechanism for such a decrease is by a decrease in osmotic potential within the guard cells themselves. The decreased osmotic potential comes from the accumulation of osmotically active solutes, either by synthesis within the guard cell or transport from adjacent cells.

Active transport of water into the guard cells is possible but thermodynamically unlikely. It is much more probable that solute accumulation within the guard cell is the mechanism for the increased turgor. Evidence supporting this probability is discussed below.

5.4.1 Physical factors affecting stomatal opening

Environmental factors affecting stomatal opening, such as light, CO_2, humidity, and temperature, were discussed in Section 5.3. In review, stomata open in the light and close in the dark, they open in response to low CO_2 partial pressures, perhaps at or near the CO_2 compensation point, and they are affected by the temperature and humidity. These factors must be included in the consideration of any mechanism explaining increases in guard-cell turgor.

As mentioned earlier, the closing response to decreased humidity probably comes about by guard cells adjusting to the decrease and losing turgor by water loss. The high-temperature effects could be caused either by water loss or perhaps by regulation of specific biochemical reactions governing guard-cell operation.

5.4.2 Oxygen and respiratory inhibitors

Anaerobiosis will prevent stomatal opening. It seems clear that oxygen is a necessary component, probably for oxidative respiration. To support this notion of the requirement for respiration, it is known that many respiratory inhibitors prevent opening and cause closing while in the light. For example, Zelitch in 1961 reported that phenolic compounds, azide, and arsenate inhibit opening. Interestingly, leading to the conclusion that the opening mechanism and the closing mechanism may be different is that azide, unlike the other compounds studied, prevented closing in the dark.

Inhibitors of photosynthesis also inhibit opening. Thus it seems reasonable to conclude that both oxidative phosphorylation and photophosphorylation (producing ATP) are necessary for stomatal opening.

5.4.3 Acidity

It has been known for a long time that treatment of epidermal strips with alkaline solutions tends to cause stomatal opening; and there is some evidence that the pH of guard cells increases when stomata are open. pH increases in the light could occur because of CO_2 depletion within guard cells during photosynthesis or by the loss of protons by proton "secretion." The possible role of a pH increase is discussed in terms of phosphorylase activation in Section 5.4.5.

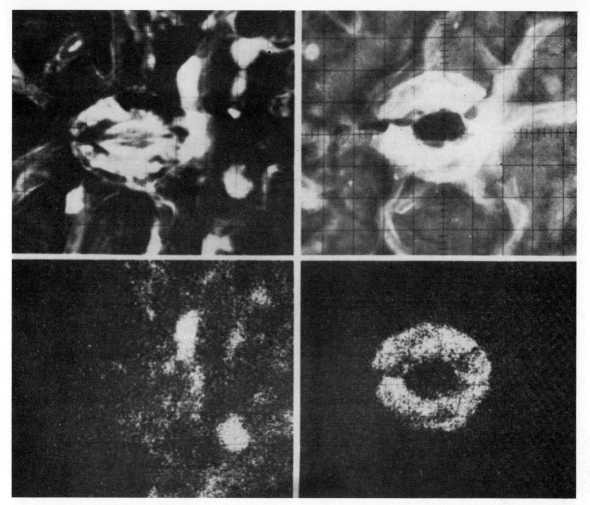

FIGURE 5.20 Photograph showing the accumulation of potassium in open guard cells (right panel). The experiment was conducted by electron-probe analysis. The upper images are secondary electron x-ray images showing a closed stoma (left) and an open stoma (right). The lower panel shows potassium images (white spots) produced from x-rays given off by the potassium during the electron-probe analysis. When the stoma is closed, the potassium is dispersed in the epidermal cells. When it is open, potassium accumulates in the guard cells. From G. D. Humble and K. Raschke. 1971. Stomatal opening quantitatively related to potassium transport. *Plant Physiol.* 48:447–453.

5.4.4 Potassium uptake

It is known that open stomata tend to accumulate potassium, perhaps from adjacent epidermal cells. If potassium is supplied to epidermal strips, stomata will open (Fig. 5.20). There is a definite relationship between the extent of potassium accumu-lation and the extent of stomatal opening. Endogenously produced organic anions such as citrate and malate may be the "counter" ions that balance the charge of potassium. There is some data suggesting that, in addition to the production of organic anions to maintain ionic balance, some chloride is taken up by the guard cells along with the potassium but this varies with species. In the absence of chloride uptake, protons from the organic acids could exchange with potassium, resulting in an increase in guard-cell pH. It is known, of course, that stomatal opening is correlated with an increase in pH.

It can be assumed that the decreased osmotic potential of guard cells results because of the ATP-dependent active uptake of K^+, Cl^-, and the endogenous production of organic anions. The role of light here is perhaps for ATP generation through photosynthesis. The effect of reduced CO_2 causing opening is not too clear, although perhaps an effect on pH is the explanation.

5.4.5 The hypothesis for stomatal opening

The classical hypothesis introduced and developed by Scarth and others in the 1930s for stomatal opening takes into consideration the increase in pH by CO_2 reduction, the light effect that reduces CO_2 by photosynthesis, and the known decrease in guard-cell starch and concomitant increase in solute content during opening.

The classical hypothesis mentioned above states that light, as an energy source for photosynthesis, causes a reduction in CO_2 within the guard-cell cytoplasm. This reduction in CO_2 is accompanied by an increase in pH, which in turn activates the enzyme phosphorylase. Phosphorylase is more active at high pH than at low pH. Phosphorylase hydrolyzes starch or other polysaccharides to glucose-1-phosphate, which is hydrolyzed to glucose and inorganic phosphate, decreasing the osmotic potential. The decreased osmotic potential decreases the guard-cell water potential, resulting in water uptake and increased turgor. The stomata open in response to the increased turgor.

There are some inconsistencies in this hypothesis. First, the small decrease in CO_2 may not alter the pH significantly. Second, all guard cells do not have starch and perhaps not phosphorylase. The hypothesis does not account for the potassium effect and it does not account for night opening of stomata in succulent plants. The night stomatal opening of succulents is a response to the reduced CO_2 during dark CO_2 fixation. The most serious objection seems to be the potassium effect, because even during night opening of succulent plant stomata, potassium accumulates.

Even though the precise mechanism for stomatal opening is not known, some facts are well recognized. Most stomata of green plants open in the light and in the absence of CO_2. The physical mechanism comes about by increased turgor of the guard cells relative to adjacent epidermal cells. Turgor is a response to the accumulation of solutes that decreases the osmotic potential. There is substantial evidence that the solutes include sugars, organic acids, and potassium. The accumulation of potassium is most certainly by active processes, requiring ATP generated through phosphorylation. Water enters passively in response to water-potential gradients established by the energy-dependent solute accumulation.

It should be clear that general water status will affect the degree of opening independently of the general process. If the plant wilts, guard cells will become flaccid and stomata will close. The plant hormone abscisic acid increases in plants that are water-stressed, and this same hormone will close stomata when applied to leaves. Thus there is evidence that abscisic acid may control stomata, particularly in plants that are exposed to drought conditions.

5.5 Methods to Measure Stomatal Opening and Gas-Transfer Resistance

There are two basic means to estimate stomatal opening, direct observation and porometry. In the direct method, observation is made of the aperture by microscopic measurement. This direct-observation method gives an indication of stomatal opening but only limited information about the capacity for diffusion. Figure 5.15, illustrating the relationship between gas diffusion and stomatal aperture, shows why knowledge of aperture opening does not necessarily indicate knowledge of capacity for diffusion. A 50% decrease in stomatal aperture may only cause a 20% reduction in gas diffusion if the stomata are wide open, whereas if the stomata are nearly closed, such a reduction would be accompanied by an equal reduction in diffusion capacity.

Porometric methods more nearly measure stomatal capacity for gas transfer by forcing gas or fluid through open stomata. The amount of pressure required to force the gas or fluid into the leaf is a measure of stomatal opening. In addition, the diffusive capacity of a leaf can be measured with a porometer. In this situation the diffusion of water vapor can be estimated directly and used to calculate diffusive capacity.

5.5.1 Direct observation

In all of the direct-observation methods stomata are observed with a high-power microscope. If the latter is equipped with a calibrated ocular micrometer, the aperture can be measured in micrometers (μm). The limitation here is the resolution of the microscope, usually about 2 μm. These methods are of limited quantitative use but have value in experiments requiring some knowledge of open or closed stomata. Direct observation of tissue surfaces is possible with top-illuminating compound microscopes.

Lloyd in 1908 developed a method by which thin strips of epidermis could be fixed by plunging them into absolute alcohol. If the strips are thin, allowing immediate replacement of the tissue water by the alcohol, the stomata will be fixed in the position they were in prior to sampling. Figure 5.21 is a reproduction of a 24-hour stomatal opening period from some of Lloyd's work, clearly showing the usefulness of the technique.

Direct surface replication can also be used to view stomata. An impression is prepared of the tissue surface and is then viewed to observe sto-

matal condition. The simplest technique is to paint the surface with cellulose acetate, allow it to dry for a minute or two, and strip off the negative impression. If the impression is gently tapped onto a drop of water on a microscope slide, there will be an excellent representation of the surface. This method damages the tissue surface and is not useful for repetitive estimations of one area.

A more elegant method, introduced by Sampson in 1961, is to prepare a silicone-rubber negative on the surface, followed by a cellulose-acetate positive impression of the silicone rubber mold.

In this method, silicone rubber of the proper viscosity is allowed to polymerize on the leaf or other surface. An exact impression of the leaf will be obtained. Certain advantages inherent in the method are little to no injury to the surface and a permanent mold. Provided that no injury occurs, the same area can be repeatedly sampled.

5.5.2 Porometry

Infiltration

In 1954 Alvim and Havis introduced a method based on the rate of infiltration of organic solvents of differing viscosity into leaves through open stomata. They prepared a series of decreasing viscosity by diluting nujol with xylol. The percentage of nujol solution that penetrates the leaf through open stomata is the index of stomatal opening. Advantages of this method are that penetration is easily observed by leaf discoloration; thus open or closed stomata can immediately be determined. The surface, of course, is permanently injured.

Viscous-flow porometry

The first viscous-flow porometer was introduced in 1911 by Darwin and Pertz. The device was attached to a leaf by means of wax and a water column within the manometer tube was adjusted with a bulb and three-way stopcock to a predetermined level. As air is pulled into the leaf through

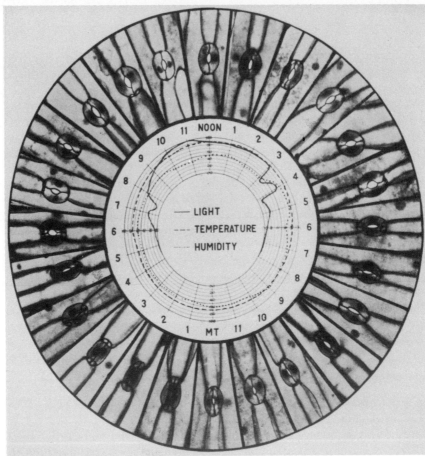

FIGURE 5.21 Clock diagram illustrating diurnal stomatal opening. Epidermal strips from onion were quickly removed and plunged into absolute alcohol. The alcohol fixes the stomata by rapidly dehydrating the tissue. After staining with Congo red, they can easily be viewed with a light microscope. From J. V. G. Loftfield. 1921. *The Behavior of Stomata.* Carnegie Institution of Washington Publ. No. 314.

open stomata by a vacuum created by the water column, the water column will drop. The greater the rate of drop, the greater the stomatal opening.

This method is most useful for amphistomatous surfaces but can be used for hypostomatous leaves since air will be pulled in from the region outside of the cup. This method is similar to other porometer methods in that it gives an indication of resistance to flow. The drop in water pressure will be directly proportional to the resistance of the open stomata to air flow.

A Wheatstone bridge porometer was introduced by Gregory and Pearse in 1934 in the form of a vacuum porometer that pulled air through a leaf.

The entire system was attached to a double-manometer system, one responding to the aspirator and one to the stomata. The difference in manometric readings between the two is an indication of stomatal resistance to flow.

In 1951, Heath and Russell introduced their Wheatstone bridge modification of the Gregory–Pearse instrument. Figure 5.22 shows a simplified version of such a Wheatstone bridge porometer. In principle, flow from a split air stream is balanced between the leaf and a needle valve by adjustment with a single manometer. The needle value is opened or closed to just balance the pressure required to force air through the leaf. Greater read-

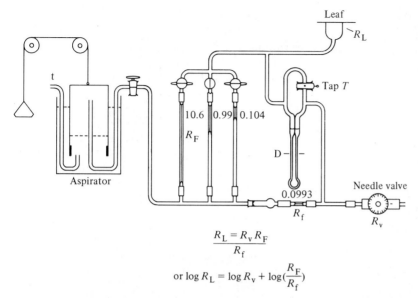

$$R_L = \frac{R_v R_F}{R_f}$$

$$\text{or} \log R_L = \log R_v + \log\left(\frac{R_F}{R_f}\right)$$

FIGURE 5.22 Diagram of the Wheatstone bridge porometer of Heath and Russell. The aspirator pulls (or pushes) air both through the leaf and simultaneously through the needle valve. The U-tube manometer is balanced using settings of the needle valve. The capillary-tube resistors (R_F = 10.6, 0.99, and 0.104) give a wide range for the instrument. R_L (leaf resistance) is calculated from the needle-valve resistance (R_v) and the ratio of R_F to standard resistance (R_f = 0.0993). From O.V.S. Heath and J. Russell. 1951. The Wheatstone bridge porometer. *J. Exp. Bot.* 2:111–116.

ings on the needle valve indicate lower resistance of the leaf to air flow.

Since these porometric methods rely on viscous flow of air through small pores, the relationship between rate of flow and pore size obeys Poiseuille's law, in which flow is proportional to the fourth power of the pore radius:

$$F = kpr^4,$$

where

F = pressure-driven viscous flow,

k = a constant inversely proportional to the length of flow path and viscosity,

p = pressure required for flow, and

r = radius of the pore.

Diffusion porometry

A variety of porometers using the diffusion of gas into or out of leaves has been developed to estimate stomatal resistance. Gas-diffusion porometers are useful because the process of gas exchange by plants is a diffusion process and the measurements relate directly to the phenomenon being studied.

Diffusion porometers using hydrogen and nitrous oxide have been developed but because of the difficulty of handling these gases their use has not been extensive. Water-vapor-diffusion porometers have proved to be the most useful instruments available today. A simple device first introduced by Wallihan in 1964 and further developed by van Bavel in 1965 consists of a microammeter and a lithium chloride relative-humidity sensor. The humidity sensor is housed in a leaf cup, which can be clamped onto a leaf. As water diffuses from the leaf the increase in humidity is measured on the ammeter as the conductivity of the lithium chloride increases with moisture. In practice, the sensor is dried to a predetermined humidity level by passing dry air through the leaf cup. The time required for transpiration to bring the humidity to some point is measured. The transit time is an indication of transpiration rate and, as well, an indication of leaf resistance to water-vapor transfer.

Calibration of the instrument in leaf-resistance units is based on the following expression:

$$R_s = \frac{S\Delta t - L_o}{D},$$

where

R_s = leaf resistance in s cm^{-1},

$S =$ a leaf-cup temperature-dependent sensitivity factor,

$t =$ transit time required for transpiration to change the relative humidity in cup from one level to another,

$L_o =$ leaf-cup geometry factor independent of temperature, and

$D =$ diffusion coefficient for water.

The instrument is calibrated by measuring t as water diffuses through cylinders of varying length ($R_s = L/D$) or through multipored plates of known R_s. Curves of R_s versus t must be determined at several temperatures in order to obtain the relationship between S and temperature.

These instruments are extremely useful for experimental use since leaf resistance (R_s) and transpiration can be estimated in a period of a few seconds with little or no disturbance to a leaf.

Isotope porometers using either tritiated water or $^{14}CO_2$ can be used to estimate transpiration and photosynthesis, with subsequent calculation of leaf resistance, R_s.

A simple porometer such as that designed by Shimshi in 1969 is composed of a high-pressure gas bottle containing $^{14}CO_2$, a needle to regulate gas flow, and a leaf cup to allow exposure of the leaf to $^{14}CO_2$. $^{14}CO_2$ of a known specific activity is pumped over the leaf at a constant rate of flow, and the amount of ^{14}C in the leaf after a given time is

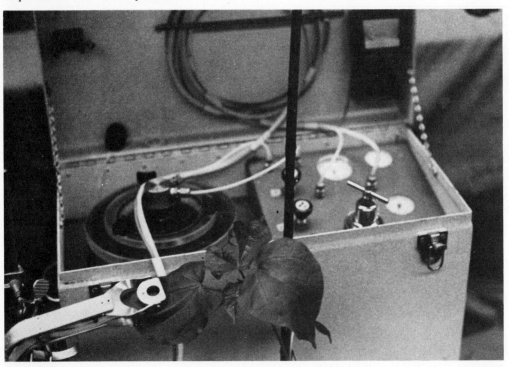

FIGURE 5.23 Photograph of a double-isotope porometer, used to measure photosynthesis and transpiration simultaneously on the same sample. $^{14}CO_2$ from a high-pressure gas bottle is pumped through a water reservoir containing tritiated water and then supplied to the leaf by the leaf gun. Data can be used to calculate rates of photosynthesis, rates of transpiration, and the conductances for CO_2 and water transfer. The porometer was built in 1978 at the University of California, Riverside.

an indication of the photosynthetic rate. Estimation of total resistance (R) to CO_2 uptake (P) is based on the gas-transfer equation:

$$P = \frac{\Delta}{R},$$

where

$P = {}^{14}CO_2$ fixation in units of radioactive disintegrations per minute (dpm) cm^{-2} s^{-1},

$\Delta = {}^{14}CO_2$ supplied to the leaf in dpm cm^{-3}, and

$R = $ resistance to $^{14}CO_2$ uptake in s cm^{-1}.

Thus

$$R = \frac{\Delta}{P} = \frac{\text{dpm cm}^{-3}}{\text{dpm cm}^{-2}\text{ s}^{-1}} = \text{s cm}^{-1}.$$

Transpiration and photosynthesis can be estimated simultaneously by pumping the $^{14}CO_2$ first through a tritiated-water (THO) reservoir at known temperature in order to bring the tritiated-water vapor in the $^{14}CO_2$ airstream to a known level. Both $^{14}CO_2$ and THO are passed over the leaf and the amount of ^{14}C and tritium uptake are determined.

Since water loss from a leaf is a physical process and is limited by the leaf resistance to water-vapor transfer, the instantaneous rate of THO uptake will encounter the same resistance. Thus the uptake resistance will equal the transpiration resistance provided that the measurement is completed prior to THO equilibrium with the water of the leaf.

A double-isotope porometer such as the one described above is shown in Fig. 5.23.

Review Exercises

5.1 Would a tub of water maintained at 30°C lose water by evaporation more rapidly on a cold, clear winter day (temperature 0°C) or on a warm summer day (temperature 35°C)? Assume 40% R.H. at both temperatures. Make calculations here and below to support your answers.

5.2 Two similar tobacco plants are growing in glass culture chambers, one at 20°C, the other at 30°C. The atmospheric moisture in each chamber is adjusted to give a Δe of 12 mbar. When sunlight strikes the chambers, the leaf temperature rises to 5°C above air temperature in each chamber. Will the rates of transpiration also increase by the same amount in each of the two chambers?

5.3 A small bean plant is growing under a bell jar in a saturated atmosphere at 25°C. As the sunlight strikes the leaf, its temperature rises 10°C above the air temperature. A similar bean plant is growing under a cloth shade nearby. Air temperature is 25°C, R.H. is 70%, and leaf temperature is 25°C. Which plant can be expected to have the more rapid transpiration rate? Compute the rates, assuming $R_s = 1$ s cm^{-1} and $R_a = 0.5$ s cm^{-1}.

5.4 What would be the influence on transpiration of raising the air temperature without increasing the temperature of the leaf? Why does an increase in the temperature of the air usually increase the rate of water loss from a leaf?

5.5 From the following data, compute the transpiration ratios in 21% and 0% oxygen.

Parameter	Atriplex species			
	1		2	
O_2 treatment (%)	21	0	21	0
Transpiration (g cm^{-2} s^{-1})*	16	16	15	16
Photosynthesis (g cm^{-2} s^{-1})*	0.038	0.061	0.073	0.071

*× 10^{-6}

From the data and using the gas-transfer equations for transpiration and photosynthesis, can you guess what it is about low oxygen that causes the reduced transpiration ratio in species 1? Inspect each term of the gas-transfer equation and decide if the absence of oxygen would be a factor.

References

BANGE, G. G. J. 1953. On the quantitative explanation of stomatal transpiration. *Acta Bot. Neerlandica* 2:255–297.

COOK, G. D., and R. VISKANTA. 1968. Mutual diffusional interference between adjacent stomata of a leaf. *Plant Physiol.* 43:1017–1022.

COWAN, I. R. 1977. Stomatal behaviour and environment. *Adv. Bot. Res.* 4:117–228.

JACOBS, M. H. 1967. *Diffusion Processes.* Springer-Verlag, New York.

JOHNSON, H. B., P. G. ROWLANDS, and I. P. TING. 1979. Tritium and carbon-14 isotope porometer for simultaneous measurements of transpiration and photosynthesis. *Photosynthetica* 13:409–418.

LEE, R., and D. M. GATES. 1964. Diffusion resistance in leaves as related to their stomatal anatomy and microstructure. *Amer. J. Bot.* 51:963–975.

MEIDNER, H., and T. A. MANSFIELD. 1968. *Physiology of Stomata.* McGraw-Hill, New York.

PENMAN, H. L., and R. K. SCHOFIELD. 1951. Some physical aspects of assimilation and transpiration. *Symposia Soc. Exp. Biol.* 5:115–129.

RASCHKE, K. 1975. Stomatal action. *Ann. Rev. Plant Physiol.* 26:309–340.

ROSE, C. W. 1966. *Agricultural Physics.* Pergamon Press, Oxford.

TING, I. P., and W. E. LOOMIS. 1965. Further studies concerning stomatal diffusion. *Plant Physiol.* 40:220–228.

VAN BAVEL, C. H. M., F. S. HAKAYAMA, and W. L. EHRLER. 1965. Measuring transpiration resistance of leaves. *Plant Physiol.* 40:535–540.

VERDUIN, J. 1949. In J. Franck and W. E. Loomis (eds.), *Photosynthesis in Plants.* Iowa State College Press, Ames, Iowa.

ZELITCH, I. (ed.). 1963. Stomata and water relations in plants. *Connecticut Agricultural Experiment Station Bull.* 664. New Haven.

ZELITCH, I. 1969. Stomatal control. *Ann. Rev. Plant Physiol.* 20:329–350.

ZELITCH, I., and P. E. WAGGONER. 1962. Effect of chemical control of stomata on transpiration and photosynthesis. *Proc. Natl. Acad. Sci. (U.S.)* 48:1101–1108.

CHAPTER SIX

Transport Processes in Plants

The study of transport processes in plants includes salt and water uptake, the short-distance transport of salts and other solutes from cell to cell, and the long-distance transport of water and solutes throughout the plant by the vascular system. Water and salt uptake are treated in Chapters 4 and 13, respectively. In this chapter, the short-distance transport of solutes from cell to cell and the long-distance transport of solutes and water through the vascular system are discussed.

The short-distance transport of solutes from cell to cell is important in cell–cell interaction and probably takes place mostly by diffusion through the cellular connecting tubules, which are called plasmodesmata. The entire living portion of the plant is considered to be a single unit called the symplasm. The nonliving portion, i.e., the non-symplastic portion of the plant, is called the apoplasm. The symplasm is connected from cell to cell throughout the plant body by the plasmodesmata. Transport from cell to cell through the plasmodesmata is termed symplastic transport, whereas

131

transport from cell to cell in which solutes leave one cell and enter another by going through cell walls and intercellular spaces is called apoplastic transport. Apoplastic transport is transport through the nonliving portion of the plant.

Long-distance transport in vascular plants is a function of the complex vascular system. It was, in fact, the evolution of the vascular system, the xylem and phloem, that ensured the success of land plants. Not only does the vascular system result in efficient transport throughout the plant body, but also it gives physical support to the plant. Without the supporting and transporting vascular system, land plants would not be very large in size and would most likely be restricted to damp or wet environments.

6.1 Short-distance Transport

Although apoplastic transport from cell to cell is feasible, probably much of the cellular transport and communication between and among cells is symplastic, occurring through the plasmodesmata. The term plasmodesma was introduced in 1901 by Strasburger and refers to a protoplasmic connection between cells. The actual process of transport through plasmodesmata is not understood, but more than likely it is by diffusion. Transport and solute movement within any one cell are by diffusion and cytoplasmic streaming (cyclosis). The latter is aided by protein microfilaments in the cytoplasm. It can be visualized, therefore, that symplastic transport from cell to cell is by means of diffusion aided by cytoplasmic streaming.

In some cases, cell-to-cell transfer of solutes is an active process, requiring the energy of ATP. It is especially true of vein loading and perhaps unloading during sugar transport in the phloem. It is not entirely clear, however, whether such energy-dependent, cell-to-cell transport is apoplastic or symplastic.

In some tissues with an active-transport function, there are specialized cells called transfer cells. The transfer cells present a large surface for transport that is primarily associated with apoplastic movement. Both plasmodesmata and transfer cells are discussed below.

6.1.1 Plasmodesmata—symplastic transport

Plasmodesmata are cytoplasmic tubules that connect adjacent cells (Fig. 6.1). They occur in most higher-plant cells except that possibly they do not occur between mature guard cells of stomata and adjacent subsidiary cells. They are also lacking between adjacent reproductive cells of gametophyte and sporophyte generations. They are not found in mature tracheids and vessel elements. There can be as many as one million plasmodesmata per mm^2. As mentioned above, it has been proposed that the entire living portion of the plant is connected from cell to cell through the plasmodesmata. According to the symplasm hypothesis, there is symplastic transport throughout the entire plant through the plasmodesmata. Apoplastic transport is aided by transfer cells.

The structure of a plasmodesma as interpreted by Robards in 1975 is shown in Fig. 6.1 next to the electron micrograph. It appears to be lined by the plasma membranes from the adjacent cells. Within the channel of the plasmodesma is a central tubule called a desmotubule. There is some evidence that the desmotubule is composed of endoplasmic reticulum from the adjacent cells. It has a central rod. Thus it appears that not only is the plasma membrane of cells connected, but also the endoplasmic reticulum is connected.

It is generally considered that the plasmodesmata present little resistance to transport of small, nonionic solute molecules, exhibiting little or no selectivity. Even if soluble, large molecules such as proteins and nucleic acids probably do not pass through the plasmodesmata. Transport of charged molecules would be limited by electrical gradients. There is evidence, however, that viruses are transported symplastically through plasmodesmata.

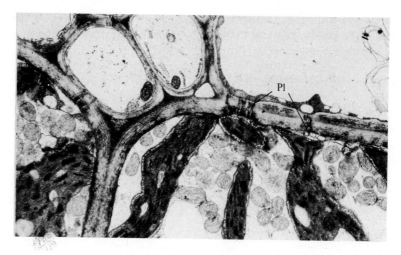

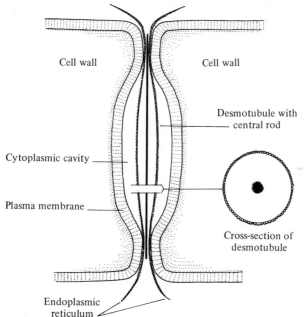

FIGURE 6.1 Electron micrograph showing plasmodesmata connecting a chloroplast-containing leaf cell and an adjacent phloem parenchyma cell. Presumably there is symplastic transport from the photosynthetic tissue into the vascular tissue. Below the micrograph there is a drawing of a plasmodesma connecting two adjacent cells. Note that the plasma membrane is continuous between the two cells and that the endoplasmic reticulum goes through the center of the plasmodesma that forms the tubule. The central desmotubule with its rod is shown in cross section in the insert drawing. Electron micrograph courtesy of Dr. W. W. Thomson, University of California, Riverside.

Viruses have been observed with the electron microscope in the tubules of plasmodesmata, but when viruses are present the desmotubule appears to be absent.

The actual evidence for transport from cell to cell through plasmodesmata is rather slight. The symplastic hypothesis is based on the fact that plasmodesmata are very abundant between cells that actively transport solutes. In addition, there is some experimental evidence for chloride transport through plasmodesmata. Experimentation has shown that chloride supplied to cells can be precipitated in the plasmodesmata channel with silver nitrate.*

The exact pathway for transport through the plasmodesmata is not clear, but most evidence suggests that it is through the desmotubule, aided by the endoplasmic reticulum. This seems likely because the channel at the neck is restricted, closing off the cytoplasmic cavity of the plasmodesmata from the adjacent cells. It was mentioned above that when viruses are seen in plasmodesmata channels the desmotubule appears to be absent. We can only speculate on this observation, but perhaps the desmotubule maintains some degree of selectivity. When it is absent or destroyed, foreign bodies (the viruses) can pass. Alternatively, the desmotubule may be an artifact of preparation of tissue for electron microscopy.

In summary, the symplasm hypothesis states that the entire living portion of the plant (the symplasm) is connected from cell to cell through channels (tubules) called plasmodesmata. Much of the short-distance transport of solutes that include inorganic ions, sugars, organic and amino acids, and regulatory compounds (hormones, for example) takes place through the plasmodesmata. This hypothesis does not preclude the possibility of cell-to-cell transport across adjacent cell plasma membranes and cell walls as must take place between those adjacent cells lacking plasmodesmata. A good example of such transport between adjacent cells lacking plasmodesmata is the transport of potassium from epidermal and subsidiary cells to the guard cells of stomata during stomatal opening (see Chapter 5). Such apoplastic transport may allow for greater selectivity than would be possible with symplastic transport through the nonselective plasmodesmata. In the case of the stomata, perhaps apoplastic transport allows only potassium to leave subsidiary cells and enter the guard cells during opening and also prevents the transport of other soluble salts.

6.1.2 Transfer cells—apoplastic transport

Transfer cells are specialized cells frequently found at the juncture of symplasm and apoplasm in regions of active transport. They are found in the xylem and phloem, in reproductive tissues, and in secreting glandular tissue. It is believed that they function by gathering and subsequently secreting solutes. An example is the salt glands that receive and secrete high concentrations of NaCl. It is visualized that the transfer cell gathers NaCl and then secretes it by active transport. Figure 6.2 shows a transfer cell.

The distinguishing anatomical feature of transfer cells is the presence of cell-wall ingrowths that greatly increase the surface area of the plasma membrane. It appears that the transfer cell has an increased surface area that aids in transport. Transport is normally against a concentration gradient and requires ATP energy. Because transport is apoplastic (going across the plasma membrane into the apoplast) rather than symplastic (through plasmodesmata), there is a greater possibility of selectivity. This seems to be true because the plasmodesmata channels should allow virtually all small, soluble molecules to pass, whereas it is visualized that transport from transfer cells would be aided by specific transporters acting on specific solutes.

The concept of a transfer cell is based mostly on anatomical considerations. There are epidermal transfer cells, xylem transfer cells, phloem transfer cells, and others named for their function or location. Evidence for the transport role of transfer

*The reaction is: $AgNO_3 + Cl^- \rightarrow NO_3^- + AgCl\downarrow$.

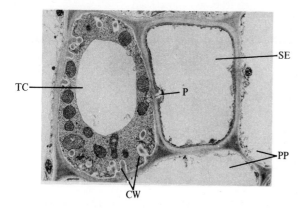

FIGURE 6.2 Phloem transfer cell and an adjacent sieve element ($\times$15,700). The transfer cell has very distinct cell-wall ingrowths on the walls next to the parenchyma cells but not on the wall adjacent to the sieve element. The connecting cell wall between the transfer cell and the sieve element has plasmodesmata. Numerous mitochondria are present in the transfer cell. TC = transfer cell. SE = sieve element. PP = parenchyma cells. CW = cell wall of transfer cell with ingrowths. P = plasmodesmata. From B. J. Bentwood and J. Cronshaw. 1978. Cytochemical localization of adenosine triphosphate in the phloem of *Pisum sativum* and its relation to the function of transfer cells. *Planta* 140:111–120. Electron micrograph courtesy of Dr. Cronshaw.

cells is very circumstantial. They have a greatly increased surface area of the plasma membrane because of the cell-wall ingrowths lined with membrane, and they occur in regions of known high-transport activity. It appears quite logical, therefore, to assign a transport function to these highly modified cells.

Consistent with the symplasm hypothesis, transport involving transfer cells would be apoplastic since the substances being transported leave the symplasm and enter the apoplast (cell walls, for example). It should be clear, however, that there is

much apoplastic transport in plants that does not involve transfer cells directly. Water and salt uptake by roots in which the water and ions pass through the cell walls and intercellular spaces is apoplastic transport. Similarly, transport within the nonliving xylem elements and vessels is apoplastic. However, it is entirely possible that at some point upon entering or leaving the apoplast, passage may be aided by transfer cells. Conclusive statements about these questions must await further experimentation.

6.2 Long-distance Transport—Water

The long-distance transport of water and solutes throughout the plant body is one of the most important physiological processes of plants. Despite the fact that there are no pumps comparable to the mammalian heart, transport is quick and efficient. The long-distance transport of both water and solutes is a physical process. Water appears to be pulled up the plant by forces created at the leaf surface during evaporation (transpiration), and solutes are transported by being carried along in the xylem with the transpiration stream or by being transported in the phloem under pressure. These two processes for long-distance transport are discussed in the remainder of this chapter.

6.2.1 Water transport

Water transport, from soil to roots to leaves and ultimately out of the plant by a process termed transpiration, takes place on a water-potential gradient. The entire system is governed by physical processes from soil to atmosphere and is referred to as the soil–plant–air continuum (SPAC). Provided that there is a decrease in water potential from the soil to the air, water will pass through the plant spontaneously on a water-potential gradient. The rate and path of water transport is largely a function of the anatomy of the plant and the physical mechanisms of transport.

The water-potential gradient from soil to atmosphere through the SPAC is shown in Fig. 6.3. As stated in Chapter 4, the ψ of soils with available water will fall within the range of -0.3 to -15 bars. Thus the water potential of a well-watered soil will be in the vicinity of 0.

A slight drop in potential is expected from soil to root tissues. A root in a soil close to field capacity would have a water potential of not much less than -5 bars. As the soil dries and the water potential approaches the wilting point of -15 bars, the water potential of the root will also decrease.

Roots have some capacity to adjust potential depending on the soil solution. In one experiment, Bernstein in 1961 measured root and leaf osmotic potential of plants growing in differing osmotic solutions. As the osmotic potential of the solution was depressed, the osmotic potential of the plant tissue decreased, apparently the result of salt uptake (see Table 6.1). The depressed osmotic potential and water loss by the root tissue causes a reduction in water potential.

There is a potential drop from roots to leaves primarily because of solute accumulation in the leaves but perhaps also because of a reduced pressure potential. Scholander in 1965 measured water potential at different heights in trees and found the expected drop in potential with height. Within the crown of Douglas firs, there was always a drop in water potential from the base of the crown at 30 meters to the top of the crown at 79 meters. During midday, the top of the crown had a water potential of about -22 bars (these data are further discussed under the section "Tensions within the xylem" later in this chapter).

The greatest drop in water potential occurs at the

TABLE 6.1   Osmotic adjustment of cotton-plant tissue in sodium chloride solutions

NaCl solution	Bars	
	ψ_π—roots	ψ_π—leaves
0	-5.6	-10.2
-3	-8.6	-12.5
-6	-8.2	-17.9
-12	-12.4	-21.3

Data from Bernstein (1961).

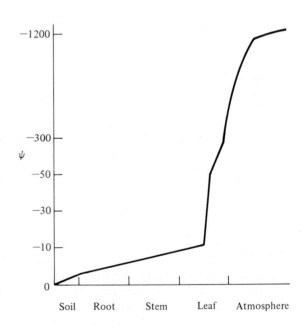

FIGURE 6.3 Diagram to show how the water potential decreases through the soil-plant-air continuum (SPAC). The water potential decreases fairly steadily through the SPAC until the phase change between the leaf mesophyll and substomatal chamber. Provided that the relative humidity is less than 100%, the water potential may decrease to well below –1000 bars. Because of the very sharp change in water potential at the leaf-air interface, it is assumed that the greatest control of transpiration is at the stomata. Modified from J. R. Philip. 1957. The physical principles of soil water movement during the irrigation cycle. *Proc. III International Congr. Irrig. and Drainage* 8:125–154.

phase change as the SPAC goes from the liquid water of the leaf tissue to the gaseous water vapor of the atmosphere. In dry air, this drop in potential may be as much as 1000 bars.

It is generally assumed that the mesophyll cell surfaces have a fairly uniform water potential and that the relative humidity approximates 100%, being nearly equal to the saturation vapor pressure of the temperature of the tissue. There have been some measurements of reduced relative humidity at the mesophyll cell walls that have resulted in estimations of water potential far below the usually expected values of about −10 to −20 bars. In some desert plants, for example, water potentials of the mesophyll cell walls were estimated at less than −100 bars. Regardless of any incipient drying at mesophyll cell surfaces, the greatest regulation of water balance is at this point in the SPAC because of the substantial drop in water potential from leaf tissue to air.

Of the total water taken up by the plant during a growing season, about 99% is lost through transpiration. Roughly 0.9% is retained as free water within the tissue, and about 0.1% enters into the plant's metabolism as a reactant in chemical reactions. The latter is the chemically bound water that is no longer identifiable as water.

It seems most logical to discuss water transport in the order of soil to root, leaf to air, and finally transport from root to leaf. This is because water uptake by roots is largely osmotic in nature and relatively simple to understand. The forces for upward transport of water from root to leaf are created to a large extent by transpiration, the evaporation of water from the leaves to the air. In order to understand water transport within the plant, it is necessary to understand first how water enters and second how water is lost, creating the forces for upward pull.

Water uptake

As a generalization, water is taken up by plant roots from the soil by osmosis involving water-potential gradients. Provided that the water poten-tial of the soil is greater than the water potential of the water in the root tissues, water will move into the root. The principles of water uptake are discussed below.

THE PATH OF WATER UPTAKE

A cross-sectional diagram of a young root is shown in Fig. 1.20 (Chapter 1). Most water uptake occurs behind the root tip in the vicinity of the root hairs. There appears to be resistance to water uptake at the tip of the root in the regions of cell division and cell elongation. Beyond the root-hair zone, the root becomes highly suberized and water uptake is prevented, although breaks and cracks in the epidermis are common and would allow water to enter freely.

Figure 6.4 illustrates a young root tip showing the presence of root hairs just beyond the region of elongation. It is generally assumed that the ephemeral root hairs arising from the epidermis of the root greatly enhance the absorbing surface of the root. It can be calculated that a single rye plant has about 14 billion root hairs with a combined surface area of almost 400 m^2. The total surface of the root plus root hairs is nearly 600 m^2. The presence of the root hairs increases the surface of the root by 67%. The root hairs increase the root–soil-particle contact by at least 20 times and allow penetration into a greater soil volume.

Water will follow the path of least resistance when passing from the soil through the root into the transporting stele (the portion of the root inside the endodermis that has the vascular tissue). The long-distance water transport is mostly apoplastic, passing through cell walls and intercellular spaces, although water does enter the symplasm within each cell and there is undoubtedly much symplastic water transport through plasmodesmata. The highly suberized endodermis between the outer cortex and inner stele makes an effective waterproof barrier to apoplastic water transport across the endodermis. Thus water passing from the cortical region of the root to the stele must pass through the living cells of the endodermis via the symplasm.

FIGURE 6.4 Photograph showing root hairs on young root of rye seedling (X28). Root hairs may increase the surface area of the root by as much as 100% and increase the root-soil contact by as much as 20 times. Micrograph courtesy of the New Zealand Department of Scientific and Industrial Research. From J. Troughton and L. A. Donaldson. 1973. *Probing Plant Structures.* McGraw-Hill, New York.

Whereas water can pass entirely through cell walls and intercellular spaces throughout the cortex, in order to enter the stele it must pass through the living cell membranes of the endodermis. Some of the water transport must be symplastic. Figure 1.19 shows the relationship between the endodermis with its suberized Casparian strips and the inner stele.

Some roots have passage cells in the endodermis and there may be breaks that allow some apoplastic water transport across the endodermis. But the usual pathway for water uptake and transport from the soil to the stele is through root hairs either directly into the epidermal cells or into cell walls and intercellular spaces. Once in the root, water may pass either through the apoplast (cell walls and intercellular spaces) or from cell to cell through the symplasm. In order to pass into the stele from the

cortical region, the water must be in the symplasm of the endodermis because the suberized axial walls present a waterproof barrier.

The structural and chemical composition of the plant cell wall is important with respect to water and mineral transport. The cell wall is considered to be extremely permeable to water and ions. It is composed largely of cellulose, hemicelluloses, and pectins. Cellulose is a β-D-glucose polymer, and the pectins are polymers of α-galacturonic acid (pectic acid). Carboxyl groups of the latter are frequently esterified, but both cellulose and the pectin are hydrophilic. Hemicelluloses are polymeric compounds of sugars such as mannose, arabinose, and others and are also hydrophilic (see Chapter 10 for a further discussion of these carbohydrates).

Lignin, the material giving cell walls strength and resistance is a hydrophilic complex polyflavone.

The cell walls have associated with them hydrophilic proteins and hydrophilic carbohydrate polymers of gums and mucilages. It is only the hydrophobic, lipoidal fatty acids and waxes of the cutin and suberin that resist water transport.

The chemical nature of the plasma membrane and tonoplast alter water permeability but, by and large, do not account for much resistance to water transport. In summary, it is safe to say that plant cells are very permeable to water. Selectivity of transport only becomes important with respect to solute molecules.

ACTIVE VERSUS PASSIVE WATER UPTAKE

Two types of water uptake can be delineated, active and passive. Active uptake refers to water uptake in which the driving forces originate in the root. This does not necessarily imply metabolic uptake per se but may refer to an osmotically produced gradient from soil solution to root tissue.

Passive uptake occurs because of forces originating in the atmosphere or in leaf tissue. Here the potential drop is from soil to atmosphere or from soil to leaf tissue. Water is truly pulled up the plant body.

Active water uptake

Active water uptake can take two possible forms: (1) metabolic uptake, which refers to a pumping mechanism in which water is pumped into the tissues, and (2) uptake that is osmotically generated by forces called root pressure. These two forms are discussed below. It should be mentioned here that metabolic uptake of water probably does not exist according to the definition given below and that uptake by root pressure is only important under rather restricted conditions.

METABOLIC UPTAKE   The question of active metabolic uptake of water has intrigued plant physiologists for years. Levitt (1967) has given a rather strict definition of active metabolic water uptake that is useful for discussion. Active uptake

dependent on metabolic energy is uptake not accounted for by diffusion or by hydrostatic flow; it is dependent on a pump that is operated with metabolic energy. The pump acts directly on the substance being transported, which in this case is water.

Possible evidence for such an active pump is severalfold. First, it was shown long ago that the osmotic potential of a solution had to be considerably lower than the root-tissue osmotic potential to prevent water uptake, meaning that water was apparently taken in against a water-potential gradient. It was also shown that respiratory inhibitors such as dinitrophenol, azide, and arsenic prevented water uptake. Furthermore, anaerobiosis will impair water uptake by roots. Such experiments directly implicate respiration in water uptake. They do not, however, necessarily prove that a pump in the sense of Levitt's definition is involved in water uptake. In fact, it is more likely that respiration maintains cellular integrity and balance without which water uptake would not occur.

Experiments have shown that treatment of root with auxin-type growth regulators such as indoleacetic acid (IAA) and naphthalene acetic acid (NAA) is accompanied by enhanced water uptake. Several hypotheses have been proposed to explain the enhanced water uptake in response to auxin treatment. First, it is known that auxin causes starch hydrolysis. Starch hydrolysis produces soluble sugars that will decrease the cellular osmotic potential, resulting in lowered water potentials and enhanced water uptake. Measurements, however, have not shown a depressed osmotic potential in response to auxin treatment.

In addition, auxin-induced increases in membrane permeability have been invoked. However, it is clear that increases in membrane permeability will only alter the rate of uptake and not the total amount of uptake.

Most researchers now agree that auxin alters cell-wall extensibility, causing a lowered pressure potential. A decrease in turgor will necessarily decrease the water potential of the cell, and water will move into the tissue on a water-potential gradient. The above evidence suggests that active transport of water in plants is unlikely.

ROOT PRESSURE If certain plant species (*Coleus,* for example) are decapitated, sap will ooze from the cut stump. The force of exudation is usually in the range of 1 to 2 bars but may be as high as 5 bars. The plant root system is acting like an osmometer, and the pressure created is called root pressure. It can be shown that the volume flow (*F*) caused by root pressure is

$$F = L_p \, (\psi_{\pi s} - \psi_{\pi r}),^*$$

where

L_p = conductivity of the root in g water s^{-1} bar^{-1},

$\psi_{\pi s}$ = osmotic potential of the soil in bars, and

$\psi_{\pi r}$ = osmotic potential of the xylem vessels in the root in bars.

In root pressure, the driving force for water uptake is the difference in osmotic potential between root and soil solution. The gradient is maintained by solute production within the root and hence falls within the definition of active transport.[†] Metabolism is necessary to maintain solute concentration; the solute concentration is perhaps even maintained by an active solute pump. It is clear here that the pump would act on the solute molecules and not on the water.

6.2.2 Transpiration

Transpiration or water loss from plant surfaces has been studied in detail for over one century. It is a remarkable process in that it is easily studied but plays no major direct role in plant growth and development insofar as is known. It is a grand paradox in biology because any function it may have is quite subtle and not generally accepted by all. Transpiration occurs largely through the stomata that open in response to light. When open, CO_2 enters during photosynthesis and vast quantities of water are simultaneously lost, frequently as much as 1000 times the volume of CO_2 taken in. If ex-

cessive quantities of water are lost by leaves, stomata will tend to close because of a water deficit and CO_2 uptake and photosynthesis will be reduced. This paradox is so great that some have termed transpiration a "necessary evil." It is even more astounding when one realizes that the only significant pathway for CO_2 uptake is the stomata and that a water deficit is the factor most limiting to plant growth the world over.

The magnitude of transpiration

The usual daytime rates of transpiration are in the range of 0.1 to 2.5 g water loss dm^{-2} h^{-1}; nighttime rates are usually less than 0.1 g dm^{-2} h^{-1}. The high diurnal rates are the result of open stomata and the high evaporative demand during the daylight hours. Water loss through the cuticle, or cuticular transpiration, is usually 1 to 10% of the open-stomata rate. Thus daytime rates of cuticular transpiration are in the range of 0.001 to 0.25 g dm^{-2} h^{-1}.

These high rates of water loss by transpiration can be expressed in rather startling terms. Turrell (1934) estimated that a *Catalpa* tree 21 years old and 10 meters tall had 26,000 leaves. After estimating the area of single leaves, he calculated that the 26,000 leaves had a combined surface area of 390 meters². At an average transpiration rate of 1.0 g dm^{-2} h^{-1}, this tree would lose 390 kilograms of water in a 10-hour day.

Kozlowski (1964) mentioned an example in which it was estimated that a single 47-foot maple tree with 177,000 leaves amounting to a surface area of 675 m² lost 220 kilograms of water per hour in the summer. Furthermore, it is estimated that hardwood forests may lose as much as 75,000 kilograms per hectare per year.

The daily course of transpiration

Many researchers have followed the course of transpiration during a 24-hour period. The pattern follows pretty closely the vapor pressure deficit (Δe) of the atmosphere, at least during the day when stomata are open. Water loss is reduced at night

---

*Compare with 4.3.1.

†Note that this is not Levitt's definition of active transport. In Levitt's definition, the active pump acts directly on the water.

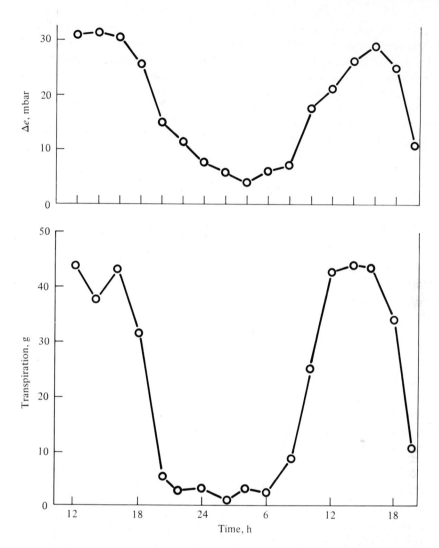

FIGURE 6.5 Experiment by P. J. Kramer showing how transpiration follows the vapor-pressure deficit (Δe) of the atmosphere. The vapor-pressure deficit (upper curve) is high during the daylight hours when it is warm and low at night when it is cool. The lower curve shows transpiration by a sunflower plant expressed in grams of water loss. During the cycle of the first day there is a midday decrease in transpiration that is not related to the Δe. Evidently this is because of stomatal closure in response to temporary wilting. Water uptake by the roots did not keep up with water loss by the leaves. The very low transpiration at night is primarily because stomata are closed. Redrawn from P. J. Kramer. 1937. The relation between rate of transpiration and rate of absorption of water in plants. *Amer. J. Bot.* 24:10–15.

because of closed stomata. Figure 6.5 shows the daily course of transpiration for sunflower plants along with the measured vapor-pressure deficit (Δe) of the atmosphere. It is not uncommon to observe a midday depression of transpiration as shown on the first day in the figure. This results because of temporary or incipient wilting when water loss exceeds water uptake. Stomata will close for a short period and will not reopen until leaf turgor is regained.

Throughout the day a cyclic variation in water loss can be observed in many plants. This cyclic variation, shown in Fig. 6.6, is accompanied by a cyclic variation in stomatal opening as explained in Chapter 5. It probably occurs as a result of slight water deficits created because of water loss that exceeds water uptake. Evidently, the roots represent a resistance to water uptake.

In actively transpiring plants, water loss almost always exceeds water uptake during the day (see Fig. 6.7). If water loss is excessive as is frequently the case, there will be incipient wilting during the day. Permanent wilting occurs only when the soil water is depleted.

Leaf structural considerations

A cross section of a typical dicotyledonous leaf is shown in Fig. 1.16. The entire surface is covered by an epidermis that is waterproofed by the deposition of cutin and wax that forms the cuticle. Figure 5.1A is a scanning electron micrograph showing the surface of a cactus with heavy deposition of wax around the stomata. Cutin deposition even extends into the substomatal chamber such that all exposed surfaces are protected by cuticle. Evidently, the cuticle within the substomatal chamber is not as waxy as it is on the epidermal surface since

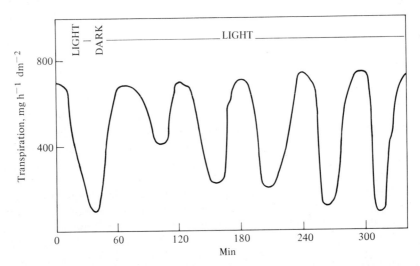

FIGURE 6.6 The cycling of transpiration by cotton plants after induction by a 20-minute dark period. The low points on the graph are attributed to stomatal closure. The stomatal closing comes about because of slight wilting of the leaves when water loss exceeds water uptake. After stomatal closure the leaves regain turgidity and the stomata open again. In addition, when the stomata are closed in the light, CO_2 depletion during photosynthesis will tend to cause stomatal opening. From H. D. Barrs and B. Klepper. 1968. Cyclic variations in plant properties under constant environmental conditions. *Physiologia Plantarum* 21:711–730.

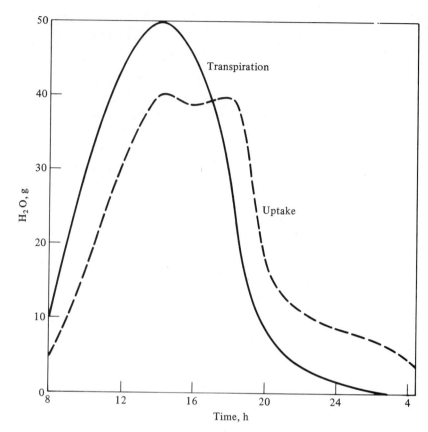

FIGURE 6.7 Graph to show transpiration and water uptake by a sunflower plant. During the morning and midday, transpiration exceeds water uptake by the roots. In the evening and at night, water uptake is faster then transpiration and the plant rehydrates. Redrawn from P. J. Kramer. 1937. The relation between rate of transpiration and absorption of water in plants. *Amer. J. Bot.* 24:10–15.

a large portion of the water evaporation must take place through these mesophyll cell surfaces. The upper mesophyll of the dicotyledonous leaf is arranged in a palisade and is thus called palisade parenchyma. Numerous chloroplasts are present. In many dicotyledons, the lower parenchyma is more loosely arranged and is called spongy parenchyma. The internal-to-external surface ratio is quite variable, ranging from about 7 to over 30 for leaves developing in the sun and shade, respectively. In leaves that develop in the sun, the cells are more compact.

According to our present understanding, the internal surface area is not a significant factor in transpiration since it is assumed that the internal atmosphere around the mesophyll cells and within the substomatal chamber is at 100% relative humidity.* The important surface area for transpiration is the external leaf surface area.

In some types of plants, there may be some incipient drying at the mesophyll, resulting in a vapor-pressure depression within the leaf. Under these circumstances of incomplete saturation, internal surface area may become an important factor.

The fine structure of the vascular bundles within the leaf is important in water transport. Bulk water is evidently transported within the xylem. Water then moves from the open xylem vessel tips to the mesophyll cell walls, where it evaporates into the

*Internal surface area does, however, become important in CO_2 uptake because it evidently is a component of R_m (see Chapters 5 and 16).

intercellular spaces. It is this evaporation from mesophyll cell walls that brings about the forces for upward water pull.

The morphology of the leaf surface influences transpiration. A thin cuticle will allow greater rates of cuticular transpiration than will a thick, waxy cuticle. The presence of sunken stomata or stomata buried below leaf hairs and other appendages, will increase diffusion path lengths, and the stomatal resistance (R_s) will be greater than predicted from actual stomatal dimensions.

The presence of hairs, bladders, and glands tends to increase the boundary-layer resistance (R_a), further reducing transpiration. Moreover, leaves with white, scurfy (loose scales) depositions on their surfaces (such as occur in many desert plants) will absorb less radiation and will therefore be cooler and transpire less water.

Factors affecting transpiration

Many factors influence water loss from plants. Perhaps most important are environmental factors that directly affect the vapor pressure of the water in the leaf and the vapor pressure of the water in the atmosphere. But in addition to atmospheric factors (most important are light, temperature, humidity, and wind), plant physiological factors become important. These include the stomatal mechanism, solutes, hormones, wax depositions on leaves, and anatomical and morphological features of plant surfaces. Many of these factors are discussed below.

SOIL MOISTURE

The effect of soil moisture on transpiration has been studied extensively. In the simplest sense, since the greatest change in water potential occurs at the leaf–air interface of the soil–plant–air continuum provided that the soil is not at the permanent wilting percentage, transpiration should be nearly independent of water-absorption rates. This suggests that soil moisture does not have a marked effect on transpiration. As illustrated in Fig. 6.7, water loss can far exceed water uptake, even to the

extent of temporary or incipient wilting. Experiments such as the one shown in the figure support the conclusion that soil moisture is not an important factor for transpiration.

Nevertheless, some careful measurements of transpiration rates with variable soil moisture do indicate some influences. As soil water decreases toward the permanent wilting percentage, transpiration rates are lower. Although water availability at the root–soil interface may influence transpiration directly, it is more likely that the reduced soil water potentials cause reduced leaf water potentials and consequent increases in stomatal resistances because of lack of guard-cell turgor. Transpiration rates will decrease if stomatal resistances increase.

Because the driving force for water uptake from the soil by the roots is $\triangle \psi$, water will not be equally available over the range of soil water potentials from field capacity (−0.3 bars) to the permanent wilting percentage (−15 bars). Of course, as the soil dries the root water potential will tend to decrease, partially compensating for the decrease in soil water potential.

ATMOSPHERIC FACTORS

A detailed analysis of the effects of light, temperature, humidity, and wind on the resistances to transpiration was given in Chapter 5 and should be reviewed. Light has a direct effect on transpiration because of its effect on stomatal opening. Most stomata open in response to light. When stomata are open, transpiration occurs. The rate of transpiration when stomata are open is a function of other environmental factors such as temperature, humidity, and wind.

Other than the stomatal opening response to light, temperature is by far the most important environmental factor affecting transpiration. Transpiration can be shown to be a direct function of the vapor pressure of water at the surface of the leaf mesophyll. The only important factor governing the vapor pressure of this water is the temperature of the water. As temperature increases, the vapor pressure (or vapor density) increases exponentially (refer to Fig. 5.8).

Relative humidity of the air is a factor in transpiration, but only in relation to air temperature. At any one air temperature, as the relative humidity decreases, transpiration will increase because it is the vapor-pressure difference between the water in the leaf and the water in the air that is the driving force for transpiration. This concept is explained in detail in Chapter 5. As the leaf temperature increases or the relative humidity of the air decreases, the driving force (Δe) will increase and transpiration will increase.

Leaf temperature is much more important than air temperature in affecting the vapor pressure. In the leaf, vapor pressure will increase exponentially with temperature. In the air, vapor pressure will increase exponentially as well, but the air also expands, compensating for the actual increase in vapor pressure; thus there is little net change.

Wind plays a dual role in transpiration. First, wind may increase transpiration, reducing the boundary layer (R_a) over the leaf. Second, wind will tend to alter the temperature of the leaf, bringing it closer to the temperature of the wind. If the leaf is warmer than the air mass passing over it, the leaf will cool. This is the usual situation that causes a slight decrease in transpiration. The decrease in transpiration is ordinarily not as great as the increase in transpiration because of the reduction in the boundary-layer resistance. Therefore, wind usually is accompanied by an increase in transpiration. Of course, if the wind is warmer than the leaf, the leaf temperature will be brought closer to the wind temperature, causing an increase in the vapor pressure of the water in the leaf and increasing transpiration.

An indirect effect of wind on transpiration may result because of the exchange of air masses. If the wind brings in air with a different water content, transpiration will be affected accordingly. If the new air mass is wetter, transpiration will decrease; if drier, transpiration will increase.

Provided that there is not a change in air mass, the water content of the air stays fairly constant throughout the day–night period, but the relative humidity does change. As the temperature of the air increases during the day, the air can hold more water and the relative humidity will decrease. The actual amount of water in the air stays fairly constant. Thus it is frequently possible to calculate the relative humidity throughout the day, knowing the air temperature and one measurement of relative humidity from which the vapor pressure of the air can be determined.

Nonstomatal water loss

Water can be lost from any surface of the plant that is not entirely waterproofed. And although water from nonstomatal surfaces rarely accounts for more than 10% of the total loss, small losses may be important from the standpoint of physiology.

CUTICULAR TRANSPIRATION

Cuticular transpiration can account for as much as 10% or more of the total water loss from plant surfaces depending on the chemical and physical nature of the cuticle. The cuticular transpiration of a variety of plants measured under standard conditions may vary by several orders of magnitude (see Table 6.2).

From Table 6.2, it can be seen that the rates of water loss from the cuticle are in the realm of expected adaptations to xerophytic (dry) environments. *Impatiens,* a plant of very mesic (moderate water) habitats, loses much more water on a per-unit-area basis than do oaks or pines, plants that

TABLE 6.2 Cuticular transpiration rates of several plants

Genus	mg H_2O h^{-1} g^{-1} (fresh wt.)
Impatiens	130
Caltha	47
Quercus	24
Pinus	1.5
Opuntia	0.1

Data from Pisek and Berger (1938).

are much more xerophytic. The *Opuntia* cactus, the most xerophytic of all, loses very little water through the cuticle. In fact, cuticular resistance to water loss in cactus can be as high as 1000 s cm^{-1}. Figure 5.1B shows the surface of a cactus plant with a heavy deposition of wax. The waxy cuticle accounts for the very high cuticular resistance to water transport.

LENTICULAR TRANSPIRATION

Woody plants have definite structures of loosely arranged cells within the bark that permit gas exchange (see Fig. 1.17). These structures are called lenticels. Depending on the arrangement of the cells of the lenticels, water loss can be rather significant. However, there are few data on lenticular transpiration.

GUTTATION

Under certain conditions of good water balance and low water loss as frequently occurs just before dawn, root pressure may force liquid water out through leaf hydathodes (large stoma-like structures of leaves that are fixed open). Open xylem vessels may terminate at hydathodes, and if the liquid of guttation (the water forced out the hydathodes by root pressure) does not evaporate, droplets can be observed on leaves (see Fig. 6.8). The total amount of water loss from plants by guttation is rather low inasmuch as it only occurs under special conditions of ample soil water and high relative humidity of the air.

WATER LOSS FROM GLANDS, HAIRS, AND APPENDAGES

Epidermal appendages may represent a significant pathway of water loss to plants, although such structures are frequently highly cutinized. Glands that secrete fluids, salts, resins, terpenes, and other organic materials will necessarily lose some water concomitantly. By and large, water loss from ap-

FIGURE 6.8 Photograph of a young corn plant losing water by guttation.

pendages is probably not too significant from the standpoint of the overall water balance of the plant.

Hormonal effects on transpiration

There have been several studies on the effects of plant hormones on water balance and on transpiration. Auxins and gibberellins are growth hormones known to influence water balance. Gibberellic acid regulates starch synthesis in some tissues (see Chapter 18). During starch hydrolysis, soluble sugars are produced that will decrease the osmotic potential. Thus hormones may regulate osmotic properties in some plant tissues.

Abscisic acid increases in plants that are water-stressed. Treatment of leaves with abscisic acid causes stomata to close; thus abscisic acid imparts some degree of drought tolerance. Whether the increase in abscisic acid in plants subjected to water stress is related to the fact that abscisic acid closes stomata is not known. Nevertheless, such an idea represents an attractive hypothesis and deserves further experimentation.

The significance of transpiration

It was previously mentioned that some plant physiologists believe that transpiration is a "necessary evil." Stomata open in response to light, CO_2 enters, and water is necessarily lost. Still, it seems that a process so complicated and of such magnitude would have some "biological meaning." Obviously, it is beyond the scope of science to look for "meaning" in a process (in a teleological sense), but there are many consequences of transpiration that can be discussed.

First, the evaporation from leaf surfaces is accompanied by a significant amount of heat loss. For each gram of water transpired, about 2.3 kilojoules are lost. Thus active transpiration enters into heat-load considerations; it tends to cool leaves. It is not uncommon to measure leaf temperatures below air temperature on bright, sunny days. Given these considerations, it is reasonable to assume that evaporative cooling of leaves is an important consequence of transpiration. The evaporative cooling of leaves is treated in Chapter 7 in a more quantitative manner.

An actively transpiring plant will also move large quantities of water from roots to leaves. The process will carry minerals and other substances from roots to stems and leaves. By and large, however, most minerals can be transported in the plant independently of transpiration, and transpiration is not considered to be the primary mode of mineral transport (but see the section in this chapter titled "The Daily Course of Transpiration").

Overall, it is probably best not to attempt rationalizations of transpiration but to simply treat it as a process occurring in virtually all vascular plants.

Antitranspirants

Conservation of water is of prime importance for efficient agriculture. The use of plants highly efficient in the use of water, and the control of transpiration without affecting photosynthesis, have been proposed as means to conserve water. Antitranspirants are materials that reduce transpiration and that have been used in attempts to conserve water.

Certain chemicals when applied to leaves will close stomata, increase R_s, and reduce transpiration. Because transpiration is limited just by R_s, and CO_2 uptake during photosynthesis is limited by both R_s and R_m, partial stomatal closure is apt to reduce transpiration more than it will photosynthesis. A good antitranspirant would be a compound that increases R_s but does not affect R_m so that transpiration is reduced with a minimal effect on photosynthesis.

Phenylmercuric acetate is a compound that has been used as an antitranspirant, but there is some evidence that it affects the chemistry of photosynthesis and thus directly increases R_m. If growth and agricultural yield are not important factors, as may be the case for ornamentals, ground covers, and other utility plants, increases in R_m may not be important considerations.

Sealants that prevent water loss without interfering with CO_2 uptake have been proposed and investigated, but none is known that is very satisfactory. Of course, many fruits, root stocks, and certain other propagules (structures that result in propagation) are routinely sealed with waxes to prevent excessive water loss during storage and transportation for which CO_2 uptake is not an important consideration.

6.2.3 Internal water transport

Virtually all of the evidence for how water is transported from roots to leaves supports the hypothesis that water is pulled under tension rather than being pushed under pressure (by a hydraulic system). When it is remembered that large trees can exceed

100 meters in height, the tremendous forces required are apparent. Since 1 bar is sufficient to raise water approximately 10 meters, it would require a force comparable to 10 bars to raise water to the top of a tall tree. This amounts to a gradient of 0.1 bar per meter.

It can be calculated on the basis of Poiseuille's law (refer to Chapter 2), which describes flow through a capillary, that for a velocity of 4.5 meters per hour up a 30-meter tree with conducting vessels 1 mm in diameter, it would require a pressure gradient of 0.1 bar per meter. Vessels 0.5 mm in diameter would require about 0.5 bars per meter, and for vessels 0.1 mm in diameter (the more usual size of xylem vessels), 1 bar per meter would be required. Because of the problem of air embolisms forming in vessels (cavitation; discussed in a subsequent section), the upper limit for vessel diameter has been calculated to be about 0.5 mm. Questions that must be accounted for by any theory include where water is transported, the rapidity of transport, and whether or not such high forces are present.

In the first experiments designed to ascertain the path of water and food transport in plants, tissues within woody stems were cut to block the transport paths. In a typical experiment such as the one shown in Fig. 6.9, bark and wood were cut in separate trials and the plants allowed to sit for observation. In the case of the cut-wood trial it was necessary to separate the bark and wood from each other with waterproof paper to prevent lateral water transport. When the bark alone was cut, leaves maintained their turgidity and no wilting occurred. After several weeks, excessive growth indicative of the accumulation of soluble food materials occurred above the cut. When the wood was cut, leaves wilted within a short period. It was concluded from these kinds of experiments that food material is largely transported in the bark (the phloem) and water is transported in the wood (the xylem).

Of course, lateral transport must occur, but for water this is largely by cell-to-cell transport on water-potential gradients as outlined in Chapter 4.

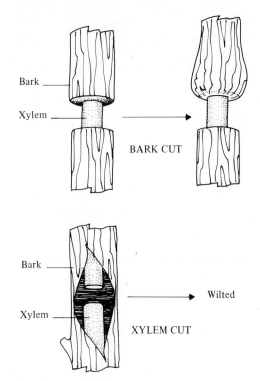

FIGURE 6.9 Experiment in which bark (with the phloem) is removed, leaving the wood (xylem) intact, and a parallel experiment in which the wood is cut, leaving the bark with phloem intact. The plant with the xylem cut and phloem intact will wilt, indicating that upward transport of water is through the xylem. When the bark is cut and the xylem is intact, downward translocation of food material stops at the cut. The accumulation of sugars results in excessive growth just above the cut.

The path of water transport

Cross sections of herbaceous monocotyledonous and dicotyledonous plants are shown in Fig. 1.18. Upward transport of water is through the xylem tissue of the vascular bundle. Although the bundles seem to be arranged in a definite pattern, any one bundle seems to follow a tortuous path within the stem (Fig. 6.10).

Dicotyledonous woody plants are perhaps more interesting from the standpoint of their xylem

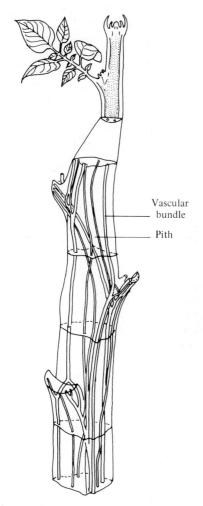

FIGURE 6.10 Drawing to illustrate the tortuous path that vascular bundles take in the stem of a herbaceous plant. Adapted from E. F. Artschwager. 1918. Anatomy of the potato plant, with special reference to the ontogeny of the vascular system. *J. Agr. Res.* 14:221–252.

transport system. Stem cross sections illustrating two wood types of woody dicotyledonous shrubs are shown in Fig. 6.11. In the ring-porous wood of *Artemisia filifolia,* water moves in the outer annual rings of the large vessels, which occur in definite bands. Ring-porous wood is also characteristic of oak, ash, and chestnut. In diffuse-porous wood

such as found in *Artemisia kauaiensis,* the large transporting vessels are more diffuse and water moves within several outer rings. The transport of water in diffuse-porous wood takes a very spiral path (Fig. 6.12). Kozlowski (1964) has shown a spiral path of upward water movement to exist within trees (poplar) by injecting dyes at the base into the wood and then preparing stem sections at various heights above the point of injection. The dye, as shown in Fig. 6.12, is transported with the water in the xylem and appears to radiate out in a spiral.

The tracheids and vessel elements are structured to be good conduits for water transport (see Fig. 1.13). They are elongated and have thick secondary walls and pits or openings on their radial walls. Water moves up the stem through tracheids and from tracheid to tracheid through the pits. The more complex xylem elements (vessel elements) are also elongated cells with thick secondary walls and pits, but they have perforated end walls. Vessel elements thus line up end to end, forming vessels that may be up to several meters long. Although water moves largely through the vertically arranged tracheids and vessel elements, most lateral transport occurs from cell to cell. A water column within a single conduit the length of a tall tree would not be very stable; the lateral connections between cells give stability to the conducting system.

Lateral transport is easily demonstrated with two transverse cuts, one above the other but on opposite sides of the trunk. If upward transport of water is maintained with such cuts, lateral flow must have occurred.

Transport velocity in the xylem

The velocity of water movement can be estimated by a variety of techniques. Estimations have been made using either radioactive tracers or dye transport within the xylem and also heat-pulse transport* between two parts on a stem. Velocities are

---

*A point along the stem is warmed, and the time required for the heat to travel upward is a measure of transpiration velocity.

FIGURE 6.11 Photomicrographs of wood from two species of *Artemisia* showing ring-porous wood (left; *A. filifolia*) and ring-diffuse wood (right; *A. kauaiensis*) (×100). For the most part, the water transport in ring-porous wood occurs in the youngest, outer vessels. In the ring-diffuse wood water transport is throughout the xylem. From S. Carlquist. 1966. Wood anatomy of Anthemideae, Ambrosiede, Calenduleae, and Arctotideae (Compositae). *Aliso* 6:1–23. Photographs courtesy of Dr. Sherwin Carlquist.

as high as 45 meters per hour in certain trees, but the usual velocities are 1 to 5 meters per hour. Transport velocities in ring-porous wood with large vessel diameters (up to 0.3 mm) are greater than in ring-diffuse wood that usually has smaller vessel diameters (about 0.1 mm). The larger-sized vessels are readily observed by comparing the ring-porous wood with the diffuse-porous wood of the two species of *Artemisia* shown in Fig. 6.11. Conifers with small tracheids (about 0.05 mm in diameter) have the slowest water-transport velocities.

The velocity of water transport within the xylem is partially governed by the rate of transpiration, because evaporation at the mesophyll surface creates forces for the upward pull of water. In addition, the rate of uptake of water and the resistance within the xylem elements to water flow will influence the rate of water movement. As explained earlier, transpiration frequently exceeds the rate of water uptake during the day if evaporative conditions are optimal.

The mechanism of water transport in the xylem

Much time and effort has been devoted by plant physiologists to formulate theories to explain how water moves up plant stems. Many have attempted to incorporate the living system of the plant (the so-called vital theories), but none has gained any favor. The most important theory is the transpiration-pull hypothesis in which water is pulled up the

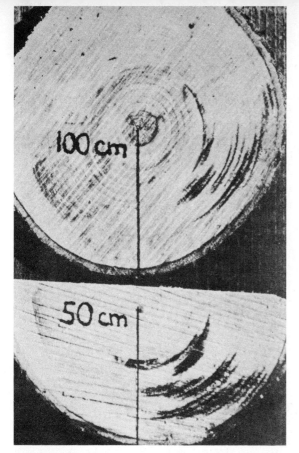

FIGURE 6.12 Experiment to illustrate the path of upward movement of water through the many vessels in the diffuse-porous wood of poplar. Dye was injected in a radial hole in the trunk, and sections were then cut 50 cm and 100 cm above the point of dye injection. The pattern of dye in the two sections indicates a spiral upward transport of water. From M. H. Zimmermann. 1971. Transport in the xylem. In M. H. Zimmermann and C. L. Brown (eds.), *Trees: Structure and Function.* Springer-Verlag, New York.

plant by the forces of evaporation created at the leaf surface during transpiration. Some of the older ideas about water transport are discussed below before the currently accepted transpiration-pull hypothesis is described.

SOME OLD HYPOTHESES FOR WATER TRANSPORT

Since plants are living organisms, it seems a reasonable assumption that water transport must be governed by living processes within the plant. And because it was known at the end of the last century that each xylem conducting cell was in association with at least one living parenchyma cell, the living parenchyma cells were believed to be involved in the transport process. Perhaps they functioned as pumps or osmotic generators for transport. Some even proposed that water was forced up large trees when wind caused trunks to sway in a pumping motion.

However, in 1890 Strasburger performed a simple experiment that seemed to eliminate the necessity of involving living parenchyma cells of the xylem in water transport. He cut a 22-meter-tall oak tree at the base and placed the cut stump in a solution of picric acid. The picric acid solution was transported to the leaves within the sap, surely killing the living parenchyma cells on its way to the leaves. Transpiration and upward picric acid movement occurred until the acid solution reached the living leaves. Acid destruction of the leaves stopped transpiration and further upward movement of the water.

This simple experiment yielded virtually all the information needed to formulate the hypothesis for water uptake. First, since the roots were removed, root pressures could be eliminated as the main mechanism of water transport up the xylem. This is not to say that, in small plants and even in some trees, root pressure could not account for some transport. But root pressure was not the main force for upward transport of water in the xylem. Second, living tissue of the xylem within the stem was not a requirement; water simply moved through the nonliving tracheids and vessel elements. Third and perhaps most important is that the water must have been pulled up by forces created in the leaves, presumably by transpiration. Once the leaves were killed and transpiration ceased, water flow stopped.

THE TRANSPIRATION-PULL HYPOTHESIS

The transpiration-pull hypothesis (the cohesion of water theory) for water transport states that water is pulled under tension up stems by forces created in leaves through transpiration. Evidence for such a hypothesis is rather overwhelming, and there are virtually no serious objections at present. It can be shown that sufficient forces can be generated at the leaf mesophyll surface and that the tensile strength of water is sufficient to maintain continuous water columns against a 20-bar force in order to lift water 100 meters. The most serious objection is cavitation, i.e., air embolism formation, that should tend to break water columns. Cavitation is treated below.

Tensions within the xylem

If water is truly pulled up stems during active transpiration, the xylem fluids must be under tension and not pressure. Many simple experiments support such a notion. The early Strasburger experiment was mentioned in the previous section. One of the simplest ways to measure transpiration is with a potometer. A cut stem is placed in a reservoir that is attached to a capillary tube. During transpiration, water will be pulled along the capillary. In addition, dendrographic measurements indicate that stems decrease in diameter when plants are actively transpiring. These three kinds of experiments suggest that the xylem fluids are under tension and not pressure during active upward transport.

In some plants, particularly prior to leaf development in the spring, it is possible to measure positive pressures in the xylem. Scholander *et al.* (1965) measured positive pressure gradients in vines prior to leaf formation. The collection of sap from maple trees for syrup depends on positive pressures. When trees were actively transpiring, however, Scholander was able to measure pressure drops with the pressure bomb (see Chapter 4). In Douglas fir trees, a pressure drop always occurred from the bottom of the crown to the top of the crown (Fig. 6.13). Thus there seems to be adequate evidence that in actively transpiring plants the xylem fluid is under tension because it is being pulled up by transpiration.

The tensile strength of water

Water has high tensile strength because of the cohesive forces of hydrogen bonding (see Chapter 2). It can be calculated that tremendous forces bind water molecules to each other. Recall that for lift up to the leaves of a 100-meter-tall tree, forces of at least 10 bars are required; if resistances are considered, perhaps the tensile strength of water would have to be much greater. Simple centrifugal experiments in which capillary tubes containing pure water are centrifuged until the water columns break have indicated tensile strengths of water up to several hundred bars. Thus the tensile strength of water is sufficient to allow it to be pulled up trees 100 meters tall.

Mesophyll forces

The forces necessary to pull water against an energy gradient of 20 bars or more are believed to be created by evaporation at the mesophyll surface. As explained earlier, at this interface between mesophyll and atmosphere there can be a drop in water potential of several hundred bars (refer to Fig. 6.3).

Assuming that there are capillaries within the mesophyll cell walls, it can be calculated that a capillary 0.1 mm in diameter will generate sufficient capillary forces to raise water 30 meters. A capillary of 0.1 μm will generate forces sufficient to raise water 300 meters (the forces involved are discussed in Chapter 2).

A model describing this concept is shown in Fig. 6.14. Evaporation from the capillary meniscus results in transpiration. A continuous water column, from the roots through the xylem elements to the capillaries of the mesophyll cell walls, is maintained by capillary forces.

Cavitation

A major problem with the transpiration-pull hypothesis for water transport is that at such low pressures and high tensions, dissolved gases readily

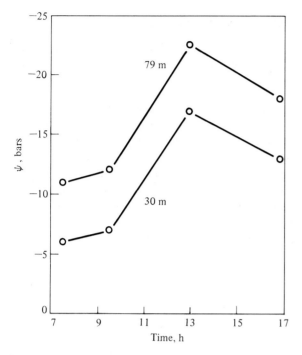

FIGURE 6.13 Measurement of water potential (ψ) at two heights in the crown of a Douglas fir tree. The potential was less at 79 m than at 30 m throughout the daylight period, indicating that the water in the tree was under tension. This is consistent with the hypothesis that water is pulled up the tree by forces created during transpiration. From early morning to midafternoon the ψ decreased; it then recovered in the evening. Redrawn from P. F. Scholander, H. T. Hammel, E. D. Bradstreet, and E. A. Hemmingsen. 1965. Sap pressure in vascular plants. *Science* 148:339-346. Copyright 1965 by the American Association for the Advancement of Science.

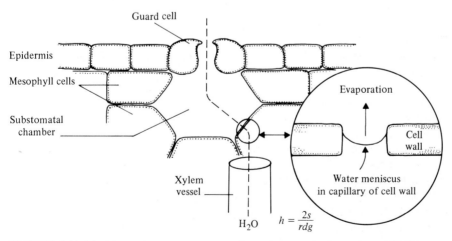

FIGURE 6.14 Model to show how evaporation of water from the fine capillaries of the mesophyll cell walls creates the force for the upward pull of water. The equation (explained in Chapter 2) is for the height of fluid in a capillary (h) as a function of the surface tension energy (s), and as an inverse function of the radius (r) of the capillary, the density of the fluid (d—here water), and the acceleration due to gravity (g).

tend to come out of solution. Gases such as CO_2, oxygen, and nitrogen in the xylem fluids will form bubbles if they come out of solution. Such gas formation at low pressure is called cavitation. Once a bubble forms and enlarges to the extent that the liquid–vapor interface is lost, excessive pressures will be necessary to pull or push the water column. Thus cavitation will prevent water transport in the xylem. In addition to cavitation at low pressure, freezing also presents a serious obstacle to xylem flow because of the low solubility of gases in ice.

Because of the extensive lateral transport between adjacent xylem elements, water continuity is maintained. Pits connecting xylem elements are sufficiently small that air bubbles cannot pass except perhaps in conifers. In conifers, tori (thickenings in the membrane) between bordered pits may act as valves, preventing gas bubbles from passing from one tracheid to another. It appears, therefore, that a xylem element with an air embolism could be effectively isolated from the main xylem stream. Furthermore, reestablishment of the continuous xylem stream may occur at night (when it is usually cool) or during periods of low transpiration. In the event of cavitation after freezing, thawing and then warming may cause the gas to go back into solution, thereby reestablishing a continuous water stream in the xylem.

Even though cavitation appears to be a serious problem for the acceptance of the transpiration-pull hypothesis, there do appear to be natural ways in which the gas can be redissolved. Moreover, as explained above, the effects of air embolism may be minimized by the isolation of individual xylem elements in which the cavitation has occurred.

6.3 Long-distance Transport—Solutes

The transport of inorganic minerals and organic compounds throughout the plant body has been studied for many decades, yet we still do not have a comprehensive theory to account for all aspects. The primary questions being asked by plant physiologists are: where does the translocation occur in the plant, how does it occur, and what factors, internal and external, influence it? Most evidence indicates that the process of solute transport is a physical phenomenon.

The transport of solutes is properly termed translocation and will be referred to as such here. As a generalization, mineral nutrients are taken up from the soil and translocated upward, whereas organic products of photosynthesis (photosynthates) are produced in the green leaves and transported downward. Thus the bidirectional translocation of substances is apparent from the very first consideration.*

6.3.1 Solutes of the vascular system

An unambiguous determination of the various solutes found in the vascular system is not possible, but a variety of studies has given much information about the materials transported in the plant vascular system. Samples from the xylem contents are usually collected by obtaining exudate from decapitated stumps (the exudate is forced out by root pressure) or by extracting a sample directly from the xylem fluid. By such methods, it is rather difficult to be sure that other fluids are not mixed with those of the xylem. Similarly, sampling from the phloem is more complex with even less certainty. The primary method of sampling the phloem takes advantage of the fact that the phloem is under positive pressure. Cutting into the phloem allows the contents to be pushed out by pressure. However, as explained earlier, once the phloem cells are damaged a slime may form and prevent leakage. The slime is most likely the protein of the phloem (P-protein, discussed later). The aphid-stylet method is perhaps the most elegant method of sampling the

*Some authors use translocation to refer only to transport in the phloem. Translocation is used here in the more general sense.

phloem and has been used to obtain reliable information about phloem contents. It is also discussed later.

One reason for wanting to know the composition of the phloem fluids is to learn where materials are transported. Second, such information frequently gives insight into the mechanism of translocation. Such knowledge becomes important for agriculture and horticulture in the application of growth regulators, systemic pesticides, and other materials taken up by the plant. Whether substances are transported in the xylem or in the phloem are important considerations.

Both the xylem and phloem have solutes that are translocated. The phloem is much more concentrated than is the xylem, with about 50 to 300 mg dry matter ml^{-1} in the phloem and 1 to 20 mg dry matter ml^{-1} in the xylem. The phloem is relatively alkaline with a usual pH of 8.0 to 8.4, whereas the xylem fluids have a low pH ranging from 5.2 to about 6.5.

Even though the xylem fluid is relatively dilute there is a wide variety of organic and inorganic materials present, and undoubtedly xylem transport is important. Many nitrogenous compounds, especially those that are synthesized in the roots, are found in the xylem and appear to be translocated from the roots to the leaves. Amino acids, amides, and amines may be present in the xylem, but usually only one such compound will be present in any appreciable concentration. Common amino acids present in the xylem are aspartic acid, asparagine, and glutamine. These nitrogenous compounds may be present in quantities of 0.1 to 5 mg ml^{-1}. In addition to the organic nitrogen compounds the xylem will also have some sugars, organic acids, and even some of the plant hormones. A common organic acid of the xylem is citrate, which may be chelated with iron.

Of the total solids found in the xylem, about one-third is usually inorganic. Potassium, calcium, magnesium, sodium, phosphate, chloride, and sulfate are the most mobile in the xylem. Some of the phosphate and sulfate occur in organic compounds.

With such a wide variety of compounds found in the xylem, it is reasonable to assume that the xylem plays a major role in solute transport in vascular plants. The phloem, however, is much more concentrated than the xylem and is the solute transport system of most importance in vascular plants. Of the 50 to 300 mg ml^{-1} dry weight in the phloem contents, about 80 to 90% is sugar. The most common sugar of the phloem is sucrose, and it is sucrose that is most frequently the major organic compound translocated in plants. Other nonreducing sugars such as raffinose, stachyose, and verbascose may be transported as well as sucrose, and occasionally these other sugars will be translocated exclusively. In addition, the sugar alcohols such as sorbitol and mannitol may be the predominant organic compounds translocated in some plants. The significance of nonreducing sugars being translocated is not known; however, it has been suggested that nonreducing sugars are less reactive and therefore more amenable to long-distance transport.

Amino acids are common constituents of the phloem contents. Most frequently, the same amino acids that predominate in the xylem will also be the most common in the phloem, but they are more concentrated in the phloem, occurring in concentrations of 20 to 80 mg ml^{-1}. Other nitrogenous compounds may be present, even protein. Most of the protein occurring in the phloem is P-protein. It is assumed that the P-protein is not translocated. There is not much nitrate in the phloem, but ammonia may be present.

Organic acids such as malate and citrate may be present in the phloem. A whole variety of other organic compounds are present in the phloem including the plant hormones. The phloem probably plays a major role in the transport of many of the plant hormones, but auxin is transported polarly (i.e., in one direction) through parenchymatous cells.

Many of the same inorganic ions found in the xylem also occur in the phloem but frequently in different concentrations. Calcium, for example, is usually less concentrated in the phloem than in the xylem. Phloem may have relatively high ratios of potassium to sodium and magnesium to calcium. Inorganic anions such as chloride, carbonate, sulfate, and phosphate are common phloem constituents. The latter two may occur in organic compounds such as sugar phosphates and the amino acids methionine, cysteine, and cystine.

The phloem can transport fairly large polymers, including proteins and complex lipids. There is much evidence as well that many viruses are transported throughout the plant body in the vascular system. When a plant leaf is inoculated with tobacco mosaic virus, the virus appears in other leaves because of transport in the phloem. It is not entirely clear if the virus is transported as the intact virus or if the nucleic acid core is transported alone. In any case, the pores of the sieve plates are apparently sufficiently large to allow passage of large particles along with the usual materials that are translocated. There is evidence as well for xylem transport of viruses.

6.3.2 Xylem transport

For the most part, the transport of solutes within the xylem is a passive process; the solutes are carried up with the transpiration stream. Water transport within the xylem was discussed in Section 6.2.3. Because solute transport is a passive process in the xylem stream, it is to be expected that the rate and velocity of transport would be governed by the rate of transpiration. There is some evidence from experiments to support this.

Some inorganic minerals appear to be largely transported in the xylem and thus appear to be dependent on transpiration for movement from the roots to the leaves. Calcium and boron fall into this category. There is some evidence that calcium accumulates in leaves that are transpiring rapidly since calcium is not very mobile once transported. Similarly, a calcium deficiency might be expected in the absence of transpiration. Although being transported upward from the roots into the leaves through the xylem in the transpiration stream, certain other inorganic minerals such as potassium

are rather mobile and are readily redistributed by phloem translocation.

6.3.3 Phloem transport

As early as 1679 Malpighi knew of the experiment shown in Fig. 6.9, but he believed translocation took place largely in the wood (xylem). Hartig in 1837 discovered sieve-tube cells in plants and knew that cut bark would exude material that was nearly 30% sugar from the phloem. But because the morphology of sieve-tube cells involved small sieve pores, it was not believed that much transport could occur. However, with the experiments of Mason and Maskell in the 1930s it became apparent that phloem functioned in translocation. In extensive experiments with cotton plants they were able to correlate sugar content in leaves with that in the phloem. Coupled with the earlier ringing experiments (Fig. 6.9) and those of their own, it was then apparent to Mason and Maskell that phloem tissue functioned in translocation.

Now much more sophisticated experiments using isotopes have proven that bulk transport of solutes is largely in the phloem. One type of experiment that has yielded much information on phloem transport is the aphid-stylus experiment of Zimmermann. When aphids feed, they insert their stylet into single sieve-tube elements, withdrawing the contents. By removing the aphid body from the stylet after it is inserted into a sieve element, Zimmermann was able to collect directly from the phloem elements. Figure 6.15 shows an aphid stylet in a single sieve-tube cell.

Phloem anatomical considerations

Like xylem, phloem is a complex tissue composed of four cell types: fibers, parenchyma cells, the conducting sieve-tube cells, and companion cells that are associated with the sieve-tube cells.* The basic

---

*Gymnosperms do not have companion cells but rather comparable albuminous cells that are associated with the sieve elements.

Aphid

Sieve-tube cells

Fibers

Stylet tip inserted
into sieve-tube cell

FIGURE 6.15 Photograph of an aphid feeding on *Tilia americana* (linden). The stylet of the aphid is inserted through the epidermis into the phloem. The tip of the stylet has actually penetrated into a sieve element. After the aphid is removed, phloem exudate continues to ooze out of the stylet, indicating that the phloem contents are under pressure. From M. H. Zimmermann. 1961. Movement of organic substances in trees. *Science* 133:73–79. Copyright 1961 by the American Association for the Advancement of Science.

anatomy of phloem is discussed in Chapter 1; diagrams of sieve elements and companion cells are shown in Fig. 1.14. Sieve elements are the basic transport cells of the phloem and can be fairly long, up to 50 μm long in angiosperms and over 1 mm long in gymnosperms. They are variable in width but average about 5 μm. Most characteristic of sieve elements, however, is the presence of the sieve plates with sieve pores. In the angiosperms, the sieve plates occur on the end walls and sieve areas occur on the lateral walls. The sieve cells of the gymnosperms do not have sieve plates on the end walls but just sieve areas on lateral walls. These sieve plates and sieve areas connect adjoining sieve elements and form the passage for solute transport from cell to cell.

Any considerations of function for the sieve elements must take into account the fact that they lose their nuclei when they become functional after maturation. Companion cells, which are always associated with the sieve elements, do have nuclei and perhaps function for the sieve elements. There are many plasmodesmata connecting companion cells and adjacent sieve elements. The sieve elements lack definite vacuoles and thus do not have tonoplasts, but they do have plasma membranes and cytoplasm with the usual cytoplasmic inclusions. Figure 6.16 is an electron micrograph

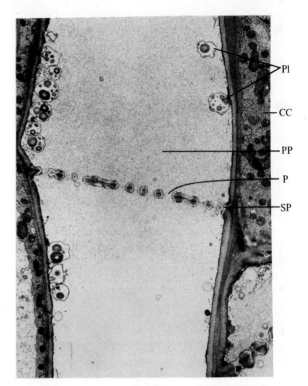

FIGURE 6.16 Portion of a sieve tube from tobacco (*Nicotiana tabacum*) showing two contiguous sieve elements and adjacent cells (X4400). The P-protein (phloem-protein) has accumulated in the lower portion of the upper sieve element and is beginning to form a slime plug. It is believed that the slime plug is an artifact of preparation because of the extreme sensitivity of the sieve elements and contents. The sieve plate and pores are distinctly visible between the two sieve elements. There is a thin layer of cytoplasm along the cell walls of the sieve elements that contains characteristic plastids. The adjacent companion cells have many mitochondria. Pl = plastids. CC = companion cells. PP = P-protein. P = pore of sieve plate. SP = sieve plate. From R. Anderson and J. Cronshaw. 1970. Sieve-plate pores in tobacco and bean. *Planta* 91:173–180. Electron micrograph courtesy of Dr. J. Cronshaw.

showing the sieve-plate ultrastructure and portions of two continuous sieve elements.

The cell lumen of the sieve element appears open except for the presence of transcellular strands going from sieve element to element. These transcellular strands are composed of protein called phloem-protein (P-protein), shown in the electron micrograph in Fig. 6.17. As mentioned previously, the P-protein tends to precipitate easily, forming slime bodies and slime plugs in the sieve elements. In fact, it is believed that such slime formation of P-protein after sieve-element injury forms plugs that prevent exudation from the injured cells. Thus slime-plug formation is presumably a protective mechanism to prevent the loss of translocation contents.

The P-protein has been an enigma for years, tending to complicate the study of phloem structure and the theories for the mechanism of trans-

location. In many sections prepared for electron microscopy the P-protein is precipitated, forming slime plugs in the vicinity of the sieve plates. In some preparations, the plugs are present on the downstream side of the sieve plate and appear to be pulled away below on the upstream side. It has been concluded from carefully prepared sections that the slime precipitation is an artifact of preparation and that it does not occur in functional sieve elements *in vivo*. Evidently the P-protein is very sensitive. As mentioned above, it may serve to protect plants by plugging punctured sieve elements, preventing loss of the contents. A puncture in a sieve element should be accompanied by exudation since the contents are under pressure. Even though the presence of the P-protein within the sieve-element lumen would appear to interfere with translocation, it is generally believed that P-protein somehow aids in translocation.

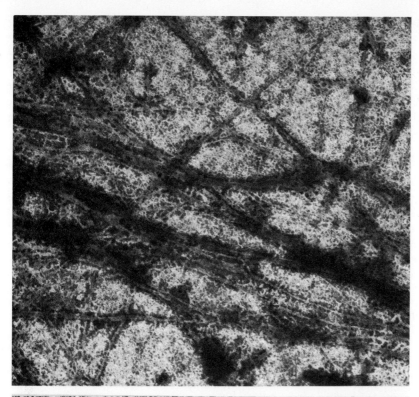

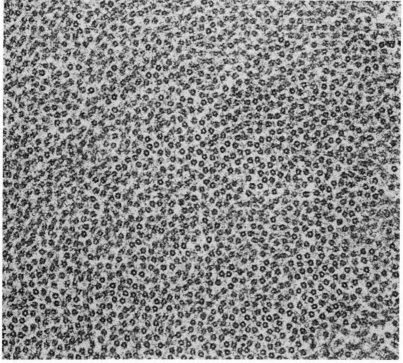

FIGURE 6.17 Electron micrographs of phloem-protein (P-protein). The upper plate is P-protein from the phloem exudate of *Cucurbita*. The fibrils were stained with sodium phosphotungstate. The smallest P-protein fibrils in the photograph are about 8 nm in diameter. The larger tubules of P-protein are about 23 nm in diameter. The lower plate shows a transverse section through a P-protein body in a young sieve element of tobacco. Here the P-protein occurs as definite tubules that run in the same direction as the long axis of the sieve elements. There appear to be connecting strands between the hexagonally shaped P-protein tubules. This figure should be compared with the P-protein shown in the sieve elements of Fig. 6.16. From James Cronshaw. 1974. P-proteins. In S. Aronoff, J. Dainty, P. R. Gorham, L. M. Srivastava, and C. A. Swanson. *Phloem Transport*. Plenum, New York.

The plasma membrane is continuous from sieve element to sieve element. There has been controversy about plasmolysis of sieve elements that lack a tonoplast. However, careful experiments seem to indicate that the sieve element will plasmolyze, which indicates that the plasma membrane is differentially permeable. It therefore appears that the plasma membrane of the sieve element is comparable to the plasma membranes of other cells.

At maturity a complex carbohydrate, callose, forms around the sieve-plate pores. After the functional life of the sieve element ends, which may be from a few days to an entire growing season, callose may completely cover the sieve pores.

Rate and velocity of transport in the phloem

If a pulse of radioactive $^{14}CO_2$ is given to photosynthesizing leaves, the rate (flux) and velocity of phloem translocation can be estimated as the radioactivity passes points along the stem. Rates of transport range from 0.5 to 5 g dry material cm^{-2} phloem tissue h^{-1}. The velocity* of transport is also variable, ranging from 10 to 100 cm h^{-1}, some 5 times slower than water transport in the xylem. However, in certain species of cucurbits that have massive translocation to large fruits, translocation velocities have been measured to be as great as 3 m h^{-1}.

Translocation of the short-lived isotope of carbon, ^{11}C, has been used to estimate translocation velocities. A short pulse of radioactive isotope is given to leaves, and then because of radioactive decay the experiment can be repeated without the problem of residual radioactivity. Such experiments have shown velocities of about 250 cm h^{-1}.

Bidirectional flow

Bidirectional transport was studied by supplying radioactive CO_2 ($^{14}CO_2$) to leaves and radioactive phosphate (^{32}P) to roots. The phosphate went up

*Note that rate is an indication of the number of molecules passing a point and velocity is an indication of how fast the molecules are moving.

and the carbon went down, evidence for simultaneous flow in opposite directions. Of course, the upward flow of phosphate was probably in the xylem and the downward flow of carbon in the phloem.

In later research using the aphid-stylet technique, radioactive urea (^{14}C) was supplied above the inserted stylet and a marker dye was supplied below the stylet. By collecting exudate from the aphid stylet, it was shown that there was simultaneous bidirectional flow to a single sieve element. It is not possible, of course, to completely rule out transport to the single sieve element after translocation occurred, so the experiment does not necessarily imply bidirectional transport in a single sieve tube. The experiment does show transport in two directions in the phloem, however.

Experiments in which two or more substances were measured for simultaneous velocities of transport indicated that materials are translocated at different velocities. For example, it was shown that radioactive sucrose (^{14}C) moved at a velocity of 200 to 500 cm h^{-1} and radioactive potassium (^{42}K) supplied at the same time moved with a velocity of 30 to 60 cm h^{-1}. In such experiments, however, it is difficult to know if the uptake into the phloem is the same for both substances. The different velocities may reflect differences in loading rather than actual transport velocities. In the case of the experiment mentioned above, the isotopes were injected directly into the sieve elements, eliminating the loading step; thus the evidence for differential velocities of transport in the phloem is quite good. Such differential velocities can be the result of chromatographic phenomena. Different chemical species all follow the mass-flow direction but at different rates, depending on individual properties.

However, other experiments in which foreign materials such as viruses and herbicides were supplied do not support bidirectional transport. Such foreign substances seem to move just with the main direction of flow. It could be argued that foreign matter would not be subjected to the same controls as would naturally occurring substances and could simply just flow with the main direction of the translocation stream.

Overall, the experiments that were designed to ascertain if bidirectional flow occurs in the phloem are complicated and difficult to interpret unequivocally. It does seem that materials travel at different velocities and that, at least with some substances, there can be simultaneous bidirectional flow in the phloem.

Factors affecting solute translocation in the phloem

Environmental factors as well as endogenous factors influence translocation in the phloem. In the case of environmental factors, the effects may be rather indirect. Some of these factors are discussed below.

LIGHT

There is increased translocation in the presence of light. Such translocation is correlated with sugar production and starch accumulation. It is more likely that the correlation of translocation activity with light is the result of sugar production by photosynthesis rather than the result of energy production. The latter, of course, cannot be completely eliminated as a cause.

An interesting question concerns the effect of photosynthetic rate on the velocity of solute translocation in the phloem. It seems reasonable to assume that the faster sugar is produced by photosynthesis, the faster it would be transported out of the leaves. Many studies indicate that translocation is related to photosynthesis, but the relationship is indirect in that translocation is more directly a function of the concentration of sugar in the leaf. Thus the greater the concentration of sugar in the leaf, the greater the translocation velocity.

TEMPERATURE

Since virtually all processes are governed to some extent by temperature, it is not surprising to learn that translocation is temperature-dependent. Furthermore, any temperature dependence of a particular process such as translocation is complicated by temperature effects on related processes.

Cold temperatures are expected to decrease translocation fluxes, if for no other reason than by directly increasing the viscosity of the phloem fluids. Specific experiments to overcome confounding effects have been conducted in which cold jackets have been placed over small regions of the stem. When only petioles are chilled, translocation out of leaves is affected by a marked reduction in transport. These data seem to indicate a direct temperature effect on translocation.

It is worth noting here that over the temperature range of about 20 to 30° C, the Q_{10} for translocation is about 1.3. A Q_{10} of 1.3 is consistent with processes that are limited by physical phenomena. At a low temperature in the range of chilling temperatures, the Q_{10} for translocation may be as high as 6.

METABOLIC FACTORS

The light and temperature effects on translocation that are mentioned above are a direct indication that metabolic activity is correlated with translocation velocities. Other experiments that have shown respiratory inhibitors such as dinitrophenols, azide, arsenite, iodoacetate, fluoride, and cyanide to inhibit translocation seem to implicate respiratory processes as well; hence the requirement for ATP.

We do not know, however, if such experiments reflect processes taking place within the sieve elements or directly on the solute loading process. The respiratory inhibitors could also interfere directly with sugar synthesis. But, in summary, the data are clear and indicate that metabolic processes are important for translocation in the phloem, and although the energy requirement may not be for actual transport at least it is required for phloem loading.

HORMONES

Since hormones and growth regulators influence growth processes, they are expected to alter patterns of translocation. One group of hormones, the cytokinins, seem to have a direct influence on translocation. Cytokinin treatment of a senescing leaf that would ordinarily be translocating carbohydrate out and becoming chlorotic (chlorosis = chlorophyll deficiency) causes it to remain green and prevents the outward translocation of solutes. The cytokinins seem to prevent senescence by preventing the mobilization of reserves.

BORON DEFICIENCIES AND TRANSLOCATION

Studies have linked boron to carbohydrate translocation. Boron deficiencies cause a reduction in translocation. The exact reason for the boron effect is not known, but boron does affect some of the enzymes involved in carbohydrate metabolism, such as those involved in the synthesis of UDP-glucose. Altered sugar metabolism would be expected to affect translocation. In addition, there is some evidence that sugars may be translocated as borate complexes. Sugar complexes with boron readily, but there does not seem to be sufficient boron in plants to complex with all the sugar in the phloem. Further studies are necessary before the exact role of boron in translocation can be stated.

The mechanism for solute transport in the phloem

As discussed by Peel (1974), the transport system has three major components: (1) the source for the solutes, (2) the conduits through which the transport occurs, and (3) the sink, or destination, of the solutes. The important considerations are: how solutes are loaded into the sieve elements at the source, the actual mechanism of transport along the sieve elements, and how the solutes are unloaded (removed) from the sieve elements once they have arrived at the sink. These considerations are taken into account in the discussions below.

For a valid explanation of the translocation mechanism, the facts and observations that must be taken into account are as follows.

1. The living-cell requirement

2. The bidirectional flow

3. The high velocities and large fluxes

4. The daily periodicity

5. The metabolic requirement

6. The boron requirement

7. The phloem anatomy

Thus any theory must be consistent with the energy-dependent process of translocation that occurs in living cells in two directions simultaneously and at high velocities. Although the translocated fluid flows through living cells, much of the energy dependence may be for loading and unloading of solutes.

PHLOEM LOADING

The metabolic requirement for phloem translocation is very well documented. There is evidence that ATP is required, that temperature affects the velocity beyond the expected effects on viscosity, and that metabolic inhibitors of respiration reduce translocation. Because it is generally believed that the actual transport within the sieve elements is a physical process, more than likely most of the energy requirement is for the loading of sieve elements with solute from the adjacent cells and perhaps also for unloading at the sink.

Loading refers to the uptake of solutes by sieve elements from the adjacent parenchyma cells, companion cells, or transfer cells. There have been some experiments directed toward understanding the loading process, but less is known about loading than about actual transport.

The loading process shows much selectivity. For example, organic acids such as malic and citric acids are not readily taken up by sieve elements. Certain amino acids such as threonine, serine, and alanine are taken up, but others such as aspartic acid are not. Inorganic ions also show differential degrees of uptake. The monovalent cations, sodium and potassium, are readily taken up, but some divalent cations such as calcium are not. Interestingly, whereas the loading process shows great selectivity, virtually all substances are readily translocated once they are in the sieve elements. Perhaps one of the most important aspects of the loading process is the selectivity of sugar uptake. Only nonreducing sugars are taken up in any appreciable quantity, and in most plants it is sucrose that is the primary sugar transported.

Two questions asked about the loading process involve whether it is symplastic and whether the sucrose is loaded (and unloaded) as sucrose *per se*. Evidence for or against symplastic loading is meager. The photograph in Fig. 6.2 would suggest that there is symplastic uptake of solutes by the sieve element from the adjacent transfer cell since there are plasmodesma connections. Such connections are not always seen between sieve elements and adjacent cells, however. The electron micrograph in Fig. 6.18 shows the relationship between photosynthetic bundle sheath cells in crabgrass and a vascular bundle with xylem and phloem. These chloroplast-containing cells have many plasmodesmata connecting to the adjacent vascular bundle. It would appear that solute transport from the photosynthetic cells to the vascular bundle would be symplastic through the plasmodesmata. Sugar could enter the apoplast (cell walls and intercellular spaces) from the vascular parenchyma cells during subsequent transport to the sieve elements. Evidence for such a notion was obtained by Evert (1977), who found that the main solute of the apoplast of corn vascular bundles was in fact sucrose. Some degree of apoplastic involvement would appear to impart a greater degree of selectivity because it is assumed that passage through the symplasm shows very little selectivity. Of course, for the loading process all of this may be idle conjecture, and selectivity could exist for symplastic transport as well as for apoplastic transport.

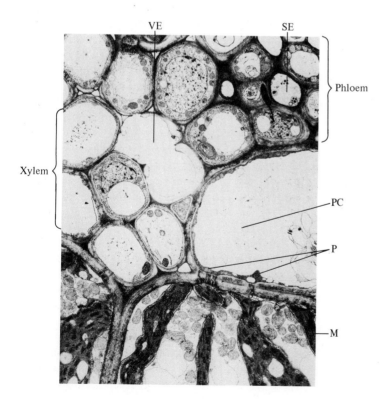

FIGURE 6.18 Electron micrograph prepared from a portion of a vascular bundle of crabgrass, showing plasmodesmata connecting a chloroplast-containing leaf cell and a vascular parenchyma cell (×9300). Within the phloem there is a sieve element visible, as can be identified from the presence of the characteristic plastid (refer to Fig. 6.16). The xylem has at least one apparent vessel element (tracheid). Transport from the photosynthetic mesophyll cells to the vascular parenchyma appears to be symplastic through the plasmodesmata. Once sugar is in the vascular bundle, subsequent sugar transport to the sieve elements could be either apoplastic (through cell walls and intercellular spaces) or completely symplastic. SE = sieve element. VE = vessel element (tracheid). PC = parenchyma cell. M = chloroplast-containing mesophyll cell. P = plasmodesmata. Electron micrograph courtesy of Dr. W. W. Thomson.

In addition, the evidence for sucrose hydrolysis to form glucose and fructose (inversion) during loading is very equivocal and confusing. In plants such as sugarcane there is evidence of inversion during loading. It appears that during the loading process the sucrose is hydrolyzed to glucose and fructose and then resynthesized after entering the sieve elements. In plants such as sugar beet it appears as though the sucrose is loaded without inversion. Most experiments to test the hypothesis of inversion during loading are done with sucrose-^{14}C. Sucrose with a ^{14}C label in the fructose and not in the glucose is supplied to the tissue and then recovered in the phloem. If the ^{14}C label remains in the fructose of sucrose and does not appear in the glucose, it is taken as evidence against inversion. Because of the presence of enzymes that can convert fructose to glucose (tautomerization—see Chapter 10 for a discussion of this and of inversion), randomization of the label between fructose and glucose would be expected if inversion occurs. The data are, however, quite equivocal because it is possible that the loading process prevents randomization even if inversion occurs. There is some evidence for inversion during unloading.

In summary, we know that loading during translocation is an active process requiring metabolic energy. It is uncertain if the process occurs sym-plastically or if it is partially apoplastic. Furthermore, it is not possible to make generalizations about sucrose inversion during loading. It is likely that there are species differences in the loading mechanism.

TRANSPORT HYPOTHESES

The main hypothesis for solute transport in the phloem is the mass-flow concept, but there are some other ideas having some experimental evidence. The mass-flow hypothesis is discussed first, and then some of the other hypotheses are discussed from the standpoint of their consistency with known experimental facts.

The mass-flow hypothesis

Mass or pressure flow as a mechanism for translocation was first proposed in 1860 by Hartig to account for the pressure exudation from tissues. The theory was expanded by Münch in 1930 for transport in all cells but is now restricted to translocation in the phloem.

A model describing the mass-flow hypothesis is shown in Fig. 6.19. Two osmometers are each in a different beaker of a dilute water solution and are connected to each other by a capillary. In one,

FIGURE 6.19 Osmometer model to illustrate the mass-flow hypothesis for solute transport within the phloem. Two osmometers are connected by a capillary tube. The osmometer at the left is assumed to be the source (leaves, for example) and the osmometer at the right is assumed to be the sink (roots, for example). Each is immersed in a dilute water solution. The osmometer at the left (source) has a high sugar concentration because of photosynthesis, and the osmometer at the right (sink) has a lower sugar concentration because of consumption.

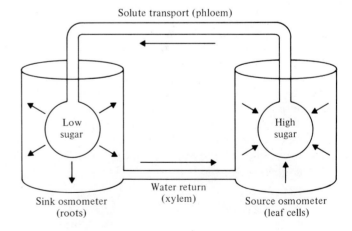

called the source, there is a high sugar concentration because of photosynthesis, while in the other, called the sink, there is a low sugar concentration because of consumption (by respiration, for example). As was true for the osmometer described in Chapter 4, water will flow into the osmometer at the source from the beaker and create a hydrostatic pressure head. This excess hydrostatic pressure will force the source osmometer fluid with sugar contents through the capillary connecting tube to the osmometer at the sink. Provided that there is a water return from sink to source and that sugar is consumed or removed at the sink, the process will continue indefinitely. Except for the loading and unloading of sugar the process is purely physical.

The flux (F) of such a mass-flow system can be described as follows:

$$F = \frac{k \, r^4 \, (P_1 - P_2)}{\eta \cdot \ell},$$

where

F = flux in g (dry matter) cm^{-2} s^{-1};

k = conductivity coefficient in units of g-poise cm^{-5} bar^{-1} s^{-1};

r = radius of conduit in cm;

P = hydrostatic pressure (P_1 is the pressure at the source and P_2 is the pressure at the sink, both in bars);

ℓ = length of conduit in cm; and

η = viscosity of phloem fluid in units of poises.

The flux of mass transport within a sieve element, described in the above equation, is a direct function of the pressure differential between source and sink (ΔP). The flux is directly proportional to the fourth power of the radius of the conduits (r^4), inversely proportional to the length of the conduits (ℓ) and to the viscosity of the fluid (η). Some of these terms can be combined into a resistance term, R. The latter would be

$$R = \frac{\eta \cdot \ell}{k \cdot r^4}.$$

Thus the mass flow through sieve-element conduits can adequately be expressed by the following equation:

$$F = \frac{\Delta P}{R}.$$

The flux equation for pressure flow within the sieve elements is written as a function of the fourth power of the radius of the sieve elements (r^4) consistent with Poiseuille's law for flow through capillaries with small diameters (see Chapter 2). Because water can enter and leave any one sieve element along the length of the sieve tube, Poiseuille's law may not be valid and flux may not vary with the fourth power of the radius but rather with the square of the radius (r^2).

Although there usually is a pressure gradient from source to sink such that ($P_1 - P_2$) is positive, there are instances of reverse-flow translocation. Photosynthesis in leaves (usually considered to be the source region) will contribute solutes, and osmotic potentials should be much lower in the leaves than in the roots. For this reason, pressures should be higher in the leaves than in the roots so that positive pressure gradients can exist from leaves to roots. Yet we know that bidirectional flow occurs, contradicting the notion of flow on positive pressure gradients. We do not, of course, know that bidirectional flow necessarily occurs in any one sieve tube, but it is still difficult to visualize significant pressure differences existing in adjacent sieve tubes. Furthermore, the bulk-flow theory described here does not seem to account for differential velocities of translocation in one direction.

Perhaps a more serious inconsistency is the nature of the resistances, R, in the translocation path. The sieve elements are filled with P-protein, and transcellular strands of P-protein can be seen through the tubes in electron micrographs. In addition, the sieve pores seem to be quite small relative to the size of the lumen. All this adds up to high resistances to flow, so high that it seems the pressure gradient would have to be excessive to push material at the high rates measured.

Because of the high apparent resistances to flow within the sieve-element lumen, it was proposed that bulk flow was through the walls of the sieve elements rather than in the lumen. When it was discovered that the plasma membrane of the sieve elements was differentially permeable, the notion of flow through the cell walls was abandoned because there could not be much transfer of sugar from the lumen to the cell wall. It seems to be an unavoidable conclusion that bulk flow is through the lumen of the sieve element.

Because of the problems associated with the concept of mass flow, there have been a few other theories proposed that are worthy of serious consideration. Some of these are discussed below.

Electroosmosis

Because of the high pressures required to force material through the sieve pores, Spanner (1958) proposed a theory involving mass flow of solutes through the sieve-element lumens and then electroosmosis across the sieve plates, the region of the greatest apparent resistance. In the simplest sense, electroosmosis occurs when ions move across a membrane in response to an electric gradient that pulls along water and other contents because of solvent drag. In the sieve-tube element, it is visualized that the pores of the sieve plate are negatively charged and have associated with them many positive ions. There are thus more soluble, mobile cations than there are anions. If an electric potential exists across the sieve plate, the mobile cations will move toward the cathode and the mobile anions will move toward the anode, both carrying water and other materials with them. Since there are more mobile cations than there are mobile anions, the net flow of water will be toward the cathode. In addition, the dipolar water molecule will migrate on an electric-potential gradient, carrying solutes.

A model for electroosmotic (i.e., potassium-mediated) phloem transport in plants is illustrated in Fig. 6.20. An electrogradient is maintained upstream of the direction of flow by a potassium pump

associated with the companion cells. Potassium is maintained at a high level on the upstream side of the sieve plate and a low level on the downstream side. Cations including the dipolar water molecule move downstream (toward the cathode) on the electric-potential gradient, carrying the phloem contents along. Hence mass flow tends to push bulk material along the sieve tubes, and transport across the sieve plates is aided or actually caused by electroosmosis set up by a potassium pump.

Although the electroosmotic system for phloem translocation is an attractive physical process and is supported by observations that potassium deficiencies inhibit translocation, there are many serious objections. First, potassium itself tends to be translocated in the phloem contents. Furthermore, there is reason to doubt that sufficient potentials can be created across the sieve plate to drive transport. On the basis of experiments with large, single-celled algae it was computed that a pressure differential across the sieve plate of 1000 bars per volt could be created, provided that water does not leak back. However, the resistance through the electroosmotic pores is estimated to be 10^5 times greater than it is for other pathways available for movement of water backward against the electroosmotic gradient. Thus even electric potentials across the sieve plate of 0.1 volt would only be sufficient to create a pressure differential of about one millibar, insufficient for transport across the sieve plate.

Streaming and contractile-protein hypotheses

Older theories for solute transport in the phloem supposed that cytoplasmic streaming may account for translocation. Young sieve elements have active cyclosis, but this seems to cease at maturity when translocation begins. Recently, active bidirectional streaming along the transcellular strands has been reported, but this observation has not received much support.

In 1932 van den Honert proposed a translocation mechanism based on interfacial flow along transcellular strands. This idea is analogous to a

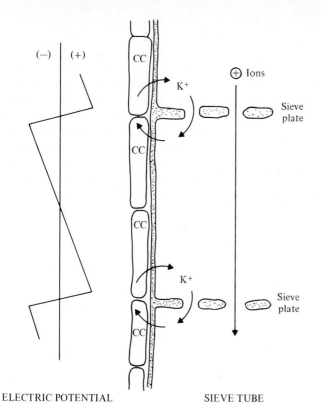

(−) (+)

CC

K⁺

⊕ Ions

Sieve plate

CC

CC

K⁺

Sieve plate

CC

ELECTRIC POTENTIAL   SIEVE TUBE

FIGURE 6.20 Model to illustrate Spanner's hypothesis for electroosmosis-driven translocation. Potassium (K^+) is pumped upstream in the sieve tube by a potassium pump associated with the companion cells (CC). The K^+ distribution across the sieve plate maintains an electric-potential gradient across the sieve plate. The slightly positively charged water (δH_2O) moves on an electric-potential gradient, carrying along the solute contents.

drop of oil spreading out on a water surface. The process would seem to be rapid enough to account for the high velocities, and the transcellular strands present a large surface area, but the fact that there is bulk flow seems to preclude the hypothesis. Interfacial flow has not received much support.

Other notions for transport in the phloem assume that the P-protein of the transcellular strands is contractile. Contraction and relaxation somehow aids in the transport. There is no evidence for contraction of P-protein; however, there is some evidence that it has ATPase activity (that is, it can catalyze the hydrolysis of ATP). So an active role for P-protein in translocation is attractive.

Overall, the mass-flow hypothesis for the translocation of solutes in the phloem seems to be the one most consistent with the available evidence. Only time will tell if the few remaining inconsistencies such as the apparent high resistances to flow in the sieve tubes and the bidirectional flow will be explained or if a new theory will replace the mass-flow hypothesis.

6.4 Concluding Remarks about Transport

Despite the fact that transport in plants is one of the most obvious phenomena being studied by plant physiologists, we still have little in the way of unequivocal hypotheses. We know little about symplastic transport through plasmodesmata and virtually nothing about the functioning of transport cells. In fact, the evidence that transfer cells actually function in transfer processes is quite conjectural.

With respect to long-distance transport of water and solutes, we know that it occurs but are quite uncertain of the mechanisms. The transpiration-pull hypothesis for water transport up stems was proposed by Dixon in 1914 and still stands as the best idea despite its problems and criticisms from contemporary investigators. Solute transport in the phloem is equally perplexing. The mass-flow hypothesis was first proposed in 1860 and then in its present form by Münch in 1930; despite numerous problems and apparent inconsistencies, it remains as the single best account of the mechanism for solute translocation in the phloem.

It should be reemphasized that both hypotheses for transport, the transpiration-pull hypothesis for water transport in the xylem and the mass-flow hypothesis for solute transport in the phloem, are based purely on physical principles. It is more than likely in the years to come that future investigators will modify existing ideas; perhaps with sufficient work and insight someone will propose a theory that accounts for all the data and observations about transport.

Review Exercises

6.1 Compare and contrast apoplastic cell-to-cell transport with symplastic cell-to-cell transport. In terms of selectivity of transport, how do the two methods compare? Why do they differ? What are the factors that would govern transport between cells by diffusion?

6.2 How high could atmospheric pressure alone be expected to push water up a tree? Assuming a maximum root pressure of 3 bars and that root pressure is the only force involved in water transport, what would be the upper limit for plant growth?

6.3 If the mean diameter of a xylem vessel is 100 μm, what would be the maximum height that capillarity could pull water up a tree?

6.4 Compute the flux (F) for transport in the phloem if the pressure differential between source and sink ($\triangle P$) is 10 bars. Use a resistance of 10^4 cm^2 s bar g^{-1}. Show that the units of the resistance (R) as given in this review question are consistent with the definition of R for flux in the phloem. How does flux differ from velocity of transport?

6.5 Criticize in some detail the two presently accepted theories for transport in plants: transpiration-pull for water in the xylem, and mass flow for solutes in the phloem.

References

ARNOLD, N. W. 1968. The selection of sucrose as the translocate of higher plants. *J. Theor. Biol.* 21:13–20.

BERNSTEIN, L. 1961. *Amer. J. Bot.* 48:909–918.

BIDDULPH, O., and R. CORY. 1960. Demonstration of two translocation mechanisms in studies of bidirectional movement. *Plant Physiol.* 35:689–695.

BOWLING, D. J. F. 1969. Evidence for the electroosmosis theory of transport in the phloem. *Biochim. Biophys. Acta* 183:230–232.

BRIGGS, G. E. 1967. *Movement of Water in Plants.* F. A. Davis, Philadelphia.

CANNY, M. J. 1973. *Phloem Translocation.* Cambridge University Press, New York.

CRAFTS, A. S., and C. E. CRISP. 1971. *Phloem Transport in Plants.* W. H. Freeman, San Francisco.

CRONSHAW, J., and K. ESAU. 1968. P-protein in the phloem of *Cucurbita.* *J. Cell Biol.* 38:293–303.

EVERT, R. F. 1977. Phloem structure and histochemistry. *Ann. Rev. Plant Physiol.* 28:199–222.

FERRIER, J. M., and M. T. TYREE. 1976. Further analysis of the moving strand model of translocation using a numerical calculation. *Canad. J. Bot.* 54:1271–1282.

GUNNING, B. E. S., and A. W. ROBARDS. 1976. *Intercellular Communication in Plants: Studies on Plasmodesmata.* Springer-Verlag, Berlin.

HODGES, T. K. 1976. ATPase associated with membranes of plant cells. In U. Lüttge and M. G. Pitman (eds.), Transport in Plants II. Part B. *Encyclopedia of Plant Physiology.* New Series 2. Springer-Verlag, Berlin.

KOZLOWSKI, T. T. 1964. *Water Metabolism in Plants.*

Biological Monographs. Harper & Row, New York.

LEVITT, 1967. *Physiol. Plantarum* 20:263–264.

LÜTTGE, U., and N. HIGINBOTHAM. 1979. *Transport in Plants.* Springer-Verlag, Berlin.

LÜTTGE, U., and M. G. PITMAN (eds.). 1976. Transport in plants II. Part B. *Encyclopedia of Plant Physiology.* New Series 2. Springer-Verlag, Berlin.

MACROBBIE, E. A. C. 1971. Phloem translocation. Facts and mechanisms; a comparative survey. *Biol. Rev.* 46: 429–481.

PEEL, A. J. 1974. *Transport of Nutrients in Plants.* Wiley, New York.

PISEK, A., and E. BERGER. 1938. *Planta* 28:124–155.

POOLE, R. J. 1978. Energy coupling for membrane transport. *Ann. Rev. Plant Physiol.* 29:437–460.

ROBARDS, A. W. 1975. Plasmodesmata. *Ann. Rev. Plant Physiol.* 26:13–29.

SCHOLANDER, P. F., H. T. HAMMEL, E. D. BRADSTREET, and E. A. HEMMINGSEN. 1965. Sap pressure in vascular plants. *Science* 148:339–346.

SPANNER, D. C. 1958. Translocation of sugar in sieve tubes. *J. Exp. Bot.* 9:332–342.

STOCKING, C. P., and U. HEBER (eds.). 1976. Transport in plants III. *Encyclopedia of Plant Physiology.* New Series 2. Springer-Verlag, Berlin.

TURRELL, F. M. 1934. Leaf surface of a twenty-one year old *Catalpa* tree. *Iowa Acad. Sci.* 41:79–84.

VAN DEN HONERT, T. H. 1948. Water transport in plants as a catenary process. *Discussions of the Faraday Society* 3:140–153.

WARDLAW, I. F., and J. B. PASSIOURA (eds.). 1976. *Transport and Transfer Processes in Plants.* Academic Press, New York.

ZIMMERMANN, M. H., and J. A. MILBURN (eds.). 1975. Transport in plants I. *Encyclopedia of Plant Physiology.* New Series 2. Springer-Verlag, Berlin.

CHAPTER SEVEN

The Energy Budget of Plants

The sun is the ultimate source of all the energy on earth available for biological processes. Ordinarily, the physiologist thinks in terms of radiant energy being the driving force for the photochemical reactions of plants, such as the primary processes of photosynthesis. There is, however, much more to the interaction of plants with the radiant energy of the sun than the photochemical reactions *per se*. In terms of direct interaction of the plant with radiant energy, the thermal balance of the plant is perhaps the most important. All of the physical and chemical processes of the plant have temperature optima, and they all have temperature minima and maxima.

From the standpoint of the thermal energetics of the plant, an evaluation of the energy budget is one of the most useful tools. When radiant energy strikes the leaf, which because of its flattened shape is a good absorber, much of the energy is absorbed. Of course, some is reflected and some is transmitted. The amount of each depends on the properties of the leaf and the wavelength of radiant energy. Of that which is absorbed by the leaf, some is consumed by the photochemical reactions and some is consumed by the thermal reactions of metabolism. Some will heat the leaf. Once the leaf has come to thermal equilibrium, i.e., when it is neither heating nor cooling, the quantity of incoming radiation that is absorbed balances exactly that which is used in the photochemical and thermal reactions and lost by the leaf through radiation, conduction and convection, and the latent heat of evaporation. In this chapter, the terms of the energy-budget equa-tion are described and a simple method is developed to balance the energy budget on the basis of ab-sorbed radiation and that lost by radiation, con-duction/convection, and evaporation.

In addition to a discussion of the energy-budget equation of a plant, the consequences of high and low temperatures for plants are discussed.

7.1 The Energy Budget

As stated above, of the total incoming radiation that strikes the green plant, a certain portion is absorbed, a certain portion is transmitted directly through the tissues, and the rest is reflected. Of that which is absorbed, some will enter into biochemical reactions such as photosynthesis and photo-morphogenic reactions (growth phenomena de-pendent on light) and some will heat the plant. The shortwave radiation in the ultraviolet may be dam-aging and the long-wave thermal radiation in the far-red and infrared regions of the spectrum acts as heat waves. The radiation with wavelengths of 400 to 700 nm is active in photosynthesis and is fre-quently called photosynthetically active radiation (PAR). In this chapter, however, it is just the thermal radiation that is considered. Chapter 3 in-cludes the important aspects of visible radiation (that is, PAR) as it applies to plants.

7.1.1 Heat load

A plant or plant leaf, if not heating or cooling, will be in thermal equilibrium with its environment. If more energy is absorbed than is lost it will heat, and if more energy is lost than is absorbed it will cool. The overall energy budget is dependent on the total amount of radiation absorbed either directly from the sun or reflected from clouds and other surfaces. The direct radiation from the sun can be referred to as solar (sun) radiation and the indirect thermal radiation can be referred to as sky radiation. Heat energy can also be absorbed by convective or con-ductive processes.

Heat (energy) loss from the plant can be by infra-red thermal radiation (that is, reradiation), by conductive and convective processes, and by evaporation of water during transpiration. The radiant energy striking the surface, i.e., the heat load, is depicted in Fig. 7.1.

Of the total radiation coming to the earth from the sun (insolation), some is absorbed by the atmo-sphere, some is scattered, and some is reflected by clouds. That which strikes the earth may be ab-sorbed by the surface or objects such as plants or it may be reflected. The reflection by the ground is called the albedo. Some of the absorbed energy is reradiated as long-wave radiation.

Because the average temperature of the earth is about 15°C (288°K), the wavelength maximum of reradiation is about 10 μm (discussed later). Much of this reradiation is absorbed by water vapor sur-rounding the earth, which results in heating of the atmosphere. Subsequently, the radiation can be re-turned as sky radiation. The remainder of the energy absorbed is stored or lost by convection, conduction, or evaporation.

The heat load of the plant is a complex function of the incoming radiation, the physical nature of the environment, the air temperature, wind, and humidity. Characteristics of the plant also become important, as will be discussed below.

7.1.2 The sun's output

The radiation emitting from the sun is generated by the fusion reaction of hydrogen to form helium.

$$4_1H^1 \rightarrow {}_2He^4 + \text{Energy}$$

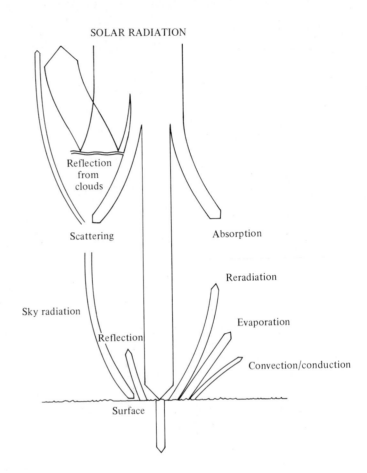

SOLAR RADIATION

FIGURE 7.1   Diagram to illustrate the heat load at the surface of the earth. Drawn after D. M. Gates. 1962. *Energy Exchange in the Biosphere.* Harper & Row, New York.

The loss in mass accompanying this reaction produces vast quantities of energy as determined by the familiar expression, $E = mc^2$. Of course, of the total energy produced by the sun and emitted as electromagnetic energy, only an extremely small fraction reaches the earth's surface.

The emission spectrum of the sun is shown in Fig. 7.2. The maximum emission at 480 nm in the green portion of the spectrum is a function of the temperature of the surface of the sun. Its emission spectrum and intensity are comparable to a blackbody emission with a temperature of 6000°K.

According to the Wein's displacement law, the maximum wavelength of emission (λ_{max}) is related inversely to temperature:

$$\lambda_{max} = \frac{2864}{T},$$

where

T = absolute temperature in °K, and

2864 = a constant, in units of μm °K.

From the above expression it can be calculated that the surface temperature of the sun is 6000° K when the λ_{max} of emission is 0.48 μm or 480 nm.

From Fig. 7.2 it can be seen that the radiation striking the outer atmosphere ranges from about 250 nm in the ultraviolet to 3000 nm in the infrared.

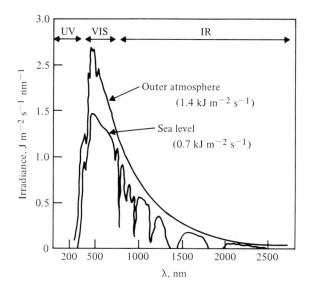

FIGURE 7.2 Spectral distribution of the energy of the sun reaching the outer atmosphere of the earth and that of the energy reaching the surface of the earth. The decrease, or attenuation, is due to scattering, reflection, and absorption by atmospheric constituents. Much of the ultraviolet (UV) is absorbed by the ozone layer of the outer atmosphere. The major absorption peaks accounting for the infrared (IR) attenuation are due to absorption bands of water, carbon dioxide, and ozone. Much of the visible (VIS) radiation passes through the "atmospheric window." Only about 50% of the incoming radiation actually reaches the surface of the earth. Redrawn and modified from S. L. Valley. 1965. *Handbook of Geophysics and Space Environments.* Air Force Cambridge Research Laboratories. Office of Aerospace Research. United States Air Force.

The actual amount of energy striking the earth's outer atmosphere is about 1.35 kilojoules (kJ) m^{-2} s^{-1} (1.94 calories cm^{-2} min^{-1}). This value of 1.35 kJ m^{-2} s^{-1} is referred to as the solar constant and has been measured by satellite above the earth's atmosphere. Because of attenuation by absorption, reflection, and scattering as the radiation passes through the atmosphere, only about 47% actually strikes the earth's surface. Of course, depending on cloud cover, humidity, dust, and other objects, the average can vary from near 0% to as high as about 70%.

As a rule, it is useful to remember a solar constant of 1.4 kJ m^{-2} s^{-1} (2 cal cm^{-2} min^{-1}) striking the earth's outer atmosphere and an average of 0.7 kJ m^{-2} s^{-1} striking the earth's surface and vegetation. Furthermore, if we assume that there is about 5% consumed by green-plant tissue during photosynthesis, then there is about 70 J m^{-2} s^{-1} of the total solar constant available for use by living organisms for energy production. The energy absorbed by the plant that is not used in photochemical reactions contributes directly to the heat load.

The atmosphere is composed of about 79% nitrogen, 21% oxygen, 0.03% carbon dioxide, and small amounts of other gases such as argon. The average amount of water vapor in the atmosphere the world over ranges from about 0.02 to 0.4%. In addition to these important gases, the outer limit of the atmosphere, at an altitude of 25 km (18 miles), is composed of an ozone layer. Because of the presence of oxygen, ozone, carbon dioxide, and water vapor in the atmosphere, which differentially absorb radiation, the quality of radiation striking the earth differs significantly from the sun's emission spectrum (see Fig. 7.2). Water vapor absorbs radiation in the infrared region of the spectrum and tends to reduce the heat load on plants. Water, oxygen, and ozone absorb in the visible region of the spectrum, altering it, and ozone absorbs in the ultraviolet. Because of the presence of the ozone protective layer at the outer limits of the earth's atmosphere, much of the incoming ultraviolet radiation is absorbed. Without the protective layer, there would surely be much injury to living organisms.

The spectral distribution of energy emitted from the sun is shown in Table 7.1. Even after attenuation by the earth's atmosphere, the greater portion of the energy falls within the long-wave, thermal

TABLE 7.1 Energy associated with the sun's output for three spectral ranges

Range (nm)	kJ m^{-2} s^{-1}	%
< 400	0.16	12
400–700	0.49	36
> 700	0.70	52
Total	1.35	100

Modified from **Seliger and McElroy (1965) (see References, Chapter 3).**

region of the spectrum. For this reason, a consideration of the heat load of plants is an important aspect of the physiology of plants.

7.1.3 Energy absorbed

The first consideration for an understanding of the thermal relations of the plant is the amount of energy absorbed. The actual amount of energy absorbed by the plant is called the Q-absorbed or Q_a.

Q-absorbed (Q_a)

Sky or solar radiation striking a leaf may be absorbed by pigments, water, or other molecules, it may be transmitted through the leaf, or it may be reflected. The amounts of absorption, transmission, and reflection are functions of the wavelength of radiation and the characteristics of the leaf.

Virtually all of the ultraviolet radiation reaching the leaf surface is absorbed by the upper tissues. Little penetrates into the mesophyll and still less is transmitted. Of the visible radiation that impinges on the leaf, up to 70% is absorbed by chlorophyll pigments and some is absorbed by other pigments such as carotenoids and anthocyanins. There is some reflection and transmission, particularly in the green. It is, of course, for this reason that leaves appear green. Short-wavelength infrared radiation with wavelengths up to about 1.5 μm is not absorbed to any appreciable extent and is about equally transmitted and reflected. The longer-wavelength infrared radiation, particularly that

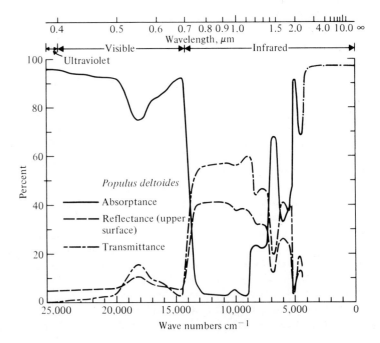

FIGURE 7.3 Graph to show percent absorption, reflection, and transmission of radiation by a typical leaf (*Populus deltiodes*). Wavelength is expressed as wave number (cm^{-1}) below and micrometers (μm) above. From D. M. Gates. 1965. A. Heat, radiant and sensible. *Meteorological Monographs* 6:1–26.

greater than 2–3 μm, is absorbed as heat radiation. There is much less radiant energy in this range than in the visible (see Fig. 7.2). Figure 7.3 shows absorption, reflection, and transmission of radiation by a typical leaf.

The total incoming radiation (Q) can be divided into that absorbed (Q_a), that transmitted (Q_t), and that reflected (Q_r):

$$Q = Q_a + Q_t + Q_r.$$

Thus the Q-absorbed (Q_a) of a leaf is

$$Q_a = Q - (Q_t + Q_r).$$

From the standpoint of the energy budget of the plant, it is the Q_a that is important and in particular the Q_a of the thermal infrared radiation (heat).

The energy-budget equation

The overall change in heat load (H) of the leaf changes with time according to the energy absorbed (Q_a) and the energy dissipated. Thus the rate of heat-load change on the plant is

$$\frac{dH}{dt} = Q_a + R + C + LE + S,$$

where

$\dfrac{dH}{dt}$ = rate of heat-load change, in units of $kJ\,m^{-2}\,s^{-1}$;

Q_a = heat (energy) accumulated as absorbed radiation ($kJ\,m^{-2}\,s^{-1}$);

R = heat loss as thermal radiation ($kJ\,m^{-2}\,s^{-1}$);

C = heat loss or gain from convection and conduction ($kJ\,m^{-2}\,s^{-1}$);

LE = heat loss by evaporation during transpiration ($kJ\,m^{-2}\,s^{-1}$); and

S = heat-storage term ($kJ\,m^{-2}\,s^{-1}$).

In the assessment of the above expression, it should be noted that any of the terms can be positive or negative; that is, heat can be gained or lost. For most purposes, however, it can be considered that Q_a is the positive term and that R, C, LE, and S are negative terms accounting for heat loss. S, the storage term, can be ignored because it is a small part of the total (less than 5%). Furthermore, we can visualize that S may include photosynthesis, which contributes to storage (a gain in heat), or respiration, which causes a decrease in S.

If the leaf is neither heating nor cooling, the rate of heat change will be zero:

$$\frac{dH}{dt} = 0.$$

It is now possible to easily solve the heat-load (energy-budget) equation for any factor:

$$\frac{dH}{dt} = Q_a + R + C + LE + S = 0.$$

If we assume that R, C, and LE are negative terms (that is, they account for losses) and we ignore S because it is small, the equation reduces to:

$$Q_a = R + C + LE.$$

Given any three of the above quantities, the other may be estimated. As an example, if the energy lost by radiation (R), conduction/convection (C), and the heat lost by evaporation (LE) is known, the energy absorbed (Q_a) can be computed. Similarly, if Q_a can be estimated and R and C calculated, transpiration can be computed from a knowledge of LE since the latent heat of evaporation (L) is 2.4 kilojoules per gram of water. The specific details of solving the energy-budget equation are discussed below. Each term is treated separately and the equations for computation of each term are developed.

RADIATION (R)

The quantity of thermal radiation emitted from any body is proportional to the fourth power of the absolute temperature of the body:

$$R = e\,k\,T^4,$$

where

R = thermal radiation, in units of $kJ\,m^{-2}\,s^{-1}$,

e = emissivity (a dimensionless factor),

k = the Stefan-Boltzmann constant $(5.7 \times 10^{-11}$ kJ m^{-2} s^{-1} °K^{-4}), and

T = absolute temperature in °K.

For a blackbody that is a perfect absorber or radiator of thermal radiation, e would be unity. For plant leaves, the value of e is approximately 0.95 for infrared radiation greater than 3 μm. Leaves are thus good absorbers and radiators of thermal radiation, coming close to being blackbodies in the infrared.

The most important consideration for the energy-budget equation is that leaves will lose energy by thermal radiation as a function of the fourth power of the temperature of the leaves. The greater the temperature, the greater the loss of energy.

CONVECTION (C)

Heat lost by convection is first lost by conduction to the surrounding air and then dissipated by convective processes. Convective heat exchange is proportional to the absolute temperature difference between the leaf and the air and is inversely proportional to a boundary-layer resistance term (refer to Chapter 5) that limits the rate of heat exchange. The expression defining convection is

$$C = \frac{T_1 - T_a}{R_a},$$

where

C = convective heat loss (or gain), in units of kJ m^{-2} s^{-1};

T_1 = temperature of the leaf in °K;

T_a = temperature of the air in °K; and

R_a = m^2 s °K kJ^{-1} units of boundary-layer resistance.

The boundary-layer resistance (R_a) may be redefined as a heat-transfer coefficient (h_c) by simply taking the reciprocal of R_a:

$$h_c = R_a^{-1}.$$

The units of h_c are kJ m^{-2} s^{-1} °K^{-1}. The heat-transfer coefficient is a complex function and rather difficult to assess. According to Gates (1962), in still air h_c can be approximated with the expression

$$h_c = 9.1 \times 10^{-2} \left(\frac{\Delta T}{d}\right)^{\frac{1}{4}}$$

where

ΔT = the temperature difference between the leaf and air, and

d = the average diameter of the leaf.

When the leaf is in moving air, d is the diameter in the direction of the wind, and the function for h_c becomes much more complicated. The greater the air movement, the larger h_c will be. However, as the effective leaf diameter (d) increases, h_c decreases. Thus the greater the wind velocity, the greater the heat loss by convection, and the larger the leaf, the less heat loss by convection. Table 7.2 gives some values of h_c as a function of leaf diameter and wind velocity.

Convective heat losses or gains can be predicted by knowing the temperature difference between the leaf and the air and by using values of h_c determined from Table 7.2. The following expression is the most useful.

$$C = h_c \, \Delta T$$

If the leaf temperature is greater than the air temperature, convection will cause heat loss. If the air

TABLE 7.2 Heat-transfer coefficients (h_c) in wind for three different leaf diameters

Diameter, m	Wind		
	2 m s^{-1}	5 m s^{-1}	13 m s^{-1}
0.01	0.06	0.08	0.15
0.05	0.03	0.04	0.06
0.10	0.02	0.03	0.05

The units of h_c are kJ m^{-2} s^{-1} °K^{-1}.
Data from Gates (1962).

FIGURE 7.4 Schlieren (shadow) photograph showing the rising convection currents during convective heat loss from a pine branch. The convective heat loss is seen as a dark shadow above the branch. As the warm air rises, cool air descends to the branch, and the process is repeated. From D. M. Gates. 1965. Heat transfer in plants. *Sci. Amer.* 213: 76–84.

temperature is greater than the leaf temperature, heat will be gained by the leaf.

Using schlieren (shadow) photography, Gates (1968) was able to study convective heat exchange by pine needles (Fig. 7.4). In the figure, the rising convection currents can be viewed as dark shadows.

EVAPORATION (*LE*)

The energy lost by evaporation (i.e., transpiration) can be estimated by knowing the transpiration rate in units of g cm^{-2} s^{-1} and the energy lost by the evaporation of one gram of water (2.4 kJ g^{-1}). The rate of transpiration as defined in Chapter 5 is

$$T = \frac{\triangle e}{R},$$

where

$T =$ transpiration, in units of g (H$_2$O) m^{-2} s^{-1},

$\triangle e =$ vapor-pressure difference between leaf and air (g m^{-3}), and

$R =$ resistance to transpiration, in units of s m^{-1}.

The transpiration rate (T), multiplied by the latent heat of evaporation (L), 2.4 kJ g^{-1}, will represent the heat lost by transpiration (LE). The units of LE are kJ m^{-2} s^{-1}, consistent with the units of Q_a, $R,$ and C.

The term LE in the energy-budget equation is one of the most important from the standpoint of physiology since it has a physiological factor (R_s) that is a function of the stomata (recall that $R = R_a + R_s$, where R is the total resistance to transpiration,

R_a is the boundary-layer resistance, and R_s is the stomatal resistance). R_s, being a function of the stomata, is a variable resistance that depends on the degree of stomatal opening.

In contrast to LE, the terms R and C are purely physical. Q_a has a physiological component dependent on pigments in the leaf and surface characteristics of the leaf, but for the most part it is a constant for any one leaf. LE, however, will vary daily, depending on environmental conditions.

STORAGE (S)

The storage term is difficult to assess but can be considered to be a function of the biochemistry of the plant that is dependent on energy. Photosynthesis and respiration are the most important, but even these amount to less than 5% of the total energy budget. S is usually not a consideration for heat load except perhaps for highly sophisticated research.

7.1.4 Leaf heating rates

Most leaves are thin with little mass and tend to respond very rapidly to slight changes in heat load. Small wind gusts, cloud-cover changes, and slight shading from other leaves or objects bring about rapid adjustments in energy balance.

The instantaneous rate of heating

In 1959 Loomis and his coworkers evaluated the instantaneous rate of heating by plant leaves by considering the total quantity of heat absorbed (Q_a), the mass of the leaf (M), and the specific heat capacity (K). In their considerations, they ignored heat dissipation in the form of R, C, and LE. The expression that they used is

$$\frac{dH}{dt} = \frac{Q_a}{M \cdot K},$$

where

$\dfrac{dH}{dt} =$ the instantaneous rate of heating, in °K g^{-1} s^{-1};

$Q_a =$ the heat absorbed, in units of kJ m^{-2} s^{-1};

$M =$ the mass of the leaf, in g m^{-2}; and

$K =$ the specific heat capacity, in kJ °K^{-1}.

Assume a Q_a of about 0.7 kJ m^{-2} s^{-1}, absorbed from all sources. A typical leaf will have a mass about 200 g m^{-2} and a heat capacity of about 3.7 J °K^{-1}. Calculation of the rate of heating from these data,

$$\frac{dH}{dt} = \frac{700 \text{ J m}^{-2} \text{ s}^{-1}}{(3.7 \text{ J °K}^{-1})(200 \text{ g m}^{-2})},$$

gives an instantaneous rate of heating of 0.95 °K s^{-1} g^{-1} tissue, or about one degree per second per gram.

A succulent leaf with a greater mass would heat much more slowly, but because the heat capacity of the succulent would be closer to 4.2 (because of more water) it would heat to a higher temperature.

Figure 7.5 illustrates the rate of heating by a succulent leaf and a nonsucculent leaf under laboratory conditions. The succulent leaf has a greater mass and heats more slowly but to a higher temperature.

Heat dissipation

It is instructive at this point to compute the transpiration necessary to dissipate a Q_a of 0.7 kJ m^{-2} s^{-1}. Since the latent heat of evaporation is 2.4 kJ g^{-1}, a transpiration rate of 0.29 g m^{-2} s^{-1} would be necessary to remove 0.7 kJ m^{-2} s^{-1} of absorbed energy:

$$\frac{0.7 \text{ kJ m}^{-2} \text{ s}^{-1}}{2.4 \text{ kJ g}^{-1}} = 0.29 \text{ g m}^{-2} \text{ s}^{-1}.$$

Converting to the usual units for transpiration, this would be about 10.5 g (H$_2$O) dm^{-2} h^{-1}, a rate 5 to 10 times greater than is usually encountered.

If we assume an average transpiration of about 0.07 g m^{-2} s^{-1}, the heat dissipation would be 0.17 kJ m^{-2} s^{-1}, or about 24% of the 0.7 kJ m^{-2} s^{-1} Q_a. This means that approximately 76% of the Q_a must be dissipated by convective and radiation processes if the leaf is to stay in thermal equilibrium.

To take this reasoning further, we can assume a wind condition of about 2 m s^{-1} over a leaf 0.05 m

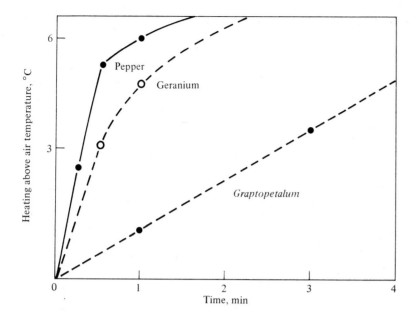

FIGURE 7.5 Experiment illustrating that thin leaves of plants such as pepper and geranium heat much faster than thick, succulent leaves such as *Graptopetalum*. The thick leaf, however, heats to a much higher temperature. From A. Q. Ansari and W. E. Loomis. 1959. Leaf temperatures. *Amer. J. Bot.* 46:713–717.

in diameter. From Table 7.2, the heat-transfer coefficient (h_c) from these data is 0.03 kJ m^{-2} s^{-1} °K^{-1}. If there is a 3° difference between the leaf and air temperature, the heat lost by convection would be 0.09 kJ m^{-2} s^{-1}:

$$C = h_c \, \Delta T = (0.03 \text{ kJ m}^{-2} \text{ s}^{-1} \text{ °K}^{-1}) \, (3°K).$$

Thus under these average conditions about 13% of the total Q_a would be lost by convection, leaving 0.44 kJ m^{-2} s^{-1} to be lost by radiation if the leaf stays in thermal equilibrium. The energy lost by radiation from a leaf with a temperature of 27° is 0.44 kJ m^{-2} s^{-1}.

We can conclude from the above analysis that substantial quantities of heat are lost by transpiration, convection, and radiation. Under the usual conditions, the greater proportion is lost by radiation. As a rule, leaves are good absorbers of radiation and hence are good radiators. Unless leaf and air temperatures are substantially different, convection is of lesser importance than is transpiration. Under the usual conditions, the order of importance for heat dissipation is radiation, transpiration, and then convection.

Leaf temperature

When in direct sunlight, leaves usually heat above air temperature. In an experiment with pepper (*Capsicum*) leaves, direct sun caused leaf heating 9° C above air temperature in about one minute. Shading the leaf resulted in a return to the original temperature just above air temperature in about one minute. The rate of heating was about the same as the rate of cooling and thermal equilibrium was reestablished about one minute after a change in the radiation (see Fig. 7.6).

The mass of the pepper leaves was 190 g m^{-2}. The succulent leaves of *Graptopetalum* have a mass of 5540 g m^{-2} and heat much more slowly than do the pepper leaves but to a higher temperature. As shown in Fig. 7.7, even after 20 minutes in full sun with a leaf temperature 17°C above air temperature, the succulent leaf was still heating.

Loomis calculated that the rate of heating of the pepper leaf in full sunlight was about 0.15°C s^{-1}. The succulent *Graptopetalum* leaf heated at a rate of about 0.007°C s^{-1}, some 21 times slower than the thin pepper leaf.

FIGURE 7.6 Experiment to show the rapid rate of heating of thin leaves (pepper) when placed in the sun. When shaded again, the rate of cooling was as rapid as the rate of heating. The experiment illustrates that the thin leaves of plants are good absorbers of radiation and also that they are good radiators. From A. Q. Ansari and W. E. Loomis. 1959. Leaf temperatures. *Amer. J. Bot.* 46:713–717.

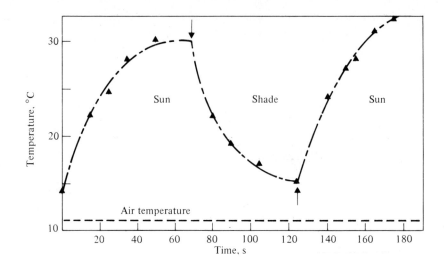

FIGURE 7.7 Experiment showing the slow heating by a succulent leaf of *Graptopetalum*. After 20 min in the sun, the thick leaf was still heating. When shaded, the rate of cooling was similar to the rate of heating, requiring about 25 min to return to the original shade temperature. The maximum temperature reached after 20 min in the sun was 17° above air temperature. From A. Q. Ansari and W. E. Loomis. 1959. Leaf temperatures. *Amer. J. Bot.* 46:713–717.

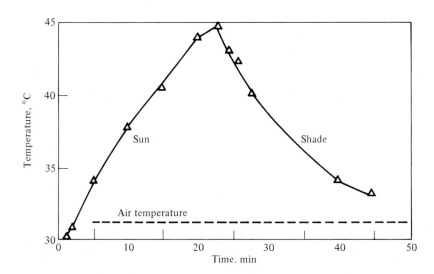

7.2 Temperature Responses

At the molecular level, temperature influences the kinetic energy of molecules. Because increases in temperature increase kinetic energy, temperature increases the probability of molecular collisions. Thus virtually all reactions and physiological processes respond to temperature. An exception, of course, is a simple photochemical reaction in which the energy source is a photon. Photochemical processes, however, are complex and have some components that are thermal as well as light-driven.

Most processes of living organisms are controlled by enzyme reactions. Enzyme reactions are limited by low temperatures, and at high temperatures the protein enzymes are denatured. The general tem-

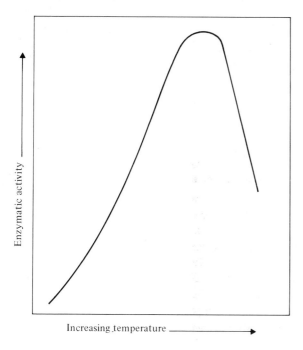

Enzymatic activity

Increasing temperature

FIGURE 7.8   Idealized temperature-response curve for a typical enzyme. At low temperature the activity is low. As the temperature increases, the activity increases to an optimum and then decreases at higher temperatures. The increase with increasing temperature is most likely the result of activation, whereas the high temperature inactivation is the result of protein denaturation.

perature response of an enzyme reaction is illustrated in Fig. 7.8. As a consequence of the temperature-response curves of enzyme reactions, growth phenomena of plants tend to respond similarly. Of course, each process, whether it is germination, root growth, stem growth, flowering, or leaf expansion, will have different temperature ranges, but the shapes of the temperature-response curves will be similar. All will have minimum temperatures at which the process is slow or does not proceed, all will have optimum temperatures, and all will have maximum temperatures that limit the process.

A temperature–growth-response curve for tomato growing in a controlled-temperature environment is shown in Fig. 7.9. For this experiment the day temperature was held constant at 26° while the night temperature was varied. Since most plant growth is assumed to be at night when water balance is more favorable, it is the night temperature rather than the day temperature that is fre-

quently most important for plants. Such a conclusion was established by Sachs as early as 1880 and fully supported by the experiments of Fritz Went in 1944 in the plant-growth facility he developed at the California Institute of Technology.

7.2.1 Cardinal temperatures

The cardinal temperatures are the minimum, optimum, and maximum temperatures at which the response occurs. Each kind of process will have its own unique cardinal temperatures, and for any one response the cardinal temperatures may change when other factors that control the process change. As a generalization for plant growth, the cardinal temperature ranges are the following.

MINIMUM	5–10°
OPTIMUM	25–30°
MAXIMUM	35–40°

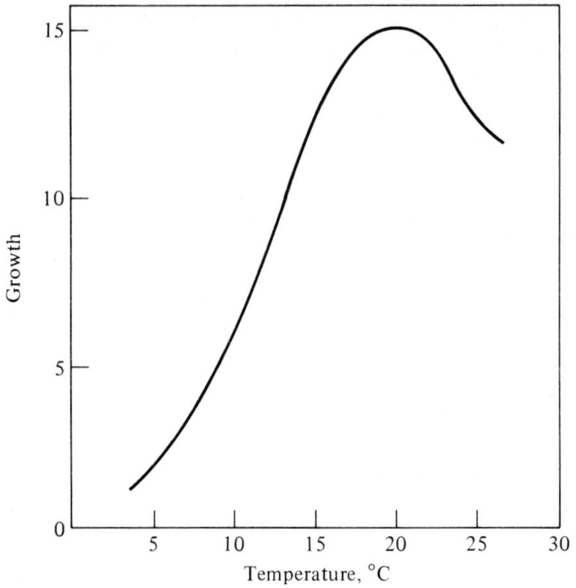

FIGURE 7.9   Growth of a tomato plant at constant day temperature (26°) and variable night temperature. The temperature-response curve is quite similar to the enzyme temperature-response curve. Redrawn from F. W. Went. 1944. Plant growth under controlled conditions. *Amer. J. Bot.* 31:597–618.

These temperature ranges are valid mostly for land plants of temperate environments. Tropical plants may "chill" at temperatures far above 10°, and some desert plants do better at temperatures in the range of 35 to 40°. Some thermophilic blue-green algae exist at temperatures approaching 100°.

7.2.2 The Q_{10}

The temperature response of a process can be assessed by determining the increase in the rate of the process with a 10° increase in temperature. This index, called the Q_{10}, is defined as the ratio of the rate at one temperature, T, to the rate at a temperature 10° higher, $T + 10°$. The defining equation is

$$Q_{10} = \frac{R_2}{R_1}^{\left(\frac{10}{T_2 - T_1}\right)}$$

or

$$\log Q_{10} = \frac{10}{T_2 - T_1} \; \log \frac{R_2}{R_1}.$$

It should be noted here that when the temperature difference between the two is 10 (that is, when $T_2 - T_1 = 10$), the expression reduces to

$$Q_{10} = \frac{R_2}{R_1}.$$

For physical processes, the Q_{10} is in the range of 1.2 to 1.4. Because the energy comes from a photon and not from thermal energy, photochemical reactions should have a Q_{10} of 1.0. Enzymatic and physiological processes have Q_{10} estimates of 2 to 3; however, the Q_{10}'s are variable and can range as much as 1 to 10 or more.

The curve for an enzyme reaction as a function of temperature, shown in Fig. 7.8, and the growth curve, shown in Fig. 7.9, illustrate the variable nature of Q_{10}. Since these processes have cardinal temperatures (with a definite lag at the minimum, an optimum, and then a decline), the Q_{10} has little meaning unless assessed where log Q_{10} is linear with log R_2/R_1 between the lag and the optimum. If Q_{10} is computed during the lag phase, it may be considerably greater than 3. At the optimum, the Q_{10} will be 0, and beyond the optimum the Q_{10} may be negative. Thus it is very important to assess Q_{10} over the proper response range; otherwise misleading results will be obtained.

7.2.3 Thermal consequences

As explained in the previous section, high temperatures tend to disrupt physiological processes by thermal denaturation of enzymes and perhaps alteration of important cellular and subcellular structure. Low temperatures interfere with controlling enzyme reactions as well and most certainly disrupt other processes. In the discussion that follows, some aspects of cold injury and heat injury are presented.

Cold injury

Cold injury can take two basic forms. First, cold can injure at temperatures above freezing by disruption of important reactions. Such injury is called chilling injury. In addition, cold can injure by causing actual freezing of the tissues. Aspects of chilling injury and freezing (frost) injury are discussed below along with winterkill and winter hardiness.

CHILLING INJURY

Low temperatures disrupt the metabolic activity of plants, and growth and development are retarded. In many respects, however, cold is a relative term and depends on the genetic composition as well as the past history of the plant. Thus temperate plants can usually withstand lower temperatures than can tropical plants, but even plants normally growing in cold environments will be cold-injured if the cold is experienced during the warm growing period. Table 7.3 includes the winter-injury temperatures for tropical, subtropical, temperate, and cold-winter (e.g., boreal/subarctic) plants in the winter. It can be seen from these data that tropical plants can be injured by above-freezing temperatures, whereas the more cold-adapted plants can withstand temperatures far below freezing.

Chilling injury best describes the phenomenon of cold injury above freezing temperatures. It may

result because of either reduced metabolic activity or structural alterations, the most important of which are phase changes in membrane lipids. Processes such as respiration are highly temperature-dependent, and reductions because of low temperature will lead to less energy available for growth. Any other reactions that are limited at low temperatures are apt to interfere with growth processes, causing chilling injury. Being a photochemical reaction, photosynthesis is less affected by cold temperatures, but all of the associated carbon-metabolism reactions are ordinary thermal reactions and are sensitive to temperature.

The graph in Fig. 7.10 shows how temperature affects respiration and photosynthesis. Photosynthesis continues at low temperature when respiration is drastically reduced. The reduction in respiration causes a reduction in the energy available for plant processes, even though carbohydrates are still synthesized. Any excess photosynthate such as carbohydrate will go into storage and perhaps form some toxic end products. Accumulation of the latter may contribute to chilling injury.

Plants growing in warm climates tend to have a greater proportion of saturated fatty acids in comparison to unsaturated fatty acids. Furthermore, chilling-sensitive plants tend to have more saturated fatty acids than unsaturated fatty acids when compared to plants that are not sensitive to chilling. There is evidence that lipids will go from a liquid-crystalline form (having some flexibility) to a solid-gel form when chilled to the temperature range of 9 to 12°. Such a transition corresponds to changes in physiological and biochemical phenomena associated with membranes. For example, in 1970

TABLE 7.3   Temperature ranges in which woody plants are injured in winter and summer

	Winter (°C)	Summer (°C)
Tropics	+5 to −2	45 to 55
Subtropics	−8 to −12	50 to 60
Temperate	−6 to −15	55 to 65
Cold-winter	−40 or lower	44 to 50

Data from Larcher (1975).

Lyons and Raison studied the oxidation of succinate by mitochondria as a function of temperature in chilling-sensitive and chilling-insensitive plants (Fig. 7.11). When an Arrhenius graph (a plot of log values of the reaction against the reciprocal of the absolute temperature) was used to analyze the data, it was observed that in the chilling-sensitive plants, there was a definite break in the curve at about 9 to 12°. There was no such break in the slope of the curve with the chilling-insensitive plants. The change in the graph at 9 to 12° was attributed to a phase change in membrane lipids inasmuch as the succinate dehydrogenase enzyme is membrane-bound.

The Q_{10} for the reaction over the temperature range of 25° to 12° was 1.3 to 1.6, but from 9° to 1.5° it was 2.2 to 6.3. These data were taken as evidence for structural changes in the membranes at low temperature. A conclusion from these data is that chilling injury results because of phase changes of membrane lipids that bring about detrimental alterations in physiological function. These alterations could involve either a reduction in enzyme activity (as is the case here for succinate oxidase) or general changes in membrane permeability. The chilling injury may result from imbalances in metabolism, the accumulation of toxic substances, or alterations in the structure of lipids.

FROST INJURY

Many plant tissues can withstand subfreezing temperatures provided that ice crystals do not form and physically disrupt cellular structure. Perhaps as important to cellular death by freezing is the drying that accompanies ice formation. Ice crystals

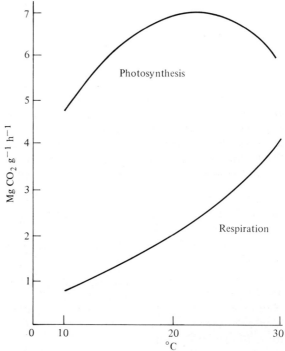

FIGURE 7.10   Graph to show the effect of temperature on both photosynthesis and respiration in *Artemisia tridentata*. Photosynthesis reaches an optimum at about 22° and then declines. Respiration continues to increase with increasing temperature. The result is that carbohydrate supplies will be depleted and injury will result. The graph is plotted from data in Table 3 of H. A. Mooney and M. West. 1964. Photosynthetic acclimation of plants of diverse origin. *Amer. J. Bot.* 51:825–827.

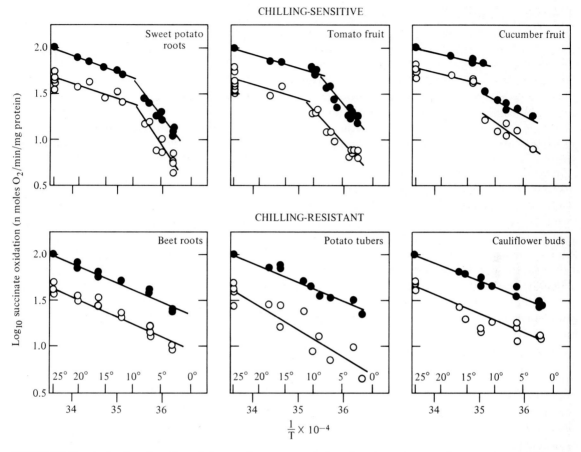

CHILLING-SENSITIVE

CHILLING-RESISTANT

FIGURE 7.11 Arrhenius plots (log of the reaction rate graphed against the reciprocal of the absolute temperature, $1/T$). The upper three panels show the graph for three plants sensitive to chilling while the lower three panels show the response for three plants resistant to chilling temperatures. The sensitive plants show a break in the slope of the curve at temperatures in the range of 9 to 12°C. The break is taken as evidence for a phase change in the lipids of the membranes that results in an altered and lowered activity. From J. M. Lyons and J. K. Raison. 1970. Oxidative activity of mitochondria isolated from plant tissues sensitive and resistant to chilling injury. *Plant Physiol.* 45:386–389.

forming within intracellular spaces will draw water from cells, causing desiccation. Severe desiccation ultimately causes death of cells and tissues. Support for the desiccation theory of cold (frost) injury comes from observations that drought resistance and cold resistance are parallel phenomena in many plants. Hence those cellular features associated with cold resistance are similar to or the same as those of drought resistance.

WINTERKILL

Reduced soil temperatures interfere with water uptake by roots. Winterkill occurs when warm days are present during the winter such that transpiration rates far exceed the potential for water uptake from cold soils by chilled roots. Winterkill is similar to frost injury in that both are caused by desiccation.

WINTER HARDINESS

It is well known that plants can be "hardened" to killing frosts by pretreatment with cold temperatures. Freezing temperatures experienced during the growing season may result in the rapid death of plants. Few plants that are actively growing in temperate climates can withstand temperatures more than a few degrees below 0°. However, some plants that have previously experienced cold temperatures can survive air temperatures as low as −10° to −20°.

In a 1972 experiment designed to determine the factors associated with winter hardiness, Kohn and Levitt exposed cabbage to successively decreasing temperatures, from 5° to −3°, over a 6-week period. Prior to treatment, when the plants were at 25° during the day and 15° at night, −2° to −4° caused frost killing. After the 6-week treatment, the plants were not killed with exposure to −20°. The investigators noted in this experiment that frost hardiness was correlated with a decrease in osmotic potential and was better developed under conditions of active photosynthesis.

It was mentioned in Chapter 2 that protoplasm would freeze at a few degrees below 0°, but frequently intact tissue could withstand much lower temperatures because of supercooling. How supercooling of protoplasm is related to winter hardiness is not known but represents a fruitful area for research in the future.

The theory for winter hardiness

Levitt (1972) developed a theory for cold hardiness in plants. His extensive studies indicated that there was a relationship between sulphydryl (SH) concentration of protoplasm and frost resistance. Most likely the sulphydryls are groups on proteins. In the theory, freezing causes oxidation of SH to form disulphide bridges (S–S). Since the disulphide bridges are covalent bonds, the free SH groups do not form upon thawing, and there is irreversible damage.

There have been some attempts to prevent cold injury in plants by chemical treatment. For example, treatment of bean seeds with decenylsuccinic acid prevents cold injury and also injury from desiccation. Further studies with such chemicals may prove useful for the treatment of frost injury.

High-temperature injury

High temperatures cause plant death in several different ways. It is interesting that the upper limit for most plant processes is 50° to 60°, whereas the low-temperature limit for survival is much more variable. In fact, if they are quick-frozen, some plants can withstand temperatures near absolute zero.

High-temperature injury is probably caused mostly by irreversible protein destruction, although metabolic disturbances independent of protein destruction may be important. Desiccation is considered to be less important, but clearly drying at high temperatures will harm plant activities.

An excellent example of the disruption of metabolic activity is the exponential increase in respiration with increasing temperature, shown in Fig. 7.10. The curve also shows injury to photosynthesis at about 22° (here probably largely due to stomatal closure but perhaps also because of other factors affected by temperature). The decrease in photosynthesis while respiration is increasing will rapidly cause depletion of carbohydrate stores and ultimately cellular injury. In addition, translocation of foodstuffs is disrupted at high temperatures, reducing carbohydrate availability.

Some plants can be hardened to prevent high-temperature injury. In one such experiment, Schroeder in 1963 exposed avocado tissue to 50° for a brief period and noted that the tissue could then tolerate 55°. Similar experiments were conducted by Yarwood in 1961 with leaves of a variety of plants. A short exposure to high temperature predisposes plants to increased heat tolerance.

It has also been shown that certain types of high-temperature lesions can be overcome by leaf treatment with chemicals. For example, leaf treatment with vitamin C, nicotinic acid, B vitamins, and even sucrose tends to minimize heat injury.

7.3 An Overview of Thermal Balance in Plants

Although the plant may have a temperature somewhat different than its environment, for the most part plant and air temperatures will be similar to within a few degrees. Under conditions of high insolation, leaf temperatures will tend to be above air temperature, although cooling by transpiration may reduce leaf temperatures below air temperature. As a generalization, in direct sunlight leaf temperatures will be above air temperatures. In a wind or in the shade, leaf temperatures will be about equal to or somewhat below air temperatures. On a clear, cool night, leaf temperatures will be a little below air temperatures, occasionally resulting in dew deposition. On a cloudy night leaf temperatures would be expected to approximate air temperatures. For these reasons, excessively high or low air temperatures will frequently result in thermal injury to plants. Resistance to such thermal injury is largely by tolerance and not avoidance.

Chilling injury that occurs to some plants in the vicinity of 10° was discussed above. Virtually all plants unless hardened will be injured when temperatures reach about −5°. Such injury results because of tissue and membrane changes resulting from ice formation that physically disrupts and dehydrates tissue. Support for the dehydration notion comes from observations that plant tissue can be cooled at rates of 10^5 to 10^6 degrees per second to liquid-nitrogen temperatures without injury. It is assumed that the water of the tissue is supercooled in a noncrystalline state; i.e., it is vitrified. Since water in the tissue is not dislocated and ice crystals do not form, the tissue remains intact when warmed. Dry tissue can withstand much lower temperatures than can hydrated tissues. An excellent example would be a dormant seed.

At "cold" temperatures above −5° cold injury may still occur, but this is the result of more indirect effects such as starvation, protein or nucleic acid breakdown, respiratory imbalance, permeability changes, specific biochemical lesions, or toxin accumulations.

High temperatures as well as low temperatures will result in injury. Unlike low temperatures, however, high-temperature effects involve more of a time factor. Some thermophilic algae can withstand temperatures up to 98°, but there is little growth above about 80° to 85°. More often even the thermophilic organisms do not survive at temperatures much above 65°. At these high temperatures the longer the exposure, the greater the probability of injury. In a pattern similar to low-temperature injury, dry tissue can withstand much higher temperature than hydrated tissue. For example, dried seeds can withstand treatment in ovens, but moist seeds will be quickly killed.

The general mechanisms that cause high-temperature injury are quite similar to those that cause low-temperature injury. The important factors are starvation due to metabolic imbalances such as high respiration and low photosynthesis, biochemical lesions of specific enzymes, protein and nucleic-acid denaturation, and the accumulation of toxins.

As stated above, the plant is very tightly coupled to the environment through the energy-budget equation, and thus high-temperature avoidance is not very likely. The only obvious exception is that the leaves of some species do not heat above air temperature as much as do others. In this sense, there is a high-temperature avoidance. High-temperature resistance is thus largely by tolerance. Any mechanism that results in the prevention of the thermal disruptions mentioned above will result in high-temperature tolerance.

There are many plants that can be made to tolerate high or low temperatures by exposure to sub-threshold injury temperatures. This process is called hardening. The hardening process in plants is poorly understood but must involve changes that tend to resist those imbalances such as membrane changes, metabolic imbalances, protein and nucleic-acid disruption, formation of biochemical lesions, and toxin accumulation, all of which cause injury.

Review Exercises

7.1 Compute the average temperature of the earth in °C if the wavelength maximum of emission is 0.001 cm.

7.2 Calculate the instantaneous rate of heating of a leaf with a mass of 500 g m^{-2}. Assume a Q_a of 1.0 kJ m^{-2} s^{-1} and a heat capacity of 3.3 J per °C.

7.3 How much heat would a leaf lose by convection if it is exposed to a wind of 5 m s^{-1} over an effective diameter of 0.1 m? Assume that the air temperature is 5° below the leaf temperature. If the wind were to increase to three times the original speed and the

leaf temperature were reduced to 1° above air temperature, what would be the convective heat loss?

7.4 What would be the Q_{10} for respiration if at 20° the rate were 2 mg CO_2 evolved per g tissue per hour and at 24° the rate increased to 2.7 mg?

7.5 Compute the expected rate of transpiration from a leaf that is absorbing 1.5 kJ m^{-2} s^{-1} of incoming radiation. Assume that the leaf temperature is 30°, the air temperature is 25°, and there is a wind of 2 m s^{-1} over an effective leaf diameter of 0.05 m.

References

ALDEN, J., and R. K. HERMANN. 1971. Aspects of the cold hardiness mechanism in plants. *Bot. Rev.* 37: 37–142.

BURKE, J. J., L. V. GUSTA, H. A. QUAMME, C. J. WEISER, and P. H. LI. 1976. Freezing and injury in plants. *Ann. Rev. Plant Physiol.* 27:507–528.

CLAYTON, R. K. 1970. *Light and Living Matter,* Vol. 1, *The Physical Part.* McGraw-Hill, New York.

CLAYTON, R. K. 1971. *Light and Living Matter,* Vol. 2, *The Biological Part.* McGraw-Hill, New York.

GATES, D. M. 1962. *Energy Exchange in the Biosphere.* Biological Monographs. Harper & Row, New York.

GATES, D. M. 1968. Energy exchange between organisms and environment. *Austral. J. Sci.* 31:67–74.

GATES, D. M. 1968. Transpiration and leaf temperature. *Ann. Rev. Plant Physiol.* 19:211–238.

GATES, D. M., and L. E. PAPAIN. 1971. *Atlas of Energy Budgets of Plant Leaves.* Academic Press, New York.

JOHNSON, H. B. 1975. Plant pubescence: an ecological perspective. *Bot. Rev.* 41:233–258.

LARCHER, W. 1975. *Physiological Plant Ecology.* Springer-Verlag, Berlin.

LEVITT, J. 1972. *Responses of Plants to Environmental Stress.* Academic Press, New York.

LI, P. H., and A. SAKAI (eds.). 1978. *Plant Cold Hardiness and Freezing Stress.* Academic Press, New York.

LYONS, J. M. 1973. Chilling injury in plants. *Ann. Rev. Plant Physiol.* 24:445–466.

MONTEITH, J. L. 1973. *Principles of Environmental Physics.* University Park Press, Baltimore.

PROSSER, C. L. (ed.). 1967. Molecular mechanisms of temperature adaptation. *Amer. Assoc. Advancement Sci.* (AAAS) *Publ.* 84. Washington, D.C.

TRANQUILLINI, W. 1964. The physiology of plants at high altitudes. *Ann. Rev. Plant Physiol.* 15:345–362.

WEISER, C. J. 1970. Cold resistance and injury in woody plants. *Science* 169:1269–1278.

ii

Biophysical Processes: Exchanges with the Physical Environment

PROSPECTUS The preceding six chapters covered the important aspects of biophysical physiology. It probably comes as a great surprise to students that so much physiology can be interpreted on physical grounds alone. Equally surprising, however, is that despite the fact that phenomena such as water uptake and loss, gas exchange, water and solute transport within the plant, and transport between cells obey physical principles, we have not been able to formulate unequivocal theories. Much still needs to be learned before acceptable hypotheses are developed. It should be readily apparent that much of our limitation is not due to an inadequate knowledge of physiological functions but rather is because of our limited understanding of physics and mathematics. Thus serious students of plant physiology would do well to concentrate much effort in understanding the physical sciences. Major advances in plant water relations, gas exchange, photobiology, and transport will only come from those well versed in physics, mathematics, and chemistry.

Like the obvious questions concerning water and solute transport, our knowledge of photobiology is also meager. Whereas we have a reasonably good knowledge of how radiant energy interacts with pigments, just how such interaction is translated into physiological function is quite obscure. These questions will be discussed in more detail in Parts IV and V.

Ultimately, the physical principles discussed in this part will be applied to plant metabolism and the growth and development of plants. But before this is possible, the chemistry of plants must be understood. The basic properties of the major plant constituents—nucleic acids, proteins, carbohydrates, lipids, and natural products—must be studied. The next part, on plant biochemistry, covers the details necessary to appreciate plant metabolism and plant development.

Biochemical
Constituents
of Plants

III

PROLOGUE The primary chemical constituents of plants fall into five natural groups. These are the nucleic acids, proteins, carbohydrates, lipids, and the "natural products," including the phenolics and alkaloids. The nucleic acids are among the most important plant compounds and include DNA, RNA, and the metabolic nucleotides that play a central role in cellular energetics. Recent advances in molecular biology that elucidated the role of DNA and RNA in protein synthesis and cellular regulation have had a profound influence on plant physiology. Although there are many important differences between the nucleic acid metabolism of bacteria with which most of the key discoveries were made, the studies have formed a basis for similar studies with higher plants and have contributed much to our understanding of plant metabolism. Chapter 8 summarizes our knowledge of nucleic acid metabolism in plants, including the regulation of protein synthesis.

Proteins, discussed in Chapter 9, are synthesized from amino acid building blocks using messenger RNA as a template. The proteins may be storage compounds, structural compounds, or, more importantly, regulators (as enzymes) of metabolism.

The carbohydrates are the primary carbon-storage compounds of plants, the most important of which is probably starch. Photosynthate produced by CO_2 assimilation is stored primarily as carbohydrate. The carbohydrates are subsequently used in respiration to obtain energy for cellular synthesis and maintenance. In addition to the central role of carbohydrates in cellular energetics, the main structural compounds of plant cell walls, cellulose and the hemicelluloses, are carbohydrates. The organic acids, which are derivatives of carbohydrates, act as precursors for other compounds and as the immediate substrates for respiration.

Lipids as well as proteins and carbohydrates are important carbon-storage compounds of plants. But perhaps their primary role from a metabolic viewpoint is as structural components of membranes. Membranes are lipoproteins with properties that are in large part a function of the properties of the lipids. Lipids may occur as carbohydrate complexes (glycolipids) or as phospholipids. They are present as neutral lipids (fatty acids or triglycerides), as waxes, and also include the terpenoids and sterols. The nucleic acids, proteins, carbohydrates, and lipids are constituents of all plant cells and play a central role in plant metabolism.

Their properties determine the properties of living protoplasm.

The final group discussed in this part (in Chapter 12), the "natural products," is a more heterogeneous group and cannot be defined on the basis of properties as precisely as can the nucleic acids, proteins, carbohydrates, or lipids. The natural products include phenolic compounds, alkaloids, and porphyrins. The terpenoids, even though they have lipoidal properties, are considered with the natural products because their metabolic role appears to be different from the other lipids. The natural products are sometimes called secondary products because they are synthesized from carbohydrates and amino acids.

In the last part, the importance of physics to the study of plant physiology was emphasized; chemistry is equally as important. The plant is without doubt the most complex chemical-synthesizing factory on earth. This complex chemistry along with the physics of plant life is what governs all of plant function. Thus the information in Parts II and III constitutes the very foundation of plant physiology.

CHAPTER EIGHT

Plant Nucleic Acids

Nucleic acids have been known as plant constituents for many decades and are now known to occur in virtually all living cells. These complex polymers, composed of an organic base (a substituted purine or pyrimidine—see below) and a sugar linked through a phosphate diester bond, play a central role in the life of organisms.

Nucleic acid metabolism is a main feature of the life processes of the cell. The enzymes, protein catalysts, regulate virtually all the chemistry of the cell. The myriad of chemical compounds in the cell are all synthesized through controlled chemical reactions catalyzed by enzymes. These chemicals are used for a variety of purposes, including use as building blocks for all cytoplasmic components, for which each step is catalyzed and controlled by an enzyme or group of enzymes. Much of the regulation of cellular metabolism results because the enzymes are highly specific with respect to substrates. The role of the nucleic acids is the regulation of polypeptide synthesis. Modification of the polypeptide after synthesis forms the functional protein.

The nucleic acid, deoxyribonucleic acid (DNA), contains in its linear sequence of base units the information for polypeptide assembly from amino acid building blocks. The DNA sequences that code for polypeptides (and other nucleic acids) are the genes. Much of the cellular DNA is localized in the nucleus and is complexed with basic-pH proteins, the histones. The complex of DNA and nucleoproteins (histones) is called chromatin, the structural component of chromosomes. During the process of polypeptide synthesis, the information contained in the sequence of bases making up DNA is transcribed to form ribonucleic acid (RNA). The RNA is the immediate gene product. A portion of the newly formed RNA has the code for subsequent translation into the proper sequence of amino acids during polypeptide assembly. This RNA is called messenger RNA. The messenger RNA is transported to the cytoplasm, where it acts as the template for the polypeptide assembly. Thus one of the most important functions of the nucleus is the production of messenger RNA.

In addition to being the site of production of messenger RNA, the nucleus is the site of the synthesis of ribosomes, the small organelles that aid in decoding the information for polypeptide synthesis that is in the base sequence of the messenger RNA. The nucleolus of the nucleus is the actual site of ribosomal RNA synthesis, and the ribosomes formed from ribosomal RNA and protein are constructed in the nucleus. In addition to messenger RNA and ribosomal RNA, the nucleus produces transfer RNA, the small, soluble ribonucleic acids that act as amino acid carriers and aid in the alignment of amino acids during the assembly of the polypeptides.

Another equally important physiological role of the nucleus is the replication of DNA. Replication takes place in preparation for cell division such that identical DNA molecules with the same information can be passed on to newly formed cells.

Actual polypeptide assembly takes place in the cytoplasm on the ribosomes that ordinarily are attached to endoplasmic reticulum. The messenger RNA attaches to the ribosomes, forming structures called polysomes or polyribosomes. Translation of the code on the messenger RNA during polypeptide assembly occurs because the transfer RNAs that bond covalently to specific amino acids can recognize the base sequences of the messenger RNA. Thus assembly of the polypeptide is ordered by the transfer RNAs aligning amino acids properly according to the code of the messenger RNA. Posttranslational modification of the polypeptide produces functional proteins that, as enzymes, regulate the chemical reactions of the cell. Modification may take the form of enzymatic cleavage of the newly formed polypeptide, adding nonprotein groups (prosthetic groups), coiling and folding, or combination with other peptides. Of course, some polypeptides are functional without any posttranslational modification. The enzymes themselves are regulated by the timing of their synthesis and by small metabolite molecules or other proteins that act as effectors of activity.

A summary of the current central dogma of cell biology is shown in the following diagram:

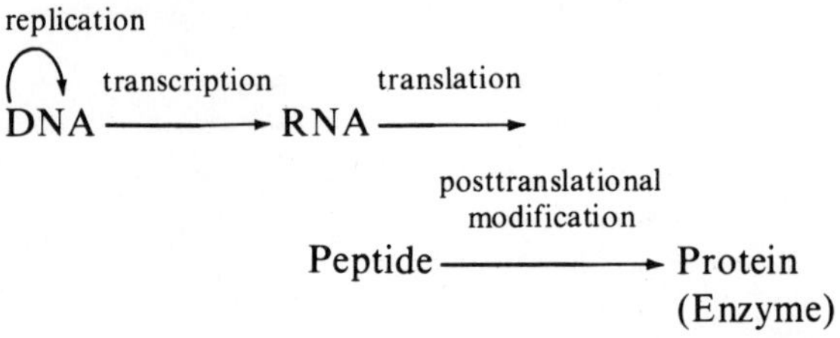

replication

DNA $\xrightarrow{\text{transcription}}$ RNA $\xrightarrow{\text{translation}}$

Peptide $\xrightarrow[\text{modification}]{\text{posttranslational}}$ Protein (Enzyme)

It is not only the nucleus and cytoplasm that are involved in polypeptide synthesis but the complex organelles, chloroplasts and mitochondria as well. Both chloroplasts and mitochondria have DNA, RNA, ribosomes, and the necessary enzymes for transcription and translation. Thus both kinds of these organelles are capable of protein synthesis and self-replication. However, not all of the protein of these organelles is the result of their own biosynthetic activity. In fact, much of the protein in the chloroplasts and mitochondria is synthesized from nuclear messenger RNA that is translated on cytoplasmic ribosomes. Thus there is much regulation and coordination of chloroplasts and mitochondria by the nucleus.

Although the various nucleotides are the substrates for nucleic acid synthesis, they do play many other important roles in the physiology of cells. Compounds such as adenosine triphosphate (ATP) and nicotinamide adenine dinucleotide (NAD) are important in biochemical reactions that require or produce energy. The compound coenzyme A is a complex nucleotide that functions as an acetate carrier in respiratory metabolism and in many bio-

syntheses. Therefore, nucleotides are common and ubiquitous constituents of cellular metabolism.

In this chapter, important aspects of nucleic acid metabolism are discussed, including biosynthesis of the purines and pyrimidines; replication, transcription, and translation during polypeptide synthesis; and the role of the metabolic nucleotides. Actual protein synthesis is discussed in the next chapter (9).

8.1 The Structure of Nucleic Acids

The repeating units of the nucleic acid sequence are the nucleotides. These are composed of a nitrogenous base, a sugar, and a phosphate. The unit minus

the phosphate is called a nucleoside. Units without the sugar are the free bases. The structural organization of nucleic acids is shown in Fig. 8.1.

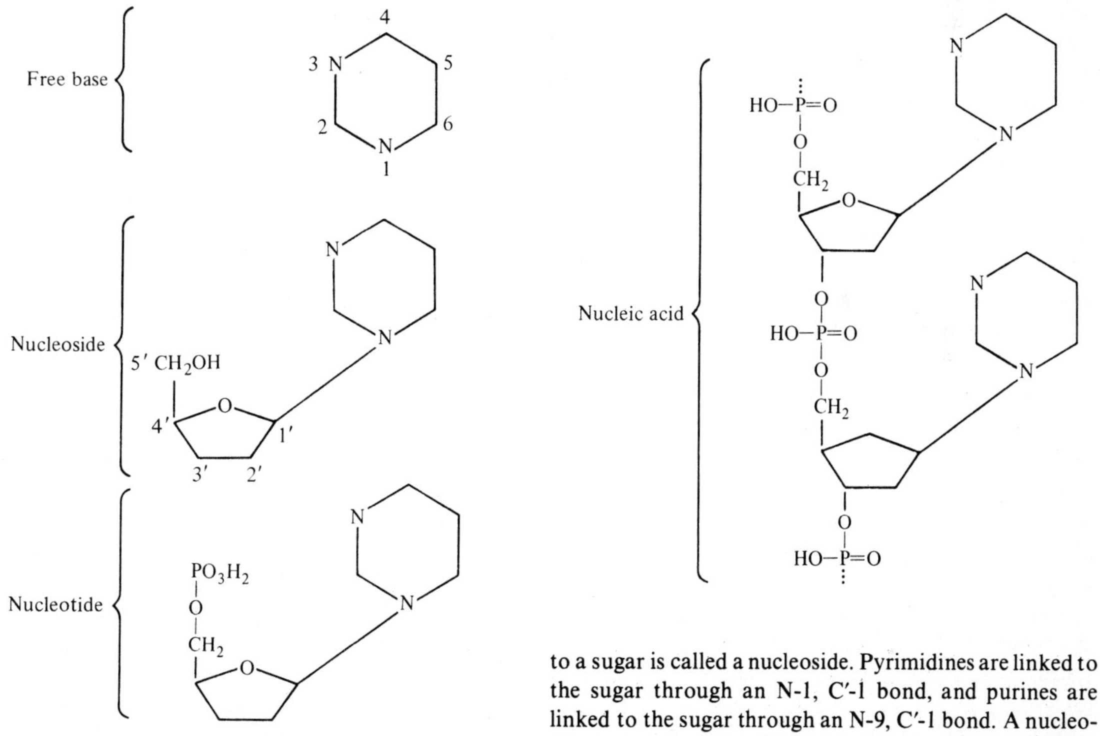

FIGURE 8.1   Models to illustrate the organization of a nucleoside, a nucleotide, and a nucleic acid. The nitrogenous ring compound, which in nucleic acids is a purine or pyrimidine, is called the free base. The base attached

to a sugar is called a nucleoside. Pyrimidines are linked to the sugar through an N-1, C'-1 bond, and purines are linked to the sugar through an N-9, C'-1 bond. A nucleotide is a nucleoside with a phosphate bonded in the C'-5 position. The nucleic acid polymer is formed by linking nucleotides through phosphodiester bonds from the C'-3 of one nucleotide to the C'-5 position of the adjacent nucleotide.

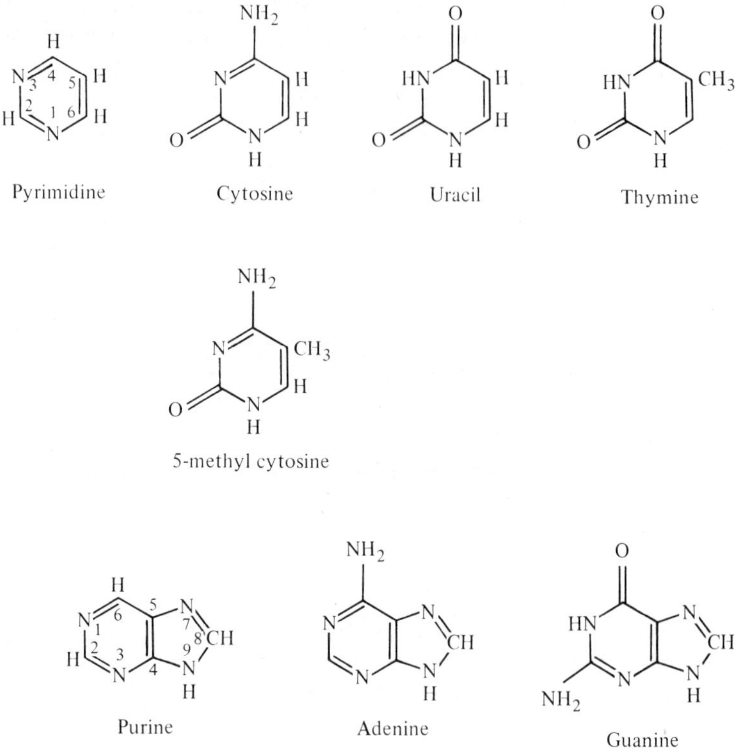

Pyrimidine   Cytosine   Uracil   Thymine

5-methyl cytosine

FIGURE 8.2 Basic structures of pyrimidines and purines, and the six most common free bases found in plant nucleic acids.

Purine   Adenine   Guanine

The common organic bases usually found in nucleic acids are depicted in Fig. 8.2. There are two basic chemical types: the pyrimidines, with a single-ring structure, and the purines, with a double-ring structure. Bases occurring in DNA are adenine, guanine, cytosine, 5-methyl cytosine, and thymine. RNA also has adenine, guanine, and cytosine but has uracil in place of thymine. In addition, many RNA polymers such as those directly involved in amino acid transfers during protein synthesis—the transfer RNAs—have a variety of other substituted bases such as hypoxanthine, 1-methyl hypoxanthine, N^2-dimethyl guanine, 1-methyl guanine, 5-hydroxymethyl cytosine, pseudouracil, dihydrouracil, and 5-methyl cytosine.

Sugars found in RNA and DNA are the 5-carbon sugars, D-ribose and 2-deoxy-D-ribose. RNA has ribose and DNA has deoxyribose, as their names

indicate. The structures of ribose and deoxyribose are as follows.

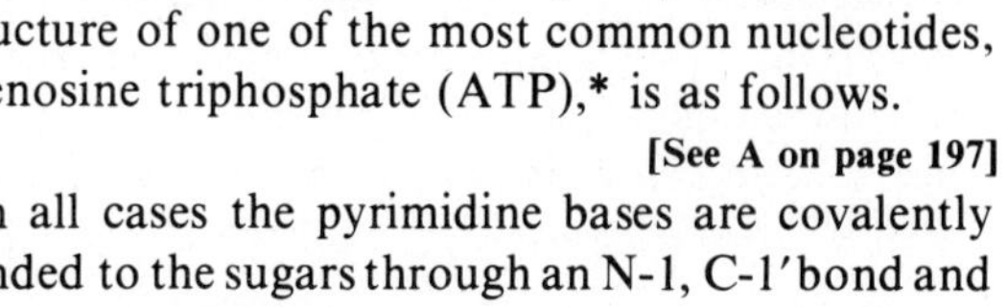

D-ribose   2-deoxy-D-ribose

A complete nucleotide is composed of a nitrogenous base and a sugar that is phosphorylated. The structure of one of the most common nucleotides, adenosine triphosphate (ATP),* is as follows.

[See A on page 197]

In all cases the pyrimidine bases are covalently bonded to the sugars through an N-1, C-1′ bond and

*At physiological pH the hydroxyls of the phosphates would be ionized, forming an anion with a charge of −4.

NH₂

(structure diagram)

Adenosine triphosphate (ATP)

A

TABLE 8.1 The names of the common nitrogenous bases and corresponding ribosides found in nucleic acids

BASE	RIBOSIDE	DEOXYRIBOSIDE
Adenine	Adenosine	Deoxyadenosine
Guanine	Guanosine	Deoxyguanosine
Uracil	Uridine	Deoxyuridine
Cytosine	Cytidine	Deoxycytidine
Thymine	Thymine riboside	Thymidine

Note that uracil is not an ordinary component of DNA and thymine is not an ordinary component of RNA. The deoxyribotide of thymine was named thymidine and not deoxythymidine because it does not occur in DNA. The nondeoxyriboside of thymine is called thymine riboside.

the purine bases are linked through an N-9, C-1′ bond.* In this nomenclature the nonprime numbers (N-1 and N-9) refer to the bases, and the prime numbers (C-1′ and C-5′) refer to the sugars. Phosphates are attached at the C-5′ position of the sugar. Using ATP as a model, the other nucleotides can be constructed if their names are known.

The names of the common nucleosides along with their bases are given in Table 8.1. To name a nucleotide, simply add phosphate to the name. Thus the nucleotide with adenine as the base is adenosine phosphate. For one phosphate the term adenosine monophosphate (AMP) is used, whereas for two or three in ester linkage the names adenosine diphosphate (ADP) and adenosine triphosphate (ATP) are used.

As well as being the substrates for nucleic acid assembly, nucleotides play a central role in cellular energetics. Hydrolysis of the terminal two phosphates of the di- and triphosphate nucleotides yields substantial free energy that can be used in many cellular syntheses and processes. Nucleotides such as ATP, GTP, UTP, and CTP play important roles in cellular energetics.

In addition, the nucleotide diphosphates have an important role in transferring sugar residues during carbohydrate biosynthesis and hydrolysis. These important metabolic events are discussed more fully in Chapter 10.

The nucleotides making up the nucleic acid polymer are linked in a linear sequence through phos-

*The bond is a N-glycosidic bond (see Chapter 10).

phate diester bonds from the 3′-hydroxyl of one sugar to the 5′-hydroxyl of another. A sequence of four nucleotides is shown in Fig. 8.3. The base sequence of this single strand of RNA can be indicated as pApGpCpU from 5′ to 3′, indicating an adenosine–guanosine–cytidine–uridine sequence. Note that in this nomenclature convention, Ap indicates a phosphate ester bond attached to the 3′-hydroxyl of adenosine and pG indicates a phosphate ester at the 5′-hydroxyl of guanosine.

Similarly, a four-nucleotide sequence of DNA could be pAGpCpT, indicating a sequence of deoxyadenosine — deoxyguanosine — deoxycytidine —deoxythymidine.

Single-stranded RNA can have a primary structure (nucleotide sequence) of any base combination. It could be polyadenosine (all adenosine), polyguanosine, or any combination of the four common nucleotides: adenosine, guanosine, cytodine, and uridine. Similarly, the primary structure of DNA can be any sequence of the four bases, deoxyadenosine, deoxyguanosine, deoxycytidine (or 5-methyl deoxycytidine), or deoxythymidine.

The structure of DNA was determined in 1953 through the efforts of many workers, including Watson, Crick, Wilkins, and Chargaff. Important discoveries in their work showed that native DNA exists *in vivo* as a double-stranded molecule

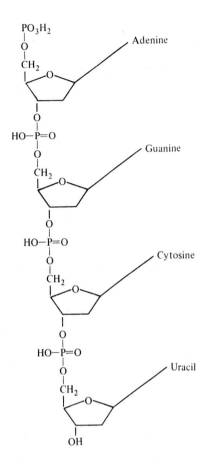

FIGURE 8.3 Sequence of four nucleotides linked as they would be in a nucleic acid polymer. The sequence shown is pApGpCpU-OH.

oriented in a right-handed helical form; a double helix, in fact, because of the double strand. The double helix is shown in Fig. 8.4.

An important observation was that the ratio of adenine to thymine and the ratio of cytosine to guanine was unity. In the case of DNA in plants, in which 5-methyl cytosine occurs along with cytosine, the ratio of cytosine plus 5-methyl cytosine to guanine is near unity (Table 8.2). Although there is similarity in the ratios, the other possible combinations of bases in DNA have no apparent consistency.

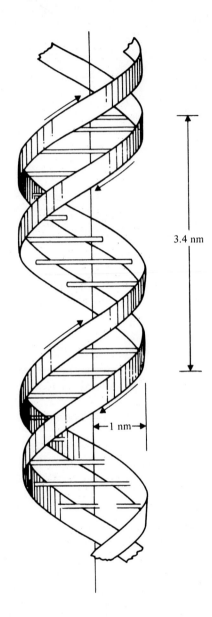

FIGURE 8.4 Diagram to illustrate the double-helical organization of DNA. The two complementary strands run in opposite directions (from C′-3 to C′-5 in one and from C′-5 to C′-3 in the other) and are held together by hydrogen bonding between complementary bases, adenine–thymine and cytosine–guanine.

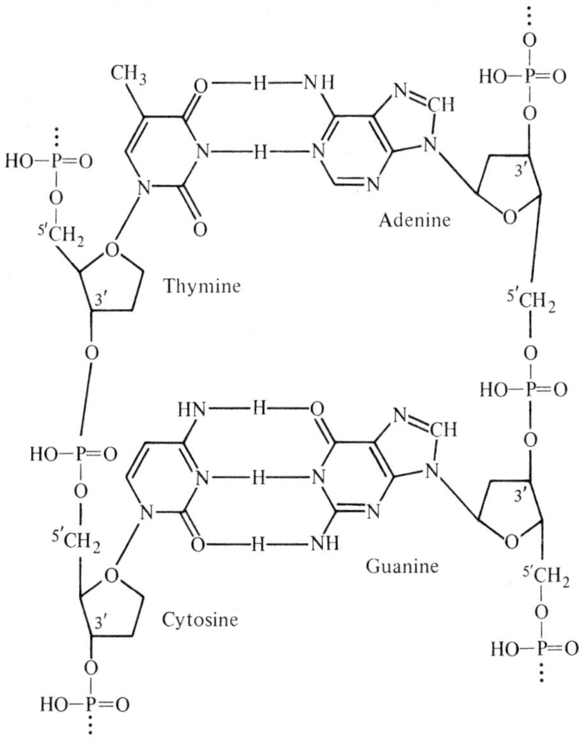

FIGURE 8.5   Diagram to illustrate hydrogen bonding between complementary bases (adenine–thymine and cytosine–guanine) in the double helix of DNA. The cytosine–guanine pair has three hydrogen bonds and the adenine–thymine pair has two hydrogen bonds.

The consistency of adenine to thymine and cytosine to guanine, along with x-ray diffraction studies, led to the present concept that the double-stranded DNA is formed of two complementary strands running in opposite directions and in which an adenine of one strand is paired with a thymine of the other and a guanine is always paired with a cytosine or 5-methyl cytosine. This fact places much restriction on the structure of DNA. Some similar base pairing may also exist in RNA but is not as restrictive. Much of the DNA stability comes about because of the perfect fit for hydrogen bonding between the complementary bases, adenine–thymine and cytosine–guanine (refer to Fig. 8.5). Adenine and thymine form two hydrogen bonds, whereas guanine and cytosine (or 5-methyl cytosine) will form three.

There is great stability of the secondary and tertiary structures of DNA because of the base pairing between complementary strands and the helical coiling. However, at high temperatures the strands will part. The breaking of hydrogen bonds at a high temperature to form single strands is called "melting." Melting temperatures are unique features of DNA molecules from different species. The reassociation upon cooling yields a double-stranded struc-

TABLE 8.2   The average base composition of DNA from 10 plant species, expressed in percentage

ADENINE	THYMINE	GUANINE	CYTOSINE + 5-METHYL CYTOSINE
29.9	30.0	20.2	18.4 (13.2 + 5.2)
$\dfrac{A}{T} = 0.99$		$\dfrac{G}{C} = 1.09$	

Data from Bonner in Bonner and Varner (1976).

ture identical to the original, which indicates restrictive base pairing. The base pairing is so restrictive that it is used to ascertain degrees of similarity between DNAs from different sources by determining the ease with which hybrid DNAs are formed.

Melting and reassociation is easily measured since the absorbance at 260 nm increases with melting. Apparently the double-helical structure with base pairing by means of hydrogen bonds results in reduced ultraviolet light absorption. Melting results in greater UV absorption, a phenomenon called hyperchromicity.

In summary, DNA is composed of two nucleic acid strands made of linear sequences of nucleotides. The two strands are complementary and run in opposite directions; that is, one strand runs from the 5' to the 3' end while the complementary strand runs from 3' to 5'. The complementary strands, coiled in a helix, are held together by hydrogen bonds between the complementary bases, adenine–thymine and cytosine–guanine.

8.2 The Biosynthesis of Nucleic Acids

8.2.1 Deoxyribonucleic acid (DNA)

The replication of DNA is a fundamental process of plant nucleic acid metabolism. During or just prior to cell division, DNA replicates to form identical DNA strands that are subsequently passed on to successive generations. For the most part, replication is without error. Mistakes are mutations and are mostly detrimental.

Polymerizing enzymes, called DNA polymerases, catalyze the synthesis of DNA using preexisting DNA molecules for templates. These polymerase enzymes require a DNA primer (DNA-3'-OH) to begin polymerization, a DNA template for proper sequencing, a metal ion (magnesium or manganese), and the deoxyribotide triphosphate substrates (dATP, dCTP, dGTP, and dTTP). The reaction is as follows.

$$\text{DNA-3'-OH} + \begin{matrix} \text{dATP} \\ \text{dGTP} \\ \text{dCTP} \\ \text{dTTP} \end{matrix} \xrightarrow[\text{DNA template}]{\text{Mg}^{+2}} \text{DNA} + \text{P}_i\text{P}_i$$

Polymerization begins from the 5'-hydroxyl end toward the 3'-hydroxyl end and begins with a DNA-3'-OH primer. The terminal product is a 5'-triphosphate. Polymerization initiation can begin with either a DNA-3'-OH or an RNA-3'-OH. The substrates are the triphosphate derivatives of the deoxyribonucleotides.

Initially replication begins with the double-stranded DNA forming single strands. Then a DNA polymerase catalyzes the synthesis of a complementary strand from one of the preexisting strands that functions as the template. A second polymerase replicates the other strand. The DNA template is neither lost nor consumed during the replication process. The net result of the replication process is two new double strands of DNA, each with both an original strand and a newly formed strand. Because one of the original strands is conserved in each of the two new strands, the process is termed semiconservative. The newly formed double helix is composed of complementary strands, each running in opposite directions and paired through the complementary bases, adenine–thymine and guanine–cytosine. An adenine of one strand is always paired with a thymine of the complementary strand and a guanine is always paired with a cytosine.

There are several DNA polymerases, each with different properties, known to exist in any one plant. DNA polymerase I, the first enzyme discovered that can catalyze DNA polymerization, requires a DNA-3'-OH or RNA-3'-OH, a metal ion, and deoxyribotide triphosphates. Such an enzyme is present in plant tissues, but because of very slow rates of DNA polymerization it may function primarily as a repair enzyme. Another polymerase, DNA polymerase II, with similar characteristics and requirements, is more likely to be the *in vivo* polymerizing protein since it catalyzes DNA syn-

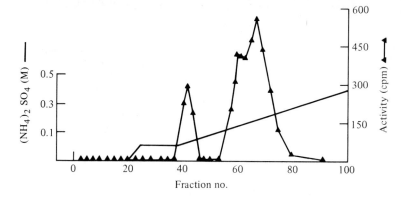

FIGURE 8.6 Two different DNA-dependent RNA polymerases present in the nucleus of maize. The polymerase isozymes were separated by column chromatography. There is yet another different DNA-dependent RNA polymerase in the chloroplast. Drawn from data of G. C. Strain, K. P. Mullinix, and L. Bogorad. 1971. RNA polymerases of maize: nuclear RNA polymerases. *Proc. Natl. Acad. Sci.* (U.S.) 68:2647–2651.

thesis semiconservatively at rates comparable to *in vivo* rates.

An enzyme, DNA ligase, is known to form 3′,5′-phosphodiester bonds between DNA pieces with 3′-hydroxyl and 5′-phosphate ends. This enzymic protein, which in plants requires ATP as the energy donor, is involved in DNA synthesis, linkage of preformed DNA fragments, or repair of broken strands. This ligase enzyme functions in both repair and in replication.*

Much existing evidence suggests that the DNA associated with a single chromosome may be a single piece; that which is isolated from cells by present techniques is probably fragmented. Native DNA occurring *in vivo* may have a size as great as 10 million daltons, with 10,000 to 15,000 base pairs.

Nuclear DNA exists as a nucleic acid–protein complex. The nucleoproteins (histones) that bind to DNA have a high pH since they are rich in lysine and arginine.

8.2.2 Ribonucleic acid (RNA)

In most respects, RNA synthesis is similar to DNA synthesis. DNA acts as the template, and an RNA

polymerase that uses ribonucleotide triphosphate precursors produces RNA.

$$\begin{matrix} GTP \\ ATP \\ CTP \\ UTP \end{matrix} + \xrightarrow[\text{DNA template}]{Mg^{+2}} RNA + P_iP_i$$

Synthesis proceeds as with DNA polymerase except that uracil matches adenine. The new RNA is complementary to the DNA template. DNA is completely conserved.

In higher plants, there are at least three different forms of RNA polymerase (Fig. 8.6). Such multiple forms of an enzyme are called isozymes and are discussed in more detail in Chapter 9. Two of the RNA polymerases of plants (called RNA polymerase I and RNA polymerase II) are found in the cell nucleus. The other is most likely localized in chloroplasts. The presence of multiple forms of RNA polymerase is consistent with the fact that there are different kinds of RNA in cells (these are discussed in the following section). Evidently, the different polymerases synthesize the different kinds of RNA.

8.3 RNA Types

The cell has several different kinds of nucleic acid. If a preparation is made from plant tissue, the nucleic acid fraction can be separated into several

different components by chromatographic procedures (Fig. 8.7). The figure shows, in addition to DNA, at least five additional peaks of RNA. The RNA molecules are the direct gene products. These and others are discussed below.

---

*DNA ligase is the enzyme used in gene-splicing experiments.

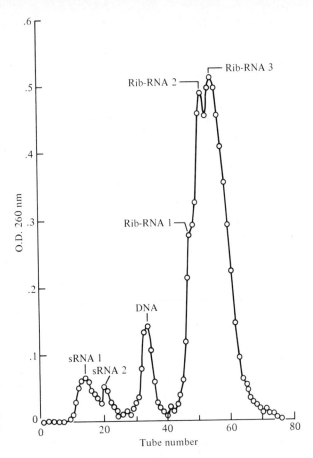

FIGURE 8.7 Column chromatography of a nucleic acid preparation obtained from wheat seeds. The nucleic acids were extracted from the seeds with a series of solutions containing ethanol, ether, and water and were dried and then resuspended in a tris-Cl buffer. The nucleic acids were extracted from the buffer with water-saturated phenol and precipitated with absolute ethanol. The preparation was applied to a silica column (a methylated albumen kieselguhr column called an MAK column) and eluted with a salt gradient. The various fractions eluted from the column were detected by their absorption at 260 nm. Two transfer RNA (sRNA in figure) and three ribosomal RNA peaks were obtained in addition to the DNA peak. From an experiment by B. S. Vold and P. S. Sypherd. 1968. Changes in soluble RNA and ribonuclease activity during germination of wheat. *Plant Physiol.* 43:1221–1226.

8.3.1 Messenger RNA

Most messenger RNA (mRNA) polymers are short-lived, single-stranded ribonucleic acids that function in transcribing the genetic code of DNA. Their lifetime is on the order of hours. The base sequence of mRNA is complementary to the corresponding DNA. Messenger RNA synthesis takes place in the nucleus, catalyzed by RNA polymerase. This polymerase requires metal ions (Mg^{+2} or Mn^{+2}), ATP, GTP, CTP, and UTP plus DNA as a template and is inhibited by the antibiotic actinomycin-D.

As will be explained in a subsequent section, mRNA migrates from the nucleus and attaches to the ribosomes prior to protein synthesis. Because mRNA makes up only about 5% of the RNA within the cell and because it is quite unstable and short-lived, little is known about its properties. Molecular sizes are about 1 million daltons.

Mature seeds apparently have long-lived mRNA that persists until germination. Research has shown that during embryo development, there is transcription of DNA to form mRNA, which persists in the mature seed. During germination the preexisting mRNA is translated for protein synthesis. Evidence for the long-lived mRNA comes from the observation that seeds treated with actinomycin-D still have mRNA. Evidently, the mRNA present prior to treatment persists. If it were not present in the seed prior to treatment with actinomycin-D, it would be expected that the actinomycin would inhibit mRNA synthesis and that mRNA would be totally absent.

8.3.2 Transfer RNA

The transfer ribonucleic acids (tRNA), which constitute up to 15% of the cellular RNA, are the smallest nucleic acid polymers found in cells and have masses on the order of 25,000 daltons, with approximately 90 nucleotides in sequence. These single-stranded nucleic acids, which are synthesized directly from chromosomal DNA, are water-soluble. Because of their small size and water solubility, they used to be referred to as soluble RNA. Now the term transfer RNA is preferred. They are known to have up to 10% of the uncommon bases, mostly methylated derivatives of the more common bases. The base pseudouridine is also found in tRNA. Transfer RNAs function as the amino acid carriers during protein synthesis.

Many different kinds of tRNA have been isolated from cells and studied. They differ from each other in amino acid specificity. For each of the amino acids known to occur in proteins there is at least one tRNA type. For some amino acids such as leucine there are several tRNAs. Although each of these leucine tRNAs will bond to leucine and transfer it

to a growing polypeptide, other properties are different. An exact role for these multiple forms of tRNA is not known, but they may function in protein synthesis in different subcellular compartments such as in chloroplasts, mitochondria, or the cytosol. Perhaps they function in the synthesis of different kinds of proteins.

Because the actual base sequence of many tRNAs is known, x-ray diffraction studies of tRNA crystals have allowed prediction of the secondary and tertiary structure. Figure 8.8 is a diagrammatic representation of a tRNA molecule, and Fig. 8.9 is a stereoscopic pair drawing of phenylalanine tRNA showing the three-dimensional structure. Insofar as is known, all of the tRNA molecules occur in the "cloverleaf" form. The end positions of the two lateral arms shown in the figure are loops formed because there is no pairing of the uncommon substituted bases that occur principally in the loop

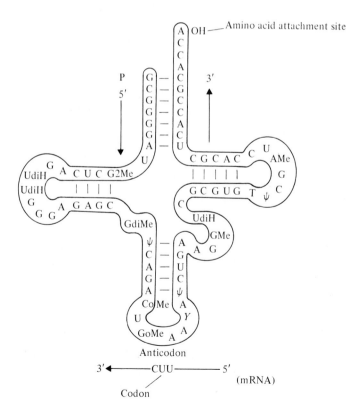

FIGURE 8.8 Diagrammatic representation of the cloverleaf form of phenylalanine transfer RNA isolated from wheat germ. The cloverleaf structure is formed from a combination of base pairings between the complementary bases, adenine–uracil and cytosine–guanine, and the lack of pairings or hydrogen bonding between a modified base and an adjacent base. The nucleic acid polymer runs from the 5′ end to the 3′ end, where the amino acid attaches. The arm at the opposite end of the amino acid attachment has the anticodon that recognizes the proper codon on mRNA for amino acid sequencing during polypeptide assembly. (AMe = 1-methyl adenosine; ψ = pseudouridine; UdiH = 5,6-dihydrouridine; GMe = 7-methyl guanosine; Y = unknown base; GOMe = O-methyl guanosine; COMe = O-methyl cytosine; GdiMe = N^2-dimethyl guanosine; G2Me = N^2-methyl guanosine.) From B. S. Dudock, G. Katz, E. K. Taylor, and R. W. Holley. 1969. Primary structure of wheat germ phenylalanine transfer RNA. *Proc. Natl. Acad. Sci.* (U.S.) 62:941–945.

FIGURE 8.9 Stereoscopic pair drawing of phenylalanine tRNA isolated from yeast. The molecule has 76 nucleotide residues and has an L shape. The amino acid attachment site is the small extension at the upper right (74–76). The anticodon is at the bottom (34–36). The figure is by A. Rich from S. H. Kim *et al.* (see S. H. Kim *et al.* 1974. *Proc. Natl. Acad. Sci.* (U.S.) 71:4870–4974) in Margaret O. Dayhoff. 1976. *Atlas of Protein Sequence and Structure,* Suppl. 2. National Biomedical Research Foundation (Georgetown University Medical Center), Washington, D.C.

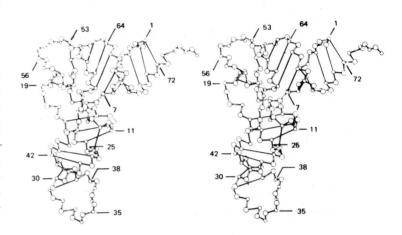

regions. It is believed that one loop may be involved in binding to ribosomes during protein synthesis and that the other may function as a recognition site for amino acid-activating enzymes that attach the amino acid to the tRNA molecule. The 3′-terminus of all known tRNA molecules has the same base sequence: pCpCpA. Amino acids attach to the adenosine terminus to form an amino acyl (adenylic acid) derivative.

Activation of an amino acid is enzymatically catalyzed and requires ATP as an energy source:

Amino acid + ATP + tRNA →

Amino acid-tRNA + AMP + P_iP_i.

The lower arm of each tRNA (see Fig. 8.8) has a unique triplet (three-base) sequence that is apparently complementary to a triplet of messenger RNA. This unique triplet is called an anticodon because it is complementary to the three-base codon sequence of RNA that specifies the amino acid. Thus the tRNA can recognize the proper site on the mRNA for amino acid sequencing during protein assembly.

8.3.3 Ribosomal RNA and ribosomes

Ribosomal RNA (rRNA) is transcribed from DNA associated with the nucleolus. It is a component of the ribosomes and evidently functions in translation of the message on messenger RNA during polypeptide assembly from amino acids. Ribosomal RNAs are large molecules with masses of 0.9 to 1.3 million daltons.

Ribosomes are small subcellular particles containing about 50% rRNA and 50% protein. The molecular dimensions are approximately 10 to 30 nm on a side, and they are about 70 or 80 *S*.* The two sizes of ribosomes are known from higher plants, 70 *S* and 80 *S*. The 80 *S* ribosomes are located in the cytoplasm and are composed of a 40 *S* and a 60 *S* subunit (Fig. 8.10). The 70 *S* ribosomes

*Recall from Chapter 2 that the Svedberg unit (*S*) is a sedimentation coefficient. When two small units are put together, their *S* estimates do not add numerically.

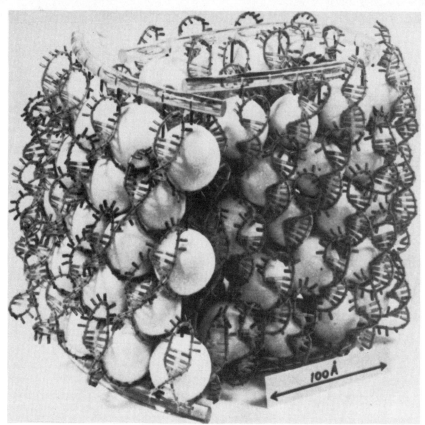

FIGURE 8.10 Molecular model of an 80 *S* ribosome. According to the model, which may not necessarily be the way it occurs *in situ,* the smaller subunit (40 *S*) fits over the larger subunit (60 *S*). The light-colored, spherical structures are the proteins. After R. A. Cox and S. A. Bonanou. 1969. A possible structure of the raddit reticulocyte ribosome. *Biochem. J.* 114:769–774.

are components of chloroplasts and of mitochondria. These somewhat smaller ribosomes are composed of 30 *S* and 50 *S* subunits. Although the 70 *S* ribosomes of chloroplasts and mitochondria are similar in size, it has been shown that they are in fact different.

Each of the small subunits of the ribosomes are in themselves composed of two or more smaller (in terms of *S* units) rRNAs. As an example, the large subunits of the ribosomes (50 and 60 *S*) contain a 23 to 28 *S* rRNA and a 5 *S* rRNA molecule. The others are also composed of smaller-sized rRNA molecules. During the synthesis of the rRNA molecules, a large precursor is first produced within the nucleus (nucleolus) and is then cleaved to form the active rRNA, which is a component of the intact ribosome.

Within the intact ribosome the rRNA may be folded onto itself to form double-stranded sections. Associated with the rRNA of the ribosomes are a variety of different kinds of proteins. The 70 *S* and 80 *S* ribosomes contain different kinds of proteins, further substantiating their differences. The exact role of the ribosomal proteins is not known, but it is more than likely that they partially contribute to the structure of the ribosomes and have enzymatic activity that aids in polypeptide assembly.

The 80 *S* ribosomes are found free in the cytoplasm and are associated with the endoplasmic reticulum. Endoplasmic reticulum with ribosomes is called rough endoplasmic reticulum (rough ER). During protein synthesis the ribosomes are attached to messenger RNA. The units formed by the association of ribosomes and mRNA are called polysomes or polyribosomes and can be isolated from cell homogenates by centrifugation. Polysomes are shown *in vivo* in the electron micrograph of part of a young cotton fiber in Fig. 8.11.

Several polypeptides can be synthesized simultaneously from one mRNA, depending on the number of ribosomes and the length of the mRNA. This is further discussed in Chapter 9.

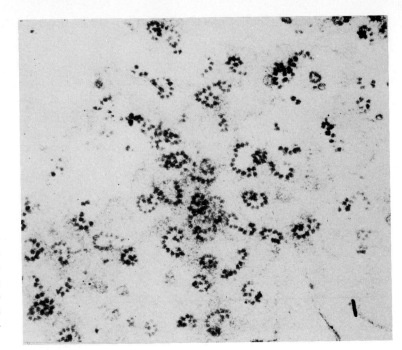

FIGURE 8.11 Electron micrograph showing polysomes (polyribosomes) from a young, developing cotton-fiber cell ($\times$75,000). The presence of the polysomes is indicative of active protein-synthesis activity. Electron micrograph courtesy of Dr. J. Berlin.

8.4 The Genetic Code

Through the efforts of many researchers it is now known how information is stored in DNA and transferred to RNA for ultimate protein synthesis. In the simplest sense, DNA sequences are composed of four different nitrogenous bases: adenine, guanine, cytosine, and thymine. These may be considered the DNA alphabet, a four-letter alphabet. Each unit, or word, of the code is made up of a three-base sequence that forms triplets, or codons. Each three-base sequence of DNA is a codon for a specific amino acid. Since there are a minimum of four bases, the possibility of 64 different triplet permutations exists, far more than needed to code for the 20 to 22 amino acids occurring in proteins. Table 8.3 shows the genetic code as presently understood.

Many of the amino acids are specific for multiple triplets or codons. Thus the codons for alanine are GCU, GCA, GCC, and GCG, and for leucine UUA, CUU, CUA, UUG, CUC, and CUG, but for trypto-

phan and methionine, for example, the codons are UGG and AUG respectively. Three of the possible codons (of the 64 possible) do not code for any of the amino acids, UAA, UGA, and UAG, but function as terminators or ends of the code for a polypeptide. There are no known starting codons or punctuations within the code. Thus the information for assembly of amino acids in the proper sequence for protein synthesis is contained in a linear sequence of base triplets.

During transcription of the DNA code by mRNA, the mRNA is synthesized such that it is exactly complementary. The codon for alanine GCU on mRNA would be CGA on DNA, and similarly the UUA codon for leucine would be AAT on DNA.

The overall sequence consists of transcription of the DNA to form mRNA with the complementary bases and migration of the mRNA to the cytoplasm, where it forms polysomes with the ribosomes. The anticodon base sequence of tRNA (see Fig. 8.8)

TABLE 8.3 **The triplet codons making up the genetic code for various amino acids found in proteins**

Alanine	GCU	GCC	GCA	GCG		
Arginine	CGU	CGC	CGA	CGG	AGA	AGG
Asparagine	AAU	AAC				
Aspartic acid	GAU	GAC				
Cysteine	UGU	UGC				
Glutamic acid	GAG	GAA				
Glutamine	CAA	CAG				
Glycine	GGU	GGC	GGA	GGG		
Histidine	CAU	CAC				
Isoleucine	AUU	AUC	AUA			
Leucine	UUA	UUG	CUU	CUC	CUA	CUG
Lysine	AAA	AAG				
Methionine	AUG					
Phenylalanine	UUU	UUC				
Proline	CCU	CCC	CCA	CCG		
Serine	UCU	UCC	UCA	UCG	AGU	AGC
Threonine	ACU	ACC	ACA	ACG		
Tyrosine	UAU	UAC				
Tryptophan	UGG					
Valine	GUU	GUC	GUA	GUG		
Terminators	UAA	UAG	UGA			

A = adenine   C = cytosine   G = guanine   U = uracil

Note that these triplets are those that occur in messenger RNA.

then recognizes the proper codon of mRNA, allowing amino acid sequencing during protein assembly. More details of protein synthesis are presented in Section 8.6.

Evidence that DNA is the genetic material is now overwhelming. With the use of breeding experiments, it can be shown that the genetic information is contained on the chromosomes of the nucleus. Furthermore, most of the cellular DNA exists as a structural component of the chromosomes. Given these two observations, coupled with the fact that cellular DNA is quite constant and comparable within a species and that it is largely unaffected by environmental conditions, it is a reasonable hypothesis that the DNA polymer contains the genetic information. There are now quite conclusive experiments with bacteria involving genetic alterations through DNA transfer between different strains that leave no doubt about the role of DNA.

8.5 Organelle Genomes

A unique feature of plants is that they have three different genomes and three different protein-synthesizing systems. The DNA of the mitochondria and chloroplasts is circular and seems to be similar to procaryote DNA. Chloroplast DNA has a mass of about 90×10^6 daltons and mitochondrial DNA is about 10×10^6 daltons. These special circular DNAs have the code for organelle ribosomal RNA, transfer RNAs, and messenger RNA for some of their own proteins. Despite the fact that the transfer RNA molecules are in themselves specific for amino acids, both the mitochondria and the chloroplasts have unique transfer RNA molecules that differ from the nuclear-transfer RNA molecules that function in protein synthesis in the cytoplasm. In addition, the aminoacyl-tRNA synthetases (ligases) of each of the three compartments differ in certain properties. That the RNAs are organelle-specific is further supported by the observation that the ribosomal RNA produced in the nucleus, in the chloroplasts, and in the mitochondria all differ in sedimentation coefficients (*S*) and in their proteins.

The mitochondria code for and synthesize some of their membrane proteins but no soluble protein. Over 90% of the protein in the mitochondria is the result of nuclear codes and is synthesized on cytoplasmic ribosomes. The transport of the protein from the cytoplasm into the mitochondria has not been studied extensively in plants.

The chloroplasts probably synthesize more of their own protein than do the mitochondria (the DNA is almost 10 times as large). Many of the membrane-bound proteins are from the chloroplast-DNA code, including some cytochromes. Unlike the mitochondria, the chloroplasts produce at least one soluble peptide, which is the large subunit of the complex ribulose bisphosphate carboxylase, the enzyme responsible for CO_2 fixation in the chloroplast. The other, smaller subunit is coded for by nuclear DNA and is synthesized in the cytoplasm on cytoplasmic ribosomal RNA. This observation that the two subunits of one functional protein are the result of two different genomes, the nuclear and the chloroplast, clearly indicates much cooperation and coordination between the nucleus and the organelles.

There is still much to be learned about the DNA of the mitochondria and the chloroplasts. Methods such as nucleic acid hybridization and the *in vitro* study of the translation products (peptides) of organelle messenger RNA will lead to the furhter clarification of the role of the three different genomes in plants.

8.6 *In Vitro* Polypeptide Synthesis

An important goal of the molecular biologist is to develop a system for the *in vitro* synthesis of polypeptides from DNA sequences that represent genes. Once a well-defined system is developed it will be possible to gain much insight into the method of protein synthesis by plants and the means by which protein synthesis is regulated. The first step is the development of a system for the transcription of DNA to form messenger RNA, the direct product of the gene. For this step to be accomplished, there is a requirement for the proper DNA sequence, RNA polymerase, and the proper substrates (that is, nucleotides) and cofactors for RNA synthesis. Second, once the messenger RNA is formed, polypeptides can then be produced on ribosomes using amino acyl transferases, amino acids, and other cofactors. Other factors, perhaps even proteins, are necessary for posttranslational modification to form the functional proteins.

With certain bacteria, it has been possible to meet many of the goals stated above, although the preparations are not completely defined. With higher plants, some progress has been made with wheat germ and a few other tissues. Production of the gene product (messenger RNA) can be accomplished, but it is rather difficult to detect and assay quantitatively. Much progress has been made, however, because messenger RNA is high in adenosine monophosphate. The AMP tends to be in sequence, forming regions of 50 to 250 AMP nucleotides. These polyadenylate regions will bind to cellulose columns in a high-salt medium. Thus a mixture of ribonucleic acids containing messenger RNA with long sections of poly A can be separated by column chromatography. Using such techniques the gene products can be isolated and studied directly.

Since the messenger RNA has a base sequence complementary to the DNA code, DNA–RNA hybrids can be formed. In this manner the amount of DNA coding for ribonucleic acid can be determined.

Once messenger RNA has been purified it can be used for subsequent study of the initial translation products, the polypeptides. As stated above, the best system prepared from higher plants is that prepared from wheat germ. The cell-free wheat germ system for translation of messenger RNA contains ribosomes, transfer RNAs for the various amino acids, the proper enzymes for catalysis, an ATP generating system, magnesium, potassium, and the necessary amino acids. At least one of the amino acids must be labeled with a radioisotope (either ^{14}C or ^{3}H) so that the translation product of the messenger RNA can be detected easily.

8.7 Metabolic Nucleotides

Many nucleotides are energy-rich compounds functioning as cofactors in enzyme reactions. Perhaps the best known and understood is adenosine triphosphate, ATP. The structure of ATP is given on page 197. The two terminal phosphates yield a substantial amount of free energy on hydrolysis. This potential free energy can be used to drive energy-requiring metabolic reactions by direct phosphate transfer.

Although not entirely understood, the free energy comes about in part from the negative repulsion of the adjacent phosphate groups. Hydrolysis of both the first and second phosphates but not the third yields about 30 kilojoules of free energy per mole under standard conditions.

Other energy-rich phosphate compounds are phosphoenolpyruvate (-53.6 kJ mol^{-1}), acetyl phosphate (-42.3 kJ mol^{-1}), and glucose-1-P (-21 kJ mol^{-1}). Compounds such as glucose-6-P and glycerol-1-P yield substantially less than 20 kilojoules on hydrolysis.

The dinucleotides, nicotinamide adenine dinucleotide (NAD) and nicotinamide adenine dinucleotide phosphate (NADP), are ubiquitous and important oxidation–reduction cofactors in intermediary metabolism.

In the older literature, NAD is referred to as coenzyme I or diphosphopyridine nucleotide (DPN) and NADP is referred to as coenzyme II or triphosphopyridine nucleotide (TPN). The original names first proposed by Warburg in 1934 have been changed to be more consistent with their actual structure. The structures of NAD and NADP are as follows.

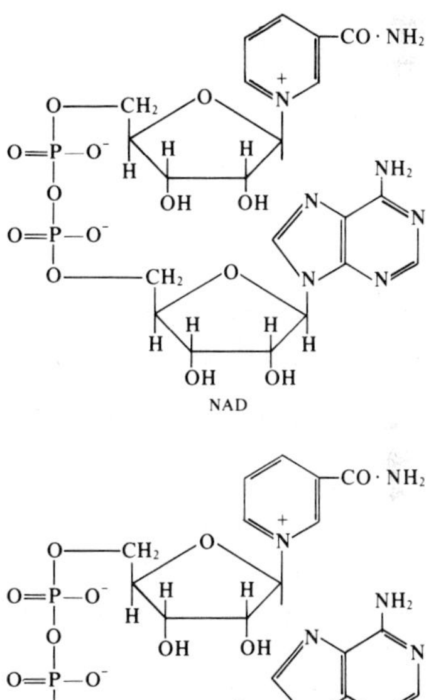

NAD

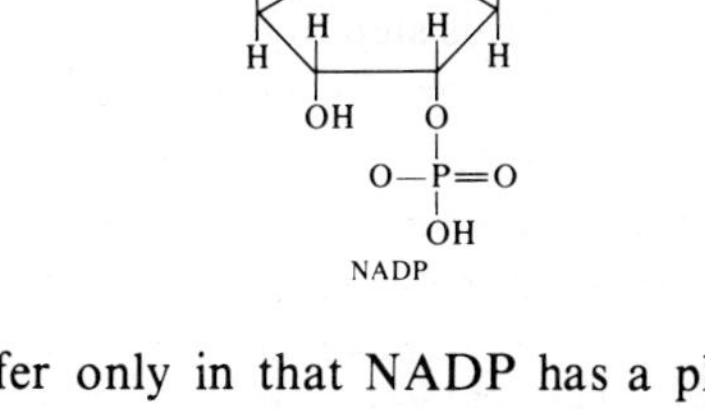

NADP

They differ only in that NADP has a phosphate group at the C′-2 position of adenine. NAD and NADP are cofactors, transferring hydrogen (with electrons) in reactions catalyzed by dehydrogenase enzymes. These enzymes that catalyze oxidation–reduction steps are very important in intermediary metabolism and are specific for either NAD or NADP. Very few dehydrogenases can use both NAD and NADP effectively, most being quite specific for one or the other.

It is the pyridine ring that becomes oxidized or reduced during the redox reactions. NAD is reduced to NADH by hydride ion acceptance at the 4 position. The addition of hydrogen is stereospecific for one side of the ring. In both NAD and NADP the oxidation and reduction are at the same position. NADP is reduced to NADPH in the same way that NAD is reduced to NADH. The phosphate of NADP attached to the C′-2 position of the ribose moiety of adenosine functions in maintaining specificity for those reactions that require NADP.

The reversible oxidation–reduction reaction of NAD is illustrated below.

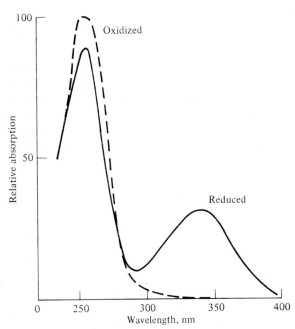

During reduction, both NAD and NADP change in absorbancy at 340 nm. As shown in Fig. 8.12, both NADH and NADPH increase in ultraviolet absorption, so the oxidation or reduction is easily followed with a spectrophotometer. Enzyme reactions can be monitored by the change in absorbance at 340 nm, or NADH and NADPH can be assayed quantitatively by the absorbance at 340 nm.

Two other redox nucleotide cofactors that are important in intermediary metabolism are flavin mononucleotide (FMN) and flavin adenine dinucleotide (FAD). Flavin mononucleotide is considered to be a nucleotide because it is formed from ribitol phosphate that is bonded to the nitrogenous base, dimethylisoalloxazine. Riboflavin (vitamin B_2) is ribitol linked to the dimethylisoalloxazine ring. The structure of flavin mononucleotide is as follows.

Flavin mononucleotide (FMN)

FIGURE 8.12 Absorption spectrum for the reduced form (NADH) and the oxidized form (NAD) of nicotinamide adenine dinucleotide. The increase in absorbance at 340 nm as NAD is reduced to NADH can be used to follow enzymatic reactions that use NAD (or NADH) as a cofactor. The change in absorbance is quantitative, following Beer's law. Nicotinamide adenine dinucleotide phosphate (NADP) shows a similar increase in absorbance at 340 nm upon reduction.

Flavin adenine dinucleotide is constructed of an FMN molecule linked to AMP through a phosphate ester bond. Since two nucleotides are present, it is named a dinucleotide. Its structure follows.

[See B on page 211]

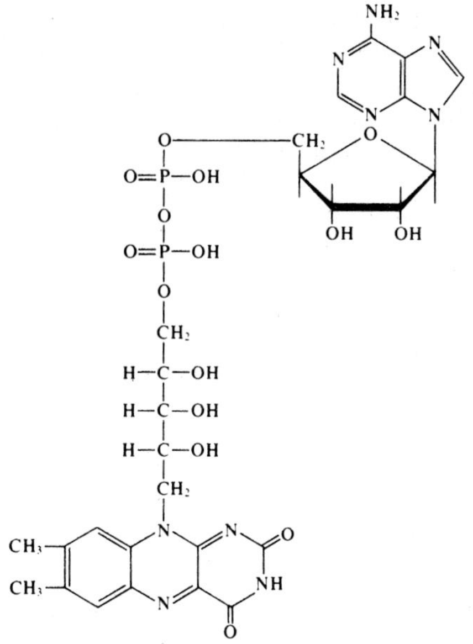

B   Flavin adenine dinucleotide (FAD)

These and other nucleotide diphosphates may be sugar carriers during syntheses.

Uridine diphosphate glucose and similar sugar derivatives of the nucleotide diphosphates function primarily in polysaccharide biosynthesis. Polysaccharide biosynthesis by means of the nucleotide diphosphate carriers is discussed in more detail in Chapter 10. The structure of uridine diphosphate glucose follows.

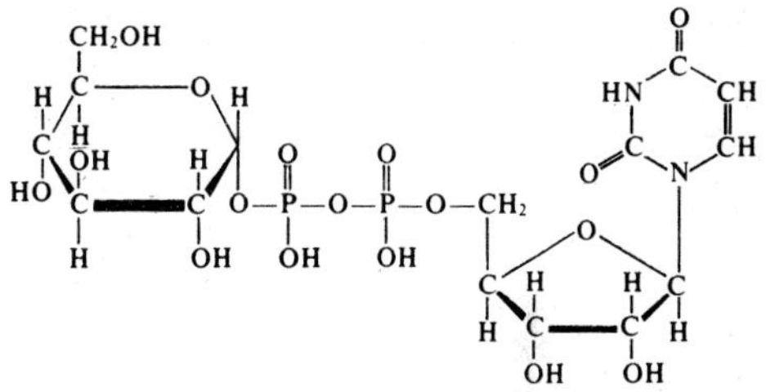

Uridine diphosphate glucose

During oxidation–reduction reactions the dimethylisoalloxazine ring structure becomes either oxidized or reduced by hydride-ion transfer.

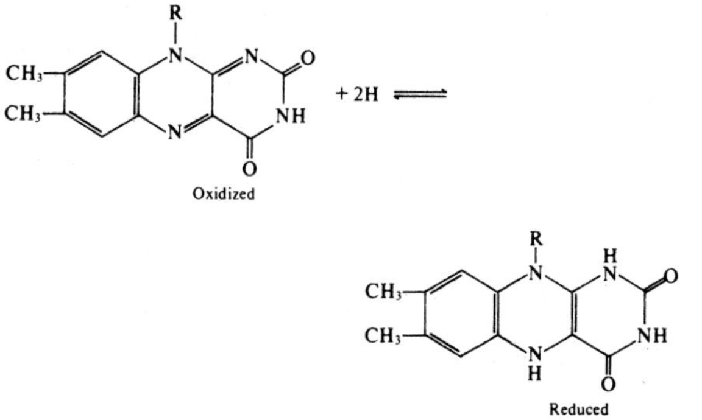

Coenzyme A is a mononucleotide with phosphopantotheine, a derivative of the vitamin pantothenic acid. Coenzyme A functions as an acyl carrier with the acyl group linked to coenzyme A through a thio ester bond. The acyl-S-CoA derivatives function in a variety of degradation and biosynthetic reactions. Perhaps the most central acyl-S-CoA derivative is acetyl-S-CoA. The structure of coenzyme A follows.

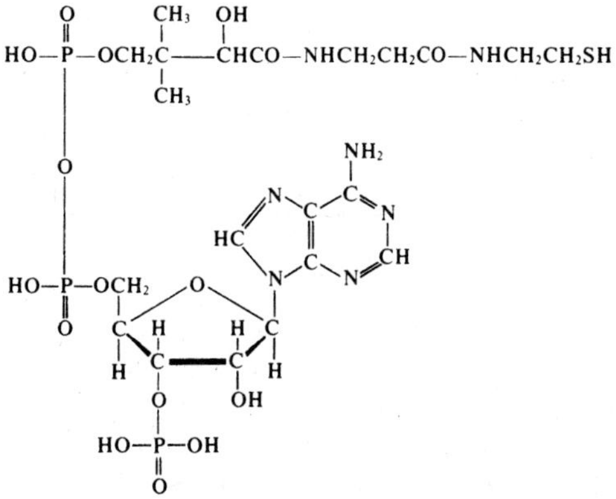

Coenzyme A

Other nucleotide cofactors that are extremely important in intermediary metabolism are coenzyme A and the sugar derivatives of the nucleotide diphosphates. The latter include uridine diphosphate glucose (UDPG), guanosine diphosphate glucose, and adenosine diphosphate glucose.

The biosynthesis of acetyl CoA takes place through a complex reaction requiring ATP and catalyzed by the enzyme acetate thiokinase. The reaction is:

$$CH_3-COO^- + ATP + CoA-SH \rightarrow$$

(Acetate)

$$CoA-S-\overset{\overset{O}{\|}}{C}-CH_3 + AMP + P_iP_i$$

(Acetyl CoA)

The ATP is necessary to drive the reaction because the hydrolysis of acetyl CoA takes place with a high standard-free-energy change ($\triangle G^{\circ\prime} = -31$ kJ mol^{-1}), making the synthesis from acetate and coenzyme A unlikely. ATP with a $\triangle G^{\circ\prime}$ of about -30 kJ per mole makes the reaction favorable.

8.8 Pyrimidine and Purine Biosynthesis

Because of the importance of pyrimidines and purines to nucleic acid metabolism, protein synthesis, and cellular energetics, their biosynthesis is discussed here. The biosynthesis takes place by a series of complex metabolic reactions, each catalyzed by a specific enzyme. Studies with radioactive tracers have allowed the elucidation of the pathways in which the various carbon, nitrogen, and oxygen atoms are inserted into the rings. The simplest is the synthesis of the pyrimidine ring, inasmuch as it is constructed from only three molecules: an aspartate molecule, NH_3, and CO_2.

8.8.1 Pyrimidine biosynthesis

Figure 8.13 illustrates the source of atoms for the biosynthesis of the basic pyrimidine ring and the metabolic pathway for the synthesis of uridine monophosphate (UMP). The sequence begins with the condensation of carbamyl phosphate and aspartate to form carbamyl aspartate ureidosuccinic acid. Ring closure proceeds through a dehydration to form dihydroorotic acid, which is subsequently reduced to orotic acid. The nitrogen and carbon at positions 2 and 3 are donated by carbamyl phosphate that was synthesized previously from NH_3, CO_2, and ATP. Phosphoribosyl pyrophosphate (PRPP) then reacts with the orotic acid to form the N-1,C-1' bond of orotidine phosphate, liberating inorganic pyrophosphate. Subsequent decarboxylation forms uridine monophosphate.

Cytidine triphosphate is then synthesized through uridine triphosphate by ammonia displacement at the C-4 position of the pyrimidine ring. The deoxyribotides are synthesized by reduction of the C-2' position through the uridine derivative. Thymidylic acid is synthesized by methyl addition to the C-5 position of the ring through tetrahydrofolic acid and a formate donor.

8.8.2 Purine biosynthesis

Figure 8.14 illustrates the source of atoms for the biosynthesis of the basic purine ring and the metabolic pathway for the synthesis of inosine phosphate from which the other ribosides are made. Beginning with phosphoribosyl pyrophosphate (PRPP), the amide nitrogen from glutamine is added to the C-1' position to form phosphoribosyl amine, liberating the pyrophosphate. The phosphoribosyl amine reacts with glycine to form glycinamide ribotide through a peptide-like bond. Formate is then added to the amino group of the glycine by donation from tetrahydrofolic acid, forming N-formylglycinamide ribotide. Another glutamine donates its amide nitrogen to form formylglycinamidine ribotide, which subsequently goes through a dehydration for ring closure to form 5-aminoimidazole ribotide. Carboxylation results in aminoimidazole carboxylic acid ribotide. Addition of nitrogen from aspartate and a formate from tetrahydrofolic acid followed by dehydration to close the ring completes the synthesis of the purine ring. The final form in this sequence is inosine phosphate.

Adenosine monophosphate is synthesized from

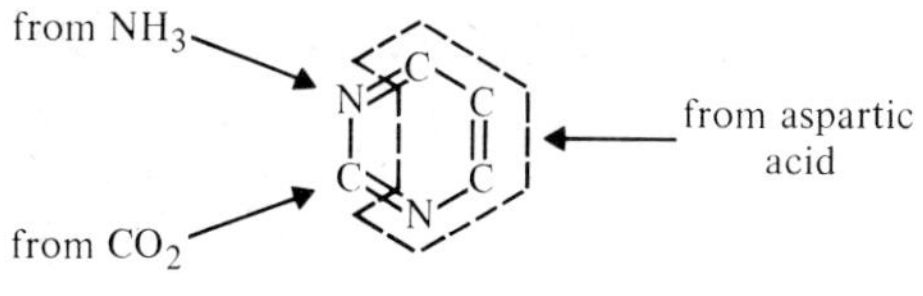

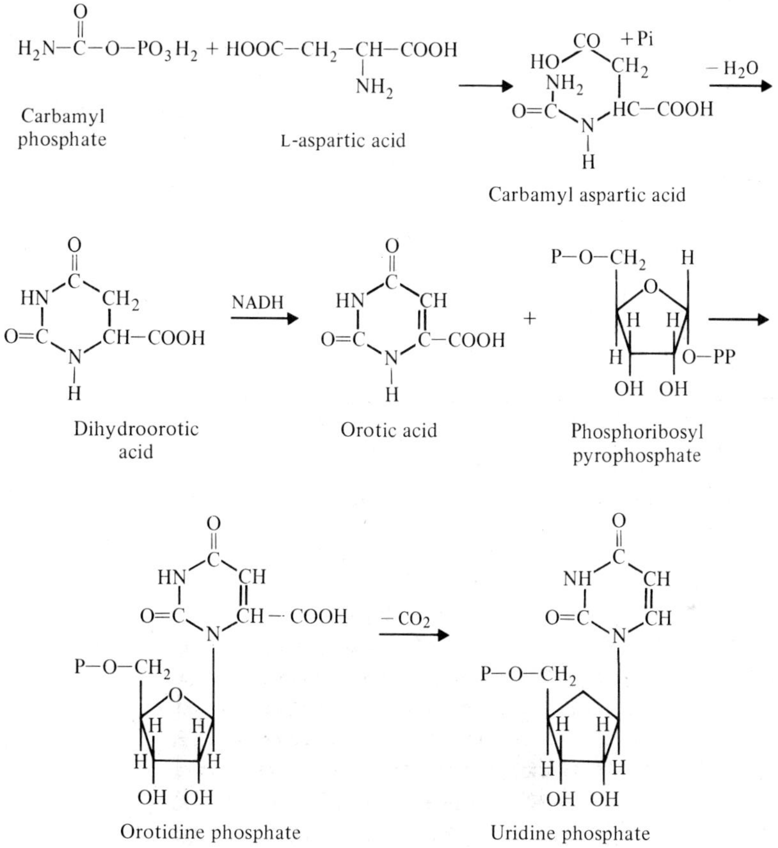

METABOLIC PATHWAY
FOR SYNTHESIS
OF PYRIMIDINE RING

FIGURE 8.13 Metabolic pathway for the biosynthesis of pyrimidines. Shown at the top is the source of atoms for the biosynthesis of the pyrimidine ring. Atoms come from aspartic acid, ammonia, and CO^2. The first base synthesized is uracil, which is produced as the nucleotide uridine phosphate. The other pyrimidine bases are produced through transformations of uridine phosphate.

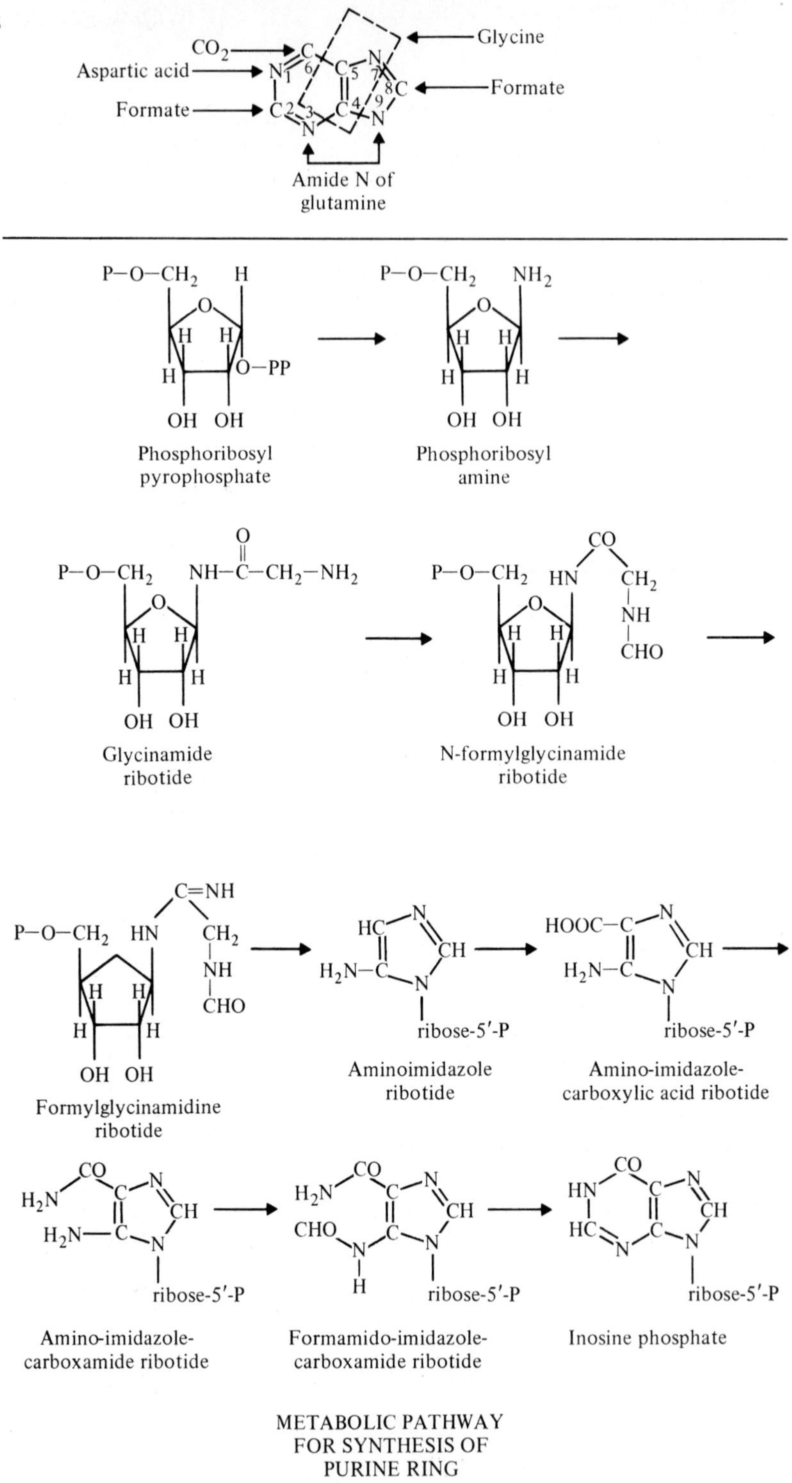

FIGURE 8.14 Metabolic pathway for the biosynthesis of purines. Shown at the top is the source of atoms for the biosynthesis of the purine ring. Atoms come from aspartic acid, formate, the amide nitrogen of glutamine, glycine, and CO_2. The first to be produced is the nucleotide phosphate, inosine phosphate. From this purine compound the other common purines are formed by transformations.

METABOLIC PATHWAY
FOR SYNTHESIS OF
PURINE RING

inosine phosphate by amination from the amino group of aspartate, and guanosine monophosphate is produced by amination from the amide group of glutamine. Like the pyrimidine derivatives, the deoxyribotides are synthesized by reduction of the C-2′ position of the ribose sugar moiety.

Review Exercises

8.1 Construct a diagram showing how nucleotides are linked in nucleic acids.

8.2 What would be the exact base sequences of DNA corresponding to the proper codons for leucine?

8.3 The ribosomes of chloroplasts and mitochondria are 70 S, being composed of 30 S and 50 S subunits. What do these S numbers mean and what information is gained from a knowledge of them? How are S numbers determined?

8.4 The following reaction is common in plant tissues.

$$PEP + ADP \rightleftharpoons PYR + ATP$$

On the basis of the free energies of hydrolysis of PEP and ATP, in which direction do you suppose the reaction usually proceeds?

8.5 The extinction coefficient for NADH is 6.2 A μmol^{-1}. If the decrease in absorbance (A) of the reaction

$$OAA + NADH \longrightarrow MAL + NAD$$

is 0.1 A min^{-1}, what would be the rate of the reaction in terms of μmol min^{-1}? The assay was conducted in a 1-cm cell.

References

ADAMS, R. L. P., R. H. DURDON, A. M. CAMPBELL, and R. M. S. SMELLIE (revisers). 1976. *Davidson's The Biochemistry of the Nucleic Acids.* 8th ed. Academic Press, New York.

BOGORAD, L., and H. H. WEIL (eds.). 1977. *Nucleic Acids and Protein Synthesis in Plants.* Plenum Press, New York.

BONNER, J., and J. E. VARNER (eds.). 1976. *Plant Biochemistry.* 3d ed. Academic Press, New York.

DILLON, L. S. 1973. The origins of the genetic code. *Bot. Rev.* 39:301–345.

DUDA, C. T. 1976. Plant RNA polymerases. *Ann. Rev. Plant Physiol.* 27:119–132.

GIVAN, C. V., and R. M. LEECH. 1971. Biochemical autonomy of higher plant chloroplasts and their synthesis of small molecules. *Biol. Rev.* 46:409–428.

GOODENOUGH, U., and R. P. LEVINE. 1970. The genetic activity of mitochondria and chloroplasts. *Sci. Amer.* 223:22–29.

HALL, T. C., and J. W. DAVIES (eds.). 1980. *Nucleic Acids in Plants,* Vols. I, II. Chemical Rubber Co. Press, West Palm Beach, Fla.

LOENING, U. E. 1968. RNA structure and metabolism. *Ann. Rev. Plant Physiol.* 19:37–70.

OOTA, Y. 1964. RNA in developing plant cells. *Ann. Rev. Plant Physiol.* 15:17–36.

PAYNE, P. I. 1976. The long-lived messenger ribonucleic acid of flowering plant seeds. *Biol. Rev.* 51:329–363.

POSSINGHAM, J. V., and R. J. ROSE. 1976. Chloroplast replication and chloroplast DNA synthesis in spinach leaves. *Proc. Royal Soc.* (London), Ser. B 193:295–305.

PRICE, H. J. 1976. Evolution of DNA content in higher plants. *Bot. Rev.* 42:27–52.

VAN'T HOF, J., and C. A. BJERKNES. 1979. Chromosomal DNA replication in higher plants. *Bioscience* 29:18–22.

WILSON, C. M. 1975. Plant nucleases. *Ann. Rev. Plant Physiol.* 26:187–208.

CHAPTER NINE

Amino Acids, Proteins, and Protein Biochemistry

Proteins are among the most important and interesting chemicals of living cells. The enzymes are proteins that regulate cellular metabolism. Each step of a metabolic pathway, as one chemical is transformed into another, is catalyzed by an enzyme, and it is believed that there are no chemical reactions that take place in living cells that are not regulated by one or more protein catalysts. As explained in the last chapter, the proteins are the ultimate translational products of ribonucleic acid. The linear base sequence of the messenger RNA is complementary to the linear base sequence of the DNA that makes up a gene (see Chapter 8). The methods by which DNA is transcribed to form RNA and RNA is translated to form polypeptides were discussed in the last chapter. The actual steps of polypeptide synthesis during translation and how the translation products (the polypeptides) are modified to form functional proteins are discussed here.

In addition, the mechanism of enzyme function is discussed in this chapter from the standpoint of the kinetics of the reaction and regulation of the reaction. Whereas the nucleic acids are directly involved in the regulation of protein synthesis, it is the proteins that regulate cellular metabolism. But even the enzymatic proteins are regulated by mechanisms involving their own products. There will be further discussion explaining just how these small metabolite molecules (which include products of reactions) regulate their own synthesis through feedback mechanisms.

Recently, it has become apparent that, in addition to genetic regulation of protein synthesis and the metabolite regulation of enzymic activity, much of intermediary metabolism is regulated by compartmentation of different metabolic pathways in different organelles of the cell. Thus the reactions of photosynthesis are kept separate from those of respiration by the former taking place in chloroplasts and the latter occurring in mitochondria or in the cytosol. However, in many of the metabolic pathways of plant metabolism there are common metabolic steps or sequences. The enzyme malate dehydrogenase occurs in respiration, in photosynthesis in some plants, and in dark CO_2 fixation. How is it that the same enzyme can function in such diverse metabolic sequences in different subcellular compartments? What regulates the synthesis and the localization of the enzyme? In partial answer to these questions, it has been discovered that even though the enzyme malate dehydrogenase is common to several metabolic pathways occurring in different cellular organelles, the actual proteins differ in properties. They have different kinetic properties and even have different amino acid sequences, giving strong evidence that they are translation products of different messenger RNA molecules. This also means that they are coded for by different genes. Such groups of enzymes that catalyze the same reaction but have different properties and are frequently localized in different organelles are called isozymes. Virtually all enzymes are known to exist as isozymes. Of course, some are just the result of heterozygote organisms,

and others are similar to the malate dehydrogenase example given above and function in different ways. Isozymes are discussed in more detail in this chapter.

As well as being catalysts, proteins serve additional important functions in the cell. Some are structural as components of membranes and cell walls. But unlike the structural protein of animals (collagen), most of the structural proteins of plants have regulatory roles as well. Most, if not all, of the proteins of cellular membranes are enzymes. Some, such as the cytochromes, are structural components of mitochondrial and chloroplast membranes and also function in electron transport. Others are transport proteins, giving specificity to transport across membranes. Still others simply act as catalysts for reactions that occur on the membranes.

Much of the protein of certain tissues such as that of seeds is storage. One of the best-known storage proteins is zein, a component of maize seed. Zein is perhaps best known to the plant physiologist because it is an important storage compound of nitrogen and carbon in maize. It has also been demonstrated to be one of the translation products of messenger RNA isolated from maize during *in vitro* protein studies with the protein-synthesizing wheat germ system discussed in the last chapter.

In order to appreciate enzymes and protein metabolism in general, it is necessary to have a good knowledge of the biochemistry of amino acids, the building blocks of the polypeptides. Many of the properties of the individual proteins are the result of the primary structure of the polypeptides. The primary structure is the amino acid sequence. Since each amino acid has different properties, their alignment and number in a polypeptide will govern many of the properties of the polypeptide. But in addition to giving primary structure to proteins, amino acids occur both free in the cell and as components of small peptides (ten or less amino acids) or other kinds of compounds. As an introduction to this chapter on proteins, the important properties of amino acids are discussed.

9.1 The Amino Acids

Amino acids are small, water-soluble organic compounds ubiquitous in living organisms. The two diagnostic functional groups are the amino group (—NH) and the carboxyl group (—COOH). Because of these two groups, and depending on the pH, amino acids may be negatively charged, positively charged, or neutral (no *net* charge). They may exist in the zwitterion form (refer to chapter 2). Amino acids may have sulfhydryl groups (—SH), hydroxyl groups (—OH), or amide groups (—CONH) as well as carboxyl and amino groups.

Most of the naturally occurring amino acids are of the L-amino type.*

$$
\begin{array}{c}
\text{COOH} \\
| \\
\text{NH}_2\text{CH} \\
| \\
\text{R}
\end{array}
$$

As far as is known, all the plant protein amino acids are of the L configuration. These amino acids may be dextrorotatory (+) or levorotatory (−), rotating plane-polarized light to the right or left, respectively.

Amino acids can be categorized (according to the properties of their R groups) as either hydrophilic or hydrophobic, or they can be grouped according to the more general properties of their R groups. Figure 9.1 illustrates the 22 commonly occurring amino acids that fall into the following six groups.

1. Neutral amino acids in which the R group is H (glycine), alphatic (alanine, valine, leucine, isoleucine), or contains a hydroxyl functional group (serine, threonine)

*The L designation refers to the geometric position in space of the functional group associated with the penultimate carbon relative to D-glyceraldehyde (see below). The α designation indicates that the amino group is on the α carbon, or C-2. The carboxyl is at the C-1 position.

$$
\begin{array}{c}
\text{CHO} \\
| \\
\text{HCOH} \\
| \\
\text{H}_2\text{COH}
\end{array}
$$

D-glyceraldehyde

NEUTRAL AMINO ACIDS

$$
\begin{array}{ccc}
\overset{\displaystyle \text{NH}_2}{\underset{\displaystyle \text{H}}{\text{H}-\text{C}-\text{COOH}}}
&
\overset{\displaystyle \text{NH}_2}{\underset{\displaystyle \text{H}}{\text{CH}_3-\text{C}-\text{COOH}}}
&
\overset{\displaystyle \text{NH}_2}{\underset{\displaystyle \text{CH}_3\ \text{H}}{\text{CH}_3-\text{CH}-\text{C}-\text{COOH}}}
\\
\text{L-glycine} & \text{L-alanine} & \text{L-valine}
\end{array}
$$

L-leucine

$$
\text{CH}_3-\text{CH}-\text{CH}_2-\overset{\text{NH}_2}{\underset{\text{H}}{\text{C}}}-\text{COOH}
$$

L-isoleucine

$$
\text{CH}_3-\text{CH}_2-\overset{\text{H}}{\underset{\text{CH}_3}{\text{C}}}-\overset{\text{NH}_2}{\underset{\text{H}}{\text{C}}}-\text{COOH}
$$

L-serine

$$
\text{HO}-\text{CH}_2-\overset{\text{NH}_2}{\underset{\text{H}}{\text{C}}}-\text{COOH}
$$

L-threonine

$$
\text{CH}_3-\overset{\text{H}}{\underset{\text{HO}}{\text{C}}}-\overset{\text{NH}_2}{\underset{\text{H}}{\text{C}}}-\text{COOH}
$$

SULFUR AMINO ACIDS

L-cysteine

$$
\text{HS}-\text{CH}_2-\overset{\text{NH}_2}{\underset{\text{H}}{\text{C}}}-\text{COOH}
$$

L-methionine

$$
\text{CH}_3-\text{S}-\text{CH}_2-\text{CH}_2-\overset{\text{NH}_2}{\underset{\text{H}}{\text{C}}}-\text{COOH}
$$

L-cystine

$$
\begin{array}{c}
\text{CH}_2-\text{S}-\text{S}-\text{CH}_2 \\
| \qquad\qquad\quad | \\
\text{H}-\text{C}-\text{NH}_2 \ \ \text{H}-\text{C}-\text{NH}_2 \\
| \qquad\qquad\quad | \\
\text{COOH} \qquad\ \ \text{COOH}
\end{array}
$$

FIGURE 9.1 Amino acids commonly found in proteins arranged according to the following groups: neutral amino acids, sulfur amino acids, acidic amino acids (and the amides), basic amino acids, and the aromatic and heterocyclic amino acids.

ACIDIC AMINO ACIDS

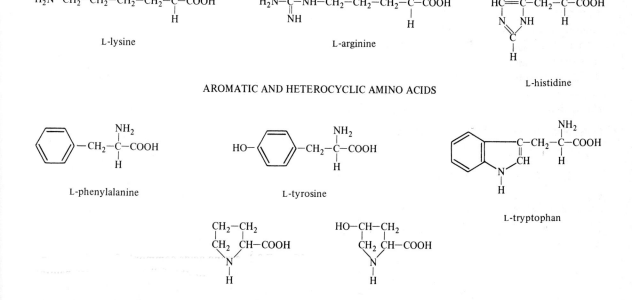

L-aspartic acid

L-glutamic acid

THE AMIDES

L-asparagine

L-glutamine

BASIC AMINO ACIDS

L-lysine

L-arginine

L-histidine

AROMATIC AND HETEROCYCLIC AMINO ACIDS

L-phenylalanine

L-tyrosine

L-tryptophan

L-proline

L-hydroxyproline

FIGURE 9.1 *Continued*

2. Sulfur amino acids containing sulfur groups (cysteine, cystine, methionine)

3. Acid amino acids in which the R group has a carboxyl, i.e., the dicarboxylic amino acids (aspartic acid, glutamic acid)

4. The basic amino acids (lysine, arginine, histidine)

5. Aromatic or heterocyclic amino acids (phenylalanine, tyrosine, tryptophan, proline) (proline is actually an imino acid with the N bonded to two carbons)

6. The amides of aspartic and glutamic acids (asparagine, glutamine)

In addition, hydroxyproline (γ-hydroxyproline) occurs in protein but is synthesized by hydroxylation of proline after proline is incorporated into protein.

These 22 amino acids making up the backbone of proteins are also found independent of protein free in the cytoplasm. Some exist as components of nonproteinaceous molecules, and many are important components of intermediary metabolism such as glycine, alanine, serine, and glutamic and aspartic acids and their amides.

In plants, there have been over 100 nonprotein amino acids isolated and identified. A few of the more common ones are illustrated in Fig. 9.2. Some of these are known to function in nitrogen metabolism (ornithine, citrulline, canavine) or in energetics (phosphoserine), whereas others such as pipecolic acid have no known function.

D-amino acids are known to occur free and in peptides of many microorganisms, but they are quite rare in higher plants. D-asparagine and D-tryptophan have been reported. A few non-α amino acids are found rather commonly in plants, such as β-alanine and γ-amino butyric acid. These are involved in intermediary metabolism.

There are several chemical reactions of the amino acids that are useful for either the qualitative or quantitative analysis of amino acids and proteins, but the most important for the plant physiologist is their reaction with ninhydrin (triketohydrindene hydrate). The α-amino acids react with oxidized ninhydrin to form α-imino acids and reduced ninhydrin.

Oxidized ninhydrin

The α-imino acid hydrolyzes and decarboxylates, forming CO_2, NH_3, and an aldehyde.

$$RC\!\!=\!\!NHCOOH + H_2O \rightarrow RCHO + CO_2 + NH_3$$

The free ammonia complexes with the oxidized and reduced ninhydrin to form a blue complex.

Blue product

The blue complex (or yellow with proline) is diagnostic for the amino acids and can be used in quantitative or qualitative tests.

Amino acids can be separated by paper (or thin-layer) chromatography and made visible by spraying the paper or plate with a solution of ninhydrin. Wherever an α-amino acid is, there will be a blue spot. In addition, solutions of amino acids can be

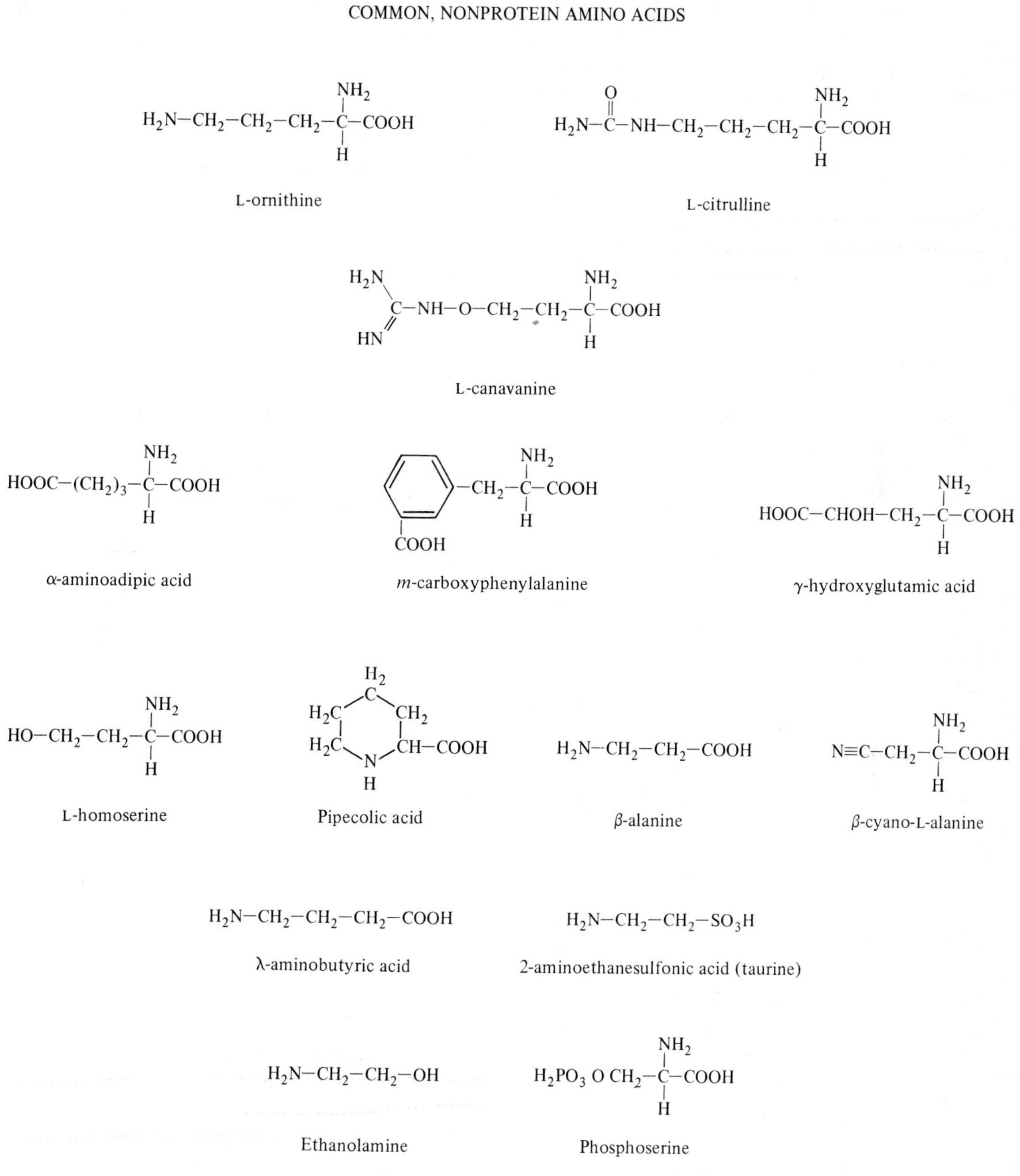

COMMON, NONPROTEIN AMINO ACIDS

$$H_2N-CH_2-CH_2-CH_2-\underset{\underset{H}{|}}{\overset{\overset{NH_2}{|}}{C}}-COOH$$

L-ornithine

$$H_2N-\overset{\overset{O}{||}}{C}-NH-CH_2-CH_2-CH_2-\underset{\underset{H}{|}}{\overset{\overset{NH_2}{|}}{C}}-COOH$$

L-citrulline

$$\underset{HN}{\overset{H_2N}{>}}C-NH-O-CH_2-CH_2-\underset{\underset{H}{|}}{\overset{\overset{NH_2}{|}}{C}}-COOH$$

L-canavanine

$$HOOC-(CH_2)_3-\underset{\underset{H}{|}}{\overset{\overset{NH_2}{|}}{C}}-COOH$$

α-aminoadipic acid

$$-CH_2-\underset{\underset{H}{|}}{\overset{\overset{NH_2}{|}}{C}}-COOH$$
COOH

m-carboxyphenylalanine

$$HOOC-CHOH-CH_2-\underset{\underset{H}{|}}{\overset{\overset{NH_2}{|}}{C}}-COOH$$

γ-hydroxyglutamic acid

$$HO-CH_2-CH_2-\underset{\underset{H}{|}}{\overset{\overset{NH_2}{|}}{C}}-COOH$$

L-homoserine

Pipecolic acid

$$H_2N-CH_2-CH_2-COOH$$

β-alanine

$$N\equiv C-CH_2-\underset{\underset{H}{|}}{\overset{\overset{NH_2}{|}}{C}}-COOH$$

β-cyano-L-alanine

$$H_2N-CH_2-CH_2-CH_2-COOH$$

λ-aminobutyric acid

$$H_2N-CH_2-CH_2-SO_3H$$

2-aminoethanesulfonic acid (taurine)

$$H_2N-CH_2-CH_2-OH$$

Ethanolamine

$$H_2PO_3\ O\ CH_2-\underset{\underset{H}{|}}{\overset{\overset{NH_2}{|}}{C}}-COOH$$

Phosphoserine

FIGURE 9.2 Some of the less commonly known amino acids found in plant tissues.

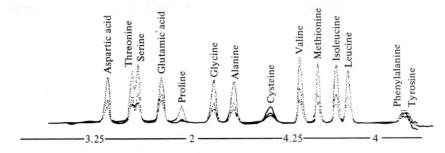

FIGURE 9.3 Tracing of chromatogram from an automatic amino acid analyzer. Amino acids are automatically separated on ion exchange columns and run through a ninhydrin reaction chamber. Whenever an amino acid appears it shows as a peak on the chromatogram. The area under each peak can be calibrated in order to quantify the amino acids. The automatic amino acid analyzer determines both the kinds and amounts of amino acids present.

measured quantitatively by reacting them with ninhydrin and measuring the amount of complex formed with a colorimeter.

Automatic amino acid analyzers have been developed in which the amino acids are separated by column chromatography, reacted with ninhydrin, and measured quantitatively. Figure 9.3 is a chromatogram from an amino acid analyzer in which a mixture of the common amino acids found in proteins has been resolved. Such amino acid analysis is common for the study of free amino acids in plants and for the determination of the amino acids in proteins. The amino acid composition of some seed proteins is shown in Table 9.1.

TABLE 9.1 Amino acid composition of some seed proteins expressed as grams of amino acid per 100 grams of nitrogen; cow's milk is presented as a comparison

AMINO ACID	BEANS	WHEAT	COTTON	MILK†
Histidine*	18	12	17	17
Isoleucine*	36	25	24	41
Leucine*	54	39	37	63
Lysine*	46	16	27	50
Methionine*	6	9	9	16
Phenylalanine*	35	29	33	31
Threonine*	27	17	22	29
Tryptophan*	6	7	7	9
Valine*	38	27	31	44
Alanine	36	20	27	22
Arginine	38	28	70	23
Aspartic acid	42	32	65	47
Cystine	6	13	10	6
Glutamic acid	100	182	114	149
Glycine	11	36	29	13
Proline	27	61	23	71
Serine	33	27	29	38
Tyrosine	24	22	17	33

*Amino acids essential for human nutrition
†Nonfat milk

Data from Altman and Dittmer (1968).

9.2 Peptides and the Peptide Bond

Amino acids are linked through a covalent bond between the α-amino group of one amino acid and the carboxyl group of another to form peptides. Two amino acids bonded through such a peptide bond is called a dipeptide, three a tripeptide, and so on. Long-chain polymers of amino acids are usually termed polypeptides. Since the average molecular weight of an amino acid is 100 to 110, even the small peptides have as many as 30 to 40 amino acid residues. Residue is a term referring to a constituent amino acid of a peptide.

The peptide bond is formed by dehydration between amino and carboxyl groups, as shown in Fig. 9.4. Polymerization through peptide bonds forms polypeptides and ultimately proteins. Peptide bonds are relatively stable but can be broken by acid- or base-catalyzed hydrolysis, which frees the individual amino acid residues.

A variety of small peptides ranging from two to twenty or more amino acid residues occur in nature, but these appear to be more prevalent in organisms other than higher plants. Perhaps the most well-known tripeptide of higher plants is glutathione,

γ-L-glutamyl-L-cysteinylglycine. Glutathione functions as a cofactor for many enzymic reactions. A number of glutamyl and polyglutamic acid polypeptides are known from higher plants. γ-L-glutamyl peptides of valine, isoleucine, leucine, methionine, phenylalanine, glycine, and alanine, among others, are known. Several of these peptides contain more than one glutamyl residue, although the polyglutamic acid polypeptides are more well known from bacteria.

Many of the better-known small polypeptides are of animal origin and function as hormones and growth regulators. For example, the polypeptides oxytocin and vasotocin, each having nine amino acid residues, are well-known animal hormones.

SOME COMMON SMALL PEPTIDES

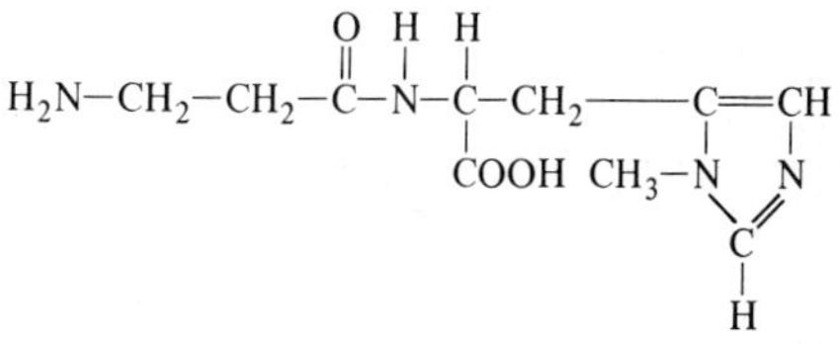

Anserine

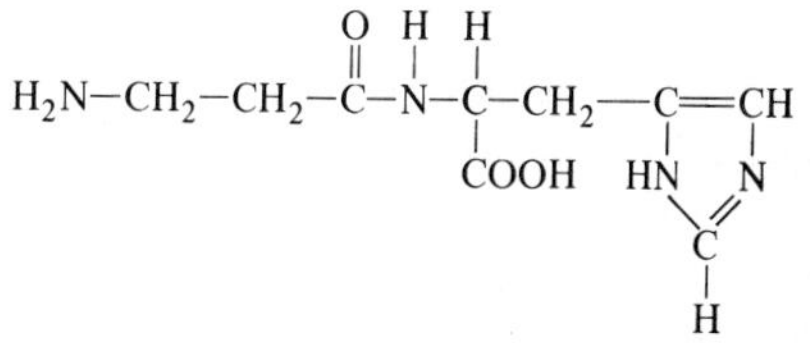

Carnosine

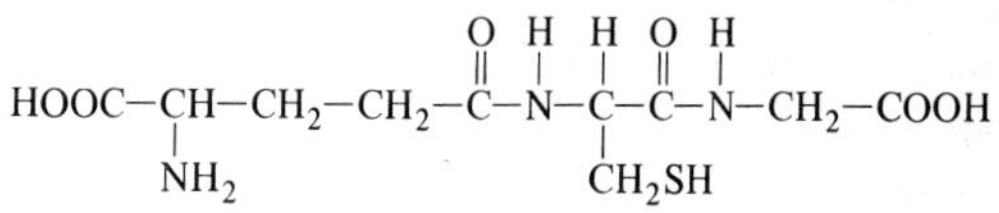

Glutathione

THE PEPTIDE BOND

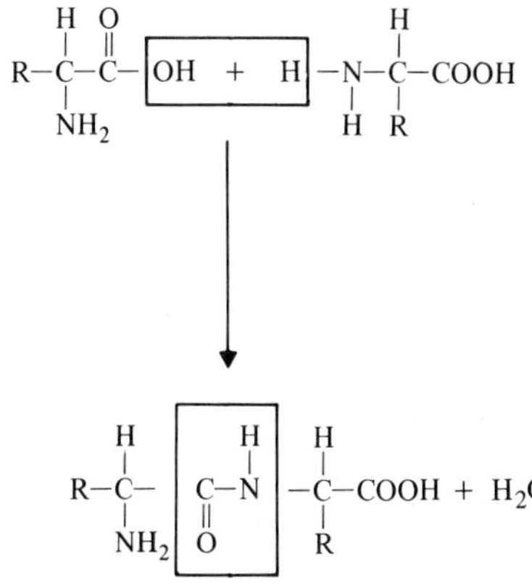

FIGURE 9.4 Diagram to illustrate the peptide bond formed between two adjacent amino acids.

FIGURE 9.5 Three small common peptides found in plant tissues.

And, of course, the large polypeptide insulin, with over 50 amino acid residues, is well known. Many of the antibiotics synthesized by microorganisms that are of therapeutic importance are small polypeptides such as penicillin, puromycin, gramicidin, the polymyxins, and streptomycin. Some of the fungi produce toxic polypeptides that are of importance to man, such as the ergot polypeptides from ergot and the toxins from the poisonous mushrooms of the genus *Amanita*. A few common peptides occurring in plants are illustrated in Fig. 9.5.

9.3 Proteins

The remainder of this chapter concerns the structure and function of proteins. As stated previously, the primary gene products resulting from the transcription of DNA sequences making up genes are ribonucleic acids. The special ribonucleic acids, the messenger RNAs, have the code for subsequent polypeptide assembly from amino acids. When attached to ribosomes, the messenger RNA−ribosome complexes called polysomes function in the translation of the code on messenger RNA and polypeptide assembly. The polypeptides, either after posttranslational modification or with no alterations, are the functional proteins of the cell. In this section, polypeptide and protein synthesis is discussed in more detail than in the previous chapter. In addition, the role of proteins in plant physiology is discussed.

9.3.1 The structure of proteins

The three-dimensional structure of a protein is a function of the kind, number, and sequence of amino acids in the polypeptide backbone and any nonprotein prosthetic groups attached to the polypeptide. The twisting, folding, and association of more than one peptide, all of which make up the functional protein, are for the most part largely predetermined by the amino acids. The overall structure almost always represents the lowest energy state of the molecule.

Four different organizational levels can be defined in the complete structure of a protein. The primary structure is defined as the number, kind, and sequence of the amino acids attached through peptide bonds. The secondary structure is defined as the conformation of the chain of amino acids resulting from the formation of hydrogen bonds between the nitrogen of the amino group and the oxygen of the carbonyl group of adjacent amino acids in the chain. Tertiary structure is a result of intramolecular interaction of the R-groups (side chains) of the amino acids in the chain that causes folding and bending of the polypeptide. To some extent, it is difficult to distinguish between the secondary and tertiary structures of the protein. Both are the direct result of the primary structure of the polypeptide. Quaternary structure is the aggregation of polypeptides.

In 1953 Frederick Sanger was the first to sequence a complex polypeptide, insulin. This was followed by the determination of the amino acid sequence of the enzymatic protein ribonuclease. The amino acid sequences of well over 1000 proteins are now known. The methods for sequencing a polypeptide and discovering the primary structure are very complex, but investigators are greatly aided by sophisticated methods and computer-assisted analysis. Basically, the procedure is to first separate the protein into component polypeptide chains if they exist and then to determine the actual sequence of the various amino acids.

First it is necessary to ascertain the total amino acid composition by (usually) using an automatic amino acid analyzer (see Fig. 9.3). Second, the N-terminal and C-terminal amino acid residues are normally identified by making specific derivatives, hydrolyzing the polypeptide, and determining which amino acids are labeled with the derivative-forming reagent. Although many reagents can be used, the first one used for determination of the N-terminal residue was 2,4-dinitrofluorobenzene, which forms yellow 2,4-dinitrophenyl derivatives of the amino acid at the N-terminal end. The C-terminal amino acid residue can be ascertained by means of either chemical reduction to an alcohol or enzymatic cleavage that specifically hydrolyzes at the C-terminal end. After the terminal amino acid residues are determined, the polypeptide chain is cleaved into small peptides by chemical means or by enzymatic means, using proteases. The component peptides are then separated and the amino acid composition and sequence of each is determined. On the smaller peptides, the sequencing is much easier; it is accomplished by means of selective hydrolysis of end groups and subsequent analysis of the freed terminal amino acids. Finally, by comparing overlapping sequences of the peptides the sequence of the entire polypeptide can be deduced.

Figure 9.6 shows the amino acid sequence of cytochrome *c* isolated from tomato. As can be seen from the figure, there are 111 amino acid residues. The amino terminus of the molecule (numbering is sequential from the amino terminus to the carboxyl terminus) is protected by an acetyl group.

As stated above, the secondary and tertiary structure of the polypeptide is dependent on the amino acid sequence. The polypeptide usually assumes a shape having the lowest energy state. The most typical secondary structure is the α-helix. The

helical shape results because the peptide bond is planar and because there are only two possible rotations about the N—C and the C—C=O bonds. The actual three-dimensional structure depends on the angles of these two bonds. In the α-helix, the angles are 132° (N—C) and 123° (C—C=O), resulting in 3.6 residues per turn of the helix, each with 13 atoms. Other helical arrangements are known, but evidently the α-helix is one of the most common, occurring in globular proteins, among others. The structure of the α-helix follows.

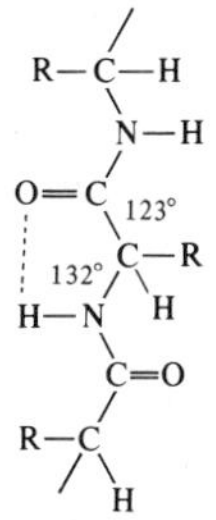

In the above structure the helix is maintained by hydrogen bonding between the nitrogen of the amino group and the oxygen of the carbonyl group of the adjacent amino acids.

Another form that can exist is the pleated-sheet structure that results from hydrogen bonding between two adjacent peptide chains. The arrangement can be either parallel, in which the adjacent chains run in the same direction, or anti-parallel, in which they run in opposite directions. A portion of

```
        1                                      10
Acetyl—Ala—Ser—Phe—Asn—Glu—Ala—Pro—Pro—Gly—Asn—Pro—Lys—Ala—Gly—Glu
               20                                      30
        Lys—Ile—Phe—Lys—Thr—Lys—Cys—Ala—Gln—Cys—His—Thr—Val—Glu—Lys
                              40
        Gly—Ala—Gly—His—Lys—Glu—Gly—Pro—Asn—Leu—Asn—Gly—Leu—Phe—Gly
               50                                      60
        Arg—Gln—Ser—Gly—Thr—Thr—Ala—Gly—Tyr—Ser—Tyr—Ser—Ala—Ala—Asn
                                    70
        Lys—Asn—Met—Ala—Val—Asn—Trp—Gly—Glu—Asn—Thr—Leu—Tyr—Asp—Tyr
               80                                      90
        Leu—Leu—Asn—Pro—Lys(Me)₃—Lys—Tyr—Ile—Pro—Gly—Thr—Lys—Met—Val—Phe
                                    100
        Pro—Gly—Leu—Lys(Me)₃—Lys—Pro—Gln—Gln—Arg—Ala—Asp—Leu—Ile—Ala—Tyr
               110
        Leu—Lys—Glu—Ala—Thr—Ala COOH
```

FIGURE 9.6 Amino acid sequence of tomato, cytochrome *c*. The residues are numbered from the NH_2 terminus to the COOH terminus. An acetyl group protects the amino terminus. From R. Scogin, M. Richardson, and D. Boulter. 1972. The amino acid sequence of cytochrome *c* (*Lycopersicon esculentum* Mill). *Arch. Biochem. Biophys.* 150: 489–492.

a parallel pleated-sheet structure is shown below.

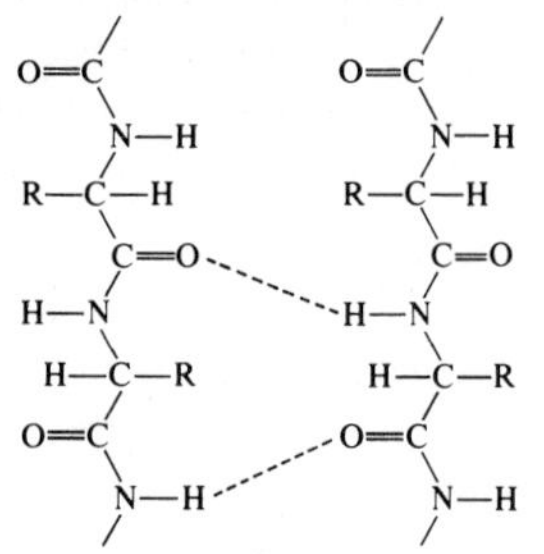

$$
\begin{array}{cc}
O{=}C & O{=}C \\
\quad N{-}H & \quad N{-}H \\
R{-}C{-}H & R{-}C{-}H \\
\quad C{=}O\cdots & \quad C{=}O \\
H{-}N & H{-}N \\
H{-}C{-}R & H{-}C{-}R \\
O{=}C & O{=}C \\
\quad N{-}H & \quad N{-}H
\end{array}
$$

The tertiary structure of a polypeptide is the result of intramolecular interaction of the R groups (side chains) of the amino acids. The most common interactions are formation of hydrogen bonds, ionic bonds, hydrophobic bonds, or disulfide co-valent bonds (see Fig. 9.7). Prosthetic groups and cofactors may aid in the formation of tertiary structure.

The three-dimensional structure of a polypeptide is ascertained by x-ray crystallography, as explained in Chapter 2. Of the many that are now known, only a few are from plants. Figure 9.8 shows the three-dimensional structures of two polypeptides when viewed as a stereoscopic pair. Cytochrome *c,* an enzyme of electron transport, has been studied widely from a number of different organisms in an attempt to ascertain evolutionary relationships on the basis of amino-acid-sequence similarities. The cytochrome *c* illustrated in Fig. 9.8 is from the photosynthetic bacterium *Rhodospirillum rubrum*. In addition, Fig. 9.8 shows the three-dimensional structure of a protein occurring in soybean that inhibits the protease enzyme trypsin.

Finally, the quaternary structure refers to the aggregation of polypeptides, each with its own primary, secondary, and tertiary structure. These aggregated units are frequently the functional proteins; however, all functional proteins will not necessarily have a quaternary structure. The mole-

SOME BONDING TYPES THAT OCCUR BETWEEN PEPTIDES

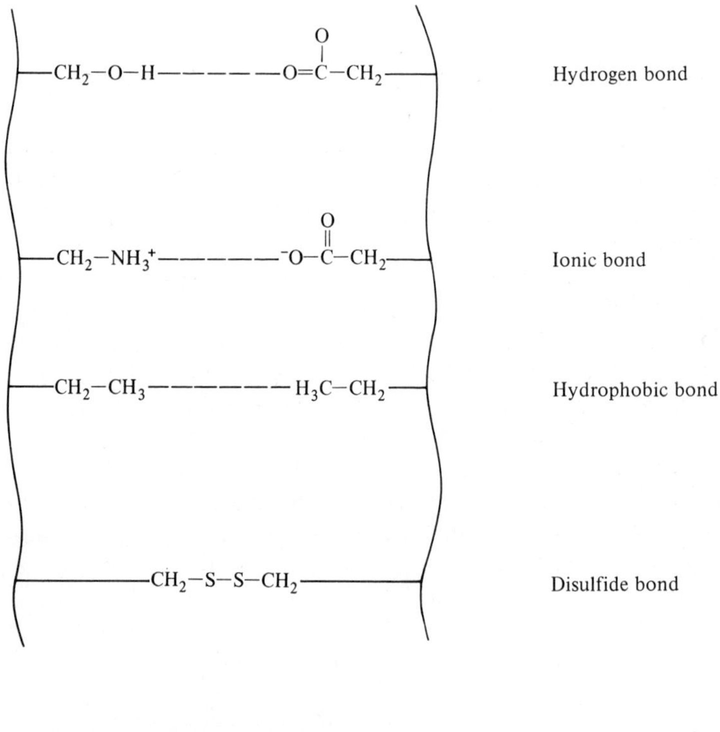

FIGURE 9.7 Various types of bonding that can occur between adjacent peptide strands that contribute to tertiary structure.

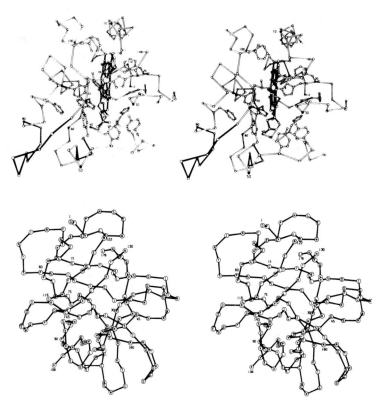

FIGURE 9.8 Stereoscopic pair drawings of two proteins, showing their three-dimensional structure. The upper pair is of cytochrome c_2 isolated from the photosynthetic bacterium *Rhodospirillum rubrum*. This particular cytochrome functions in electron transport during photosynthesis by *Rhodospirillum*. The dark portion of the structure of the center of the molecule is the prosthetic heme group. From F. R. Salemme, J. Kraut, and M. D. Kamen. 1973. Structural bases for function in cytochromes c. *J. Biol. Chem.* 248: 7701–7716. The lower pair drawing is a protein isolated from soybean that binds to and inhibits the activity of the protease enzyme trypsin. The loop at the upper left, made of residues 61 through 66, binds to trypsin. The figure is from D. M. Blow. In Margaret O. Dayhoff. 1976. *Atlas of Protein Sequence and Structure*, Suppl. 2. National Biomedical Research Foundation (Georgetown University Medical Center), Washington, D. C.

cules illustrated in Fig. 9.8 do not have a quaternary structure as such.

In the case of proteins with a definite quaternary structure, the individual polypeptide chains are called subunits and are held together by either hydrophobic or electrostatic forces and not by covalent bonds. Subunits making up the quaternary structure of proteins may be identical or dissimilar. For example, the plant protein peroxidase, with a mass of 40,000 daltons, is a single polypeptide chain. Malate dehydrogenase, on the other hand, is about 60,000 daltons and is composed of two identical subunits of 30,000 daltons each. The protein ribulose bisphosphate carboxylase, with a mass of about 568,000 daltons, is believed to have a quaternary structure of 8 large subunits, each with a mass of about 55,000 daltons, and 8 small subunits, each with a mass of 16,000.

A simple classification of plant proteins can be based on their size, solubility, and degree of basicity (number and kind of basic amino acids). These simple proteins can also be conjugated with other groups. Thus there are mucoproteins or glycoproteins with a carbohydrate moiety, lipoproteins with lipid, metalloproteins with metals, nucleoproteins with nucleic acids, and chromoproteins that have a pigment (chromophore) as a prosthetic group. Eight common classes are listed in Table 9.2.

Proteins can also be classified according to their tertiary or quaternary structure as either globular or fibrous. The globular proteins are frequently water-soluble and function in storage or as enzymes. The fibrous proteins exist as sheets or parallel polypeptides. Their main function is structural; they occur primarily in animals as components of skin, feathers, and integuments.

TABLE 9.2   A simple protein classification on the basis of size, solubility, degree of basicity, and conjugation with nonprotein groups

PROTAMINES	Small, water-soluble polypeptides with a high content of basic amino acids; frequently occur with acids, such as nucleic acids in the nucleus
HISTONES	Large, water-soluble proteins with a high number of basic amino acids; differ from the protamines largely on the basis of size
ALBUMINS	Common proteins of both plant and animals that are soluble in water and in dilute salt solutions; are readily precipitated by saturated ammonium sulfate solutions
GLOBULINS	Common proteins of plants and animals that are insoluble in water but soluble in dilute salt solutions; can be precipitated with half saturated ammonium sulfate solutions
GLUTELINS	Plant proteins that are insoluble in both water and dilute salt solutions but can be dissolved in acid and alkaline solutions
PROLAMINS	Plant proteins (largely found in seeds) that are soluble in 70% to 80% alcohol solutions but insoluble in water and salt solutions
ALBUMINOIDS	Animal proteins that are insoluble in most solvents
CONJUGATED PROTEINS	Simple proteins conjugated with a prosthetic group, including: mucoproteins with a carbohydrate moiety; lipoproteins with a lipid group; nucleoproteins conjugated to nucleic acids; phosphoproteins with a phosphate group(s); chromoproteins with a pigment (chromophore) as a prosthetic group; and metalloproteins with a metal group

9.3.2 Protein biosynthesis

Through the efforts of many research workers using procaryotic cells, the salient features of protein synthesis have been ascertained. Even though most of our knowledge has come from studies of organisms other than plants, sufficient investigations have been conducted to allow a coherent picture of protein biosynthesis in plants.

The information for protein synthesis is contained in the sequence of DNA. First the DNA is transcribed as a complementary base sequence of messenger RNA; it is then translated into the amino acid sequence of the primary structure of proteins by the biosynthetic mechanism of the cell. It is this latter translation of the base sequence of mRNA into the amino acid sequence of protein that is covered in this section.

The first important step in protein synthesis is the formation of the aminoacyl–tRNA complex from a specific tRNA and an amino acid. ATP energy is required. The transfer RNAs are specific for amino acid species. As explained in Chapter 8, there are multiple forms of tRNA for the various amino acids. For example, in higher plants there are six different tRNAs for leucine. Each will form a leucyl–tRNA complex and will insert leucine residues into growing polypeptide chains.

The formation of the aminoacyl–tRNA complex is shown in Fig. 9.9. First, the amino acid reacts with a specific aminoacyl synthetase enzyme to form an AMP derivative attached to the synthetase. The reaction liberates pyrophosphate from the ATP. Then the aminoacyl compound is transferred to the specific tRNA to form the active aminoacyl–tRNA complex plus AMP and free synthetase enzyme.

It is the tRNA, not the amino acid, that recognizes the specific codon on mRNA. Hence the tRNAs must have at least two recognition sites, one for the aminoacyl synthetase enzyme and one for the proper codon of mRNA.

Protein biosynthesis, from the formation of the ribosome–mRNA complex to amino acid poly-

FORMATION OF THE AMINOACYL-tRNA COMPLEX

$$NH_2-\underset{\underset{R}{|}}{\overset{\overset{H}{|}}{C}}-COOH + ATP + ENZYME \longrightarrow AMP-\overset{\overset{O}{\|}}{C}-\underset{\underset{R}{|}}{\overset{\overset{H}{|}}{C}}-NH_2-ENZ + P_iP_i$$

$$AMP-\overset{\overset{O}{\|}}{C}-\underset{\underset{R}{|}}{\overset{\overset{H}{|}}{C}}-NH_2-ENZ + tRNA\ C_pC_pA \longrightarrow tRNA\ C_pC_pA-O-\overset{\overset{O}{\|}}{C}-\underset{\underset{R}{|}}{\overset{\overset{H}{|}}{C}}-NH_2 + AMP + ENZYME$$

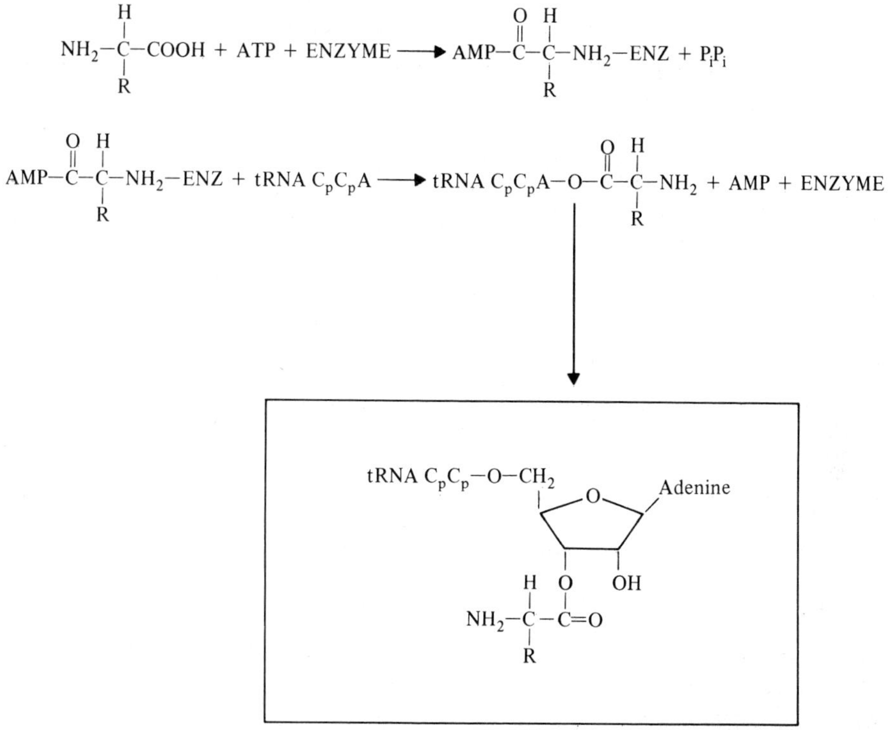

FIGURE 9.9 Formation of the aminoacyl-tRNA complex from an amino acid, ATP, enzyme, and tRNA. Each tRNA and enzyme is specific for an amino acid. The aminoacyl-tRNA complex then functions in peptide synthesis.

merization, is illustrated in Fig. 9.10. A first step is the attachment of 40 and 60 *S* ribosomes to mRNA to form the active polysomes. Chain initiation occurs by the attachment of a specific methionyl tRNA to the 3′-terminus AUG codon of the mRNA. It is worth noting here that in procaryotic cells, the chain initiator is a formylmethionyl tRNA. Chain polymerization begins with addition of an amino acid to the methionine by peptide bond formation. The amino group of the incoming amino acid and the carboxyl of the methionine form the peptide bond. In the procaryotic system mentioned above, formylation of the methionine amino acid apparently prevents initiation from the amino group. How this specificity is maintained in eucaryotes is

not known, but, as in procaryotes, chain polymerization goes from the amino terminus to the carboxyl terminus. Since the N-terminal methionyl residue does not occur in the completed polypeptide, it is removed prior to completion. All of the steps in the polypeptide chain polymerization are of course catalyzed by specific enzymes.

The incoming aminoacyl–tRNA complex binds at a specific ribosomal mRNA site called the A-site. After a peptide bond is formed, the synthesized peptidyl tRNA attaches to a site called the P-site. The migration of ribosomes is toward the 3′-terminus of mRNA. The next proper aminoacyl–tRNA molecule then attaches to the A-site of the ribosome as it moves to the next codon. Thus the

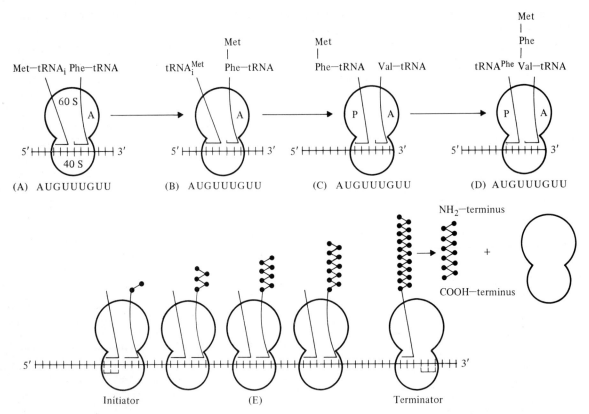

FIGURE 9.10 Model to show how proteins are synthesized on polyribosomes using tRNA's. The incoming Phe-tRNA binds to the A site of the ribosome that has a Met-tRNA at the initiator site. Peptide bonds are formed (B), and the new peptidyl-tRNA then migrates to the P site from the A site. The sequence is repeated until the peptide is completely formed, which is accomplished by reaching the terminator. From A. Marcus. 1976. Protein biosynthesis. In J. Bonner and J. E. Varner. *Plant Biochemistry*. Academic Press, New York.

ribosomes migrate, one codon at a time, from the 5' end of mRNA toward the 3'-terminus as polypeptide synthesis occurs. Because the mRNA codons are not altered by the process, a single mRNA molecule can participate in multiple polypeptide syntheses simultaneously.

Little is known about chain termination in plants. The final reaction is the addition of water (hydrolysis) to the carboxyl of the amino acid residue of the final aminoacyl–tRNA molecule to form the free carboxyl end and terminate polymerization. As of yet there are no known termination codons in plants, although terminator codons are known from other systems. Presumably, the methionyl

initiator residue is removed by a specific proteolytic enzyme for completion of the final amino acid sequence making up the primary structure.

The secondary, tertiary, and quaternary structures of protein are a function of the primary amino acid sequence. Of course, cofactors or other prosthetic groups will partially determine conformation of the final protein regardless of their nature. The fact that primary structure largely determines higher-order structure was shown through experiments in which proteins were denatured and then renatured. After denaturation, the original conformation is assumed when renatured. This is true even if covalent linkage is involved (for example, disulfide formation). Proteins that have several sulfhydryls always form the same disulfide bridges when renatured, indicating restrictive reactions. It is assumed that the higher-order structures—folding, bending, twisting, and other conformations—are largely governed by the lowest energy state of the finished molecule.

Some proteins are inactive when first synthesized and require further modification. Of course, nucleoproteins, lipoproteins, and glycoproteins require addition of their nonprotein prosthetic groups before completion. Others are synthesized in the form of inactive zymogens. These are frequently very large molecules that are reduced to smaller units by specific proteolytic enzymes. The proteolysis causes activation.

9.3.3 The regulation of protein biosynthesis

Relatively little is known about the regulation of protein synthesis in higher plants. In procaryotic cells such as bacteria, metabolite molecules are known to induce or repress the synthesis of enzymatic proteins at the transcriptional level; that is, they control protein synthesis by regulating mRNA synthesis. For example, when certain bacteria are grown in the presence of galactosides such as lactose, the enzyme β-galactosidase is induced. Thus the presence of the substrate lactose induces the synthesis of an enzyme that metabolizes it. In the terminology of cell biology, this kind of regulation of enzyme synthesis is called induction by substrate. In the presence of the amino acid tryptophan the synthesis of the biosynthetic enzyme, tryptophan synthetase, is inhibited or repressed. In the total absence of tryptophan, the enzyme is present and is said to be derepressed.

In the above examples, there is metabolite regulation of enzymes by their own substrates or products. With these regulated enzymes that are not present during all stages of the life cycle, substrates tend to induce and products tend to repress the synthesis of enzymes. The above kinds of metabolite regulation are at the level of transcription. There are other kinds known in which metabolites

regulate mRNA synthesis or enhance its breakdown.

Although the above regulatory mechanisms probably occur in higher plants, higher plant cells are usually not subjected to rapid changes in their chemical environment. In one case, however, nitrate will induce the enzyme nitrate reductase. The latter enzyme reduces nitrate to nitrite (nitrate reductase is discussed in Chapter 13). Thus plants respond in this case to the presence of a nutrient in the soil. Other examples undoubtedly occur, but probably much of the regulation of protein synthesis in plants is more complex than simple induction and repression.

There is substantial evidence that certain plant growth regulators such as the auxins, gibberellins, and cytokinins affect enzyme synthesis. Bonner and coworkers in 1962 proposed that the histone proteins associated with DNA were involved in the regulation of protein synthesis. Prior to transcription the histones are removed from specific DNA segments being transcribed. The histones are synthesized in the cytoplasm by conventional means. However, it is presently believed that histones are probably more involved in the structure of chromosomes and do not have much of a regulatory function.

9.3.4 Protein function

For convenience of discussion, proteins can be considered to have at least three primary functions in the life of the plant. Proteins are the structural components of membranes, they function in the storage of nitrogen and carbon, and they play a major role in plant metabolism as enzymes, the regulators of all cellular reactions. In addition to these major roles, proteins have a protective function at the structural level and in the immunological systems of animals. Their role in immunochemical reactions of plants is poorly understood except that immunochemical properties of proteins can be used effectively in preparation and analysis. In animals as well, some hormones are proteins (insulin, for example) and some function in movement as contractile proteins. In plants, the protein tubulin, which occurs as a component of microtubules, may also have a contraction role. The next section is a discussion of the structural, storage, and enzymatic role of plant proteins.

Structural and storage proteins

Intimately associated with membranes and cell walls are proteins. These proteins, giving structure to the plant cell particularly as components of membranes, do have other functions. As an example, the proteins of cell membranes probably all have an enzymatic function. Cytochrome *c,* one of the most thoroughly studied proteins, is an integral part of mitochondria and chloroplasts as a portion of the structure of the membranes and acting as one of the electron-transport proteins. Other proteins may function in aiding transport across the membrane.

The actual storage proteins found in seeds, tubers, and roots serve primarily as a store for nitrogen and carbon. These are rapidly consumed during the first stages of growth or regrowth. It is entirely possible that many have a catalytic or regulatory role in addition to storage. Any one protein should not be considered to have a single role in the metabolism of the plant, but more than one function (as defined here) is likely for many.

Two of the most common proteins in nature are the structural proteins of animals. Collagen makes up much of the structural protein of animals, and keratin is the hard, horny protein of appendages such as nails, horns, and shells. Such structural proteins do not exist in plants, but the protein extensin of cell walls is somewhat similar. Much of the wall structure in plants is carbohydrate—cellulose and hemicellulose.

MEMBRANE PROTEINS

As described in Chapter 1, plant cellular membranes are composed of lipid and lipoprotein. The lipids are largely glycolipids, phospholipids, sulfolipids, and isoprenoids in addition to the lipoprotein complexes. The glycoproteins are not as abundant in plant membranes as they are in animal membranes. The proteins associated with and making up the structure of the membrane can be classified into two distinct groups, depending on the ease with which they can be removed by solubilization (refer to Fig. 1.2). Peripheral proteins are those that associate with the outer surface of the membrane and that can be readily removed by washing in dilute salt solutions. Those proteins that are difficult to remove from the membrane and require harsh treatment such as detergent washing are called integral proteins. The actual structure of the membrane is in part a function of the integral proteins but not so much dependent on the peripheral proteins. The latter comprise 30% or less of the protein associated with membranes.

Because of the arrangement of the proteins and the lipids within and on the membrane, the membrane in both structure and function shows asymmetry. This asymmetry of the membrane means that the two surfaces have different properties and function differently. Therefore, it is possible for unidirectional transport to occur across the membrane and for different functions to occur on either side of the vesicles and organelles. Some of the properties of the two kinds of membrane proteins are discussed below.

Peripheral proteins

The peripheral proteins are loosely associated with the outer surface of the membrane, being bound largely by ionic attraction. These proteins are somewhat basic and have cationic sites that bind to negative sites of the lipoidal compounds of the membrane or to other proteins. There is evidence that some of the peripheral proteins are identical to the inner, more tightly bound integral proteins.

For the most part the peripheral proteins are small and water-soluble. They can be removed easily from the membranes by mild treatment with salt solutions or with solutions containing chelating agents. Their properties are similar to the common soluble proteins occurring in plants. It is generally assumed that they function at least in part as enzymes and perhaps as transport proteins.

There is evidence that the cytochrome c of the mitochondria is a peripheral protein because it is associated with the outer surface of the inner mitochondrial membrane. It is readily solubilized during the preparation of mitochondria. In addition, phytochrome, the chromoprotein that regulates morphogenesis (see Chapter 20), is known to be a peripheral protein of some membranes.

Integral proteins

Integral proteins are the most abundant proteins within the cell membrane, making up about 70% to 80% of the protein associated with the membrane. They are tightly bound and difficult to remove. Integral proteins are highly hydrophobic because they contain only small numbers of acidic and basic amino acid residues or because the tertiary structure is such that the acidic and basic amino acids are hidden within the molecule. Many of the integral proteins are lipoproteins. Because of the low number of cysteine residues, disulfide linkages are not common.

Binding of the integral proteins within the membrane structure is largely through hydrophobic bonds. Removal is rather difficult but can be done with strong detergents or protein-denaturing agents.

The integral proteins function partially as structural components and partially as enzymes. Most of the enzymes associated with membranes are integral proteins. Perhaps one of the most important in plants is the protein that occurs in the chlorophyll–protein complex of chloroplasts. This protein remains bound to chloroplasts after extraction and can only be removed by washing with detergents or other strong reagents.

CELL-WALL PROTEIN (EXTENSIN)

A glycoprotein rich in hydroxyproline is found as a structural component of plant cell walls. This protein, which is called extensin, has an arabinose oligosaccharide moiety evidently linked to the hydroxyl of the hydroxyprotein. Many other proteins are found associated with cell walls, but most are enzymatic rather than purely structural as envisioned for extensin.

STORAGE PROTEINS

Plant seeds contain an abundance of stored protein. In fact, the plant proteins are frequently divided into two groups, the seed proteins and the cytoplasmic proteins. These storage proteins that also occur in tubers and roots function as carbon and nitrogen sources during germination or regrowth and as a source of amino acids during new protein synthesis.

The storage proteins of cereals have been studied quite extensively because of their importance in human nutrition. Approximately 70% of the world's supply of protein for human consumption comes from plants; the remainder is from animals. Of the 70%, just about half is made available from cereal grains. The rest comes from the oilseeds such as pulses, nuts, and other sources. The pulses (dry, edible seeds of legumes) are mostly fed to animals. The most important cereals are wheat, rice, corn, barley, oats, rye, sorghum, and millet. They are about 10% protein on a dry-weight basis. The important oilseeds are soybeans, cottonseed, peanuts, and sunflower seeds.

The proteins of the seeds and other storage organs are mostly albumins, globulins, glutelins, and prolamins. The albumins and globulins are the enzymatic proteins, and the glutelins and prolamins are storage proteins. The latter two, which are the most predominant in seeds, are interesting because of their amino acid composition. Many of them are highly amidated such that aspartate and glutamate are replaced by asparagine and glutamine. In addition, many are rich in proline. But of more interest from the standpoint of human nutrition is that the prolamins are deficient in lysine and tryptophan, two amino acids essential for nutrition.* For this reason the prolamins of seeds have been studied in much detail.

---

*Those amino acids that cannot be synthesized by the human body to any appreciable extent are called essential amino acids. They are essential in the diet and include isoleucine, leucine, lysine, methionine, phenylalanine, threonine, tryptophan, and valine.

The seed prolamins are commonly named after the genus of plant in which they occur. Thus there is zein in corn, hordein in barley, avenin in oats, secalin in rye, but gliadin in wheat.

Zein, one of the most thoroughly studied prolamins, is the main storage protein complex of corn (*Zea mays*). Common corn is 40% to 50% zein and 20% to 30% glutelin. Because the prolamins have extremely low amounts of the essential amino acids lysine and tryptophan, and because the other groups found in seeds—albumins, globulins, and glutelins—have more, those seeds that are high in prolamins are a poor source of protein for human nutrition. Rice, for example, has about 80% glutelin and is for this reason a better source of protein, being quite sufficient in lysine and tryptophan.

There has been much effort to develop cereals with sufficient lysine. The opaque-2 gene of corn results in seed with low zein and thus high lysine. Breeding trials that introduce opaque-2 into new varieties result in seed that is enriched in lysine. Another example of plant breeding that results in progeny with a high lysine content is the cross between wheat and rye that forms the artificial genus *Triticale*. *Triticale* has a higher lysine content per unit of protein than the parent wheat.

The oilseeds are a better source of protein for human consumption since they are higher in protein on a dry-weight basis, frequently having as much as 20% to 25%. They have ample lysine but are deficient in the sulfur-containing amino acids, methionine and cysteine. Because the cereals usually contain sufficient amounts of the sulfur amino acids, a combination of cereal protein and oilseed protein is a better diet.

For the most part the storage proteins of cereals are small, with masses on the order of 30,000 to 40,000 daltons. The storage and reserve proteins of the noncereal, dicotyledonous plants such as the oilseeds are much larger and are on the order of 300,000 to 400,000 daltons.

The storage proteins of potato tubers have been studied. The potato tuber is low in protein, about 10% on a dry-weight basis. It is high in globulins called tuberin and low in albumins, glutelins, and prolamins. Because potatoes are high in globulins

and low in prolamins, lysine is sufficient. The globulins of potato tubers, however, are low in the sulfur amino acids.

The storage protein of the cereal grain is localized in the endosperm, whereas in the oilseeds the protein is found in the cotyledons. In both, the storage proteins are contained in specialized cellular organelles called protein bodies, or aleurone grains. These organelles have a single membrane and are packed with protein (refer to Fig. 11.6). Protein bodies have been identified in peanuts, cottonseed, soybeans, corn, wheat, rice, barley, beans, and peas. They are approximately 1 to 20 μm in diameter. The storage proteins are synthesized on polysomes of the cytoplasm and transported through the membrane of the protein body. In addition to the storage proteins, the protein body has associated with it hydrolytic enzymes such as phosphatases, proteases, and amylases. Evidently, during germination of the seed these enzymes hydrolyze storage compounds that are subsequently used in the germination process. The proteases hydrolyze the storage proteins of the protein body.

Enzymatic proteins

The enzymatic function of proteins is one of the most fundamental processes of living organisms. The enzymes regulate virtually all the chemical reactions that occur in the cell by virtue of their high specificity to their substrates. Because of this high specificity most enzymes will only catalyze reactions with one kind of substrate. Analogues of substrates are either ineffective as substrates or act as inhibitors because they will bind to the active site of the enzyme but cannot enter into the reaction. In 1833 Payen and Persoz noted that a preparation from malt would hydrolyze amylose (a component of starch) to simple sugar. Even though fermentation and digestion as processes were reasonably well understood, the finding of a thermolabile substance in malt extracts was the beginning of the science of enzymology. The substance they prepared, which is now named amylase by adding the suffix "ase" to the root of the substrate name (amylose), is a good example of an enzyme and also provides the basis for a systematic way of naming enzymes. In some cases there is an indication of the type of reaction catalyzed in the name of an enzyme. Malate dehydrogenase, for example, is an enzyme that catalyzes the dehydrogenation (oxidation) of malate.

There are now virtually hundreds of different enzymes known and fully described. Each catalyzes a different reaction in the cell. Any one cell may have several thousand different enzymes and each may be represented by thousands of molecules. A cell may have as many as a billion enzyme molecules. For regulation, however, along with the high substrate specificity of each enzyme, all are not necessarily active at the same time. Thus the enzymes in themselves are regulated. Introducing even more complexity is the fact that most enzymes exist in cells or tissues in different forms. These different forms of the same enzyme catalyze the same reaction but with different properties and perhaps at different rates. Enzymes that catalyze the same reaction but that differ in one or more properties are called isozymes. Isozymes are important for the regulation of metabolism.

In this section the properties of enzymes and isozymes will be discussed from the standpoint of their occurrence in the cell, their participation in metabolic reactions, the kinetics of their reactions, and the regulation of the reactions and of the enzymes themselves.

ENZYME CLASSES

There are a number of enzyme types distinguished on the basis of their catalytic function. Enzymes can catalyze hydrolyses, oxidations or reductions, group transfers involving additions or removal, and condensations. The Commission on Enzymes of the International Union of Biochemists has recommended systematic nomenclature for naming and numbering the enzymes. The major groups with some common examples are given in Table 9.3.

ENZYME KINETICS

In the study of catalysis by enzymes, it must be remembered that the enzyme is a true catalyst, only altering the *rate* of the reaction and not the *equilibrium position* of the reaction. In the reaction cata-

---

TABLE 9.3 The major groups of enzymes according to the Commission on Enzymes of the International Union of Biochemists

EC 1 OXIDOREDUCTASES: Oxidoreductases catalyze oxidation–reduction reactions in which (a) hydrogen is added or removed (dehydrogenases), (b) O_2 acts as an electron acceptor (oxidases and oxygenases), or (c) peroxide acts as an electron acceptor (peroxidases). Some examples that are discussed in the text are: alcohol dehydrogenase, malate dehydrogenase, cytochrome oxidase, nitrate reductase, peroxidase, and catalase.

EC 2 TRANSFERASES: Transferases catalyze transfers of groups such as amino, phosphate, and acetate. Some examples are: aspartate aminotransferase, glucokinase, pyruvate kinase, and adenylate kinase.

EC 3 HYDROLASES: As the name implies, hydrolases catalyze hydrolysis by water addition. Some examples are: chlorophyllase, phosphatase, amylase, cellulase, glucosidase, papain, urease, ATPase.

EC 4 LYASES: Lyases catalyze removal or addition of groups not involving water. Some important examples are: phenylalanine ammonia lyase, phosphoenolpyruvate carboxylase, aldolase, and isocitrate lyase.

EC 5 ISOMERASES: Isomerases catalyze isomerizations and include glutamate racemase, mannose isomerase, and phosphoglycerate phosphomutase.

EC 6 LIGASES: Ligases are condensing enzymes; they form bonds using the energy of ATP or its equivalent. Three examples are: asparagine synthetase, pyruvate carboxylase, and acyl-CoA synthetase.

For further information and for details consult Dixon and Webb (1964) and Boyer (1959–1976).

lyzed by malate dehydrogenase,

Oxalacetate + NADH $\rightleftharpoons$ Malate + NAD,

the equilibrium position, defined by the value of K,

$$K = \frac{[\text{Malate}]\ [\text{NAD}]}{[\text{Oxalacetate}]\ [\text{NADH}]},$$

is a constant provided that conditions that alter K do not change. The equilibrium position is defined at a specified pH, temperature, and ionic strength. The enzyme alters the rate at which the reaction proceeds, usually increasing it. Furthermore, the enzyme is neither destroyed nor consumed during the reaction and thus is a true catalyst.

A second property of enzymes is that they are highly specific for the substrates of the reactions. In one study with malate dehydrogenase by Benveniste and Munkres in 1970, the reaction with L-aspartate was only 1.5% of that with the natural substrate, L-malate, in the direction of oxidation of malate to oxalacetate. L-aspartate differs from L-malate only in that the hydroxyl of carbon-2 is replaced with an amino group. The double-hydroxy analogue of malate, mesotartrate, was oxidized at a rate that was 18% of L-malate. To further emphasize the great specificity, the isomer of L-malate, D-malate, was a completely ineffective substrate. This property of substrate specificity is important in cellular regulation of a vast number of reactions that take place in cells simultaneously.

The dehydrogenases such as malate dehydrogenase are highly specific for the oxidation–reduction substrates NAD and NADH. This particular enzyme is not reactive if NADP replaces NAD or if NADPH replaces NADH. There are several synthetic analogues of NAD that can be used to study the catalytic properties of the dehydrogenases. Commercially available analogues include the thionicotinamide analogue (TN-NAD), the deamino analogue (DA-NAD), and the pyridine aldehyde analogue (PA-NAD). In an experiment by Victor Rocha in 1971, it was found that a malate dehydrogenase of spinach-leaf mitochondria showed 21%, 28%, and 30% of the activity with the analogues TN-NAD, DA-NAD, and PA-NAD, re-

spectively, in comparison with the naturally occurring NAD. In addition, the 3-acetylpyridine analogue showed three times as much activity as did NAD. Evidently, the naturally occurring substrates are not necessarily the best when defined in terms of rapidity of reaction.

The specificity with respect to cofactors may or may not be as high as with the actual substrates. The enzyme P-enolpyruvate carboxylase that catalyzes the reaction

P-enolpyruvate + HCO_3^- $\rightarrow$ Oxalacetate + P_i

requires a metal ion as a cofactor. There is no activity in the absence of a divalent or trivalent metal ion. Monovalent metal ions such as potassium or sodium are ineffective. In an experiment by S. K. Mukerji (1971), magnesium (Mg^{+2}) was the best cofactor with the predominant P-enolpyruvate carboxylase isolated from cotton-leaf tissue. The following ions were also active but to a lesser degree: Mn^{+2} (97%), Co^{+2} (80%), Fe^{+3} (39%), Zn^{+2} (36%), and Al^{+3} (12%).

Although there are some examples of enzymes with little specificity to substrates, the rule is that there is high specificity. This high specificity is the result of the interaction of the substrate with the active site of the enzyme. The active site is defined as that portion of the enzyme that is actually involved in catalysis. It can be considered to be composed of two parts: a binding site or sites where the substrates and cofactors bind to the enzyme, and an actual catalytic site where the reaction occurs. The binding site is usually made up of a number of amino acids, whereas the catalytic site is composed of one or only a few amino acids. The geometry of the active site is a function of the entire protein and any prosthetic groups and is, therefore, ultimately dependent on the primary structure of the protein.

Since the binding of substrates to active sites is rather weak, the study of substrate–active site interaction is rather difficult, although some substrate analogues will bind tightly, allowing determination of the amino acid composition of the site. Frequently, reagents are used that will bind at the active site and prevent catalysis. Identification of the amino acid or acids involved in the binding gives information about the active site. The most common amino acids at the active sites of enzymes are cysteine, lysine, aspartate, histidine, and serine. As examples, the two proteases (enzymes that hydrolyze proteins into component amino acids) caseinase and phaseolin, isolated from *phaseolus,* have serine at the catalytic site. Papain, a protease of papaya, has cysteine at the active site. Many of the dehydrogenases such as malate dehydrogenase have cysteine at the catalytic site.

Exactly how an enzyme catalyzes a reaction is not known. It is generally believed that the substrate fits into the active site somewhat the way a key fits into a lock, with great specificity. Such a view is certainly far too simplistic. Nevertheless, the geometry of the active site and the three-dimensional structure of the substrate determine binding to the amino acids at the binding site. Distortion of the substrate due to the binding increases the free energy of the substrate, making it more reactive. If the reaction involves more than one substrate, such as water in a hydrolysis or a pyridine nucleotide (NAD or NADP) in an oxidation–reduction reaction, these may also bind at the binding site. After the reaction the products leave the site and the enzyme is unchanged.

Many enzymes have a site or sites other than the active site that bind molecules. Such sites are called allosteric sites and may bind either metal ions or metabolites. The binding of a molecule at the allosteric site will cause conformational changes in the enzyme molecule that may in turn alter the rate of the reaction. These small metabolite molecules that bind to allosteric sites and alter the rate (kinetics) of the enzyme reaction are called effectors. Effectors may enhance the reaction (positive effectors) or inhibit the reaction (negative effectors). The point here is that the allosteric sites that bind effectors are as specific in binding as are the active sites. Allosterism is discussed below in more detail.

The protein part of the enzyme is referred to as the apoenzyme. In conjunction with a prosthetic group (nonprotein component), which may be a carbohydrate, lipid, metal ion, etc., the holoenzyme is formed. The concept of an entire functional enzyme with a protein component and a nonprotein component is useful for understanding enzyme function. Some enzymes, of course, do not have prosthetic groups.

From the standpoint of understanding the metabolism of the cell and ultimately growth and development, one of the most important features of enzyme-catalyzed reactions is the kinetics of the reaction. Kinetics can be defined as the rate of change of the reaction (as substrate goes to product) and the mechanisms that alter the rates. A knowledge of the kinetic parameters of an enzyme reaction gives much useful information that can be used to make predictions about what occurs in the cell with respect to biochemistry. The pH optima, the substrate concentration necessary to bring about a reaction, whether there is any substrate inhibition of the reaction at high substrate concentrations, whether the products of the reaction inhibit or reverse the reaction, whether there are any metabolites that influence the reaction, and how temperature affects the rate are all factors that govern intermediary metabolism. In the section to follow there is a discussion of enzyme kinetics that will be useful for the study of metabolism.

Factors affecting enzyme kinetics

As mentioned above, the rate of an enzyme reaction is governed by the concentration of the enzyme protein, the concentration of substrates, products, and cofactors, the presence of activators or inhibitors (effectors), temperature, pH, and ionic strength of the medium.

PROTEIN CONCENTRATION   Provided that all other factors are optimum, an enzyme reaction rate should increase linearly with an increase in the amount of enzyme. In fact, linearity of the reaction with increasing protein concentration is one of the first criteria used to indicate that a reaction is enzyme-catalyzed.

Enzyme reaction rates are frequently expressed

on the basis of specific activity.* Most frequently the activity is that of protein. Reactions can be expressed as micromoles (μmol) of substrate consumed or product produced per minute per milligram protein. In a freshly prepared preparation from a plant there will be many proteins, and the protein base does not refer just to the protein of the enzyme being measured but to all protein in the reaction mixture. When the enzyme is purified, it will become enriched with respect to protein still present, and the specific activity will increase.

Of course, specific activity can be expressed on bases other than protein-related. For example, reaction rates of plant enzymes can be expressed on a chlorophyll basis, a fresh-weight basis, a dry-weight basis, or even on a nucleic acid basis. The most common are protein, chlorophyll, and fresh-weight.

If a pure enzyme protein is available, a turnover number can be determined. The turnover number is the number of substrate molecules reacted per molecule of enzyme per unit time, or moles of substrate per moles of enzyme per unit time. Usual turnover numbers are 2000 to 10,000 per minute. If we assume an average turnover number of 6000 per minute, there will be about one reaction every 0.01 second.

TEMPERATURE Effects of temperature on enzyme kinetics are complicated because not only are chemical reactions temperature-dependent, but also the enzyme proteins are denatured by high (and, occasionally, low) temperatures.

Usually, as temperature increases the increased thermal activity causes a greater reaction rate, but eventually, accompanying the increased temperature is a denaturation of the protein. Thus the usual temperature-dependent enzymic catalysis curve is a modified bell-shaped curve. Figure 9.11 illustrates the net effect of the temperature-dependent in-

*The Enzyme Section of the National Research Council on Biological Chemistry suggests that enzyme activity be expressed in terms of International Units (I.U.); an I.U. is 1.0 micromole of substrate reacted per minute. Specific activity would be I.U.'s per milligram of protein.

crease in the rate of the reaction and the denaturation of the protein at high temperatures.

Enzymes vary in their temperature optima, i.e., the temperature at which the net rate is at a maximum. Enzymes usually have maximum activity at about 35°, but plants from cold habitats or hot environments may have proteins with vastly different thermal stabilities.

pH AND IONIC STRENGTH Because proteins are linear sequences of amino acids that have a variety of ionic functional groups, pH will alter the ionic properties of enzymes. Functional groups associated with proteins include sulfhydryls (—SH), amino (—NH), hydroxyl (—OH), imino (—NH—), and carboxyl (—COOH), among others. Because of varying pK's associated with these groups, pH will affect their ionization differently. Furthermore, the conformation of pro-

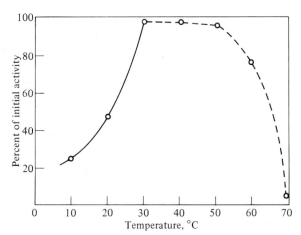

FIGURE 9.11 Graph to illustrate the effect of temperature on enzyme activity. The left curve (solid) shows how activity increases with an increase in temperature. The right curve (dashed) shows the denaturation of protein and loss of enzyme activity when incubated at the given temperatures for a few minutes. The combination of the increase in activity as a function of temperature and the denaturation at high temperature results in a bell-shaped temperature-dependence curve. From 1974 data of Russell Curry.

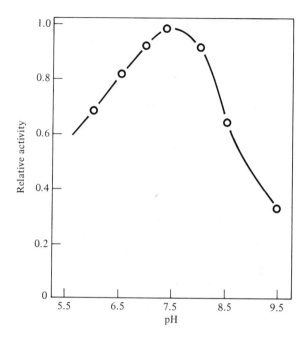

FIGURE 9.12 Relative activity of malate dehydrogenase as a function of pH. The enzyme activity reaches a maximum at a pH of about 7.5. From 1970 data of Victor Rocha.

teins based on their secondary, tertiary, and quaternary structures is in part determined by ionizable functional groups. Ionic strength and pH will, therefore, have marked effects on enzyme structure and on reaction rates. In addition, many substrates, products, effectors, and cofactors are altered by pH and ionic strength.

For these reasons, pH and ionic strength will alter enzyme rates. A typical pH-dependent rate curve is shown in Fig. 9.12. Because of the structural features of enzymes mentioned above, different enzymes will have different pH optima, and ionic strength may either increase or decrease the catalysis. Some enzymes, such as the hydrolytic enzymes including some proteases and phosphatases, have very acidic pH optima; others are more active at alkaline pH's. Acid phosphatases are most active at about pH 5 and alkaline phosphatases are more active at about pH 9. Most enzymes, however, have maximum activity near neutrality in the pH 6–8 range.

It is not pH per se that affects enzymic activity but rather the concentration of protons. Protons, as well as altering enzyme and sometimes substrate structure, may also enter into the reaction as substrates or as products. In these cases the concentration of protons will directly alter the rate of the reaction.

Although the cell tends to be highly buffered and resists changes in pH during metabolism, some cellular compartments differ from each other in pH. For example, when illuminated the chloroplast pumps protons from the outside of vesicle membranes to the inside. The outside becomes alkaline with respect to the inside. As explained in Chapter 15, this proton gradient across the chloroplast membranes is the driving force for the phosphorylation of ADP to form ATP. Changes in pH that occur during phosphorylation will alter the activity of those enzymes that are sensitive to pH. Similarly, the mitochondria establishes a proton gradient during electron transport and phosphorylation. With the mitochondria, however, the gradient is in the opposite direction; the outside becomes more acid with respect to the inside. In any case, it is quite clear that during active metabolism in the cell the concentration of protons will vary with time and within the various subcellular compartments. These changes in proton concentration, measured by pH, will have a profound effect on enzymic activity and will influence metabolism.

Michaelis–Menten kinetics

As explained above, the kinetic parameters of an enzyme reaction are fundamental properties that must be known to evaluate intermediary metabolism. In order to fully appreciate the kinetics of a reaction and be able to interpret reaction rates, substrate concentration optima, inhibition, and other aspects of enzyme function in a quantitative manner, it is necessary to be able to describe the rate of the reaction mathematically. Michaelis and his coworkers in 1913 developed a simple rate expression that is useful to describe the rate of an enzyme reaction as a function of substrate concentration. The rate expression

$$v = \frac{V \cdot [S]}{K_m + [S]}$$

is adequate to describe an enzyme reaction with a single substrate or one with more than one sub-

strate if the others are constant and at saturating levels. The equation describes a rectangular hyperbola. In the above expression,

$v =$ the rate of the reaction, for example in μmol s^{-1};

$V =$ the maximum rate obtained when the substrate concentration [S] is high enough to saturate the reaction;

$[S] =$ the substrate concentration in molarity that gives rate v; and

$K_m =$ the Michaelis constant, which is specific for the enzyme being studied and constant under uniform assay conditions. It is equal to the substrate concentration [S] when the rate of the reaction (v) is exactly equal to ½ of the maximum rate (V).

A typical rate curve that is described by the Michaelis–Menten expression is shown in Fig. 9.13. Much understanding can be obtained about enzyme catalysis by analyzing the rate curve shown in the figure with the rate expression.

Note that initially the reaction is nearly linear with

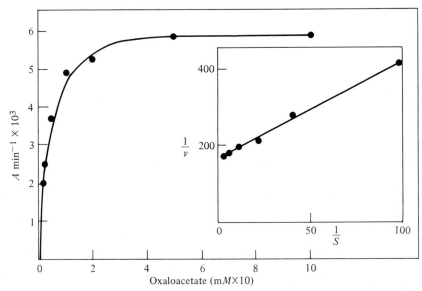

FIGURE 9.13   Hyperbolic rate curve for malate dehydrogenase with oxalacetate as the substrate. The graph is the enzymatic rate on the y-axis, expressed as change in absorbance, as NADH goes to NAD (measured at 340 nm with a spectrophotometer and expressed as A min^{-1}=absorbance per minute) against the concentration (mM) of substrate on the x-axis. NADH is held constant and at optimum concentration. The curve shows the typical saturation kinetics of many enzymatic reactions. The inset is a double-reciprocal plot of the data. From 1970 data of V. Rocha and I. P. Ting.

increasing substrate concentration. We can visualize that at low substrate concentration, when the number of substrate molecules is low relative to the number of enzyme molecules, increasing the substrate concentration will increase the reaction rate proportionally. Analytically, [S] will be low relative to K_m, and the sum of [S] plus K_m will be approximately equal to K_m. Under these conditions, the reaction will be a function of [S] alone:

$$v = \frac{V \cdot [S]}{K_m} \, ,$$

since V and K_m are constants. When [S] is low, the reaction will show first-order kinetics, that is, the rate will be dependent on [S].

When the substrate molecules are abundant with respect to enzyme (in other words, when the enzyme is saturated), increasing the number of substrate molecules will not affect the already saturated enzyme. Zero-order kinetics will prevail; i.e., the reaction rate will be independent of [S]. Analytically, [S] will be large relative to K_m, and the sum of [S] plus K_m will be approximately equal to [S]. Thus the expression reduces to

$$v = V.$$

That K_m is the concentration of substrate when the rate is ½ of the maximum can easily be shown. Let $v = \frac{1}{2} V$, and thus

$$\tfrac{1}{2} V = \frac{V \cdot [S]}{K_m + [S]}$$

$$\tfrac{1}{2} K_m + \tfrac{1}{2} [S] = [S]$$

$$K_m = [S].$$

The constants K_m and V can be evaluated either analytically or graphically. Perhaps the easiest

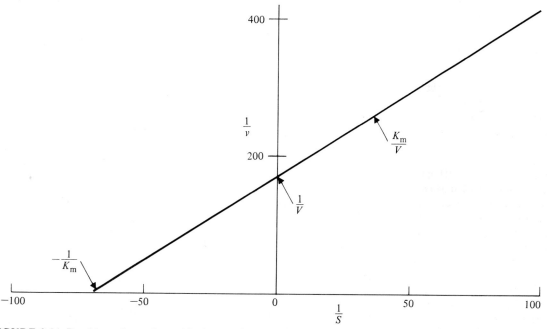

FIGURE 9.14 Double-reciprocal graphical analysis of the data shown in Fig. 9.13. According to the double reciprocal of the Michaelis–Menten equation, the slope of the line is the ratio of the K_m to V, the y-axis intercept is the reciprocal of V, and the negative x-axis is the reciprocal of the negative of K_m. Computation from this curve of K_m and V gives $K_m = 0.014$ mM and $V = 0.0059$ A min^{-1}.

method without access to a computer is a graphical analysis. A simple method is the Lineweaver–Burk graphical analysis in which the rate expression is transformed to the double-reciprocal form,

$$\frac{1}{v} = \frac{K_m}{V} \cdot \frac{1}{[S]} + \frac{1}{V},$$

and $1/v$ is plotted against $1/[S]$. This transformation converts the expression into the form of a straight line:

$$y = mx + b.$$

Here the slope (m) of the graph, shown in Fig. 9.14, will equal K_m/V, and the y-intercept (b) will be $1/V$. The reciprocal of b will be the maximum velocity, V, and the ratio of the slope of the y-intercept, m/b, will be $(K_m/V) \div (1/V)$, an estimate of K_m.

Note also that the x-axis intercept marks $-1/K_m$, and K_m can be evaluated by taking the reciprocal of the x-axis intercept and multiplying by -1.

Representative K_m's of some common enzymes are given in Table 9.4. It is worthy of note that the K_m's of enzymes, provided that they are measured under comparable conditions, are remarkably similar for the same enzyme isolated from different species.

The derivation of the Michaelis-Menten equation

The Michaelis–Menten expression for a single-substrate reaction can be derived rather easily with a few simple assumptions. If we assume that there is an enzyme–substrate complex, ES, formed that breaks down to product and free enzyme, the following will describe the sequence.

$$E + S \underset{k_2}{\overset{k_1}{\rightleftharpoons}} ES \overset{k_3}{\rightarrow} E + P$$

Here we will assume that the formation of the enzyme–substrate complex is reversible and that product formation is a nonreversible reaction. If we further assume that the breakdown of the ES complex is the rate-limiting step, then the rate of the enzyme reaction will be determined by the concentration of ES (i.e., [ES]). A rate expression for

TABLE 9.4   **Michaelis constants (K_m) for a few representative plant enzymes**

ENZYME	SUBSTRATE OR COFACTOR	K_m (mM)
P-enolpyruvate carboxylase (corn)	P-enolpyruvate	0.4
	Bicarbonate	0.02
(PEP + HCO$_3^-$ $\overset{Mg}{\rightarrow}$ OAA + P$_i$)	Magnesium*	0.5
Malate dehydrogenase (spinach)	Oxalacetate	0.06
(OAA + NADH → MAL + NAD)	NADH	0.02
(MAL + NAD → OAA + NADH)	Malate	0.8
	NAD	0.2
Malate enzyme (cactus)	Malate	0.15
	NADP	0.008
(MAL + NADP $\overset{Mn}{\rightarrow}$ PYR + CO$_2$ + NADPH)	Manganese*	0.02
Isocitrate dehydrogenase (corn)	Isocitrate	0.003
(ISOCIT + NADP → OXSUC + NADPH)	NADP	0.007

PEP = P-enolpyruvate, **OAA** = oxalacetate, **MAL** = malate, **PYR** = pyruvate, **ISOCIT** = Isocitrate, **OXSUC** = oxalsuccinate

*cofactors

All determinations are from student experiments.

the decrease in [ES] with time $(-d[ES]/dt)$ will describe the reaction. Note here that the rate of disappearance of ES will be proportional to the concentration of ES and to the rate constant k_3. Similarly, $-d[ES]/dt$ will be proportional to $k_2[ES]$ and inversely proportional to $k_1[S][E]$. The overall expression is

$$\frac{-d[ES]}{dt} = k_3[ES] + k_2[ES] - k_1[S][E].$$

This expression can be solved for [ES] by assuming steady state: $d[ES]/dt = 0$. If we let $[E]_o$ be the total enzyme concentration and [E] the free enzyme, then

$$[E] = [E]_o - [ES].$$

Substituting $[E]_o - [ES]$ for [E] and solving for [ES] gives

$$[ES] = \frac{[E]_o \cdot [S]}{(k_3 + k_2)/k_1 + [S]}.$$

The expression $((k_3 + k_2)/k_1)$ is the definition of K_m, and thus

$$[ES] = \frac{[E]_o \cdot [S]}{K_m + [S]}.$$

The rate or velocity of the reaction (v) is defined as the velocity of the limiting step, $k_3 \cdot [ES]$, and hence

$$v = k_3 \cdot [ES].$$

We can now substitute v/k_3 for [ES] to obtain

$$v = \frac{k_3[E]_o \cdot [S]}{K_m + [S]}.$$

The maximum velocity of the reaction (V) would be when all $[E]_o$ is bound as [ES]; that is, $[E]_o = [ES]$; hence

$$V = k_3 \cdot [E]_o.$$

Substituting V for $k_3 \cdot [E]_o$ gives

$$v = \frac{V \cdot [S]}{K_m + [S]},$$

the Michaelis–Menten expression.

Frequently, K_m is referred to as an enzyme affinity constant (or a dissociation constant). If K_m is low, the substrate is said to have high affinity for the enzyme. This would be true if k_3 is very small relative to k_1 and k_2 such that

$$\frac{k_3 + k_2}{k_1} = \frac{k_2}{k_1}.$$

Here K_m would equal k_2/k_1 and represent the affinity of [S] for the enzyme:

$$\text{ES} \underset{k_1}{\overset{k_2}{\rightleftharpoons}} \text{E} + \text{S}$$

and

$$K_m = \frac{k_2}{k_1} = \frac{[E][S]}{[ES]}.$$

In practical terms, a low K_m means that the reaction will proceed at a lower substrate concentration than it would at a higher K_m.

Simple inhibition

Competition for the active site of an enzyme between the substrate and a substrate analogue is one of the most straightforward types of enzyme inhibition known. For example, the enzyme fumarase, which catalyzes the reversible dehydration of succinate to fumarate,

$$\text{Succinate} \underset{+H_2O}{\overset{-H_2O}{\rightleftharpoons}} \text{Fumarate},$$

is competitively inhibited by malonate, a substrate analogue of succinate.

$$
\begin{array}{cc}
CO_2^- & CO_2^- \\
| & | \\
CH_2 & CH_2 \\
| & | \\
CH_2 & CO_2^- \\
| & \\
CO_2^- & \\
\text{Succinate} & \text{Malonate}
\end{array}
$$

Since the enzyme has high specificity for the substrate succinate, there is little or no catalysis with the analogue malonate, but malonate will bind to the active site because of the structural similarity.

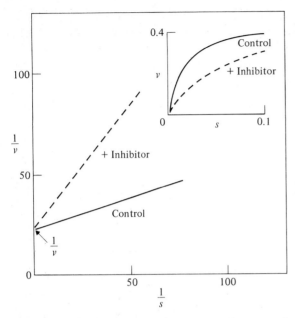

FIGURE 9.15 Double-reciprocal graph of a control curve and a curve in the presence of a competitive inhibitor. Inset shows the hyperbolic rate curves for the same data. Competitive inhibitors show an increase in K_m but no effect on the maximum velocity (V), as indicated by the same y-axis intercept.

In this case in which malonate competes for the active site, a high concentration of succinate will overcome the inhibition. This is true because both substrate and inhibitor are continually binding to and leaving the active site. A characteristic of competitive inhibition is that the maximum velocity is not reduced but that the K_m for the substrate (here succinate) is increased, indicating a requirement for a greater substrate concentration to reach half of maximum velocity.

A graphical analysis of competitive inhibition is shown in Fig. 9.15. The competitive inhibition curve crosses the y-axis at the same point as the control curve so that the maximum velocity is the same.

Allosterism

Through the recent work of Changeux, Jacob, Monod, and others, many enzymatic proteins have been shown to have sites other than the active catalytic site that bind small metabolite molecules (allosteric sites). Binding at these sites affects the kinetics of the enzyme-catalyzed reaction. The binding molecules that effect the enzyme kinetics are termed effectors. Those effectors that increase the reaction rate are called positive effectors and those that decrease the rate are called negative effectors.

It is believed that the enzymatic rate is altered because the binding of the effector at the allosteric site changes the conformation of the protein. One consequence of a conformational change is an altered fit between active site and substrate or cofactor.

Allosteric enzymes will normally have sigmoid kinetics rather than the hyperbolic kinetics depicted in Fig. 9.13. Figure 9.16 illustrates the sigmoidicity of the rate curve for an allosteric enzyme and how the usual positive and negative effectors alter the curve. Negative effectors tend to increase the sigmoidicity of the curve, increasing the K_m, whereas positive effectors reduce the apparent sigmoidicity because of alterations in K_m, with no effect on V.

Although allosteric proteins and the concept of metabolic regulation by effectors are developed to the greatest extent in procaryotic organisms, there is also substantial evidence for allosteric regulation of metabolism in higher plants.

In the simplest case, products of metabolic sequences will tend to inhibit enzymes by feedback at the beginning of biosynthetic pathways. Thus the products of reaction sequences tend to regulate their own synthesis by feedback inhibition.

$$A \xrightarrow{\text{I}} B \xrightarrow{\text{II}} C$$

Usually the end product involved in the feedback inhibition is more than one step away from the enzyme that it inhibits.

In the simple model shown above, A is a substrate of enzyme I, B is a product of I and substrate for II, and C is a product of enzyme II and a negative allosteric effector of enzyme I. Such negative effectors are common in branched metabolic pathways.

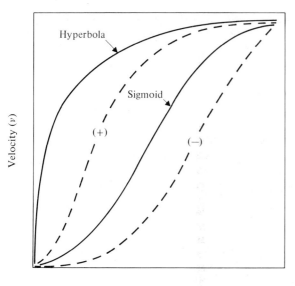

Velocity (v)

Hyperbola

Sigmoid

(+)

(−)

Substrate concentration (s)

FIGURE 9.16 Graph to show enzyme rate curves (velocity of the reaction as a function of substrate concentration) for a typical hyperbolic reaction and for an allosteric enzyme that shows sigmoid kinetics. In the presence of a positive effector (+), the curve becomes more steep with no change in maximum velocity. Positive effectors reduce the K_m. Negative effectors (−) increase the K_m with no effect on the maximum velocity.

An example of feedback inhibition is the carboxylating enzyme, phosphoenolpyruvate carboxylase,

$$\text{Phosphoenolpyruvate} + CO_2 \xrightarrow{Mg^{+2}} \text{Oxalacetate,}$$

which is an allosteric enzyme activated by glucose-6-phosphate and inhibited by malate.

$$\text{Glucose} \longrightarrow \text{G-6-P} \dashrightarrow \text{PEP} \begin{array}{c} \overbrace{}^{(+)} \\ \longrightarrow \end{array} \begin{array}{c} (-) \\ \text{OAA} \end{array} \begin{array}{c} \text{MAL} \\ \text{PYR} \quad \text{ASP} \end{array}$$

Here the flow of carbon from glucose to malate and aspartate is regulated by the concentrations of glucose-6-phosphate and malate. The negative feedback occurs just after a branch point in the metabolic sequence.

Another enzyme reported to be an allosteric protein in plants is phosphofructokinase (PFK). This enzyme, which catalyzes the phosphorylation of fructose-6-P in glycolysis to fructose-1,6-bisP, is inhibited by ATP, phosphoenolpyruvate, and citrate. Inhibition is relieved by inorganic phosphate and fructose-6-P. Thus when phosphoenolpyruvate

or citrate build up, their synthesis is prevented by allosteric feedback inhibition of PFK.

Pyruvate dehydrogenase, the enzyme that forms acetyl CoA from pyruvate prior to oxidation in the citric acid cycle, is inhibited by ATP and citrate. ATP and citrate at high levels reduce the flow of carbon to the citric acid cycle. In some plants, NAD-linked isocitrate dehydrogenase is inhibited by NADH. The ratio of NADH to NAD may therefore regulate the activity of the citric acid cycle.

There is evidence that some enzymes of polysaccharide biosynthesis and nitrogen metabolism may be regulated by allosteric mechanisms.

The energy charge

It has been proposed by Atkinson (1970) that cellular metabolism is regulated by the relative concentrations of ATP, ADP, and AMP since these nucleotides are intimately involved in cellular energetics. The above three are in equilibrium in the cell through the enzyme adenylate kinase.

$$\text{ATP} + \text{AMP} \rightleftharpoons 2\text{ADP}$$

Atkinson noted that certain enzymes were activated when ATP was high, notably those using ATP for biosynthesis. Furthermore, those enzymes involved in generating ATP were activated when ATP was low. The activation is allosteric in most cases and not because the adenosine nucleotides are used as substrates in these reactions. Enzymes such as NAD-isocitrate dehydrogenase and phosphofructokinase are inhibited by ATP and activated by low ATP and high AMP. Atkinson proposed the concept of energy charge (EC) as an index of the ATP/ADP/AMP ratio.

$$EC = \frac{ATP + \frac{1}{2}\,ADP}{ATP + ADP + AMP}$$

When all the adenine nucleotides are in the form of ATP, EC = 1.0, and when all are in AMP, EC = 0. When EC is high at about 0.85, the cell is at the crossover point between ATP generation (above 0.85) and ATP consumption (below 0.85). Because of the allosteric reactions involving the adenosine nucleotides as effectors, when the EC of the cell decreases below the crossover point of 0.85, ATP-generating systems are activated. Similarly, when the EC is above 0.85, ATP-generating systems are less functional or inhibited. The concept of energy charge is therefore useful as an indication of the energy status of the cell and gives information concerning the type of metabolism that is occurring.

In green plant tissue in the light, the usual EC is 0.8 to 0.95. For example, Heber and Santarius found in 1970 that the energy charge of chloroplasts was high in the light (about 0.95) and lower in the dark (about 0.67).

9.3.5 The multiple forms of enzymes (isozymes)

Studies by Markert and others in 1959 revealed that enzymes, despite being specific for substrates, occurred in more than one molecular form. Subsequent studies by many investigators confirmed the early work of Markert and have shown that most enzymes exist in multiple forms. These multiple forms of enzymes that catalyze the same reaction are termed isozymes. It is probably true to say that all enzymes exist as isozymes.

A broad generalization of biology is that one gene codes for one enzyme and that one enzyme catalyzes one reaction. The one gene–one enzyme–one reaction concept is useful but must be modified in the light of isozymes. There are many different metabolic situations in the cell in which the same reaction occurs. Metabolic reactions and reaction sequences occur during different developmental stages, in different organelles, in different tissues, and in different organs. Because the conditions are different for each of these situations and the kinetic properties of the isozymes are frequently different, it is logical to assume that one role of the isozymes of an enzyme is catalysis under different metabolic situations. One good example (explained below) is the organelle-specific isozymes of malate dehydrogenase. In the green plant cell, the reversible reaction that oxidizes malate to oxalacetate occurs in the cytosol during dark CO_2 fixation, in the mitochondria as a part of the citric acid cycle, and in microbodies (peroxisomes) as a part of photorespiration. Each reaction is the same—catalyzed by malate dehydrogenase—but different isozymes occur in the different compartments. Each is coded for by a different nuclear gene.

Another example of the role of isozymes in biology is the similar reactions in different metabolic sequences that occur in the same cellular compartment. Here regulation and separation are through isozymes, as shown by the different phosphatase and peroxidase reactions that occur together.

During development, the same enzymatic reaction occurs at different stages. Thus different isozymes of phosphatase, peroxidase, amylase, cellulase, and others are present at different developmental stages.

There is also some evidence that isozymes give some evolutionary stability to organisms. If there are multiple forms of the same enzyme present, a mutation in one may not be as detrimental as it would if the protein existed in only one form. Undoubtedly there are other functions of the isozymes that will be discovered in the future. But in any case

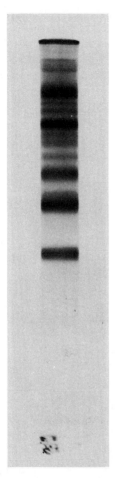

FIGURE 9.17 Acrylamide disc gel electrophoresis of the proteins extracted from cotton seeds. The proteins migrate according to their anionic charge. They are made visible by staining the gel with a protein-specific stain. Gel courtesy of Dr. B. L. Johnson.

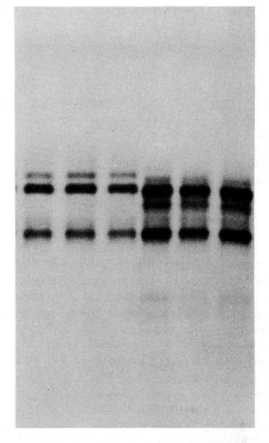

FIGURE 9.18 Peroxidase isozymes from leaf extracts of creosote bush. The isozymic proteins migrate according to their charge and separate on gels into distinct bands. The gel is flooded with peroxide and benzidine, and wherever a peroxidase is there will be a dark precipitate formed by the peroxidase reaction. Gel courtesy of Dr. Leonel Sternberg.

they are ubiquitous in plants and important to metabolism, growth, and development.

Isozymes can be detected by many different means, but the most frequent technique is that of gel electrophoresis. The medium through which the electrophoresis takes place can be of various substances; however, starch and acrylamide gels are the most commonly used. When a protein extract is electrophoresed in the gel medium and the gel stained for protein, many bands will be present (Fig. 9.17). Here the presence of proteins is indicated, but there is no notion of enzyme activity. Enzymes can be detected by staining the gel with substrates that will yield either visible products or products that can be made visible.

Figure 9.18 is an acrylamide gel stained for the enzyme peroxidase. The stain has made visible at least five prominent bands with peroxidase activity.

These are isozymes because they represent different proteins with the same enzymatic activity.

Isozymes such as those shown in Fig. 9.18 can be formed in several different ways. It is customary that the term isozyme be modified to indicate the type of isozyme. For example, many are known to be different gene products and hence are in one sense true isozymes. Isozymes formed from different genes may be either allelic or nonallelic. Nonallelic isozymes are coded for by different genetic loci, whereas the allelic isozymes are products of different alleles at the same locus. Furthermore, isozymes may be of the posttranslational type, being formed after the primary structure of the protein is assembled. Such isozymes are termed epigenetic. Epigenetic isozymes may be produced by different conformations of the identical protein, by posttranslational conjugation with carbohydrates or lipids, or by actual modification of the primary structure, usually by proteolytic activity to produce smaller units. Moreover, since active proteins may be composed of subunits, there may be subunit differences that result in isozymes. The subunit polymers may be the same, resulting in homopolymeric isozymes, or the subunits may be different, resulting in heteropolymeric isozymes.

As an example, the peroxidase isozymes (Fig. 9.18) are known to be both genetic and epigenetic. Some have different primary structures, which indicates genetic isozymes. They may be either allelic or nonallelic. Since they are glycoproteins, some of the isozymic forms result from the posttranslational addition of different carbohydrate moieties.

Although not a common enzyme in plants except in certain seeds, lactate dehydrogenase (LDH), which catalyzes the following reaction,

$$\text{Pyruvate} + \text{NADH} \rightleftharpoons \text{Lactate} + \text{NAD},$$

is one of the most thoroughly studied. LDH proteins are formed from two different subunit polymers, A and B. The A and B polymers are from different genetic loci. There can be two homopolymeric isozymes of LDH, constructed from either the A subunits or the B subunits. Since the func-tional LDH protein has four subunits, the homopolymeric types are AAAA and BBBB. Note that the A-type and B-type homopolymeric LDH isozymes are nonallelic.

Subunit binding to form the tetrameric quaternary structure is not restrictive; hence heteropolymeric isozymes form AAAB, AABB, and ABBB. In most tissues that have LDH, all five of the possible combinations are found but in different ratios.

An equally interesting group of isozymes, thoroughly studied in both plants and animals, is the malate dehydrogenase group.

$$\text{Oxalacetate} + \text{NADH} \rightleftharpoons \text{Malate} + \text{NAD}$$

Starch gel electrophoresis of a preparation from green leaves that is stained for malate dehydrogenase will show multiple forms of MDH.* The simplest case is three organelle-specific forms. These three forms are nonallelic and are known to exist in different subcellular compartments. One is found in the microbodies, one in the mitochondria, and the third in the cytosol of the cell (note that Fig. 9.19 shows two cytosol forms). Each functions in a different metabolic pathway but catalyzes the same reaction. Furthermore, within each nonallelic group there may be allelic isozymes. The MDH protein is composed of two identical subunits; that is, its quaternary structure is homopolymeric. Thus there can be three possible allelic forms from two alleles: AA, AB, and BB.

The above examples of multiple forms of enzymes illustrate some of the metabolic roles for the isozymes. In the case of esterases, they are probably specific *in vivo* for different kinds of ester linkages. The different peroxidases are frequently specific for different kinds of phenolic compounds. MDH

---

*The MDH staining procedure can be used to illustrate the general method. MDH oxidizes malate to oxalacetate (OAA), reversibly using NAD as the redox cofactor. NAD is reduced to NADH when malate is oxidized to OAA. NADH can reduce the dye nitroblue tetrazolium (NBT) by transfer of electrons through phenazine methosulfate (PMS). NBT is a blue insoluble compound when reduced and is easily seen on gels as a blue bland. A gel flooded with a mixture of malate, NAD, PMS, and NBT at the proper pH (about 8.0) will show blue bands wherever MDH is present.

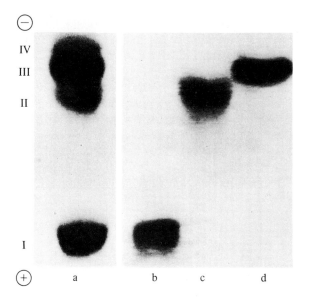

FIGURE 9.19 Isozymes of malate dehydrogenase separated on a starch gel by electrophoresis. After migration in the electric field, the gel is flooded with the reaction mixture that includes nitroblue tetrazolium. A blue precipitate that is formed when the nitroblue tetrazolium is reduced marks the presence of an isozyme of malate dehydrogenase. Band I is a microbody-specific form of the enzyme, Band II is the mitochondrial form, and Bands III and IV are soluble (cytosol) forms. Column a is a total homogenate from spinach leaf tissue, column b is the malate dehydrogenase from microbodies, column c is from mitochondria, and column d is from a soluble fraction. Each of the organelle-specific isozymes functions in a different metabolic pathway. From a 1970 experiment by Victor Rocha and I. P. Ting.

occurs in different subcellular compartments and hence has a role in metabolic regulation. Different RNA polymerases that occur in nuclei, mitochondria, and chloroplasts were discussed in the previous chapter on nucleic acids. The function of the isozymes discussed here is the same: they catalyze the same reaction but occur in different cellular compartments. The many other isozymes known include those with different thermal sensitivities and different pH optima, and in the case of the allelic forms, they probably do not have different functions.

There is speculation that different allelic forms of the same enzyme give some stability to evolution. Mutations can occur without dramatic detrimental effects. The allelic forms could be beginning points for evolution as well.

9.3.6 Protein and enzyme preparation

For metabolic and biochemical purposes it is frequently desirable and indeed necessary to separate one protein from another, and on occasion it is useful to purify to homogeneity. Many of the physical properties of proteins can be used to advantage in their preparation.

Since many proteins exist in solution as colloids, they can be precipitated by the same procedures that precipitate colloids (refer to Chapter 2). Organic solvents such as acetone or alcohol will remove hydration shells from protein and cause precipitation. High ionic strength as produced by solutions of ammonium sulfate will cause protein precipitation, a procedure called salting out. Similarly, proteins can be precipitated by pH treatment depending on their isoelectric points. If advantage is taken of differing physical properties among proteins they can be selectively precipitated by proper adjustment of solvent concentration, ionic strength, and pH.

Column chromatography is a useful method for protein preparation; it takes advantage of charged groups (ion exchange chromatography) or molecular size (gel filtration chromatography). In ion exchange chromatography, charged proteins are adsorbed to an oppositely charged matrix of an exchange resin contained within a cylinder or column and are then eluted from the column with

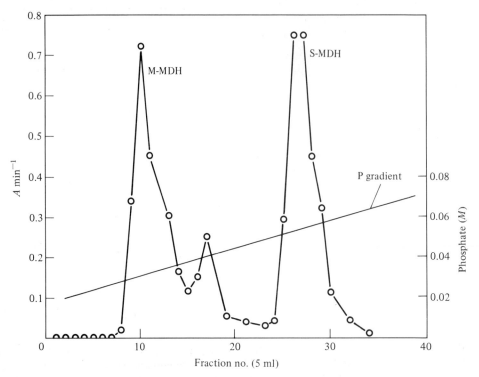

FIGURE 9.20 Fractionation of three forms of malate dehydrogenase from maize seed by DEAE-cellulose column chromatography. The extract is placed on the cellulose column (anion-exchange cellulose) and eluted with a linear phosphate gradient. The different malate dehydrogenase isozymes elute from the column according to their charge. The mitochondrial forms (M-MDH) elute before the soluble form of the enzyme (S-MDH). From I. P. Ting, 1968. Malic dehydrogenase in corn root tips, *Archives of Biochem. Biophys.* 126:1–7.

either a pH or ionic-strength gradient (see Fig. 9.20). Either anion or cation exchange columns can be used depending on the electric properties of the proteins. Since individual proteins differ in charge, selective elution is possible and allows for separation and purification. Proteins can also be separated on the basis of charge by electrophoresis.

Gel filtration chromatography takes advantage of gels with pores of molecular dimensions such that proteins are eluted according to their size. In the usual procedure, a protein mixture is sieved through a sieving gel, which allows small proteins or molecules to enter the interstices. However, larger proteins flow around the gel beads and elute faster than the smaller proteins (refer to Chapter 2).

Figure 9.21 shows an experiment in which the size

(mass) of isocitrate dehydrogenase isolated from *Zea mays* was estimated by gel filtration. The graph is a plot of the volume of buffer required for elution of the various proteins from the column, expressed as a ratio of elution volume to the void volume of the column (buffer volume passed through the column before any protein elutes) against the logarithm of the mass expressed in daltons. The standards used in this particular experiment are cytochrome *c* (12,500 daltons), pig heart isocitrate dehydrogenase (68,000 daltons), malate dehydrogenase purified from maize (60,000 daltons), and glucose-6-phosphate dehydrogenase (102,000 daltons). The elution of the four marker proteins graphs as a straight line against their mass (expressed in daltons). The mass of the isocitrate de-

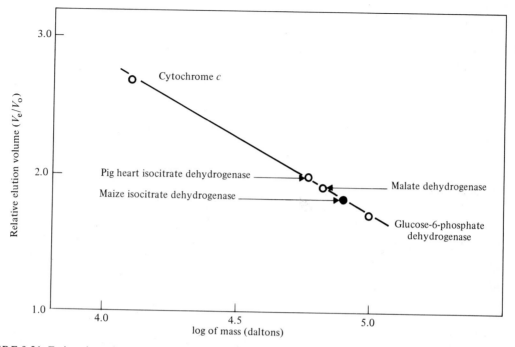

FIGURE 9.21 Estimation of the mass of isocitrate dehydrogenase isolated from *Zea mays*. The graph is an elution profile of marker proteins from a gel filtration column. The marker proteins are cytochrome *c* (12,500 daltons), pig heart isocitrate dehydrogenase (60,000 daltons), maize malate dehydrogenase (68,000 daltons), and glucose-6-phosphate dehydrogenase (102,000 daltons). The y-axis is the relative elution volume of each protein, where V_e is the actual volume of buffer passed through the column to elute a particular protein and V_o is the void volume of the column. The void volume is the buffer required to elute a protein (or other substance) that is completely excluded from the interstices of the gel beads. It represents the actual volume around the beads in the column. When V_e/V_o for globular proteins is graphed against the logarithm of their mass in daltons a straight line is obtained. From this relationship the mass of any protein can be obtained knowing its V_e/V_o. In this particular experiment, the *Zea* isocitrate dehydrogenase has a calculated mass of 83,000 daltons. Redrawn from R. A. Curry and I. P. Ting. Purification, properties, and kinetic observations on the isoenzymes of NADP isocitrate dehydrogenase of maize. 1976. *Arch. Biochem. Biophys.* 176:501–509.

TABLE 9.5   The purification of isocitrate dehydrogenase from maize root tissue

STAGE OF PURIFICATION	PROTEIN (mg)	YIELD (%)	SPECIFIC ACTIVITY (I.U. mg^{-1})
Fresh homogenate	—	100	—
Acetone precipitation	7.0	57	0.1
DEAE*-cellulose column chromatography	3.6	56	0.9
Ammonium sulfate precipitation	0.9	29	2.7
Hydroxylapatite column chromatography	0.12	25	11.5
Acrylamide gel electrophoresis	0.3	3	33.5

*DEAE = diethylaminoethyl

The data are for the purification of an isozyme of NADP isocitrate dehydrogenase. From Curry (1975).

hydrogenase isolated from maize is estimated from this graph to be 83,000 daltons. It should be recognized that this is an estimate of mass and not molecular weight because the proteins eluting from the column will be hydrated and may have ions bound to them. Furthermore, additional experiments with this isocitrate dehydrogenase have shown that it is composed of two identical subunits of 41,500 daltons.

An efficient and specific method for the separation of enzymatic proteins from each other takes advantage of the specificity of the active site with the substrate or a substrate analogue. This method, called affinity chromatography, is conducted by binding a substrate or analogue to an inert column matrix and then passing the protein mixture through the column. Because of the affinity of the active site of the enzyme for the substrate, the enzyme protein will bind to the substrate–column matrix. After all unbound protein is removed, the protein to be purified is removed from the substrate complex by altering the substrate–enzyme complex. Frequently, a protein can be purified in a single step by affinity chromatography.

Table 9.5 illustrates the purification of NADP isocitrate dehydrogenase by common laboratory procedures. The degree of purity and enrichment at each step is expressed as enzyme activity per unit of protein, or specific activity. As the specific activity increases during purification, it means that the protein is being concentrated relative to the other proteins in the original mixture or extract.

Another property of proteins that is useful to physiologists is that of antibody formation. If a foreign protein (antigen) is injected into an animal, the globulin fraction of the animal will form antibodies that are specific for the foreign protein. Although the antibodies represent an immune reaction by the animal, they can be purified from serum and used to advantage by the protein chemist or physiologist.

If pure protein (i.e., antigen) is available, very specific antibodies can be obtained. The Ouchterlony double-diffusion technique can be used to ascertain the purity of protein. As shown in Fig.

9.22, antiserum is placed in the center well of an agar plate and protein is placed in perimeter wells. The antibody from the antiserum will diffuse radially, as will the protein. At the point of contact, a sharp precipitin band will be formed if the protein is pure.

The antiserum with antibody can be used to assay for the presence of a specific protein, and with the proper procedures the assay can be made quantitative.

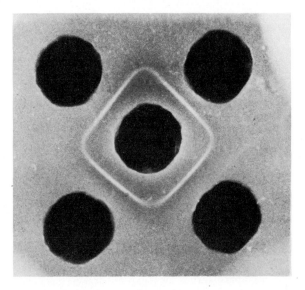

FIGURE 9.22 Ouchterlony double-diffusion technique to show precipitation of an enzyme in the presence of antiserum made against the enzyme. The antiserum, containing antibodies made by white rabbits injected with purified enzyme, is placed in the center well. The enzyme is placed in the outer four wells. The antiserum and enzyme diffuse from the wells into the agar until they meet. On contact, a white precipitin band forms in the agar that marks the presence of antiserum and enzyme. Sharp precipitin bands indicate a high purity of enzyme. The specific reaction of enzyme and antibody can be used to detect the presence of a specific enzyme (in this case malate dehydrogenase) and can be made quantitative. From W. C. Zschoche and I. P. Ting, 1973. Purification and properties of microbody malate dehydrogenase from *Spinacea oleracea* leaf tissue, *Arch. Biochem. Biophys.* 159:767–776.

Review Exercises

9.1 Draw the structure of aspartic acid when in an acid solution and when in a basic solution.

9.2 In a particular leaf in the light, the ratio of ATP to ADP to AMP is 5 to 3 to 2. Compute the energy charge and indicate what the value means about the metabolic state of the tissue.

9.3 What would be the approximate molecular weight of a protein that has 500 amino acid residues? State the difference between mass, molecular weight, and molar mass.

9.4 The following data were obtained in an experiment with malate dehydrogenase,

$$OAA + NADH \longrightarrow MAL + NAD,$$

in which the rate of the reaction (change in A per minute, measuring NADH at 340 nm) was determined at different oxalacetate concentrations. By graphical analysis, compute the Michaelis constant for oxalacetate.

$\triangle A$ min^{-1}	OAA mM
0.96	0.23
0.73	0.12
0.44	0.05
0.21	0.02
0.13	0.01

9.5 A common method to estimate the concentration of soluble protein in extracts is by the technique of Lowry in which the protein amino acids tyrosine and tryptophan reduce phosphomolybdate and phosphotungstate ions in the presence of copper. The complexes formed are blue and follow Beer's law. The following data were obtained from a standard series of solutions containing known concentrations of bovine serum albumin as a protein standard.

A at 750 nm	BSA mg ml^{-1}
0.09	0.05
0.18	0.10
0.27	0.15
0.37	0.20
0.46	0.25

A 25-ml extract from a pea cotyledon had an A of 0.25. How many mg of protein are there in a cotyledon?

References

ALTMAN, P. L., and D. G. DITTMER. 1968. *Metabolism.* Biological Handbooks, Federation of American Societies for Experimental Biology, Bethesda, Maryland.

ASHTON, F. M. 1976. Mobilization of storage proteins of seeds. *Ann. Rev. Plant Physiol.* 27:95–117.

ATKINSON, D. E. 1970. Enzymes as control elements in metabolic regulation. In P. D. Boyer (ed.), *The Enzymes,* Vol. 1. Academic Press, New York.

BEIL, E. A. 1976. "Uncommon" amino acids in plants. *FEBS Letters* 64:29–35.

BENDER, M. L., and L. J. BRUBACHER. 1973. *Catalysis and Enzyme Action.* McGraw-Hill, New York.

BOGORAD, L., and J. H. WEIL (eds.). 1977. *Nucleic Acids and Protein Synthesis in Plants.* Plenum Press, New York.

BOSCH, L. 1972. *Mechanism of Protein Synthesis and Its Regulation.* American Elsevier, New York.

BOULTER, D., R. J. ELLIS, and A. YARWOOD. 1972. Biochemistry of protein synthesis in plants. *Biol. Rev.* 47:113–175.

BOYER, P. D. (ed.). 1959–1976. *The Enzymes.* 13 vols. Academic Press, New York.

CHANGEUX, J. P. 1965. The control of biochemical reactions. *Sci. Amer.* 212:36–45.

COMMISSION ON ENZYMES (International Union of Biochemistry). 1973. *Enzyme Nomenclature.* American Elsevier, New York.

CURRY, R. A. 1975. Purification and properties of NAD malate dehydrogenase and NADP isocitrate dehydrogenase from maize, *Zea mays.* Thesis, University of California, Riverside.

DAYHOFF, M. O. 1968–1978. *Atlas of Protein Sequence and Structure,* Vols. 1–5. National Biomedical Research Foundation, Georgetown University Medical Center, Washington, D.C.

DIXON, M., and E. D. WEBB. 1964. *Enzymes.* Academic Press, New York.

EISENBERG, D. 1970. X-ray crystallography and enzyme structure. In P. D. Boyer (ed.), *The Enzymes,* Vol. 1. Academic Press, New York.

HARBORNE, J. B., and C. F. VAN SOUMERE (eds.). 1975. *The Chemistry and Biochemistry of Plant Proteins.* Academic Press, New York.

INGLETT, G. E. (ed.). 1972. *Seed Proteins.* The Avi Publ. Co., Westport, Conn.

KOSHLAND, D. E. 1973. Protein shape and biological control. *Sci. Amer.* 229:52–64.

MARCUS, A. 1971. Enzyme induction in plants. *Ann. Rev. Plant Physiol.* 22:313–336.

MARKERT, C. L. (ed.). 1975. *Isozymes,* Vols. 1–4. Academic Press, New York.

MEISTER, A. 1965. *Biochemistry of the Amino Acids,* Vols. 1–2. 2d ed. Academic Press, New York.

MIFLIN, B. J., and P. J. LEA. 1977. Amino acid metabolism. *Ann. Rev. Plant Physiol.* 28:299–329.

MILLERD, A. 1975. Biochemistry of seed proteins. *Ann. Rev. Plant Physiol.* 26:53–72.

PIRIE, N. W. (ed.). 1971. *Leaf Protein.* IBP Handbook No. 20. Blackwell Scientific Publ. Co., Oxford.

SMITH, H. (ed.). 1977. *Regulation of Enzyme Synthesis and Activity in Higher Plants.* Academic Press, New York.

STADTMAN, E. R. 1970. Mechanisms of enzyme regulation in metabolism. In P. D. Boyer (ed.), *The Enzymes,* Vol. 1. Academic Press, New York.

ZALIK, S., and B. L. JONES. 1973. Protein biosynthesis. *Ann. Rev. Plant Physiol.* 24:47–68.

CHAPTER TEN

Carbohydrates and Organic Acids

Carbohydrates are the most abundant chemical constituents of plants, making up from 60% to 90% of their dry matter. From a physiological viewpoint they are extremely important. The simple, water-soluble carbohydrates with a sweet taste are the sugars involved in intermediary metabolism that act both as substrates for the synthesis of more complex compounds and as respiratory intermediates. The sugars are found either singly or in polymeric form as components in nucleic acids, lipids, proteins, and a variety of other natural products.

The simple sugars glucose and fructose are important respiratory compounds. Once they are phosphorylated, making them more reactive, they are oxidized by glycolysis and go through the tricarboxylic acid cycle in the mitochondria to CO_2 and water, yielding energy in the form of reduced pyridine nucleotides (NADH and NADPH) and nucleotide triphosphates such as ATP. The sugar transformations and oxidations to organic acids during respiration are fully discussed in Chapter 14.

255

The simple sugars are also among the first products of photosynthesis. Even though the very first compound of photosynthesis is an organic acid, phosphoglycerate, the first reduced compound is a simple sugar, glyceraldehyde. Subsequent to the formation of glyceraldehyde in photosynthesis other simple sugars such as glucose and fructose are produced.

The most abundant soluble sugar in plants and certainly the best-known sugar is the disaccharide, sucrose. It is a simple chemical composed of a glucose and fructose and serves as the primary carbohydrate of transport in plants. Many tissues accumulate vast quantities of sucrose, such as the roots of sugar beets and the stems of sugar cane. Sucrose is common table sugar and is therefore what comes to mind when the word "sugar" is mentioned.

Equally as important as the soluble sugars and their derivatives are the abundant polymeric sugars that function as both storage and structural components of the plant. The most important storage compound of plants is starch, a complex polymer of glucose. And although some plants store carbohydrates as other kinds of polymers, such as inulin (a polymer of fructose), and some tissues store carbon compounds other than carbohydrate, such as protein and lipid, starch is the most common of all the storage compounds.

Perhaps one of the most abundant compounds in nature is cellulose, a complex polysaccharide similar to starch composed of glucose units. Both cellulose and starch are glucose polymers, and the linkage between adjacent glucose residues is by attachment of the same carbons. Carbon-1 of one glucose unit is linked to carbon-4 of the adjacent glucose. The primary difference is the arrangement of the bond around carbon-1. This slight difference, which will be explained in detail in this chapter, makes starch an easily metabolized molecule and cellulose extremely resistant to metabolism (e.g., decay). There are many enzymes in plants and animals that can hydrolyze starch. Some of these enzymes, such as the amylases, are primarily involved in digestion processes, and others, such as the phosphorylases, are respiratory enzymes. But

because of the high specificity of enzymes they will not hydrolyze cellulose even though starch and cellulose are both 1→4 polymers of glucose. Plants and, of course, some bacteria do have cellulase enzymes that can metabolize cellulose, but for the most part these enzymes are highly specialized. The cellulases of plants are involved both in maturation processes such as fruit ripening and in leaf and fruit drop. The gut of ruminants contains bacteria that can metabolize cellulose so that cows and other related animals can use cellulose to obtain soluble sugars.

Cellulose is important in plants because it is a major structural component of cell walls. But cell walls also have many other complex polymers of carbohydrates. Hemicelluloses are polymers of sugars such as arabinose and mannose. These polymers are abundant in both primary and secondary cell walls and make up a large part of the primary cell wall of plants. The pectins and pectic compounds, which are complex polymers of uronic acids such as galatouronic acid (an oxidation product of the sugar galactose), and other sugars make up the third important group of polymeric carbohydrates that give many of the important structural and physiological properties to plant cell walls. In addition to these polysaccharides, cell walls have protein, lipid, and the very complex, aromatic polymer lignin. Lignin is a component of secondary cell walls and is discussed more fully in Chapter 12.

The sugars and carbohydrates in general are of extreme historical importance to the study of biochemistry. Much of our knowledge of stereochemistry and optical isomerism comes from the early studies of carbohydrates. The constitution of a particular chemical compound is the nature and sequence of bonding. In addition, the individual atoms or groups can be arranged about a central atom in different configurations. Such configurational differences between molecules having even the same constitution have profound effects on metabolism and physiology. Enzymes, being highly specific, can detect differences in configuration. Sugars with the molecular formula $C_6H_{12}O_6$ are abundant in plants and include glucose, galactose, and mannose. Despite having the same molecular formula, the chemical properties of these sugars are quite different. The enzyme hexokinase readily catalyzes the phosphorylation of glucose to form glucose-6-phosphate but will not catalyze the addition of phosphate to galactose or mannose. The basic principles of these configurational arrangements are discussed in this chapter.

Although not carbohydrates by definition, the water-soluble organic acids are intimately asso-

ciated with carbohydrate metabolism. Reduction of the carboxyl group of an organic acid forms an aldehyde, one of the diagnostic functional groups of the carbohydrates. Similarly, oxidation of the aldehyde functional group of sugars forms organic acids. Such oxidations and reductions are extremely important steps in many biochemical sequences, including those of both respiration and photosynthesis. Water-soluble organic acids and carbohydrates are discussed in this chapter.

10.1 Properties of the Carbohydrates

The term carbohydrate comes from the initial observation that carbohydrates structurally are carbon atoms hydrated with water (CH_2O). Of course, many carbon compounds that fall into the carbohydrate classification are known that do not have an atomic ratio of 1:2:1 for carbon to hydrogen to oxygen. For the most part, however, carbohydrates do have this ratio or they are derived from compounds that had it.

More precisely, carbohydrates can be defined as polyhydroxy aldehydes or polyhydroxy ketones. Two of the most common carbohydrates occurring in plants, glucose and fructose, are shown below.

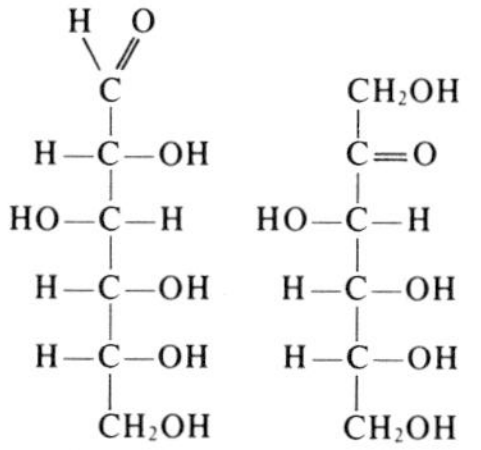

Glucose   Fructose

Glucose is an aldehyde sugar and fructose is a ketose sugar. They are structural isomers with the same molecular formula, $C_6H_{12}O_6$. The aldehyde sugar, glucose, is an aldose and has the same molecular formula as fructose, a ketose carbohydrate. They are structural isomers with the same molecular formula, $C_6H_{12}O_6$.

Carbohydrates can be classified as monosaccharides, oligosaccharides, or polysaccharides, de-

pending on the number of identifiable sugars in the molecule. Any or all may have functional groups containing sulfur, nitrogen, or phosphorus. The monosaccharides are the simplest and may have three to many carbons. Trioses have 3 carbons, tetroses have 4, pentoses have 5, hexoses have 6, heptoses have 7, and so on.

Oligosaccharides contain up to six monosaccharides. They are named according to the number of monosaccharide residues in the molecule. There are disaccharides, trisaccharides, tetrasaccharides, pentasaccharides, and hexasaccharides. For the most part, the monosaccharides and oligosaccharides are white, crystalline, water-soluble compounds that are sweet and are commonly called sugars.

Polysaccharides yield many monosaccharides when hydrolyzed. Because of their large size and complex secondary structure, many of the polysaccharides are relatively insoluble in water.

10.1.1 Chemical properties

If a sugar has a free or potentially free aldehyde group, it will act as a reducing agent in alkaline solution and will be oxidized to a sugar acid. A number of reagents have been designed that will indicate the presence of a reducing sugar. One reagent, called Fehling's solution, is a mixture of Cu^{+2} tartrate in a basic solution that will react with reducing sugars to form cuprous (Cu^+) ions. The Cu^+ forms a brick-red precipitate of Cu_2O, indicating the presence of a reducing sugar. The only common nonreducing sugars found in plants are sucrose, trehalose, and raffinose. Most sugars, including glucose, fructose, mannose, galactose, ribose, and xylose, are reducing sugars with a free aldehyde group or a group that can readily convert to an aldehyde group (fructose, for example).

The free sugars will react with hydrazine to form hydrazones and osazones. The osazone crystals are characteristic for the individual sugars, and their physical properties can be used as diagnostic tests for identification. The hydrazine reacts with the top two carbons of the free sugar. For this reason, glucose, fructose, and mannose, which differ only in the arrangement of groups around carbons 1 and 2, form identical osazones.

Aldose and ketose sugars will tautomerize in the presence of base through their common enediol salts. For example, if glucose is placed in an alkaline solution, mannose and fructose as well as glucose will result. As can be seen from the following diagram, mannose, glucose, and fructose differ only in the arrangement of groups around the two top carbons.

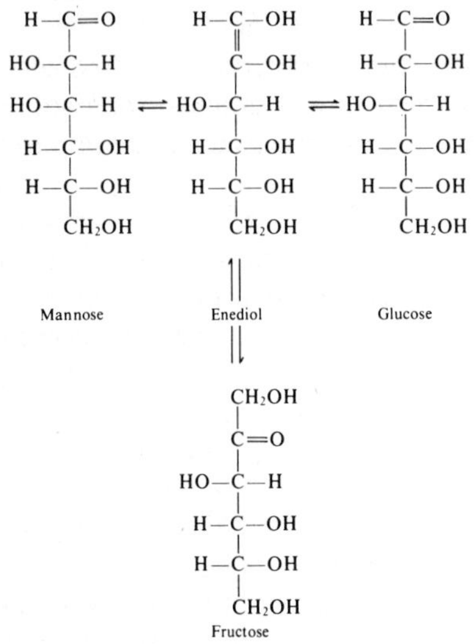

Mannose   Enediol   Glucose

Fructose

The aldose sugars may be mildly oxidized to form aldonic acids (carboxyl at C-1), or, if the aldehyde group at C-1 is protected, more rigorous oxidation will form a uronic acid, with the terminal carbon oxidized to the carboxyl. Oxidation of the aldose sugars with a strong oxidizing agent such as nitric acid will form the dicarboxylic acid derivatives.

Upon treatment with strong mineral acids, the aldopentoses and the aldohexoses form furfural and furfural derivatives that will in turn form colored complexes with reagents, such as α-naphthanol, resorcinol, and orcinol. The colored complexes can be used to identify these sugars. The saccharides also form a blue color when treated with the reagent anthrone. The latter reagent has been used extensively to quantify saccharides by means of glucose residues. It can be used to assay glucose directly. More commonly, starch and other glucose polymers (all called glucans) are hydrolyzed into component glucose residues that are measured. The anthrone reagent is not specific for glucose and will react with other free saccharides.

Other useful reactions of the sugars are the formation of cyanohydrins when reacted with hydrogen cyanide and the formation of oximes when reacted with hydroxylamine. The student should consult a comprehensive biochemistry textbook for further information about these reactions and their uses.

10.1.2 Optical properties

Perhaps one of the most interesting features of carbohydrates is that of isomerism. Isomers of carbohydrates may be of two general types, structural (structural isomerism) and optical (stereoisomers). Structural isomers have the same molecular formula but are arranged differently. They may have branched or straight chains, cis–trans isomers in which the arrangement around a double bond differs, or functional group isomers such as fructose and glucose.

Stereoisomers have the same molecular formula but differ in the configuration of atoms or groups around an asymmetric center. As is well known, the carbon atom is assumed to form a tetrahedron, with the nucleus in the center and the four asymmetric bonds forming 109° angles with each other. If bonding is to four different groups, the molecule will be asymmetric and there will be two possible arrangements of the four groups around the carbon nucleus center. The two isomers formed are exact mirror images of each other and are called in this case enantiomers. Each is an enantiomer of the other. To help you visualize the arrangement, observe the fingers of each of your hands. They are essentially identical but are exact mirror images of each other.*

Glyceraldehyde, the simplest sugar, has an asymmetric carbon center, C-2, since four different groups are attached, —CHO, —CH_2OH, —OH, and —H. As shown below, there are two possible arrangements. Also shown is a structural isomer of glyceraldehyde, dihydroxyacetone.

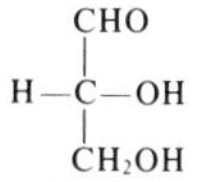

```
        CHO               CHO            CH₂OH
         |                 |               |
   H—C—OH           HO—C—H           C=O
         |                 |               |
       CH₂OH             CH₂OH          CH₂OH
  D-glyceraldehyde    L-glyceraldehyde  Dihydroxyacetone
```

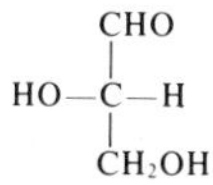

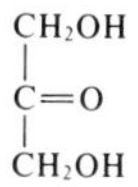

All of these compounds have the same molecular formula, $C_3H_6O_3$. The chemical and physical properties of glyceraldehyde and dihydroxyacetone differ in many respects, but the only significant difference between the glyceraldehyde enantiomers (other than biological) is the direction in which they rotate plane-polarized light. From a biological viewpoint the asymmetry becomes extremely important, because enzymes are specific for only one enantiomer.

The atomic groups of the above diagrams are arranged according to the Fischer projection method. Those groups that are vertical to the center carbon are by convention said to be in front of the plane of the paper, and those to the left and right are oriented behind the plane of the paper.

Emil Fischer assigned the designation D to the glyceraldehyde enantiomer with the —OH group to the right of the asymmetric center when drawn

*If you have never constructed molecular models to illustrate an asymmetric carbon atom, you should try now so that you may appreciate the subtle differences between enantiomers.

in the projection formula. Its enantiomer was designated as L.

Carbohydrates that have their penultimate asymmetric carbon center (the next to the last numbering from the functional group) arranged similarly to D-glyceraldehyde are called D-compounds; those that are arranged as L-glyceraldehyde are called L-compounds.

The D and L designates stand for dextra (D)—"to the right"—and levo (L)—"to the left." In the case of glyceraldehyde, D-glyceraldehyde rotates plane-polarized light to the right and L-glyceraldehyde rotates plane-polarized light to the left. However, this is not necessarily true in all other enantiomer pairs because optical rotation is a property of the entire molecule. Thus the symbols (+) and (−) are used to indicate dextrorotatory (+) and levorotatory (−) molecules. The D and L designations refer only to the arrangement of the penultimate carbon relative to the arrangement of the glyceraldehyde enantiomers. As an example, D(+)-glucose has its penultimate carbon arrangement the same as D-glyceraldehyde and rotates plane-polarized light to the right. D-fructose, however, rotates plane-polarized light to the left, and thus its proper designation is D(−)-fructose.

The number of possible optical isomers can be determined from 2^n, where n is the number of asymmetric carbons. The aldotrioses, such as glyceraldehyde, have two possible optical isomers ($2^n = 2^1 = 2$), aldotetroses have 2^2 or 4, aldopentoses have 2^3 or 8, and aldohexoses have 2^4 or 16. Dihydroxyacetone, the only ketotriose sugar, does not have an asymmetric carbon. The ketotetroses have one asymmetric carbon and thus two isomers, the ketopentoses have two asymmetric carbons and four isomers, and the ketohexoses have three asymmetric carbons and eight isomers. Figure 10.1 illustrates the possible isomers of the aldoses and ketoses through the hexoses.

From Fig. 10.1 it should be apparent that in the case of the isomers formed from more than one asymmetric center (the tetroses and larger), all are not mirror images of each other and are therefore not enantiomers. Those optical isomers that are not

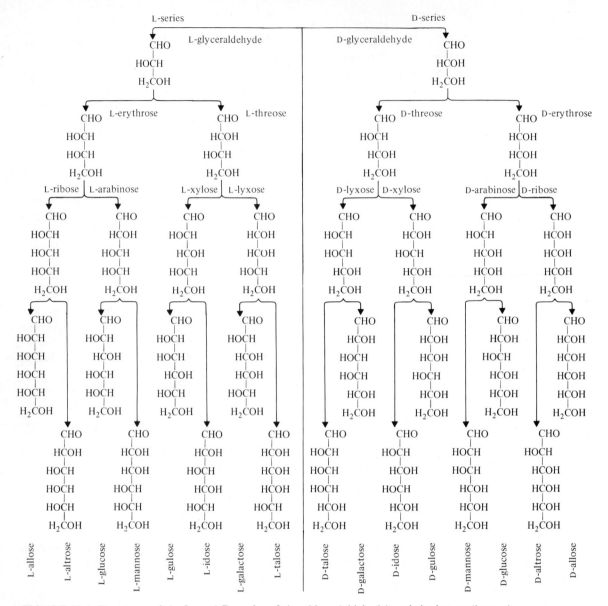

FIGURE 10.1 Structures of the L- and D- series of the aldose (aldehyde) and the ketose (ketone) sugars.

enantiomers are called diastereomers and will have different chemical and physical properties. Referring to the figure, D- and L-glyceraldehyde are optical isomers and enantiomers since they are mirror images of each other. Of the tetroses, erythrose and threose are optical isomers but are not enantiomers. They are diastereomers. L-erythrose and D-erythrose are enantiomers, and L-threose and D-threose are enantiomers of each other. L- and D-erythrose will have the same physi-

KETOSE SUGARS

```
                          CH₂OH
                          |
                          C=O
                          |
                          CH₂OH
L-series                                            D-series
                        Dihydroxyacetone
```

$$
\begin{array}{cc}
\text{L-series} & \text{D-series}
\end{array}
$$

```
        CH₂OH                                   CH₂OH
        |                                       |
        C=O                                     C=O
        |                                       |
    HO—C—H                                   H—C—OH
        |                                       |
        CH₂OH                                   CH₂OH
     L-erythrulose                           D-erythrulose
```

```
   CH₂OH          CH₂OH              CH₂OH          CH₂OH
   |              |                  |              |
   C=O            C=O                C=O            C=O
   |              |                  |              |
HO—C—H         H—C—OH            HO—C—H         H—C—OH
   |              |                  |              |
HO—C—H         HO—C—H            H—C—OH         H—C—OH
   |              |                  |              |
   CH₂OH          CH₂OH              CH₂OH          CH₂OH
 L-ribulose     L-xylulose         D-xylulose     D-ribulose
```

```
 CH₂OH    CH₂OH      CH₂OH    CH₂OH      CH₂OH    CH₂OH      CH₂OH    CH₂OH
 |        |          |        |          |        |          |        |
 C=O      C=O        C=O      C=O        C=O      C=O        C=O      C=O
 |        |          |        |          |        |          |        |
HO—C—H   H—C—OH     HO—C—H   H—C—OH     HO—C—H   H—C—OH     HO—C—H   H—C—OH
 |        |          |        |          |        |          |        |
HO—C—H   HO—C—H     H—C—OH   H—C—OH     HO—C—H   HO—C—H     H—C—OH   H—C—OH
 |        |          |        |          |        |          |        |
HO—C—H   HO—C—H     HO—C—H   HO—C—H     H—C—OH   H—C—OH     H—C—OH   H—C—OH
 |        |          |        |          |        |          |        |
 CH₂OH    CH₂OH      CH₂OH    CH₂OH      CH₂OH    CH₂OH      CH₂OH    CH₂OH
L-psicose L-fructose L-sorbose L-tagatose D-tagatose D-sorbose D-fructose D-psicose
```

FIGURE 10.1 *Continued*

cal and chemical properties except for optical rotation. L- and D-threose will also have the same chemical and physical properties except for optical rotation. Erythrose and threose have different physical and chemical properties. Similar statements can be made for the pentoses and hexoses.

Those compounds that differ from each other by the arrangement around a single asymmetric carbon center are termed epimers. Thus D-glucose and D-mannose are epimers, as are D-glucose and D-galactose. D-galactose and D-mannose differ at two carbon centers and are not epimers.

In solution, many of the simple sugars (pentoses, hexoses, heptoses) form internal hemiacetals, which result in 5-carbon or 6-carbon ring structures.

```
   H   OH          H    O           HO   H
    \ /             \  //            \ /
     C               C                C
     |               |                |
  H—C—OH          H—C—OH           H—C—OH
     |               |                |
 HO—C—H    O   ⇌  HO—C—H    ⇌   HO—C—H    O
     |               |                |
  H—C—OH          H—C—OH           H—C—OH
     |               |                |
  H—C              H—C              H—C
     |               |                |
     CH₂OH           CH₂OH            CH₂OH
  α-D-glucose       D-glucose       β-D-glucose
```

The five-membered rings similar to furan are termed furanose forms, and the six-membered rings similar to pyran are called pyranose forms.

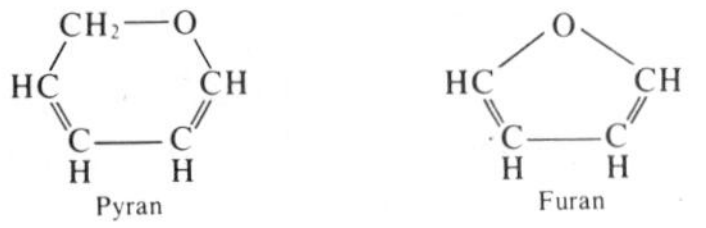

It is immediately apparent that there are two possible ring structures for glucose since a new asymmetric carbon is created at carbon-1. In the Fischer projection, if the hydroxyl appears to the right, α is used as the designation, and if it appears to the left, β is used. The α and β isomers are called anomers. Thus there are 32 different isomers of the aldohexoses when in the furanose form ($2^5 = 32$).

One piece of evidence for the equilibration of α- and β-glucose through hemiacetal formation is that pure α-D-glucose has an optical rotation of $+113°$ (α-L-glucose would have a rotation of $-113°$). When α-D-glucose is allowed to dissolve in water, the rotation shifts from $+113°$ to $+52.5°$. If pure β-D-glucose is dissolved in water, the optical rotation shifts from $+19°$ to $+52.5°$. Thus the α and β forms are in equilibrium through hemiacetal formation with 37% α and 63% β. Optical rotation shift through hemiacetal formation is called muto-rotation.

Because many of the sugars in nature exist as ring compounds, Haworth in 1929 proposed a ring formula to depict the sugars. In the Haworth formulas, the ring is oriented perpendicular to the plane of the paper.

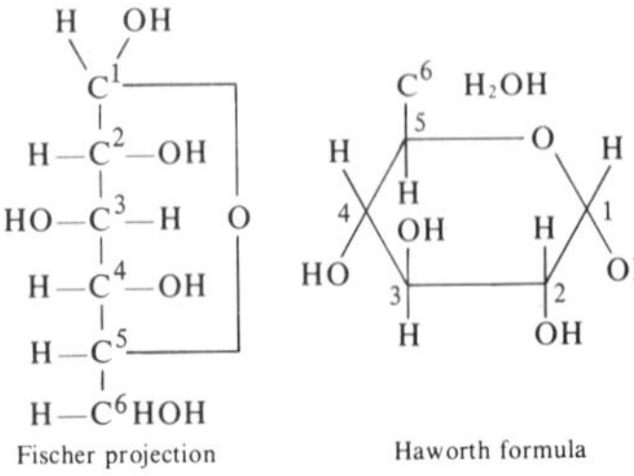

Those groups to the right in the Fischer projection are drawn below the plane of the ring and those to the left in the Fischer projection are drawn above the plane of the ring in the Haworth formula. Haworth formulas are commonly used because they more properly depict the structure of the sugars and can be drawn readily without the necessity of indicating all of the carbons, hydrogens, and hydroxyls of the sugar.

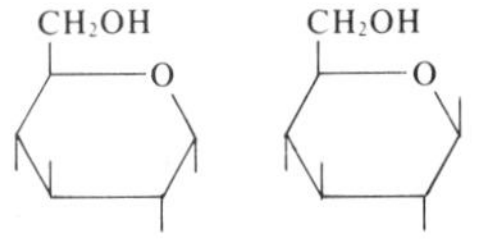

10.2 The Monosaccharides

The monosaccharides are the simplest of the sugars. They function as beginning substrates for respiration (such as glucose and fructose); when phosphorylated, they serve as intermediates in respiratory pathways. Many are the important building blocks for structural polymers of cell walls. The pentose sugars arabinose and xylose are important constituents of the hemicelluloses, and, of course, glucose is the primary if not only sugar residue of starch and cellulose. The properties of the poly-meric carbohydrates are in large part a function of the properties of their constituent sugar residues. In this section, the important monosaccharides of plants and their properties will be discussed.

10.2.1 D-sugars

Almost without exception, the monosaccharides occurring in higher plants are D-sugars. The D-aldoses and D-ketoses up through the six-carbon hexoses are illustrated in Fig. 10.1. The ketose D-sedoheptulose is the only seven-carbon sugar of note in plants.

$$
\begin{array}{c}
CH_2OH \\
| \\
C=O \\
| \\
HO-C-H \\
| \\
H-C-OH \\
| \\
H-C-OH \\
| \\
H-C-OH \\
| \\
CH_2OH \\
\end{array}
$$

D-sedoheptulose

The monosaccharides, being involved in cellular energetics, participate in a variety of reactions. Some of the most important reactions of intermediary metabolism are discussed below. These reactions will be encountered in subsequent sections on plant metabolism.

Phosphorylation

Before being metabolized, virtually all of the monosaccharides are phosphorylated. α-D-glucose is phosphorylated, forming α-D-glucose-6-phosphate by catalysis with the ubiquitous enzyme hexokinase.

$$
\text{Glucose} + \text{ATP} \xrightarrow{\text{hexokinase}} \text{Glucose-6-P} + \text{ADP}
$$

Fructose and mannose are similarly phosphorylated by specific kinases.

Hexose-6-P can be mutated by phosphoglucomutase to form hexose-1-P. For glucose-6-P, the reaction is as follows.

$$
\text{Glucose-6-P} \underset{\text{phosphoglucomutase}}{\rightleftharpoons} \text{Glucose-1-P}
$$

Isomerization

Isomerase enzymes can catalyze the isomerization of phosphoketoses to phosphoaldoses reversibly.

For example, phosphoglucoisomerase catalyzes the isomerization of glucose-6-P to fructose-6-P.

$$
\text{Glucose-6-P} \underset{\text{phosphoglucoisomerase}}{\rightleftharpoons} \text{Fructose-6-P}
$$

Ribose and xylose are isomerized to ribulose and xylulose by specific isomerases. In metabolism, these isomers are thus readily interchangeable.

Epimerization

There are specific epimerase enzymes that catalyze the epimerization of galactose and glucose, glucose and mannose, and arabinose and xylose. Most of the epimerizations take place with the sugars bound as sugar-nucleotide derivatives rather than as free sugars.

Aldol condensations

Enzymes termed aldolases, transaldolases and transketolases, catalyze condensations of aldo or keto groups. For example, transaldolase transfers a ketotriose group from one sugar to another, and transketolase transfers a 2-carbon keto group to an aldose to form a new ketose. The aldol condensation reactions are some of the most important carbon reactions of respiration and photosynthesis.

Oxidations and decarboxylations

The sugars can be oxidized by enzyme catalysis to form sugar acids. Aldonic acids are sugar acids with carboxyls at C-1, and uronic acids are sugar acids with terminal carboxyls. Two common sugar acids are shown below.

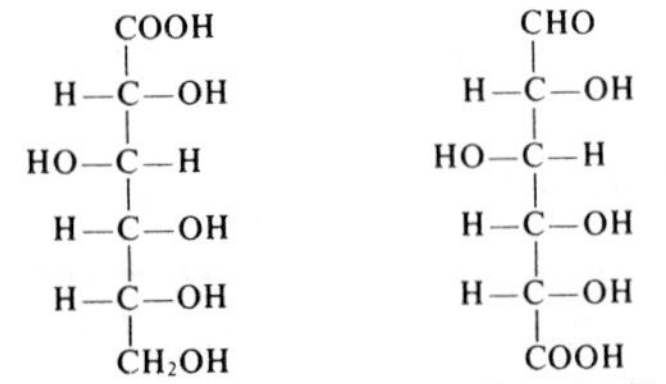

```
   COOH              CHO
    |                 |
H—C—OH            H—C—OH
    |                 |
HO—C—H            HO—C—H
    |                 |
H—C—OH            H—C—OH
    |                 |
H—C—OH            H—C—OH
    |                 |
  CH₂OH             COOH
D-gluconic acid   D-glucuronic acid
```

Gluconic acid may be decarboxylated to form a pentose.

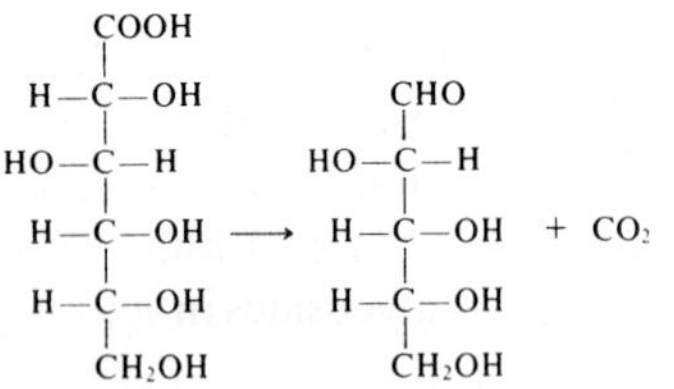

```
   COOH                    CHO
    |                       |
H—C—OH                  H—C—OH
    |                       |
HO—C—H                  HO—C—H
    |          ⟶          |
H—C—OH                  H—C—OH   + CO₂
    |                       |
H—C—OH                  H—C—OH
    |                       |
  CH₂OH                   CH₂OH
D-gluconic acid        D-arabinose
```

The most common reaction that occurs *in vivo* (catalyzed by an NADP-requiring dehydrogenase) decarboxylates 6-phosphogluconic acid to ribulose-5-P and CO₂.

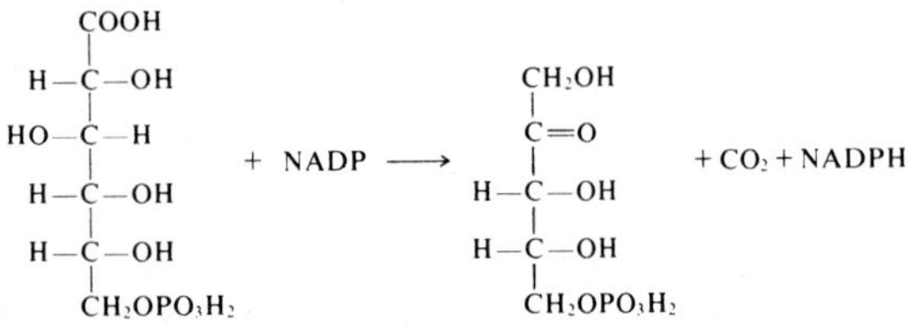

```
   COOH                      CH₂OH
    |                         |
H—C—OH                     C=O
    |                         |
HO—C—H     + NADP ⟶       H—C—OH      + CO₂ + NADPH
    |                         |
H—C—OH                     H—C—OH
    |                         |
H—C—OH                     H—C—OH
    |                         |
  CH₂OPO₃H₂                CH₂OPO₃H₂
```

10.2.2 Amino sugars and sugar alcohols

Although not too common, the most frequent naturally occurring sugar is α-D-glucosamine. It may be found in plants as an acetyl derivative, N-acetyl-D-glucosamine. Their structures follow.

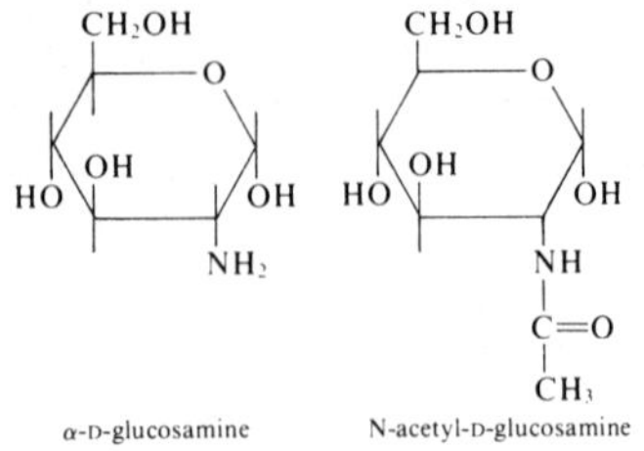

α-D-glucosamine   N-acetyl-D-glucosamine

The amino sugars are frequently found in glycoproteins.

The sugar alcohols are quite common in plants. Three of the most common, D-mannitol, D-sorbitol, and D-ribitol, are derivatives of mannose, glucose, and ribose. Their structures follow.

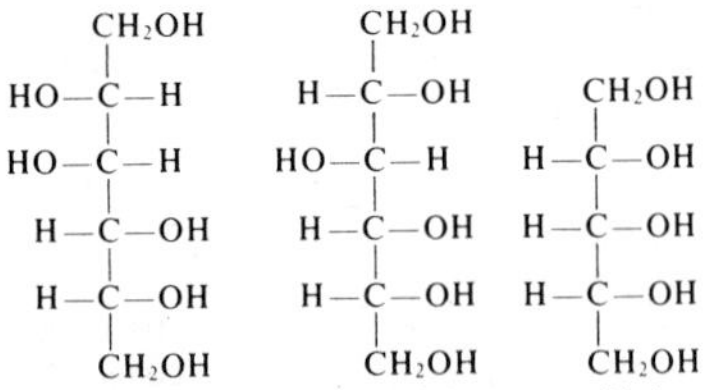

```
   CH₂OH           CH₂OH
    |               |
HO—C—H          H—C—OH            CH₂OH
    |               |               |
HO—C—H          HO—C—H          H—C—OH
    |               |               |
H—C—OH          H—C—OH          H—C—OH
    |               |               |
H—C—OH          H—C—OH          H—C—OH
    |               |               |
  CH₂OH           CH₂OH           CH₂OH
D-mannitol       D-sorbitol       D-ribitol
```

The cyclohexanehexols, or inositols, are common in plant tissues. There are nine possible isomers of inositol, of which about four are commonly found. The most common and perhaps the most important is the optically inactive isomer, myoinositol (meso-inositol). The structure of myoinositol is as follows.

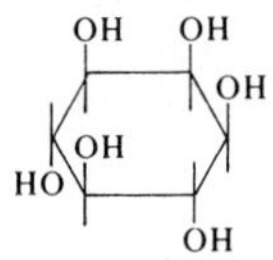

Myoinositol

The hexaphosphate derivative of inositol, phytic acid, is also quite common and perhaps functions as a phosphate storage compound.

10.2.3 Sugar acids

As stated above, sugars can be oxidized to form acids. Aldonic acids are sugar derivatives oxidized in position C-1, forming a carboxyl. Uronic acids are oxidized to the carboxyl in the terminal carbon (farthest from the functional group), and aldaric acids are dicarboxylic sugar derivatives.

Glucuronic and galacturonic acids are probably the most common plant constituents of the sugar acids. Glyoxylic acid could be considered the simplest uronic acid.

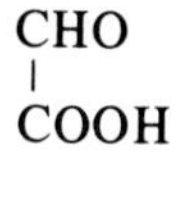

$$
\begin{array}{c}
CHO \\
| \\
COOH
\end{array}
$$

Glyoxylic acid

Uronic acids are easily decarboxylated and play an important role as intermediates in metabolism. Galacturonic acid and some others are components of the pectins and pectic acids found in cell walls.

Probably the most important sugar acid found in plants is L-ascorbic acid. It is important because it functions in terminal oxidation during some types of respiration through the enzyme ascorbic acid oxidase. It is known to be a cofactor in some enzyme reactions and is of course an important vitamin (vitamin C) in human nutrition.

The biosynthesis of L-ascorbic acid is poorly understood in plant tissues; it may be formed by more than one metabolic pathway. A glance at the structure of ascorbic acid suggests that it could be synthesized from D-glucose if the glucose molecule were inverted, the C-1 reduced to hydroxyl, and the C-6 oxidized to uronic acid. The latter is supported by some evidence found in plant tissues. However, other evidence from using ^{14}C labels in positions C-1 and C-6 of D-glucose indicates that the molecule is not inverted; biosynthesis proceeds by inversion

of the penultimate carbon (C-5) of D-glucose to form the L-configuration.

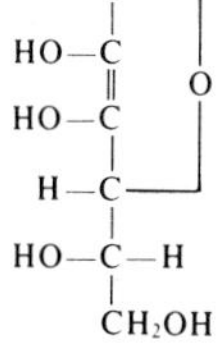

Some of the plant acids are derived from the sugars and sugar acids. Tartaric acid, a common acid of some plant tissues (see Section 10.5), could be considered to be the aldaric acid derivative of erythrose.

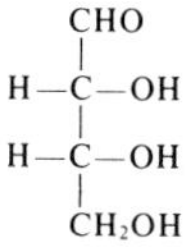

D-erythrose 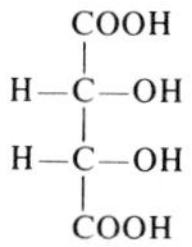   Tartaric acid

10.2.4 Deoxysugars

Some deoxysugars occur in plants. The pentose sugar 2-deoxyribose is a constituent of DNA. Some of the 6-deoxysugars, 6-deoxy-D-glucose, 6-deoxy-L-galactose (L-fucose), and 6-deoxy-L-mannose (L-rhamnose), occur in plants. Fucose is found in certain marine plants (algae), and deoxyglucose and rhamnose occur as glycosides in some higher plants.

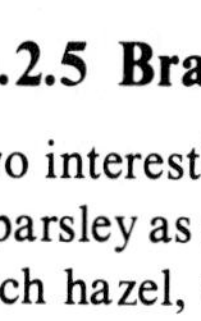

2-deoxyribose 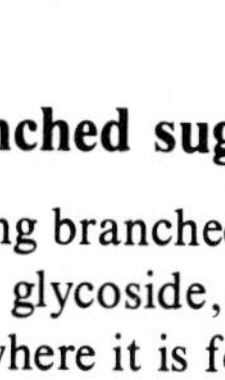   6-deoxy-D-glucose 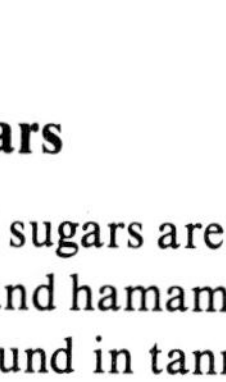   L-fucose 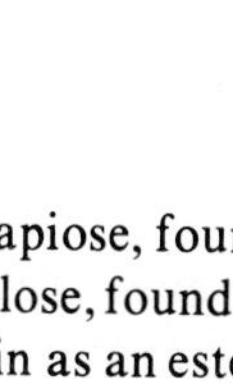   L-rhamnose

10.2.5 Branched sugars

Two interesting branched sugars are: apiose, found in parsley as a glycoside, and hamamelose, found in witch hazel, where it is found in tannin as an ester.

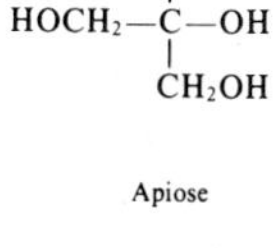

Ascorbic acid

Apiose 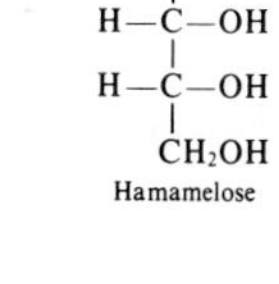   Hamamelose

10.2.6 Glycosides

The anomeric carbon of sugars can form ether linkages and together with other alcoholic compounds form O-glycosides. Although such glycosides are not quite as stable as other ether linkages, they may play a role as carbohydrate storage compounds, intermediates in biosyntheses, or as end products of metabolism.

Glycosides are named by naming the R group or aglycone first and then changing the -ose ending of the carbohydrate to -oside to indicate the glycosyl portion. Thus a simple O-glycoside would be alkyl-O-β-glucoside. It is interesting that most naturally occurring glycosides have β linkages and that the most common glycosyl group by far is glucose. Thus most glycosides are β-glucosides. Many of the common glycosides are known by their trivial (common) names.

The structure of dhurrin, a common cyanogenic glucoside, is as follows.

Dhurrin

In addition to the O-glycosides, N-, S-, and C-glycosides also exist in plants. Nucleosides are N-glycosides in which the base is the aglycone. The S-glycosides, or thioglycosides, are abundant in certain plants such as mustards. A common S-glycoside of black mustard is sinigrin.

Sinigrin

Many plant glycosides are discussed in Chapter 12 with reference to their aglycone moieties.

10.3 Oligosaccharides

Oligosaccharides are very common and important carbohydrates of plant metabolism. Some of the common oligosaccharides occurring in plants are depicted in Fig. 10.2. The oligosaccharides of plants can be readily separated from each other and identified by chromatographic techniques. Figure 10.3 shows a chromatogram of some oligosaccharides that were synthesized by cotton fibers supplied with UDP-glucose-^{14}C. As a rule, the migration of oligosaccharides on chromatograms is in ascending order according to the number of monosaccharide residues. The fewer the residues, the farther the migration.

The most abundant oligosaccharide of plants is the disaccharide sucrose. Sucrose, or cane sugar, occurs in virtually all plants and usually serves as the primary carbohydrate of transport. The sucrose molecule is formed from a glucose linked to a fructose through an α-$(1 \rightarrow 4)$ O-glycosyl (or glycosidic) bond.* Its proper name is α-$(1 \rightarrow 4)$-D-glycopyranosyl-β-D-fructofuranoside. The common and systematic names of several plant oligosaccharides are listed in Table 10.1.

The biosynthesis of sucrose is poorly understood. There is more than one enzyme that can catalyze the syntheses. In all probability the enzyme sucrose phosphate synthase is the *in vivo* catalyst. The reaction proceeds by formation of an O-glycosyl bond between the glucose of UDP-glucose and fructose-6-P to produce sucrose-P. Hydrolysis of the phosphate bond by a specific phosphatase pulls the reaction in the direction of sucrose synthesis.

UDP-glucose + Fructose-6-P $\rightleftharpoons$ Sucrose-P + UDP

Sucrose-P $\longrightarrow$ Sucrose + P_i

*The linkages between the oligosaccharides and polysaccharides are O-glycosidic bonds.

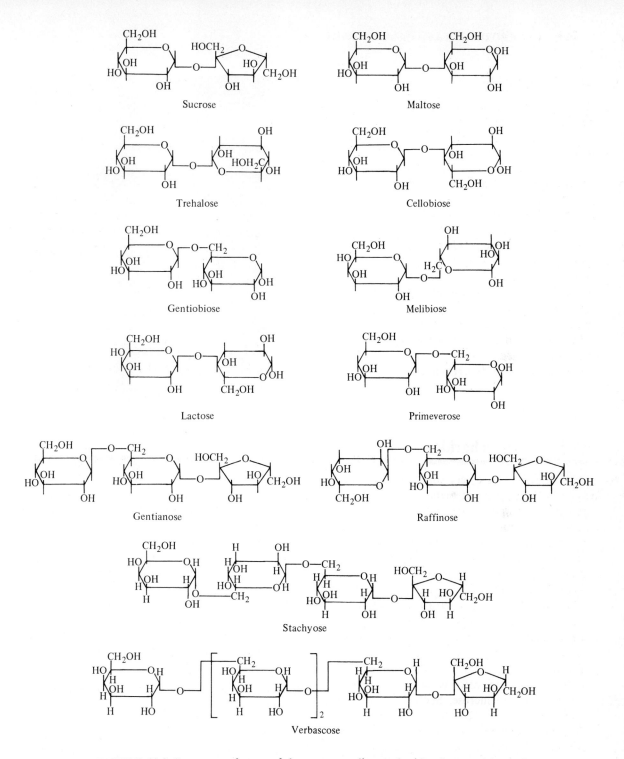

FIGURE 10.2 Structures of some of the common oligosaccharides that occur in plants.

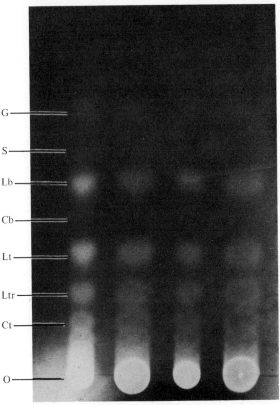

FIGURE 10.3 Thin-layer chromatography of the radio-active products produced in young cotton fibers that were supplied with UDP-glucose-¹⁴C. The young cells are actively synthesizing cell-wall material and producing many oligosaccharides. The migration on the thin-layer plate tends to be an inverse function of the size of the sugar. G = glucose, S = sucrose, Lb = laminaribose (β-(1→3) O-glucosyl bond), Cb = cellobiose, Lt = laminaritriose, Ltr = laminaritetrose, Ct = cellotriose. The spots close to the origin are longer oligosaccharides (O). From an experiment by R. L. Palmer and W. M. Dugger. 1978. University of California, Riverside.

A different enzyme, sucrose synthetase, is also capable of the catalysis of the O-glycosyl bond during sucrose synthesis, but it probably functions largely in the formation of nucleoside diphosphate glucosides for subsequent biosynthesis of glycosidic bonds. The enzyme sucrose synthetase acts best with UDP-glucose, but other nucleotide diphosphates such as ADP-glucose, GDP-glucose, CDP-glucose, and TDP-glucose can also be used. The reversible reaction catalyzed by sucrose synthetase follows.

$$\text{Sucrose} + \text{UDP} \rightleftharpoons \text{UDP-glucose} + \text{Fructose}$$

Another common enzyme prevalent in storage tissues that can catalyze the hydrolysis of sucrose is invertase. Invertase catalyzes sucrose hydrolysis to glucose and fructose, a reaction termed inversion because during the reaction the optical rotation shifts from +66.5° to −28°. The −28° is an average

of glucose (+52.5°) and fructose (−92°). Invertase is also called sucrase. Since the hydrolysis is not reversible, sucrase cannot catalyze the synthesis of sucrose.

The enzyme sucrose phosphate synthase catalyzes the synthesis of sucrose, and the enzymes invertase and sucrose synthetase both catalyze the hydrolysis of sucrose, the latter producing the nucleotide diphosphate derivatives of glucose for subsequent biosynthetic reactions.

Although not occurring in plants to any extensive degree, maltose is a major product of starch hydrolysis (see Section 10.4). Maltose is composed of two glucose residues linked in an α-(1→4) O-glycosyl bond.

Similarly, cellobiose is the disaccharide hydrolytic product of cellulose and is composed of two glucose residues but is linked through a β-(1→4) O-glycosyl bond. Trehalose is composed of two

TABLE 10.1   Some common oligosaccharides

COMMON NAME	SYSTEMATIC NAME
Sucrose	α-D-glucopyranosyl-β-D-fructofuranoside
Maltose	4-α-D-glucopyranosyl-D-glucopyranose
Trehalose	1-α-D-glucopyranosyl-α-D-glucopyranoside
Cellobiose	4-β-D-glucopyranosyl-D-glucopyranose
Gentiobiose	6-β-D-glucopyranosyl-D-glucopyranose
Melibiose	6-α-D-galactopyranosyl-D-glucopyranose
Lactose	4-β-D-galactopyranosyl-D-glucopyranose
Primeverose	6-β-D-xylopyranosyl-D-glucopyranose
Gentianose	β-D-glucopyranosyl-(1$\rightarrow$6)-sucrose
Raffinose	α-D-galactopyranosyl-(1$\rightarrow$6)-sucrose
Stachyose	α-D-galactopyranosyl-(1$\rightarrow$6)-O-α-D-galactopyranosyl-(1$\rightarrow$6)-sucrose

glucose residues linked through an α-(2$\rightarrow$1) O-glycosyl bond, and gentiobiose is two glucose units linked in a β-(1$\rightarrow$6) arrangement. Melibiose and lactose are both galactopyranosyl-glucopyranose disaccharides, with the former linked α-(1$\rightarrow$6) and the latter linked β-(1$\rightarrow$4). Primeverose is a xylo-pyranosyl-glucopyranose molecule linked in a β-(1$\rightarrow$6) O-glycosyl bond.

The two common nonreducing trisaccharides gentianose and raffinose can both be considered as addition products of sucrose. Gentianose has an additional glucose linked in an α-(1$\rightarrow$6) O-glycosyl bond to the sucrose moiety and raffinose has a galactose unit linked in an α-(1$\rightarrow$6) O-glycosyl bond to sucrose. The tetrasaccharide stachyose has galactose units linked to sucrose.

10.4 Polysaccharides

Polysaccharides are important because they play both a structural and a storage role. Perhaps the most ubiquitous carbohydrate in nature is the structural polymer of cell walls, cellulose. Cellulose is a glucan, or polymer, of glucose. Starch, also a polymeric glucan, is the primary storage polysaccharide of plants. Other kinds of polysaccharides discussed below are fructosans, mannans, and a variety of pentosans such as xylan and arabinan.

10.4.1 Starch

Starch, a complex mixture of straight-chain and branched-chain glucans, is the most important reserve carbohydrate of plants. Starch is most frequently stored in plants as starch grains. The grains, or granules, are of various shapes and are frequently unique for a species. Starch is deposited on the grain in concentric layers that may be either uniform or of different densities, depending on the time of the year or stage of the growing season. Grains are synthesized and stored in plastids: either chloroplasts or rather amorphous plastids called amyloplasts. The chloroplasts of many plants are particularly good starch-grain-synthesizing organelles (Fig. 10.4). A mature starch grain is insoluble in water primarily because of a lipid coat. Removal of the coat will solubilize the starch.

Starch can be separated into two compounds: a straight-chain, mostly water-soluble component called amylose and a large, relatively insoluble, branched molecule called amylopectin.

Amylose is a straight-chain glucan with α-(1$\rightarrow$4) glucosidic linkages with a molecular weight range of 10,000 to 100,000. It has about 60 to 600 glucose residues, depending on the molecular weight. The molecular structure of amylose is shown in Fig. 10.5.

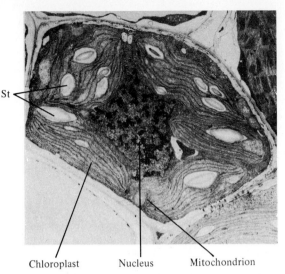

St

Chloroplast   Nucleus   Mitochondrion

FIGURE 10.4 Electron micrograph of a bundle sheath cell from crabgrass, showing chloroplasts with starch grains. St = starch grains. Micrograph courtesy of Dr. W. W. Thomson.

Amylopectin is a highly branched, large poly-saccharide with a molecular weight as great as 500,000 and 3000 or more glucose residues. Like amylose, the straight-chain portion is composed of α-D-glucose residues linked in α-(1→4) glucosidic bonds. At intervals of about 20 to 25 glucose residues, a branch occurs through an α-(1→6) glucosidic linkage. The structure of amylopectin is shown in Fig. 10.5 along with amylose.

Starches isolated from different plants will have a variable content of amylose. Usually amylose makes up from 10% to 25%, the rest being the branched polymer amylopectin. Waxy cereals get their names from the fact that their starch is almost exclusively amylopectin. The high content of amylopectin in the grain makes the endosperm feel waxy. In the grains referred to as starchy or wrinkled the starch is predominantly insoluble amylose.

The helical secondary structures of amylose and amylopectin can bind iodine molecules, with the gyres forming colored complexes. Amylose develops a deep blue iodine complex, and amylopectin forms a reddish purple complex. The iodine reaction is diagnostic for the starchlike glucans.

The metabolism of starch

The metabolism of starch can be separated into hydrolysis and biosynthesis. Hydrolyses of starch are catalyzed by two groups of enzymes, the amy-lases and the phosphorylases. The amylases occur in reserve and storage tissues and apparently function largely in starch degradation for sugar mobilization during growth. Phosphorylases are more closely associated with respiratory processes; they hydrolyze starch to form the highly active phosphorylated sugars that act as substrates for respiration.

HYDROLYSIS

As stated above, starch hydrolysis functions in mobilization of reserves such as occurs during seed germination or for hexose-phosphate production during respiration and biosyntheses. The main starch-hydrolyzing enzymes of starchy seeds are the amylases. The amylases hydrolyze starch to maltose.

$$\text{Starch} + n \text{ H}_2\text{O} \longrightarrow n \text{ Maltose}$$

Maltose is then further hydrolyzed to glucose units by the action of maltase.

$$\text{Maltose} + \text{H}_2\text{O} \longrightarrow 2 \text{ Glucose}$$

The amylases only hydrolyze the α-(1→4) O-glucosidic bonds of amylose, amylopectin, and smaller, straight-chain glucan polymers called dextrins. Two types of amylases are known, α-amylase and β-amylase. The α-amylases catalyze the hydrolysis of the α-(1→4) bonds seemingly at random to produce maltose and larger molecules that have

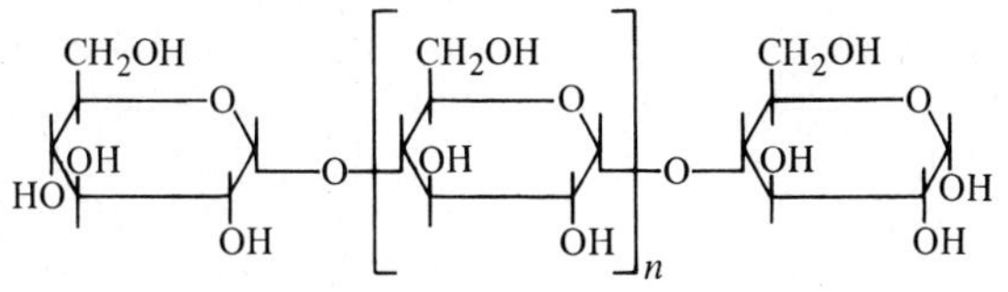

Amylose

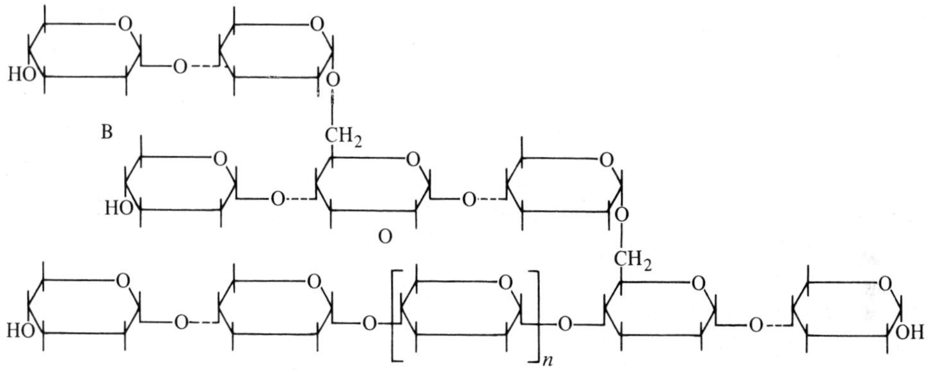

Amylopectin

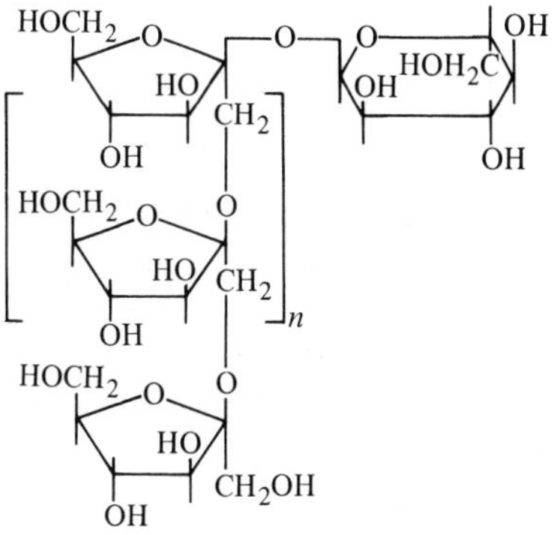

Inulin

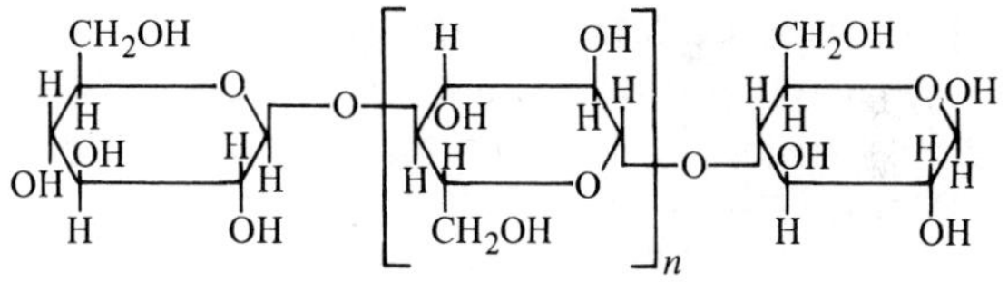

Cellulose

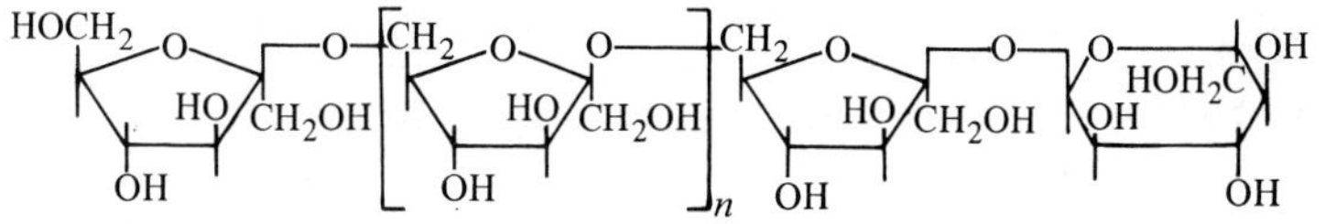

Levan

FIGURE 10.5 Molecular structures of some of the common polysaccharides in plants.

the α-(1→6) glucosidic bonds. These larger units are called limit dextrins. β-amylase removes maltose units, starting from the nonreducing end of the glucan, and will completely degrade amylose to maltose. Amylopectin is hydrolyzed to maltose units up to the α-(1→6) glucosidic bonds. Therefore, amylopectin yields maltose and limit dextrins from the hydrolytic action of β-amylase.

Another similar enzyme known as isoamylase will hydrolyze amylopectin to form short-chain amylose molecules. Isoamylase is specific for the α-(1→6) O-glycosidic bonds.

The phosphorylases hydrolyze starch by attacking the α-(1→4) linkages, forming glucose-1-phosphate. Complete hydrolysis of starch by phosphorylase to glucose-1-P is not possible because of the α-(1→6) linkages of amylopectin.

Active phosphorylases are produced from inactive phosphorylases by phosphorylation with ATP as the phosphate donor. The phosphorylases catalyze the hydrolysis of starch, producing glucose-1-P.

$$\text{Starch} + n\,P_i \longrightarrow n\,\text{Glucose-1-P}$$

It is worthy of note that *in vitro* phosphorylase can catalyze the formation of α-(1→4) O-glucosidic bonds from glucose-1-P and a glucan primer. Thus phosphorylase may be able to synthesize amylose. However, thermodynamic considerations suggest that it is primarily a degradative enzyme rather than a biosynthetic one.

BIOSYNTHESIS

The mode of starch biosynthesis is only partially resolved and represents one of the interesting complicated questions remaining for plant biochemists studying carbohydrates. Not only is the method of polymerization of the glucose units to form the straight-chain α-(1→4) O-glucosidic bonds in doubt, but also how the complicated branching occurs in amylopectin biosynthesis is not totally resolved.

Basically, it seems that polymerization proceeds through starch synthetase (ADP-glucose- or UDP-glucose–starch transglucosylase), which transfers glucose to an existing glucan by donation from a nucleotide diphosphate-glucose, ADP-glucose, or to a lesser extent from UDP-glucose. These dinucleotide-glucose–starch transglucosylases require a primer (at least maltose). The reaction can be depicted as follows.

$$n\,\text{ADP-glucose} + \alpha\text{-}(1\text{–}4)\text{-glucan (primer)} \rightarrow$$
$$\text{Starch} + n\,\text{ADP}$$

Because sucrose is the primary mobile saccharide of photosynthesis and is also the primary transport sugar of plants, most starch synthesis probably begins with sucrose. The ADP-glucose substrate can be formed from sucrose by two means, either directly by catalysis with sucrose synthetase, as explained in Section 10.3, or by sucrose inversion by the enzyme invertase to form glucose and fructose, followed by phosphorylation and ADP-glucose synthesis.

$$\text{Sucrose} + H_2O \longrightarrow \text{Glucose} + \text{Fructose}$$

$$\text{Glucose} \xrightarrow[\text{Hexokinase}]{\text{ATP}} \text{Glucose-6-P}$$

$$\text{Glucose-6-P} \longrightarrow \text{Glucose-1-P}$$

$$\text{Glucose-1-P} + \text{ATP} \longrightarrow \text{ADP-glucose} + P_iP_i$$

The process of branching to form amylopectin is more complicated. An α-(1→6)-ADP-glucose–starch transglucosylase has been isolated from plant tissues and is believed to catalyze the formation of the α-(1→6) linkages in amylopectin.

Another enzyme called branching enzyme or Q-enzyme catalyzes the formation of α-(1→6) bonds from the α-(1→4) bonds. These enzymes have the name amylo-(1,4→1,6)-transglycosylase.

Thus starch biosynthesis in plants is believed to proceed, starting with sucrose, by the forming of ADP-glucose or UDP-glucose and then by transglucosylation of glucose from the nucleotide diphosphate-glucose to a glucan primer, first forming the α-(1→4) linkages of amylose. The α-(1→6) glucosyl bonds are subsequently formed by catalysis with a specific α-(1→6)-ADP-glucose–starch transglucosylase or by the catalysis of Q-enzyme, transferring the α-(1→4) to an α-(1→6) bond.

10.4.2 Fructosans: inulin and levan

As a carbohydrate reserve, many plants store polymers of fructose rather than glucose. The best known is the polyfructosan inulin, abundant in many monocots such as grasses and also in dahlia and Jerusalem artichoke. Inulin is a polymer of fructose linked by β-(2→1) linkages (β-(2→1)-D-fructofuranose) terminated with a sucrose molecule. Another fructosan, levan, also found in grasses such as *Lolium* and *Dactylis,* is β-(2→6)-D-fructofuranose. Levan is also terminated by a sucrose molecule. Inulin and levan are shown in Fig. 10.5.

The biosynthesis and degradation of the polyfructosans have not been completely elucidated, but transfructosidase enzymes have been detected in plants, and the biosynthesis is more than likely similar to that of the other polymeric carbohydrates.

10.4.3 Cellulose

Cellulose is a straight-chain polymeric glucan with β-(1→4) O-glucosidic linkages (see Fig. 10.5). Complete hydrolysis yields mostly β-D-glucose and trace quantities of other sugars. Cellulose thus differs from amylose only in that it is a polymer of β-D-glucose, whereas amylose is an α-D-glucose polymer. The molecular weight of cellulose has been estimated to be between 200,000 and 2,000,000 and perhaps varies considerably *in situ.* The cellulose molecules found in primary walls are smaller than those found in secondary walls, and there is some evidence that they may be synthesized by a different mechanism. Since glucose has a molecular weight of 180, cellulose is composed of about 1000 to 10,000 glucose residues.

Cellulose is not only one of the most abundant polysaccharides in nature but it is also one of the most resistant to decay. Few organisms can metabolize cellulose to any significant extent. Some bac-

teria and fungi have cellulases (β-glucosidases) that hydrolyze cellulose to form either glucose or cellobiose. The bacteria inhabiting the gut of ruminants have cellulases and can hydrolyze cellulose to small units that are subsequently metabolized by the animal.

Plant tissues have cellulases, but these are special-purpose enzymes that function in leaf and organ abscission and cell-wall breakdown during such processes as ripening rather than functioning in the mobilization of carbohydrate reserves. Cellulase function is discussed in Chapter 19.

The biosynthesis of cellulose has not been completely resolved. Several enzymes have been isolated and studied that can catalyze the β-transglucosylation of glucose from nucleoside diphosphate glucose derivatives. Whether UDP-glucose or GDP-glucose is the *in vivo,* natural donor is perhaps the most interesting question remaining. Several experiments have shown that both UDP-glucose and GDP-glucose (but not ADP-glucose) can donate glucose during β-glucan biosynthesis, but a β-(1→4) glucan was not always synthesized. In some experimental situations β-(1→3) glucans are produced, and there is some evidence that GDP-glucose functions in the biosynthesis of β-(1→4)-mannan or glucomannan polymers. Rates of β-(1→4) glucan or cellulose biosynthesis from UDP-glucose seem to be faster than from GDP-glucose, and it is reasonable to conclude at this time that UDP-glucose is the natural *in vivo* glucosyl donor during cellulose biosynthesis.

Complicating the question of cellulose biosynthesis is that the polymeric structure of cellulose in primary walls is somewhat different from that found in secondary walls. In some experiments with developing cotton fibers, GDP-glucose was the best glucosyl donor for primary-wall biosynthesis. It is possible that different donors function at different periods of development.

10.4.4 Other plant polysaccharides

All of the sugars can exist in polymeric form either as homopolymers (one kind of sugar) or as mixed heteropolymers. Many of these polymeric saccharides, such as the hemicelluloses and pectins, are structural components of cell walls.

Hemicelluloses: mannans and xylans

The group of polysaccharides called hemicelluloses were named because they are celluloselike, can be extracted from cell walls with a strong base, and were at one time believed to be precursors for cellulose biosynthesis. The hemicelluloses, including polymannose and polyxylose, are named according to their sugar constituents. Thus there are mannans and xylans and mixed polymers such as glucomannans and galactomannans. The xylans (polymers of xylose) are, next to cellulose, the most abundant cell-wall polysaccharides.

Mannopyranoses are known that are linked both through α-$(1 \rightarrow 4)$ and α-$(1 \rightarrow 6)$ O-glycosyl bonds. As stated above, many of these mannans have galactose and glucose residues as well. Since both ADP-mannose and GDP-mannose are known from plants, either ADP-mannose or GDP-mannose may function as the donors during mannan biosynthesis. The mannans, like starch, may function as carbohydrate reserves.

Pectins

Pectins and pectic acids are common polysaccharide constituents of primary cell walls. They are mixtures of araban, galactan, and galactouronic acid. The arabans are small, branched polymers with α-$(1 \rightarrow 3)$- and α-$(1 \rightarrow 5)$-L-arabinofuranose units, whereas the galactans are larger, straight chains of β-$(1 \rightarrow 4)$-D-galactopyranose units. The acidic portion comes from polymers of α-$(1 \rightarrow 4)$-D-galactopyranosyluronic acid.

The substances called pectic acid are small polymers with 100 or fewer residues. They are common in the middle lamella and, along with the larger pectins, function somewhat as a cementing material to hold adjacent cells together. The pectic acids are water-soluble but may form insoluble salts with calcium. The pectins are usually larger than pectic acid and have the carboxyls methylated.

They occur in the middle lamella, within the primary cell wall, and as cytoplasmic constituents.

A still larger polymer with properties similar to pectin and having 200 or more residues is called protopectin. Protopectin has methylated and free carboxyl groups. It is abundant in primary cell walls. During the ripening of some fruits protopectin is converted to pectin.

Pectinases such as polygalacturonase are common enzymes of plants and microorganisms. The decay of plant material and the breakdown or overripening of fruit frequently occur because of the activity of pectinases that hydrolyze the pectic substances to their constituent monosaccharide residues.

Pectinases have recently become useful for the preparation of protoplasts from plant cells. Protoplasts are cells without the cell walls. Pectinases are used to hydrolyze the middle lamella, freeing individual cells. Cellulases are then used to remove the cell wall and form the free protoplast. Protoplasts have been used in a variety of physiological studies, such as the determination of the mechanism for ion uptake. The protoplasts are useful for ion uptake studies because the highly charged cell wall is removed and transport is more a function of the cell membrane. Protoplasts prepared enzymatically are shown in Fig. 1.11.

Gums and mucilages

The plant gums and mucilages are hydrophilic (not hygroscopic) polymers of rhamnose, arabinose, xylose, galactose, glucuronic acid, and galacturonic acid. They are abundant in many seeds and in the leaves and stems of succulent plants. They may occur free throughout cells or may be concentrated in specialized mucilage cells such as occur just below the epidermal surface in some cacti (Fig. 10.6).

Because of their large water-holding capacity there is the belief that they may function in seed germination by absorbing water during imbibition and also as water-holding molecules in plants adapted to arid conditions, such as the succulents.

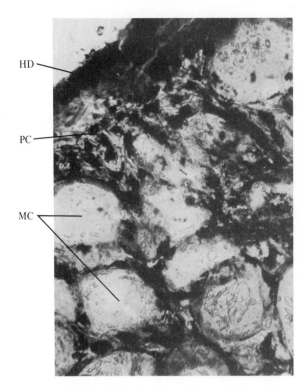

HD

PC

MC

FIGURE 10.6 Cross section through stem tissue from cactus showing hypodermis (HD), chloroplast-containing parenchyma cells (PC), and mucilage-containing cells (MC). The mucilage-containing cells occur in groups just below the hypodermis. From an experiment by Dr. B. G. Sutton. 1976. University of California, Riverside.

10.4.5 The chemical structure of the cell wall

The primary cell wall is composed of cellulose, various hemicelluloses, pectic substances, and protein. In 1976 Peter Albersheim, using primary cell walls isolated from suspension cultures of sycamore cells, has developed a model for the arrangement of the polymers that make up the structure of the primary cell wall. The evidence suggests that the polymers are bonded to each other through either hydrogen bonds or covalent bonds. A simplified version of the Albersheim model, based on his own work and that of others, is shown in Fig. 10.7. The model is primarily based on the chemical structure of the known cell-wall constituents, their properties, and selective enzymatic degradation of the cell wall.

The hemicelluloses are bonded directly to the cellulose chains, evidently through hydrogen bonds. The evidence largely indicates that the hemicellulose xyloglucan will bind to purified cellulose and to the cellulose of cell walls. Pectins are covalently bonded to the hemicellulose and to the cell-wall protein, extensin, through serine residues of the protein.

The structure of the secondary wall is not as well understood, but it is known to contain the same substances plus lignin. Organization of the secondary wall is not known, but the cellulose molecules are larger and laid down in layers of microfibrils. The cellulose organization of the secondary wall seems to be more compact than that of the primary cell wall.

Although the cell wall appears to be highly permeable to most substances and shows little if any selectivity with respect to transport, it seems quite evident that there will be some influence. There is much structure and organization and many charged groups that will affect transport.

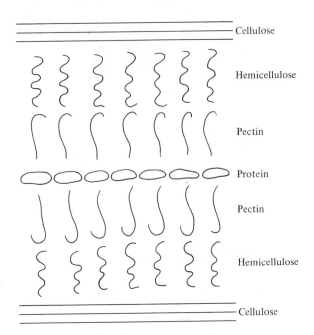

FIGURE 10.7 Arrangement of chemicals in the primary cell wall. The diagram is a simplification of a model proposed by Albersheim. The model is based on the known chemical composition of the primary cell wall, the properties of the various constituents, and on chemical and enzymatic-degradation studies of the cell wall. The hemi-celluloses are bonded to the cellulose probably through hydrogen bonds. The pectins are covalently linked to hemicelluloses on one side and to the cell wall protein on the other. From P. Albersheim in J. Bonner and J. E. Varner. 1976. *Plant Biochemistry*. Academic Press, New York.

10.5 The Organic Acids

The organic acids play a central role in plant metabolism. They can be defined as small mono-, di-, and tricarboxylic acids that are soluble in water. They do not include the fatty acids. The latter are discussed in Chapter 11. Some of the important and interesting organic acids of plants are shown in Fig. 10.8.

Organic acids are important in plant metabolism because of their role in intermediary metabolism in the tricarboxylic acid cycle (see Chapter 14). They are also integral components of photosynthesis and dark CO_2 fixation, and they play a role in the storage of carbon and in ionic and osmotic balance.

It is not uncommon for plants to accumulate one or more of the organic acids in excess. In fact, many plants have names consistent with the organic acids that they accumulate. Thus *Oxalis* accumulates oxalic acid, *Citrus* accumulates citric acid,

Malus (apple) accumulates malic acid, and *Fumaria* accumulates fumaric acid.

The organic acids may exist in plant tissues as free acids or as anions. In the plants mentioned above, the free acid accumulates. For example, oxalic acid accumulates in *Oxalis*. The ending "ic" in oxalic acid designates the protonated form (COOH—COOH), whereas the ending "ate" of oxalate indicates the ionized, anionic form (COO⁻—COO⁻). The names of the free acids end in "ic" (e.g., oxalic acid, citric acid, etc.) and the names of their anions end in "ate" (e.g., oxalate, citrate, etc.). With the exception of the kinds of examples given above and the accumulation of malic acid during crassulacean acid metabolism (see Chapter 16), it is not the free acid that occurs but the anion. This is because at physiological pH the acids are ionized, forming anions. For this reason, it is usually more proper to talk about the

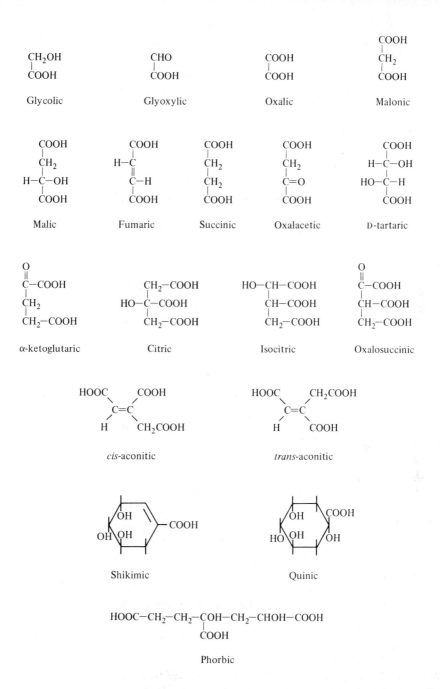

FIGURE 10.8 Structures of some of the common organic acids found in plants.

anions of the organic acids rather than about the free acids. The most important anions are citrate, *cis*-aconitate, isocitrate, oxalosuccinate, α-ketoglutarate, succinate, fumarate, malate, and oxalacetate, all components of the tricarboxylic acid cycle of respiration. As part of the respiratory cycle they occur in mitochondria in catalytic amounts.

With the possible exception of oxalosuccinate, all of the abovementioned organic acids occur outside the mitochondria in other cellular compartments. Except for the very reactive keto acids, oxalacetate and α-ketoglutarate, many may be in concentrations exceeding 100 mM. Depending on the cellular compartment in which they accumulate, the acids will be either free acids or anions. As a rule, those occurring in active metabolic compartments such as chloroplasts, microbodies, the cytosol, and in mitochondria will be anions, whereas those accumulating in high concentrations in vacuoles may be free acids depending, of course, on the pH of the vacuolar contents.

The first product of photosynthesis is an organic acid, phosphoglycerate, and it is not until reduction to an aldose sugar, glyceraldehyde phosphate, that sugar is formed. Furthermore (as will be explained in Chapter 16) the two two-carbon acids, glyoxylate and glycolate, play a major role in the energetics of photosynthesis.

The simple monocarboxylic acids (which can also be considered the simplest fatty acids) formate ($HCOO^-$) and acetate (CH_3COO^-) are important metabolic intermediates. Formate is an important component of biosyntheses in which one carbon is added or removed to other groups through the tetrahydrofolic acid derivative (see purine biosynthesis in Chapter 9). Acetate is the beginning organic acid of respiration in the tricarboxylic acid cycle and is both a substrate for fatty acid synthesis and a product of fatty acid oxidation (see Chapter 11). Malonate is also involved in fatty acid synthesis in catalytic amounts.

The organic acids discussed in this section are those that tend to accumulate to high concentrations and that play a role other than in cellular energetics. Many, of course, have more than one role. Malate, citrate, isocitrate, and fumarate, as previously mentioned, are metabolic intermediates and also accumulate outside the mitochondria. Many studies have shown conclusively that many of the plant acids occurring in cellular storage pools were not in equilibrium with their counterparts in the mitochondria. The integral organic acids of the citric acid cycle, such as malate, isocitrate, and citrate, do not equilibrate with the pools of malate, isocitrate, and citrate outside of the mitochondria. It was shown, for example, that less than 30% of the malate was in an active turnover pool in the mitochondria. The remainder was probably stored in cellular vacuoles.

Malate, perhaps the most important organic acid, may be synthesized by carboxylation reactions. The most common reaction is the carboxylation of P-enolpyruvate to form oxalacetate, which is rapidly reduced to malate. The carboxylating enzyme is the cytoplasmic P-enolpyruvate carboxylase (PEP carboxylase), and the enzyme that reduces oxalacetate to malate is an NADH malate dehydrogenase localized in the cytosol. The overall reactions follow.

$$PEP + HCO_3^- \longrightarrow OAA + P_i$$

$$OAA + NADH \longrightarrow MAL + NAD$$

The enzyme actually uses bicarbonate, not free CO_2. The product, malate or malic acid, accumulates in vacuoles up to concentrations of 100 to 200 mM in certain plants, especially the succulents that have crassulacean acid metabolism (CAM). In the CAM plants malic acid accumulates, not malate. In plants other than those having CAM, its major role is in ionic and osmotic balance, where it accumulates primarily as malate. There is some evidence that the buffer of the xylem is a malate buffer.

Malate can be metabolized by oxidative decarboxylation in the cytosol to pyruvate and CO_2 by NADP malate enzyme (there is also an NAD-specific form of this enzyme found in mitochondria).

$$MAL + NADP \rightleftharpoons PYR + CO_2 + NADPH$$

There is one other important plant enzyme that can synthesize or decarboxylate oxalacetate during malate metabolism: P-enolpyruvate carboxykinase, a mitochondrial enzyme.

$$PEP + CO_2 + ADP \rightleftharpoons OAA + ATP$$

These enzymes—P-enolpyruvate carboxylase, P-enolpyruvate carboxykinase, and malate enzyme—are important catalysts in C_4-photosynthesis and crassulacean acid metabolism (see Chapter 16).

Oxalic acid is an important acid of plant tissues, accumulating as the acid in *Oxalis, Begonia,* and spinach, and as the insoluble calcium salt in tobacco, rhubarb, and cactus. The synthesis of oxalate is somewhat uncertain; however, it probably results from oxidation of glycolate and glyoxylate.

$$\underset{\text{Glycolate}}{COO^- \ CH_2OH} \longrightarrow \underset{\text{Glyoxylate}}{COO^- \ CHO} \longrightarrow \underset{\text{Oxalate}}{COO^- \ COO^-}$$

Oxalate can be consumed by oxidative decarboxylation to form CO_2, but the degradative reactions are usually much slower than the biosynthetic reactions, and either oxalate or oxalic acid accumulates.

Malonate is known to accumulate in many legumes such as beans and in some species of the *Umbelliferae.* Its metabolic role is unknown other than in fatty acid biosynthesis as malonyl CoA (see Chapter 11), but it could function in ionic balance and osmotic regulation or as a carbon reserve since it can readily be metabolized to acetate.

Its synthesis was shown to be by oxidative decarboxylation of oxalacetate.

$$Oxalacetate + \tfrac{1}{2} O_2 \longrightarrow Malonate + CO_2$$

Tartaric acid is one of the better known plant acids; it accumulates in grapes as well as in geraniums and legumes. Both (+) and (−) tartrate occur in plants. It exists as the insoluble salt of potassium and as the free acid. Its path of biosynthesis is not known.

Aconitate accumulates in many grasses and it is the predominant organic acid of maize and sugarcane. Interestingly, the active acid of the tricarboxylic acid cycle in mitochondria is the *cis*-isomer,

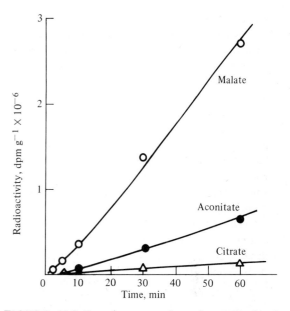

FIGURE 10.9 Experiment to show the synthesis of organic acids from $^{14}CO_2$ (dark CO_2 fixation) by maize roots. $^{14}CO_2$ was supplied to the roots for the times indicated on the abscissa. Organic acids were extracted and identified by chromatography. The predominant organic acid synthesized by the carboxylation is malic acid, but *trans*-aconitic acid is the most abundant organic acid in the roots. From a student experiment. 1970. University of California, Riverside.

whereas the *trans*-isomer is the form that accumulates in vacuoles. Its biosynthesis is not known, although in maize it is the main end product of dark CO_2 fixation. Malate is the first stable organic acid formed from CO_2 fixation in the dark, but it is *trans*-aconitate that accumulates. Figure 10.9 shows the accumulation of malate, *trans*-aconitate, and citrate during $^{14}CO_2$ uptake by maize roots. After 60 minutes of dark CO_2 fixation, malate has most of the label, but after several hours it is aconitate that accumulates.

In many succulent plants of the Crassulaceae, isocitric acid tends to accumulate because it is the most predominant organic acid in the leaves. Unlike malic and citric acids, it is not synthesized in the Crassulaceae through the reactions of dark CO_2 fixation (that is, by crassulacean acid metabo-

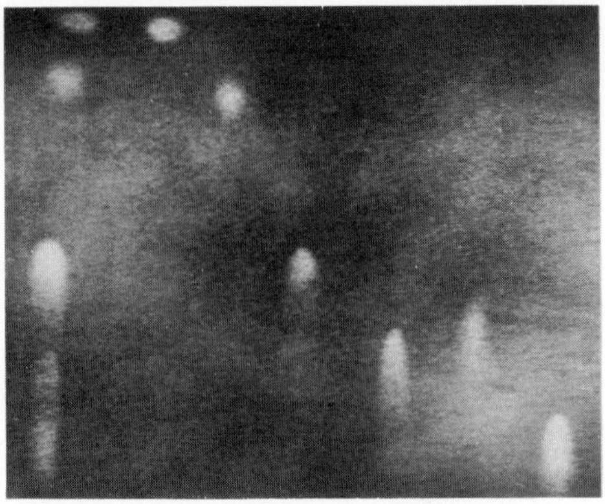

FIGURE 10.10 Thin-layer chromatography of water-soluble organic acids in plant tissue. Column 1 is the mixture obtained from the plant, 2 = fumaric acid standard, 3 = succinic acid standard, 4 = malic acid standard, 5 = citric acid standard, 6 = isocitric acid standard, and 7 = tartaric acid standard. To visualize the acids, the silica gel chromatography plate was sprayed with a blue solution of bromophenol blue that turns yellow when it encounters acid. If an organic acid is present, it will appear as a yellow spot on the blue background. From a student experiment. 1970. University of California, Riverside.

lism). Moreover, it is not in equilibrium with the isocitrate of the tricarboxylic acid cycle. Its function is unknown.

Other plants accumulate the acids found in the tricarboxylic acid cycle as well. In addition to *Fumaria,* fumaric acid also accumulates in *Tagetes* and *Helianthus.* Succinic acid accumulates in *Vinca* and in alfalfa.

Many other plants accumulate a variety of little-known mono-, di-, and tricarboxylic acids. Acids such as phorbic acid (an 8-carbon tricarboxylic acid) in *Euphorbia* and certain other succulents, and piscidic acid (*p*-hydroxybenzyl tartaric acid) in *Agave,* legumes, and lilies have been isolated and identified.

It is clear from many studies that the same acids found in the tricarboxylic acid cycle tend to accumulate in plant vacuoles in large quantities. Their exact metabolic role is uncertain since they tend to accumulate out of active metabolic pools. Under some circumstances, such as during stress, they may be mobilized and metabolized. For example, if malate production is inhibited in the tricarboxylic acid cycle, storage malic acid may be transported to the mitochondria and oxidized. Thus storage is at least one function for some of the organic acids that accumulate. In other tissues, however, the storage organic acids do not change in concentration even

under severe starvation such as experienced in total darkness.

Since organic acids accumulate in developing fruit, a role in fruit maturation is likely. In fruits such as grapes, which become sweet, tartaric acid decreases during ripening. However, in fruits such as citrus, citric acid remains high throughout the ripening process. In young fruits, citramalic, quinic, and shikimic acids may accumulate as well as citric acid, malic acid, and tartaric acid.

There is some evidence that organic acids such as malic and oxalic acids may detoxify heavy metals such as zinc and copper.

Most measurements and assays of the organic acids take advantage of the carboxyl groups. They can be readily separated and identified by paper and thin-layer chromatographic procedures. Detection on paper and on thin-layer plates of silica gel or cellulose is usually accomplished by spraying acid–base indicators to give a reaction wherever acids are present. Figure 10.10 is a photograph of a one-dimensional separation of the organic acids of the tricarboxylic acid cycle on silica gel. The acids were located by spraying the plate with bromophenol blue, which turns yellow if acid is present.

In addition, organic acids can be separated by silicic acid chromatography. A mixture of acids prepared from plant tissue can be eluted from the

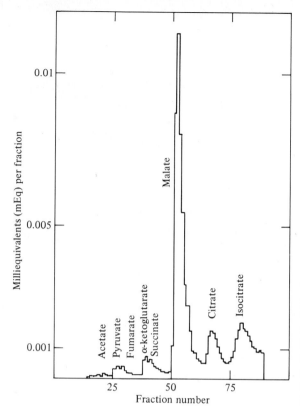

FIGURE 10.11 Silicic acid column chromatography of the organic acids extracted from *Kalanchoe*, a succulent plant. The organic acids are eluted from the column with ether, and each fraction (containing several milliliters) is dried and titrated with NaOH to measure the amount of acid in the fractions. The acids are identified by the sequence in which they elute from the column. From an experiment by Dr. Orlando Queiroz. Cited in M. Kluge and I. P. Ting. 1978. *Crassulacean Acid Metabolism.* Springer-Verlag, New York.

column with ether. Sequential samples are dried and titrated with base to quantify the acids present. The acids are identified by the sequence in which they come off the column. Figure 10.11 shows such an elution profile of the organic acids from *Kalan-* *choe.* This particular sample has acetate, pyruvate, fumarate, α-ketoglutarate, succinate, malate, citrate, and isocitrate. Malate and isocitrate are present in the highest concentrations.

Review Exercises

10.1 How could you differentiate between glucose and sucrose on the basis of their chemical properties?

10.2 Draw the Fischer projection formulas and the Haworth formulas for the D and L forms of the possible hexoses. How many optical isomers of a heptose sugar could there be? Can you draw them?

10.3 With the use of diagrams differentiate between transaldolase and transketolase reactions.

10.4 What is the structural difference between starch and cellulose? Illustrate your answer using a diagram. Do you think that cellulose and starch are sufficiently similar so that the amylases (enzymes that hydrolyze starch) would hydrolyze cellulose? Why?

10.5 With the use of chemical diagrams, indicate the structural relationships among aspartate, oxalacetate, malate, fumarate, and succinate. Do the same for citrate, isocitrate, α-ketoglutarate, and glutamate.

References

BAILEY, R. W. 1965. *Oligosaccharides*. Macmillan, New York.

BANKS, W., and C. T. GREENWOOD. 1975. *Starch and Its Components*. Halsted Press, New York.

GUTHRIE, R. D. 1974. *Introduction to Carbohydrate Chemistry*. 4th ed. Clarendon Press, Oxford.

HASSID, W. Z. 1967. Transformation of sugars in plants. *Ann. Rev. Plant Physiol.* 18:253–280.

HAWORTH, W. N. 1929. *The Constitution of Sugars*. Longmans, New York.

LOEWUS, F. 1971. Carbohydrate interconversions. *Ann. Rev. Plant Physiol.* 22:337–364.

LOEWUS, F. (ed.). 1973. *Biogenesis of Plant Cell Wall Polysaccharides*. Academic Press, New York.

NORDIN, J. H., and S. KIRKWOOD. 1965. Biochemical aspects of plant polysaccharides. *Ann. Rev. Plant Physiol.* 16:393–414.

PRESTON, R. D. 1974. *The Physical Biology of the Plant Cell Wall*. Chapman and Hall, London.

PRIDHAM, J. B. (ed.). 1974. *Plant Carbohydrate Biochemistry*. Academic Press, London.

TURNER, J. F., and D. H. TURNER. 1975. The regulation of carbohydrate metabolism. *Ann. Rev. Plant Physiol.* 26:159–186.

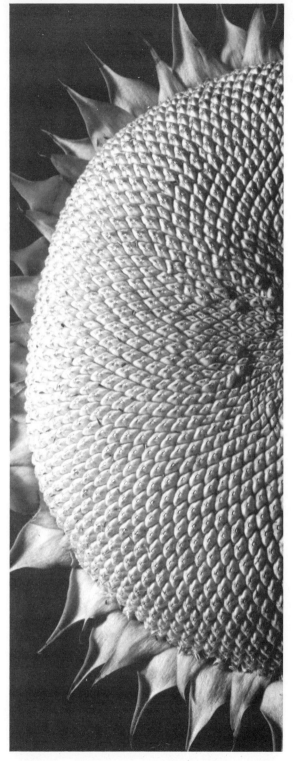

CHAPTER ELEVEN

Lipids and Lipid Biochemistry

Lipids are a rather mixed group of compounds that have the characteristic of being hydrophobic. The best definition for lipids is that they are soluble in organic solvents such as acetone, chloroform, and ether and are only slightly soluble in water. Of course, within the other two major groups of organic compounds, proteins and carbohydrates, there are water-insoluble compounds such as the prolamine proteins and the carbohydrate cellulose. The lipids have a greater proportion of hydrophobic groups than do the other classes of biochemicals.

Lipids can be grouped into neutral lipids, phospholipids, glycolipids, terpenoids, and waxes. Lipids also include some vitamins and pigments. Because of their heterogeneity lipids are difficult to classify on the basis of their chemical properties.

Lipids are important in plant physiology and metabolism because they are storage compounds and components of membranes and act as protective substances for exposed surfaces. The neutral lipids, including the fats and oils, function primarily as storage molecules, and the phospholipids

283

and glycolipids are membrane components. The terpenoids include the sterols and a variety of secondary products such as rubber, chicle, and the essential oils. Protective lipids, mostly waxes, are largely long-chain fatty acids and long-chain esters of fatty acids and alcohols.

Many of the lipids are quite well known either in pure form or as mixtures. Examples of well-known plant lipids are olive oil, linseed oil, peanut oil, castor oil, lecithin, vitamin A, rubber, and carnauba wax.

The lipids of plants are discussed below except for the terpenoids and sterols, which are discussed as natural products in Chapter 12.

11.1 Neutral Lipids

The neutral lipids include the fats and oils. They have no charged functional groups. When hydrolyzed they yield only fatty acids and the trihydroxy alcohol glycerol. Waxes are basically neutral lipids that yield long-chain fatty acids and alcohols upon hydrolysis. They are discussed in detail in Section 11.4 because of their unique role in physiology.

The fats and oils are almost exclusively triglycerides (or, more properly, triacylglycerides)composed of fatty acids linked to glycerol by means of ester bonds. Fatty acids may be straight-chain carboxylic acids, branched-chain molecules, or may have heterocyclic ring groups. They may be saturated or unsaturated, with double or triple carbon-carbon bonds.

Oils, which are the main triglycerides of plants, are liquid at room temperatures because of the preponderance of unsaturated fatty acids. Fats are generally solids at room temperatures and contain high proportions of straight-chain, saturated fatty acids.

The basic structure of the triglyceride follows.

$$
\begin{array}{l}
CH_2OOCR \\
| \\
CHOOCR \\
| \\
CH_2OOCR \\
\text{Triglyceride}
\end{array}
$$

R denotes the hydrocarbon portion of a fatty acid.

A monoglyceride has one fatty acid, and a diglyceride has two fatty acids. The free alcohol group(s) of these glycerides may be substituted with other groups, forming other classes of lipids, which are discussed in subsequent sections.

Acid hydrolysis of triglycerides produces glycerol, a three-carbon trihydroxy alcohol, and free fatty acids in the acid form. Free fatty acids and alcohols will readily esterify such that the hydrolysis is reversible.

$$
\begin{array}{l}
CH_2OH \\
| \\
CHOH \\
| \\
CH_2OH \\
\text{Glycerol}
\end{array}
$$

Base hydrolysis (with KOH, for example) is not reversible because the salts of the fatty acids have charged groups and thus little tendency to reform ester bonds. Base hydrolysis of fats and oils is termed saponification, the process that produces soaps. Soaps are the salts of fatty acids, and because of their solubility in water and hydrophobic side chains, they will bind to grease and oil.

The neutral lipids (and also phospholipids, glycolipids, and waxes) can be saponified, but the terpenoids cannot be. Saponification of a triglyceride can be illustrated as follows.

$$
\begin{array}{l}
CH_2OOCR \\
| \\
CHOOCR \quad +3KOH \longrightarrow \\
| \\
CH_2OOCR \\
\text{Triglyceride} \qquad \text{Base}
\end{array}
\quad
\begin{array}{l}
CH_2OH \\
| \\
CHOH \quad +3RCOOK \\
| \\
CH_2OH \\
\text{Glycerol} \quad \text{Fatty acid} \\
\qquad\qquad \text{salts}
\end{array}
$$

For the most part, fats and oils are mixtures of triglycerides and not pure chemicals in the usual sense. Very few triglycerides with three identical fatty acids exist in nature. Most of the naturally

TABLE 11.1   Names and formulas for some of the common fatty acids

SATURATED FATTY ACIDS

BUTYRIC ACID	$CH_3(CH_2)_2COOH$
CAPROIC ACID	$CH_3(CH_2)_4COOH$
CAPRYLIC ACID	$CH_3(CH_2)_6COOH$
CAPRIC ACID	$CH_3(CH_2)_8COOH$
LAURIC ACID	$CH_3(CH_2)_{10}COOH$
MYRISTIC ACID	$CH_3(CH_2)_{12}COOH$
PALMITIC ACID	$CH_3(CH_2)_{14}COOH$
STEARIC ACID	$CH_3(CH_2)_{16}COOH$
ARACHIDIC ACID	$CH_3(CH_2)_{18}COOH$
BEHENIC ACID	$CH_3(CH_2)_{20}COOH$
LIGNOCERIC ACID	$CH_3(CH_2)_{22}COOH$
CEROTIC ACID	$CH_3(CH_2)_{24}COOH$

UNSATURATED FATTY ACIDS

CROTONIC ACID	$CH_3CH = CHCOOH$
PALMITOLEIC ACID (9)*	$CH_3(CH_2)_5CH = CH(CH_2)_7COOH$
OLEIC ACID (9)	$CH_3(CH_2)_7CH = CH(CH_2)_7COOH$
LINOLEIC ACID (9,12)	$CH_3(CH_2)_3(CH_2CH = CH)_2(CH_2)_7COOH$
ELEOSTEARIC ACID (9,11,13)	$CH_3(CH_2)_3(CH = CH)_3(CH_2)_7COOH$
LINOLENIC ACID (9,12,15)	$CH_3(CH_2CH = CH)_3(CH_2)_7COOH$
ARACHIDONIC ACID (5,8,11,14)	$CH_3(CH_2)_3(CH_2CH = CH)_4(CH_2)_3COOH$

*Double bond between carbons 9 and 10.

occurring triglycerides are mixed triglycerides and have two or three different fatty acids esterified to the glycerol moiety.

11.1.1 Fatty acids

The common straight-chain fatty acids found in nature are illustrated in Table 11.1. Not including formic and acetic acids, which are quite water-soluble because of their small size (only a small portion of the molecule is hydrophobic), the common fatty acids range from butyric acid with four carbons to cerotic acid with 26 carbons. The most common straight-chain saturated fatty acid in plants is the 16-carbon palmitic acid, although 12-carbon lauric acid, 14-carbon myristic acid, and 18-carbon stearic acid are all quite abundant in plant material.

Most fatty acids are even-numbered (an even number of carbon atoms), and only trace amounts of the odd-numbered fatty acids are detected in plant tissues. The even numbers are probably the result of the mode of biosynthesis (see Section 11.5.1).

By far the most common fatty acid in plants is the 18-carbon monounsaturated fatty acid known as oleic acid. The double bond occurs between carbons 9 and 10. By convention, unsaturation is indicated by designating the first carbon of the pair with an unsaturated bond, with numbering starting from the carboxyl.* Thus oleic acid is 9-octadecenoic acid. The common unsaturated fatty acids are depicted in Table 11.1.

*It is common to indicate the degree of unsaturation and the number of carbon atoms in a fatty acid as follows: C 16:0 = 16-carbon fatty acid with no double bonds; C 16:1 = 16-carbon fatty acid with 1 double bond; C 16:2 = 16-carbon fatty acid with 2 double bonds, and so on.

TABLE 11.2　Examples of fatty acids found in some seeds commonly used as sources of vegetable oils

Seed	Oleic	Linoleic	Palmitic	Stearic
Olive	83%	7%	6%	4%
Sunflower	34	59	4	3
Cotton	31	45	22	2
Peanut*	61	22	6	4

*Also has arachidic acid and trace quantities of others.

Data are reported in Altman and Dittmer (1968).

Linoleic acid, with 18 carbons and double bonds at carbons 9 and 12, and linolenic acid, with 18 carbons and double bonds at carbons 9, 12, and 15, are the next most common unsaturated fatty acids. Table 11.2 shows some examples of the proportions of fatty acids in seeds.

The *cis* isomers of the unsaturated fatty acids are the most common in nature, although the *trans* isomer of oleic acid is known to occur in plants. The *trans* isomer of oleic acid has its own name—elaidic acid.

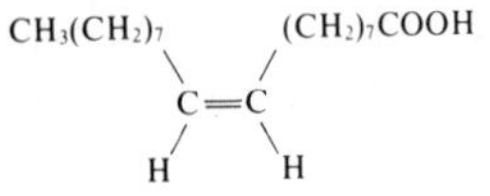

Oleic acid

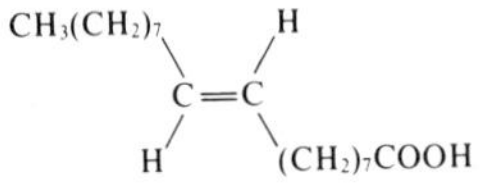

Elaidic acid

Several other fatty acids have been isolated and identified from plant sources. These include tariric and ximenynic acids with triple bonds, ricinoleic acid with a hydroxyl group, sterculic and chaulmoogric acids with ring components, vernolic acid with an epoxy group, and licanic acid with a ketone group. All of these fatty acids have 18 carbons. The only 20-carbon fatty acid of importance is japanic acid, found in *Rhus*. Japanic acid is also a dicarboxylic acid. A sulfhydryl-containing fatty acid, lipoic acid, is a common constituent in plants and functions in metabolism as a cofactor in enzymatic reactions. These complex fatty acids are illustrated in Fig. 11.1.

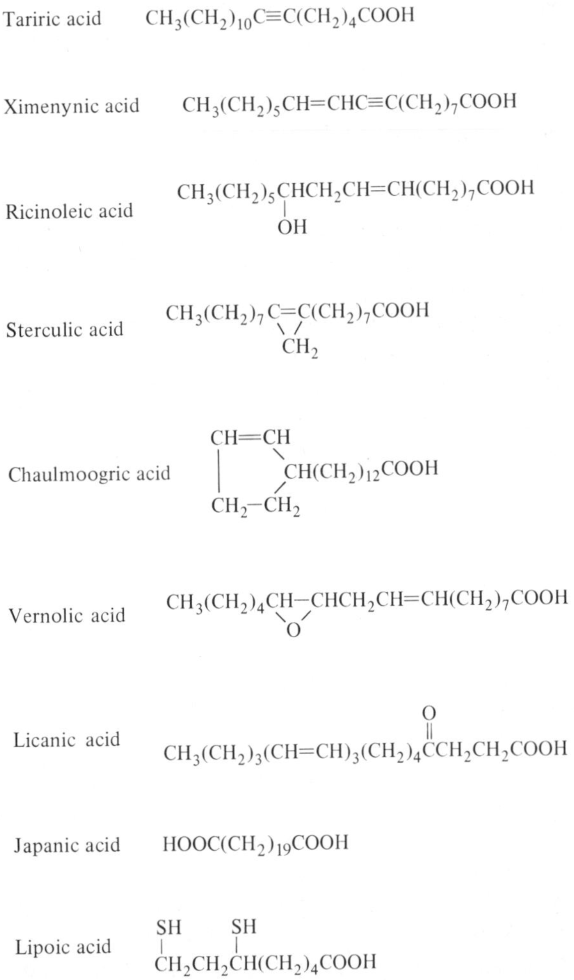

Tariric acid　　$CH_3(CH_2)_{10}C\equiv C(CH_2)_4COOH$

Ximenynic acid　　$CH_3(CH_2)_5CH=CHC\equiv C(CH_2)_7COOH$

Ricinoleic acid　　$CH_3(CH_2)_5\underset{\underset{OH}{|}}{C}HCH_2CH=CH(CH_2)_7COOH$

Sterculic acid　　$CH_3(CH_2)_7\underset{\underset{CH_2}{\diagdown\diagup}}{C}=C(CH_2)_7COOH$

Chaulmoogric acid　　$\begin{matrix}CH=CH\\ |\qquad\diagdown\\ \quad CH(CH_2)_{12}COOH\\ CH_2-CH_2\diagup\end{matrix}$

Vernolic acid　　$CH_3(CH_2)_4\underset{\underset{O}{\diagdown\diagup}}{C}H-CHCH_2CH=CH(CH_2)_7COOH$

Licanic acid　　$CH_3(CH_2)_3(CH=CH)_3(CH_2)_4\overset{\overset{O}{\|}}{C}CH_2CH_2COOH$

Japanic acid　　$HOOC(CH_2)_{19}COOH$

Lipoic acid　　$\underset{CH_2CH_2CH(CH_2)_4COOH}{\overset{SH\quad\ SH}{\ |\qquad\ |}}$

FIGURE 11.1 Structures of some of the complex fatty acids found in plants. See Table 11.1 for the common saturated and unsaturated fatty acids.

There are several important chemical properties of the fatty acids that are useful for analysis. Hydrolysis by saponification was previously mentioned and is frequently the first step in the analysis of fats and oils. An index known as the saponification number represents the milligrams of KOH required to neutralize the fatty acids from one gram of fat. It can be used to estimate the average molecular weight of the fatty acids isolated from fats or oils. The Reichert–Meissel number, determined by titration of volatile fatty acids (fatty acids with 12 carbons or less will steam-distill), gives an indication of the proportion of short- and long-chain fatty acids in fats and oils. The double or triple bonds of fatty acids can be reduced by hydrogenation or oxidized by manganate or ozone. Oxidation results in cleavage, forming smaller aldehydes and carboxylic acids that can be readily identified by chromatographic procedures.

Paper and thin-layer chromatography of fatty acids is easy and quick and provides an excellent analytical tool for analysis. Furthermore, many gas-liquid chromatographic techniques are available for the quantitative analysis of fatty acids. Figure 11.2 is a gas–liquid chromatogram showing fatty acids isolated from maize leaves. The most

predominant fatty acid is linolenic acid with three double bonds.

11.1.2 Triglycerides

The triglycerides existing in nature are of the mixed type, having different fatty acids rather than being homogeneous with a single kind of fatty acid. One or more of the fatty acids of triglycerides may be characteristic of a plant species, but they are always present in mixtures. Triglycerides of plants are rich in unsaturated fatty acids, meaning that they are liquids at room temperatures. In contrast, most animal triglycerides (except fish) have a higher proportion of saturated fatty acids. Table 11.2 is an indication of the preponderance of unsaturated fatty acids in plant triglycerides.

The triglycerides are abundant in certain seeds and are deposited throughout cells. Oil droplets can frequently be observed in plastids, particularly in chloroplasts. Many chloroplasts deposit oil droplets called plastoglobuli rather than starch grains. A cross section through the palisade parenchyma of a citrus leaf showing extensive deposition of oil droplets in plastoglobuli within the chloroplasts is shown in Fig. 11.3.

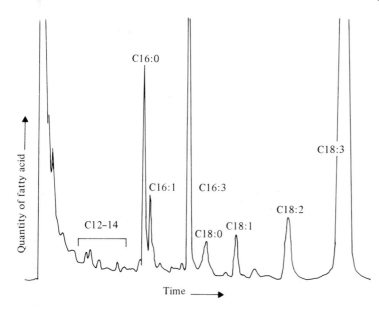

FIGURE 11.2 Tracing of chromatogram obtained by the gas-liquid chromatographic separation of fatty acids isolated from maize leaves. The beginning portion represents fragments and fatty acids with 12 to 14 carbons. The 16- and 18-carbon fatty acids are the most predominant. Chromatogram courtesy of R. L. Heath, University of California, Riverside.

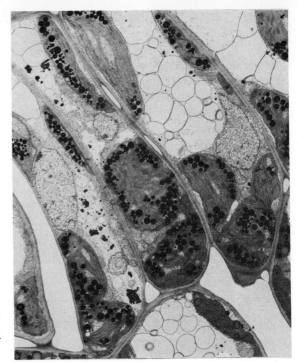

FIGURE 11.3 Electron micrograph through the palisade parenchyma of a citrus (navel orange) showing extensive deposition of oil droplets (dark spots) within chloroplasts. The oil droplets are called plastoglobuli and are believed to be mostly lipid. Photograph courtesy of Dr. W. W. Thomson, University of California, Riverside.

11.2 Phospholipids

Phospholipids are common components of cellular membranes and are characterized by the yielding of free inorganic phosphate upon hydrolysis. The parent compound derived from glycerol-3-phosphate is phosphatidic acid, a diglyceride with phosphate linked by an ester bond to glycerol.

$$CH_2COOR$$
$$CHCOOR$$
$$CH_2OPO_3H_2$$
Phosphatidic acid

In many, the diglyceride moiety has a saturated fatty acid in position 1(α) and an unsaturated fatty acid in the central position (2 or β).

Phosphatidic acid itself is rare in plant tissues and probably is present to any appreciable extent only during active metabolism of phospholipids. Phosphatidic acid can be synthesized in plants either by phosphorylation of a diglyceride through ATP or by acylation of glycerol-3-P with fatty acid-CoA substrates. The various phosphatidic acid derivatives found in plants are called phosphatides.

The structure of the phospholipids relates directly to membrane form and function. Hydrophobic chains of the phospholipids readily form hydrophobic bonds with adjacent nonpolar molecules, resulting in lipid micelles with exposed hydrophobic groups. Monolayers over liquids can form sheets of exposed hydrophobic groups as well. Examples of such structures are depicted in Fig. 11.4.

11.2.1 The common phosphatides

Perhaps the most common phosphatide of leaf tissue is phosphatidyl glycerol.

$$CH_2OOCR$$
$$CHOOCR$$
$$CH_2OPO_3-CH_2-CHOH-CH_2OH$$
Phosphatidyl glycerol

Phosphatidyl glycerol is a phosphatidic acid linked to glycerol through a phosphate ester bond.

In seeds and certain other tissues, the two most

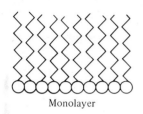

Monolayer

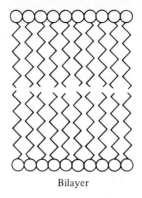

Bilayer

Micelle

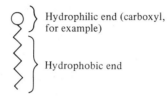

} Hydrophilic end (carboxyl, for example)

Hydrophobic end

Monomer

FIGURE 11.4 Diagrammatic representation of the orientation of lipid molecules that form a monolayer, a bilayer, and micelles. The monomer represents fatty acids with hydrophobic tails (hydrocarbon portion) and hydrophilic heads (carboxyl group). Hydrophobic bonding causes the orientation. Refer also to the membrane-structure diagrams in Fig. 1.3.

common phosphatides are phosphatidyl ethanolamine and phosphatidyl choline. The latter is also known as lecithin, a common preservative.

$$CH_2OOCR$$
$$|$$
$$CHOOCR$$
$$|$$
$$CH_2OPO_3—CH_2—CH_2—NH_3^+$$
Phosphatidyl ethanolamine

$$CH_2OOCR$$
$$|$$
$$CHOOCR$$
$$|$$
$$CH_2OPO_3—CH_2—CH_2—N^+—(CH_3)_3$$
Phosphatidyl choline

Other phosphatides that are particularly abundant in seeds are phosphatidyl serine and phosphatidyl inositol.

$$CH_2OOCR$$
$$|$$
$$CHOOCR$$
$$|$$
$$CH_2OPO_3—CH_2—CH—COO^-$$
$$|$$
$$NH_3^+$$
Phosphatidyl serine

$$CH_2OOCR$$
$$|$$
$$CHOOCR$$
$$|$$
$$CH_2OPO_3—O$$
OH OH
HO
OH
OH
Phosphatidyl inositol

During the biosynthesis of the phosphatides, the glycerol, serine, ethanolamine, choline, and other groups are probably added as phosphate derivatives directly to diglycerides. Phosphatidic acid is probably not an intermediate; it is found free only during hydrolysis of the phosphatides.

11.2.2 Sphingolipids

The sphingolipids are complex membrane components found in plants. Sphingosine-phosphate, the parent compound, is derived from serine and palmitic acid.

$$(CH_2)_{14}CH_3$$
$$|$$
$$CHOH$$
$$|$$
$$CHNH_2$$
$$|$$
$$CH_2OPO_3H_2$$
Sphingosine

The sphingosine of animal membranes is similar except that the hydrophobic side chain is unsaturated. Phosphate diester bonds can be formed with sugars, choline, and inositol, forming complex sphingolipids.

11.3 Glycolipids

Glycolipids are important components of chloroplast membranes. They do not contain phosphate. By definition, glycolipids are glycerides with a carbohydrate substitution, the most common being galactosyl derivatives. Unsaturated fatty acids such as linolenic acid are the most common side chains of the glycolipids. Two of the most common glycolipids of chloroplasts are monogalactosyl and digalactosyl diglyceride.

Monogalactosyl diglyceride

Digalactosyl diglyceride

11.4 Waxes

Waxes differ from the other saponifiable lipids in that glycerol is replaced by a long-chain alcohol. Waxes are defined as esters of long-chain fatty acids and alcohols and may include free long-chain fatty acids, primary and secondary alcohols, ketones, and even alkanes, in addition to the esters. Chain lengths are 36 or more carbons.

Plant waxes are deposited on most exposed surfaces and probably function largely in protection and prevention of excessive water loss. The electron micrographs shown in Fig. 5.1 show extensive deposition of wax on plant surfaces. The protective depositions, cutin and suberin, have chemical structures comparable to the simple plant waxes. Cutin is the waxlike lipid of the cuticle, and suberin is the lipid material of cork.

11.5 Lipid Metabolism

The primary concern in the study of lipid metabolism is the biosynthesis of fatty acids and their subsequent degradation during respiratory metabolism. Although lipids function in part as structural components of membranes, from a metabolic viewpoint their role in energy storage is equally important. In this section, the synthesis and degradation of lipids will be considered from the standpoint of fatty acid metabolism. In addition, the metabolic pathway linking storage lipids with carbohydrate metabolism (the glyoxylate cycle) will be discussed.

11.5.1 The biosynthesis of long-chain fatty acids

The biosynthesis of fatty acids in plants is complex and not completely understood. There is evidence that it occurs by more than one pathway, especially with unsaturated fatty acids. The synthesis of saturated and unsaturated fatty acids is treated separately below because of this uncertainty and complexity.

Saturated fatty acid biosynthesis

The biosynthesis of fatty acids in plants has been studied extensively. In the simplest and most straightforward biosynthetic scheme, acetyl units are condensed to ultimately form palmitic acid in a *de novo* biosynthetic pathway. This simple pathway does not result in fatty acids with carbons greater than 16 (that is, palmitic acid). For example, subsequent chain elongation to stearic acid (C-18), proceeds by a chain-elongating sequence that requires an existing fatty acid, usually palmitic acid.

The substrate for palmitic acid biosynthesis is acetyl CoA; however, the actual carrier during chain elongation is not coenzyme A but rather a small protein called acyl carrier protein (ACP). This sulfhydryl protein has a molecular weight on the order of 10,000.

An acetyl transacylase enzyme catalyzes the formation of acetyl-S-ACP from acetyl CoA and carrier protein.

$$CH_3-\overset{O}{\underset{\parallel}{C}}-SCoA + ACP-SH \longrightarrow$$

$$CH_3-\overset{O}{\underset{\parallel}{C}}-S-ACP + CoASH$$

The acetate of the acetyl-S-ACP is subsequently transferred directly to an enzyme, β-ketoacyl-ACP synthetase, to form acetyl-S-enzyme(-ENZ).

$$CH_3-\overset{O}{\underset{\parallel}{C}}-S-ACP + ENZ \longrightarrow$$

$$CH_3-\overset{O}{\underset{\parallel}{C}}-S-ENZ + ACP$$

The acetyl group then displaces CO_2 from malonyl-S-ACP to form acetoacetyl-S-ACP.

$$CH_3-\overset{O}{\underset{\parallel}{C}}-S-ENZ + CO_2-CH_2-\overset{O}{\underset{\parallel}{C}}-S-ACP$$

$$\longrightarrow CH_3-\overset{O}{\underset{\parallel}{C}}-CH_2-\overset{O}{\underset{\parallel}{C}}-S-ACP + CO_2$$

The malonyl derivative is synthesized from acetyl CoA by carboxylation with a biotin-requiring enzyme, acetyl CoA carboxylase. ACP-SH displaces the CoA from malonyl CoA to yield malonyl-S-ACP.

Subsequent reactions, all associated with acyl carrier protein, involve the reduction of the β-keto group to a methylene carbon, using primarily NADPH as the reductant.

First, acetoacetyl-S-ACP is reduced to D-β-hydroxybutyryl-S-ACP.

$$CH_3-\overset{O}{\underset{\parallel}{C}}-CH_2-\overset{O}{\underset{\parallel}{C}}-S-ACP + NADPH \longrightarrow$$

$$CH_3-\overset{H}{\underset{\underset{OH}{|}}{C}}-CH_2-\overset{O}{\underset{\parallel}{C}}-S-ACP + NADP$$

Removal of water by the enzyme enolyl-ACP-hydrase forms *trans*-crotonyl-S-ACP.

$$CH_3-\overset{H}{\underset{\underset{OH}{|}}{C}}-CH_2-\overset{O}{\underset{\parallel}{C}}-S-ACP \longrightarrow$$

$$CH_3-CH=CH-\overset{O}{\underset{\parallel}{C}}-S-ACP + H_2O$$

Subsequent reduction of crotonyl-S-ACP by NADPH-enoyl-ACP reductase produces butyryl-S-ACP.

$$CH_3-CH=CH-\overset{O}{\underset{\parallel}{C}}-S-ACP + NADPH \longrightarrow$$

$$CH_3-CH_2-CH_2-\overset{O}{\underset{\parallel}{C}}-S-ACP + NADP$$

Further chain elongation proceeds by repeating steps outlined above until eight acetates are polymerized to form palmitic acid. These subsequent steps in chain elongation proceed along the newly formed chain, adding to malonyl-S-ACP as discussed above. Carbon dioxide is displaced after the first condensation so that chain elongation is effectively the addition of acetate molecules. The overall biosynthetic sequence of *de novo* chain elongation is shown in Fig. 11.5 along with the scheme for fatty acid degradation, β-oxidation.

Palmitic acid is the end product of *de novo* fatty acid biosynthesis. A second biosynthetic scheme, not unlike the *de novo* pathway, catalyzes chain

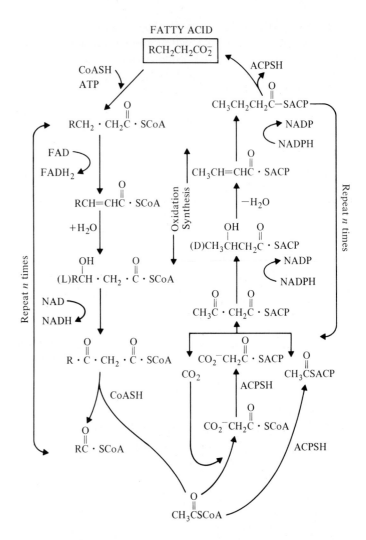

FIGURE 11.5 Biochemical pathways for synthesis and oxidation of fatty acids. The oxidation sequence is called β-oxidation because it is the beta carbon (second from the functional group) that becomes oxidized. The biosynthesis of fatty acids takes place for molecules up to the size of palmitic acid by a process that is similar to the oxidation step except that different carriers and cofactors are involved. In addition, the β-hydroxy intermediate of oxidation is the L form, whereas it is the D form in the biosynthetic sequence. See text for further details.

elongation to produce fatty acids with more than 16 carbons.

Fatty acid biosynthesis occurs in several cellular compartments within the plant cell, including chloroplasts, other plastids, and probably in association with the endoplasmic reticulum, the cytosol. The chloroplast is a particularly good fatty acid biosynthetic compartment but produces little if any triglycerides. The electron micrograph in Fig. 11.3 emphasizes the fact that chloroplasts produce extensive lipoidal materials. Chloroplasts from some species, however, may store carbohydrates rather than lipids.

Unsaturated fatty acid biosynthesis

The biosynthesis of unsaturated fatty acids—oleic (9), linoleic (9,12), and linolenic (9,12,15)—is not as well understood in plants as is the biosynthesis of the saturated fatty acids.

An oxygen-dependent NADP-stearyl-ACP desaturase is known from plants that can form oleyl-S-ACP from stearyl-S-ACP. The reaction also requires the nonheme iron protein, ferredoxin (Fd).

$$\text{Stearyl-S-ACP} + \text{NADP} \xrightarrow{\text{Fd}}$$
$$\text{Oleyl-S-ACP} + \text{NADPH}$$

There is evidence for subsequent desaturation of oleic acid to produce linoleic acid, but the reaction appears to be rather slow. Linolenic acid does not appear to be produced by further unsaturation of linoleic acid; a different pathway is involved.

11.5.2 Fatty acid degradation

Lipids are important storage compounds of many plant tissues. As seen in Fig. 11.3, some chloroplasts store lipid rather than starch. From a human nutritional point of view, the oil-storing seeds of most nut crops, the cucurbits, safflower, soybean, cotton, peanut, sunflower, castor bean, and, to some extent, wheat and corn are the best known among the oil-storing tissues. They are high in unsaturated fatty acids and may contain the essential fatty acids (those that the human body cannot synthesize) such as linoleic and arachidonic acids.

When seeds that store oil germinate (see Fig. 11.6), the lipid is mobilized to ultimately form carbohydrate. Since most of the storage lipid is in the form of triglycerides, the first step is hydrolysis to form glycerol and free fatty acids. The latter are oxidized to acetate by a process that forms reduced pyridine nucleotides. The acetate is converted to carbohydrate that is mostly sucrose, which is the transport sugar during the initial stages of seedling growth. In addition, the acetate is used as a starting point for other kinds of biosyntheses and as a major substrate for respiration in the tricarboxylic acid cycle (see Chapter 14). The processes of triglyceride hydrolysis and fatty acid oxidation are discussed below.

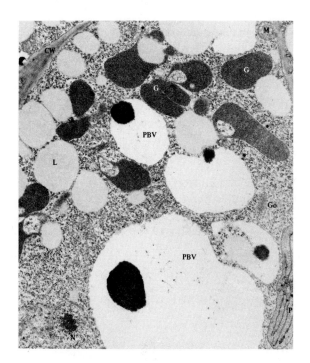

FIGURE 11.6 Electron micrograph (X21,000) showing a portion of cucumber cotyledon cell 4 days after germination. The glyoxysomes (microbodies) that metabolize the lipid that ultimately forms sucrose are closely associated with lipid bodies. The large, empty vacuoles are protein-body vacuoles, left after storage protein was mobilized during the first days of germination. G = glyoxysomes. L = lipid body. PBV = protein body vacuole. M = mitochondrion. N = nucleus. Go = Golgi body. P = plastid. CW = cell wall. Electron micrograph courtesy of Dr. R. N. Trelease, Arizona State University, Tempe.

The hydrolysis of triglycerides

Enzymes called lipases are capable of hydrolyzing the triglycerides to glycerol and free fatty acids. These enzymes are abundant in oil-storing seeds and are present where triglycerides are metabolized.

$$
\begin{matrix}
CH_2\!-\!O\!-\!R \\
| \\
R\!-\!O\!-\!CH \qquad + 3H_2O \longrightarrow \\
| \\
CH_2\!-\!O\!-\!R
\end{matrix}
$$

$$
\begin{matrix}
CH_2OH \\
| \\
CHOH + 3ROOH \\
| \\
CH_2OH
\end{matrix}
$$

Other lipases are specific for phospholipids and glycolipids and yield, upon hydrolysis, free fatty acids, glycerol-galactosyl and glycerol-phosphate derivatives. In addition, phospholipases are known that produce free choline, ethanolamine, and serine plus phosphatidic acid. Phosphatidic acid can be further metabolized by phosphate hydrolysis through the action of specific phosphatases and lipases to produce free fatty acids, glycerol, and phosphate.

Perhaps the most important aspect of triglyceride hydrolysis by the lipases is the production of the free fatty acids. These are further metabolized to produce sucrose, the compound of carbon translocation, and acetate, which acts either as the substrate of aerobic respiration or as the beginning for a variety of biosyntheses.

β-oxidation of fatty acids

Free fatty acids are largely oxidized to acetate (actually acetyl CoA) by a repeating sequential series of reactions called β-oxidation (see Fig. 11.5). β-oxidation apparently occurs in the cytosol and in seeds that store oil in the microbodies, called glyoxysomes.

The acetate produced in the form of acetyl CoA is further metabolized in respiration via the tricarboxylic acid cycle, or in many cases (particularly if the fatty acid was a storage compound) converted to sucrose via the series of reactions coupling the glyoxylate cycle (see Section 11.6) and a reversal of glycolysis.

Fatty acid oxidation by β-oxidation begins with coenzyme A to form acyl CoA. This activation reaction, catalyzed by the enzyme thiolase, requires ATP.

$$RCO_2H + CoA + ATP \longrightarrow$$

$$
\overset{\displaystyle O}{\overset{\displaystyle \|}{R\!-\!C}}\!-\!SCoA + ADP + P_i
$$

The first oxidative reaction is catalyzed by a flavin adenine nucleotide (FAD)-requiring enzyme, acyl-CoA dehydrogenase, producing a double bond between carbons 2 and 3. Carbon-3 is the β carbon that is oxidized from the methylene form to the carbonyl form (hence the name β-oxidation). Reduced FAD is produced that can subsequently be used for the production of 2 ATP during electron transport in the mitochondria (see Chapter 14).

$$
R\!-\!CH_2\!-\!CH_2\!-\!\overset{\displaystyle O}{\overset{\displaystyle \|}{C}}\!-\!SCoA + FAD \longrightarrow
$$

$$
R\!-\!CH\!=\!CH\!-\!\overset{\displaystyle O}{\overset{\displaystyle \|}{C}}\!-\!CoA + FADH_2
$$

Water is subsequently added across the newly formed double bond to produce the L-β-hydroxyl derivative of the fatty acid. The reaction is catalyzed by an enzyme called enoyl-CoA hydrase.

$$
R\!-\!CH\!=\!CH\!-\!\overset{\displaystyle O}{\overset{\displaystyle \|}{C}}\!-\!SCoA + H_2O \longrightarrow
$$

$$
R\!-\!CHOH\!-\!CH_2\!-\!\overset{\displaystyle O}{\overset{\displaystyle \|}{C}}\!-\!SCoA
$$

Next the hydroxyl is oxidized by an NAD-specific β-hydroxy acyl-CoA dehydrogenase to form the β-keto derivative of the fatty acid. NADH is produced, which can yield 3 ATP during respiration (see Chapter 14).

$$R—CHOH—CH_2—\overset{\overset{\displaystyle O}{\|}}{C}—SCoA + NAD \longrightarrow$$

$$R—\overset{\overset{\displaystyle O}{\|}}{C}—CH_2—\overset{\overset{\displaystyle O}{\|}}{C}—SCoA + NADH$$

Finally nucleophilic displacement of acetyl CoA from the β-keto fatty acid by CoASH yields a fatty acid with two less carbons plus free acetyl CoA. This reaction is catalyzed by β-ketoacyl-CoA thiolase.

$$R—\overset{\overset{\displaystyle O}{\|}}{C}—CH_2—\overset{\overset{\displaystyle O}{\|}}{C}—SCoA + CoASH \longrightarrow$$

$$R—\overset{\overset{\displaystyle O}{\|}}{C}—SCoA + CH_3—\overset{\overset{\displaystyle O}{\|}}{C}—SCoA$$

The entire β-oxidation sequence is repeated until the final step, in which β-keto-butyryl CoA is cleaved to form two acetyl CoA units. The latter will occur provided that the substrate fatty acid had an even number of carbon atoms. If it was odd-numbered the β-oxidation sequence will stop with propionyl CoA. Odd-numbered fatty acids may be oxidized by the α-oxidation sequence discussed in Section 11.5.2.

Approximately 40% of the total energy of the fatty acid can be converted to ATP or ATP equivalents through oxidation. As an example, the complete oxidation of butyric acid through the citric acid cycle will result in a net yield of 28 ATP. A single ATP is required to form the initial butyryl CoA derivative, and the production of the two acetyl CoA units results in one reduced FAD molecule and one NADH molecule. FAD will yield two ATP and NADH will yield three, assuming maximum efficiency in the electron transport system of the mitochondria. Thus a total of four net ATP will be produced during the formation of the two acetates from the butyrate. Since the complete oxidation of acetate in the tricarboxylic acid cycle yields 12 ATP, the net gain from the oxidation of butyrate is 28 ATP. Since the heat of combustion of butyrate is about 2200 kilojoules per mole and the hydrolysis of ATP yields about 30 kilojoules per mole, the conservation of energy in the yield of 28 ATP is about 38%. A similar analysis with caproic acid, a six-carbon fatty acid, results in a net yield of 45 ATP and an energy conservation of 38%.

It is instructive to compare fatty acid β-oxidation with the biosynthetic pathway for fatty acid production. Although these processes are basically similar, several important features differ. First, the oxidation redox cofactors are NAD and FAD in the β-oxidation sequence whereas the reductant for the biosynthetic pathway is largely NADPH. The actual substrates for the oxidative pathway are acyl-CoA derivatives whereas the acyl carrier for the biosynthetic process is a small sulfhydryl protein, acyl carrier protein (ACP). Finally, the β-hydroxy product of the β-oxidation sequence is the L isomer whereas the D isomer is the intermediate in the biosynthetic pathway.

It seems reasonable to assume that these three major differences keep the oxidative and biosynthetic functions separate within cells that may be carrying on both processes simultaneously.

α-oxidation of fatty acids

It is known that under certain circumstances long-chain fatty acids can be oxidized to produce CO_2, reducing the fatty acid chain length by one carbon. Stumpf (1970) and others have extensively studied this degradation pathway and have partially elucidated the intermediate steps. Both O_2 and H_2O_2 appear to be involved, and an NAD dehydrogenase is required. The overall reaction follows.

$$R—CH_2—COOH + NAD \longrightarrow$$
$$R—COOH + CO_2 + NADH$$

The best fatty acid substrate is palmitic acid. Lauric acid is an ineffective substrate.

Unlike β-oxidation, the role of α-oxidation does not appear to be the complete oxidation of fatty acids. Stumpf (1970) proposed two possible roles for α-oxidation. First, it may function as a bypass of a substitution group that could prevent the operation of the β-oxidation sequence:

$$
\begin{array}{c}
R' \\
| \\
R—CH_2—CH—CH_2—COOH
\end{array}
\xrightarrow{\text{(}\alpha\text{-oxidation)}}
$$
$$
\begin{array}{c}
R' \\
| \\
R—CH_2—CH—COOH + CO_2,
\end{array}
$$

followed by

$$
\begin{array}{c}
R' \\
| \\
R—CH_2—CH—COOH
\end{array}
\xrightarrow{\text{(}\beta\text{-oxidation)}}
$$
$$
\begin{array}{c}
R' \\
| \\
R—COOH + CH_2—COOH.
\end{array}
$$

Second, a long, even-chain fatty acid could be reduced by one carbon to produce an odd-chain fatty acid. Subsequent β-oxidation would ultimately yield propionic acid rather than acetic acid:

$$CH_3(CH_2)_{13}CH_2COOH \xrightarrow{\text{(}\alpha\text{-oxidation)}}$$
$$CH_3(CH_2)_{13}COOH + CO_2 +$$

followed by

$$CH_3(CH_2)_{13}COOH \xrightarrow{\text{(}\beta\text{-oxidation)}}$$
$$6CH_3COOH + CH_3CH_2COOH.$$

11.6 Lipid–Sugar Conversion: The Glyoxylate Cycle

During germination of seeds that store lipids, such as castor bean and cucurbits, stored triglycerides are mobilized and sucrose is produced. Sucrose is subsequently used as an energy source for growth and development processes. Sucrose is the main product of the conversion because it is sucrose that acts as the compound for carbon translocation. Thus stored lipids are converted to sucrose, which is translocated to sites requiring energy.

The stored triglycerides are mobilized first by lipase activity producing glycerol and free fatty acids. The free fatty acids are oxidized to acetate via the β-oxidation sequence described in Section 11.5.2. The acetate is the direct substrate for sucrose synthesis. Acetate is first converted to malate through a series of reactions known as the glyoxylate cycle. In this cycle an acetate and a glyoxylate molecule are condensed to form malate. The malate is then converted to phosphoenolpyruvate (PEP) through oxalacetate and ultimately to hexoses and then sucrose. The sequence of β-oxidation and the glyoxylate cycle takes place within the seed in the single-membrane microbody organelles, glyoxysomes. Figure 11.6 shows the close association of glyoxysomes and lipid bodies during germination of an oil-storing seed. The photograph shows cotyledons four days after germination.

The glyoxylate cycle is depicted in Fig. 11.7. First, an acetyl CoA is condensed with oxalacetate through the catalysis of an enzyme called condensing enzyme to form citrate and free coenzyme A. Citrate is dehydrated through the action of the enzyme aconitase to form aconitate, which is converted to isocitrate by hydration. Catalysis is by the same enzyme, aconitase.

The subsequent step in the glyoxylate cycle is a central reaction that produces succinate and glyoxylate. Isocitrase (also called isocitric lyase) is the enzyme catalyzing the reaction and is unique to the glyoxylate cycle, as far as is known. The reaction

GLYOXYLATE CYCLE AND GENERATION OF MALATE
FOR SUGAR SYNTHESIS

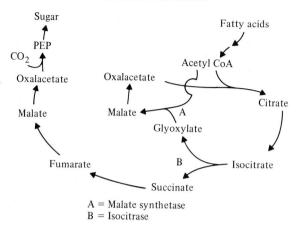

A = Malate synthetase
B = Isocitrase

FIGURE 11.7 The glyoxylate cycle, which links fatty acid oxidation to carbohydrate biosynthesis. Fatty acids are oxidized by the β-oxidation sequence to form acetyl CoA. Acetyl CoA condenses with oxalacetate to form citrate. Citrate is converted to isocitrate. The enzyme isocitrase catalyzes the production of glyoxylate and succinate from the isocitrate. Glyoxylate and another acetyl CoA condense to form malate when catalyzed by the enzyme malate synthetase. The malate is oxidized to oxalacetate, regenerating the initial acceptor (oxalacetate) and completing the cycle. The succinate in mitochondria is converted to fumarate. Fumarate is converted to another malate. The malate is the substrate for carbohydrate synthesis, as discussed in the text. The β-oxidation sequence and the glyoxylate cycle occur in glyoxysomes.

catalyzed by isocitrase follows.

$$\begin{array}{c} CH_2COO^- \\ | \\ H\overset{.}{C}COO^- \\ | \\ H\overset{.}{C}COO^- \\ | \\ \overset{O}{\underset{H}{}} \end{array} \longrightarrow \begin{array}{c} CH_2COO^- \\ | \\ CH_2COO^- \end{array} + CHO\!-\!COO^-$$

Isocitrate   Succinate   Glyoxylate

The second important step of the glyoxylate cycle is the condensation of glyoxylate with a second acetyl CoA to form malate. The malate synthetase step is as follows.

$$CH_3\!-\!\overset{O}{\overset{||}{C}}\!-\!S\!-\!CoA + CHO\!-\!COO^- \longrightarrow$$

$$\begin{array}{c} COO^- \\ | \\ CH_2 \\ | \\ HOCH \\ | \\ COO^- \end{array} + CoASH$$

The malate is oxidized by malate dehydrogenase, an enzyme that requires NAD as the redox cofactor, to yield and regenerate oxalacetate. The oxal-

acetate condenses with another acetyl CoA to continue the cycle. This step illustrates the significance and importance of cycles in intermediary metabolism. Such cycles are self-sustaining because they regenerate the initial acceptor molecule.*

The succinate produced from the isocitrate step is further metabolized in mitochondria by succinate dehydrogenase, a flavoprotein that produces fumarate by oxidation. Fumarate is hydrated by the enzyme fumarase to produce a second malate that is considered to be the substrate for carbohydrate synthesis.

The glyoxylate cycle uses two acetyl CoA molecules derived from fatty acid degradation to produce one net malate. During the cyclic process the initial acceptor, oxalacetate, is regenerated such that the process continues. The intermediates citrate, aconitate, isocitrate, succinate, fumarate, malate, and oxalacetate are present in catalytic amounts.

The glyoxylate cycle was first studied in plants by

*The other important cycles in plants that similarly regenerate the primary acceptor molecules are the tricarboxylic acid cycle that occurs in mitochondria during aerobic respiration and the reductive pentose cycle that occurs in chloroplasts during carbon fixation in photosynthesis.

Beevers (see Beevers, 1975) while he was investigating how the castor bean could convert fats to sugars during germination. His studies showed that the reactions of the glyoxylate cycle could accomplish the link between fatty acid oxidation and sugar conversion, accounting for the nearly quantitative production of sugar from fatty acids.

It is worth noting at this point that some of the reactions of the glyoxylate cycle are common to the tricarboxylic acid cycle. The specific enzymes catalyzing the common steps are not the same, however, but are isozymes. Isozymes, discussed in Chapter 9, are different enzymes catalyzing the same reaction. The glyoxylate cycle has a unique set of proteins that are probably coded for by specific genetic loci that are different from those that code for the comparable tricarboxylic acid cycle enzymes.

In seeds that store lipids, fatty acid conversion to carbohydrate occurs only for a brief period immediately following germination. The enzymes of the glyoxylate cycle that are contained in the glyoxysomes are presumably generated *de novo* concomitant with the germination process. In many seeds maximum activity is reached within a few days after germination, and then the glyoxylate cycle enzymes along with the glyoxysomes disappear as the young seedling becomes self-sufficient through the reactions of photosynthesis. Figure 11.8 illustrates the production of the enzyme isocitrase, which reaches a maximum activity three days after germination and is essentially completely gone by six days after germination. A second enzyme shown in the figure, glycolate oxidase, functions in photorespiration, a respiratory process coupled to photosynthesis (see Chapter 16). Glycolate oxidase does not appear until four days after germination. This process, occurring in cucumber cotyledons, illustrates the transition from dependence on stored lipids during the early stages of germination to the shift to photosynthesis as the cotyledons become green and self-sufficient.

The malate produced by the glyoxylate cycle is the substrate for carbohydrate biosynthesis. This glucogenic process is quite complicated; it requires

input of energy to overcome energy barriers. The sequence from malate to hexose sugar, the substrate for sucrose synthesis, is shown in Fig. 11.9.

The first energy barrier in the path to sucrose synthesis is the production of phosphoenolpyruvate. Pyruvate kinase, an enzyme common in plant tissues, catalyzes the synthesis of pyruvate from phosphoenolpyruvate, generating ATP from ADP. It would seem that a reversal of this reaction using the energy of ATP would be sufficient to produce the phosphoenolpyruvate. However, the hydrolysis of ATP yields about 30 kilojoules per mole under standard conditions, whereas the hydrolysis of phosphoenolpyruvate produces about 50 kilojoules per mole. Thus an input of at least 20 kilojoules would be required to produce phosphoenolpyruvate using ATP energy alone.

For the most part, this large free-energy barrier is overcome by a decarboxylation reaction that converts oxalacetate to phosphoenolpyruvate, bypassing pyruvate. First, the malate is converted to oxalacetate oxidatively by an NAD-malate dehydrogenase.

$$\text{Malate} + \text{NAD} \longrightarrow \text{Oxalacetate} + \text{NADH}$$

Phosphoenolpyruvate carboxykinase catalyzes the decarboxylation of oxalacetate, producing phosphoenolpyruvate and CO_2 and consuming ATP in the process. Apparently, the generation of CO_2 as a product tends to pull the reaction in the direction of decarboxylation despite an unfavorable energy relationship ($\triangle G^\circ = +20 \text{kJ mol}^{-1}$).

$$\text{Oxalacetate} + \text{ATP} \longrightarrow \text{PEP} + CO_2 + \text{ADP}$$

The next two reactions, the conversion of phosphoenolpyruvate to 2-phosphoglycerate (2-PGA) catalyzed by the enzyme enolase and the conversion of 2-phosphoglycerate to 3-phosphoglycerate (3-PGA) by catalysis with phosphoglyceromutase, are both reversible and pose no particular free-energy considerations.

$$\text{PEP} \longrightarrow \text{2-PGA} \longrightarrow \text{3-PGA}$$

A reaction catalyzed by phosphoglycerate kinase converts 3-phosphoglycerate to 1,3-bisphospho-

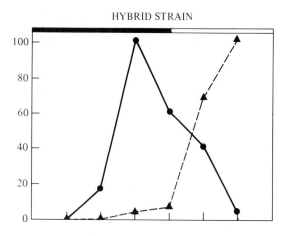

HYBRID STRAIN

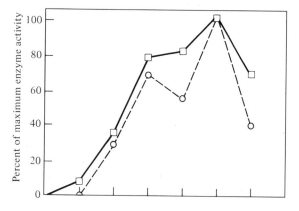

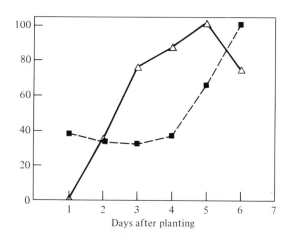

Days after planting

FIGURE 11.8 The appearance of the glyoxylate-cycle enzymes during the first few days after germination of a plant that stores lipid in the cotyledons (cucumber). Upper panel: black dots represent isocitrase and black triangles denote glycolate oxidase. Middle panel: white squares show malate dehydrogenase and white dots show catalase. Lower panel: white triangles show protein and black squares show fresh weight. In the upper panel, the glyoxylate-cycle marker enzyme, isocitrase, appears during the time in which stored lipid is being actively metabolized and converted to carbohydrate. Glycolate oxidase, an enzyme of photorespiration, does not appear until later, when the tissue becomes photosynthetic. As shown in the middle panel, two other enzymes **not** unique to the glyoxylate cycle are present during glyoxylate-cycle activity and thereafter. From Irene Wainwright. 1975. Malate dehydrogenase isozymes in developing cucumber cotyledons. Ph.D. thesis, University of California, Riverside.

glycerate, consuming ATP. This reaction is driven with the energy of ATP although the free energy is positive ($\Delta G° = +20$ kJ mol^{-1}).

$$3\text{-PGA} + \text{ATP} \longrightarrow 1,3\text{-bisPGA}$$

Once the bisphosphate derivative of glycerate is produced, it is reduced to the level of sugar by NAD-glyceraldehyde dehydrogenase, yielding glyceraldehyde phosphate (GAP). Glyceraldehyde phosphate is in equilibrium with dihydroxyacetone phosphate (DHAP) through the enzyme triosephosphate isomerase.

$$\text{GAP} \rightleftharpoons \text{DHAP}$$

Aldolase, a condensing enzyme, forms fructose-1,6-bisphosphate from glyceraldehyde phosphate and dihydroxyacetone phosphate.

$$\text{GAP} + \text{DHAP} \longrightarrow \text{Fructose-1,6-bisP}$$

A specific phosphatase hydrolyzes the fructose-1, 6-bisphosphate to fructose-6-phosphate, yielding orthophosphate; an isomerase converts the fructose-6-phosphate to glucose-6-phosphate. These hexoses are the substrates for sucrose synthesis.

Although many of the reactions in the path from malate to hexose are unfavorable from a free-energy standpoint, it should be realized from the considerations of cellular energetics in Chapter 2 that, in addition to standard free energies ($\Delta G°$), one must also consider the concentrations of the reactants and products (that is, ΔG becomes more important than $\Delta G°$). During biosynthetic sequences such as described here for the synthesis of carbohydrate from lipid, it is logical that the substrates (reactants) are in higher concentration than the products, thus tending to drive the overall biosynthetic sequence toward carbohydrate synthesis.

In summary, the overall process occurring in germinating seeds produces substrates for sucrose production from fatty acids stored as triglycerides. β-oxidation forms acetate, which is converted to malate by the glyoxylate cycle. Malate is oxidized to oxalacetate, which is decarboxylated to form phosphoenolpyruvate. The latter is converted

METABOLIC SEQUENCE FROM MALATE TO SUCROSE

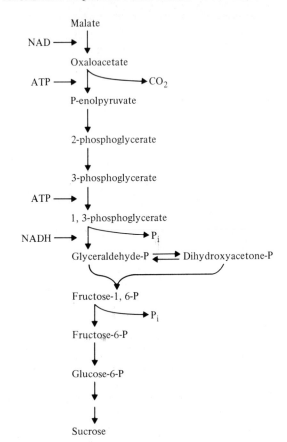

FIGURE 11.9 The metabolic pathway that produces sucrose from the malate formed through the reactions coupling fatty acid oxidation and carbohydrate synthesis. The sequence is essentially the reversal of glycolysis, which is discussed in Chapter 14.

to hexoses prior to sucrose synthesis. The important point of the entire process is that sucrose has been synthesized from carbon stored in lipids. Sucrose is the primary sugar of transport. Once it is transported to sites of high metabolic activity, sucrose is hydrolyzed to hexoses that in turn are oxidized to yield energy for metabolism, growth, and development.

Review Exercises

11.1 Define the term lipid. Differentiate between fatty acid, fat, oil, phosphatidic acid, phosphatide, and wax.

11.2 Within the fatty acids, as the degree of unsaturation increases the melting point decreases. Unsaturated fatty acids are liquids at lower temperatures than are the corresponding saturated fatty acids. Would you expect the degree of unsaturation among the fatty acids of plant tissues to change when grown under different thermal regimes? Explain.

11.3 The saponification number is defined as the milligrams of KOH required to saponify 1 gram of a fat or lipid. The triglyceride tristerin has a molecular weight of 891. What would be the saponification number of tristerin? Would a triglyceride with a lower molecular weight than tristerin have a smaller or higher saponification number?

11.4 The degree of unsaturation of a lipid can be estimated by the "iodine number," the grams of iodine taken up by 100 grams of lipid. According to data published in Altman and Dittmer (1968) (see References), coconut oil has an iodine number of 6–10, that of castor oil is 84, olive oil is 79–88, and cottonseed oil is 103–111. Rank these oils according to their degree of unsaturation. Do you think you can rank them according to their melting points? During iodination, two moles of iodine are absorbed per double bond. What would be the iodine number of pure linoleic acid (molecular weight = 280.44)?

11.5 Compute the total ATP or equivalents that could be obtained from the complete oxidation of stearic acid through the β-oxidation pathway and the tricarboxylic acid cycle.

References

ALTMAN, P. L., and D. S. DITTMER (eds.). 1968. *Metabolism.* Federation of American Societies for Experimental Biology, Bethesda, Md.

BEEVERS, H. 1975. Organelles from castor bean seedlings: Biochemical roles in gluconeogenesis and phospholipid biosynthesis. In T. Gallaird and E. I. Mercer (eds.), *Recent Advances in the Chemistry and Biochemistry of Plant Lipids.* Academic Press, London.

BEWLEY, D., and M. BLACK. 1977. *Physiology and Biochemistry of Seeds in Relation to Germination,* Vol. 1. Springer-Verlag, New York.

CHRISTIE, W. W. 1973. *Lipid Analysis.* Pergamon Press, Oxford.

DURE, L. S. 1975. Seed formation. *Ann. Rev. Plant Physiol.* 26:259–278.

EISENBERG, M., and S. MCLAUGHLIN. 1976. Lipid bilayers as models of biological membranes. *BioScience* 26:436–443.

GALLAIRD, T., and E. I. MERCER (eds.). 1975. *Recent Advances in the Chemistry and Biochemistry of Plant Lipids.* Academic Press, London.

HITCHCOCK, C., and B. W. NICHOLS. 1971. *Plant Lipid Biochemistry.* Academic Press, London.

MARTIN, J. T., and B. E. JUNIPER. 1970. *The Cuticles of Plants.* St. Martin's Press, New York.

STUMPF, P. K. 1970. Fatty acid metabolism in plants. In S. J. Wakil (ed.), *Lipid Metabolism.* Academic Press, New York.

TEVINI, M., and H. K. LICHTENTHALER (eds.). 1977. *Lipids and Lipid Polymers in Higher Plants.* Springer-Verlag, Berlin.

CHAPTER TWELVE

Natural Products

Natural products are those chemicals produced by plants by means of secondary reactions resulting from primary carbohydrates, amino acids, and lipids. The alternate name for this group of compounds is secondary products, but it is perhaps less useful since it carries the connotation of lesser importance. Natural products are frequently end products of metabolism, but they do have many important roles in the physiology of plants as well.

The chemistry of natural products is rather sophisticated, requiring the use of analytical tools such as thin-layer chromatographic instrumentation and gas–liquid chromatographs equipped with diagnostic mass spectrometry.

In the simplest sense, we can separate natural products into three main groups: phenolics, alkaloids, and terpenoids. Natural products, of course, are not restricted to these three groups, and it is common to include other kinds of compounds such as pigments and porphyrins within the broad category of natural products.

In many cases grouping is not entirely logical, even though there may be structural similarities. For example, phenolics are those compounds with an aromatic ring substituted with a hydroxyl group or those compounds derived from the latter. Alkaloids, a more diverse group, are nitrogen-containing products mostly synthesized from amino acids. Terpenoids are those compounds considered to be derived from isoprene. These latter lipoidal compounds were discussed briefly in Chapter 11.

The natural products are used by chemotaxonomists to classify and understand relationships among various taxa of plants. Because they are considered to be end products of metabolism and are to a large extent characteristic for a species, they are very useful diagnostic chemicals for phylogenetic studies. The volumes by R. Darnley Gibbs (1974) on the chemotaxonomy of flowering plants covers much of our knowledge about the chemical basis for the taxonomy of flowering plants.

The entire group of natural products produced by plants has had a profound influence on the development of human populations. They include such diverse substances as medicines, poisons, narcotics, stimulants, essential oils, phenolics that are used as tanning agents, resins for preservatives, dyes, and rubber, among others. Some of the most important medicinals that are in common use are cardiovascular drugs: reserpine from *Rauwolfia serpentina* and digitoxin from *Digitalis purpurea*. Quinine from cinchona is still widely used as an antimalarial drug, and even the synthetics, chloroquine and atabrin, are structurally related to quinine. Curare, the well-known poison, is an extract from either *Strychnos toxifera* or *Chondodendron tomentosum* that contains tubocurarine as the active ingredient. And opium, the alkaloid that has contributed such important drugs as morphine and codeine but has also been responsible for much misery, is a natural product of poppy.

Any discussion of useful natural products includes the alkaloids that are stimulants: caffeine from coffee, tea, and cocoa; and nicotine of tobacco. The latter is also used as an insecticide. Cocaine, also an alkaloid, is a stimulant and an anesthetic when properly used.

More recently, there has been much interest in locating and studying natural products with teratogenic and carcinogenic properties. Not only do some plants have tumor-and cancer-causing chemicals, but also some may contain compounds that arrest the growth of tumors. Amygdalin (laetrile), a cyanogenic glucoside (see Chapter 10), has been suspected of being such a compound. Other natural products are neurotoxins or cause contact allergies, and some are suspected of having antimicrobial properties. These are called phytoalexins. Most are phenolic compounds.

Those chemicals produced by a plant that affect the growth, health, or behavior of other organisms are called allelochemics. The allelochemics can be divided into two kinds: (1) those that appear to have a beneficial effect on the producer, such as animal and insect repellents, growth suppressants used against competing plants, and poisons that deter eating; and (2) those that seem to have a beneficial effect on the recipient. The latter could include food attractants given off by a plant that attract herbivores. In addition, there are examples of plants producing insect hormones that coordinate the growth of the insect with that of the plant.

Allelopathy is the suppression of growth of plants by other plants because of the production of allelochemics. These allelochemics can be leached from leaves by rainwater, secreted by roots, and be formed from decaying plant matter. For the most part they are phenolics, flavonoids, alkaloids, and terpenes.

The coevolution of plants and animals (particularly insects) has been studied from this point of view. It is assumed that as plants evolve the mechanisms for the biosynthesis of allelochemics, insects evolve mechanisms to cope with them. An example is the phenolic compound hypericin produced by *Hypericum* that causes photosensitivity in animals, which may lead to simple external irritation or to blindness. Beetles of the genus *Chrysolina* have enzymes that can detoxify hypericin.

The physiological and metabolic roles of the various natural products in plants other than that of allelopathy are quite obscure. Many appear simply to be end products of metabolism. The plant has

SIMPLIFIED SCHEME FOR THE BIOSYNTHESIS OF NATURAL PRODUCTS

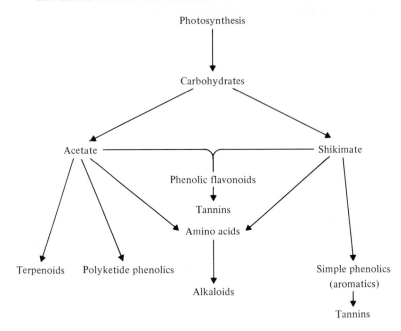

FIGURE 12.1 A simplified scheme for the biosynthesis of the several classes of natural products.

little in the way of an excretory system, and waste products are frequently stored in vacuoles. In some cases the secondary products may function as storage compounds that are mobilized during specific stages of growth and development. Products such as the cinnamic acids (phenolics) are precursors for lignin biosynthesis. Those such as the flavonoids are colored and involved in insect–plant interaction processes including pollination. The tannins formed by polymerization of phenolics are probably involved in resistance to insect and pathogen attack; they are formed most frequently when plant tissues are injured. The terpenoid compounds function in some aspects of metabolism. The carotenes, which are terpenoids with 40 carbons, play a role in photosynthesis. Porphyrins are integral components of chlorophyll, the enzymes catalase and peroxidase, and the cytochrome electron-transport proteins of photosynthesis and respiration.

In this chapter the various natural products found in plants are discussed from the standpoint of their occurrence, structure, and synthesis.

Figure 12.1 is an abbreviated scheme for the biosynthetic relationships among the various natural products. All are ultimately derived from the carbohydrates of photosynthesis. Phenolics, including the flavonoids and other aromatics, are formed from acetate and shikimate. Shikimate is a main intermediate in the biosynthesis of the aromatic amino acids tryptophan, phenylalanine, and tyrosine, from which the aromatic phenolics are formed. Alkaloids are also synthesized from acetate and shikimate, and most have amino acids as intermediate precursors. Other alkaloids are synthesized from glutamate, ornithine, and lysine, which are derived from acetate. Terpenoids are formed from acetate through mevalonic acid, the precursor for isopentenyl pyrophosphate. The latter is called active isoprene.

12.1 Phenolics

The plant phenolics can be defined as compounds having an aromatic ring with at least one hydroxyl functional group or derivatives of aromatic hydroxyl compounds.

12.1.1 Occurrence

The plant phenolics are perhaps the most abundant of all the chemicals that fall into the category of natural products. They include a variety of end products of metabolism, active plant hormones, redox cofactors such as coenzyme Q (ubiquinone), the flavonoid pigments, and one of the most abundant natural chemicals on earth, lignin. Humus, the main organic material of soil, is composed of phenolic polymers and other materials such as peptides.

Many exist in nature as glycosides rather than being in the free state, and thus many of the plant phenolics are aglycones. The sugar moiety of the O-glycosides may be glucose, galactose, xylose, rhamnose, arabinose, and a number of other less common sugars. Many uncommon disaccharides such as gentiobiose and primeverose are found in phenolic glycosides. Some C-glycosides are known as well as the common O-glycosides.

The most common phenolics found are phenol, catechol, hydroquinone, and phloroglucinol.

Phenol

Catechol

Hydroquinone

Phloroglucinol

These may occur free but are usually a part of the phenolic aglycones. Of these phenolics, hydroquinone is probably the most widely distributed.

12.1.2 The shikimic acid pathway

Biosynthesis of phenolics is quite complex and not entirely understood. Perhaps the best way to understand the biosynthesis of phenolics begins with the biosynthesis of the aromatic amino acids tryptophan, phenylalanine, and tyrosine. These amino acids are the immediate precursors of many of the phenolics. The aromatic amino acids are synthesized through the shikimic acid pathway, shikimate being a central intermediate. An abbreviated scheme is illustrated in Fig. 12.2.

In the first step, phosphoenolpyruvate and erythrose-4-phosphate are condensed to form a 7-carbon intermediate, 3-deoxy-D-arabino-heptulosonic-acid-7-phosphate (DAHP). The DAHP intermediate is converted to 5-dehydroquinic acid by reactions that ultimately form a ring structure. The 5-dehydroquinic acid is in equilibrium with quinic acid. The exact role of quinic acid in aromatic biosynthesis is not clear, but it is a major compound in plants, accumulating in many tissues such as acidic fruits. Next the 5-dehydroquinic acid is converted to dehydroshikimic acid, which is reduced to shikimic acid by catalysis with an NADPH dehydrogenase.

Shikimate is converted to an intermediate, chorismic acid, which is the precursor for either prephenic acid or anthranilic acid. Prephenic acid is the precursor for phenylalanine and tyrosine, and anthranilic acid is the precursor for tryptophan.

These three aromatic amino acids, phenylalanine, tyrosine, and tryptophan, are the main precursors for the remainder of the phenolics.

12.1.3 Phenolic types

The group referred to here as phenolics is a large and diverse assemblage of compounds that all have in common an aromatic ring with a hydroxyl substitution. They may only represent an artificial group, but they are considered here as a group for

SHIKIMIC ACID PATHWAY FOR THE BIOSYNTHESIS OF AROMATIC AMINO ACIDS

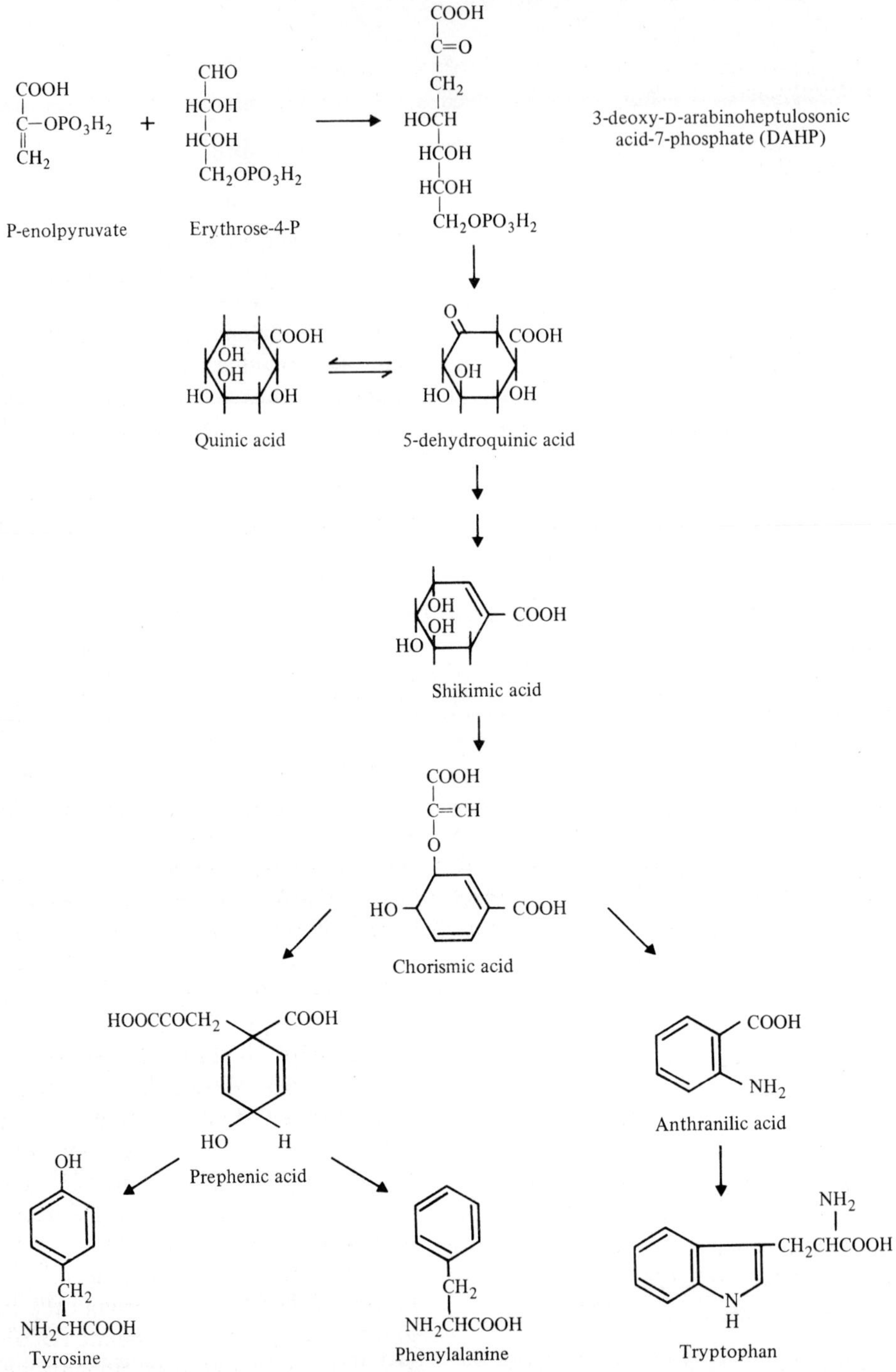

FIGURE 12.2 The shikimic acid pathway for the biosynthesis of the aromatic amino acids tyrosine, phenylalanine, and tryptophan. Biosynthesis begins with P-enolpyruvate and erythrose-4-P and proceeds via several steps. A major intermediate is shikimic acid. The aromatic amino acids are subsequently used in the biosynthesis of a variety of nitrogenous and aromatic natural products.

convenience of classification and because other possible groupings are not substantially better.

Simple phenolics

CINNAMIC ACIDS

Cinnamic acid is synthesized directly from phenylalanine by oxidative deamination (refer to Fig. 12.3) by the enzyme phenylalanine ammonia lyase (PAL). Hydroxylation of cinnamic acid in the para position produces p-coumaric acid. The latter can also be synthesized directly by oxidative deamination of tyrosine by the enzyme tyrosine ammonia lyase (TAL). PAL seems to be the more important biosynthetic enzyme (except perhaps in some grasses) that can produce coumaric acid.

Subsequent hydroxylation of p-coumaric acid in the meta position produces caffeic acid. Further steps from the latter two acids form ferulic and sinapic acids.

These common phenolics rarely occur free but rather occur as glycosides or as esters. Caffeic acid is the most common of the cinnamic acids and perhaps even the most abundant phenolic in plants (other than lignin). It occurs most commonly as an ester of quinic acid to form the common chlorogenic acid.

Chlorogenic acid

Chlorogenic acid has been implicated in the resistance of plants to infection. Caffeic acid also forms esters in combination with numerous sugars and organic acids such as tartaric acid.

COUMARINS

Oxidation and ring closure of cinnamic acid form coumarin, a common bicyclic compound

of plants.

Cinnamic acid   Coumarin

Several steps are required for the synthesis of coumarin from cinnamic acid, the entire process going through a glucoside derivative. Coumarin itself is not properly a phenolic, of course, since it lacks a hydroxyl, but it is in the phenolic family by virtue of its biosynthetic relationship to true phenolics.

Other common coumarins found in plants are umbelliferone (see below) from p-coumaric acid, aesculetin from caffeic acid, scopoletin from ferulic acid, and isofraxetin from sinapic acid.

Umbelliferone

LIGNIN

Lignin is a structural component of plant secondary cell walls. Next to cellulose it is the most abundant compound of plants. Its synthesis begins shortly after secondary-wall biosynthesis commences. Synthesis proceeds from the middle lamella toward the plasma membrane so that both the primary and secondary walls are impregnated with lignin. Through covalent bonds it evidently binds the cellulose microfibrils together. Because of its chemical structure and the fact that it is hydrophobic, it is highly resistant to decay. It gives chemical and biological resistance to the cell wall and much mechanical strength.

Lignin is a large, complex polymer of the alcoholic derivatives of the common cinnamic acids, coumaryl alcohol, coniferyl alcohol, and sinapyl alcohol. Biosynthesis is through the shikimic acid pathway, involving the formation of phenylalanine and tyrosine, followed by cinnamic acid synthesis (refer to Figs. 12.2 and 12.3). The alcoholic derivatives of the cinnamic acids are then produced by reduction.

BIOSYNTHESIS OF THE CINNAMIC ACID FAMILY

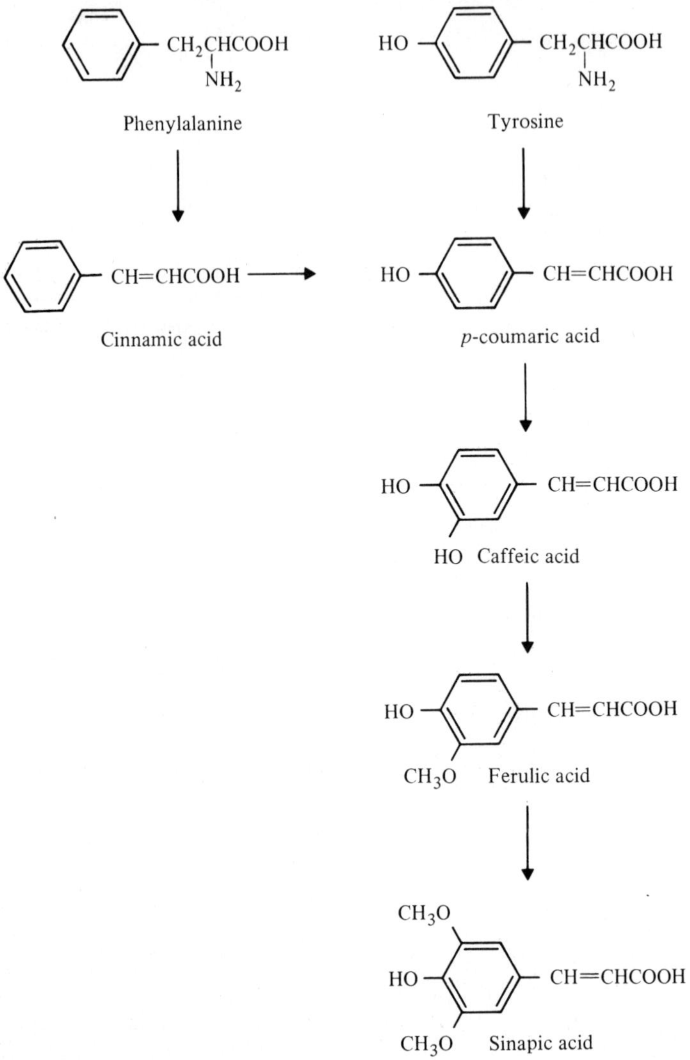

FIGURE 12.3 Biosynthetic scheme for synthesis of acids in the cinnamic acid family. Biosynthesis begins with the aromatic amino acids phenylalanine and tyrosine.

Polymeric linkages are by carbon-carbon (C—C) and ether (C—O—C) bonds in several different positions. Because there is not a constant ratio of the component residues within the lignin polymer due to variation even within a species during de-velopment, lignin is not a readily described chemi-cal compound.

Reduction of the precursor cinnamic alcohols takes place by reduction of the cinnamyl CoA derivatives, using NADPH-specific enzymes. Polymerization begins with the O-glucosides of the alcohols through the action of β-glucosidases to remove the glucose moieties; polymerization then occurs through the catalysis of polyphenol oxidases and peroxidases.

Flavonoids

The flavonoids constitute a large group of complex, water-soluble compounds widespread in the plant kingdom. Their name comes from the fact that many are yellow (flavus). However, they may range in color from white through blue, depending on the substitution of the aromatic rings. The basic ring with its numbering system follows.

Basically the structure is two benzene rings separated by three carbons. The latter are cyclized by means of oxygen to form the heterocyclic ring. Most exist as glycosides in nature, the sugar moiety being attached to hydroxyls at the C-5 or C-7 positions of the A ring or at C-3 of the heterocycle. The anthocyanins are always found as glycosides.

The various flavonoid derivatives differ in the variation of the oxidation or substitution of the heterocycle. The most common variations are the following.

[See A above]

Chalcones, of course, are not technically flavonoids since there is no heterocycle but are included because of their obvious similarity and because of the fact that chalcone is an intermediate in the biosynthesis of the various flavonoids.

Biosynthesis of the flavonoids begins with condensation of malonyl CoA and cinnamyl CoA. From the condensation product, several additional steps produce a chalcone. Ring closure and substitution yield flavanone. Flavanone is converted to flavone, flavonol, and anthocyanidin.

The various kinds of chalcones, flavanones, fla-

A

vonols, and anthocyanin are produced by different R groups on the B ring. Properties differ depending on the kind of glycoside that is ultimately formed. A brief biosynthetic scheme is outlined in Fig. 12.4.

The flavonoids are common, water-soluble pigments in plants; they are abundant in flowers. They range in color from white through yellow, red, and blue. The common red and blue pigments are anthocyanins.

Blues become deeper with more hydroxylation, and methylation results in red coloration. *In vitro*, anthocyanins are red if in an acidic medium and blue if in a basic medium. *In vivo*, blue complexes result by Al^{+3} and Fe^{+3} chelation through the hydroxyls of the A ring. The structures of four common flavonoids are illustrated in Fig. 12.5.

There are a variety of other compounds found in plants related to flavonoids but having more complex ring structures. These will not be discussed here. The texts listed at the end of this chapter have more detailed information.

Tannins

Tannins are polymerization products of phenolic compounds. They account for the brown coloration

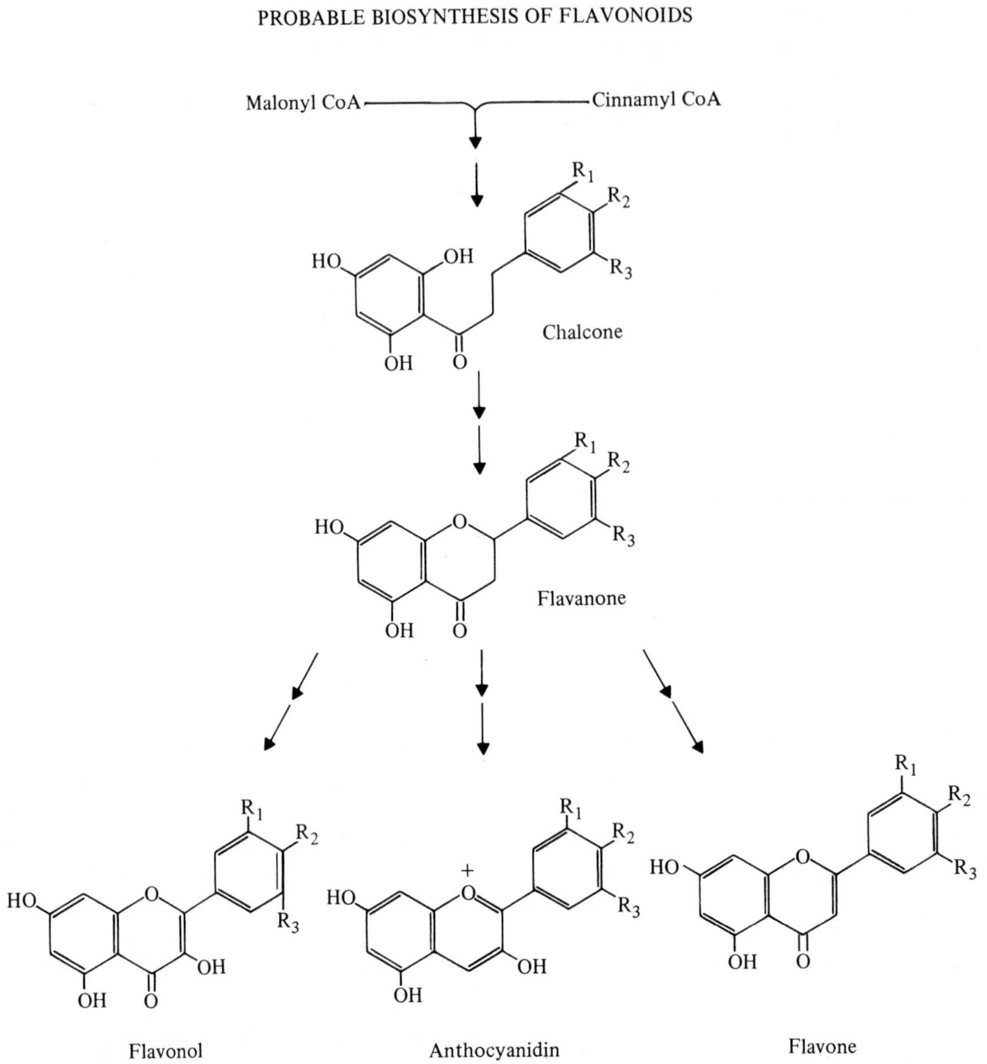

PROBABLE BIOSYNTHESIS OF FLAVONOIDS

FIGURE 12.4 Possible biosynthesis of flavonoids, beginning with malonyl CoA and cinnamyl CoA (refer to Fig. 12.3). Only the major intermediates are shown.

of some leaves in the fall (e.g., oak) as well as the discoloration of injured plant tissues. An excellent example of the latter is the browning of potato tubers or apples when they are cut and exposed to air.

These little-understood polyphenolic compounds will link to animal-skin proteins and render them resistant to decay, a process called tanning. The tanning process converts raw hides to leather.

Many healthy tissues have tannins present, but more frequently they are synthesized when injury causes cellular disruption, which brings into con-

SOME COMMON FLAVONOID GLYCOSIDES FOUND IN PLANTS

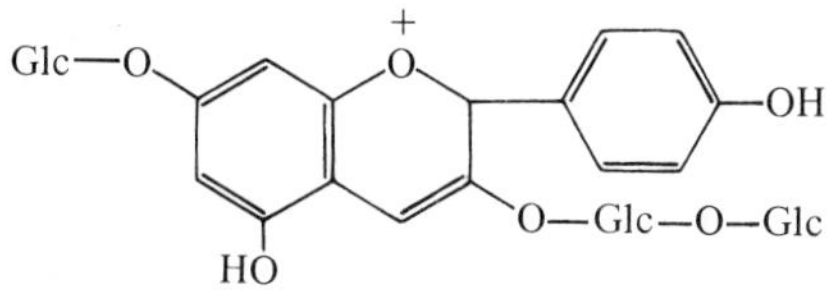

Pelargonidin 3-sophoroside, 7-glucoside

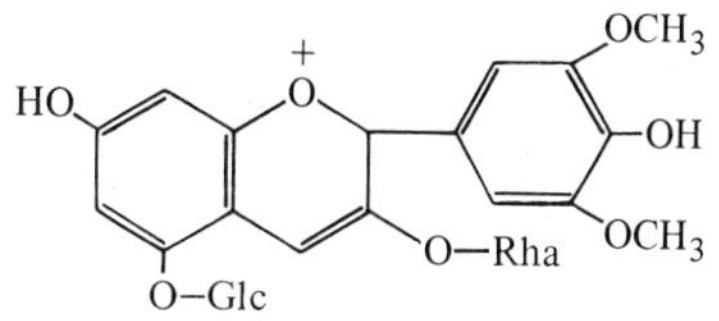

Malvidin 3-rhamnoside, 5-glucoside

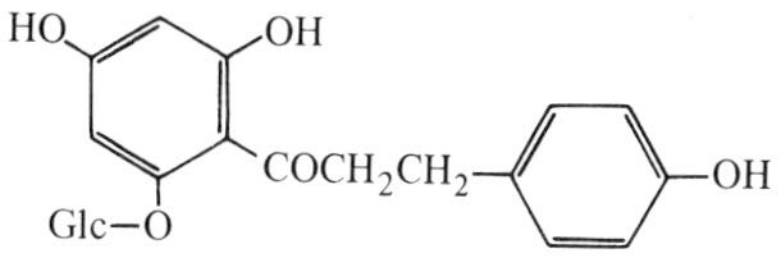

Phloridzin

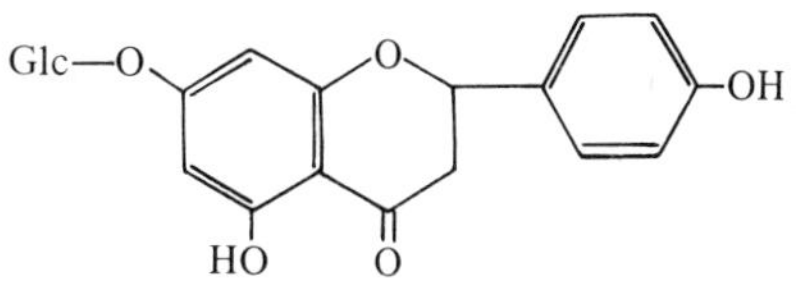

Narigenin 7-glucoside

FIGURE 12.5 Structures of four common flavonoid glycosides found in plants.

tact polyphenol oxidases* and phenolic substrates such as gallic acid, chlorogenic acid, caffeic acid and the flavonoids. Oxidation of the latter phenolic compounds by polyphenol oxidases produces quinone products, which polymerize and form tannins.

Tanning may function as a protective device to prevent infection of injured tissues. However, it is an interfering complication for the study of plant proteins since protein extraction of tissue is frequently accompanied by tanning. The tannins will bind to proteins, and if the proteins are enzymatic, activity is reduced or completely inhibited.

Tanning can be prevented by anaerobic extraction or by inhibiting the polyphenol oxidases with

*Polyphenol oxidases include a large group of copper-containing enzymes that oxidize phenolics to quinones.

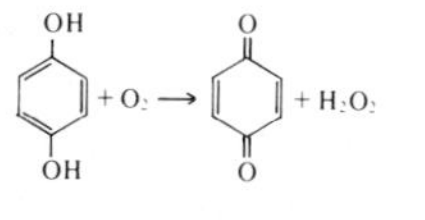

The quinones polymerize, forming the tannins.

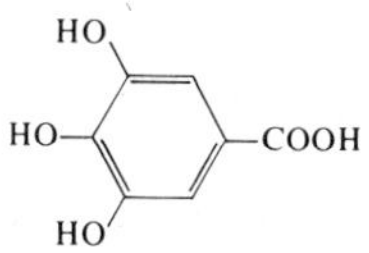

copper-chelating agents. For the latter, diethyl-dithiocarbamic acid (DIECA) is commonly used. If tannins are present in the cell they can be scavenged during extraction with tannin-binding agents such as polyvinylpyrrolidone (PVP), which is used to clarify beer by removal of tannin.

There are two main groups of tannins, hydrolyzable tannins and condensed tannins.

HYDROLYZABLE TANNINS

The hydrolyzable tannins can be hydrolyzed with hot hydrochloric acid into component phenolics and sugars, of which they are polymeric esters. Tara tannin is composed of an ester of gallic and quinic acids.

Gallic acid

It is probably produced from shikimic acid through the shikimic acid pathway, as is quinic acid (Fig. 12.2).

Commercial tannin (tannic acid) is a mixture of gallic acid and gallic acid esters of glucose.

CONDENSED TANNINS

The condensed tannins are largely polymeric condensation products of catechin (flanan-3-ol) and flavan-3,4-diols.

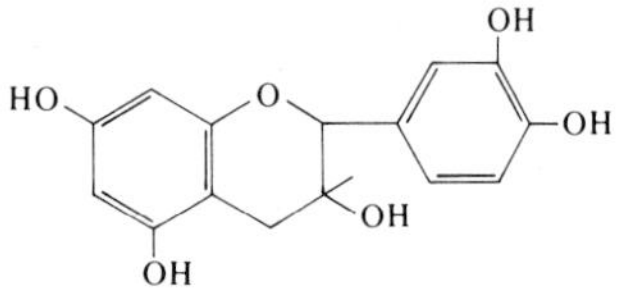

Catechin

The exact mode of polymerization is not completely understood, but it is believed to be through both spontaneous condensations and enzymatic catalysis by polyphenol oxidases.

Because of the binding of phenolics and tannins to proteins, there has been much interest in the nature of the interaction. As mentioned above, the tannins interfere with protein extraction from plant tissues. They also interfere with the preparation of viruses from plants because of binding. There is much of interest to food technologists because tannins cause hazes in beer and wines by binding to proteins; they may also result in unwanted flavors in certain foods, particularly those made from fruits. The high astringency of some foods is due to the presence of tannins.

The interaction of the tannins and polypeptides is through both hydrogen bonding and covalent bonding. The condensed tannins bind strongly to protein up to a pH of 7 or 8 but not above. This is taken as evidence that the binding is through the hydrogen of the phenolic hydroxyls. Above pH 8 phenolic groups are ionized. Hydrolyzable tannins bind to proteins very strongly at pH 3 to 4 but much less at higher pH values. This suggests that strong hydrogen bonds are formed with the carboxyl groups of the phenolics and weak bonds with phenolic hydroxyls.

There is evidence that some proteins have ferulic acid covalently bonded to them. Although this could be an artifact of preparation, phenolics could be normal constituents of proteins.

Polyketide phenolics

Many of the phenolic compounds found in nature are not synthesized through the shikimate pathway but rather are derived from the cyclization of linear poly-β-keto chains with the following structure.

The poly-β-keto chain is called a polyketide and is synthesized through the acetate–malonate pathway described in Chapter 11 for fatty acid biosynthesis. The primary difference between acetate-chain elongation during fatty acid biosynthesis and polyketide biosynthesis is that in polyketide-chain elongation the carbonyl groups are not reduced. After polyketide synthesis comes cyclization, reduction of the carbonyls, and substitution primarily by alkalation, producing the various polyketide phenolic compounds.

The polyketide phenolics are common in fungi and occur in ferns and some higher-plant families. Two polyketide phenolics that have been identified in plants are sorigenin, found in *Rhamnus japonicus,* and eleutherinol, found in *Eleutherine bulbosa.* Their structures follow.

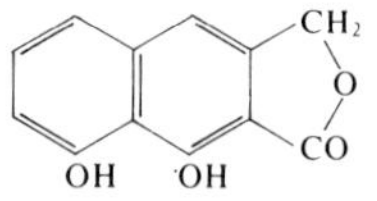

β-sorigenin

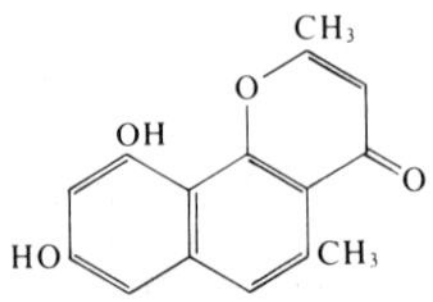

Eleutherinol

Sorigenin occurs in nature as a glycoside of primeverose. The latter is a disaccharide formed from xylose and glucose.

Without complex experimentation it is rather difficult to ascertain if a phenolic is synthesized from the acetate–malonate pathway or from the shikimate pathway. As a rule, it seems that those phenolics produced by cyclization of polyketides have their phenolic hydroxyl groups in the meta position. An example would be phloroglucinol shown on page 305. Those phenolics derived from the shikimate pathway have hydroxyl groups in the ortho position (e.g., catechol).

12.2 Alkaloids

The alkaloids are a large group of heterogeneous compounds that have basic (cationic) properties because of the presence of a nitrogenous heterocyclic ring. Like the other groups of natural products, there are compounds occurring naturally in plants that are considered to be alkaloids and that do not strictly fit the above definition. Some compounds classed with the alkaloids do not have a nitrogenous heterocycle and others may not even be basic, e.g., nicotinic acid. Such compounds are usually classed as alkaloids because of structural or biosynthetic similarities.

Of course, not all plants have alkaloids; however, some may have several different kinds of alkaloids. They are of interest primarily because of their pharmaceutical and medicinal uses. Alkaloids are used as emetics, anesthetics, narcotics, tranquilizers, aphrodisiacs, muscle paralyzers, and stimulants.

Their biological function in plants is quite obscure, but they are suspected of playing a role in storage of waste nitrogen, cationic balance, and protection against parasites, infectious microbes, and predators. They are growth regulators and are perhaps end or waste products of metabolism. Since the alkaloids fall within a mixed group of substances, it is likely that they have multiple functions.

12.2.1 Biosynthesis

With few exceptions, the alkaloids are synthesized from the amino acids tryptophan, phenylalanine, glutamate (glutamic acid), ornithine, and lysine. The nitrogen component of the diagnostic heterocycle comes mostly from amino nitrogen. As shown in Fig. 12.1, the carbon of alkaloids comes from acetate and shikimate. The amino acids glutamate, ornithine, and lysine are in the aspartate and glutamate family of amino acids, and tryptophan and phenylalanine are in the shikimate group of amino acids.

Examples of alkaloid biosynthesis from glutamate, ornithine, and lysine are illustrated in Fig. 12.6. Examples of biosynthesis from tryptophan and phenylalanine are shown later in Fig. 12.8.

12.2.2 Alkaloid classes

Piperidine and related alkaloids

The relatively simple alkaloids common in the Leguminosae (e.g., lupines, peas, and beans) and Solanaceae (e.g., tobacco) are largely derived from glutamate, ornithine, and lysine. They contain simple heterocylic rings: pyrrolidine, piperidine, and pyridine.

Pyrrolidine   Piperidine   Pyridine

Perhaps the most common and well known of this group are nicotine and anabasine.

Nicotine   Anabasine

Nicotine is the main alkaloid of commercial to-

EXAMPLES OF ALKALOID BIOSYNTHESIS FROM GLUTAMIC ACID,
ORINITHINE, AND LYSINE

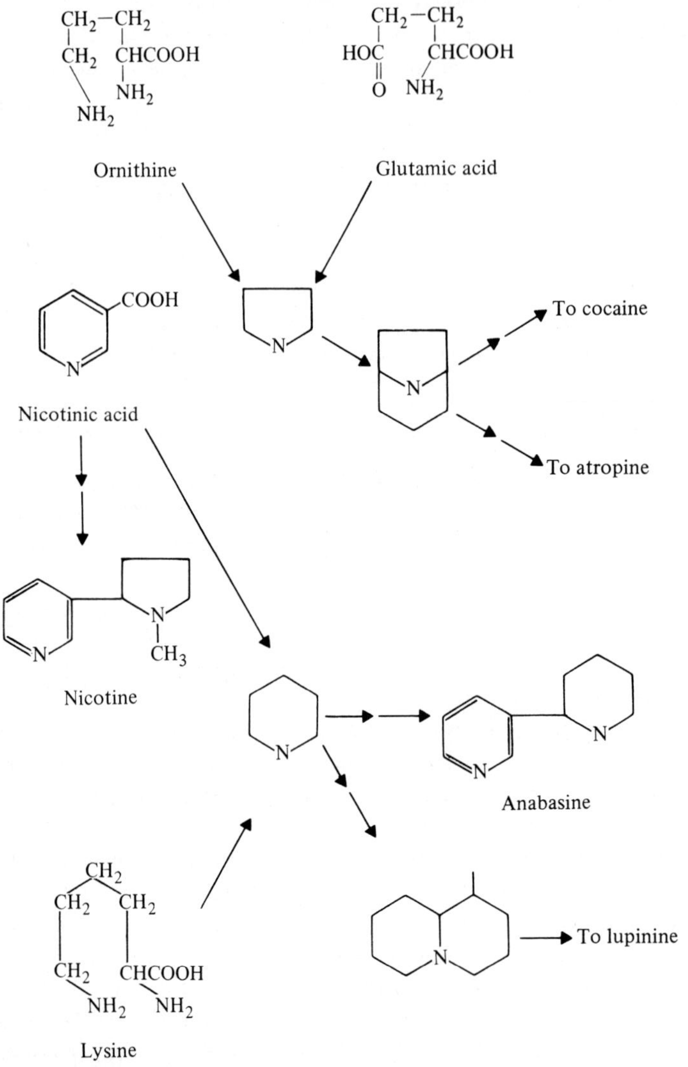

FIGURE 12.6 Examples of the biosynthesis of alkaloids from the amino
acids ornithine, glutamic acid, and lysine. Only the beginning and major
intermediates are shown.

bacco, *Nicotiana tabacum,* and anabasine is most prevalent in tree tobacco, *N. glauca.*

Aromatic derivatives

The aromatic derivatives include alkaloids composed of the tropane ring and other related forms.

Tropane

Tropane is synthesized from ornithine or glutamate through the pyrrolidine ring (see Fig. 12.6). The aromatic portions also come from tryptophan or phenylalanine.

Cocaine, from *Erythroxylon,* is an aromatic tropane alkaloid used as an anesthetic and as a stimulant.

Cocaine

Another tropane alkaloid of great commercial importance is atropine, found in *Atropa belladonna* and *Datura stramonium.* Atropine has anticholinergic properties and is used as a morphine antagonist.

Colchicine, from *Colchicum autumnale,* is synthesized from tyrosine and phenylalanine through cinnamic acid. It has the interesting property of inhibiting cytokinesis during cell division but allowing karyokinesis to proceed, resulting in the formation of polyploids.

Colchicine

As explained in Chapter 1, colchicine binds to

tubulin, the protein of microtubules, preventing assembly during cell division.

Betacyanins

In the Centrospermae, which includes the Chenopodiaceae, water-soluble red pigments occur that are named betacyanins because of their color similarity to anthocyanins. Perhaps most abundant in red beets, they are glycosides of a pyridine ring derivative plus indole derivatives. The aglycone, i.e., the pyridine and indole moiety, is called betanidine.

Closely related yellow pigments, the betaxanthins, are similar in structure.

Isoquinoline and indole alkaloids

The isoquinoline alkaloids form an important group that contains the alkaloids from *Papaver,* including papaverine, thebaine, morphine, codeine, and other opiates. They are constructed of isoquinoline condensed with a benzene ring derived from phenylalanine.

The indole alkaloids, like the isoquinoline alkaloids, are complex cyclic structures. They are found in a variety of higher plants in the families Apocyanaceae, Euphorbiaceae, and Convolvulaceae, among others. In the latter family the ergot alkaloids are found. The indole nucleus of these alkaloids comes from tryptophan, and the remainder comes primarily from isopentenyl pyrophosphate and mevalonic acid.

The base ring structures of the indole and isoquinoline alkaloids are illustrated in Fig. 12.7. Some of the common indole alkaloids are reserpine, strychnine, quinine, and lysergic acid.

Purine alkaloids

The purine alkaloids form an important group that includes caffeine and theobromine. They are somewhat related at least structurally to the cytokinins. The latter are important growth regulators in plants. Biosynthesis of the purine ring was de-

EXAMPLES OF ALKALOID BIOSYTHESIS FROM TRYPTOPHAN AND PHENYLALANINE

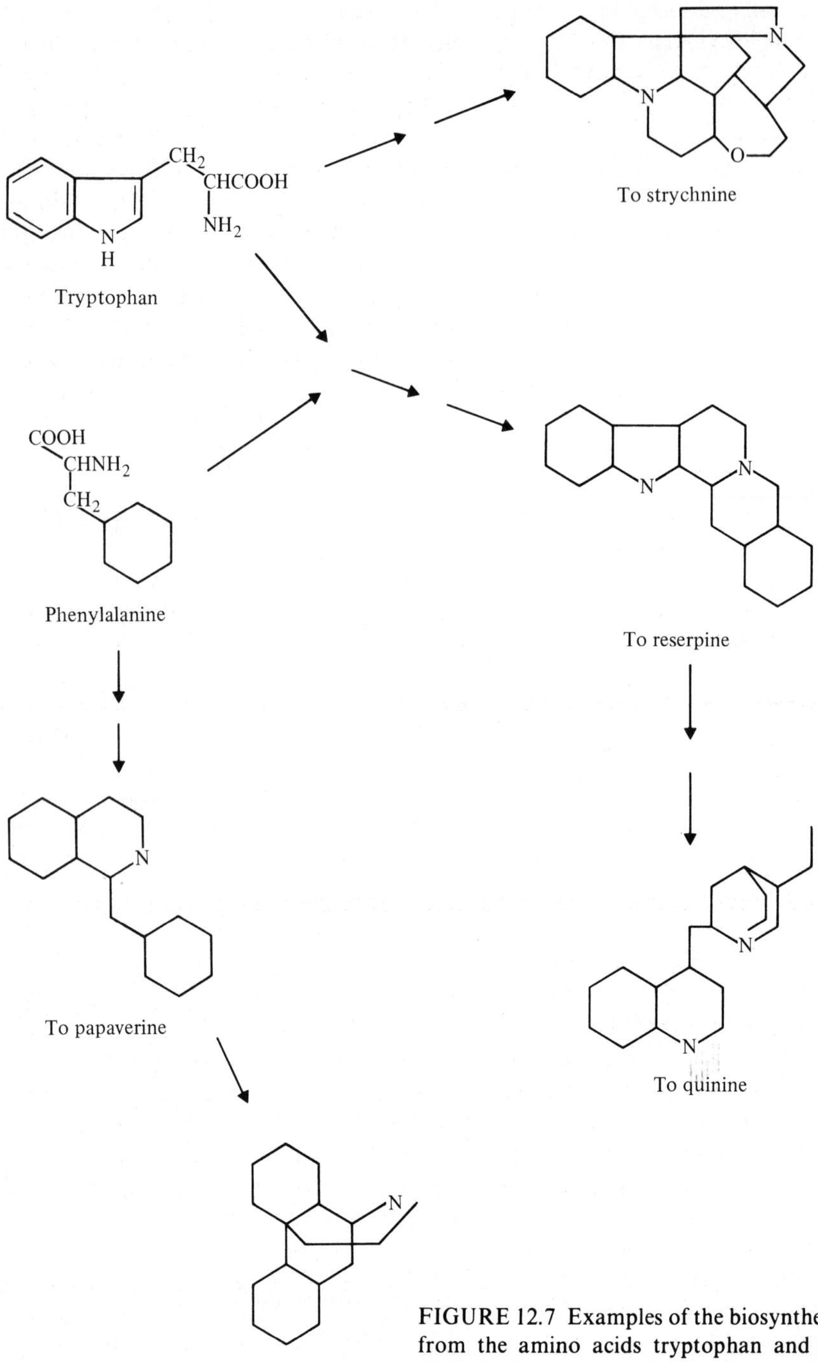

CH2
CHCOOH
NH2
N
H
Tryptophan
N
N
O
To strychnine
COOH
CHNH2
CH2
Phenylalanine
N
N
To reserpine
N
To papaverine
N
N
To quinine
N
To morphine

FIGURE 12.7 Examples of the biosynthesis of alkaloids from the amino acids tryptophan and phenylalanine. Only the beginning and end steps are shown.

scribed in Chapter 8, and the initial steps of the synthesis of the purine alkaloids take place similarly.

The best-known alkaloid of this group is caffeine, found in coffee, tea, and cocoa.

Caffeine

The purine alkaloids are structurally related to the bases of the nucleic acids.

Aliphatic alkaloids

Certain basic compounds of pharmaceutical importance that contain an aromatic ring but no nitrogenous heterocycle can be considered as alkaloids. These are included in the alkaloids because of their functional and structural similarity. Two of the best examples are mescaline and ephedrine, both of which have profound effects on humans. Mescaline induces psychotic alterations in personality and behavior (that is, it is psychotomimetic), and ephedrine causes narrowing of the lumens of blood vessels (vasoconstriction) because of its effect on sympathetic nerves. Mescaline has been isolated from cacti such as *Lophophora williamsii*, and ephedrine is from *Ephedra*.

Mescaline   Ephedrine

12.3 Terpenoids

The terpenoids, falling into the general class of lipids, were discussed briefly in Chapter 11. They are largely water-insoluble acyclic or cyclic compounds with five to several hundred carbons. Included in this group are the essential oils, the steroids, and large polymers such as rubber.

Biosynthesis of the terpenoids proceeds from acetyl CoA to form acetoacetyl CoA; then, with a second condensation with acetyl CoA, hydroxymethyl-glutaryl CoA is formed. The latter is converted to mevalonic acid, the precursor of the primary intermediate of terpenoid biosynthesis, isopentenyl pyrophosphate. Isopentenyl pyrophosphate synthesis from acetyl CoA is outlined in Fig. 12.8. Isopentenyl pyrophosphate is the active "isoprene" for terpenoid biosynthesis.

In living cells, isopentenyl pyrophosphate and dimethylallyl pyrophosphate (see Fig. 12.9) are in equilibrium. Condensation and polymerization produce the various groups of terpenoids in carbon multiples of five.

As a generalization, condensation of the isopentenyl pyrophosphate is usually "head to tail":

$$\underset{\displaystyle C}{\overset{\displaystyle C}{|}} \qquad \underset{\displaystyle C}{\overset{\displaystyle C}{|}}$$
$$C-C-C-C \text{———} C-C-C-C,$$

but, as the student might expect, there are exceptions to this "isoprene rule."

A brief scheme for the biosynthesis of terpenoids by isoprene (isopentenyl pyrophosphate) condensation is shown in Fig. 12.9. The first condensation product of isopentenyl pyrophosphate and dimethyllallyl pyrophosphate is geranyl pyrophosphate. A second addition of isopentenyl pyrophosphate to geranyl pyrophosphate forms the 15-carbon compound farnesyl pyrophosphate. Subsequent condensations produce the various groups of terpenoids.

The terpenoids are usually studied and measured in plants by chromatographic processes. They are readily separated from each other by paper or thin-layer chromatography. However, gas chromatography is a much more quantitative procedure

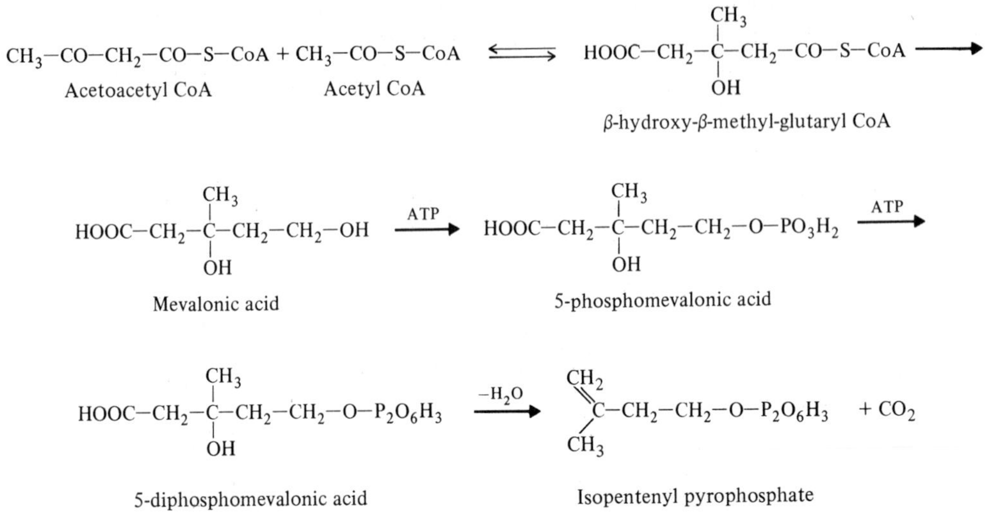

$$CH_3-CO-CH_2-CO-S-CoA + CH_3-CO-S-CoA \rightleftharpoons HOOC-CH_2-\overset{\overset{\displaystyle CH_3}{|}}{\underset{\underset{\displaystyle OH}{|}}{C}}-CH_2-CO-S-CoA \longrightarrow$$

Acetoacetyl CoA　　　　Acetyl CoA

β-hydroxy-β-methyl-glutaryl CoA

$$HOOC-CH_2-\overset{\overset{\displaystyle CH_3}{|}}{\underset{\underset{\displaystyle OH}{|}}{C}}-CH_2-CH_2-OH \xrightarrow{\text{ATP}} HOOC-CH_2-\overset{\overset{\displaystyle CH_3}{|}}{\underset{\underset{\displaystyle OH}{|}}{C}}-CH_2-CH_2-O-PO_3H_2 \xrightarrow{\text{ATP}}$$

Mevalonic acid　　　　　　　　　　　　5-phosphomevalonic acid

$$HOOC-CH_2-\overset{\overset{\displaystyle CH_3}{|}}{\underset{\underset{\displaystyle OH}{|}}{C}}-CH_2-CH_2-O-P_2O_6H_3 \xrightarrow{-H_2O} \overset{\overset{\displaystyle CH_2}{\|}}{\underset{\underset{\displaystyle CH_3}{}}{C}}-CH_2-CH_2-O-P_2O_6H_3 \quad + CO_2$$

5-diphosphomevalonic acid　　　　　　Isopentenyl pyrophosphate

FIGURE 12.8 The biosynthesis of active isoprene (isopentenyl pyrophosphate) from acetoacetyl CoA and acetyl CoA. The active form of isoprene is the main precursor for terpenoid biosynthesis.

and can be used to resolve large numbers of terpenoids in plant samples. Such studies are used to ascertain the diversity of terpenoids, to study their metabolism and their distribution in plants, and for taxonomic purposes. The gas chromatogram shown in Fig. 12.10 shows the extreme diversity and complexity of the terpenoids called essential oils in *Citrus*.

12.3.1 Hemiterpenoids

The hemiterpenoids are direct derivatives of isoprene. Isoamyl alcohol is a hemiterpenoid and occurs as an ester in essential oils of *Eucalyptus* and mints.

$$\begin{array}{c} CH_3 \\ \diagdown \\ \diagup \\ CH_3 \end{array} CHCH_2CH_2OH$$

Isoamyl alcohol

12.3.2 Monoterpenoids

The monoterpenoids are compounds made of two isoprene residues. They may be straight-chained or

ring structures with a variety of functional groups. Characteristically, they are volatile, insoluble in water, and frequently fragrant. Usually they have ten carbons because they are derived from two isoprene units, but as with all isoprenoids, post-biosynthetic modification may change the number of carbons. The monoterpenoids may be acyclic, monocyclic (with one ring), or bicyclic (with two rings). Two examples are geraniol and menthol.

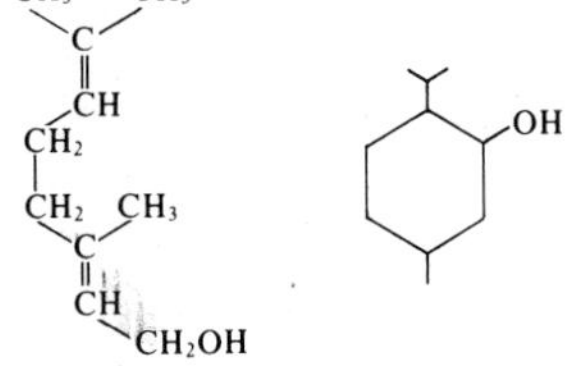

Geraniol　　　　　Menthol

12.3.3 Sesquiterpenoids

Sesquiterpenoids are volatile terpenoids composed of three isoprene units having 15 carbons. They may be acyclic, monocyclic, or bicyclic. Three examples

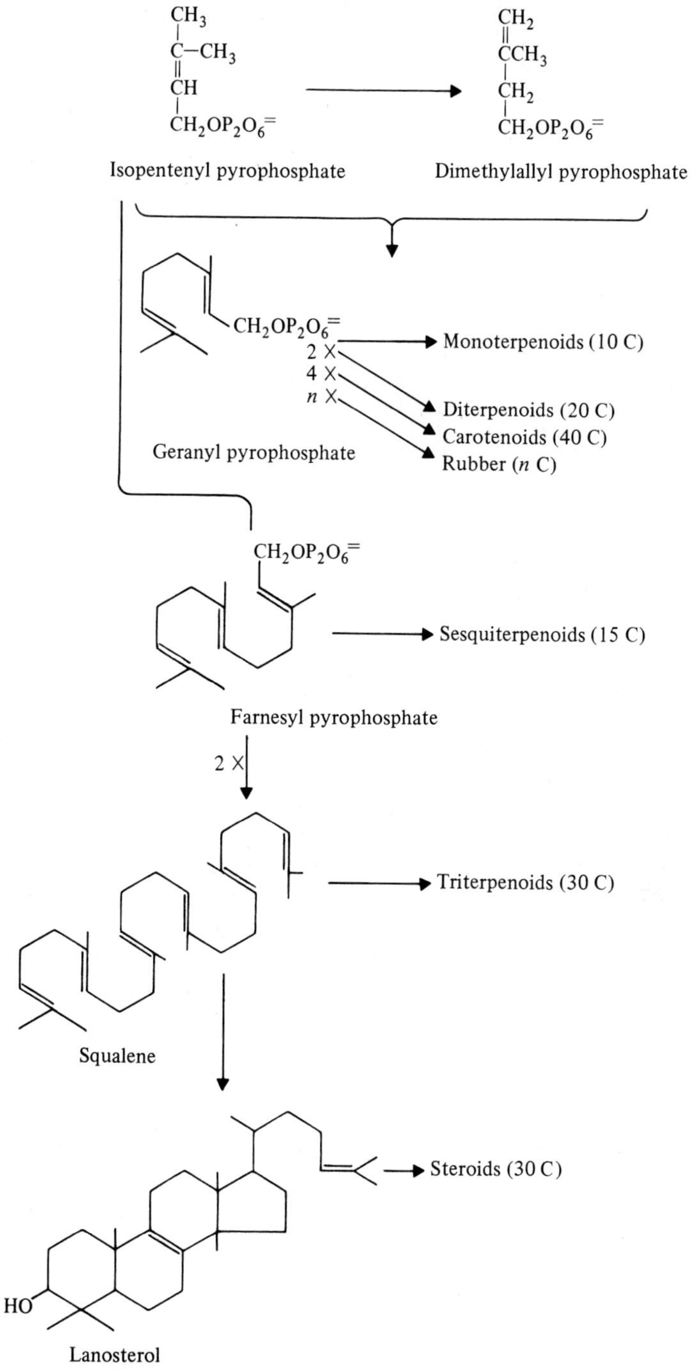

BIOSYNTHESIS OF TERPENOIDS

FIGURE 12.9 Possible mode of biosynthesis of the various groups of terpenoids, beginning with active isoprene (isopentenyl pyrophosphate).

FIGURE 12.10 Gas chromatograph-mass spectroscopic analysis of the volatile essential oils of cotton, mostly monoterpenes. 1 = α-pinene, 2 = β-pinene, 3 = limonene, 4 = γ-terpinene, 5 = β-phellandrene (all monoterpenes); and 6 = an unidentified sesquiterpene. From an experiment by J. Kumamoto, J. G. Waines, J. L. Hollenberg, and R. W. Scora. 1979. Identification of the major monoterpenes in the leaf oil of *Gossypium sturtianum* Var. *nandewdrense* (Der.) Fryx. *J. Agr. Food Chem.* 27: 203-204.

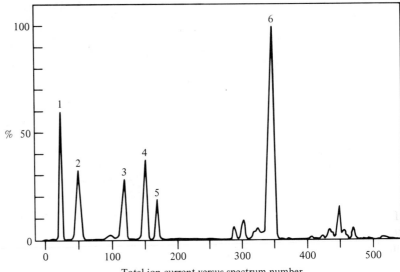

Total ion current versus spectrum number

are the acyclic farnesol, the monocyclic zingiberene from *Zingiber officinale,* and the bicyclic cadinene from cedar.

Farnesol

Zingiberene

Cadinene

12.3.4 Diterpenoids

The diterpenoids are 20-carbon compounds considered to be derived from four isoprene residues. They are not volatile and may be either acyclic or ring compounds. Two examples are the acyclic phytol, which forms the "tail" of the chlorophyll molecule, and camphorene from *Cinnamonum camphora.*

Phytol

Camphorene

The gibberellins are plant growth hormones that are diterpenes. They are discussed in Chapter 18.

12.3.5 Triterpenoids

The triterpenoids are complex 30-carbon compounds that may be acyclic or tetracyclic. Few tricyclic terpenoids exist, and there are no known mono- or bicyclic structures. Squalene, shown below, occurs in some seed oils, such as olive oil, and is a precursor in steroid biosynthesis. The tetracyclic lanosterol is shown below with squalene.

Squalene

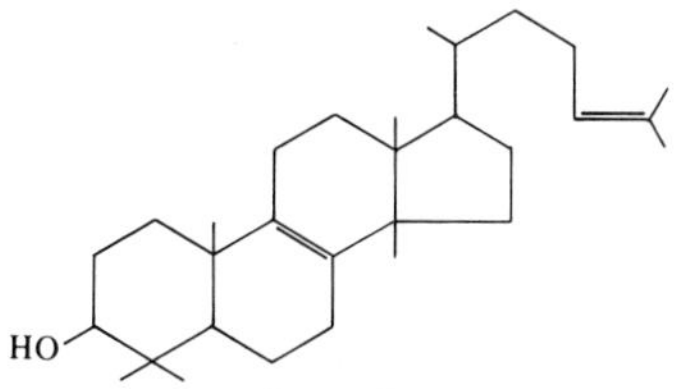

Lanosterol

Triterpenoid glycosides fall within the group that includes saponins and cardiac glycosides. They are frequently surface-active compounds, and many are poisonous to vertebrates. Perhaps those from *Digitalis purpurea* are the best known.

Plant sterols are triterpenoids with an additional methyl group. They are quite common in the plant kingdom but are poorly understood functionally. Virtually all have a hydroxyl group at position C-3. Ergosterol from wheat is shown below. Other common sterols found in plants are cholesterol, stigmasterol in soybean, and spinasterol in spinach.

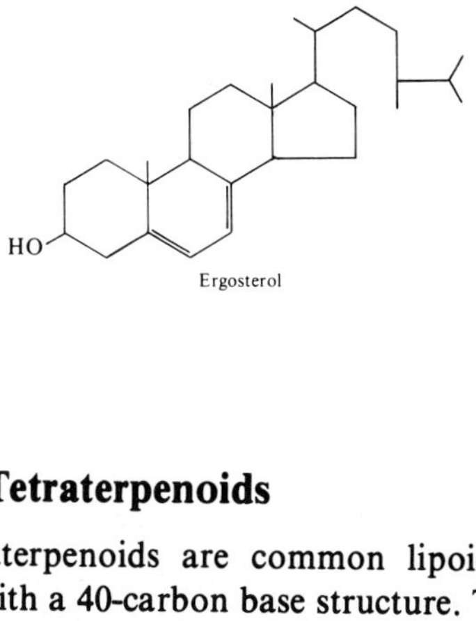

Ergosterol

12.3.6 Tetraterpenoids

The tetraterpenoids are common lipoidal compounds with a 40-carbon base structure. They may be acyclic, monocyclic, or bicyclic. Higher-order cyclization is rare. The best known are the carotenes, such as β-carotene.

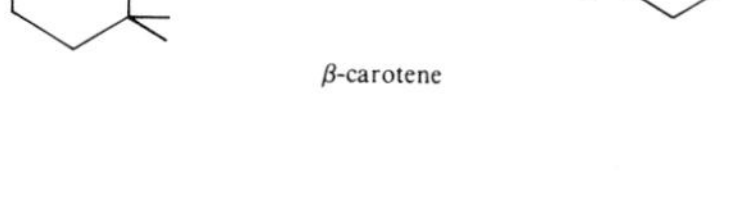

β-carotene

Their role is not entirely understood, but they may function to some extent as light- and oxidation-protective agents in photosynthesis. In addition, there is evidence that they function as accessory pigments (to chlorophyll) by gathering light during photosynthesis.

12.3.7 Polyterpenoids

Although only a few polyterpenoids are known, from an economic viewpoint the polyterpenoid rubber is perhaps the most important. Its structure is illustrated below.

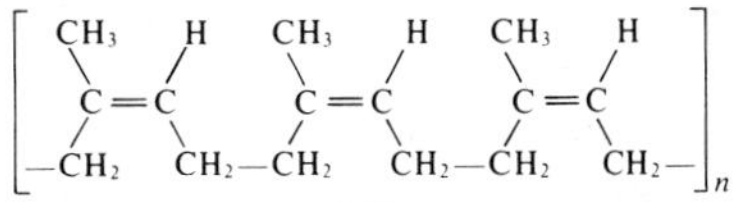

Rubber

The repeating unit contains from 3000 to 6000 isoprene units and all double bonds are arranged in the *cis* configuration. The all-*trans* isomer, gutta, is also common in certain plants. Rubber is somewhat larger than gutta and more elastic. Balata is another polyterpenoid with a *trans* configuration that has properties similar to rubber and gutta. The main component of chewing gum, chicle of *Achras sapota,* is a polyterpenoid.

These polyterpenoids are frequently found in plants within the branched, continuous lactiferous ducts. The liquid component of the lactiferous ducts is called latex and is composed of a variety of substances such as lipids, sugars, salts, and terpenoids. Not all of the latex ducts contain rubber or other polyterpenoids, but in those species which have polyterpenoids they are suspended within the latex. The rubber found in the rubber tree (*Hevea*) is in the latex ducts. In other species, the polyterpenoids may be found in specialized cells. The rubber of guayule, for example, is not in latex ducts but in separate cells in the stem.

The polyterpenoids are found mostly in the families Euphorbiaceae, Moraceae, Asclepidaceae, Apocynaceae, Compositae, and Sapotaceae. The rubber of commerce is obtained mostly from *Hevea brasiliensis* but is also obtained from *Parthenium argentatum* (guayule) and *Taraxacum kok-saghyz* (Russian dandelion).

12.3.8 Mixed terpenoids

There are a variety of compounds that are structurally similar to terpenoids but have additional groups. These include the tocopherols and vitamin K, the benzoquinones, plastoquinone, and ubiquinone (coenzyme Q), the insecticide pyrethrin, and the psychotogen cannabinol. These are mostly aromatic compounds with isoprenoid tails.

The structure of plastoquinone, a photosynthetic pigment, is a good illustration of a mixed terpenoid.

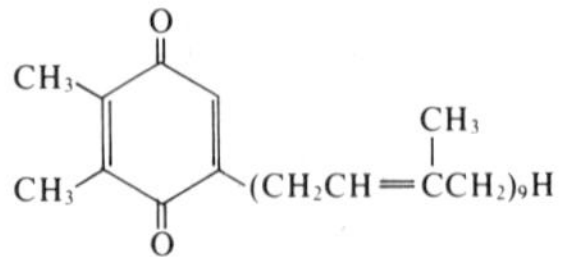

Plastoquinone

12.4 Porphyrins

Porphyrins are closed tetrapyrrole structures that have associated metals when functional. For example, chlorophylls are magnesium porphyrins. The enzymes catalase and peroxidase and the cytochromes of electron transport are iron porphyrins (or hematins).

The basic structural unit is the pyrrole ring.

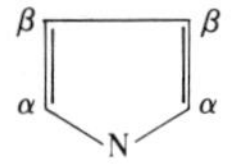

Cyclization of four pyrroles to form the closed tetrapyrrole occurs through carbon linkages between the α positions of the pyrrole ring.

The various porphyrins are distinguished on the basis of the substitutions at the β positions of the pyrroles. Magnesium or iron is bound by chelation within the ring by the nitrogens of the pyrrole heterocycle.

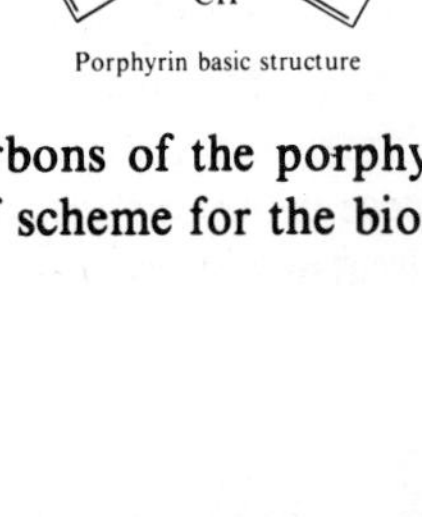

Porphyrin basic structure

All of the carbons of the porphyrin come from acetate. A brief scheme for the biosynthesis of the tetrapyrrole structure is illustrated in Fig. 12.11. The δ-amino levulinic acid precursor of the basic porphobilinogen may come from the condensation of glycine and succinate. However, it has been shown that for chlorophyll synthesis the δ-amino levulinic acid comes directly from glutamate. The C-1 carboxyl, after reduction to a keto group, and the C-2 amino group are rearranged to form δ-amino levulinic acid.

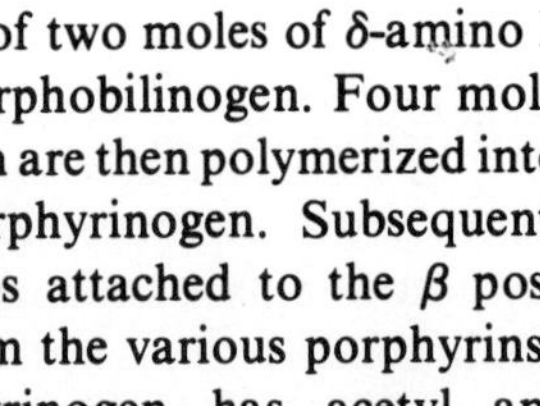

Cyclization of two moles of δ-amino levulinic acid produces porphobilinogen. Four moles of uroporphobilinogen are then polymerized into a ring structure, uroporphyrinogen. Subsequent changes of the R-groups attached to the β positions of the pyrroles form the various porphyrins.

Uroporphyrinogen has acetyl and propionyl groups attached to each of the pyrroles. Unsaturation of the methylene bridges gives the uroporphyrin. Protoporphyrin is produced by R-group substitution at the β positions of the pyrroles. Insertion of magnesium or iron ultimately forms either chlorophyll or the iron-porphyrin hematins.

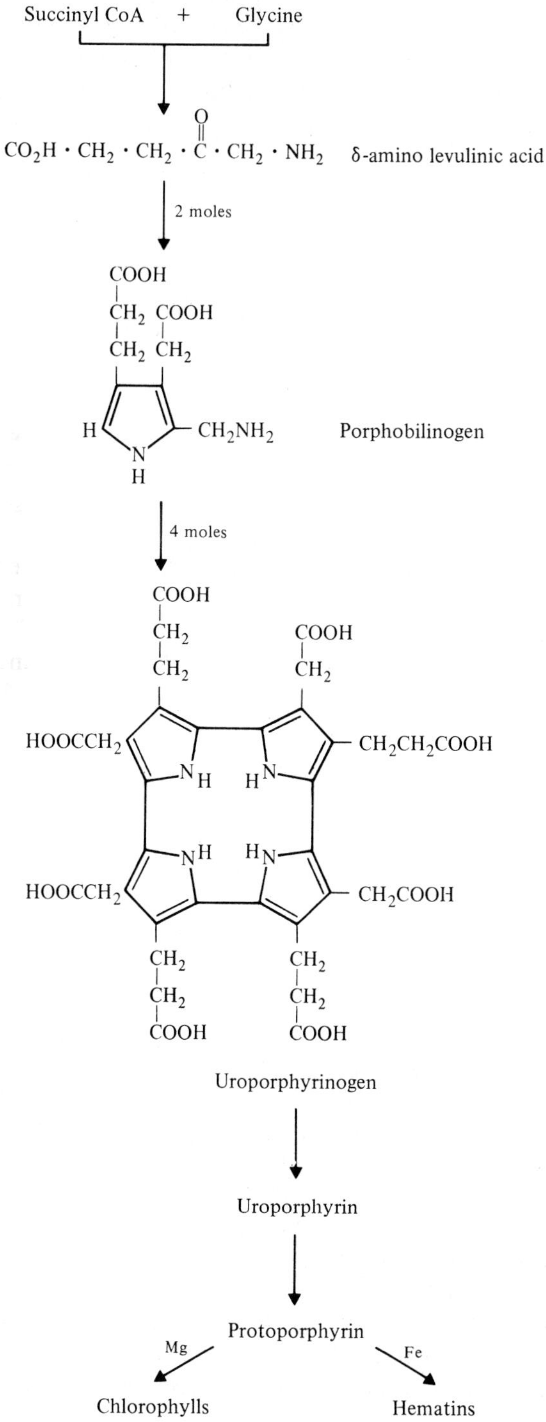

Succinyl CoA   +   Glycine

$CO_2H \cdot CH_2 \cdot CH_2 \cdot \overset{O}{\underset{}{C}} \cdot CH_2 \cdot NH_2$ δ-amino levulinic acid

2 moles

Porphobilinogen

4 moles

Uroporphyrinogen

Uroporphyrin

Protoporphyrin

Mg   Fe

Chlorophylls   Hematins

Substitution on the pyrrole rings is somewhat restricted and includes methyl (—CH_3), formyl (—CHO), acetyl (—CH_2COOH), ethyl (—CH_2CH_3), allyl (—CH=CH_2), and propionyl (—CH_2CH_2COOH). The porphyrins are numbered I through IV on the basis of the pyrrole substitutions.

An important precursor of chlorophyll and hematin is protoporphyrin IX; it is produced from uroporphyrin III. Subsequent introduction of magnesium forms Mg-protoporphyrin IX and introduction of iron forms Fe-protoporphyrin IX. Further steps produce chlorophyll from the magnesium intermediate and hematin from the iron intermediate.

In chlorophyll biosynthesis in angiosperms, Mg-protoporphyrin IX or protochlorophyllide *a* is converted to chlorophyllide *a* through a light-dependent reaction. Chlorophyllide is the immediate precursor of chlorophyll.

The various porphyrins can be distinguished on the basis of absorption spectra, and the metal-free porphyrins fluoresce with an intense red coloration in neutral or acid solution.

The function of the porphyrins is light absorption in the case of the chlorophylls and electron transport by oxidation–reduction reactions in the case of the iron porphyrins. When cytochromes function in electron transport during respiration, the iron goes reversibly from the ferric to the ferrous state during reduction.

The hematin pigments differ from the chlorophylls in that the iron prosthetic group of the hematins is tightly bound, whereas the magnesium of the chlorophylls is readily removed.

Chlorophyll *a* and *b* are the common green pigments of higher plants. Chlorophyll *a* is blue-green and differs from the yellow-green chlorophyll *b* by having a methyl group rather than a formyl group at one of the *β* positions of a pyrrole moiety.

FIGURE 12.11 Biosynthetic pathway for the synthesis of porphyrins, beginning with succinyl CoA and glycine.

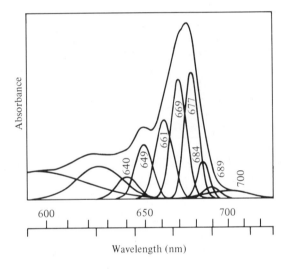

Chlorophyll *a*

Chlorophyll function is discussed more fully in Chapter 15.

The chlorophylls are hydrophobic molecules because of the presence of the phytol chain ($C_{20}H_{39}OH$) attached to the propionyl group of a pyrrole ring through an ester bond. Chlorophyll *in vivo* is most frequently found bound to small hydrophobic proteins. These proteins that have masses on the order of 10,000 daltons are high in leucine and isoleucine and have relatively small amounts of acidic and basic amino acids. The chlorophyll–protein complex is formed by hydrophobic bonds. It is believed that the different absorption and emission spectra of chlorophyll *a*-protein complexes are the result of different proteins creating a different kind of environment for the chlorophyll molecule. As will be explained in Chapter 15 with regard to the primary processes of photosynthesis, the primary light traps of photosynthesis have absorption spectra in the red (photosystem I at about 700 nm and photosystem II at about 680 nm). Because of the nature of the proteins complexed with other chlorophyll *a* molecules within the same photosynthetic unit, many have absorption spectra shifted toward the blue, that is, they have maxima at lower wavelengths. These accessory chlorophylls absorb radiant energy and transfer it to the chlorophylls that absorb at longer wavelengths by excitational energy transfer. Ultimately the energy is transferred to the primary trap, either the pigment 680 of photosystem II or the pigment 700 of photosystem I.

C. S. French and his colleagues (Brown, 1972), using computer-assisted analysis, have studied the absorption spectra of chlorophyll in great detail. Figure 12.12 shows the absorption spectrum of a chlorophyll-containing fragment of chloroplast grana believed to have photosystem I. The spectrum, measured over the range of just less than 600 nm to just beyond 700 nm, shows several peaks and shoulders. The various shoulders and peaks were analyzed by assuming that they are the sum of a series of individual peaks from chlorophyll–protein complexes with different absorption spec-

FIGURE 12.12 Curve analysis of the absorption spectrum from chloroplast grana fragments isolated from spinach. The fragments have mostly photosystem I activity (reduction of NADP to NADPH). The absorption spectrum (upper curve) with several shoulders and peaks is assumed to be the sum of several component absorption bands from chlorophyll-protein complexes that abosrb at different wavelengths. The absorption spectrum was resolved by computer-assisted analysis. From J. S. Brown, R. A. Gasanov, and C. S. French. 1973. A comparative study of the forms of chlorophyll and photochemical activity of system 1 and system 2 fractions from spinach and *Dunaliella*. *Carnegie Institution Yearbook* 72:351–359.

tra. The sum of the individual peaks will add to produce the absorption spectrum of the fragments. In this particular fraction, which contained photosystem I, there were chlorophyll *a*–protein complexes that had absorption peaks at 640 nm, 649 nm, 661 nm, 669 nm, 677 nm, 684 nm, 689 nm, and 700 nm.

Although there are pigments in addition to chlorophyll *a* that act as energy gatherers ("antennae chlorophylls"), chlorophyll *a* is the primary pigment of photosynthesis. It must be reemphasized that even though the chlorophyll *a*–protein complexes differ in their absorption spectra depending on the protein structure, all the evidence indicates that the chlorophyll *a* molecule is the same. It is conceivable, of course, that side-chain substitution could also alter absorption spectra, but nuclear magnetic resonance studies of isolated chlorophyll *a*–protein complexes do not suggest that the chlorophyll *a* structure differs.

The open-chain tetrapyrroles of the phycobiliproteins are important in plant metabolism. They are pigments of lower plants, for example, the phycocyanins and phycoerythrins of algae. The important growth and development pigment, phytochrome, is a tetrapyrrole.

Review Exercises

12.1 In your own mind, what do you visualize to be the metabolic or physiological role of the various compounds classified as natural products? Consider in your answer products such as rubber, alkaloids, and porphyrins.

12.2 Define a glycoside. Some common glycosides that are not usually thought of as glycosides are nucleic acids. What would be the aglycone portion of the nucleic acid molecule? Draw the glycosidic bond of the nucleic acids. The nucleic acids can be classified as purine and pyrimidine alkaloids. Why?

12.3 When plant tissues are injured, it is quite common to observe a rapid brown coloration develop because of the polymerization of quinones. How could you prevent such tanning from occurring? You should be able to think of several different ways.

12.4 What do you think is the significance of the fact that many of the compounds falling within the group called natural products have such a profound effect on the physiology of mammals such as humans? What are the main products obtained from plants that have the capacity to alter human physiology?

12.5 Prepare a list of useful and beneficial natural products that are obtainable from plants. Classify them as phenolics, alkaloids, or terpenoids.

References

BEYTIA, E. D., and J. W. PORTER. 1976. Biochemistry of polyisoprenoid biosynthesis. *Ann. Rev. Biochem.* 45:113–142.

BROWN, J. S. 1972. Forms of chlorophyll in vivo. *Ann. Rev. Plant Physiol.* 23:73–86.

DEVERALL, B. J. 1977. *Defence Mechanisms of Plants.* Cambridge University Press, Cambridge.

FREUDENBERG, K., and A. C. NEISCH. 1968. *Constitution and Biosynthesis of Lignin.* Springer-Verlag, Berlin.

GEISSMAN, T. A., and D. H. G. GROUT. 1969. *Organic Chemistry of Secondary Plant Metabolism.* Freeman, Cooper, & Co., San Francisco.

GIBBS, R. D. 1974. *Chemotaxonomy of Flowering Plants,* Vols. 1–4. McGill Queens University Press, Montreal.

GOODWIN, T. W. (ed.). 1976. *Chemistry and Biochemistry of Plant Pigments,* 2d ed., Vol. 1. Academic Press, London.

GOODWIN, T. W. 1979. Biosynthesis of terpenoids. *Ann. Rev. Plant Physiol.* 30:369–404.

HARBORNE, J. B. (ed.). 1972. *Phytochemical Ecology.* Academic Press, New York.

HARBORNE, J. B., T. J. MABRY, and H. MABRY (ed.). 1975. *The Flavonoids.* Academic Press, New York.

KEELER, R. F. 1975. Toxins and teratogens of higher plants. *Lloydia* 38:56–86.

LUCKNER, M., and K. SCHREIBER (eds.). 1979. *Regulation of Secondary Products and Plant Hormone Metabolism.* Pergamon Press, Oxford.

MARINI-BETTOLO, G. B. (ed.). 1978. *Natural Products and the Protection of Plants.* American Elsevier, New York.

REBEIZ, C. A., and P. A. CASTELFRANCO. 1973. Protochlorophyll and chlorophyll biosynthesis in cell-free systems. *Ann. Rev. Plant Physiol.* 24:129–172.

RICE, E. L. 1979. Allelopathy—an update. *Bot. Rev.* 45:15–109.

ROBINSON, T. 1975. *The Organic Constituents of Higher Plants: Their Chemistry and Interrelationships,* 3d ed. Cordus Press, North Amherst, Mass.

SARKANEN, K. V., and C. H. LUDWIG (eds.). 1971. *Lignins.* Wiley, New York.

SCHULTES, R. E. 1970. The botanical and chemical distribution of hallucinogens. *Ann. Rev. Plant Physiol.* 21:571–598.

STAFFORD, H. A. 1974. The metabolism of aromatic compounds. *Ann. Rev. Plant Physiol.* 25:459–486.

WALLACE, J. W., and R. L. MANSELL (eds.). 1975. Biochemical interaction between plants and insects. *Recent Advances in Phytochemistry,* Vol. 10. Plenum Press, New York.

WALLER, G. R., and E. K. NOWACKI. 1978. *Alkaloid Biology and Metabolism in Plants.* Plenum Press, New York.

iii

Biochemical Constituents of Plants

PROSPECTUS With the obvious exception of the natural products, the biochemicals found in plants differ little from those found in animals and microbes. The major chemical constituents of life are quite universal. All plants and animals have nucleic acids, proteins, carbohydrates, and lipids. Nevertheless, there are some striking differences between plant and animal biochemicals, which to some extent reflect their basic differences in metabolism and structure. Plants undergo photosynthesis and photorespiration, and although most of the intermediates are common to both plants and animals, the protein catalysts and the lipids associated with plant cell membranes differ. In addition, plants have a cell wall with cellulose and hemicelluloses and a middle lamella composed of pectins.

There are some marked differences between plant and animal nucleic acids. Plant DNA is high in the uncommon base 5-methyl cytosine, and plant tRNA is rich in cytokinins. But both have DNA, mRNA, tRNA, and ribosomal RNA. The lipids of plants and animals differ in kind. Both galacto- and sulfur lipids are common in plants and uncommon in animals.

There is still much to learn about the biosynthesis of the chemical constituents of plants. We do not understand cell-wall biosynthesis. *In vitro* studies using different nucleotide sugar donors produce varying results including 1,3 polymers as well as 1,4 cellulose polymers. The biosynthesis of many of the organic acids that accumulate in plants such as isocitrate, malonate, and oxalate is poorly understood. Studies concerning the biosynthesis of unsaturated fatty acids have been confusing, suggesting that more than one metabolic pathway exists. Other areas of interest include the DNA and RNA metabolism of chloroplasts and mitochondria. And there is much interest in cell-free peptide synthesis from mRNA isolated from plants.

Without doubt, the entire topic of natural product chemistry is the least understood. We know very little about the biosynthesis of the various natural products and even less about their metabolic function. Are they storage products, end products of metabolism, waste products, or metabolic intermediates? Since many of the natural products are important industrial or pharmaceutical compounds, they will continue to remain of interest.

Metabolic Processes
of Plants

I V

P R O L O G U E The study of plant metabolism is interdisciplinary, relying heavily on knowledge gained in other fields. Biophysics is important to ion-transport studies and for an understanding of the light reactions of photosynthesis. Chemistry is important for plant nutrition studies, and biochemistry is an integral part of understanding respiration and carbon metabolism of photosynthesis. Despite the fact that some of the reactions of photosynthesis are common to most living organisms, the process of photosynthetic carbon reduction is unique to plants. Thus both carbon metabolism and the photochemistry of photosynthesis are of central interest to plant physiologists. Mineral nutrition is also a field that is centrally within the realm of plant physiology. For the most part, respiration of plants is similar to the respiration of animals, and many studies and findings are common to both plant and animal physiology.

Although it has been known for centuries that some form of fertilization is necessary for adequate plant growth and development, especially under agricultural conditions, serious studies to ascertain the elements essential for plant growth and development did not begin until the investigations of the German botanists Sachs and Knop in the mid-1800s. Their studies, along with those that followed, used aqueous cultures, which led to the discovery of the 16 elements essential for plant growth. Subsequently, mineral

329

nutrition studies have concentrated more on ion uptake and transport. Recently, specific ATPase enzymes implicated in the transport of ions have been isolated from plants. The most extensively studied is a potassium-dependent ATPase assumed to be involved in potassium uptake.

The important research on plant respiration to a large extent followed along with the studies of Hans Krebs (1937) and others, who elucidated the pathways of oxidative respiration in animal tissues. These workers discovered the metabolic pathways of glycolysis and the citric acid cycle and the mechanism of electron transport during oxidative phosphorylation. Basically the same processes occur in plants as in animals. Recent investigations of plant metabolism include studies of the cyanide-resistant electron transport pathway and the mechanism by which phosphorylation occurs in mitochondria. Peter Mitchell (1966) developed the hypothesis of proton gradients driving phosphorylation, which has largely been investigated in plants with photophosphorylation in chloroplasts rather than those with oxidative phosphorylation in mitochondria.

Perhaps one of the most exciting discoveries of plant metabolism during the past several decades was that certain plants of tropical origin (C_4 plants) have a modified carbon-assimilation pathway that results in high photosynthetic efficiency. These plants have a special leaf anatomy (Kranz anatomy), fall into limited taxonomic groups, and have high rates of photosynthesis. The initial discovery

was made in the 1960s and stimulated the interest of physiologists, anatomists, taxonomists, biochemists, and ecologists because of the wide range of implications. The discovery resulted in much renewed interest and research on the carbon metabolism of plants. Subsequent to the discovery of C_4 photosynthesis it was realized that the crassulacean acid metabolism occurring in certain succulent plants was also a modification of the basic C_3 photosynthetic cycle elucidated by Calvin and his associates. There is presently much research to clarify these alternative pathways and to search for additional modifications that may exist in nature. Furthermore, there is much interest in developing desirable agricultural crops with attributes of either C_4 photosynthesis or the CAM pathway.

Photosynthesis represents one of the most distinct metabolic events of plants, and although much is now known about the carbon metabolism of plants, we still lack information about much of the photochemistry of photosynthesis. The Z scheme, which proposes how electron transport drives ATP synthesis and NADP reduction, is consistent with most of the data, but the oxidized and reduced intermediates are not known. The photochemistry of photosynthesis is an active research area of plant metabolism.

In this part there is an overview of plant metabolism, including mineral nutrition and ion uptake and transport, respiration, and photosynthesis.

CHAPTER THIRTEEN

Plant Mineral Nutrition and Ion Uptake

Some form of organized agriculture dates to ancient times, but there are few records of activity such as soil fertilization or soil mineral nutrition until the time of the Greeks and Romans. The early Egyptians cultivated their fields, but there is evidence that the Greeks actually added organic matter to soil prior to sowing in order to improve crop productivity. Roman agriculturists used crop rotation, added lime and organic matter, and grew legumes to improve soils for plant growth.

Perhaps the beginning of scientific agriculture occurred in the 17th century with the European observations that minerals containing nitrogen would enhance plant growth. The earliest kinds of experiments to study plant mineral nutrition and soil improvement were done in the field. In 1843 the oldest existing experiment station, The Rothamsted Experiment Station, was established in England by Sir John Lawes and John Gilbert. Together they established the beginnings of the extensive experimentation that subsequently followed on plant mineral nutrition.

Subsequent to the establishment of field stations there was much interest in studying nutrition using aqueous culture solutions. It was generally believed that many of the principles that applied to mineral nutrition of plants growing in soils could be discovered through growth in culture solutions. It was easier to produce well-defined conditions and reproduce experiments in culture. Much of the early work is attributed to Hoagland, Stout, and Arnon, working at the University of California in Berkeley (see Hoagland, 1923). Their research showed that minerals were taken up primarily in the ionic form and that different ions were absorbed by roots at different rates. Furthermore, such research showed that one ion influenced the uptake of another. For example, potassium from KCl was taken up at a greater rate than potassium from K_2SO_4. Ions such as sodium also interfered with the uptake of potassium and calcium. Ions are usually absorbed by plant roots at a greater rate from dilute solutions than they are from concentrated solutions. This information, with the additional knowledge that the rates of ion uptake were not necessarily correlated with water uptake and transpiration, led to many questions about the mechanisms of ion uptake.

The above phenomena can be partially explained by hypothesizing specific transport mechanisms for the various ions. But the question of ion transport and mineral nutrition is much more complex than just uptake and subsequent use in the physiological processes of the plant. Many other factors influence the availability of ions in the soil solution. Soil properties, exchange capacity, pH, and the presence of different kinds of minerals and salts are all important. Thus the problems of plant nutrition vary depending on the kind of soil, the kind of environment, and even the presence of various plants.

13.1 Mineral Uptake

In this chapter two aspects of the mineral metabolism of plants are discussed. First the principles of ion uptake and transport are covered. Second, the actual use of the ions and minerals by the plant during growth and developmental processes are discussed.

13.1.1 The problem

Plants live in a dilute and highly variable environment with respect to the minerals necessary for their growth and development. The main question is how plants accumulate minerals against concentration gradients ranging from one to five orders of magnitude in excess of environmental concentrations.

An example is shown with the marine alga, *Halicystis*. As shown in Table 13.1, sea water has more sodium and less potassium than does the plant vacuole whereas chloride is distributed equally, indicating that the plant has a concentrating mechanism for potassium and an exclusion mechanism for sodium. This implies expenditure of energy to maintain concentrations above or below those predicted by electrochemical gradients. The distribution of ions across plant cell membranes can be studied with the Nernst equation, which relates the electrochemical potential across a membrane to the distribution of ions across the membrane.

13.1.2 The Nernst equation

Electrochemical potentials are established across membranes because of unequal charge distribution that is brought about by a variety of circumstances,

TABLE 13.1 Distribution of K^+, Na^+, and Cl^- in seawater and in the vacuole of *Halicystis*

Element	Seawater, m*M*	Vacuole, m*M*
Na^+	488	257
K^+	12	337
Cl^-	523	543

From Blount and Levedahl (1960).

including nondiffusible and membrane-bound charged species. For example, in an equilibrium situation potassium ions will distribute across a membrane as a function of the electrochemical potential difference across the membrane. If the ions are not in electrochemical equilibrium, it is implied that energy is required to maintain the nonequilibrium situation. The difference between the membrane potential and the actual potential created by the nonequilibrium distribution is a measure of the quantity of energy required. A modification of the Nernst equation that can be used for such calculations is

$$\epsilon = -\frac{R \cdot T}{Z \cdot F} \cdot \ln \frac{a_i}{a_o},$$

where

ϵ = electrochemical potential between cells of root and external solution, in millivolts (mv);

R = gas constant (8.3 J mol^{-1} K^{-1});

T = absolute temperature (K);

Z = net charge on ion (dimensionless);

F = Faraday constant (96,400 J mol^{-1});

a_i = activity of ion on inside of tissue; and

a_o = activity of ion outside of tissue.

For quick calculation, it is convenient to remember that

$$\frac{R \cdot T}{F} = 26 \text{ mV}.$$

For pea roots the electrochemical potential between root cells and the medium was found to be -110 mV. Because the activities of each ion in the solution bathing the roots were known and since it was assumed that they distributed according to the Nernst equation (that is, on an electrochemical gradient), the expected activities in the root were ascertained (Table 13.2).

It is apparent that potassium uptake by the root followed electrochemical gradients. But sodium and magnesium were excluded against the gradient and nitrate and chloride were accumulated against

TABLE 13.2 Actual activity and calculated activity from the Nernst equation for various ions in pea roots

Ion	Actual	Calculated
K$^+$	75 μEq g^{-1}	74 μEq g^{-1}
Na$^+$	8	74
Mg^{+2}	3	2700
Ca^{+2}	2	10,800
NO$_3^-$	28	0.027
Cl$^-$	7	0.014

From an experiment by Higinbotham et al. (1967).

the gradient. Theories to explain accumulation and exclusion against potential gradients are discussed below.

13.1.3 The path of mineral uptake

A cross section of a typical root is illustrated in Fig. 1.20. Like water, both ions and nonionic materials can move into roots (through free intercellular spaces or cell walls) until they reach the endodermis. Because of the suberization of endodermal lateral walls (the Casparian strips), all substances that enter the root must pass through living membranes before transport, eventually reaching the conducting tissues of the stele.

Transport can be visualized in two ways. Transport may be either through the cytoplasm of cells and from cell to cell through plasmodesmata, or throughout much of the plant through nonliving spaces and within cell walls.

The symplasm hypothesis was discussed in Chapter 6. According to the symplasm/apoplasm concept, the plant body is divided into two physiological regions, the symplasm (living portion) and the apoplasm (nonliving portion). The symplasm is composed of the living cytoplasmic portion of the plant body including all regions bounded by selective membranes, such as vacuoles, whereas the apoplasm includes nonliving regions, such as cell walls and the nonliving xylem elements. This hypothesis led to concepts of free space as opposed to inner space within the living portion.

If a root or other tissue is placed in a solution containing a solute such as mannitol, which will not pass freely through the plasma membrane, the mannitol solution will occupy a volume (outside the symplasm) called the water free space (WFS). It is explicitly implied here that the WFS is within the apoplasm.

However, in cases involving ionic material, if tissue is bathed in a solution the ions will freely occupy a certain space that does not necessarily correspond to the WFS. This space in which the ions can be quickly removed by washing is called the apparent free space (AFS). It is termed apparent free space because it may not necessarily correspond to the apoplast but may include some of the symplast or even exclude some of the apoplast. Furthermore, each substance is apt to have its own AFS.

AFS is defined in a physiological sense and not in the anatomical sense of symplasm and apoplasm. However, for most purposes the apoplast and AFS can be considered synonymous.

13.1.4 General rules for uptake

There are some rules that apply to the uptake of solutes by plants. First, nonionic hydrophilic substances are generally taken up as an inverse function of their size. Thus small molecules or ones with little hydration will be taken up more freely than larger molecules. It is implied here that uptake is through pores (holes) in hydrophilic portions of membranes and that uptake would follow concentration gradients.

Second, hydrophobic substances will be transported as a function of their lipid solubility. Those hydrophobic substances most soluble in lipid will be most readily taken up. Again, it is implied here that uptake would follow concentration gradients and be through lipid portions of membranes.

Some ionic substances enter roots easier when they are in the form of ions, while others enter more readily when not in ionic form. Acidic and basic substances will be influenced by pH.

Finally, many materials are excluded either partially or completely, and many are accumulated against potential gradients. This latter observation implies selectivity of uptake by cellular membranes and strongly suggests active-transport processes.

13.1.5 Simple uptake

Simple uptake of ions can be described with the following flux equation:

$$J = L_a \frac{d\mu}{dx},$$

where

J = flux in mol cm^{-2} s^{-1},

L_a = permeability in cm^{-1} s^{-1}, and

$\frac{d\mu}{dx} = \triangle\mu$ = chemical potential gradient, or driving force, in mol cm^{-1}.

A positive flux results if the driving force, $\triangle\mu$, is positive such that there is a concentration or potential gradient across the membrane.

The time span of simple ion uptake is shown in Fig. 13.1. If roots are bathed in a solution containing K^+ there will be a rapid influx of ions followed by a steady, linear uptake. If the roots are subsequently washed to remove free-space K^+ ions and the amount of K^+ remaining in the roots is determined, a graph will show a straight line from 0 time with a slope (or rate) equal to the linear portion of the first curve. The linear uptake curve is shown in Fig. 13.1 below the actual uptake curve.

The K^+ that freely washes out is the K^+ that entered the AFS, whereas the linear uptake portion represents ions transported to the inner space. Inner-space ions do not readily exchange and are not washed out.

The upper curve in Fig. 13.1 can be described with the following expression:

$$C_t = C_i + C_e (1 - e^{-kt}) + vt,$$

where

C_t = ions taken up at time t;

C_i = ions initially present in tissue (at zero time, $t = 0$);

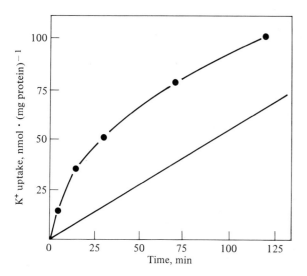

FIGURE 13.1 The time course of potassium uptake by roots. The upper curve shows the actual uptake with time. The lower, linear curve depicts the expected steady-state uptake after the free space is in equilibrium with the potassium in the medium. Data are from Ion transport in isolated plant protoplasts, a thesis by Dr. I. J. Mettler. 1977. University of California, Riverside.

C_e = equilibrium concentration of exchangeable ions in AFS;

k = diffusion coefficient for ion movement into AFS;

v = rate or flux of transport into inner space (that is, $\Delta C / \Delta t$); and

t = time.

In an experiment in which a radioactive tracer or other marker is used, C_i is zero and the expression reduces to

$$C_t = C_e \left(1 - e^{-kt}\right) + vt.$$

The term $C_e \left(1 - e^{-kt}\right)$ is a decreasing function that describes the diffusion into the AFS. As time increases, the rate decreases until final equilibrium is reached. When equilibrium is reached at infinite time, this term approaches C_e:

$$C_e - C_e \left(e^{-kt}\right) = C_e.$$

Thus when equilibrium is established with the outside medium and the AFS, it is only vt that is important:

$$C_t = vt.$$

Therefore, the linear rate of uptake is

$$v = \frac{C_{t_2} - C_{t_1}}{t_2 - t_1} = \frac{\Delta C}{\Delta t},$$

where the rate (v) is described as ions taken up during some time interval ($C_{t_2} - C_{t_1}$) divided by the time interval ($t_2 - t_1$).

13.1.6 The carrier concept

Ion uptake (v) shows saturation kinetics. This means that as the concentration of ion in the external medium increases from near zero, there is a linear increase in uptake until high concentrations in the medium are reached. Then further increases in the medium do not result in further increases in the uptake. Figure 13.2 illustrates saturation kinetics. In the figure, the rate of K^+ uptake, nmol h^{-1} (mg protein)$^{-1}$, is graphed against KCl concentration in the medium.

Figure 13.2 should be carefully compared with Fig. 13.1 to note the differences. In Fig. 13.1 the cumulative uptake of ions is graphed against time, whereas in Fig. 13.2 the rate of uptake is graphed

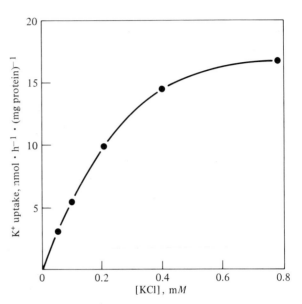

FIGURE 13.2 Graph illustrating the saturation kinetics of ion uptake by plants cells. The rate of potassium uptake is graphed against the concentration of potassium supplied to isolated tobacco-cell protoplasts. The curve is hyperbolic, showing a rapid increase in the rate of uptake at low potassium concentrations and a maximum rate at high potassium concentrations. From Ion transport in isolated plant protoplasts, a thesis by Dr. I. J. Mettler. 1977. University of California, Riverside.

against the concentration of ions in the medium. Each datum point in Fig. 13.2 is obtained from the slope of the linear portion of the curve in Fig. 13.1 at different concentrations of ions. Thus for each datum point of Fig. 13.2 a separate experiment comparable to the one shown in Fig. 13.1 is necessary.

A simple interpretation of the saturation kinetics shown in Fig. 13.2 is that there are carrier sites for potassium on the transporting membranes. When there are few ions present, there are sufficient carrier sites such that increasing the amount of ions in the medium simply increases the uptake rate in direct proportion. However, when the carrier sites are saturated, an increase in ions in the external medium has no effect on the rate of uptake because the transport system is already at full capacity. The transport sites are filled and the transport system is saturated.

13.1.7 Evidence for active uptake of ions

There is much evidence for the hypothesis that ions are taken up by an energy-dependent process dependent on specific carriers. First, interference with respiration will reduce uptake of ions. The respiratory inhibitors oligomycin and dinitrophenol

(DNP) decrease the uptake of potassium (Fig. 13.3). In addition, photosynthetic inhibitors will interfere with ion transport in green tissues in the light. Anaerobiosis and low temperatures will also reduce ion uptake. This evidence is taken as proof that ion transport is linked to, and dependent on, ATP produced from respiration and photosynthesis.

Second, once ions reach the inner space there is very little exchange or leakage, further evidence that uptake and ion accumulation are active processes. Third, different ions will interfere with each other by a process known as antagonism. The latter evidence suggests specific carrier sites for the ions and will be discussed in more detail later. Finally, as mentioned previously, frequently ions are accumulated or excluded against electrochemical potential gradients, a condition only likely if energy is expended.

13.1.8 Ion transport kinetics

Epstein (1972) developed a simple method to describe ion uptake. The curve in Fig. 13.2, which is a rectangular hyperbola, can be described as follows:

$$v = \frac{V \cdot I}{K_s + I},$$

where

v = velocity or rate of uptake in millimoles per unit time;

V = maximum rate of uptake;

I = concentration of ions in the external medium; and

K_s = constant identical to the concentration of ions (I) when the rate of uptake (v) is equal to ½V.

Thus if $V = 1.0$ and $v = 0.5$, then

$$0.5 = \frac{(1.0)\,(I)}{K_s + I}$$

and

$$(0.5)\,K_s + 0.5\,I = I$$

and

$$K_s = I.$$

An alternative way to describe the shape of the curve in Fig. 13.2 is through an analysis of the uptake at both low and high ion concentrations. Such an analysis is shown in Fig. 13.4. At low I concentrations, the value of $K_s + I$ is approximately equal to K_s, and the entire expression reduces to

$$v = \frac{V \cdot I}{K_s},$$

and the rate of uptake is linear with I. An increase in ion concentration (I) will be accompanied by a linear increase in the rate of uptake (v).

However, if the concentration of ions in the external medium is high, $K_s + I$ will be approximately equal to I, and the expression reduces to

$$v = \frac{V \cdot I}{I} = V.$$

Here the rate of uptake is independent of the con-

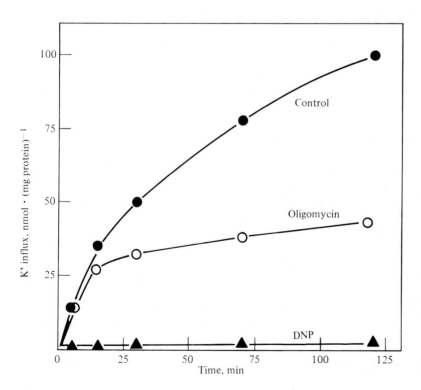

FIGURE 13.3 Graph showing how the respiratory inhibitors oligomycin and dinitrophenol (DNP) inhibit the uptake of potassium by isolated tobacco-cell protoplasts. From Ion transport in isolated plant protoplasts, a thesis by Dr. I. J. Mettler. 1977. University of California, Riverside.

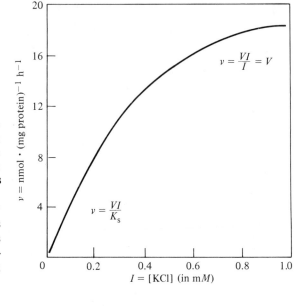

FIGURE 13.4 Analysis of the hyperbolic uptake curve showing saturation kinetics. The actual curve is shown in Fig. 13.2. The analysis is done with the expression for ion uptake rate as a function of ion concentration in medium $[(v = V \cdot I/(K_s + I)]$. At low ion concentration (I), I is much smaller than K_s, and the sum of $I + K_s$ reduces to K_s. Thus the rate of uptake (v) is directly proportional to I $(v = V \cdot I/K_s)$. At high concentrations of I, however, the sum of $I + K_s$ reduces to I since I is much greater than K_s. In this case the expression reduces to $v = V$, and the rate of uptake is at a maximum and independent of ion concentration.

centration of ions, and the rate is at the maximum (V).

The expression for saturation kinetics can be analyzed quite readily by taking the double reciprocal of the hyperbolic expression to obtain

$$\frac{1}{v} = \frac{K_s}{V} \cdot \frac{1}{I} + \frac{1}{V}$$

and then graphing $1/v$ against $1/I$. Since the expression has been transformed to a linear equation of the form

$$y = mx + b,$$

the slope of the graph (m) will equal K_s/V and the y-intercept (b) will equal $1/V$. Therefore, the reciprocal of b (or $1/V$) will equal V, and the ratio of the slope to the y-intercept (m/b) will equal the half-saturation constant K_s.

Shown in Fig. 13.5 is such a double-reciprocal plot, using the data of Fig. 13.2. Calculating from the intercept and slope, the K_s for potassium in this experiment is 0.4 mM.

The similarity between enzyme kinetics (Michaelis–Menten) discussed in Chapter 9 and the

saturation kinetics of ion uptake should be apparent.

Competition

Similar ions will frequently compete for the same carrier sites as shown by analysis using saturation kinetics. For example, it has been determined by Epstein (1972) and others that, in barley roots, K^+ will compete with Rb^+ but Na^+ will not. Thus it appears that Rb^+ and K^+ are transported at a common site and Na^+ by a different site.

K^+, Cs^+, and Rb^+ are apparently transported at the same site and will compete with each other. Na^+ and Li^+ are both transported at another site and will compete with each other but not with K^+, Cs^+, or Rb^+. Ca^{+2} and Mg^{+2} are each transported at different sites.

The expected double-reciprocal curves for competing ions are shown in Fig. 13.5. Interference from a competing ion can be overcome by increasing the concentration of the main ion. This condition is true because competition for the carrier sites depends on relative concentrations. In practice, it

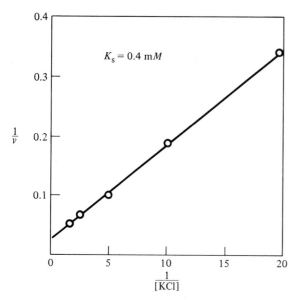

FIGURE 13.5 A double-reciprocal graph of the data shown in Fig. 13.2. The reciprocal of the rate of uptake $(1/v)$ is graphed against the reciprocal of the ion concentration in the medium $(1/KCl)$. The K_s calculated from these data is 0.4 mM potassium.

means that competing ions only interfere at low concentrations of I and that the maximum rate of uptake is not affected, as shown by the same y-intercept of the two curves in Fig. 13.6. In the presence of a competing ion, the K_s will increase such that it requires more of I to reach one-half of the maximum velocity.

Other kinds of interference, such as those that are not competitive for the carrier site, will alter V or both K_s and V. An example would be a poison (such as dinitrophenol, arsenic, cyanide, or azide) that destroys the carrier or interferes with required energy production.

Antagonism

In some circumstances, a very small amount of one ion will interfere with the uptake of another, un-related ion. Interference that is not of the competitive type and in which the interfering ion is effective at very low concentrations is called antagonism.

Calcium is antagonistic with respect to potassium and sodium. In addition, K^+ and Na^+ will antagonize Ca^{+2}, but Na^+ will not antagonize K^+. Ca^{+2} will not antagonize Ba^{+2}. The mechanism of antagonism is not clear but is most certainly related to direct effects on membranes.

13.1.9 The active uptake mechanism

The exact mechanism of active transport is poorly understood. Evidence for an energy-dependent, site-selective transport of ions was presented above.

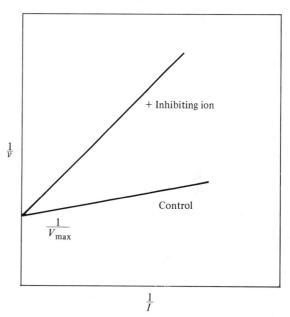

FIGURE 13.6 Idealized graphic analysis of ion uptake, showing the common y-axis intercept for uptake in the presence of a competing ion. Competing ions interfere with uptake at low concentrations of the ion being taken up but not at high concentrations. The different slopes mean that the K_s is reduced in the presence of a competing ion (it takes more ion to get the same rate), but the common y-axis intercept means that the maximum velocity of uptake has not been altered.

It seems clear from these data and many other observations that there is in fact a highly selective, energy-dependent ion transport system in plants.

Anion respiration

In the older literature, the concept of anion respiration was developed to account partially for active-transport processes. Here electron transport in respiration produced positive charges that resulted in anion uptake. Thus anion uptake is energy-dependent, i.e., dependent on the energy of respiration.

Organelles have active proton pumps. The exclusion of protons by an organelle may be accompanied by cation uptake that maintains ionic balance. Thus proton pumps are involved in ion transport.

Transfer proteins

Studies with microbial systems have revealed the presence of specific membrane proteins that bind organic molecules. Sugar- and amino acid-binding proteins have been isolated and characterized. It is implied, since binding is specific for metabolites, that the proteins with binding specificity mediate transfer through membranes.

In higher plants such proteins have not been characterized, but there are specific ATPase enzymes associated with membranes. And since there is a vast amount of evidence showing ATP hydrolysis upon ion uptake, it is strongly suggested that these membrane ATPase proteins function directly in active transport. The active research in this area, particularly with regard to specific ATPase potassium pumps, should soon clarify active uptake of ions.

Physical uptake—the Donnan theory

A Donnan equilibrium system can account for ion uptake against an apparent concentration gradient.

Of course, the uptake is not against an electrochemical potential gradient. In Donnan theory, if a membrane separates a solution from diffusible ions, the diffusible ions will diffuse across the membrane into the solution. Any time a cation diffuses, an anion will diffuse, thus maintaining electric balance. Similarly, cations will balance anions. In the event that there are nondiffusible ions on one side of the membrane, the diffusible ions will move by diffusion across the membrane until the products of anion and cation concentration on each side are equal.

Assume that on one side of the membrane there is a negatively charged protein (nondiffusible) and K^+ counter-ions. On the other side, a KCl solution is present.

K^+	K^+
Cl^-	P^-
(1)	(2)

Initially,

$$[K^+]_1 = [Cl^-]_1 \quad \text{and} \quad [K^+]_2 = [P^-]_2;$$

otherwise, there would be electric imbalance.

Chloride will diffuse from (1) to (2) and K^+ will accompany it. At equilibrium, the products of the diffusible ions must be equal:

$$[K^+]_1[Cl^-]_1 = [K^+]_2[Cl^-]_2.$$

If C_1 is the initial concentration of $K_1^+ = Cl_1^-$, C_2 the initial concentration of K_2^+, and x the amount of $K^+ = Cl^-$ diffusing from (1) to (2), then

$$(C_1 - x)\,(C_1 - x) = x(C_2 + x).$$

Thus there will be more K^+ on side (2) in the presence of the nondiffusible anion (that is, P^-), and it will appear that K^+ is accumulated against a concentration gradient. The accumulation of K^+ on side (2) in the presence of the nondiffusible, negatively charged protein takes place, meeting the requirement of electrochemical balance.

In Chapter 5, organic anions (malate) and K^+ were linked to stomatal opening. The energy-dependent production of the negatively charged malate anion would be accompanied by the uptake of K^+ in a Donnan equilibrium. This is an example of active transport of potassium—active in the sense that it is energy-dependent. It is not active in the strictest sense, however, since K^+ enters passively, thereby maintaining electric neutrality.

POTASSIUM TRANSPORT MODEL

The hydrolysis of ATP by a specific ATPase drives the uptake of potassium. Electronic neutrality is maintained by actual transport of a proton or other cation, e.g., sodium. Similar cation ATPase transport proteins are visualized for other cations. In addition, anion carriers could drive anion transport.

The pH gradient established by the proton exchange drives hydroxyl (OH$^-$) transport and allows for anion exchange. Thus it can be visualized that ions are transported either by "pumps" that require ATP or by pH gradients established by ATP hydrolysis.

The evidence for an ATP-driven cation transport mechanism resides largely in the observation of the presence of cation-stimulated ATPase activity associated with the plasma membrane. Potassium stimulates the hydrolysis of ATP,

$$ATP + H_2O \xrightarrow{K^+} ADP + P_i,$$

and the kinetics of hydrolysis are similar to the kinetics of potassium uptake by plant tissues. Here a cause-and-effect relationship is indicated. At this point, however, it cannot be considered as proven that there are actually ATP-dependent cation pumps associated with plant membranes.

Transport across the tonoplast (vacuolar membrane) into the vacuole is even less understood. In some cases the uptake of potassium shows different kinetics, depending on the concentration of potassium. It is assumed that uptake at low potassium concentration is transport across the plasma membrane and that uptake at high potassium concentration, with different kinetics, represents transport into the vacuole across the tonoplast.

Figure 13.7 shows a model for ATPase-dependent transport of potassium. Hydrolysis of ATP drives protons across the membrane (a proton pump), and potassium ions move in, passively maintaining electric balance. If hydroxyl ions diffuse out along with the protons, they may be replaced by other

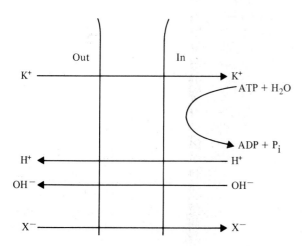

FIGURE 13.7 Model to show how an ATPase-dependent ion pump might operate in plant tissues. Hydrolysis of ATP results in protons being "pumped" out. Potassium moves in, maintaining ionic balance. Hydroxyl ions may be transported out with the protons and balanced by the influx of anions (chloride, for example) as counter-ions for the potassium.

anions (X^-). ATP-driven proton pumps may be accompanied by the transport of both cations and anions to maintain electrochemical balance.

Much needs to be studied to confirm such a model for active transport, but the evidence is quite overwhelming that ATP and proton pumps are involved. There are many specific cases in which proton pumps have been implicated. The mechanism of phosphorylation (formation of ATP from ADP and P_i) in both mitochondria and chloroplasts involves proton transport across membranes (this is the Mitchell hypothesis for phosphorylation discussed in Chapter 15). Stomatal opening may involve an ATP-dependent proton pump associated with the potassium uptake. In succulent plants (CAM plants), which accumulate large quantities of malic acid, it has been suggested that malate is transported passively into the vacuoles following the active transport of protons across the tonoplast.

It can be concluded that the selectivity of ion transport across membranes and the accumulation against apparent concentration gradients is in large part a function of ATP-dependent transporters that may act initially on protons, establishing gradients across membranes. The transport outward of counter-ions along with protons may be passive just as is the transport of like ions inward. In addition, there may be other specific ATPase transporters that act directly on other ions or metabolites.

13.2 The Essential Elements

There are sixteen different elements known to be essential for the growth and development of plants. If one were to purchase a container of "plant food" at a local nursery, the mixture is apt to contain either nitrate or some form of ammonia, phosphate, and potash. These three substances containing nitrogen, phosphorus, and potassium (NPK) are the most commonly used and, when added to soils, are most frequently beneficial for growth and development. One is also likely to purchase sulfate of ammonia $(NH_4)_2SO_4$, iron sulfate, or some other form of sulfur. Iron compounds such as $FeSO_4$ or chelated iron are common "plant foods" added to soils.

In point of fact, these elements should not properly be called foods since a food is an organic constituent produced by living organisms and subsequently oxidized by living organisms, releasing energy. Food materials ordinarily are not supplied to plants and are never energy sources except under quite extraordinary circumstances, such as in tissue culture.

The adding of minerals to soils in which plants are growing is for the purposes of supplying essential elements. This is true even if the elements are supplied in organic form, such as nitrogen in the form of urea or manure. Mineralization in the soil (conversion to inorganic form) will occur by means of microbial activity before use by higher plants. Of course, mineral and organic constituents are also added to soils to improve soil structure.

The 16 elements known to be required (essential) for all plant growth and development are listed in their order of abundance in Table 13.3. The elements are: carbon, oxygen, hydrogen, nitrogen, potassium, calcium, magnesium, phosphorus, sulfur, chlorine, iron, manganese, boron, zinc, copper, and molybdenum. Other elements, such as sodium, cobalt, and silicon, may be essential for all or some plants but as yet have not been shown to meet the necessary criteria for all plants as established in the following text.

13.2.1 Criteria for essentiality

Using techniques of water culture (that is, hydroponics), Arnon and Stout in 1939 established three criteria that they felt an element must meet in order to be properly called an essential element. Even if an element aids or enhances growth or a fundamental

TABLE 13.3 The usual concentrations of essential elements in higher plants

Element	% dry weight
Carbon	45
Oxygen	45
Hydrogen	6
Nitrogen	1.5
Potassium	1.0
Calcium	0.5
Magnesium	0.2
Phosphorus	0.2
Sulfur	0.1
Chlorine	0.01
Iron	0.01
Manganese	0.005
Boron	0.002
Zinc	0.002
Copper	0.0001
Molybdenum	0.0001

Data of Stout in Rains (1976).

process, it would not be considered essential unless the three criteria are met.

The first criterion for an element to be essential is that "a deficiency of the element prevents completion of the plant's life cycle." All of the essential elements listed above meet this criterion. Thus there are 16 elements known which must be present for a plant to germinate, grow, flower, and set seed.

Second, "for an element to be essential, it cannot be replaced by another element with similar properties." Even though sodium has properties similar to potassium, sodium cannot replace potassium completely. Trace concentrations of potassium are essential. This is equally true for the other 15 essential elements.

The final criterion that must be met is that "the element must be involved directly in the metabolism of the plant and must not be beneficial only by virtue of improving soil characteristics, aiding in the growth of beneficial microflora, or the like."

It is perhaps tempting to simplify the above three criteria by saying that an element is essential if it is necessary for the completion of the plant's life

cycle or if it is shown to be a constituent of an otherwise essential component. One such example would be cobalt, a component of the cobalamines of which vitamin B_{12} is an example. Even though vitamin B_{12} is not known to occur in plants, since the cobalamines are important in intermediary metabolism of other organisms we might infer that cobalt is essential. Furthermore, cobalt is essential for species of *Rhizobium,* symbiotic nitrogen-fixation bacteria living in the nodules of legumes. It is clear here that the cobalt is essential for the bacteria and not the legume. Therefore, cobalt is not at this time considered to be an essential element for plants.

For the reasons stated above, it is perhaps better to apply the three criteria of Arnon and Stout. They will hold under all circumstances and, since they are more rigorous, they will prevent errors.

Because some elements are required in such low concentrations (for example, molybdenum is required at about 0.1 ppm), it is difficult to obtain solutions free from trace contamination. Thus there may be elements such as cobalt required in such low concentrations that determination of essentiality has so far not been possible.

The presence of an element within a plant in high concentration is surely not indicative of essentiality. There are plants such as *Astragalus* and *Stanleya* that are selenium indicators. These plants grow in soils containing high levels of selenium and, as a consequence, accumulate selenium. There is no indication that they require selenium; in fact, the evidence suggests that they simply tolerate it.

There are many sodium-tolerant plants (halophytes) such as species of the Chenopodiaceae, the mangroves, and others, but there is little evidence that these plants require sodium. Recently, however, it was shown that C_4-photosynthesis plants (see Chapter 15) require sodium.

13.2.2 *In vivo* concentrations

Table 13.3 lists some representative concentrations of the essential elements in plants. Carbon and oxygen make up the bulk of the dry weight of the plant body by far, approximately 45% each. Thus two elements make up about 90% of the dry weight of plants. Hydrogen makes up about 6% of the dry weight; thus carbon, hydrogen, and oxygen account for over 95% of the total. This, of course, is not surprising when it is recalled that these three are the major elemental components of organic compounds.

The next most abundant element is nitrogen, at about 1.5%. Nitrogen is also found mostly as an organic constituent. Other elements that can be expressed conveniently in terms of percentage dry weight are potassium (1%), calcium (0.5%), phosphorus and magnesium (0.2% each), and sulfur (0.1%). These nine elements, C, O, H, N, K, Ca, Mg, and S, make up the macroelements. They are called macroelements because they occur in relatively high concentrations and not because of any relative importance to plant nutrition.

However, iron, which is present at approximately 100 ppm (0.01%) is sometimes considered with the macroelements because of its frequent deficiency and importance to plant nutrition.

Another six elements, chlorine, manganese, zinc, boron, copper, and molybdenum, are present in low concentrations and are more easily expressed in parts per million (ppm) than in dry weight. These trace elements are termed micronutrients because of their relatively low concentrations in plants.

13.2.3 The role of the elements

In this section there is a cursory look at the elements essential for plant growth and development. Each element is considered separately, and its main physiological or metabolic roles are mentioned. Plant symptoms denoting a deficiency are discussed. Nitrogen and sulfur are treated in more detail at the end of the chapter.

With respect to the deficiency symptoms, it should be noted carefully that, perhaps with the exception of nitrogen and iron, it would be nearly impossible to guess an elemental deficiency from foliar symptoms without extensive knowledge of both the plant species or variety and the geographical area. Even though there are certain characteristic deficiency symptoms for each element, species vary considerably, and many disturbances alter or mask classical symptoms.

Careful studies of mineral nutrition using aqueous solutions were first conducted by Hoagland, Arnon, Stout, and their co-workers. In 1938 Hoagland and Arnon published their now widely used "Hoagland's solution" for the aqueous culture of plants. This nutrient solution includes all the essential elements in a ratio conducive to good growth for most plants. Certain plants may, of course, require more or less of any one element for optimal growth. Hoagland's solution is listed in Table 13.4. Abbreviated information concerning nutrient-deficiency symptoms is given in Table 13.5.

Carbon

Carbon, which forms bonds in the shape of a tetrahedron, is the backbone of organic compounds. By definition, those compounds that have carbon as their basic molecular component are technically organic compounds. In biology, we usually ex-

TABLE 13.4 Contents of a complete nutrient solution as devised by Hoagland*

Compound	Concentration
$Ca(NO_3)_2$	5mM
KNO_3	5
$MgSO_4$	2
KH_2PO_4	2
Fe	1
Micronutrients	
H_3BO_3	0.029 ppm
$MnCl_2 \cdot 4H_2O$	0.018 ppm
$ZnSO_4 \cdot 7H_2O$	0.0022 ppm
$CuSO_4 \cdot 5H_2O$	0.008 ppm
$H_2MoO_4 \cdot H_2$	0.002 ppm

*See, for example, Machlis and Torrey (1956).

TABLE 13.5   Abbreviated key to mineral deficiency symptoms

1. YOUNG TISSUE AFFECTED FIRST—ELEMENT NOT TRANSLOCATED
  2. Dieback of growing points
    3. Necrosis of leaves—deformation and abscission   **BORON**
    3. Chlorosis of leaves—curled, crinkly, necrotic   **CALCIUM**
  2. No dieback of growing points
    3. Chlorosis of leaves
      4. Leaf necrosis—striped appearance   **MANGANESE**
      4. No necrosis
        5. Veins chlorotic—leaves almost white   **SULFUR**
        5. Veins not chlorotic—leaves white except veins   **IRON**
    3. No leaf chlorosis—may be wilted   **COPPER**
1. OLDER TISSUE AFFECTED FIRST—ELEMENT TRANSLOCATED
  2. Whole plant affected—stunted
    3. Lower leaves chlorotic—plant faded   **NITROGEN**
    3. Plant dark green to purple—some fading later   **PHOSPHORUS**
  2. Effects localized—chlorosis, necrosis, mottling, striping
    3. Interveinal chlorosis—striping   **MAGNESIUM**
    3. Marginal or general chlorosis
      4. Marginal necrosis   **POTASSIUM**
      4. Necrosis over entire leaf—spotty   **ZINC**

Without experience with a particular plant, it is almost impossible to accurately guess a mineral deficiency from symptoms alone. The above table is meant to be a guide in learning the most general symptoms associated with some of the major elements when deficient.

clude such simple compounds as CO_2 and HCN by treating them as inorganic compounds. There is a tendency to consider only the reduced carbon compounds as organic.

Virtually all of the complex molecular components of living organisms are organic in the carbon sense, although polyphosphates do occur and undoubtedly are important. The basic carbon compounds of living organisms are carbohydrates, proteins, lipids, and nucleic acids. These are discussed in detail in Part III.

Little need be said about a carbon deficiency aside from the fact that a reduction of carbon (that is, CO_2) to plants results in reduced growth. And, as explained in Chapter 16, photosynthesis is limited in many plants by ambient levels of CO_2. An increase in CO_2 above the ambient 0.3 mbar will result in more photosynthesis.

Oxygen

Oxygen is somewhat like carbon in that it occurs in virtually all organic compounds of living organisms: carbohydrates, proteins, lipids, nucleic acids, and the natural products. There are some terpenoid compounds (such as rubber) and perhaps some halogen compounds that are oxygen-free, but these are quite rare in living organisms.

The role of free oxygen is primarily as an electron acceptor in terminal respiration.

$$O_2 + 4e^- + 4H^+ \longrightarrow 2H_2O$$

Relative to ambient—about 21% O_2—the concentration of O_2 required for full respiratory activity is low, on the order of 1%.

Oxygen functions as a substrate in certain reactions involving the terminal oxidases: ascorbic acid oxidase and the polyphenoloxidases. Some hydroxylation reactions use free O_2 rather than OH^-. In addition, the main carboxylating enzyme of photosynthesis, ribulose bisphosphate carboxylase, can accept O_2 as well as CO_2 (see photorespiration in Chapter 16). Hence O_2 competes with carbon assimilation.

Aerobic organisms such as the green plants cannot live under anaerobic conditions, but since O_2 is a product of photosynthesis, there is little likelihood of anaerobiosis in the illuminated atmosphere of plants. In certain waterlogged soils, O_2 deficiency does occur.

The role of O_2 as an inhibitor of metabolism is perhaps more interesting than the potential problem of a deficiency. It is well known that reduced atmospheric O_2 will result in enhanced photosynthesis of many green plants. This condition is in part the consequence of the oxygenase activity of ribulose bisphosphate carboxylase mentioned above. Furthermore, the high oxidation potential of oxygen makes it an inhibitor of reactions that require reduced substances or reducing conditions. The role of oxygen in photosynthesis and respiration is discussed in more detail in Chapters 14, 15, and 16.

Hydrogen

Hydrogen plays a central role in plant metabolism. With the exception of CO_2, all organic compounds have hydrogen. In its oxidized atomic form as a proton, it plays as important a role as any other element. Protons are virtually ubiquitous and are important in ionic balance. The familiar pH scale and its importance in expressing proton or hydrogen ion concentration attests to the central role of hydrogen.

In its reduced form, H^-, it is the main reducing agent (reductant) of living organisms. Such familiar reducing agents as NADH and NADPH transfer hydrogen in oxidation–reduction reactions. Thus hydrogen plays a key role in energy relations of cells. It is inconceivable to think of hydrogen being deficient because it is so abundant in water.

Nitrogen

Nitrogen is a component of many important organic compounds ranging from proteins to nucleic acids. These and other nitrogenous compounds are discussed in Part III; nitrogen metabolism is discussed in more detail at the end of this chapter.

Nitrogen deficiencies are among those most frequently encountered in agriculture and ornamental horticulture. Since nitrogen is readily translocated throughout plants, deficiency symptoms show first on older leaves. As the plant becomes nitrogen-deficient, nitrogen is mobilized from old leaves and transported to young leaves.

Because of the role of nitrogen in protein and chlorophyll synthesis, deficiency symptoms appear first as chlorosis. Frequently, anthocyanins develop along veins or in other areas. Figure 13.8 illustrates typical chlorosis of old leaves due to nitrogen deficiency.

Phosphorus

Phosphorus is a common component of organic compounds in plants. It occurs as a component of activated carbohydrates (for example, glucose-6-P, fructose-6-P, phosphoglycerate, etc.), nucleic acids, and phospholipids. There are also phospho-amino acids (e.g., phosphoserine) that may occur in the phosphoproteins.

The central role of phosphorus is in energy transfer. As stated before, for activation prior to metabolism, carbohydrates are phosphorylated. The presence of phosphorus in the molecular structure of sugars makes them more reactive. The highly reactive nucleotides—adenosine triphosphate (ATP), adenosine diphosphate (ADP), guanosine triphosphate (GTP), guanosine diphosphate (GDP), uridine triphosphate (UTP), uridine diphosphate (UDP), cytosine triphosphate (CTP), and cytosine diphosphate (CDP)—play an essential role in energy transfer by phosphorylation.

Despite the central role of phosphorus in energetics and protein metabolism, the deficiency symptoms are rather nondescript. If there is a marked deficiency, there will be reduced growth. The symptoms of small deficiencies are a dark purplish coloration largely because of anthocyanin production. Plants may be stunted or dwarf, and fruits mature slowly if phosphorus is deficient.

Sulfur

Sulfur is somewhat like phosphorus in that it is involved in plant cell energetics. Activated sulfate occurs in compounds analogous to ATP: adenosine phosphosulfate (APS) and phosphoadenosine phosphosulfate (PAPS). It is also an integral part of amino acids such as cysteine and methionine and thus functions partially in protein metabolism both by maintaining protein structure through disulfide bonds and by being the active site (—SH) for many enzymatic proteins. As coenzyme A (CoA-SH) it is involved in energy relations, along with APS and PAPS. More details of sulfur metabolism are given in a subsequent section.

Although soils are generally not deficient in sulfur, sulfur deficiency symptoms are known and characterized for plants. Because of its role in protein metabolism, sulfur deficiencies result in marked chlorosis, sometimes even to the extent that leaves appear nearly white. At first, yellowing may appear along veins before general chlorosis appears.

Organic sulfur is not easily translocated (it is translocated mostly as SO_4^-), hence deficiency symptoms occur first on younger leaves.

Calcium

Calcium plays an important role in plant metabolism. It is intimately involved in cell division because it is, as calcium pectate, an integral component of the middle lamella. Pectins are polymeric galacturonic acids that form salts with calcium. In addition, calcium is found in vacuoles and as a counter-ion for negative charges on proteins.

Calcium is not readily translocated once taken up; thus deficiency symptoms occur in young tissues. Perhaps more important, though, is that because of its structural role in the middle lamella of cells, cell division cannot proceed in its absence. Thus growing points die in the absence of calcium.

Calcium also plays a major role in the maintenance of membrane integrity, and deficiencies cause malfunction of transport selectivity. Other symptoms include general chlorosis except along veins, which remain green, and a crinkly appearance of the leaves. Roots are almost always poorly developed and may be gelantinous. Fruiting is impaired.

Magnesium

Magnesium occurs as a component of chlorophyll and as a divalent cation cofactor for many enzymatic reactions. Kinases and other enzymes involved in phosphate transfers frequently have a requirement for magnesium.

For the most part magnesium occurs as a free ion in cell solutions, although it may be associated with negatively charged components such as proteins and nucleotides through ionic bonds. Magnesium is an integral component of the chlorophyll molecule.

Deficiency symptoms include general chlorosis, particularly in interveinal regions. Such a chlorotic pattern frequently gives the magnesium-deficient leaf a striped appearance, especially in monocotyledons. Formation of brown spots and dying of tips are common deficiency symptoms of magnesium. Magnesium-deficient leaves may be quite brittle and frequently abscise. Severe deficiencies result in much leaf necrosis. Because magnesium is not too mobile in the plant, symptoms most frequently occur first on older leaves.

Soil deficiencies, although not too common, are usually corrected by the addition of magnesium sulfate.

Potassium

Unlike the other macroelements mentioned above, potassium is not known to occur in plants in organic form or in an organic complex. Being present in

A

D

C

FIGURE 13.8 Photographs of leaves with mineral deficiency symptoms. (A) Nitrogen deficiency symptoms on the leaves of lemon. At left are leaves having sufficient nitrogen, and at the right are leaves showing the typical chlorosis of nitrogen deficiency. Photograph courtesy of Dr. Robert G. Platt, University of California, Riverside. (B) Phosphorus deficiency symptoms on leaves of walnut tree. Upper is a control that received triple superphosphate fertilizer. From E. F. Serr. 1960. Walnut orchards on volcanic soils deficient in phosphorus. *Calif.*

E

Agr. 14:6–7. (C) Potassium deficiency symptoms on leaves of potato, showing marginal and interveinal necrosis. From J. W. Oswald, O. A. Lorenz, T.

B

Bowmand, M. Snyder, and H. Hall. 1958. Potato fertilization and internal black spot in Santa Maria Valley. *Calif. Agr.* 12:8–10. (D) Magnesium deficiency symptoms on avocado, showing interveinal chlorosis and necrosis. Photograph courtesy of Dr. Frank Bingham, University of California, Riverside. (E) Calcium deficiency symptoms on the leaves of grapefruit, showing chlorosis on young leaves. Photograph courtesy of Dr.

F

G

H

Robert G. Platt, University of California, Riverside. (F) Iron deficiency symptoms on orange leaves, showing typical very white chlorosis of leaves, but not veins, on young leaves. The leaf at the right has sufficient iron. Photograph courtesy of Dr. T. W. Embleton, University of California, Riverside. (G) Copper deficiency symptoms on an orange leaf. From A. R. C. Haas and J. N. Brusca. 1956. Nitrate levels for Valencias. *Calif. Agr.* 10:8–9. (H) Plate comparing zinc (upper), manganese (middle), and magnesium (lower) deficiency symptoms on the leaves of orange. Photographs courtesy of Dr. T. E. Embleton, University of California, Riverside.

abundance, its primary role seems to be in osmotic and ionic regulation. It is abundant in plant vacuoles and is bound through ionic bonds to proteins and other components with negative functional groups. Its role as an osmotic agent in stomatal opening was discussed in Chapter 5.

Potassium functions as a cofactor or activator for many enzymes of carbohydrate and protein metabolism. One of the most important, pyruvate kinase, is a central enzyme of glycolysis and respiration. In most instances, the concentration of potassium for effective activity is 10 mM or greater.

Since it is readily translocated throughout the plant body, deficiency symptoms occur first on older leaves as a slight yellowing. Leaf edges subsequently become brown and necrotic. In grasses, interveinal areas turn brown.

Iron

Iron is intimately involved in oxidation–reduction reactions of living organisms. In electron transport, ferrous (Fe^{+2}) iron is oxidized to ferric (Fe^{+3}) iron. When in the cytochrome chains, it is an integral portion of heme, making up iron porphyrins. Like the cytochromes, iron porphyrins are a part of the hydrogen peroxide-metabolizing enzymes, catalase and peroxidase. In other metabolic sequences, iron occurs in a nonheme form, for example, in ferredoxin. Ferredoxin is one of the most electronegative components in plants and plays an important role in electron transport in photosynthesis (reduction of NADP to NADPH) and in the reduction of nitrogen during nitrogen fixation.

Because of its intimate association with organic compounds, iron is not readily translocated, and deficiency symptoms show first in young tissue. Young leaves show deficiency symptoms by becoming extremely chlorotic, sometimes appearing virtually white.

Along with nitrogen deficiency, iron chlorosis is one of the most frequently encountered mineral-deficiency symptoms of plants. This situation occurs primarily because the solubility of iron salts is very pH-dependent, and although it is usually abundant in soils it is frequently unavailable.

Iron can be supplied as ferrous or ferric salts of halides or sulfate, but is most commonly supplied in chelated form to ensure availability. This makes good sense since the deficiency almost always results from the insolubility of iron salts, although addition of the salt $FeSO_4$ may correct unfavorable pH and allow an availability of iron.

Manganese

Manganese, one of the micronutrients, is essential for the Hill reaction of photosynthesis in which water is "split," yielding electrons and oxygen.

$$2H_2O \xrightarrow{\quad Mn^{+2} \quad} 4H^+ + 4e^- + O_2$$

In addition, it functions as an enzymatic cofactor in many reactions. Its essentiality as an enzymatic cofactor, however, is not entirely clear, because other divalent cations such as Mg^{+2}, Co^{+2}, Ni^{+2}, and Zn^{+2} can substitute for Mn^{+2}. P-enolpyruvate carboxylase, for example, can use any of the above cations.

Manganese may also function as a counter-ion for anionic groups and in general ionic balance.

Because the micronutrients are required in such small amounts, the difficulty of detecting them in soils and plants thwarted an understanding of their physiological roles. As a consequence, there were many plant diseases described that later were shown to be due to micronutrient deficiencies.

Manganese deficiency is known to cause "gray speck" of oats, "marsh spot" of peas, and "speckled yellows" of sugar beets. As indicated from the names of these diseases, deficiency symptoms frequently show as spots and flecks, but as in magnesium deficiency, chlorosis of interveinal regions is common. This may show as striping in grasses that have definite parallel venation. Unlike magnesium deficiency, symptoms usually first appear on young leaves and later on older leaves.

Zinc

Zinc, like manganese, plays an essential role in oxidation–reduction reactions of plants. It is not known to occur in organic form. Its best-known requirement is as a cofactor for the enzyme carbonic anhydrase, which catalyzes the hydration of CO_2 to form carbonic acid (H_2CO_3). The latter spontaneously hydrolyzes, depending on pH to form protons (H^+) and bicarbonate (HCO_3^-).

$$CO_2 + H_2O \xrightleftharpoons{Zn^{+2}} H_2CO_3 \rightleftharpoons H^+ + HCO_3^-$$

Carbonic anhydrase may play an important role in CO_2 transport and in forming specific CO_2 species (i.e., CO_2 or HCO_3^-) for carboxylation reactions.

Deficiency symptoms are rather nondescript, but frequently involve mottling and necroses of leaves. Zinc deficiencies are specifically associated with "little leaf" diseases of plants.

Copper

Copper, unlike manganese and zinc, commonly occurs in complex with organic compounds. The electron-transport component, plastocyanin, is a copper-containing enzyme that functions as an integral part of photophosphorylation. In addition, copper is an important component in terminal oxidases, such as polyphenol oxidase, which catalyzes the oxidation of phenolics to ketones during lignin formation and in tanning.

Other oxidases that use copper are ascorbic acid oxidase, p-diphenol oxidase, and cytochrome oxidase.

Although copper deficiencies are rare, deficiencies are known to be associated with tip-dying diseases such as "dieback of citrus." Deficiencies result in some chlorosis and reduction in carotenoids.

Molybdenum

The metabolic role of molybdenum is not entirely understood, but it is linked to nitrogen metabolism. A specific role is known for nitrogen fixation, and it functions as a cofactor for nitrate reductase.

$$NO_3^- \xrightarrow{Mo^{+2}} NO_2^-$$

Molybdenum deficiency is rare, but some specific diseases such as "whiptail of cauliflower" and "yellow spot of *Citrus*" are known. There is chlorosis and stunted growth associated with molybdenum deficiency.

Chlorine

Chlorine was shown by Broyer in 1954 to be essential for photosynthesis. Like manganese, it is required for the Hill reaction.

$$2H_2O \xrightarrow{Cl^-} 4H^+ + 4e^- + O_2$$

However, bromine will substitute for chlorine when *in vitro*.

Deficiency symptoms determined in aqueous cultures are chlorosis, necrosis, and bronzing.

Boron

Boron has been implicated in carbohydrate and nucleic acid metabolism. Since sugars will form borate complexes, there is the suggestion that they may be translocated as borate complexes. There is also an implication of a role for boron in cell-wall metabolism along with calcium. A constant ratio of calcium to boron seems optimal for plant growth.

Boron deficiencies are known to be associated with several diseases, such as "heart rot" of beets, "stem crack" of celery, and "drought spot" of apples. Severe deficiencies cause apical death in a manner similar to a calcium deficiency since meristematic tissue is affected. Stems and roots may be shortened, and floral and seed development are reduced. Plants shed terminal leaves, and some form rosettes when boron is deficient.

Other elements

Deficiency symptoms of the other elements that have not conclusively been shown to be essential are either unknown or unclear. Silicon, for example, is a structural component of grass stems and some other monocots. A deficiency will clearly result in abnormal growth, but life cycles can be completed in the absence of silicon. There is no evidence that silicon is an essential element in the sense of the criteria outlined above for essential elements.

Cobalt may be an essential element for higher plants in that it is an integral component of the cobalamines. Vitamin B_{12}, a major cobalamine, is not considered to be synthesized by higher plants. Many algae, including *Euglena,* show deficiency symptoms if grown in the absence of cobalt.

13.3 The Metabolism of Nitrogen and Sulfur

Nitrogen and sulfur are two elements that are mostly taken up by the plant when in inorganic form but rapidly converted to organic form after absorption. It is their role as components of organic compounds that makes them so important to the growth and development of plants. In this section, some of the important aspects of nitrogen and sulfur metabolism are discussed.

13.3.1 Nitrogen fixation

Nitrogen is the single most important soil-based element for plant growth and development. Except for water availability, it is nitrogen that limits plant growth most frequently. The fact that nitrogen is frequently deficient is somewhat enigmatic since nitrogen is the most abundant element in the atmosphere, making up approximately 80%. Before the nitrogen of the atmosphere, which occurs as N_2, can be used by living organisms, it must be either oxidized or reduced, usually to NO_3^- (nitrate) or NH_3 (ammonia).

Much of the nitrogen present in soils and available for plant growth is in the form of organic nitrogen, i.e., nitrogen tied up in protein, amino acids, nucleic acids, and other nitrogenous compounds. The ammonia in soils is usually bound as a cation (NH_4^+) to soil clays; some nitrate is frequently present, but much is lost by leaching. Soils have on the order of 0.01 to 0.25% nitrogen.

The term "nitrogen fixation" refers to the oxidation or reduction of atmospheric N_2 to a form that can be used by living organisms. The usable forms are most frequently NO_3^- or NH_4^+, but some small organic molecules such as urea and amino acids can

be taken up and used. From the standpoint of industrial use, which primarily means fertilizer, nitrogen fixation is done on a very large scale. Although there are now attempts to develop catalytic, nonenzymatic approaches to N_2 fixation through the use of reduced organic salts, virtually all of the nonbiologically produced, commercial nitrogen comes from the Haber process. In the Haber process, atmospheric N_2 is reduced directly to ammonia with molecular hydrogen under high pressure and at high temperature.

$$N_2 + 3H_2 \longrightarrow 2NH_3$$

Of the estimated 237 million metric tons of available nitrogen formed on earth each year (Table 13.6), approximately 57 million tons are produced commercially by the Haber process and 40 million tons of this are used for agriculture. The remainder is used for other industrial purposes. Approxi-

TABLE 13.6 Summary of nitrogen-fixation processes estimated for 1974

SOURCE	METRIC TONS PER YEAR, MILLIONS
NONBIOLOGICAL	88
Industrial	57
Combustion	20
Lightning	10
Ocean	1
BIOLOGICAL	149
TOTAL	237

Data from Pratt (1977).

TABLE 13.7 Examples of free-living (asymbiotic) and symbiotic nitrogen-fixing organisms

Asymbiotic

Bacteria
    Aerobic (*Azotobacter*)
    Anaerobic (*Clostridium*)
    Anaerobic—photosynthetic (*Chromatium*)
Blue-green algae (*Nostoc*)

Symbiotic

HOST PLANTS	N$_2$-FIXING ORGANISMS
Angiosperms	
Legumes	(*Rhizobium*)
Nonlegumes	(Actinomycetes)
Tropical grasses	(*Spirillium, Klebsiella*)
Gymnosperms	
Cycads	(Blue-green algae)
Ferns, lichens	(Blue-green algae)

mately 63%, or 149 million tons, are the direct result of biological nitrogen fixation. The remainder comes from natural processes such as electrical storms and combustion.

The importance of nitrogen fertilizer to crop growth is illustrated by the realization that developed countries in 1974 used an estimated 58.2 kg per capita and developing countries consumed 6.6 kg. per capita.

As shown in Table 13.6, approximately 149 million tons of the 237 million total are fixed through biological processes. All of the organisms capable of fixing atmospheric nitrogen are procaryotic types, either bacteria or blue-green algae. These may be free-living or symbiotic. Table 13.7 illustrates the kinds of organisms known to fix nitrogen. Free-living aerobic and anaerobic bacteria of the genera *Azotobacter* and *Clostridium* and photosynthetic bacteria of the genus *Chromatium* fix atmospheric nitrogen. *Spirillum* and *Klebsiella* and some filamentous bacteria in the Actinomycetes also fix nitrogen. The latter have been shown to occur symbiotically with some tropical grasses. Free-living and symbiotic blue-green algae that have heterocysts may fix nitrogen. Of the symbiotic nitrogen-fixing organisms, the nodule-forming *Rhizobium* species of legumes are perhaps the best known.

Symbiotic nitrogen fixation

The symbiotic nitrogen-fixing organisms of the genus *Rhizobium* live in association with legumes. They are named according to their hosts; *Rhizobium melioti* lives with alfalfa, *R. trifoli* with clover, *R. phaseoli* with beans, *R. japonicum* with soybeans, and so on.

The infection process forming the symbiosis has been studied in much detail. The free bacteria are attracted to young root hairs (Fig. 13.9). During the actual infection process that must involve the production of hormones and digestive enzymes, the root hair forms a curl and then is invaded by the bacteria. Once in the root hair, migration to the root cortex takes place via a mucilaginous infection thread. After the bacteria enter the cortical cells, cell division produces polyploid cells that ultimately form nodules. Once within the cortical cells, they increase greatly in size, forming cells called bacteroids that are enclosed in membranes. The final structure, which contains bacteroids, is in intimate association with vascular tissue and is called a nodule. Nodules occurring on cowpea (*Vigna*) roots are shown in Fig. 13.10.

Although the legume nodule containing nitrogen-fixing *Rhizobium* spp. is the most well known, there are many nonlegumes that have nodules with nitrogen-fixing organisms. Most of these have the *Alnus* type of nodule, which is characterized by clusters formed by repeated branching. Genera that are known to have the *Alnus* type of nodule in addition to *Alnus* are: *Coriaria, Myrica, Casuarina, Hippophae, Elaeagnus, Shepherdia, Ceanothus, Discaria, Dryas, Purshia, Cercocarpus,* and *Arctostaphylos* (reported in Nutman, 1976). Many of these nodules occurring with nonlegumes have Actinomycetes of the genus *Frankia* as the nitrogen-fixing organism.

More recently, it was shown that certain tropical grasses including *Paspalum, Digitaria,* maize, and

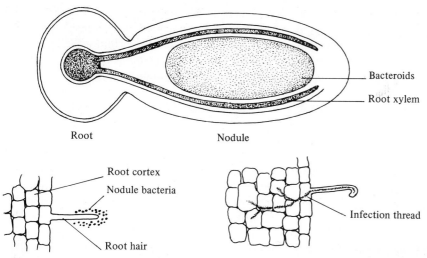

Bacteroids

Root xylem

Root

Nodule

Root cortex

Nodule bacteria

Root hair

Infection thread

FIGURE 13.9 Drawings illustrating the infection process of *Rhizobium* during nodule formation on legume roots. Bacteria are attracted to root hairs. The root hairs curl, bacteria enter and form an infection thread. Migration of *Rhizobium* down the infection thread into the cortex stimulates the formation of a nodule by division of polyploid cells of the cortex. There is intimate association between the nodule containing the bacteroid cells and the vascular tissue of the host plant. The process is described in detail by P. J. Dart in A. Quispel (ed.), 1974. The biology of nitrogen fixation. North-Holland Amsterdam and Oxford.

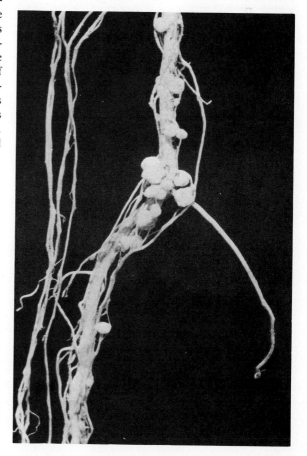

FIGURE 13.10 Photograph of the nodules on the roots of cowpea (*Vigna*). Photograph by Harrietann Joseph. University of California, Riverside.

sorghum have symbiotic, nonnodule-forming bacteria of the genera *Azotobacter*, *Spirillum*, and *Klebsiella* that fix nitrogen. These bacteria appear to live in mucilaginous plates along roots. To date, estimates of the capacity for nitrogen fixation by these organisms in association with grasses have been rather small when compared to the rhizobia of legumes (Table 13.8). Research is presently being conducted to enhance these symbionts, particularly those of the cereals.

Other plants known to have symbiotic nitrogen-fixing procaryotes include Douglas fir, with leaf bacteria, and certain ferns, with blue-green algae. An example of the latter is the water fern *Azolla*, which has a symbiotic blue-green alga of the genus *Anabaena*. Rice can be grown in paddies that have *Azolla*, even with minimal nitrogen, apparently because of the nitrogen fixation by the *Anabaena–Azolla* complex.

Nonsymbiotic nitrogen fixation

Because of the extremely high energy demand of the nitrogen-fixation process (discussed later), many of the nitrogen-fixing organisms are symbionts. Nevertheless, there are a variety of nonsymbiotic organisms capable of nitrogen fixation. The nonsymbiotic nitrogen-fixing organisms include free-living heterotrophic bacteria, photosynthetic

bacteria, and blue-green algae. The heterotrophic bacteria obtain their required energy from the environment, but almost certainly the energy for nitrogen fixation by the photosynthetic organisms comes directly from photosynthetic processes.

The free-living, heterotrophic, nitrogen-fixing bacteria fall into three groups: obligate aerobes, facultative aerobes, and obligate anaerobes. The obligate aerobes include the species of the Azotobacteraceae and some members of Pseudomonadaceae, Achromobacteraceae, Corynebacteriaceae, and Mycobacteriaceae. These organisms are somewhat peculiar since oxygen tends to inhibit the nitrogen-fixing reactions.

The facultative aerobes include species of the Enterobacteriaceae, including *Klebsiella*, *Enterobacter*, and *Escherichia intermedia*. In addition, there are some *Bacillus* facultative aerobes that fix nitrogen. Nitrogen fixation by these organisms that can live with or without oxygen is largely under anaerobic conditions, or at least with reduced partial pressures of oxygen. *Clostridium* species are the best known anaerobic bacteria that are capable of nitrogen fixation.

The photosynthetic bacteria that fix nitrogen are gram-negative organisms living in aquatic environments. They fall into three distinct groups: the purple sulfur bacteria (*Chromatium*), the purple nonsulfur bacteria (*Rhodospirillum*), and the green sulfur bacteria (*Chlorobium*). There is much evidence that photosynthesis by these organisms is directly involved in the nitrogen-fixation process.

Many species of blue-green algae have the capacity of nitrogen fixation. The nitrogen-fixing blue-green algae are mostly of the type that have heterocysts. Much of the nitrogen fixation occurs in the heterocysts, evidently because oxygen evolved by the green portion of the alga inhibits the nitrogen-fixing enzyme (nitrogenase). Thus the heterocyst may be a specialized structure in which there is reduced oxygen tensions during nitrogen fixation. It should be noted that the photosynthetic bacteria do not evolve oxygen during photosynthesis.

TABLE 13.8 Comparison of nitrogen fixation (by acetylene reduction) between a typical legume (soybean) and some common grasses

SPECIES	ACETYLENE REDUCTION*
Glycine max	11,114 ± 1010
Cynodon dactylon	244 ± 157
Paspalum vaginatum	144 ± 111
Zoysia japonica	136 ± 90
Distichlis stricta	45 ± 47

*Data are nanomoles actylene reduced per gram dry roots per hour. Experiment was conducted with excised roots. From Eskew and Ting (1977).

The biochemistry of nitrogen fixation

The product of nitrogen fixation by intact cells is ammonia. Cell-free preparations and purified proteins have shown that an enzyme called nitrogenase catalyzes the reduction of N_2 to form ammonia. Thus there is evidence that the enzyme nitrogenase is the catalyst for nitrogen fixation. The enzyme has four basic requirements for activity *in vivo*. There is a strong requirement for ATP, a requirement for Mg^{+2}, a need for a reducing agent as a source of electrons for the reduction, and a need for anaerobic conditions. The reaction as presently understood follows.

$$N_2 + 6e^- + 8H^+ + n \text{ Mg ATP} \longrightarrow$$
$$2NH_4^+ + n \text{ Mg ADP} + n \text{ P}_i$$

The multistep reduction is believed to proceed from molecular nitrogen to diimide, hydrazine, and ultimately to ammonia.

$$N_2 \longrightarrow NH{=}NH \longrightarrow NH_2{-}NH_2 \longrightarrow 2NH_3$$

Each step requires two electrons for a total of six to reduce one diatomic N_2 molecule to two ammonia molecules. The exact source of electrons is not completely understood, but most evidence suggests that reduced ferredoxin may be involved. At least in the case of the photosynthetic bacteria and blue-green algae capable of nitrogen fixation, ferredoxin is probably reduced directly in photosynthesis and is then used as the reductant in nitrogen fixation.

It is worth noting at this point that many of the organisms capable of nitrogen fixation also have an enzyme that reduces protons to molecular hydrogen, using ferredoxin as the reductant.

$$2H^+ + \text{ferredoxin (red)} \longrightarrow$$
$$H_2 + \text{ferredoxin (oxid.)}$$

This enzyme, called hydrogenase, produces hydrogen. But to complicate matters, under some circumstances (such as the absence of N_2) the enzyme nitrogenase can reduce protons to evolve hydrogen. This reaction, unlike the hydrogenase reaction, is dependent on ATP energy.

The nitrogenase reaction, reducing N_2 to ammonia, has a very high energy requirement. It appears that 12 ATP are required for the reduction. To some extent, this high energy demand is a paradox because the protein nitrogenase is extremely sensitive to oxygen, and the high energy requirement could be met by aerobic respiration. The paradox of the oxygen sensitivity and high energy requirement is solved in different ways. Many of the nitrogen-fixation organisms are anaerobic or only fix nitrogen under anaerobic conditions. ATP is evidently supplied by anaerobic respiration. In the case of the blue-green algae that produce oxygen during photosynthesis, most of the nitrogen fixation occurs in the nongreen heterocysts.

It would seem logical to conclude that symbiosis creates a situation in which energy is supplied by the host, the legume plant in the case of *Rhizobium*. In the legume nodule, which appears red if cut open, there is a protein comparable to myoglobin called leghemoglobin. The leghemoglobin has a higher affinity for oxygen than myoglobin and hemoglobin. The primary role of leghemoglobin in nitrogen fixation is assumed to be oxygen binding in the vicinity of the oxygen-sensitive enzyme nitrogenase. Alternatively, it could play a role in supplying oxygen to meet high energy demands during aerobic respiration.

Experiments have shown that the plant host synthesizes the globin portion of the leghemoglobin molecule and that most of the heme is produced by the bacteroids. The structure of the globin is independent of the species of infecting *Rhizobium*, and it is concluded that the information for globin synthesis is within the genome of the host. During the symbiosis, the heme portion is made and the entire molecule is assembled outside the bacteroids.

Nitrogenase is a large, complex enzyme composed of two major components. Component I contains both molybdenum and iron and is called Mo-Fe protein. Depending upon the species, it has a molecular weight of about 100,000 to 300,000, 15 to 20 iron atoms, and 1 or 2 molybdenum atoms. The Mo-Fe protein is composed of four subunits of two different types. Component II contains just iron and is called Fer-protein. It is smaller, with a molecular weight of about 50,000 to 70,000, and contains 4 irons. Component II has two identical subunits and requires ATP. The nitrogenase enzyme is believed to be constructed of two Fer-proteins for each Mo-Fe protein.

Neither component I nor component II exhibits nitrogenase activity by itself. The two in complex produce nitrogenase with full activity.

Nitrogenase, as well as catalyzing the reduction of nitrogen to ammonia, also catalyzes a variety of other reactions.

$$N_2O \longrightarrow N_2 + H_2O$$
$$N_3^- \longrightarrow N_2 + NH_3$$
$$2H^+ \longrightarrow H_2$$
$$C_2H_2 \longrightarrow C_2H_4$$

The fourth reaction, the reduction of acetylene to ethylene, forms the basis for one of the most common assays for nitrogen fixation. Direct determination of N_2 to NH_3 can be done with ^{15}N, but it is difficult, cumbersome, and expensive. Estimation by the conversion of acetylene to ethylene is quick and convenient with simple apparatus and a gas chromatograph. Figure 13.11 shows the time course of ethylene production from acetylene by root nodules of a legume. The inset shows a trace from the gas chromatograph. The reduction in the area below the acetylene peak or the increase in the area of the ethylene peak is taken as an indication of nitrogen fixation.

Since acetylene to ethylene reduction requires two electrons,

$$HC\equiv CH + 2e^- + 2H^+ \longrightarrow H_2C\equiv CH_2,$$

and N_2 reduction to NH_3 requires six, one-third the rate of acetylene reduction would be an appropriate estimate of the potential rate of nitrogen fixation. Because of N_2 competition in the acetylene reduction assay, the presence of other reductive activities such as hydrogen production, and the uncertainty of the reaction mechanism, the 1-to-3 conversion is only an estimate.

Because of the importance of nitrogen to agriculture, there is much interest in the genetics of nitrogen fixation. Recently, experiments by Streicher and Valentine (1973) showed that it was possible to transfer DNA with nitrogen-fixation genes (called *nif* for *ni*trogen *f*ixation) from a nif⁺ (nif-positive) strain of *Klebsiella* to a nif⁻ strain. Because of our knowledge of molecular biology and the capability of genetic transfer by the techniques of transduction and conjugation, there is now the possibility of introducing nif genes into higher plants.

Of course, even if the nif⁺ gene were introduced into the plant genome, there would still remain problems of regulation. Nevertheless, when it is realized that legumes can be grown agriculturally without the addition of nitrogen fertilizer, the importance of plants that can fix nitrogen is apparent.

Little is actually known about the regulation of nitrogenase, but ammonia represses its synthesis. The ammonia effect, however, seems to be indirect and operates through the enzyme glutamine synthetase. Glutamine synthetase produces glutamine from ammonia and glutamate (discussed later). Apparently, ammonia converts glutamine synthetase to a form that acts on the nif genes, preventing transcription and the ultimate synthesis of nitrogenase.

Once ammonia or another fixed form of nitrogen is taken up by the plant, it is converted to organic nitrogen. The assimilation of nitrate and ammonia is discussed below.

NITRATE ASSIMILATION

The nitrogen in the soil is in the form of organic compounds, ammonium ions, and nitrate ions. Through processes of mineralization, organic nitrogen is converted to inorganic form by microbial activity. Ammonia that is produced is usually rapidly converted to nitrate. The nitrification of ammonia is very efficient in most well-aerated soils. For most plants it is nitrate that is preferentially taken up, but some plants, particularly in acid soils, absorb ammonia and ammonium ions.

The first step in nitrate assimilation is the reduc-

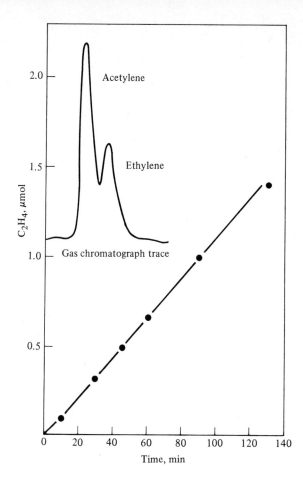

FIGURE 13.11 Graph showing the linear production of ethylene from acetylene by nitrogen-fixing bacteria in the nodules of legume roots. The inset graph shows a trace from a gas chromatograph depicting the separation of acetylene and ethylene. The area under the curves is proportional to the concentration of the substrate and can be used to quantify nitrogen fixation. From data of David Eskew, University of California, Riverside.

tion of nitrate to nitrite by the enzyme nitrate reductase.

$$NO_3^- + 2H^+ + 2e^- \longrightarrow NO_2^- + H_2O$$

The enzyme has a requirement for molybdenum. Nitrate reductase is very complex and not too well understood. A variety of reductants are used including NADPH, NADH, FADH, and FMNH. In addition, some artificial electron donors can be used, such as methyl viologen (MV). Specificity of the electron donors evidently depends on the nitrate reductase and the species of plant.

The localization of nitrate reductase activity in the plant is poorly understood. More than likely it is present in roots and leaves, although in some plants it appears to be lacking in roots. Nitrate is evidently translocated to leaves prior to reduction to nitrite in those plants that lack nitrate reductase in roots.

An interesting feature of nitrate reductase activity in roots is that it is induced by the presence of nitrate, and the activity tends to disappear in the absence of nitrate. The induction of the enzyme activity in the presence of nitrate (the substrate) appears to be an example of enzyme induction by *de novo* protein synthesis.

Once nitrite is formed by reduction through the catalysis of nitrate reductase, it is rapidly reduced to ammonia by nitrite reductase. Most plants contain only small quantities of nitrite, it being a toxic chemical. The process of nitrite reduction is complex and poorly understood. The overall reaction, requiring six electrons, is as follows.

$$NO_2^- + 6e^- + 6H^+ \longrightarrow NH_3 + H_2O + OH^-$$

The intermediates in the reaction are not known but probably include hyponitrite ($N_2O_2^{-2}$) and hydroxylamine (NH_2OH). The intermediates are evidently bound to the enzyme during the reaction.

The reductants are not too well known but may include NADPH, NADH, FMNH, or FADH. Inasmuch as nitrite reductase is frequently found as a component of chloroplasts, there is the possibility that reduced ferredoxin supplies electrons used in nitrite reduction.

As with nitrate reductase, there is some evidence that nitrite reductase is induced by its substrate, nitrite. The enzyme is also subject to control by ammonia in some tissues.

The final product of nitrate reduction, ammonia, does not accumulate in plant tissues to any appreciable extent. Ammonia is rapidly incorporated into amino acids and other nitrogenous organic compounds. Some of the reactions of ammonia assimilation are discussed below.

AMMONIA ASSIMILATION

Of the various forms of inorganic nitrogen readily available to plants (that is, nitrate, ammonia, and to a very limited extent, nitrite), it is only ammonia that is incorporated into organic form. The first step is normally the incorporation of ammonia into the amide nitrogen of the amino acid glutamine and to a lesser extent into aspartate to form asparagine. Subsequent reactions can transfer the nitrogen into other nitrogenous compounds.

As explained in the section above, most of the ammonia available for the synthesis of nitrogenous organic compounds comes from the reduction of nitrate by nitrate reductase. Of course, the degradation of many nitrogen-containing compounds in the plant body will produce ammonia that may be reassimilated. One additional source of ammonia that is worthy of note is urea. Urea can be taken up by plant roots (and leaves). It is metabolized by the enzyme urease, the first protein demonstrated to have enzymatic activity. The reaction involving

urease follows.

$$\underset{\text{(urea)}}{NH_2-\overset{\overset{\displaystyle O}{\|}}{C}-NH_2} + H_2O \longrightarrow 2NH_3 + CO_2$$

Ammonia in water will form ammonium ions, depending on the pH.

$$NH_3 + H_2O \longrightarrow NH_4^+ + OH^-$$

Probably the ammonium ion acts as the substrate in the reactions that incorporate ammonia into organic compounds. Perhaps the most important enzyme involved in the assimilation of ammonia is glutamine synthetase, which catalyzes the following reaction.

$$\text{Glutamate} + NH_3 + ATP \longrightarrow$$
$$\text{Glutamine} + ADP + P_i$$

Using the energy of ATP, ammonia is incorporated into the amide bond of the amino acid glutamate. Subsequently, the enzyme glutamate synthase produces glutamate from the glutamine and α-ketoglutarate.

$$\text{Glutamine} + \alpha\text{-ketoglutarate} \longrightarrow 2 \text{ Glutamate}$$

These two enzymes in series produce one net glutamate from α-ketoglutarate (a keto acid), ATP, and ammonia.

$$\alpha\text{-ketoglutarate} + ATP + NH_3 \longrightarrow$$
$$\text{Glutamate} + ADP + P_i$$

The above series of reactions, sequentially using glutamine synthetase and glutamate synthase, is probably the most important step in the assimilation of ammonia by plant tissues.

An enzyme analogous to glutamine synthetase, asparagine synthetase, catalyzes the following reaction.

$$\text{Aspartate} + NH_3 + ATP \longrightarrow$$
$$\text{Asparagine} + ADP + P_i$$

It has been described from microbial systems, but

its presence in higher plants is still somewhat questionable. This plant enzyme may use glutamine as the ammonia donor and thus would not be directly involved in ammonia assimilation.

Once ammonia is incorporated into glutamate, other amino acids can be synthesized by transamination reactions. The reaction of aspartate aminotransferase can be used to illustrate the basic reaction.

Glutamate + Oxalacetate $\longrightarrow$

Aspartate + α-ketoglutarate

These reversible reactions transfer the α amino group from an amino acid to an α-keto acid, forming the corresponding amino acid of the α-keto acid acceptor and the α-keto acid of the amino acid donor. In addition, the amide nitrogen of glutamine is frequently the substrate for the incorporation of nitrogen into other kinds of nitrogenous compounds. The biosynthetic discussions in Chapters 8, 9, and 12 should be reviewed.

Another interesting reaction that incorporates nitrogen into amino acids is the production of asparagine from cyanide. Free cyanide is incorporated into serine or cysteine to form β-cyanoalanine.

Serine + HCN $\longrightarrow$ β-cyanoalanine

Hydrolysis of the β-cyanoalanine forms asparagine.

β-cyanoalanine + H_2O $\longrightarrow$ Asparagine

Although the above reactions could contribute to the assimilation of nitrogen, it is more probable to assume a detoxification mechanism for the highly poisonous cyanide. Cyanide is produced by many different kinds of plants, primarily by hydrolysis of cyanogenic glycosides. The metabolic role of the latter in plants is not known.

13.3.2 Sulfur metabolism

Sulfur, like nitrogen, is taken up by plants mostly in inorganic form, reduced, and incorporated into organic compounds. In the recycling of sulfur, it is returned in the organic form to soil, where it must be mineralized by microorganisms before being available to higher plants. For the most part, sulfur is taken up as sulfate.

Some sulfur is made available from the weathering of parent rock (i.e., the source of the inorganic portion of the soil). Furthermore, pollution from fossil fuel burning liberates sulfur dioxide (SO_2), hydrogen sulfide (H_2S), and other gaseous forms of sulfur compounds. Forms such as SO_2 can be taken up through open stomata and used in plant metabolism. But in any case, regardless of the mode of uptake, sulfate and sulfur dioxide are first reduced prior to incorporation into organic compounds. There are some sulfate esters found in plants that apparently use sulfate directly.

The main organic, metabolic forms of sulfur are the amino acids cysteine, cystine, and methionine, as well as the activated sulfur compounds analogous to ATP, adenosine 5'-phosphosulfate (APS) and 3'-phosphoadenosine 5'-phosphosulfate (PAPS). In addition, sulfur occurs in a variety of sulfate esters such as choline sulfate, mustard oil glycosides, and polysaccharide sulfates.

For assimilation into organic compounds, the following reductive sequence occurs.

Sulfate (SO_4^{-2}) $\longrightarrow$ Sulfite (SO_3^{-2}) $\longrightarrow$

Sulfide (S^{-2})

First, active sulfate is produced in the form of PAPS. The following reaction sequence occurs.

$$ATP + SO_4^{-2} \longrightarrow APS + P_iP_i$$
$$APS + ATP \longrightarrow PAPS + ADP$$

The former reaction is catalyzed by the enzyme ATP-sulfurylase and the latter by APS-kinase. Because the reaction catalyzed by ATP-sulfurylase in the direction of APS formation is endergonic ($\triangle G^{o'} = +46$ kJ mol^{-1}), it favors ATP formation. Sulfate activation only occurs because of the APS-kinase reaction ($\triangle G^{o'} = -25$ kJ mol^{-1}), which is favorable. In addition, the hydrolysis of the pyrophosphate by a pyrophosphatase pulls the reaction toward PAPS synthesis.

$$P_iP_i + H_2O \longrightarrow 2P_i$$

The hydrolysis of pyrophosphatase releases about 21 kJ mol^{-1}. Because of these energy considerations, little APS occurs in plants, and PAPS is the active useful form of sulfate.

It is known that the reduction of sulfate to sulfide involves both APS and PAPS, but the exact steps and mechanism are not understood. It seems that APS and PAPS are directly involved in the reduction of sulfate to sulfite but not in the reduction of sulfite to sulfide. In photosynthetic tissue there is evidence that reduced ferredoxin is involved in the reduction of sulfite to sulfide.

Sulfide is incorporated into organic form through the enzyme sulhydrylase with L-serine as the acceptor, forming cysteine.

$$\text{L-serine} + S^{-2} \longrightarrow \text{Cysteine}$$

It is known that acetyl CoA is involved in the above reaction. Subsequently, the other two sulfur amino acids, cystine and methionine, are synthesized from cysteine.

The sulfate esters are mostly formed from the reaction between PAPS and alcohols. The general reaction is as follows.

$$\text{ROH} + \text{PAPS} \longrightarrow \text{ROSO}_3^- + \text{PAP}$$

Review Exercises

13.1 What is the molarity of a solution that has 5 g of $Ca(NO_3)_2$ made up to 100 ml with water? What are the equivalents of calcium in the solution? What are the equivalents of nitrate in the solution? If 10 ml of this solution is diluted to one liter, what will be the molarity? What will be the ppm of calcium?

13.2 The electrochemical potential between roots and the medium in which the roots are immersed was measured to be -100 mV. If the medium has 50 μEq K^+ per ml, what would you predict the potassium to be in the root tissue if the distribution between medium and roots followed that predicted by the Nernst equation? If the actual potassium in the root was estimated to be 500 μEq g^{-1} fresh weight of roots, what information would you have about the mechanism of uptake?

13.3 From the following data obtained during an experiment in which potassium uptake was measured

using plant-cell protoplasts, compute the K_s for potassium.

KCl, mM	Uptake (nmol mg^{-1} protein h^{-1})
0.1	5
0.2	9
0.5	15
0.8	17
1.0	18

13.4 Calculate the estimated rate of nitrogen fixation per g fresh weight if 15 g of root tissue reduced 1.25 μmol of acetylene to ethylene per hour. What are the various assumptions necessary when using the acetylene reduction assay for an estimation of nitrogen fixation?

13.5 How would you prepare a nutrient solution for growth of sunflowers if you wanted to estimate the effect of iron deficiency? What would you observe or measure to demonstrate iron deficiency?

References

ANDERSON, W. P. 1972. Ion transport in the cells of higher plant tissues. *Ann. Rev. Plant Physiol.* 23:51–72.

ANDERSON, W. P. (ed.). 1973. *Ion Transport in Plants.* Academic Press, New York.

BEEVERS, L. 1976. *Nitrogen Metabolism in Plants.* Edward Arnold, London.

BLOUNT, R. W., and B. H. LEVEDAHL. 1960. Active sodium and chloride transport in the single-celled marine alga *Halicystis ovalis. Acta Physiol. Scand.* 49:1–9.

BOND, G. 1976. The results of the IBP survey of root-nodule formation in non-leguminous angiosperms. In P. S. Nutman (ed.), *Symbiotic Nitrogen Fixation in Plants.* Cambridge University Press, New York.

BROYER, T. C., A. B. CARLTON, C. M. JOHNSON, and P. R. STOUT. 1954. Chlorine—a micronutrient element for higher plants. *Plant Physiol.* 29:526–532.

BROYER, T. C., and P. R. STOUT. 1959. The macronutrient elements. *Ann. Rev. Plant Physiol.* 10:277–300.

BURRIS, R. H. 1978. Future of biological nitrogen fixation [plus accompanying articles]. *BioScience* 28: 563–592.

EPSTEIN, E. 1972. *Mineral Nutrition of Plants: Principles and Perspectives.* Wiley, New York.

ESKEW, D. L., and I. P. TING. 1977. Comparison of intact plant and excised root assays for acetylene reduction in grass rhizospheres. *Plant Sci. Letters* 8:327–331.

GAUCH, H. G. 1972. *Inorganic Plant Nutrition.* Dowden, Hutchinson, and Ross, Stroudsburg, Pa.

GERLOFF, G. C. 1963. Comparative mineral nutrition of plants. *Ann. Rev. Plant Physiol.* 14:107–124.

HARDY, R. W. F., R. D. HOLSTEN, E. K. JACKSON, and R. C. BURNS. 1968. The acetylene–ethylene assay for N_2 fixation: Laboratory and field evaluation. *Plant Physiol.* 43:1185–1207.

HARDY, R. W. F., and W. S. SILVER (eds.). 1977. *A Treatise on Dinitrogen Fixation.* Wiley-Interscience, New York.

HEWITT, E. J., and T. A. SMITH. 1975. *Plant Mineral Nutrition.* Halsted Press, New York.

HIGINBOTHAM, N. 1973. The mineral absorption process in plants. *Bot. Rev.* 39:15–69.

HIGINBOTHAM, N., B. ETHERTON, and R. J. FOSTER. 1967. Mineral ion contents and cell transmembrane electro-potentials of pea and oat seedling tissue. *Plant Physiol.* 42:37–46.

HOAGLAND, D. R. 1923. The absorption of ions by plants. *Soil Sci.* 16:225–246.

MACHLIS, L., and J. G. TORREY. 1956. *Plants in Action. A Laboratory Manual of Plant Physiology.* W. H. Freeman, San Francisco.

MIFLIN, D. J., and P. J. LEA. 1976. The pathway of nitrogen assimilation in plants. *Phytochemistry* 15: 873–885.

MINCHEN, F. R., and J. S. PATE. 1973. The carbon balance of a legume and the functional economy of its root nodules. *J. Exp. Bot.* 24:259–271.

NISSEN, P. 1974. Uptake mechanisms: inorganic and organic. *Ann. Rev. Plant Physiol.* 25:53–79.

NUTMAN, P. S. (ed.). 1976. *Symbiotic Nitrogen Fixation in Plants.* Cambridge University Press, Cambridge.

PRATT, P. F. 1977. Effect of increased nitrogen fixation on stratospheric ozone. *Climatic Change* 1:109–135.

QUISPEL, A. (ed.). 1974. *The Biology of Nitrogen Fixation.* North-Holland, Amsterdam.

RAINS, D. W. 1976. Mineral metabolism. In J. Bonner and J. E. Varner (eds.), *Plant Biochemistry,* 3d ed. Academic Press, New York.

SCHIFF, J. A., and R. C. HODSON. 1973. The metabolism of sulfate. *Ann. Rev. Plant Physiol.* 24:381–414.

SPRAGUE, H. B. (ed.). 1964. *Hunger Signs in Crops.* David McKay, New York.

STREICHER, S. L., and R. C. VALENTINE. 1973. Comparative biochemistry of nitrogen fixation. *Ann. Rev. Biochem.* 42:279–302.

VAN DEN DRIESSCHE, R. 1974. Prediction of mineral nutrient status of trees by foliar analysis. *Bot. Rev.* 40:347–394.

WINTER, H. C., and R. H. BURRIS. 1976. Nitrogenase. *Ann. Rev. Biochem.* 45:409–426.

CHAPTER FOURTEEN

Respiration

14.1 Significance

Respiration can be defined as the oxidative degradation of organic compounds to yield usable energy. The efficient process in higher plants involves the consumption of oxygen, oxygen being the terminal electron acceptor for the oxidation process. The overall process in the presence of oxygen begins with a simple hexose molecule such as glucose. Glucose is oxidized to pyruvate in a series of reactions called glycolysis. Pyruvate is decarboxylated to form acetate, which is oxidized to CO_2 and water within the mitochondria. Electrons from the oxidation are transported via a cytochrome electron-transport chain to oxygen. The complete combustion of glucose to CO_2 and water yields about 2900 kilojoules per mole of hexose. The combustion process,

$$C_6H_{12}O_6 + 6O_2 \longrightarrow 6H_2O + 6CO_2,$$

by aerobic respiration produces, at a maximum, 36 ATP or ATP equivalents.

In photosynthetic organisms such as green plants that can utilize radiant energy for biosynthesis, what is the significance of aerobic respiration? The answer is quite simple. Respiration is the controlled oxidation process by which all organisms ultimately obtain the energy stored in organic compounds. This is true for green plants and animals alike. The energy is conserved in the form of ATP or other equivalent compounds that are subsequently used in growth, reproduction, and maintenance processes. The principles of thermodynamics, and in particular the Second Law of Thermodynamics, which states that all systems tend toward randomness (that is, an increase in entropy), clearly tells us that an input of usable energy is an absolute prerequisite for life.

The gist of photosynthesis is the production of reduced organic compounds from CO_2, using the energy of sunlight. Once organic compounds are formed they are used by all living organisms for their own energy-dependent processes. Thus plants are little different from animals except in the source of organic compounds. Plants make their own; animals require preformed organic compounds.

In this section, the various processes by which organic compounds are oxidized to yield usable energy are discussed.

14.2 Cellular Energetics

14.2.1 High-energy phosphate compounds

As explained in Chapter 2, the terminal phosphates of ATP have a high negative standard free energy of hydrolysis ($\Delta G^{o\prime}$). Thus hydrolysis of the terminal phosphate of ATP yields about -30 kilojoules per mole when computed under standard conditions at pH 7.0. The terminal phosphate of ADP also yields a similar amount of free energy on hydrolysis, but AMP, with a single phosphate group, yields significantly less. The high negative standard free energy of hydrolysis makes ATP extremely reactive.

Other phosphate esters of biological importance that have high negative free energies of hydrolysis and participate in cellular energetics are phosphoenolpyruvate ($\Delta G^{o\prime} = -62$ kJ mol^{-1}), cyclic-AMP ($\Delta G^{o\prime} = -50$ kJ mol^{-1}), 1,3-diphosphoglycerate ($\Delta G^{o\prime} = -49$ kJ mol^{-1}), pyrophosphate ($\Delta G^{o\prime} = -33$ kJ mol^{-1}), glucose-1-P ($\Delta G^{o\prime} = -21$ kJ mol^{-1}), and glucose-6-P ($\Delta G^{o\prime} = -14$ kJ mol^{-1}). It should be clear that these values of free energy of hydrolysis are based on standard conditions at molar concentrations of the reactants. At physiological concentrations and conditions, the actual free energies of hydrolysis (ΔG) will vary. Table 14.1 lists some $\Delta G^{o\prime}$ values for some important biological compounds.

14.2.2 Other compounds with high standard free energies of hydrolysis

There are other compounds of biological interest and significance that are very reactive because of high negative standard free energies of hydrolysis. The thiol esters, of which acetyl CoA is perhaps the most important, are examples. Acetyl CoA on hydrolysis yields 31 kilojoules per mole and is thus a compound that can readily enter into acylation reactions. As is seen throughout this text, acyl esters of coenzyme A are very important in cellular energetics.

TABLE 14.1 Standard free energy ($\Delta G^{o\prime}$) of hydrolysis of some biological compounds that function in metabolism*

Phosphoenolpyruvate	-61.9
Cyclic-AMP	-50.2
1,3-diphosphoglycerate	-49.2
Pyrophosphate	-33.5
Glucose-1-phosphate	-20.9
Glucose-6-phosphate	-13.8
Acetyl CoA	-31.4

*The standard free energy ($\Delta G^{o\prime}$) of hydrolysis is the change in free energy (here expressed in kilojoules per mole) on hydrolysis when the reactants are at molar concentrations and at pH 7.0, usually at room temperature (25°C).
Data are from Sober, 1970.

It should be noted that considerations of the energetics of reactions in terms of standard free-energy changes are at specified conditions ($\triangle G^{\circ\prime} =$ pH 7.0, 25°C, and unit molar concentrations of the substrates). Recalling that $\triangle G$ is related to the reaction as y goes to x by

$$\triangle G = \triangle G^{\circ\prime} + RT \ln [x]/[y],$$

it can readily be seen that the free energy of hydrolysis will be a function of the concentration of the reactants and products. It would be necessary, therefore, to know intracellular conditions of pH and substrate and product concentrations before accurate assessments of reactions could be made on the basis of free energy.

14.3 Anaerobic Oxidation Processes

The first step in respiration is the oxidation of carbohydrate to pyruvate by a series of reactions that do not consume oxygen. The most important process in plants in glycolysis, an anaerobic fermentation in which glucose is oxidized through pyruvate to lactate or ethanol. Glycolysis, meaning lysis of sugar, was defined in 1909 by Lepine to mean the production of CO_2 and ethanol from glucose, but we can loosely refer to glycolysis as meaning the oxidation of glucose through an anaerobic sequence to produce pyruvate. Under the aerobic conditions of respiration, pyruvate is metabolized to acetate, which is oxidized in the tricarboxylic acid cycle.

14.3.1 Glycolysis

The glycolytic process occurring within the cytosol, the soluble phase of the cell, begins with phosphorylated glucose, glucose-6-P. The immediate precursor is either free glucose, sucrose, or stored polysaccharides such as starch. Hydrolysis of starch and sucrose to produce phosphorylated hexose was described in detail in Chapter 10.

It is well to review the phosphorylation of glucose to form glucose-6-P by the ubiquitous enzyme hexokinase.

Phosphorylation of glucose produces a compound with a standard free energy of hydrolysis ($\triangle G^{\circ\prime}$) of about -17 kilojoules per mole. Such a compound can be further metabolized via the glycolytic sequence. When sugars are oxidized by respiratory processes they are activated by phosphorylation. Because of the high negative free energy of hydrolysis, phosphorylation makes the phosphate compounds more reactive. After phosphorylation, subsequent transformations and oxidations yield energy in the form of ATP. The reactions of the glycolytic sequence are illustrated in Fig. 14.1.

In the first series of reactions, glucose-6-P is cleaved into two three-carbon sugars, namely glyceraldehyde-3-P and dihydroxyacetone-P. The first reaction catalyzed by phosphoglucoisomerase produces fructose-6-P. With about twice as much glucose-6-P as fructose-6-P, this reaction is nearly in equilibrium in the cell. Thus the reaction is readily reversible.

Glucose-6-phosphate ⇌ Fructose-6-phosphate

The standard free-energy change is quite low: about $+2 \, kJ \, mol^{-1}$. There is a phosphomannose isomerase known from plants that catalyzes the equilibrium between mannose-6-P and fructose-6-P so that mannose as well as glucose can serve as a substrate for glycolysis.

In the subsequent steps, fructose-6-P is phos-

D-glucose → D-glucose-6-phosphate

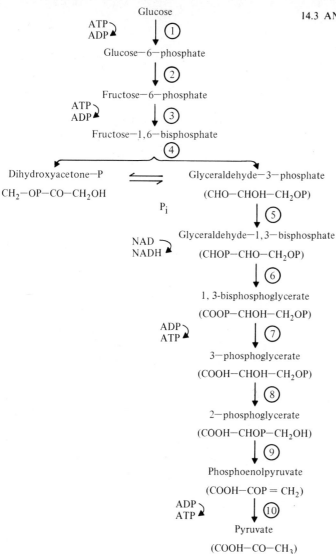

Glucose

ATP ⤵
ADP ⤴ ①

Glucose—6—phosphate

②

Fructose—6—phosphate

ATP ⤵
ADP ⤴ ③

Fructose—1,6—bisphosphate

④

Dihydroxyacetone—P ⇌ Glyceraldehyde—3—phosphate

CH_2—OP—CO—CH_2OH (CHO—CHOH—CH_2OP)

P_i

⑤

NAD ⤵
NADH ⤴ Glyceraldehyde—1,3—bisphosphate

(CHOP—CHO—CH_2OP)

⑥

1,3-bisphosphoglycerate

(COOP—CHOH—CH_2OP)

ADP ⤵
ATP ⤴ ⑦

3—phosphoglycerate

(COOH—CHOH—CH_2OP)

⑧

2—phosphoglycerate

(COOH—CHOP—CH_2OH)

⑨

Phosphoenolpyruvate

(COOH—COP = CH_2)

ADP ⤵
ATP ⤴ ⑩

Pyruvate

(COOH—CO—CH_3)

FIGURE 14.1 Glycolysis, the anaerobic oxidation of glucose to form pyruvate. 1 = hexokinase. 2 = phosphoglucoisomerase. 3 = phosphohexokinase. 4 = aldolase. 5 = phosphotriose isomerase. 6 = glyceraldehyde-3-phosphate dehydrogenase. 7 = phosphoglyceryl kinase. 8 = phosphoglyceryl mutase. 9 = enolase. 10 = pyruvate kinase.

phorylated to form the bisphosphate derivative. Phosphofructokinase is the name of the enzyme catalyzing the reaction.

$PO_3^=$—OH_2C O CH_2OH
            HO
                OH
        OH
Fructose-6-phosphate

+ ATP $\xrightarrow{Mg^{..}}$

$PO_3^=$—OH_2C O CH_2O—$PO_3^=$
            HO
                OH
        OH
Fructose-1,6-bisphosphate

+ ADP

The preceding reaction favors the formation of fructose-1,6-P because of an overall standard free-energy change of about -14 kJ mol^{-1}. Phosphofructokinase is an important regulatory enzyme of respiration.

In the subsequent reaction, fructose-1,6-P is cleaved by the enzyme aldolase into two triose-P compounds, glyceraldehyde-3-P and dihydroxyacetone-P. The reaction actually favors condensation because of a standard free-energy change of $+24$ kJ mol^{-1} but is probably pulled toward the trioses because of the unequal equilibrium between the trioses.

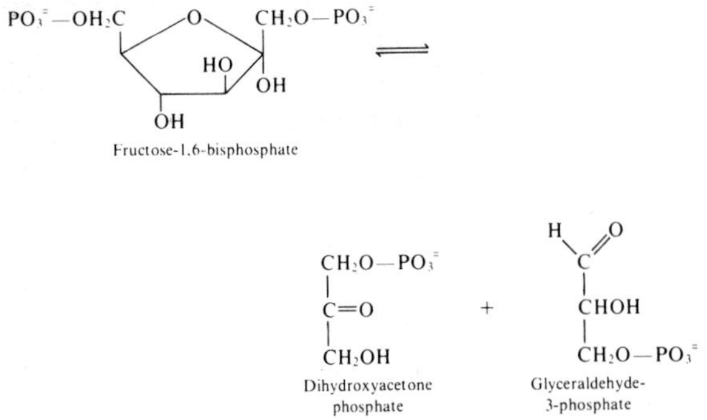

Fructose-1,6-bisphosphate

Dihydroxyacetone   Glyceraldehyde-
phosphate   3-phosphate

Dihydroxyacetone-P and glyceraldehyde-3-P are in equilibrium through the enzyme triosephosphate isomerase, with about 97% dihydroxyacetone-P.

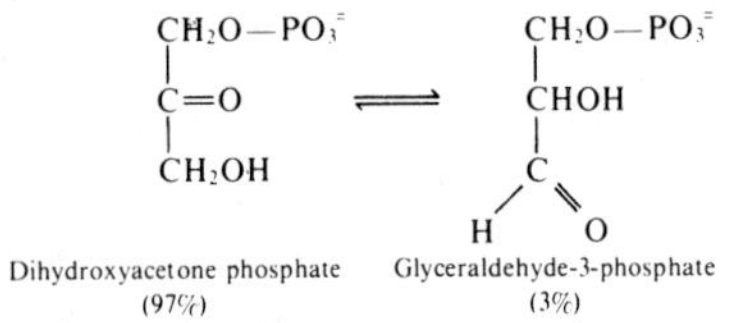

Dihydroxyacetone phosphate   Glyceraldehyde-3-phosphate
(97%)   (3%)

The standard free-energy change is about +7.5 kJ mol^{-1}. Up to this point, there has been utilization of energy. Starting with glucose, two moles of ATP are required to form the two phosphorylated trioses.

The next step in the glycolytic sequence is the oxidation of glyceraldehyde-3-P to 1,3-bisphosphoglycerate, catalyzed by the enzyme glyceraldehyde-3-P dehydrogenase. The reaction consumes inorganic phosphate and produces the reduced pyridine nucleotide NADH. The two triose molecules produced from the hexose require two moles of phosphate and yield two moles of NADH. It should be noted that this oxidation converts a sugar (glyceraldehyde) to an acid (glycerate).

[See A in Col. 2]

There is a positive free-energy change in the oxidation direction of +6.3 kJ mol^{-1}, and thus the reaction is readily reversible. Plants have another glyceraldehyde-3-P dehydrogenase that uses NADP in place of NAD. This latter enzyme, NADP

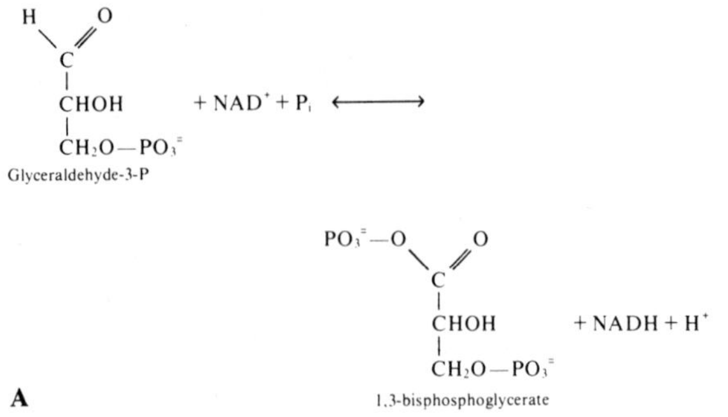

Glyceraldehyde-3-P

A

1,3-bisphosphoglycerate

glyceraldehyde-3-P dehydrogenase, is a chloroplast enzyme and can use NAD as well as NADP. The cytosol enzyme that functions in glycolysis appears to use only NAD as the redox cofactor.

The phosphate at the C-1 position of 1,3-bisphosphoglycerate is a high-energy acyl group that will yield about 49 kJ mol^{-1} on hydrolysis ($\triangle G^{o'} = -49$ kJ mol^{-1}), significantly higher than ATP ($\triangle G^{o'} = -30$ kJ mol^{-1}).

The 1,3-bisphosphoglycerate will act as a substrate for the phosphorylation of ADP, a process called substrate-level phosphorylation. The enzyme, phosphoglycerate kinase, catalyzes the phosphorylation to yield ATP and 3-phosphoglycerate.

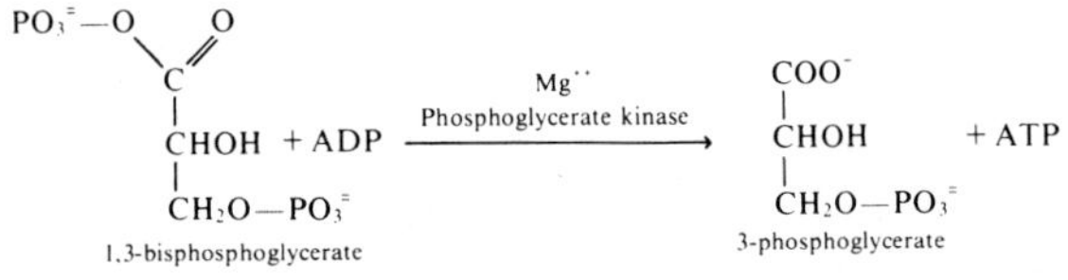

1,3-bisphosphoglycerate   3-phosphoglycerate

The standard free-energy change for the overall reaction is about -19 kJ mol^{-1}.

The reaction, catalyzed by phosphoglycerate kinase, is an excellent example of substrate-level phosphorylation. The addition of phosphate and concomitant oxidation at C-1, catalyzed by glyceraldehyde-3-P dehydrogenase, forms a high-energy acyl phosphate that on hydrolysis will yield about 49 kJ mol^{-1}. Since the hydrolysis of ATP takes place with a standard free-energy change of

-30 kJ mol^{-1}, the coupling of the two reactions,

$$1,3\text{-PGA} \longrightarrow 3\text{-PGA} + P_i \qquad -49 \text{ kJ}$$
$$\underline{ADP + P_i \longrightarrow ATP \qquad\qquad\quad +30 \text{ kJ} }$$
$$1,3\text{-PGA} + ADP \longrightarrow 3\text{-PGA} + ATP \qquad -19 \text{ kJ}$$

takes place with a $\Delta G^{\circ\prime}$ of -19 kJ mol^{-1}. Thus the coupling of the two reactions provides sufficient free-energy change to drive the phosphorylation of ADP to form ATP.

Next, 3-phosphoglycerate is mutated to 2-phosphoglycerate, catalyzed by phosphoglyceromutase.

The above is a readily reversible reaction with a standard free-energy change of $+4$ kJ mol^{-1}. Enolase then catalyzes the dehydration of 2-phosphoglycerate to phosphoenolpyruvate.

Phosphoenolpyruvate has a high standard free-energy change, yielding 62 kJ mol^{-1} on phosphate hydrolysis.

The standard free-energy change of the reaction catalyzed by pyruvate kinase is -31 kJ mol^{-1}, sufficient for the substrate-level phosphorylation that forms ATP.

The overall reaction sequence for the production of pyruvate from glucose in glycolysis is as follows.

$$\text{Glucose} + 2ADP + 2H_3PO_4 + 2NAD^+ \longrightarrow$$
$$2 \text{ pyruvate} + 2ATP + 2NADH + 2H^+ + 2H_2O$$

There are produced 2 net moles of NADH and 2 moles of ATP. As will be explained in a subsequent section (14.6), the NADH from glycolysis can be oxidized in the mitochondria to produce two moles of ATP per mole of NADH.

Overall, the standard free-energy change of the reaction from glucose to pyruvate is about -92 kJ mol^{-1}, preventing the sequence from being reversible. In Chapter 11 the reversal of glycolysis during gluconeogenesis was discussed and it was indicated that the initial kinase steps, which are not reversible since ATP would be produced rather than consumed, were bypassed, with phosphatases yielding inorganic phosphate. In the case of the pyruvate kinase step, carboxylation and decarboxylation circumvent the free-energy barrier.

The combustion of glucose to CO_2 and water will yield about 2900 kJ mol^{-1}. Oxidation of glucose to two moles of pyruvate yields 586 kJ mol^{-1}. If we assume about 30 kJ energy conserved per mole of ATP and a maximum ATP yield of 6 moles (2 from each NADH and 2 from substrate phosphorylation), there is an efficiency of conservation of 31% ($180/586 \times 100$). Thus the anaerobic oxidation of carbohydrate is reasonably efficient. If, however, we calculate the efficiency of energy conservation from glucose in glycolysis, noting that only 20% of the possible energy is actually generated ($586/2900 \times 100$), there is a yield of about 6% (20% or 31%) of the total possible. As will be discussed below, however, glycolysis is coupled with terminal oxidation (that is, respiration) in the mitochondria, yielding about 36 ATP equivalents, with an overall conservation of energy of about 37%.

14.3.2 The pentose phosphate pathway

Early research on respiration by Warburg and others demonstrated alternative routes for the oxidation of glucose to triose that did not involve the enzyme aldolase. When glucose-6-P is available it may be oxidized to pentose and CO_2, as shown by supplying glucose (1-^{14}C)-6-P. Furthermore, two enzymes that use NADP as the redox cofactor, glucose-6-P dehydrogenase and 6-phosphogluconate dehydrogenase, have been isolated from plants. These enzymes catalyze the following reactions.

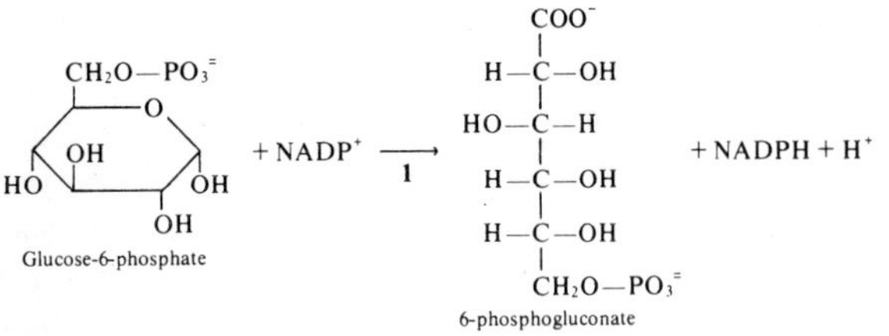

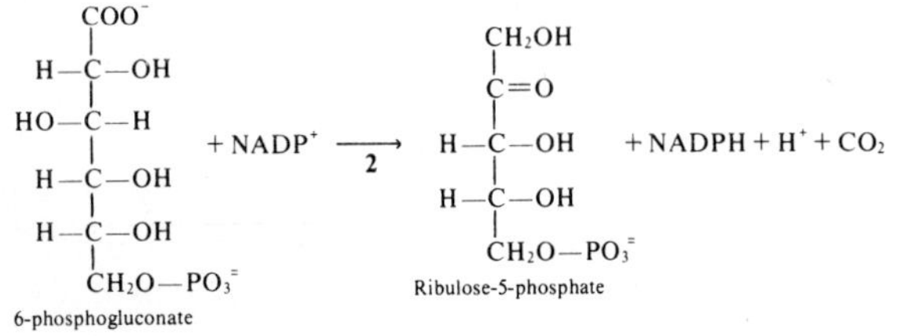

CH$_2$O—PO$_3^=$ / O / OH / HO / OH / OH
Glucose-6-phosphate

+ NADP$^+$ $\xrightarrow{1}$

COO$^-$ / H—C—OH / HO—C—H / H—C—OH / H—C—OH / CH$_2$O—PO$_3^=$
6-phosphogluconate

+ NADPH + H$^+$

COO$^-$ / H—C—OH / HO—C—H / H—C—OH / H—C—OH / CH$_2$O—PO$_3^=$
6-phosphogluconate

+ NADP$^+$ $\xrightarrow{2}$

CH$_2$OH / C=O / H—C—OH / H—C—OH / CH$_2$O—PO$_3^=$
Ribulose-5-phosphate

+ NADPH + H$^+$ + CO$_2$

Reaction 1 proceeds through 6-phosphogluconolactone and forms 6-phosphogluconate by hydration.

Subsequent to the above discoveries, a series of sugar rearrangement reactions yielding four hexose-P and two triose-P moles from the six ribulose-5-P were described. The sequence of reactions is called the pentose phosphate pathway and is illustrated in Fig. 14.2.

Oxidation of glucose-6-P yields CO$_2$, NADPH, and ribulose-5-P. Next the latter is either isomerized by phosphoriboseisomerase to ribose-5-P or epimerized by phosphoribose epimerase to xylulose-5-P.

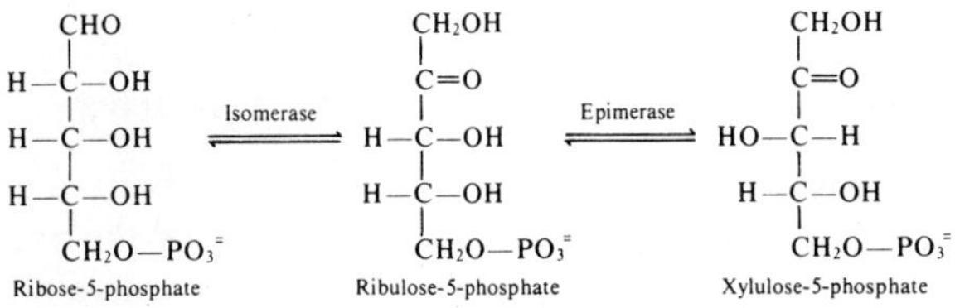

CHO / H—C—OH / H—C—OH / H—C—OH / CH$_2$O—PO$_3^=$
Ribose-5-phosphate

$\underset{\text{Isomerase}}{\rightleftharpoons}$

CH$_2$OH / C=O / H—C—OH / H—C—OH / CH$_2$O—PO$_3^=$
Ribulose-5-phosphate

$\underset{\text{Epimerase}}{\rightleftharpoons}$

CH$_2$OH / C=O / HO—C—H / H—C—OH / CH$_2$O—PO$_3^=$
Xylulose-5-phosphate

Subsequently, two xylulose-5-P donate C$_2$ fragments to two ribose-5-P acceptors to form two moles of glyceraldehyde-3-P and two moles of sedoheptulose-7-P. The enzyme catalyzing this reaction is a transketolase and was described in

Chapter 10. The transketolase reaction requires Mg^{+2} and thiamine pyrophosphate (TPP).

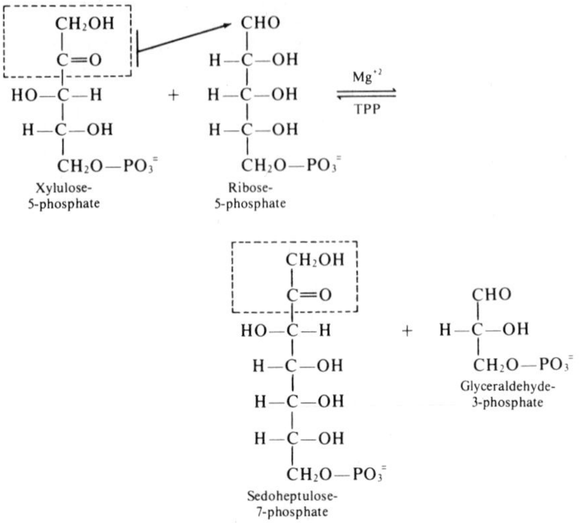

CH$_2$OH / C=O / HO—C—H / H—C—OH / CH$_2$O—PO$_3^=$
Xylulose-5-phosphate

+

CHO / H—C—OH / H—C—OH / H—C—OH / CH$_2$O—PO$_3^=$
Ribose-5-phosphate

$\underset{\text{TPP}}{\overset{\text{Mg}^{+2}}{\rightleftharpoons}}$

CH$_2$OH / C=O / HO—C—H / H—C—OH / H—C—OH / H—C—OH / CH$_2$O—PO$_3^=$
Sedoheptulose-7-phosphate

+

CHO / H—C—OH / CH$_2$O—PO$_3^=$
Glyceraldehyde-3-phosphate

Next the two sedoheptulose-7-P react with the two 3-phosphoglycerate to form two moles of erythrose-4-P and two moles of fructose-6-P. The enzyme catalyzing the transfer is a transaldolase that does not require any cofactors.

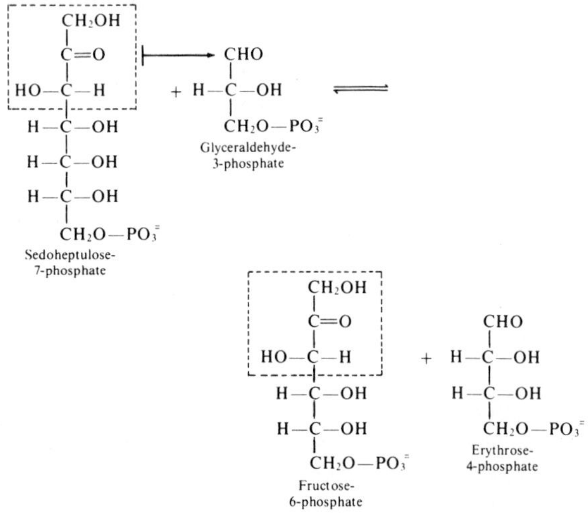

CH$_2$OH / C=O / HO—C—H / H—C—OH / H—C—OH / H—C—OH / CH$_2$O—PO$_3^=$
Sedoheptulose-7-phosphate

+

CHO / H—C—OH / CH$_2$O—PO$_3^=$
Glyceraldehyde-3-phosphate

$\rightleftharpoons$

CH$_2$OH / C=O / HO—C—H / H—C—OH / CH$_2$O—PO$_3^=$
Fructose-6-phosphate

+

CHO / H—C—OH / CH$_2$O—PO$_3^=$
Erythrose-4-phosphate

The two moles of erythrose-4-P then react with two additional xylulose-5-P formed from ribulose-5-P to form two moles of 3-phosphoglycerate and two moles of fructose-6-P. The enzyme cata-

PENTOSE PHOSPHATE PATHWAY

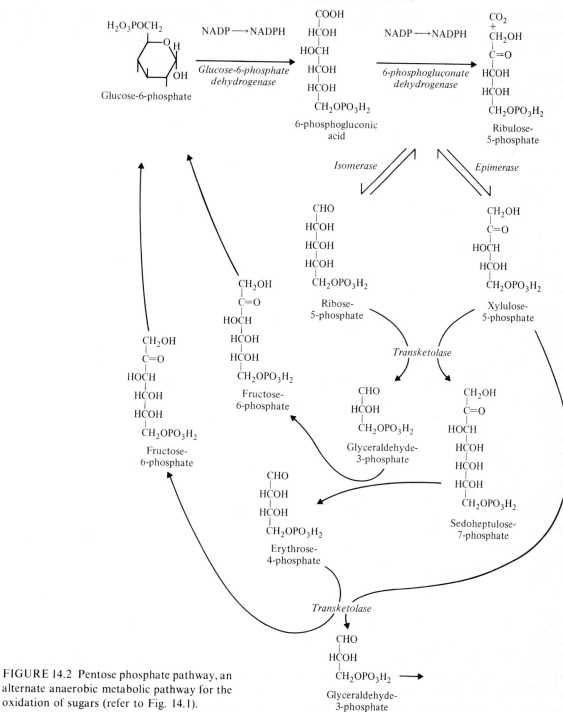

FIGURE 14.2 Pentose phosphate pathway, an alternate anaerobic metabolic pathway for the oxidation of sugars (refer to Fig. 14.1).

lyzing the reaction is a transketolase.

Thus there are four moles of hexose-6-P and two moles of triose-3-P produced from the six moles of pentose-5-P. The two moles of triose can be considered as products and enter the glycolytic sequence at a point after the aldolase reaction. Alternatively, the two moles of triose can condense, catalyzed by aldolase, to yield an additional hexose-6-P. By continual recycling, hexose can be completely oxidized to CO_2, yielding 12 NADPH per mole of hexose. Assuming that NADPH can yield a maximum of two ATP equivalents, there could be 30 kJ mol^{-1} $\times$ 25 mol = 720 kJ energy conserved from the 2900 kJ mol^{-1} possible from the hexose, or approximately 25% efficiency.

Although the pentose phosphate pathway could conserve energy in the form of ATP as explained above, an additional important metabolic role would be the production of reducing power in the form of NADPH for cellular biosyntheses. In many cellular biosyntheses, such as fatty acid biosynthesis, NADPH is the reducing agent and not NADH. Thus the pentose phosphate pathway may also play a role in the production of reducing power for biosyntheses.

In terms of energy conserved, the $\triangle G^{o\prime}$ of NADPH hydrolysis is about 220 kJ mol^{-1}. Twelve moles produced from the complete oxidation of

hexose through the pentose phosphate pathway would conserve over 90% of the possible energy (220 kJ mol^{-1} $\times$ 12 mol/2900 kJ mol^{-1}).

14.3.3 An assessment of anaerobic carbohydrate metabolism

Bloom and Stetten in 1955 developed a method to assess the relative importance of the glycolytic and pentose phosphate pathways in the oxidation of hexose. In the oxidation of hexose through glycolysis, triose is produced with no loss of carbon as CO_2. Subsequent to aerobic respiration in the mitochondria, C-1 carbons of triose are removed by decarboxylation. These carbons were the original 3- and 4-carbons of the hexose and are indistinguishable in the triose. In the pentose phosphate pathway, the first carbon removed is C-1 when hexose is decarboxylated to pentose.

Thus Bloom and Stetten proposed that the ratio of $6\text{-}^{14}C$ to $1\text{-}^{14}C$ from CO_2 would be an indication of the relative contribution of each pathway. If the ratio were unity, the only oxidative pathway would be glycolysis. If the ratio $6\text{-}^{14}C/1\text{-}^{14}C$ was less than one, there must have been an alternative pathway for the production of CO_2. The simplest interpretation is that some hexose was oxidized via the pentose phosphate pathway. Of course, many metabolic events occur that complicate the analysis, but since the respiratory processes are very active, initial rates of $1\text{-}^{14}CO_2$ and $6\text{-}^{14}CO_2$ production from hexose are a reasonable estimate of the relative contribution of the two pathways.

When plant tissues were assessed, it was found that most plant tissues (with the exception of young meristematic tissues) had $6\text{-}^{14}C/1\text{-}^{14}C$ production ratios less than unity. It was concluded that both the glycolytic and pentose phosphate pathways are operative in plant tissues.

14.3.4 Terminal electron acceptors of fermentation

As stated previously, in aerobic respiration oxygen is the terminal electron acceptor. In fermentation, oxidized organic compounds function as terminal electron acceptors. Acetaldehyde and pyruvate are the usual acceptors reduced by NADH. As explained in Section 14.3.1, oxidation of hexose through the glycolytic pathway yields two moles of

ATP and two of NADH. The latter must enter into reductive reactions, supply energy for ATP synthesis, or be dissipated in some other manner.

Anaerobic organisms or aerobic organisms under anaerobic conditions accumulate either lactate or ethanol. The latter two compounds are the usual terminal electron acceptor products in fermentation processes.

In the production of ethanol, pyruvate is decarboxylated to acetaldehyde through pyruvate decarboxylase, followed by acetaldehyde reduction to ethanol by alcohol dehydrogenase.

Alternatively, pyruvate can be reduced directly to lactate by lactate dehydrogenase.

Both reactions consume NADH. Thus in the fermentation of hexose, there will be two moles of ATP formed but no NADH yield.

14.4 Respiration

Respiration is the oxidation of organic compounds with the concomitant consumption of oxygen. The primary reaction is the oxidation of acetate in mitochondria, oxygen being the terminal electron acceptor. The complete oxidation of acetate,

$$CH_3COOH + 2O_2 \longrightarrow 2CO_2 + 2H_2O,$$

yields 870 kJ mol^{-1}. Assuming a maximum ATP production of 12, the efficiency is about 40% $(350 \text{ kJ mol}^{-1})/(870 \text{ kJ mol}^{-1})$. Because the sub-

strate is combusted completely to CO_2, respiration is considered more efficient than anaerobic carbohydrate metabolism, which yields organic compounds as end products.

During anaerobic carbohydrate metabolism through the glycolytic sequence, two moles of pyruvate are produced per mole of hexose oxidized. As explained above, under anaerobic conditions the pyruvate is usually converted to either lactate or ethanol. However, if tissue is aerobic, acetate is produced but not free acetate. The acetate produced has been called active acetate, and was shown to be the coenzyme A derivative of acetate, acetyl CoA. It is the active acetate or the acetate moiety of acetyl CoA that is oxidized to CO_2 and water under aerobic conditions within the mitochondria through the reactions of the tricarboxylic acid (TCA) cycle.

Before proceeding with the details of the TCA cycle it is perhaps useful to discuss a method by which carbon–carbon bonds are oxidized by cells. In the usual sequence, the carbon–carbon bond,

$$-\overset{|}{\underset{|}{C}}-\overset{|}{\underset{|}{C}}-,$$

is oxidized by a flavoprotein to the carbon–carbon double bond.

$$-\overset{}{\underset{|}{C}}=\overset{}{\underset{|}{C}}-$$

Subsequently, water is added across the newly formed double bond to form a hydroxy derivative.

$$\overset{H}{\underset{|}{}}\quad\overset{OH}{\underset{|}{}}$$
$$-\overset{|}{\underset{|}{C}}-\overset{|}{\underset{|}{C}}-$$

Then the hydroxyl group is oxidized to a keto group with a pyridine nucleotide (NAD or NADP) as the oxidant.

$$\overset{O}{\underset{|}{}}\quad\overset{\|}{}$$
$$-\overset{|}{\underset{|}{C}}-\overset{}{C}-$$

The oxidation sequence produces two moles of reducing equivalents. This overall sequence occurs in fatty acid oxidation and in the TCA cycle.

14.4.1 Acetate formation

The initial step of respiration can be considered as the decarboxylation of pyruvate to form acetate, for it is acetate that is oxidized to CO_2 and water within the mitochondria. A single enzyme does not catalyze the decarboxylation; the catalyst is the pyruvate dehydrogenase complex. This complex is known to be constructed of multiple units of three different enzymes. The individual enzymes catalyzing the three different steps are: (1) pyruvate decarboxylase, (2) lipoate acetyl transferase, and (3) lipoate dehydrogenase. The entire sequence requires five cofactors: Mg^{+2}, coenzyme A, NAD, lipoate, and thiamine pyrophosphate (TPP). The overall reaction of the pyruvate dehydrogenase complex is as follows.

Pyruvate + CoA + NAD $\longrightarrow$

Acetyl CoA + NADH + CO_2

The first reaction catalyzed by pyruvate decarboxylase requires thiamine pyrophosphate and Mg^{+2} and forms the TPP derivative of acetate, hydroxyethyl thiamine pyrophosphate.

$$\underset{COO^-}{\overset{CH_3}{\underset{|}{\overset{|}{CO}}}} + TPP \longrightarrow \underset{TPP}{\overset{CH_3}{\underset{|}{\overset{|}{CHOH}}}} + CO_2$$

In this step, pyruvate is decarboxylated and CO_2 is released. Subsequently catalyzed by lipoate acetyl transferase, lipoate displaces thiamine pyrophosphate to form acetylhydrolipoate and free TPP.

$$\underset{TPP}{\overset{CH_3}{\underset{|}{\overset{|}{CHOH}}}} + \underset{\underset{COO^-}{\overset{|}{(CH_2)_4}}}{\overset{S-CH_2}{\underset{|}{\overset{|}{\underset{S-CH}{CH_2}}}}} \longrightarrow CH_3-CO-S-\underset{\underset{COO^-}{\overset{|}{(CH_2)_4}}}{\overset{HS-CH_2}{\underset{|}{\overset{|}{\underset{CH}{CH_2}}}}} + TPP$$

Lipoate   Acetylhydrolipoate

Finally, the acetate from the lipoyl acetate complex, acetylhydrolipoate, is displaced by coenzyme A to form free reduced lipoate and acetyl CoA. The enzyme catalyst is lipoate dehydrogenase. It should be noted that the energy of decarboxylation is now conserved within the reduced lipoate molecule.

$$CH_3-CO-S-\underset{\underset{COO^-}{\overset{|}{(CH_2)_4}}}{\overset{HS-CH_2}{\underset{|}{\overset{|}{\underset{CH}{CH_2}}}}} + CoASH \longrightarrow CH_3CO-S-CoA + HS-\underset{\underset{COO^-}{\overset{|}{(CH_2)_4}}}{\overset{HS-CH_2}{\underset{|}{\overset{|}{\underset{CH_2}{CH_2}}}}}$$

However, the ultimate reduced compound formed in the series of reactions is not reduced lipoate but rather NADH. The enzyme catalyzing the reduction of NAD to NADH through the transfer of electrons from reduced lipoate is called dihydrolipoate dehydrogenase. This enzyme is a flavoprotein with tightly bound flavin adenine dinucleotide (FAD). The FAD associated with the protein is first reduced to $FADH_2$, and the $FADH_2$ then reduces NAD to NADH. The double-step reaction is shown below.

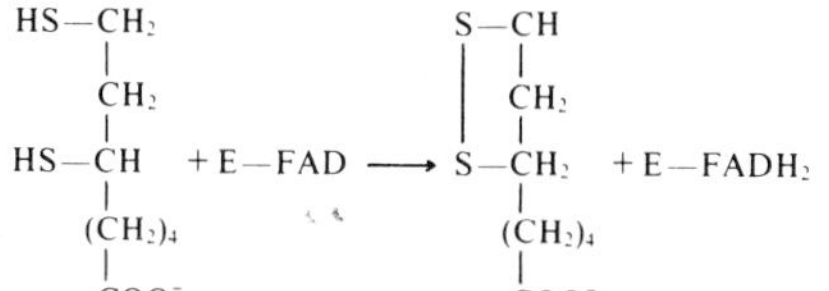

$$HS-CH_2 \atop | \atop CH_2 \atop | \atop HS-CH \atop | \atop (CH_2)_4 \atop | \atop COO^- \quad +E-FAD \longrightarrow \quad S-CH \atop | \atop CH_2 \atop | \atop S-CH_2 \atop | \atop (CH_2)_4 \atop | \atop COO^- \quad +E-FADH_2$$

$$E-FADH_2 + NAD^+ \longrightarrow E-FAD + NADH + H^+$$

Thus the decarboxylation of pyruvate to form acetyl CoA releases energy that is conserved in NADH. Two moles of NADH are produced from each mole of hexose that is oxidized to acetate.

The overall reaction of the pyruvate dehydrogenase complex is not readily reversible because of the large negative standard free energy of the decarboxylation step. The process takes place with a $\triangle G^{o\prime}$ of about -35 kJ mol^{-1}.

In the next section the oxidation of acetate to CO_2 and water through the reactions of the tricarboxylic acid cycle will be described. Of course, glycolysis is not the only source of acetyl CoA. In many tissues, β-oxidation of fatty acids produces acetyl CoA as an immediate substrate of respiration. The details of β-oxidation to produce acetyl CoA were discussed in Chapter 11.

14.4.2 The tricarboxylic acid (TCA) cycle

The tricarboxylic acid cycle was first described by Hans Krebs in 1937 as a cyclic process whereby acetate is oxidized to CO_2 and water, yielding reduced NAD and FAD. The latter two compounds in the form of NADH and $FADH_2$ ultimately transfer electrons to oxygen through the electron transport chain, conserving energy in the form of ATP.

The tricarboxylic acid cycle is also called the citric acid cycle, citrate being the first intermediate, or the Krebs cycle, after Hans Krebs. Because of the commonly used acronym TCA, the name tricarboxylic acid cycle is frequently preferred and will be used throughout this text.

In the 1930s biochemists studying biological oxidations noted that animal tissues would oxidize certain organic acids such as malate, succinate, and citrate, and that catalytic amounts of organic acids would maintain oxygen-consumption levels by aerobic tissues.

Szent-Gyorgi in 1935 showed experimentally that pyruvate could be oxidized by animal tissues if catalytic amounts of organic acids such as malate and succinate were also supplied. Krebs discovered that citrate, one of the acids that could be oxidized by tissues, was synthesized from acetate and oxalacetate. It was further shown that malonate, a specific inhibitor of succinic dehydrogenase (the enzyme that oxidizes succinate to fumarate) will not block the synthesis of succinate from fumarate. He proposed on the basis of such evidence the cycle illustrated in Fig. 14.3 for the oxidation of acetate and other, more complex organic acids. Aerobic tissues do not oxidize free acetate carbon directly but oxidize acetate using organic acids as carriers.

In the late 1940s Kennedy and Lehninger demonstrated that isolated mitochondria would oxidize pyruvate with the consumption of oxygen. These experiments showed that the mitochondria are the subcellular organelles that function in aerobic respiration. Mitochondria contain all the enzymes of the tricarboxylic acid cycle, including the pyruvate dehydrogenase complex. With the exception of succinate dehydrogenase, all are localized in the soluble matrix of the mitochondria. Succinate dehydrogenase is known to be a component of the inner mitochondrial membrane and is only solubilized with difficulty. It was thus shown con-

THE TRICARBOXYLIC ACID CYCLE

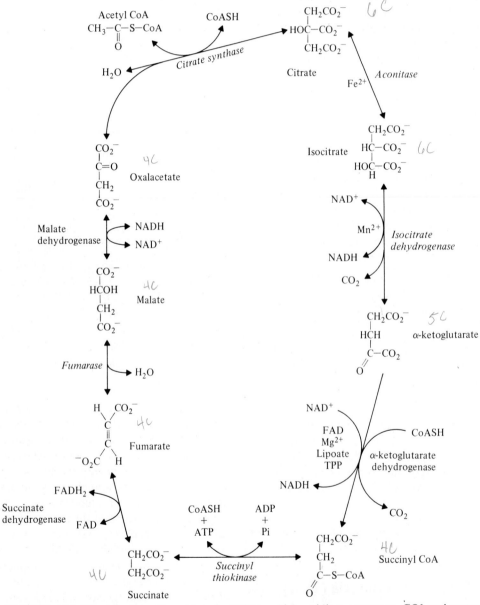

FIGURE 14.3 The tricarboxylic acid cycle (TCA), which oxidizes acetate to CO^2 and water, generating NADH and ATP.

clusively that the TCA cycle is the major metabolic component of the mitochondria.

In the first step of the TCA cycle, active acetate (acetyl CoA) is condensed with oxalacetate to form citrate through catalysis by the condensing enzyme, citrate synthetase.

Oxaloacetate + C—SCoA → Citrate + CoASH

The tricarboxylic, six-carbon citrate is then oxidized in a series of oxidation-reduction reactions to yield NADH and FADH², regenerating the initial acetate acceptor, oxalacetate. The equilibrium of the citrate synthetase reaction favors citrate formation.

Citrate is metabolized first by the enzyme aconitase, which isomerizes citrate to isocitrate. The reaction goes through *cis*-aconitate, which is enzyme-bound and does not occur free within the mitochondria. Citrate is dehydrated to form the double-bonded compound *cis*-aconitate, and the latter is then hydrated to form isocitrate. The net result is a shift in the position of the hydroxyl group. The reaction is readily reversible.

Citrate ⇌ *cis*-aconitate + H₂O ⇌ Isocitrate

However, there is some evidence that the iron-dependent reaction of aconitase does not go through aconitate but rather that aconitate is in equilibrium with citrate and isocitrate.

Isocitrate is then decarboxylated to α-ketoglutarate in a complex two-step reaction that generates reduced NAD. This is the first oxidation step that conserves energy in the form of NADH.

[See B in Col. 2]

isocitrate + NAD⁺ → α-ketoglutarate + CO₂ + NADH + H⁺

B

The decarboxylation of isocitrate proceeds through oxalosuccinate, the actual oxidation step. Oxalosuccinate is then decarboxylated to form α-ketoglutarate. Plant tissues have a second isocitrate dehydrogenase that requires NADP rather than NAD, but this is not the enzyme of the TCA cycle. The reaction is not readily reversible because of the decarboxylation.

Next, α-ketoglutarate is decarboxylated with an enzyme complex similar to the pyruvate dehydrogenase complex. Cofactors required are coenzyme A, lipoate, TPP, and NAD. NADH is ultimately formed, and the organic acid product is succinyl CoA.

α-ketoglutarate + NAD⁺ + CoASH —(TPP, lipoate, FAD)→ Succinyl CoA + NADH + H⁺ + CO₂

The reaction catalyzed by the α-ketoglutarate dehydrogenase complex is not readily reversible. Succinyl CoA is converted to succinate through the catalysis of succinyl CoA synthetase, generating ATP from ADP. The enzyme can also use GDP and IDP, but probably ADP is the usual substrate.* The succinyl CoA synthetase reaction is reversible ($\Delta G^{\circ\prime} = -3$ kJ mol^{-1}).

Succinyl CoA + ADP + P$_i$ ⇌ Succinate + ATP + CoASH

*The enzyme in animal tissues evidently is specific for GDP.

In a third oxidation, succinate is oxidized by a membrane-bound enzyme, succinate dehydrogenase, to fumarate. Succinate dehydrogenase is a flavoprotein that utilizes FAD as the redox cofactor and generates $FADH_2$. Although it requires iron, it is not a heme-containing protein. It is this enzyme that is inhibited by the competitive substrate analogue malonate.

$$
\begin{array}{ccc}
COO^- & & COO^- \\
| & & | \\
CH_2 & & CH \\
| & + FAD \longrightarrow & \| \\
CH_2 & & HC \\
| & & | \\
COO^- & & COO^- \\
& & \text{Fumarate} + FADH_2
\end{array}
$$

The enzyme fumarase catalyzes the hydration across the double bond of fumarate to form the hydroxydicarboxylic acid malate.

$$
\begin{array}{ccc}
COO^- & & COO^- \\
| & & | \\
CH & + H_2O \rightleftharpoons & HOCH \\
\| & & | \\
HC & & CH_2 \\
| & & | \\
COO^- & & COO^- \\
& & \text{Malate}
\end{array}
$$

The fumarase reaction is readily reversible ($\triangle G^{o\prime} \cong 0$ kJ mol^{-1}). The cycle is completed when the initial acceptor molecule, oxalacetate, is regenerated through another oxidation that is catalyzed by the enzyme malate dehydrogenase ($\triangle G^{o\prime} = +30$ kJ mol^{-1}).* The latter is an NAD enzyme, and NADH is produced as malate is oxidized to oxalacetate.

$$
\begin{array}{ccc}
COO^- & & COO^- \\
| & & | \\
HOCH & + NAD^+ \rightleftharpoons & C=O & + NADH + H^+ \\
| & & | \\
CH_2 & & CH_2 \\
| & & | \\
COO^- & & COO^- \\
& & \text{Oxaloacetate}
\end{array}
$$

Oxalacetate, the acceptor of acetate, initially formed citrate. Citrate, through a series of four oxidation–reduction reactions proceeding to oxalacetate, generated three moles of NADH, one mole of $FADH_2$, and one mole of ATP. In addition, two moles of CO_2 were formed, and ultimately two moles of water will be produced when oxygen is reduced. As will be shown in a subsequent section, since $FADH_2$ can generate two moles of ATP there can be a net gain of 12 ATP for each acetate oxidized.

Even though many of the reactions of the TCA cycle are in themselves readily reversible, the cycle is not considered to be a reversible process because of the nonreversibility of the decarboxylation steps that take place with such high negative standard free energies. As an example, the decarboxylation of α-ketoglutarate has a standard free energy of -33 kJ mol^{-1}, and overall, the free-energy change of the entire sequence ($\triangle G$) is quite negative.

14.5 Electron Transport and Oxidative Phosphorylation

The oxidative processes of glycolysis and the TCA cycle produce some ATP, but most of the energy is conserved in the form of the reduced compounds NADH or $FADH_2$. During the oxidative phosphorylation process, which produces ATP, the electron pair that reduced NAD and FAD is transported to molecular oxygen via a series of electron carriers that are firmly bound to or are components of the inner mitochondrial membrane. Collectively these redox electron-transporting proteins are known as the electron transport system (ETS).

The oxidation of NADH and NAD takes place with a $\triangle G^{o\prime}$ of about -218 kJ mol^{-1}.

$$NADH + H^+ + \tfrac{1}{2}O_2 \longrightarrow NAD^+ + H_2O$$

The drop in energy as electrons flow from NADH to O_2 is sufficient to produce several moles of ATP.

*On the basis of $\triangle G^{o\prime}$, the malate dehydrogenase reaction in the mitochondria would appear to be very unfavorable. However, it is $\triangle G$ that is important and not $\triangle G^{o\prime}$. Recall that $\triangle G$ is a function of the substrate concentrations, oxalacetate and malate (also NAD and NADH). Evidently, in the mitochondria the concentration of malate is high and oxalacetate is low, tending to drive the reaction toward oxalacetate formation.

The ETS, as presently visualized, is illustrated in Fig. 14.4. Oxidation of pyruvate (by pyruvate dehydrogenase) and of isocitrate, α-ketoglutarate, and malate within the TCA cycle produces four moles of NADH, one from each oxidation step. NADH is then reoxidized when a flavoprotein complex containing flavin mononucleotide (FMN) is reduced by electron transfer from NADH. Both the former cofactor and flavin adenine dinucleotide (FAD) can exist as free radicals and thus can accept and transfer electrons one at a time. The oxidation–reduction of NADH is, of course, a two-electron transfer. The complex, which is reduced directly by NADH, has been isolated from mitochondria and is known as Complex I. The FMN-dependent Complex I is reduced by a two-electron transfer from NADH and then becomes reoxidized as it reduces a specific quinone in the electron transport chain. The quinones can also exist as free radicals, and their oxidation–reduction reactions can be one- or two-

electron transfers. The specific quinone in the electron transport chain after Complex I is called either coenzyme Q or ubiquinone. All the subsequent electron carriers in the chain are iron-porphyrin, or heme, proteins known as cytochromes. The cytochromes are reduced as Fe^{+3} goes to Fe^{+2} in a one-electron transfer. The significance of the semiquinone forms of the flavin cofactors and the quinones is now apparent. The latter can be reduced by a two-electron transfer from NADH and can couple with the cytochromes in a one-electron transfer.

There are three major types of cytochromes in the mitochondria; they are described on the basis of their absorption spectra. They are the a, b, and c types. The order of the cytochromes in the ETS chain is based in part on their respective redox potentials ($E^{\circ\prime}$) and on the sequence in which they become reduced and oxidized as electron-donating substrates are added to mitochondrial prepara-

ELECTRON-TRANSPORT CHAIN OF MITOCHONDRIA

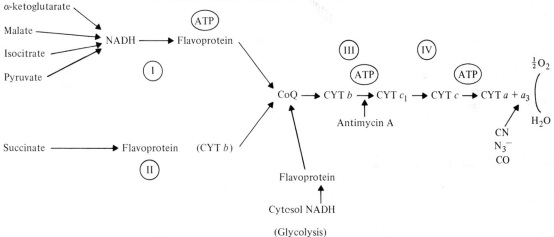

FIGURE 14.4 The electron-transport chain localized in the mitochondrial membranes. Electrons from the oxidation of organic acid substrates in the mitochondria (TCA cycle) or from glycolysis pass along the electron-transport chain in a series of oxidation–reduction steps to the ultimate acceptor, oxygen. The decrease in free energy is sufficient to phosphorylate ADP to form ATP, a process called oxidative phosphorylation. There are three "sites" along the chain where ATP is formed. Note that the NADH from the cytosol (glycolysis) enters the chain after Complex I, and only two ATP's are formed. The numbers I, II, III, and IV refer to electron-transport complexes. CYT = cytochrome. CoQ = coenzyme Q (ubiquinone). CYT $a + a_3$ is the terminal oxidase.

tions. The oxidation–reduction of the cytochromes can be followed with spectrophotometers specially designed to measure absorption spectra in complex mixtures of absorbing substances. It can be shown that cytochrome b becomes reduced before cytochrome c_1 when NADH is added to mitochondrial preparations capable of consuming O_2 when NADH is added.

Further evidence about the order of electron transport components can be gained with the use of specific inhibitors of electron transport. When the antibiotic antimycin A is added, there is an inhibition of electron flow between cytochrome b and cytochrome c_1. In the presence of NADH and O_2, antimycin A results in the reduction of cytochrome b and the oxidation of cytochrome c_1. The interpretation is that electrons from NADH reduce cytochrome b but cannot be transported to the next carrier, cytochrome c_1; hence cytochrome c_1 becomes oxidized as it transports electrons to the next carrier.

Cyanide binds very tightly to the heme group of cytochrome a, preventing electron transfer to O_2. Thus the inhibitor cyanide has been used in the study of the sequence of electron carriers of the mitochondria. Azide and carbon monoxide also inhibit cytochrome a.

There is sufficient drop in energy as electrons flow from NADH with a $E^{\circ\prime}$ of -0.32 volts to O_2 with an $E^{\circ\prime}$ of $+0.8$ volts to produce at least three ATP. Studies with isolated complexes and inhibitors have shown that ATP is produced between the electron transfer of NADH and the flavoprotein Complex I, between cytochrome b and cytochrome c_1, and between cytochrome c and cytochrome $a + a_3$.

The latter complex (cytochrome $a + a_3$) is known as the terminal oxidase or sometimes simply as cytochrome oxidase, since it transfers electrons directly to O_2.

$$2H^+ + 2e^- + \tfrac{1}{2}O_2 \longrightarrow H_2O$$

When succinate is the electron donor through succinate dehydrogenase, a reduced FAD is produced.

The latter is covalently bonded to the succinate dehydrogenase protein. This flavoprotein complex, Complex II, which includes a cytochrome b, reduces ubiquinone in a manner similar to the way Complex I is reduced by NADH. The ubiquinone then transfers electrons through the cytochrome chain, as described above. There is no ATP formed in the Complex II step, hence succinate through $FADH_2$ only produces two moles of ATP rather than three (as is the case with NADH).

The process of phosphorylation, producing ATP from ADP and inorganic phosphate, is coupled to electron flow from the substrates to O_2. However, phosphorylation can be uncoupled with certain reagents, such as 2,4-dinitrophenol (DNP).

2,4-dinitrophenol

In the presence of DNP electron flow still proceeds, since it can be shown that NADH becomes oxidized and O_2 is consumed, but there is no phosphorylation of ADP to form ATP. In fact, in the presence of ADP, inorganic phosphate, and an uncoupler, there is enhanced electron flow, as indicated by an increase in the rate of O_2 uptake. Therefore, the limiting step of the coupled sequence is phosphorylation.

Other uncouplers are the antibiotics gramicidin and valinomycin. Oligomycin will inhibit both electron flow and oxidative phosphorylation. These inhibitors of oxidative phosphorylation do not inhibit substrate-level phosphorylation. The mechanism of oxidative phosphorylation during electron transport is described in Section 14.9.

Most of the information that is available about respiration comes from studies of animals and bacteria. However, all of the enzymes of glycolysis, the pentose phosphate pathway, and the tricarboxylic acid cycle have been isolated from plant tissues and studied in some detail. In some cases the enzymes have been crystallized and fully characterized. These studies with plant enzymes have not revealed

TABLE 14.2 The yield of NADH, FADH$_2$, and ATP from the complete oxidation of glucose through glycolysis and the tricarboxylic acid cycle; the net yield of ATP or equivalents possible is also given*

STEP	NADH	FADH$_2$	ATP	NET ATP
Glycolysis	2	0	2	8
Pyruvate → acetate	2	0	0	6
TCA cycle	6	2	2	24
TOTAL	10	2	4	38

*Yield of ATP is computed on the basis of 3 moles from NADH and 2 moles from FADH$_2$.

many significant differences in enzymology between animal and plant proteins; thus most of the generalizations drawn from the extensive animal and bacterial studies are generally applicable.

Table 14.2 shows the net conservation of energy from the complete oxidation of glucose. There are produced 10 moles of NADH, two moles of FADH$_2$, and four moles of ATP. If we assume that there can be produced three moles of ATP for each

NADH and two moles of ATP from each FADH$_2$ by the electron transport system of the mitochondrial membranes, there could be a net production of at least 38 ATP from the complete oxidation of glucose. NADH produced by glycolysis only results in two ATP rather than three (see Fig. 14.4). Thus the maximum ATP production from the complete oxidation of glucose is only 36.

14.6 The Phosphate:Oxygen (P:O) Ratio

The overall reaction for phosphorylation through the electron transport system is

$$NADH + H^+ + \tfrac{1}{2}O_2 + 3ADP + 3P_i \longrightarrow$$
$$NAD^+ + 3ATP + H_2O$$

from NADH oxidation, and

$$FADH_2 + \tfrac{1}{2}O_2 + 2ADP + 2P_i \longrightarrow$$
$$FAD + 2ATP + H_2O$$

from FADH$_2$ oxidation.

In the former case, when NADH is oxidized there are three phosphates consumed by phosphoryla-

tion per oxygen atom (that is, $\tfrac{1}{2}O_2$) consumed, a ratio of phosphorus to oxygen (P:O) of 3. Similarly, oxidation of FADH$_2$ produces a P:O ratio of 2. The designation P:O is used to indicate the phosphorylation to oxygen uptake ratio.

NADH oxidation has a P:O ratio of 3. It can be shown experimentally that P:O ratios of 3 are approached when malate is the substrate and 2 when succinate is the substrate for mitochondrial metabolism. In the case of plant mitochondria, when externally added NADH is used the P:O ratio is only about 2 rather than the ratio of 3 as expected from our knowledge of electron transport. Experimentally, theoretical P:O ratios are rarely reached.

14.7 The Respiratory Quotient

The respiratory quotient (RQ) is defined as the ratio of CO$_2$ evolved to O$_2$ uptake during respiration and is an indication of the kind of substrate

being oxidized, provided that there are no significant interfering reactions that involve either CO$_2$ or O$_2$.

For example, the complete oxidation of hexose,

$$C_6H_{12}O_6 + 6O_2 \longrightarrow 6H_2O + 6CO_2,$$

will have an RQ of

$$\frac{CO_2}{O_2} = \frac{6}{6} = 1.0.$$

Oxidation of the more oxidized organic acids will result in respiratory quotients of greater than unity, since less oxygen is consumed than CO_2 liberated. For example, the complete oxidation of succinate has an RQ of about 1.14.

$$2C_4H_6O_4 + 7O_2 \longrightarrow 8CO_2 + 6H_2O$$

The complete oxidation of citrate will have an RQ of about 1.33.

$$C_6H_8O_7 + 4\frac{1}{2}O_2 \longrightarrow 6CO_2 + 4H_2O$$

Oxidation of fatty acids will have RQ's of less than one. Palmitate, for example, has an RQ of 0.68.

$$C_{16}H_{34}O_2 + 23\frac{1}{2}O_2 \longrightarrow 16CO_2 + 17H_2O$$

Trioleate has an RQ of 0.71.

$$C_{57}H_{104}O_6 + 80O_2 \longrightarrow 57CO_2 + 52H_2O$$

In many plant tissues the measured RQ will be close to unity, which is indicative of carbohydrate respiration. However, there are many notable exceptions. For example, in germinating seeds of the carbohydrate- or starch-storing type such as wheat, RQ's increase from near 1 to about 1.3, suggesting a shift from carbohydrate to organic acid oxidation. Similarly, in seeds that store fats, RQ's decrease to about 0.4 or less during germination. This is indicative of fatty acid oxidation and, to some extent, the conversion of fatty acids to sucrose. The RQ for the conversion of palmitate to sucrose is 0.36.

Another interesting example of variable RQ estimations is the gas-exchange patterns of succulent plants that have the crassulacean acid metabolism pathway. In the dark, plants such as *Kalanchoe* fix CO_2, forming malic acid concomitant with carbohydrate respiration. The uptake of virtually all available CO_2 results in RQ determinations near zero. After an extended period in the dark, CO_2 fixation will cease, and RQ's will increase to 1.3 or more, suggesting oxidation of malic acid. The complete aerobic oxidation of malic acid will have an RQ of 1.33. Roots of cactus fix substantial amounts of CO_2 and show an apparent RQ of about 0.66 when estimated from gas-exchange data.

14.8 Mitochondrial Permeability and Shuttles

The mitochondrion is limited by a double-membrane system. The outer membrane is considerably different from the inner membrane, being rather freely permeable to most metabolites of small molecular weight (10,000 or less). The inner membrane, which is made of transport proteins and the proteins of the electron transport system, is highly selective in metabolite transport. The chemiosmotic hypothesis (discussed later) for phosphorylation is based on impermeability to protons. Gases and water seem to pass freely through the inner membrane, as does phosphate and some of the tricarboxylic acid cycle intermediates. Citrate and isocitrate enter the matrix readily. Fumarate does not, but malate and succinate do, perhaps on an exchange basis. The niacin nucleotides, NAD and NADH, are not readily transported by mitochondria, but ATP and ADP are. The latter are transported on an equimolar basis; that is, an ATP is exchanged for an ADP and vice versa. Certain phosphorylated compounds are excluded from the inner mitochondrial matrix, but others, such as glycerol-P, freely enter.

Glycerol-P enters the mitochondria and is exchanged for dihydroxyacetone-P. This is the basis for one shuttle hypothesis in which glycerol-P

SHUTTLE HYPOTHESES FOR THE TRANSPORT OF NADH INTO MITOCHONDRIA

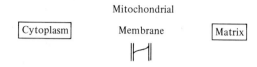

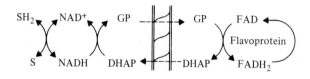

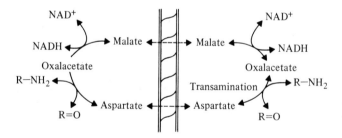

FIGURE 14.5 Two hypotheses for the transport of NADH from the cytosol into the mitochondria. Glycerol-P shuttle: NADH is produced by oxidation of a reduced substrate. The NADH reduces dihydroxyacetone phosphate (DHAP) to glycerol-P (GP), which enters the mitochondria. The GP is oxidized back to DHAP in the mitochondria, producing reduced FAD. The DHAP is transported back to the cytosol, completing the shuttle. Malate aspartate shuttle: The malate aspartate shuttle is similar to the glycerol-P shuttle except that malate is transported into the mitochondria and aspartate is transported out. For each malate transported into the mitochondria, there is one reducing equivalent of NADH transported.

enters the matrix, effectively bringing a reducing equivalent. Glycerol-P within the matrix is oxidized to dihydroxyacetone-P by an FAD flavoprotein. The FADH₂ feeds electrons into the electron transport system and the dihydroxyacetone-P returns to the cytosol, where it is once again reduced to glycerol-P by an NAD dehydrogenase. In this manner reducing equivalents of NADH are transported into the mitochondria via glycerol-P and transferred to FAD. Thus the transport of glycerol-P and dihydroxyacetone-P are each unidirectional. The glycerol-P/dihydroxyacetone-P exchange shuttle is illustrated in Fig. 14.5.

Another shuttle system in which NADH reducing equivalents are transported into the mitochondrial matrix is the malate aspartate shuttle. A cytosol NAD malate dehydrogenase reduces oxalacetate to malate, which enters the mitochondrial matrix. The mitochondrial NAD malate dehydrogenase regenerates the NADH by oxidation of the malate to oxalacetate. Because the inner mitochondrial membrane is not permeable to oxalacetate, it is postulated that a transaminase converts the oxalacetate to aspartate, which reenters the cytosol and is converted back to oxalacetate by a cytosol transaminase, completing the shuttle cycle.

It can be noted that in the case of the glycerol-P shuttle, cytosol NADH (for example, that generated by glycolysis) will only produce two moles of ATP since it effectively enters at the FAD level. However, the malate shuttle will produce three moles of ATP. The malate shuttle is illustrated in Fig. 14.5. Other shuttle systems have been proposed for the entry of substrates into the mitochondrial matrix.

The presence of transaminases within the mitochondrial matrix allows for the oxidation of both glutamate and aspartate by mitochondria. Glutamate, for example, can enter and be converted to α-ketoglutarate by aspartate aminotransferase.

Glutamate + Oxalacetate $\rightleftharpoons$

Aspartate + α-ketoglutarate

Similarly, aspartate can be converted to oxalacetate. Being intermediates, both α-ketoglutarate and oxalacetate can be readily oxidized by the tricarboxylic acid cycle. Glutamate, in fact, is an excellent substrate for mitochondria.

All of the acids of the tricarboxylic acid cycle—oxalacetate, malate, fumarate, succinate, α-ketoglutarate, isocitrate, and citrate—will act as good substrates for respiration. Succinate appears to be the best of the various acids. Pyruvate, the compound which enters into the mitochondria and is decarboxylated to acetate prior to oxidation, is in itself a rather poor tricarboxylic acid cycle substrate.

Pyruvate is a poor aerobic respiration substrate because oxalacetate is required as a carrier for the acetate product of pyruvate. When in the presence of low concentrations of tricarboxylic acid cycle intermediates, pyruvate is a much better substrate because the intermediates generate oxalacetate.

Furthermore, pyruvate is poorly transported into mitochondria, limiting the effectiveness of pyruvate as a substrate. It has been observed that a high proportion of the carbon from glucose that enters into the tricarboxylic acid cycle enters as malate and not pyruvate per se. When phosphoenolpyruvate is formed during glycolysis, much is diverted to malate through a carboxylation reaction rather

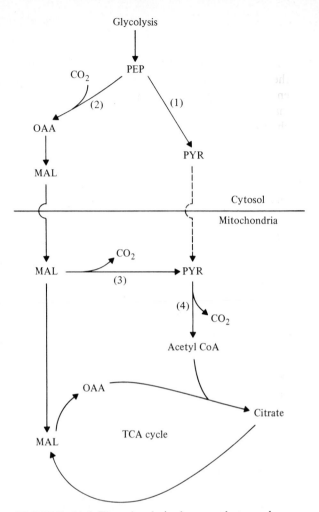

FIGURE 14.6 The glycolytic bypass that produces malate rather than pyruvate. Once phosphoenolpyruvate (PEP) is produced rather than the pyruvate kinase step (1), there is a carboxylation of PEP catalyzed by PEP carboxylase (2) to form oxalacetate (OAA). The latter is reduced to malate (MAL), which is transported into the mitochondria and enters the tricarboxylic acid (TCA) cycle. Once in the mitochondria, malate may be decarboxylated by malate enzyme (3) to form pyruvate. The pyruvate is decarboxylated by the pyruvate dehydrogenase complex (4) to form acetyl CoA, which enters the TCA cycle.

than going directly to pyruvate via the pyruvate kinase reaction (Fig. 14.6).

$$PEP + CO_2 \longrightarrow OAA$$

$$OAA + NADH \longrightarrow Malate + NAD$$

As shown in Fig. 14.6, there is a branch point in the glycolytic sequence such that some phosphoenolpyruvate goes to pyruvate and some goes to malate. Malate is readily transported into mitochondria and can be either oxidized directly or decarboxylated to pyruvate through catalysis with the malate enzyme. This enzyme may require either NADP or NAD and generates the corresponding reduced pyridine nucleotide.

$$\text{Malate} + \text{NADP (NAD)} \longrightarrow$$
$$\text{Pyruvate} + CO_2 + \text{NADPH (NADH)}$$

14.9 The Oxidative Phosphorylation Mechanism

The mechanism by which the terminal phosphate ester of ATP is made during electron transport in mitochondria has been studied extensively. It was shown prior to 1940 that tissue would form sugar phosphate esters if supplied with pyruvate and oxygen. Subsequent to this observation, others observed that mitochondrial preparations would produce ATP from ADP when in the presence of tricarboxylic acid cycle intermediates and inorganic phosphate. It was further shown that the only phosphate bond produced within the mitochondria was the terminal phosphate ester of ATP.

During the search for the actual chemical mechanism of oxidative phosphorylation two basic hypotheses have been developed, neither of which has yet been fully accepted. We have, however, much information to formulate theories, but as is true of all theories, they must be consistent with all the available data and explain all of the pertinent observations.

It is well known that the mitochondrial membranes have a transport function. During electron flow to O_2 from NADH or another electron-donating substrate, for each reducing equivalent of NADH (that is, two electrons, or an electron pair) six protons are transported from the inner mitochondrial matrix to the outside (actually to the intermembrane space between the inner and outer membranes). During oxidative phosphorylation the pH outside will decrease. Potassium is simultaneously transported into the mitochondria, maintaining ionic balance. This observation of proton pumping was very important for the formulation of the most popular hypothesis of oxidative phosphorylation, the chemiosmotic coupling hypothesis. The alternative hypothesis worthy of consideration is the chemical coupling hypothesis, in which there are intermediates of sufficient redox potential to bring about phosphorylation by displacement. Indeed, a protein called coupling factor (CF) was isolated from Complex I that actually catalyzes the phosphorylation of ADP to form ATP:

$$\text{ADP} + P_i \longrightarrow \text{ATP} + H_2O.$$

It is referred to as an ATPase but really functions in the opposite direction.

As stated previously, the oxidation of NADH is highly exergonic, yielding about 218 kJ mol^{-1}. Furthermore, it is readily shown that when pyruvate, isocitrate, α-ketoglutarate, and malate are the substrates that are producing NADH there can be up to three moles of ATP produced (a P:O of 3), and when succinate is the substrate there can be up to two moles of ATP formed (a P:O of 2). Thus it is proposed that there are three "sites" capable of phosphorylation in the mitochondria when malate, isocitrate, α-ketoglutarate, and pyruvate are the substrates and two when succinate is the substrate. The two hypotheses discussed in detail below, the chemiosmotic hypothesis and the chemical coupling hypothesis, attempt to explain the nature of these "sites" (Perhaps sites is a poor term, especially with respect to the chemiosmotic hypothesis.)

14.9.1 The chemical coupling hypothesis

The chemical coupling hypothesis, proposed by Briton Chance and others (Lehninger, 1967), is probably the more easily understood of the two, being more consistent with our general knowledge of chemical phenomena. Simply stated, electron transport produces high-energy intermediates within the inner mitochondrial membranes. It is visualized that during oxidation-reduction reactions accompanying electron flow, high-energy bonds are formed between oxidized components and a factor termed I. The high-energy bond is designated as $\sim$, and consequently the hypothesis is occasionally referred to as the "squiggle" hypothesis.

$$XH + Y + I \longrightarrow X \sim I + YH$$

In the above hypothesis the reduced component, XH, is oxidized by Y, forming the high-energy intermediate $X \sim I$ and reduced YH. This is not unlike the process of substrate-level phosphorylation at the glyceraldehyde-3-P dehydrogenase step. Recall that oxidation of glyceraldehyde-3-P by NAD in the presence of inorganic phosphate forms 1,3-diphosphoglycerate. The standard free energy of hydrolysis of the newly formed phosphate ester at the C-1 position is -470 kJ mol^{-1} as opposed to the standard free energy of hydrolysis of the phosphate ester at the C-3 position of only about -13 kJ mol^{-1}.

In the chemical coupling hypothesis, it is then visualized that I is displaced from $X \sim I$ by an enzyme to form an enzyme complex, $E \sim I$.

$$X \sim I + E \longrightarrow E \sim I + X$$

Subsequently, phosphate displaces I to form a high-energy phosphate ester, $E \sim P$.

$$E \sim I + P \longrightarrow E \sim P + I$$

Displacement of $\sim$P from the enzyme complex by ADP completes the esterification and forms ATP.

$$E \sim P + ADP \longrightarrow E + ATP$$

The chemical coupling hypothesis is consistent with known chemical observations and is a feasible hypothesis to explain high-energy phosphate ester formation during oxidative phosphorylation. With the exception of the one coupling factor, however, intermediates have not been isolated, and confirmation or refutation must await additional experimentation and discovery.

14.9.2 The chemiosmotic coupling hypothesis

The chemiosmotic coupling hypothesis, proposed chiefly by Mitchell (1966), is based on the observation that six protons are generated per electron pair transported along the electron transport system. Mitchell proposed that the inner mitochondrial membrane is not permeable to protons, that the ETS components are arranged within the inner membrane such that matrix protons are taken up by matrix-oriented proteins, and that protons are secreted into the intermembrane space by outer membrane proteins. Thus a proton gradient, which is coupled to electron flow, is created across the inner mitochondrial membrane with sufficient electrochemical energy to drive oxidative phosphorylation. A diagram of this vectorial hypothesis is shown in Fig. 14.7.

Perhaps the best evidence to support the chemiosmotic coupling hypothesis is that experimentally produced proton gradients will in fact drive phosphorylation and that actual proton gradients can be measured during phosphorylation.

Phosphorylation is visualized to occur through a reversible ATPase reaction.

$$ATP + H_2O + H^+ \longrightarrow ADP + P_i + H^+$$

We can visualize that the proton gradient from the matrix side of the inner membrane to the outside creates a proton deficiency within the matrix where phosphorylation occurs. Thus the above reaction is drawn from right to left, forcing phosphorylation. In addition, the proton gradient is visualized to cause hydrophobic conditions within the inner mitochondrial membrane, facilitating the phosphorylation of ADP rather than the hydrolysis of ATP.

The main features of the chemiosmotic coupling hypothesis are:

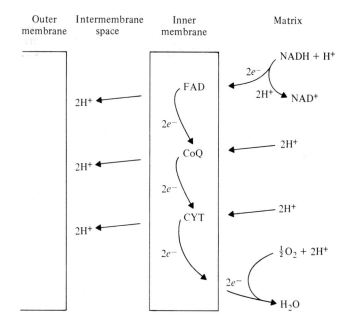

Outer membrane   Intermembrane space   Inner membrane   Matrix

FIGURE 14.7 Hypothesis for the creation of a proton gradient during electron transport in mitochondria. As electrons are transported along the cytochromes within the inner mitochondrial membrane, protons are pumped from the matrix into the space between the inner and outer membranes. As explained in the text, the gradient is sufficient to create enough electrochemical energy to drive phosphorylation. From P. Mitchell. 1966. Chemiosmotic coupling in oxidative and photosynthetic phosphorylation. *Biol. Rev.* 41:445–502.

1. There is a vectorial transport of protons across the membrane, which is relatively impermeable to protons.

2. Electron flow during respiration is coupled to proton transport.

3. A counter-ion flux across the membrane prevents positive charge accumulation.

4. The proton gradient creates a protonmotive force (pmf) that can be coupled to ATP synthesis through phosphorylation.

The electrochemical gradient across the membrane during electron flow caused by the proton gradient can be described in terms of a protonmotive force (pmf), which is a function of the electrochemical potential gradient across the membrane ($\Delta\psi$) and the pH differential across the membrane.

$$\text{pmf} = \Delta\psi + \Delta\text{pH}$$

Protonmotive force can be expressed in terms of free energy (ΔG_{H+}). The pmf expressed as $\Delta G^{o'}$ would have to be at least equivalent to -30 kJ

mol^{-1} to drive ATP synthesis.

$$\text{ATP} \longrightarrow \text{ADP} + \text{P}_i \quad (\Delta G^{o'} = -30 \text{ kJ mol}^{-1})$$

Studies with inhibitors of phosphorylation have not been too helpful in elucidating the phosphorylation mechanism. In both hypotheses, inhibitors such as cyanide that prevent electron flow would prevent phosphorylation. Uncouplers such as dinitrophenol are assumed to interfere with the high-energy intermediate formation in the chemical coupling hypothesis and to increase membrane permeability, destroying the proton gradient in the chemiosmotic coupling hypothesis.

There are some alternative hypotheses for phosphorylation that can be mentioned briefly. In one hypothesis it is visualized that electron flow causes conformational changes of the inner mitochondrial membrane, increasing the energy state of the membrane. Structural changes can be observed during active electron transport and phosphorylation, which supports such a notion. Energy states of conformational changes are then used to drive phosphorylation.

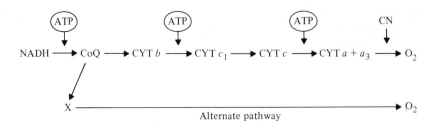

FIGURE 14.8 Hypothesis to account for cyanide-resistant respiration in which no ATP is formed. Cyanide inhibits the terminal oxidase (CYT $a + a_3$), preventing electron transport to oxygen. It is assumed that coenzyme Q transports electrons to an alternate electron acceptor (X) that can function as a terminal oxidase (i.e., that reduces O_2 to water).

14.10 Cyanide-resistant Respiration

As indicated in Fig. 14.4, cyanide will inhibit respiration by its effect on the terminal oxidase. In the presence of cyanide, oxygen uptake is inhibited and the phosphorylation of ADP to form ATP is reduced.

However, in some plant tissues there is continued respiration in the presence of cyanide. Respiration also continues in the presence of carbon monoxide and azide, which act at the same site as cyanide, and in the presence of antimycin A, an uncoupler that inhibits reactions proceeding between cytochrome b and cytochrome c_1. The respiration quotient is close to one, and ATP synthesis is reduced such that P:O ratios fall to about 1.0 from 3.0. Thus it appears that only one of three phosphorylation sites is functional in the presence of cyanide.

One possible explanation is that coenzyme Q is autooxidizable and functions as the terminal oxidase in the presence of cyanide. But there is little direct evidence for Q as a terminal oxidase. There is known an *alternate oxidase pathway* in which electron flow from Q is not to cytochrome b, as illustrated in Fig. 14.4, but rather to an alternate electron-transport chain (refer to Fig. 14.8). The alternate oxidase functions as a terminal oxidase, consuming oxygen, but has no phosphorylation sites. The only ATP formed is prior to Q at Complex I.

It is visualized that the overall oxidation–reduction potential of the alternate pathway is less positive than the usual pathway such that in the uninhibited system most electron flow is from Q to cytochrome b. When inhibited with cyanide, electron flow is through the alternate pathway.

Cyanide resistance is most evident in aged plant tissues. For example, aged potato slices show little cyanide inhibition of respiration, wheres fresh slices are inhibited 70% or more by $1 mM$ cyanide.

Perhaps the most interesting example of cyanide-resistant respiration comes from studies of the very high respiratory rates of the *Arum* spadix. Respiration rates are as high as 20,000 μL O_2 per g fresh weight per h and account for marked temperature increases within the spadix. Rather than inhibiting the respiratory rate, cyanide will frequently cause it to increase. A plausible explanation is that the alternate oxidase illustrated in Fig. 14.8 is functional in *Arum*.

14.11 Terminal Oxidases

Several enzyme systems are known in higher plants that can function as terminal oxidases in respira-

tion. Without doubt the most important is cytochrome oxidase. Cytochrome oxidase $(a + a_3)$

transfers electrons to O_2 at the final step of the electron-transport system of mitochondria. Almost all of our available evidence indicates that cytochrome oxidase is the most important terminal oxidase of plants.

There are several soluble enzyme systems other than cytochrome oxidase that can function as terminal oxidases, including polyphenoloxidase, ascorbate oxidase, glycolate oxidase, and glutathione reductase. In the basic reaction, electrons are transferred to an oxidized substrate from either NADH or NADPH, and then a terminal oxidase oxidizes the reduced substrate, transferring electrons to oxygen and forming water. In some of the reactions, hydrogen peroxide is formed, which is oxidized by catalase (see below).

The basic reaction of the terminal oxidase follows.

R is a respiratory intermediate and A is the substrate of the terminal oxidase.

Polyphenoloxidase can be used as an example of such a terminal oxidase system.

Although these oxidases are widespread in plants, they are not believed to play an important role in respiration. The role of glycolate oxidase in photorespiration will be discussed in Chapter 16.

The peroxidases oxidize reduced substrates, using hydrogen peroxide (H_2O_2) as the electron acceptor.

The basic peroxidase reaction follows.

Peroxidases are widespread in plant tissues and exist in many multiple forms. They were discussed in Chapter 9.

Catalase is an oxidase similar to peroxidase in that it uses H_2O_2 as an electron acceptor. The electron donor is also H_2O_2. The reaction that generates water and oxygen is the following.

$$H_2O_2 + H_2O_2 \longrightarrow 2H_2O + O_2$$

Catalase is a common enzyme of plant tissues associated with flavoproteins that generate H_2O_2. It very likely functions in the detoxification of H_2O_2. It is localized in the single-membrane organelles called microbodies and perhaps also in chloroplasts.

Another enzyme of interest in oxygen metabolism is superoxide dismutase, which catalyzes the production of O_2 and H_2O_2 from the superoxide anion (O_2^-).

$$2O_2^- + 2H^+ \longrightarrow H_2O_2 + O_2$$

Superoxide dismutase probably functions in the detoxification of the superoxide anion that may be generated as an intermediate during peroxide formation from oxygen.

$$O_2 + e^- \longrightarrow O_2^-$$
$$O_2 + 2H^+ + e^- \longrightarrow H_2O_2$$

14.12 The Regulation of Respiration

14.12.1 Respiratory control

When mitochondria are isolated in appropriate media it is possible to measure O_2 consumption in the presence of ADP, inorganic phosphate, and various substrates such as succinate, malate, and α-ketoglutarate. Oxygen uptake is most frequently measured with an oxygen electrode polarographically. Chance and Williams (see Lieberman and Baker, 1965) assessed the condition of isolated mitochondria by the ratio of O_2 uptake in the presence of ADP to oxygen uptake in the absence of ADP when an appropriate substrate was present. This ratio is called the respiratory control and gives an indication of the functional capacity of isolated mitochondria. They designated the rate of oxygen uptake in the presence of ADP and substrate as State 3, and the rate after ADP is consumed (i.e., phosphorylated to form ATP) as State 4. Thus respiratory control is the ratio of the O_2 uptake rate during State 3 to that of State 4. The ratio is on the order of 2 to 10 for plant mitochondria. Animal mitochondria frequently show higher degrees of respiratory control.

They further named the slow uptake of oxygen in the absence of substrate as State 1 and the rate in the presence of substrate but with an absence of ADP as State 2.

14.12.2 The Pasteur effect

It was noted by Pasteur that fermentation, which produced alcohol and CO_2, was inhibited by O_2. A switch from aerobic to anaerobic conditions results in an increase in glycolysis.

The Pasteur effect in oxygen is known to be associated with high ATP and low ADP tissue levels. Since dinitrophenol uncouples oxidative phosphorylation from electron transport, the Pasteur effect is probably the result of mitochondrial-coupled phosphorylation. Both glycolysis and oxidative phosphorylation depend on ADP availability. In the presence of O_2, oxidative phosphorylation acts as a sink for ADP, limiting that available for glycolytic substrate-level phosphorylation. Hence the rate of glycolysis is reduced in the presence of oxygen. Under anaerobic conditions, when oxidative phosphorylation is prevented, excess ADP stimulates glycolysis.

High ADP and inorganic phosphate could also be associated with control at the phosphofructokinase step of glycolysis since the latter enzyme is inhibited by ATP. When ATP is low and ADP necessarily high under anaerobic conditions, phosphofructokinase would not be inhibited and glycolysis could function at a high rate.

14.12.3 Metabolite regulation of respiration

The flow of carbon from storage carbohydrate to complete oxidation within the tricarboxylic acid cycle is regulated by specific metabolite effects on the catalytic enzymes. Metabolites may alter enzymatic rates in two primary ways. First, those metabolites that are direct substrates or products in reversible reactions regulate by mass-action. Thus a glycolytic enzyme such as phosphoglucoisomerase, which catalyzes the equilibrium between glucose-6-P and fructose-6-P, will be pushed toward fructose-6-P if glucose-6-P is high relative to fructose-6-P and vice versa. And, of course, even such poorly reversible reactions as phosphofructokinase will not proceed if fructose-6-P or ATP are not available. Another and perhaps more interesting regulatory mechanism is allosteric regulation by metabolites, which affects enzymes but not at the catalytic sites. Allosteric regulation was discussed in Chapter 9.

Within the glycolytic sequence there are two enzymatic reactions that are not readily reversible (and thus subject to regulation). Phosphofructokinase is inhibited by ATP and to a lesser extent by ADP and AMP and is stimulated by inorganic phosphate. The enzyme pyruvate kinase is also inhibited by ATP, stimulated by AMP, and inhibited by citrate. Thus one can visualize that when ATP is high in the cytosol, the two kinases will be inhibited and glycolysis will proceed slowly. However, when ATP is low and inorganic phosphate and AMP are necessarily high, the kinases will be activated. Glycolysis will function at a high rate until ATP reaches an inhibitory level.

Glucose-6-P dehydrogenase, which catalyzes the first step of the pentose phosphate pathway, is competitively inhibited by NADPH with respect to NADP. Thus the level of cytosol NADPH is governed by NADPH regulation of glucose-6-P dehydrogenase. When NADPH is low, glucose-6-P dehydrogenase will be actively producing NADPH. Therefore NADPH levels also control the flow of carbon between glycolysis and the pentose phosphate pathway.

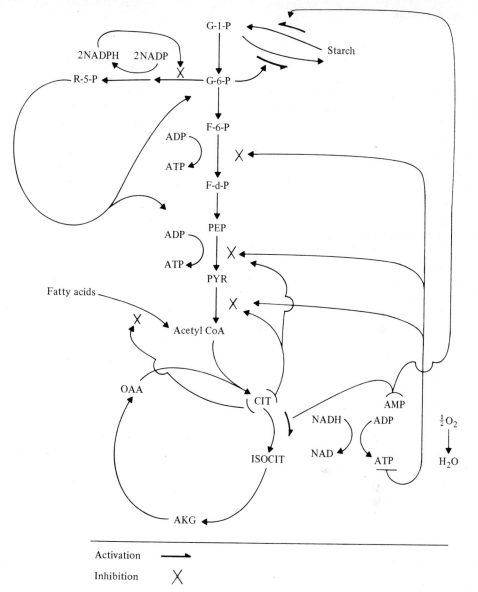

FIGURE 14.9 Possible sites of regulation of respiration by metabolites. See text for explanation and details.

Citrate levels regulate the flow of carbon through the tricarboxylic acid cycle. When citrate builds up within the mitochondria, it can feed back and inhibit its own synthesis by blocking pyruvate kinase.

Several other enzymes of the tricarboxylic acid cycle are also inhibited and activated by metabolites. Good examples are the pyruvate dehydroge-

nase complex, which is inhibited by NADH, and citrate synthetase, which is inhibited by ATP. Furthermore, it was shown that the citrate synthetase isozyme of the glyoxysomes was not inhibited by ATP, but the mitochondrial enzyme was. Isocitrate dehydrogenase is also inhibited by NADH. Once again we can visualize that the products NADH and ATP regulate their own synthesis.

When in high concentrations, they will inhibit enzymes responsible for their synthesis. A simplified regulatory scheme is outlined in Fig. 14.9.

There also is metabolite regulation of fatty acid oxidation that supplies acetyl CoA for the tricarboxylic acid cycle. Furthermore, it is known that high cellular levels of citrate inhibit the fatty acid synthesizing system; hence regulation of fatty acid oxidation and synthesis alters the amount of acetyl CoA available to the tricarboxylic acid cycle.

It should be kept in mind that even though the inhibitory properties of metabolites can be demonstrated *in vitro* with isolated and purified enzymes, the extrapolation to *in vivo* cellular conditions is speculation.

14.13 Respiration of Plant Tissues

Numerous factors affect the rate of respiration by tissues. Dormant seeds may have respiration rates so low as to be virtually nondetectable, and the flowering spadix of *Arum* can have such high rates of respiration that the tissue will heat many degrees above air temperature. The usual rates of respiration by green leaves in the dark, expressed on an area basis, are 1 to 10 mg CO_2 evolution per dm^2 per h. When expressed on the basis of dry weight of tissue, the rates are also about 1 to 10 mg CO_2 evolution per g dry weight per h. On a per-cell basis, rates can range from 1 to 20 mg $\times 10^{-8}$ per h. Of course, rates will vary depending on environmental conditions, time of day, age of tissue, disease, and on almost any factor that one can imagine.

Respiration varies greatly with age. Elongating root tips will have relatively high rates of respiration at the growing tip. Figure 14.10 illustrates oxygen consumption by young elongating corn root tips and shows that there is nearly twice as much oxygen consumption on a per-section basis at the growing tip as there is within the differentiated root zone. Shown in Fig. 14.11 is the respiration rate of strawberry leaves during the course of their development from the expansion period through senescence and death. The maximum rates of

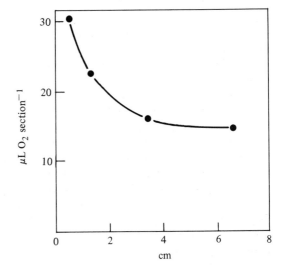

FIGURE 14.10 Respiration (oxygen consumption) by maize root tips as a function of distance from the tip. The younger (and the more active) the tissue, the greater the oxygen consumption per unit of tissue section.

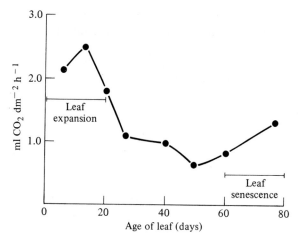

FIGURE 14.11 Change in respiration rate of a strawberry leaf during development and maturation. During
leaf expansion, the respiration rate is high. It remains
rather constant throughout much of the life of the leaf
and then increases during senescence (leaf yellowing).
Drawn from the data of S.E. Arney. 1947. Respiration of
strawberry leaves attached to the plant. *New Phytologist* 46:68–96.

respiration occur during the stage of greatest expansion of the leaf. Rates decrease throughout the
aging period and peak again during senescence
before decreasing until death.

Research shows rates of respiration correlated
with rates of growth; those tissues with the greatest
mitochondrial activity are apt to have the highest
growth rates.

Since dark respiration occurring within the mitochondria is absolutely dependent on oxygen, one
might expect that oxygen tensions would affect the
rate of respiration. This is in fact true, but the rate
of respiration is surprisingly quite uniform over the
range from ambient oxygen tensions (about 21%) to
1% or less. The affinity of the terminal oxidase for
O_2 is quite high, and large decreases in O_2 tension
from ambient levels have little or no effect. Very
high concentrations of O_2 will inhibit respiration,
presumably by oxidation of necessary respiratory
components such as substrates, products, or enzymatic proteins.

Respiration responds most dramatically to temperature. Little respiration occurs at temperatures
near freezing. As the temperature increases, there
is a nearly linear increase in the rate of respiration.
As explained in Chapter 2, for each 10° rise in temperature we expect at least a two- to threefold
increase in respiration. Most temperate plants
reach respiratory maxima near 35° and then show a
decrease beyond that. The respiration of corn root
tips as a function of temperature is illustrated in
Fig. 14.12. There is a peak at 35° followed by a
decrease. Of course, the temperature optimum is
highly dependent on the kind of plant and its adaptation to temperature. Cactus, for example, would
be expected to have a higher temperature optimum
for respiration than more mesic plants, which are
adapted to cooler environments.

The effect of light on dark respiration has been
studied extensively by many workers. Because of
the complication of photorespiration, which is a
light-dependent CO_2-evolution and O_2-consump-
tion process not directly related to dark respiration,
data obtained on O_2 uptake or CO_2 evolution by
green tissues in light are difficult to interpret.

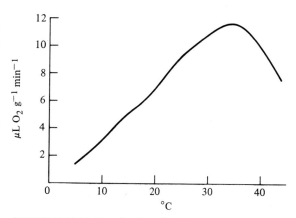

FIGURE 14.12 Respiration of maize root tips as a function of temperature. There is an increase in the rate of
respiration as temperature increases up to about 35°
followed by a decline. From a student experiment.
University of California, Riverside.

Photorespiration will be discussed in the next two chapters, but it is necessary to say here that although certain plants (namely, the C_4 plants) have little or no photorespiration, most species do. Before the process of photorespiration was fully understood, it seemed that some plants respired in the light and others did not. Most of the available data and the present interpretation of dark respiration in the light indicate that although respiration may be decreased somewhat because of the high levels of ATP produced through photosynthesis, respiration does continue in the light.

Because tricarboxylic acid cycle intermediates are oxidized in the light, it is concluded that light probably alters the rate of dark respiration but that dark respiration continued in the light. Nongreen tissues, of course, are largely independent of direct light effects on respiration.

Water stress tends to decrease plant respiration, although if the stress is transient, respiration may increase temporarily. In fact, it is not uncommon to find increases in respiration of recently injured tissue. Disease is also frequently associated with increases in respiration.

Review Exercises

14.1 Predict the maximum amount of ATP that could be produced by the complete oxidation of glucose via glycolysis and the tricarboxylic acid cycle if NADH is transported to the mitochondrial matrix by the glycerol-P shuttle.

14.2 Compute the total amount of ATP possible from the complete oxidation of caproic acid.

14.3 Speculate on the biological significance of cyanide-insensitive respiration.

14.4 The enzyme pyruvate P_i dikinase catalyzes the following reaction in plant tissues.

$$PYR + ATP + P_i \longrightarrow PEP + P_iP_i + AMP$$

In terms of energetics, do you predict that the dikinase reaction would yield much P-enolpyruvate (PEP)? The pyrophosphate (P_iP_i) is hydrolyzed *in vivo* to orthophosphate ($2P_i$) by a pyrophosphatase. If the pyrophosphatase reaction is coupled to the dikinase, would you predict the reaction to be feasible?

14.5 Why is phosphofructokinase considered to be a nonreversible enzyme? How is this step, i.e., fructose-6-P to fructose-1,6-P, reversed in cells?

References

BEEVERS, H. 1961. *Respiratory Metabolism in Plants.* Row, Peterson Co., Evanston, Illinois.

DAVIES, D. D. 1979. The central role of phosphoenolpyruvate in plant metabolism. *Ann. Rev. Plant Physiol.* 30:131–158.

DUNN, G. 1974. A model for starch breakdown in higher plants. *Phytochemistry* 13:1341–1346.

HENRY, M. F., and E. J. NYNS. 1975. Cyanide-insensitive respiration. An alternate mitochondrial pathway. *Subcellular Biochem.* 4:1–65.

IKUMA, H. 1972. Electron transport in plant respiration. *Ann. Rev. Plant Physiol.* 23:419–436.

KAHL, G. 1974. Metabolism in plant storage tissue slices. *Bot. Rev.* 40:263–314.

LEHNINGER, A. L. (ed.). 1967. Biochemistry Society Symposium. Energy coupling in electron transport. *Fed. Proc.* 26:1333–1379.

LIEBERMAN, M., and J. E. BAKER. 1965. Respiratory electron transport. *Ann. Rev. Plant Physiol.* 16:343–382.

MEEUSE, B. J. D. 1975. Thermogenic respiration in aroids. *Ann. Rev. Plant Physiol.* 26:117–126.

MITCHELL, P. 1966. Chemiosmotic coupling in oxidative and photosynthetic phosphorylation. *Biol. Rev.* 41:445–502.

NORTHCOTE, D. H. 1974. *Plant Biochemistry.* University Park Press, Baltimore, Md.

PALMER, J. M. 1976. The organization and regulation of electron transport in plant mitochondria. *Ann. Rev. Plant Physiol.* 27:133–157.

RICHTER, G. 1979. *Plant Metabolism.* University Park Press, Baltimore, Md.

SOBER, H. A. 1970. *Handbook of Biochemistry,* 2d ed. Chemical Rubber Company, Cleveland, Ohio.

SOLOMOS, T. 1977. Cyanide-resistant respiration in higher plants. *Ann. Rev. Plant Physiol.* 28:279–297.

STREET, H. E., and W. COCKBURN. 1972. *Plant Metabolism.* Pergamon Press, Oxford.

TEDESCHI, H. 1979. *Mitochondria: Structure, Biogenesis, and Transducing Functions.* Springer-Verlag, Berlin.

ZELITCH, I. 1964. Organic acids and respiration in photosynthetic tissues. *Ann. Rev. Plant Physiol.* 15:121–142.

CHAPTER FIFTEEN

The Primary Processes of Photosynthesis

Photosynthesis is by far the most important biological phenomenon on earth. It is through photosynthesis that all the useful organic material available on earth has been produced. This organic material ranges from the obvious food supply for us and other animals to our primary sources of energy, stored in fossil fuel reservoirs, and to the less obvious raw materials for the synthesis and production of synthetic fibers, plastics, polyesters, and other useful materials.

The magnitude of carbon fixed on an annual basis is staggering. It is estimated that there is about 1.55×10^{11} tons of dry matter produced annually by photosynthetic plants, with about 60% produced on land and the remainder within the oceans and continental water bodies. This represents a turnover of approximately 7% of the available atmospheric and oceanic CO_2 reserve.

15.1 The History of the Development of Modern Concepts

Some form of organized agriculture has existed for over 10,000 years, yet there were virtually no systematic investigations of photsynthesis until the

18th century. Early Greek farmers thought that plant matter came largely from the soil, and the practice of adding plant and animal debris to soil was common. Thus some form of fertilization was established early in our recorded history.

One of the first reported studies of plant growth was by Woodward, an Englishman who in 1699 reported an experiment in which he grew mint plants in rainwater, in Thames river water, and in sewage water. The plants, of course, grew better in the sewage water, and Woodward rightly concluded that some terrestrial matter in addition to water was necessary for plant growth. The rainwater, although produced by a process of distillation (evaporation), would have had some nutrients from the dirty atmosphere. River water, particularly under the scant sanitation procedures of the 18th century, would have been quite rich in organic matter but not as rich as the sewage water.

The first plant physiologist of note was perhaps Stephan Hales, who in 1727 studied the beneficial effect of air and light on plant growth. There was the suggestion at this time that plants derived some useful substances from the air and that light was a requirement for growth.

After this, a theory was developed in which combustion processes were thought to produce a toxic substance called phlogiston, the "material substance" of fire. Priestley, the discoverer of O_2, observed in 1772 that combustion, e.g., the burning of a candle, would produce "air" that would not support life. Thus mice kept in a bell jar with a burning candle would die. However, he also found that plants would somehow cleanse, or "dephlogistenize," the air contaminated with a burning candle so that it would again support life.

Priestley recognized the reversible nature of the process but was unaware of the role of CO_2 or the importance of light. It seems quite strange that the discoverer of O_2 was unaware that he was studying its physiological significance. The lesson here is that discoveries are difficult to make, and that much information and background are necessary before significant interpretation of experiments can be made. Even though we usually associate single

important events of history with one or, at most a few individuals, scientists do not work alone; many contribute to each and every discovery.

Jan Ingen-Housz, the famous Dutch scientist, recognized in 1779 that plants only dephlogistenized air in the light and furthermore that nongreen portions of plants behaved similarly to animals. Ingen-Housz thus showed the importance of light and gave us the hint that green was important.

Antoine Lavoisier's studies of the combustion process showed that O_2 was taken up and CO_2 was evolved, ending the phlogiston concept and clarifying the nature of contaminated air. On the basis of the discoveries of Lavoisier, Ingen-Housz suggested that the CO_2 in air was the source of carbon for organic matter.

Perhaps the first modern experiment of photosynthesis was that of de Saussure. In 1804, he demonstrated the uptake of CO_2 by plants when in the light and the simultaneous evolution of O_2 and that the process was reversed in the dark.

The development of the "Law of Conservation of Energy" by Robert Mayer in 1842 clearly indicated that the sun was the source of energy for plants and animals and that light energy was the energy converted to chemical energy by green plants.

At the end of the 19th century our knowledge of photosynthesis was rather slight, but some understanding allowed the rapid developments of the 20th century. Up until that time we knew that plants absorbed CO_2 and converted it to organic matter. Light was the source of energy. Oxygen was produced as a product of the process, and plants in the dark absorbed O_2 and gave off CO_2 like animals.

At the beginning of the 20th century, studies by Blackman indicated that photosynthesis consisted of two reactions: a fast, light-dependent reaction, and a slower process that could occur in the dark. His studies showed that when light was saturating and CO_2 was limiting, photosynthesis was temperature-dependent, indicating that the carbon metabolism reactions were ordinary thermal reactions. However, when CO_2 was not limiting but light was, photosynthesis was temperature-independent, indicating that the light reactions were true photochemical reactions. This observation led to studies of the photochemistry of photosynthesis (the fast reactions) and carbon metabolism (the dark, slow reactions).

In 1937 Robin Hill isolated chloroplasts from plants and demonstrated that in the presence of light, water, and an artificial electron acceptor such as potassium ferricyanide, chloroplasts would evolve O_2. To this day, oxygen evolution by chloroplasts in the light is called the Hill reaction in honor of Robin Hill's discovery. The process can be depicted as follows.

$$2H_2O \longrightarrow O_2 + 4H^+ + 4e^-$$

With the discoveries in the 1950s by Allen that properly prepared chloroplasts would fix CO_2 in the light and by Arnon that chloroplast membranes would reduce NADP to NADPH in the light, there remained for Calvin (also in the 1950s) to establish the individual steps for the flow of carbon from CO_2 to carbohydrate in photosynthesis. The discovery and elucidation of the latter pathway, the carbon reduction cycle or Calvin cycle, was the impetus for the awarding of the 1961 Nobel prize in chemistry to Melvin Calvin. These studies by Allen, Arnon, Calvin, and others led to our modern concepts of photosynthesis.

15.2 The Definition of Photosynthesis

In the most straightforward sense, photosynthesis can be defined as the conversion of light energy into useful chemical energy. Martin Kamen (1963) defined photosynthesis as a series of processes in which electromagnetic energy is converted to chemical free energy that can be used for biosyntheses. Radiant energy is trapped by pigments of green plants and used to reduce atmospheric CO_2 to sugars, which in turn can be oxidized, releasing energy in a useful form for the growth, development, and maintenance of living organisms.

Overall, photosynthesis can be depicted as follows.

$$6CO_2 + 12H_2O \xrightarrow[\text{chlorophyll}]{hv} C_6H_{12}O_6 + 6O_2 + 6H_2O$$

Electrons from the hydrogen of water are used to reduce CO_2 to sugar (glucose, $C_6H_{12}O_6$). Through a series of photochemical reactions, the light energy elevates the energy status of electrons from water to the redox potential of H_2, which is sufficient to reduce CO_2 to carbohydrate. It is known that the oxygen produced in the photosynthetic reaction comes from water; thus 12 water molecules, each with an oxygen atom, must participate as substrates to yield six diatomic O_2 molecules.

In this chapter the photochemical aspects of photosynthesis will be developed, followed by aspects of chloroplast physiology such as envelope permeability and chloroplast development. In the subsequent chapter (Chapter 16) the carbon metabolism of photosynthesis will be discussed.

15.3 The Study of Photosynthesis

Although most scientists would largely agree to the accuracy of the above definition of photosynthesis, that is, the conversion of radiant energy to useful chemical energy by green plants, the term photosynthesis may be visualized in different ways. As discussed by Kamen (1963), scientists study photosynthesis in the context of the time span of the various photosynthetic events.

During the period 10^{-15} to 10^{-9} second or, under some special circumstances, 10^{-5} second, the primary reactions of photosynthesis occur. This area is within the realm of the physicists and photochemists. During this initial period, a photon is absorbed by a pigment, usually chlorophyll, resulting in an excited pigment molecule. This excitation usually lasts for about 10^{-9} second but can be stabilized to about 10^{-5} second. A simple equation describing the excitation period follows.

$$CHL + light \longrightarrow CHL*$$

The energy associated with the excited chlorophyll molecule (CHL*) can be either transferred to another molecule and conserved or lost by radiation (heat or light).

During the next period—over the time span of 10^{-9} to 10^{-4} second—the energy of the excited chlorophyll molecule is conserved by transfer to another substance and stabilized. This reaction can be depicted as follows.

$$CHL* + X \longrightarrow CHL + X*$$

We can visualize that X* is a compound with a sufficiently negative potential to reduce NADP.

$$X* + NADP \longrightarrow X + NADPH$$

Summation of the above three equations gives the last equation.

$$CHL + light \longrightarrow CHL*$$
$$CHL* + X \longrightarrow CHL + X*$$
$$X* + NADP \longrightarrow X + NADPH$$

$$\overline{Light + NADP \longrightarrow NADPH}$$

The above reactions require a source of electrons; ordinarily it is water, as illustrated in the following balanced equation.

$$NADP^+ + H_2O + light \longrightarrow NADPH + H^+ + \tfrac{1}{2}O_2$$

The reducing agent (i.e., NADPH) formed from the photochemical events described above is used to reduce CO_2 to an organic compound such as glucose, conserving the free energy.

$$6CO_2 + 12NADPH + 12H^+ \longrightarrow$$
$$C_6H_{12}O_6 + 12NADP^+ + 6H_2O$$

Carbon dioxide reduction occurs over the time span of 10^{-4} to 10^1 seconds and is limited by the turnover time of enzymatic reactions. This is within the realm of biochemistry and is clearly identical to the slow, dark reactions first identified by Blackman as early as 1905.

For the average enzymatic reaction there are about 2000 to 10,000 molecules reacted per minute. The fastest reactions take place in about 10^{-4} second; the more complex, slower reactions may take as long as 10^1 seconds. If we assume an average turnover time of 6000 events per minute there would be 6000 per 60 seconds or 100 per second, one event for each 10^{-2} second.

The final series of processes in photosynthesis includes the conversion of the organic compounds synthesized in the above reactions into cellular constituents such as cell-wall material, membranes, and other cytoplasmic substances. These events can be visualized to occur over the time span of 10^1 to 10^3 seconds and up to the lifetime of the plant. It includes the subsequent areas of growth and development and can be depicted as follows.

$$(CH_2O)n \longrightarrow Cell\ constituents$$

15.4 Photosynthetic Pigments

There are several pigments known to be components of the photosynthetic apparatus. A pigment is defined as a molecule that absorbs light, and, of course, there can be no photochemical reactions without light absorption. Therefore, the pigments are some of the most important chemicals of photosynthesis. The light-absorbing portion of a pigment is called the chromophore.

The pigments involved in the light-trapping reactions of photosynthesis are the chlorophylls and the accessory pigments such as carotenoids. Pigments such as the flavoproteins, plastocyanin, the cytochromes, ferredoxin, and the quinones, which are also found in chloroplasts, are probably not directly involved in light gathering but rather in electron transport.

Chlorophyll Chlorophyll is the most important pigment in the light-absorption process of photosynthesis. The structure and many aspects of the photochemistry of chlorophyll are discussed in Chapters 3 and 12. In review, chlorophyll is a magnesium porphyrin attached to protein. Chlorophylls that occur in photosynthetic organisms are chlorophyll *a*, chlorophyll *b*, chlorophyll *c*, chlorophyll *d*, bacteriochlorophyll, and several other derivatives. They can be distinguished chemically by the various substitutions on the porphyrin ring and by their different absorption spectra. The chlorophyll *a* molecule is the primary absorbing pigment of green plants, although chlorophyll *b* frequently occurs with chlorophyll *a*. However, mutants are known that lack chlorophyll *b* but that can still photosynthesize. The absorption spectra of chlorophyll *a* and chlorophyll *b* are compared in Fig. 15.1.

There is much evidence, explained in Chapter 12, indicating that chlorophyll *a* occurs in more than one form. Experiments demonstrate that the different forms, primarily identified on the basis of small differences in absorption spectra, are the result of the chlorophyll *a* molecule being attached to different proteins and not because of substitutions on the porphyrin. The attachment, which is largely through hydrophobic bonding of the phytol chain to hydrophobic portions of the chlorophyll proteins, creates different environments for the chlorophyll, resulting in absorption spectrum changes.

Chlorophyll *a* is the primary energy trap in photosynthesis. As will be explained subsequently, there are two primary reaction centers of photosynthesis, called photosystem I and photosystem II. Each has chlorophyll *a* and accessory pigments, including chlorophyll *b*, carotenoids, and perhaps other pigments. The ultimate energy trap of each photosystem is a specialized chlorophyll *a* molecule that absorbs furthest into the red end of the spectrum (lowest energy). The absorption maximum

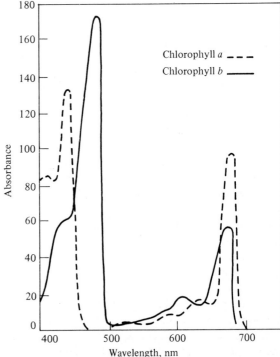

FIGURE 15.1 Absorption spectra for chlorophylls *a* and *b*. Although the spectra are somewhat different, both have major absorbance peaks in the blue and red region of the spectrum. Redrawn from data of J. H. C. Smith and A. Benitez. 1955. Chlorophylls: Analysis in Plant Materials. In K. Paech and M. V. Tracey (eds.), *Modern Methods of Plant Analysis.* Springer-Verlag, New York.

for the chlorophyll *a* of photosystem I is about 700 nm and for photosystem II about 680 nm. Other chlorophyll *a* molecules associated with these primary traps have absorption spectra shifted toward the blue (higher energy). They function as "antennae" pigments, absorbing light and ultimately transferring energy by inductive resonance to the primary traps, i.e., to the pigment 700 of photosystem I and the pigment 680 of photosystem II. The extent to which chlorophyll *b* and the other accessory pigments (carotenoids) function in light gathering and transfer to chlorophyll *a* is not known. These accessory pigments may just function as protection by preventing oxidation of the main light gatherers.

There are two main requirements for effective energy transfer after light absorption from the antennae pigments to the primary traps. First, the pigments must be in close proximity to allow interaction and transfer of energy by inductive resonance. Second, the absorption spectrum of the receiving molecule must overlap the fluorescent emission spectrum of the antennae molecules. Since the fluorescent emission spectra are shifted toward a longer wavelength (toward the red) with respect to the absorption spectra, there is a loss (decrease) of energy during the transfer. The ultimate trap, the specialized chlorophyll *a* molecules, have absorption spectra that are shifted furthest toward the red. This can be visualized to some extent by noting that the major absorption peak in the red for chlorophyll *a* is at about 663 nm (refer to Fig. 15.1) but the spectrum shows evidence of shoulders. Curve analysis of the absorption spectrum indicates that it is the result of several chlorophyll *a* molecules, each with different absorption spectra (see Fig. 12.12). According to this kind of analysis, the pigments with the peaks shifted toward the blue would be the antennae pigments and the pigment with the peak furthest toward the red would be the primary trap.

Chlorophyll in solution will readily fluoresce. However, the fluorescence from the photosystem I pigment is weak whereas the photosystem II pigment is strong and a good indicator of energy flow.

The fluorescent yield—the photons emitted by fluorescence per photons absorbed—is an indicator of the efficiency of energy transfer. When there is quenching (low fluorescent yield), there are trapping centers that receive the energy. When fluorescent yield is high, energy is lost. For this reason, studies of fluorescent yield are useful for an understanding of energy flow in photosynthesis. Fluorescent yield is also used to monitor photosynthesis during experimental procedures.

As stated above, there is evidence from mutants that lack chlorophyll *b* indicating that chlorophyll *b* is not necessary for photosynthesis to occur. Furthermore, it is not necessary for the structural organization of chloroplasts. Thus the two possible biochemical roles for chlorophyll *b* are light gathering as an antennae pigment and protection of the photosynthetic apparatus by absorption of excess light.

The chlorophylls are components of the photosynthetic membranes, but their exact orientation is not known with any certainty. It seems reasonable to assume that the phytol chain, which gives the hydrophobic properties to chlorophyll, orients in the lipid bilayer of the membrane in the same manner as the hydrophobic portions of the lipids (see Fig. 1.3). Evidently the porphyrin portion is exposed to the outside of the bilayer. Being hydrophobic, the protein portion is most likely an integral protein of the chloroplast membranes.

Carotenoids The carotenoids, which are carotenes and their oxidation products, the xanthophylls, are linear tetraterpenoids found abundantly in chloroplasts. Their chemistry was discussed in Chapter 12. Because of their association with chlorophyll in chloroplasts, they are assumed to be accessory pigments for gathering light and perhaps for transferring energy to chlorophylls. They also play a role in protection against photooxidation demonstrated by severe light-induced injury in mutants that lack carotenoids.

Biliproteins The biliproteins, or bile pigments, contain linear tetrapyrroles and are found in the photosynthetic machinery of many algae. They include the well-known pigments phycocyanin (blue) and phycoerythrin (red). They are not known to occur in higher green plants except for the non-

photosynthetic pigment phytochrome. Like the carotenoids, they apparently function as accessory light-absorbing pigments. The biliproteins of algae are usually called phycobiliproteins.

15.5 The Photochemistry of Photosynthesis

In the simplest terms, we can summarize the overall process of photosynthesis with the following unbalanced equations.

$$CHL + light + NADP + ADP + P_i \longrightarrow$$
$$ATP + NADPH + CHL$$

$$CO_2 + H_2O + NADPH + ATP \longrightarrow$$
$$CH_2O + NADP + ADP + P_i$$

The first reaction, which produces ATP and NADPH, makes up the photochemical reactions of photosynthesis, while the second reaction, which produces carbohydrate, is the thermal reaction. The remainder of this chapter will specifically cover the photochemical reactions of photosynthesis.

There are two laws of photochemistry (discussed in Chapter 2) that are important for our understanding of the photochemistry of photosynthesis. These laws are the following.

1. The Grothus−Draper law

2. Einstein's Law of the Photochemical Equivalent

The Grothus−Draper law states that for a quantum (or photon) to be used in a photochemical reaction, the quantum must be absorbed. Einstein's Law of the Photochemical Equivalent states that for each photochemical reaction that occurs, a single photon must be absorbed. This law implies that photons cannot be saved or stored to bring about a primary reaction but that the photon must have sufficient energy associated with it to cause the reaction.

If an absorption spectrum of chlorophyll is compared with an action spectrum of photosynthesis, it is observed that both have major peaks in the blue at about 430 nm and in the red at about 670 nm (see Fig. 15.2). Since light absorption by chloro-

phyll corresponds closely with the action spectrum of photosynthesis, we can assume that the light absorbed by chlorophyll is used in the primary photochemical act of photosynthesis. Thus it is quite clear that the conditions of the Grothus−Draper law are met and that photons are absorbed during photosynthesis.

To evaluate Einstein's law, we note that the electrochemical span for photosynthesis goes from the most oxidized component, molecular oxygen (O_2), to the most reduced component, equivalent to the redox potential of molecular hydrogen (H_2). The standard potential ($E^{\circ\prime}$) at the hydrogen electrode is -0.42 volts; at the oxygen electrode it is $+0.82$ volts. This represents a voltage difference of 1.2 electron volts. Since there are 96 kilojoules per electron volt, a span of 1.2 volts is equivalent to 115.5 kJ.

One mole (einstein, E) of red light (700 nm) has about 167 kJ. Thus if the process is at least 70% efficient, there is sufficient energy in red light to drive the photochemical reactions of photosynthesis. As will be explained subsequently, however, more than one photon is involved in the process through two different light systems. It should be recalled that red light is the least energetic of the light involved in photosynthesis. This simple analysis makes it clear that the criterion of Einstein's law is met in photosynthesis. That is, there is sufficient energy associated with red-light photons to bring about the reactions.

As explained in Chapter 3, when light energy interacts with an absorbing molecule, several events may occur. Fig. 3.7 illustrates some of these for the chlorophyll molecule and should be reexamined at this time. The chlorophyll a absorption spectrum has two major peaks, one in the blue at about

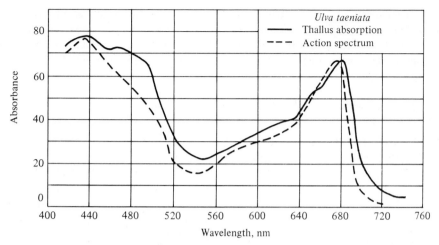

FIGURE 15.2 Absorption spectrum of an intact thallus of the green alga *Ulva taeniata* and action spectrum for photosynthesis by the same tissue. For experimental reasons the action-spectrum peak in the red was adjusted to 675 nm to correspond with the absorption-peak maximum. After adjustment the two curves correspond very well, indicating that most of the pigments including chlorophylls function in photosynthesis. Between 450 and 510 nm there is some absorption that evidently does not correspond to the photosynthetic action spectrum, indicating that some of the absorbing carotenoids do not function in the light gathering of photosynthesis. From F. T. Haxo and L. R. Blinks. 1950. Photosynthetic action spectra of marine algae. *J. Gen. Physiol.* 33:389–422. By copyright permission of the Rockefeller University Press.

430 nm and the other in the red at about 670 nm. The blue absorption band corresponds to the second singlet state. This state will return to the first singlet state (the red absorption band), with energy dissipated as heat. Thus in photosynthesis, blue light acts like red light. This process, allowing blue to act like red during the light reactions of photosynthesis, is referred to as the "blue–red transition." In the subsequent discussions of the photochemistry of photosynthesis, only red and far-red light are considered.

As was just explained, there is sufficient energy in one einstein of red light (700 nm) to drive the reaction from O_2 to H_2 (or from H_2O to O_2) but not enough to allow the synthesis of carbohydrate from CO_2.

In a series of experiments designed to study the quantum requirement for photosynthesis, Robert Emerson in 1957 found that the efficiency of monochromatic light decreased greatly as the wavelength increased above 800 nm. But if such long wavelengths were supplemented with low levels of light with wavelengths below 680 nm, the quantum efficiency could be restored. This phenomenon is referred to as the Emerson Enhancement Effect and is described in the next section.

15.5.1 The Emerson Enhancement Effect

As mentioned in the preceding section, even though there is sufficient energy associated with an einstein of red light (700 nm) to bring about the oxidation of H_2O to molecular O_2, there is insufficient energy to reduce CO_2 to the level of carbohydrate.

$$CO_2 + H_2O \longrightarrow CH_2O + O_2$$

The energy requirement here is about 494 kJ mol^{-1}, meaning that the minimum quantum requirement for reduction would be about 3 (494/167). In most studies the actual quantum requirement for CO_2 reduction is on the order of 8 to 10.

Inspection of the absorption spectrum for chlorophyll and the action spectrum for photosynthesis, both shown in Fig. 15.2, indicates that at long wavelengths (over 700 nm) there is still absorption of light but photosynthesis decreases rapidly, faster than the decrease in light absorption. During investigations of this "red drop" phenomenon, Emerson helped elucidate the quantum requirement for photosynthesis. He noticed that the quantum requirement for photosynthesis remained reasonably low below 680 nm but increased greatly above 680 nm. When plants were irradiated with low amounts of light lower than 680 nm plus low amounts of light higher than 680 nm, there was more than an additive increase in photosynthesis,

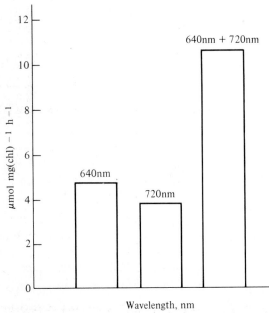

FIGURE 15.3 Experiment to illustrate the Emerson enhancement effect. There is more photosynthesis (measured by CO_2 fixation in intact chloroplasts) when 640 nm light and 720 nm light are used together than when they are used separately. 640 nm = 4.7 μmol mg^{-1} h^{-1} and 720 nm = 3.8 μmol mg^{-1} h^{-1} for a total of 8.5 μmol mg^{-1} h^{-1}. When both wavelengths are used together the rate is 10.6 μmol mg^{-1} h^{-1}, nearly 25% greater. From an experiment by D. G. Peavey and M. Gibbs. 1975. Photosynthetic enhancement studied in intact spinach chloroplasts. *Plant Physiol.* 55: 799–802.

and there was also an increase in the overall quantum efficiency.

An experiment illustrating the enhancement effect is illustrated in Fig. 15.3. In this experiment chloroplasts were irradiated with 640 nm light and 720 nm light. Photosynthesis was determined by measuring $^{14}CO_2$ fixation. When the same wavelengths of light were applied simultaneously rather than separately, total photosynthesis was increased by 25%. Thus it can be shown that low levels of red and far-red light given simultaneously result in a synergistic effect when compared with the same amount of light given separately or sequentially.

These kinds of experiments, in which light of less than 680 nm and greater than 680 nm when given together resulting in synergism, led to the suggestion that the photochemistry of photosynthesis must include at least two light systems working together, one sensitive to red light and the other sensitive to far-red light.

The two light systems for the photochemical reactions of photosynthesis, partly based on the "red drop" phenomenon and the Emerson Enhancement Effect, are described in detail in the next section. As will be seen, both red and far-red light are required to drive two cooperating photosystems that result in O_2 evolution, ATP formation, and the reduction of NADP to NADPH.

15.5.2 The Z scheme for the photochemical reactions of photosynthesis

The Z scheme for photosynthesis is outlined in Fig. 15.4. There is assumed to be two light-gathering photocenters that are an integral part of two systems, referred to as photosystem I and photosystem II. Photosystem II has the light-gathering center that functions in the Hill reaction. Here water is oxidized ("split") to form molecular oxygen as electrons are removed. Photosystem II forms a very strong oxidant and is most sensitive to red light. Photosystem I is the sequence of reactions most sensitive to far-red light, which reduces NADP to NADPH and hence forms a strong reductant. The two photosystems are coupled by a series

THE Z SCHEME FOR ELECTRON TRANSPORT IN PHOTOSYNTHESIS

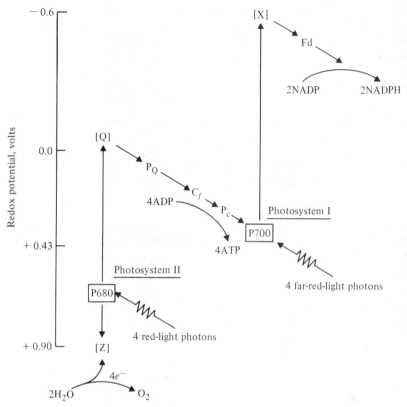

FIGURE 15.4 The Z scheme for the photochemistry of photosynthesis, showing the two photosystems acting together to produce ATP and reduce NADP to NADPH. Eight photons, four in each photosystem, produce one O_2, four ATP, and two NADPH in noncyclic electron flow. P_Q = plastoquinone. C_f = cytochrome f. P_c = plastocyanin. P680 and P700 are special chlorophyll a molecules at the reaction centers of the two photosystems. Q = the weak reductant of photosystem II that may be plastoquinone. Z = the oxidant produced in photosystem II that is sufficiently strong to oxidize water. X = the strong reductant of photosystem I that could be a ferredoxin. F_d = ferredoxin. The weak oxidant of photosystem I that oxidizes the reduced Q is believed to be P700.

of electron-transport components similar to those of the electron-transport system of mitochondria. The coupling sequence of electron carriers produces ATP.

Whereas red light is sufficiently energetic to drive both photosystem I and photosystem II, far-red light, being less energetic than red light, will only drive photosystem I.

Within the electron-transport system of the mito-

chondria there are flavoproteins, a quinone, and cytochromes of the a, b, and c type. Chloroplast membranes involved in electron transport also contain similar electron-transport components. There are flavoproteins, quinones, and cytochromes of the b and c type. The c type cytochrome of the electron-transport system of chloroplasts is called cytochrome f. The f stands for fronds (a leaf). In addition to these components there is also a copper-containing protein, plastocyanin, and nonheme iron proteins such as ferredoxin. Ferredoxin or another iron–sulfur protein is believed to be the primary electron acceptor of photosystem I.

Photosystem II (oxygen evolution)

Photosystem II is the light-gathering center that contains a short-wavelength-absorbing form of chlorophyll *a* (P680); it is most sensitive to a wavelength maximum of about 684 nm. The center also has chlorophyll *b*, but chlorophyll *a* is the primary trap and *b* functions more as an accessory pigment. In addition to the special light-trapping form (P680), there are about 200 chlorophyll molecules associated with the light-gathering center. The center also has about 50 carotenoid molecules, about four plastoquinones, two cytochrome *b*, and six manganese. In addition to the known components the center has an unidentified electron donor called Z (this is the strong oxidant) and an unidentified electron acceptor called Q (the weak reductant). Q may be plastoquinone. The photosystem II center has a requirement for chloride in addition to manganese and is inhibited by dichlorophenyl dimethyl urea (DCMU), a herbicide.

The reaction of the system, which produces charge separation, can be depicted as follows.

$$Z^+ \longleftarrow \boxed{\text{II}} \longrightarrow Q^-$$
$$\uparrow 680 \text{ nm}$$

A strong oxidant (Z^+) is produced at the center with a redox potential ($E^{\circ\prime}$) of about +0.81 volts, and a weak reductant (Q^-) is produced with a potential close to 0.0 volts. The strong oxidant is sufficiently strong to oxidize water, resulting in the evolution of molecular oxygen.

$$2H_2O \longrightarrow 4H^+ + 4e^- + O_2$$

The oxidized center may be a special kind of chlorophyll *a* with an absorption maximum at 684 nm. This chlorophyll has been studied and can be referred to as P680, the P standing for pigment. Al-though it has not been isolated or characterized in any detail, it is analogous to the more fully characterized center of photosystem I referred to as P700.

We do not know the nature of the weak reductant referred to as Q^-. It is the primary electron acceptor of photosystem II and is the compound that quenches fluorescence. The component with the lowest known potential is plastoquinone, with a redox potential of about 0.0 volts. Even though the system will not function if the plastoquinone is removed, there is some evidence that Q^- and plastoquinone are not the same. Alternatively, cytochrome b_6, with a potential of -0.06, could be Q^-. Further experimentation will be necessary before the exact nature of Q is ascertained.

Once electrons are stabilized with the weak reductant, Q, they are passed on through the electron-transport chain to the other light-gathering center of photosystem I. The electron "hole" created at the center of photosystem II is "filled" with electrons from water during the water-splitting process. Thus water is the natural electron donor for photosynthesis.

Standard redox potentials for the various components of the photochemical systems of photosynthesis are given in Table 15.1.

TABLE 15.1 Oxidation–reduction potentials ($E^{\circ\prime}$) of some compounds important in photosynthesis; potentials are in volts

Viologen dye	-0.55
Ferricyanide	-0.44
Ferredoxin	-0.43
Hydrogen	-0.42
NADPH (NADH)	-0.32
Cytochrome b_6	-0.06
Plastoquinone	0.00
Phenazine methosulfate (PMS)	$+0.08$
2,6-dichloroindophenol (DCPIP)	$+0.22$
Cytochrome f	$+0.37$
Plastocyanin	$+0.37$
P700	$+0.43$
Oxygen	$+0.82$

Photosystem I (NADP reduction)

The light-gathering center of photosystem I has a special chlorophyll *a* with an absorption maximum at 700 nm. This form of chlorophyll *a* is called P700. The center, like the center of photosystem II, has about 200 chlorophyll molecules of which one (or perhaps two) is of the P700 type. Chlorophyll *a* molecules other than P700 apparently function in light gathering and eventual transfer of energy to the P700 molecule. Transfer of energy is presumably by molecular collision. The center also has about 50 carotenoid molecules, one cytochrome *f*, one plastocyanin, two cytochrome *b*, and membrane-bound ferredoxins. The plastocyanin is a low-molecular-weight (about 21,000) copper protein with a redox potential of +0.37 volts.

Light reactions at the center of photosystem I create a weak oxidant (P700$^+$) and a strong reductant (X$^-$). The reaction can be depicted as follows.

$$P700^+ \longleftarrow \boxed{I} \longrightarrow X^-$$
$$\updownarrow 700 \text{ nm}$$

We do not know for certain that P700 is the actual weak oxidant formed with a redox potential of +0.43 volts, but such a hypothesis is consistent with much of the available evidence.

The nature of the strong reductant (primary electron acceptor, X) is not known. There is some evidence that it may be membrane-bound ferredoxin ($E^{o\prime} = -0.43$ volts). In any case, the strong reductant produced in photosystem I is a sufficiently strong reductant to reduce NADP to NADPH. The electron "hole" created at the center of photosystem I is filled by electron transfer from photosystem II via the coupling with the photosynthetic electron-transport system. Overall, the path of energy in the form of electron transfer is from water at photosystem II through the electron-transport system to photosystem I and thence to NADP to form NADPH.

One of the most important features of the photochemistry of photosynthesis is the ultimate trapping of the usable energy in the form of NADPH. We can visualize the process occurring through the reduction of ferredoxin by the unknown compound X$^-$ of Fig. 15.4. Ferredoxin, a nonheme iron protein, is one of the most electronegative of all naturally occurring biological compounds. It is a low-molecular-weight protein (about 12,000) with two moles of sulfur and two moles of iron. It is loosely bound to the chloroplast membranes and easily lost when chloroplast membranes are prepared for study. The $E^{o\prime}$ of ferredoxin is about −0.43 volts.

Actual reduction of NADP to NADPH takes place through ferredoxin and is catalyzed by a flavoprotein called ferredoxin-NADP reductase.

$$\text{Photosystem I} \longrightarrow X^- \longrightarrow$$
$$\text{Ferredoxin} \xrightarrow[\text{Fd-NADP reductase}]{\overset{\displaystyle \text{NADP}}{}} \text{NADPH}$$

The reductase is an integral component of the chloroplast membranes.

There are a variety of artificial (nonnatural) electron acceptors such as indophenol, ferricyanide, phenazine methosulfate, and several quinones that can be reduced by photosystem I. In addition, such naturally occurring compounds as cytochrome *c* can be reduced in photosystem I by ferredoxin. The latter can occur with the ferredoxin-NADP reductase.

The coupling of photosystems II and I (electron transport)

The electron-transport system that couples photosystems II and I is not fully understood but in its simplest form is as follows.

$$H_2O \longrightarrow \boxed{P680} \longrightarrow \text{Plastoquinone} \longrightarrow$$
$$\begin{matrix} \text{Cytochrome } f \\ \text{and/or} \\ \text{Plastocyanin} \end{matrix} \longrightarrow \boxed{P700} \longrightarrow \text{NADP}$$

There is a redox potential decrease from photosystem II to photosystem I of about 0.43 volts, sufficient to produce ATP. And it is, in fact, this drop in potential as electrons flow from system II to system I that brings about photophosphorylation. Recalling that the hydrolysis of ATP yields about 30 kJ mol^{-1} and that a potential of 0.43 volts is equivalent to about 41 kJ (96 kJ per electron volt $\times$ 0.43), there is sufficient energy for at least one ATP.

The evidence for the ordering of the components of the electron-transport system is somewhat controversial, but some rather elegant experiments by Gorman and Levine (see Levine, 1969) can be used to illustrate the kinds of experiments and evidence that have been used.

Gorman and Levine found two photosynthetic mutants in *Chlamydomonas,* a green alga. One lacked cytochrome *f* and the other lacked plastocyanin. Both required acetate as an energy source for growth. The designated mutants are Ac-206 (lacks cytochrome *f*) and Ac-208 (lacks plastocyanin). In many cases, electron carriers can be ordered by their respective oxidation–reduction potentials, but here cytochrome *f* and plastocyanin have redox potentials of about +0.37 volts. Of course, it would not be possible to decide if the carriers were in series or in parallel from their redox potentials since they are the same.

With the Ac-206 mutant there was no NADP reduction when water was the electron donor; that is, the Z scheme shown in Fig. 15.4 would not operate, presumably because of the block due to the absence of the cytochrome *f* component. However,

when an artificial electron donor, dichlorophenol indolphenol (DCPIP), was added to the algal suspension, NADP was reduced. It was concluded that DCPIP added electrons to the electron-transport system on the photosystem I side of cytochrome *f.* If this were not true, electrons would not be transferred to the P700 center in the absence of cytochrome *f.* In addition, light will cause reduction of cytochrome *f* in the Ac-208 mutant, indicating a block between cytochrome *f* and P700 because of the absence of plastocyanin.

With the Ac-208 mutant lacking plastocyanin, DCPIP will not result in NADP reduction. It was concluded that DCPIP electrons entered on the system II side of plastocyanin, and the absence of plastocyanin prevented transfer to P700. Moreover, light will cause reduction of cytochrome *f* in the Ac-208 mutant, indicating a block between cytochrome *f* and P700 because of the absence of plastocyanin. If cytochrome *f* were on the system I side of plastocyanin, light would cause oxidation, not reduction, of cytochrome *f.* Thus the evidence from the above experiments indicates that cytochrome *f* and plastocyanin are in series and that cytochrome *f* is located on the system II side of plastocyanin.

Actually we cannot be certain about the ordering of the electron-transport components from such experiments, and until definitive evidence about ordering becomes available it is perhaps best not to place cytochrome *f* and plastocyanin in sequence. Figure 15.4 has cytochrome *f* and plastocyanin arranged in sequence, not in parallel.

15.6 Photophosphorylation

In many respects photophosphorylation is similar to oxidative phosphorylation in that a series of electron-transport coupling factors (proteins) transfer electrons along an electrochemical gradient, yielding ATP. As with oxidative phosphorylation, we do not completely understand the mechanism of phosphorylation, although most researchers favor the chemiosmotic hypothesis of

Mitchell over the chemical coupling hypothesis (refer to Chapter 14). Indeed, much of the evidence for the chemiosmotic hypothesis comes from work with chloroplast membranes.

Unlike mitochondria, chloroplast membranes "pump" protons to the inside rather than to the outside during light-driven electron flow. Thus the pH of the medium tends to increase and become alka-

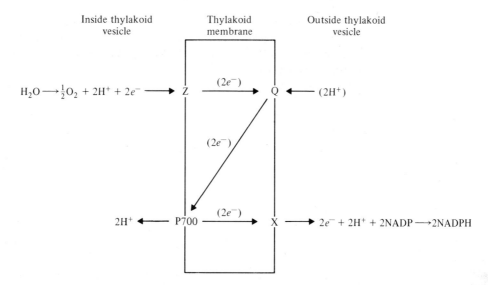

Inside thylakoid vesicle   Thylakoid membrane   Outside thylakoid vesicle

$H_2O \longrightarrow \frac{1}{2}O_2 + 2H^+ + 2e^- \longrightarrow$ Z $\xrightarrow{(2e^-)}$ Q $\longleftarrow (2H^+)$

$(2e^-)$

$2H^+ \longleftarrow$ P700 $\xrightarrow{(2e^-)}$ X $\longrightarrow 2e^- + 2H^+ + 2NADP \longrightarrow 2NADPH$

FIGURE 15.5 Hypothesis for the establishment of a proton gradient in chloroplasts during electron flow. Electron transport from water through Z to Q, then to P700, to X, and finally to NADP, forming NADPH, creates a proton gradient across the thylakoid membrane with sufficient potential to bring about phosphorylation (here called photophosphorylation). For each electron transported from inside the thylakoid to the outside there are two protons transported from outside to inside. Thus the outside becomes alkaline. Symbols are the same as in Fig. 15.4. The concept is after P. Mitchell. 1966. Chemiosmotic coupling in oxidative and photosynthetic phosphorylation. *Biol. Rev.* 41:445–502.

line rather than decrease, as is the case with mitochondria. It is visualized that the protons are transported into the spaces within the thylakoids (vesicles). Jagendorf and Uribe (1966) were able to show with isolated chloroplast membranes that if a pH gradient was established across the lamellae with the use of organic acids, the membranes would catalyze phosphorylation in the dark. This is perhaps the most convincing evidence for the chemiosmotic hypothesis.

In this experiment they equilibrated the chloroplast lamellar membranes at about pH 4 by suspension in an organic acid solution and then quickly added ADP and phosphate while increasing the pH to about 8. Significant amounts of ATP were produced in the dark.

Figure 15.5 illustrates the chemiosmotic hypothesis of Mitchell for the establishment of the proton gradient during electron transport and photophosphorylation.* Electron-transport components, including the strong oxidant Z and the weak reductant Q of photosystem II and the weak oxidant of photosystem I (P700) and the strong reductant X, are positioned such that electron flow produces uptake of protons on the outer side of the membranes and secretion of protons on the inner side. Thus there is a proton gradient established across the chloroplast membranes during electron transport.

The concept of "sideness" for chloroplast lamellar membranes is rather well established. Through studies done primarily with specific antibodies it was shown that the water-splitting process that

*Figure 15.5 should be studies in conjunction with the chloroplast lamellar diagram in Fig. 15.6.

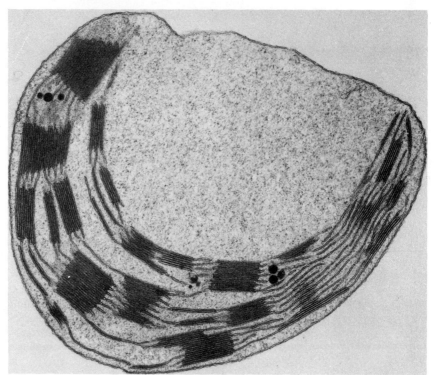

FIGURE 15.6 Detailed drawing of the probable chloroplast membrane structure.
The drawing can be com-
pared with the electron
micrograph of a tobacco-
leaf chloroplast, prepared
by B. S. Julich and W. M.
Laetsch. The envelope, or
limiting membrane, regu-
lates transport between the
cell and the chloroplast. The
photochemical reactions of
photosynthesis discussed in
this chapter occur within
and on the thylakoid mem-
branes. There is evidence
that for active functioning
of photosystem II, grana
composed of stacks of thylakoids must be present.
Photosystem I evidently
occurs on the frets (stromal
lamellae) as well as on the
granal membranes. The Hill
reaction (oxygen evolution)
occurs within the thylakoid
vesicles, and NADP reduc-
tion occurs on the outside of
the vesicles. The dark (carbon metabolism) reactions,
discussed in the next chapter
(Chapter 16), takes place
within the stromal matrix.

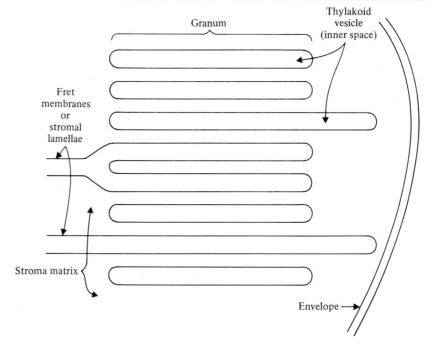

evolves O_2 occurs toward the inside of the thylakoid vesicles, and NADP reduction occurs toward the outside, or stromal, side in the presence of the carbon reduction cycle enzymes (see Figs. 15.5 and 15.6).

An as yet unresolved question is the P:O ratio or, in the case of photosynthesis, the $P/2e^-$ ratio for photophosphorylation. For a long time it was assumed that the $P/2e^-$ ratio was one, i.e., one ATP produced for each two electrons transported. This would, of course, correspond to $\frac{1}{2}O_2$ production. Many careful estimates of the $P/2e^-$ ratio result in values of about 1.3 to 1.5. If we take the chemical coupling hypothesis literally and assume specific phosphorylation sites, $P/2e^-$ ratios of other than whole numbers suggest some losses or errors in measurement. Thus a $P/2e^-$ ratio of 1.3 was interpreted as a ratio of two, that is, two ATP per two electrons, or one for one.

Fractional ratios are somewhat easier to understand with the chemiosmotic hypothesis since it does not necessarily involve specific sites. Nevertheless, our conclusion is that the correct $P/2e^-$ ratio is probably two and that there are two ATP produced for each electron pair transported, corresponding to the production of $\frac{1}{2}O_2$. This conclusion does not imply two phosphorylation sites and may mean one site requiring two electrons or perhaps no specific sites.

In support of the chemiosmotic coupling hypothesis for photophosphorylation in chloroplasts, there are estimates of a proton-motive force expressed as $\triangle G^{o\prime}$ across the chloroplast membrane of about 21 kJ mol^{-1} during electron flow. With an assumed $3H^+$ to 1ATP ratio (which is close to experimental), there would be about 63 kJ mol^{-1} ($3 \cdot 21 = 63$), sufficient to drive ATP synthesis.

$$ADP + P_i \longrightarrow ATP \ (\triangle G^{o\prime} = +30 \text{ kJ mol}^{-1})$$

A coupling factor has been isolated from chloroplasts that has the properties of an ATP synthesizing enzyme. The enzyme, which has five subunits, is believed to couple the energy formed by the establishment of proton gradients with phosphorylation. The reaction, which is driven by light or a pH gradient, requires magnesium.

$$ADP + P_i \longrightarrow ATP + H_2O$$

On treatment with trypsin, a proteolytic enzyme, the coupling factor has ATPase activity.

Overall, for each two H_2O oxidized, which yields one O_2 and four electrons, there can be a yield of four ATP. Since two electrons are required to reduce each NADP to NADPH, two NADPH can be generated by the oxidation of the two H_2O. For this there is a requirement of four photons to drive each photocenter, or a total of eight photons. Thus oxidation of two moles of water, yielding one mole of O_2, will be accompanied by the production of four ATP and two NADPH, driven by eight photons, four red and four far-red. In the following chapter it will be shown that for the reduction of a mole of CO_2 to sugar there is a requirement for three ATP and two NADPH. The overall equation follows.

$$CO_2 + 2H_2O + 3ATP + 2NADPH \longrightarrow$$
$$CH_2O + O_2 + 3ADP + 2NADP + 2H^+ + 3P_i$$

15.6.1 The quantum requirement or yield

The quantum yield of photosynthesis has been debated for some time. As explained previously, in photochemical reactions one quantum interacts with one pigment molecule, causing the excitation of one electron. Quantum yield refers to the number of quanta absorbed per reaction.

15.6.2 Noncyclic and cyclic photophosphorylation

The concept of noncyclic photophosphorylation implies electron flow from water, the natural electron donor for photosynthesis, to NADP, the ultimate acceptor. To this point we have been describing electron flow and photophosphorylation in noncyclic terms. Electrons are produced by oxidation of water and transported from photosystem II to photosystem I along the electron-transport chain. We visualize the noncyclic electron-flow pathway to be the usual one in plants because the stoichiometry of O_2 evolution to CO_2 fixation is one. Without NADPH production there can be no CO_2 reduction.

It was noted that certain artificial redox components, such as phenazine methosulfate, dichlorophenol indolphenol, and diaminodurene, when added to a suspension of properly prepared chloroplasts would catalyze the light-dependent phosphorylation of ADP in amounts greater than that predicted stoichiometrically. Hence Arnon (see Arnon, 1967) proposed that there was cyclic electron flow from the artificial donor through the electron-transport system back to the donor. Far-red light would continuously drive the process, and thus it was similar to photosystem I except that NADP was not the terminal electron acceptor and, of course, no NADPH was produced, only ATP.

Depending on the redox factor added, electrons will "loop" around photosystem I from ferredoxin to plastoquinone to cytochrome f and back to P700. The sequence for cyclic electron flow is illustrated in Fig. 15.7.

Although cyclic electron flow is illustrated as if it were a part of noncyclic electron flow, Arnon has evidence that cyclic flow may be separate and involve yet another photosystem.

There is some evidence that a cytochrome b_6 is involved *in vivo* in cyclic electron flow from ferredoxin. From studies with inhibitors and the oxidation and reduction of the various electron-transport components, it appears that when phenazine methosulfate is the added factor electrons are donated directly to P700. Despite this, high quantities of ATP are formed. When diaminodurene is the added factor, electrons enter on the system II side of plastocyanin. Dibromothymoquinone donates electrons to plastoquinone.

The site or sites for ATP production are not known, but there is sufficient energy drop during cyclic electron flow for one or two phosphorylations. Because DCMU does not inhibit cyclic electron flow, photosystem II is assumed not to be a component.

Even though cyclic electron flow can be demon-

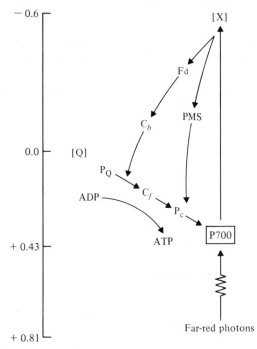

FIGURE 15.7 Diagram showing cyclic electron flow that produces ATP but no NADPH. A photosystem similar to, or identical with, photosystem I cycles electrons from P700 to X through ferredoxin and back to the electron-transport chain. The system is driven by far-red photons. When the chemical phenazine methosulfate (PMS) is used as an electron transport agent, P700 is reduced directly, yet ATP is still formed. It is believed that when cyclic electron flow occurs *in vivo*, a special b cytochrome, cytochrome b_6 (C_b), is reduced by ferredoxin and feeds electrons into the chain at cytochrome f.

strated *in vitro* with properly prepared chloroplasts, there is only limited evidence for its *in vivo* occurrence. Cyclic electron flow has gained acceptance because of the cellular requirements for ATP in excess of that produced in noncyclic flow. As explained by Kok (1976), there may be a requirement for about six ATP for the assimilation of carbon from CO_2 into cellular material. The reduction of CO_2 to carbohydrate requires three ATP (via the reductive pentose cycle of Calvin), but estimates indicate that another three ATP would be required for ultimate incorporation of the carbohydrate into cellular material. Here we can visualize the subsequent reactions of protein biosynthesis, membrane lipid biosynthesis, and structural polysaccharide biosynthesis (e.g., cellulose). In addition, as will be outlined in the subsequent chapter on carbon metabolism, C_4 plants may have cyclic electron flow in the agranal, lamellar chloroplasts of the bundle sheath cells.

15.6.3 Inhibitors of photophosphorylation

There are many known inhibitors of photophosphorylation. As in oxidative phosphorylation, many compounds uncouple phosphorylation from electron transport. These uncouplers prevent ATP formation, with the usual result of an increase in electron flow as measured by O_2 evolution. Common uncouplers of photophosphorylation are detergents, antibiotics such as antimycin A, organic amines and hydroxylamine, and ammonium salts. The latter amine and ammonium compounds are known to destroy the pH gradient when acting as uncouplers. Gramicidin, an antibiotic, is an ionophore, which alters membrane permeability such that there is a free exchange of ions resulting in uncoupling. The common oxidative phosphorylation inhibitor dinitrophenol does not uncouple photophosphorylation.

Other compounds will inhibit electron flow by interfering with electron-transport compounds. Phlorizin inhibits the phosphorylation step and does not inhibit electron flow, either the basal rate or the uncoupled rate. Cyanide will also inhibit phosphorylation, probably by preventing electron flow between cytochromes.

Perhaps the most interesting and useful inhibitors of photophosphorylation are the substituted ureas used as herbicides. Dichlorophenyl dimethyl urea (DCMU) prevents phosphorylation by inhibiting the oxidation of reduced Q of photosystem II. Thus photosystem II can be chemically separated from photosystem I with DCMU. DCMU thus inhibits the Hill reaction.

15.7 Pseudocyclic Photophosphorylation (the Mehler Reaction)

Using chloroplast membranes prepared in the proper buffers, oxygen consumption can be measured by standard manometric or polarographic techniques if CO_2 is excluded from the reaction. This reaction was first demonstrated by A. H. Mehler in 1951 and is called the Mehler reaction or pseudocyclic photophosphorylation since ATP is formed from ADP. Ordinarily, it would be virtually impossible to detect much oxygen consumption by chloroplasts that are functioning normally since the usual electron-transport pathway evolves oxygen in the process. But if carbon dioxide is removed so that there is no ultimate electron acceptor, oxygen can act as the electron acceptor. Water is still the donor and electrons are transported through the usual electron-transport chain.

$$H_2O \longrightarrow \text{photosystem II} \longrightarrow \text{plastoquinone} \longrightarrow$$
$$\text{cytochrome } f/\text{plastocyanin} \longrightarrow$$
$$\text{photosystem I} \longrightarrow O_2$$

It is usually assumed that oxygen is reduced to hydrogen peroxide directly by photosystem I, but electrons may also pass through ferredoxin. In either case, little or no NADPH is formed and oxygen is consumed. Oxygen consumption occurs because the oxidation of water by the Hill reaction in the photosystem II reaction produces $\frac{1}{2}O_2$, but the reduction of O_2 during the formation of hydrogen peroxide (H_2O_2) consumes O_2.

$$H_2O \longrightarrow \frac{1}{2} O_2 + 2e^- + 2H^+$$
$$\underline{O_2 + 2e^- + 2H^+ \longrightarrow H_2O_2}$$
$$H_2O + \frac{1}{2}O_2 \longrightarrow H_2O_2$$

Except for no NADPH formation and some oxygen consumption, electron flow here seems to be similar to noncyclic electron flow. DCMU will inhibit. The quantum requirement is the same and the same amount of ATP is formed.

It is not known if pseudocyclic photophosphorylation actually occurs *in vivo*. Light-induced oxygen consumption has been demonstrated with intact tissue using ^{18}O and mass spectrographic analysis during gas exchange, but there are other kinds of oxygen consumption reactions that complicate interpretation. However, it is reasoned that since more ATP than NADPH is required for the carbon metabolism of photosynthesis, once NADPH reaches saturation ATP could still be produced by pseudocyclic electron flow (in addition to cyclic electron flow). Moreover, if stomata close and limit the supply of CO_2 to chloroplasts, electron flow and ATP synthesis may continue through the Mehler reaction.

Another interesting reaction catalyzed by some algae is light-dependent hydrogen production. In 1940 Hans Gaffron showed that green algae such as *Scenedesmus* consume hydrogen. There is a photoreduction of CO_2 that uses hydrogen as the electron donor. If carbon dioxide is not present, NADPH will become saturating in the light and there will be production of hydrogen. Thus certain algae can both use hydrogen and produce hydrogen during photosynthetic electron transport.

The significance, if any, of hydrogen production by algae is not known, but because hydrogen is a potential useful fuel source there is much interest in a thorough understanding of this process. Evidently, electron flow is through the photosynthetic electron-transport system of chloroplasts to ferredoxin. Water is the substrate and the electron donor. Because electron flow is through the electron-transport system, ATP is produced.

$$H_2O \longrightarrow \text{photosystem II} \longrightarrow \text{plastoquinone} \longrightarrow$$
$$\text{cytochrome } f/\text{plastocyanin} \longrightarrow \text{photosystem I} \longrightarrow$$
$$\text{ferredoxin} \longrightarrow H_2$$

Hydrogen production has been demonstrated in many different green algae and in some red and brown algae but not in any higher plants.

15.8 The Photosynthetic Unit

The concept of a photosynthetic unit is that there is a minimum-sized particle associated with the chloroplast membranes, which carry out the functions of photosynthesis. As early as 1932 Emerson and Arnold approached this question by using very short flashes of light just sufficient to bring about CO_2 fixation. From these experiments and knowledge of the amount of chlorophyll they determined that there were about 2500 chlorophyll molecules for each CO_2 molecule fixed.

If we assume a quantum requirement of 8 to 10 quanta (photons) per reaction, there would be about 250 chlorophyll molecules per quantum unit (2500/8–10). This would be a functional or biochemical unit.

Chloroplasts were fractionated into units about 100 Å in diameter, which would still complete the Hill reaction. With the use of the electron microscope, repeating structures on chloroplast membranes in the same size range have been observed. After determining the density of the repeating units through sucrose density-gradient centrifugation, it was possible to calculate that each unit would have about 230 chlorophyll molecules. Thus there appears to be a structural unit comparable to the functional unit described above. These function/structure units were called quantasomes and may be equivalent to the elementary particles of mitochondria.

Much of the evidence obtained through chloroplast-membrane fractionation indicates that the photosystem II reactions require thylakoid stacking (i.e., grana), whereas photosystem I is a part of the stromal lamellae or fretwork (refer to Fig. 15.6). In this context, C_4-photosynthetic plants that have agranal, lamellar chloroplasts in bundle sheath cells have a reduced photosystem II and hence reduced oxygen evolution.

In further experiments, researchers separated chloroplast membrane particles treated with digitonin (a detergent) into small units each of which would reduce NADP to NADPH and a residue of larger particles capable of the Hill reaction. Thus it is possible that photosystem I is separable from photosystem II.

Perhaps the significance of a photosynthetic unit with about 200–250 chlorophylls and other accessory pigments is that it allows more efficient use of absorbed energy. After absorption by a pigment, the excitation energy can migrate from pigment molecule to pigment molecule until it reaches the reaction center, where it is rapidly converted into chemical energy. At the reaction center there is a charge separation to form a negative and positive charge. The primary electron acceptor becomes reduced by the negative site and the center is oxidized. The center returns to ground state by receiving an electron from the primary electron donor.

The concept of a photosynthetic unit is not universally accepted. For one reason, the chemiosmotic hypothesis for photophosphorylation involves a thylakoid membrane with an inner and outer region in order to physically separate protons and maintain a gradient. Nevertheless, the evidence that a minimum unit for photosynthesis involves 200 to 250 chlorophyll molecules is quite overwhelming and is useful for further study and understanding of the photochemical processes of photosynthesis.

15.9 The Permeability of the Chloroplast Envelope

The chloroplast envelope is relatively impermeable to ATP and NADPH, yet there is substantial evidence that much of the cytoplasmic ATP is the direct result of photophosphorylation. Sucrose and the hexoses are not readily transported across the chloroplast envelope. Nevertheless, the strong requirement for ATP by the cell and the fact that many ATP-dependent processes such as cytoplasmic streaming occur in the light at higher rates than in the dark suggest that the chloroplast is a direct source for cellular ATP.

Shuttle schemes have been proposed in which reduced substances produced through photosynthesis are transported into the cytoplasm, where they are used for ATP and NADH production. In one such scheme, malate is transported from the chloroplast to the cytoplasm, where malate dehydrogenase oxidizes it to oxalacetate, generating NADH. The oxalacetate is shuttled back to the chloroplast, perhaps as aspartate, where it is reductively converted back to malate to complete the shuttle. Thus through the malate shuttle, chloroplast-generated NADPH can be used to produce cytoplasmic NADH. This shuttle is similar to the mitochondrial shuttle outlined in Fig. 14.6 in which malate and aspartate are transported and produce NADH in the mitochondria.

A phosphoglycerate–glyceraldehyde shuttle as proposed by Stocking and Larson in 1969 is illustrated in Fig. 15.8. The chloroplast membrane is permeable to triose phosphate (GAP) and phosphoglycerate (PGA) but not NADPH. This scheme accounts for transport of chloroplast ATP and NADPH from the chloroplast to the cytoplasm. Glyceraldehyde phosphate, produced from phosphoglycerate in the chloroplast, is transported to the cytoplasm where it is reduced back to phosphoglycerate by triosephosphate dehydrogenase, generating NADH and ATP. As mentioned previously, the triosephosphate dehydrogenase of the cytoplasm is an NAD enzyme, whereas the chloroplast enzyme is an NADP enzyme. The triosephosphates and phosphoglycerate are transported across the chloroplast envelope on an exchange basis for inorganic phosphate.

In addition to the above, other shuttle schemes have been proposed for the transport of ATP and "reducing power" (NADPH) from the chloroplast

FIGURE 15.8 Hypothesis for the transport of NADPH out of chloroplasts. Phosphoglycerate from photosynthesis (PGA) is reduced by NADPH to glyceraldehyde phosphate (GAP). GAP is transported into the cytosol where it is oxidized back to PGA by an NAD glyceraldehyde phosphate dehydrogenase, producing NADPH. PGA is transported back into the chloroplast, effectively completing a cycle in which NADPH is converted to NADH through the catalysis of two different glyceraldehyde phosphate dehydrogenase enzymes. From C. S. Stocking and S. Larson. 1969. A chloroplast cytoplasmic shuttle and the reduction of extraplastid NAD. *Biochem. Biophys. Res. Commun.* 37:278–282.

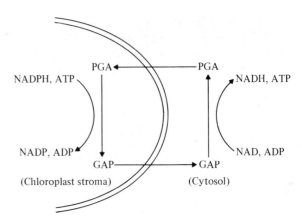

into the cytoplasm. These will not be discussed here.

The chloroplast envelope is relatively impermeable to many common cations such as hydrogen ions (protons), potassium, sodium, and magnesium. Transport of these is through energy-dependent processes that frequently involve ATPase enzymes. Of course, the chemiosmotic hypothesis for phosphorylation depends on chloroplast membrane impermeability to protons.

The chloroplast envelope is not permeable to all compounds of biological interest. This was dis-cussed to some extent in Chapter 4 under reflection coefficients. Sucrose is essentially excluded, whereas the permeability of amino and organic acids differ, depending on their structure. Sugar alcohols are more readily taken up than are the latter compounds.

The chloroplast envelope is quite permeable to CO_2 but probably isn't to the bicarbonate ion. Since much of the carbonic anhydrase of the cell is in the chloroplast, CO_2 and HCO_3^- are virtually interchangeable.

15.10 Chloroplast Development

Being the characteristic organelle of plant cells, the chloroplast (plastid) is quite interesting not only from its physiological role in photosynthesis but also because it has some degree of autonomy. The chloroplast has DNA, RNA, and the capacity for protein synthesis; however, the extent of the cooperation between nuclear DNA and chloroplast DNA during chloroplast development is still not fully understood.

Functional chloroplasts develop from proplastids that are inherited maternally through the egg cells. The proplastid is a relatively undifferentiated plastid with little internal structure. After germination the proplastid develops into an etioplast, which is characterized by the presence of an internal, tubular structure called a prolamellar body. Upon illumination, this body forms the thylakoids of the functional chloroplast. Since some species do not have etioplasts with prolamellar bodies, transition from proplastid through the etioplast during chloroplast development is evidently not obligatory. Figure 15.9 shows a young developing plastid with a prolamellar body.

The lipid components of the chloroplasts are not much different from those of the etioplast except for chlorophyll. Chlorophyll synthesis in angiosperms is light-dependent. Upon illumination of a dark-grown, etiolated plant, green chlorophyll

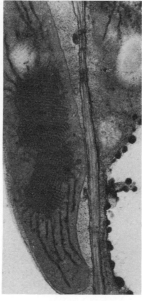

FIGURE 15.9 Stereoscopic pair photograph of a young developing plastid, showing the crystallinelike prolamellar body. Typical thylakoids develop from the prolamellar body. Electron micrographs courtesy of K. Platt-Aloia, University of California, Riverside.

appears with a short lag time (see Fig. 15.10 on page 418). The precursor of chlorophyll synthesis is protochlorophyllide, which is synthesized from protoporphyrin IX (see Chapter 12) by magnesium insertion into the porphyrin ring. Protochlorophyllide is photoconvertible to chlorophyllide *a*, the immediate precursor of chlorophyll. Photoconversion of protochlorophyllide to chlorophyllide *a* is accompanied by a spectral shift toward the red, which is more characteristic of chlorophyll. Chemically the conversion is a photoreduction (Fig. 15.10).

Chlorophyllide is converted to chlorophyll by phytylation (addition of phytol) to form chloro-

phyll *a*, as shown in Fig. 15.10. Chlorophyll *a* is oxidized to chlorophyll *b*. During the initial stages of greening the ratio of chlorophyll *a* to chlorophyll *b* is quite high, but after several hours the ratio decreases to three chlorophyll *a* to one chlorophyll *b*, which is characteristic of most green plants.

The chloroplast becomes photochemically competent after a few minutes of illumination. Evidence suggests that the photosystems develop sequentially, with photosystem I activity appearing prior to photosystem II activity. In barley, photosystem II (measured by oxygen evolution) begins after about 30 minutes of illumination and reaches the mature-leaf level of activity in two or three hours.

Review Exercises

15.1 In the Gorman and Levine experiment with mutant alga cells, how could reduction of NADP to NADPH be measured experimentally? What kinds of problems would be encountered in such measurements?

15.2 Explain the rationale for assuming that ATP is produced by cyclic electron flow. If cyclic electron flow is not a real physiological process, where do you suppose the ATP comes from for the synthesis of cellular components? Estimate the quantum re-

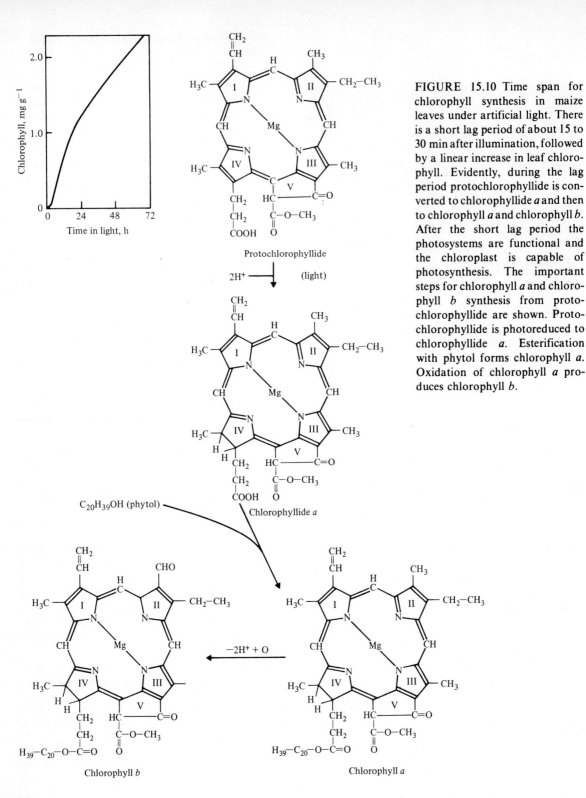

Protochlorophyllide

$2H^+ \longrightarrow$ (light)

Chlorophyllide a

$C_{20}H_{39}OH$ (phytol)

$-2H^+ + O$

Chlorophyll b

Chlorophyll a

FIGURE 15.10 Time span for chlorophyll synthesis in maize leaves under artificial light. There is a short lag period of about 15 to 30 min after illumination, followed by a linear increase in leaf chlorophyll. Evidently, during the lag period protochlorophyllide is converted to chlorophyllide a and then to chlorophyll a and chlorophyll b. After the short lag period the photosystems are functional and the chloroplast is capable of photosynthesis. The important steps for chlorophyll a and chlorophyll b synthesis from protochlorophyllide are shown. Protochlorophyllide is photoreduced to chlorophyllide a. Esterification with phytol forms chlorophyll a. Oxidation of chlorophyll a produces chlorophyll b.

quirements for such biosyntheses. Can you think of any experiments to demonstrate *in vivo* cyclic electron flow?

15.3 The herbicide DCMU can be used to experimentally separate photosystem II from photosystem I by preventing reoxidation of Q. What are other ways in which the two photosystems can be separated for experimental purposes?

15.4 Do you think chloroplasts with intact outer envelopes, when prepared for *in vitro* studies, would make ATP from ADP and inorganic phosphate better than chloroplasts lacking the outer envelope and stroma? Explain your answer. Would the chloroplasts lacking outer envelopes and stroma fix CO_2?

15.5 If the ratio of NADPH to $NADP^+$ in chloroplasts in the light is 10, compute the actual oxidation–reduction potential. The $E^{o'}$ for the NADPH/ $NADP^+$ couple is -0.32 mV.

References

ARNON, D. I. 1967. Photosynthetic activity of isolated chloroplasts. *Physiol. Rev.* 47:317–358.

AVRON, M. 1977. Energy transduction in chloroplasts. *Ann. Rev. Biochem.* 46:143–155.

AVRON, M., and J. NEUMANN. 1968. Photophosphorylation in chloroplasts. *Ann. Rev. Plant Physiol.* 19:137–166.

BARBER, J. (ed.). 1977. *Primary Processes of Photosynthesis.* Elsevier, Amsterdam.

BEARDEN, A. J., and R. MALKIN. 1975. Primary photochemical reactions in chloroplast photosynthesis. *Quart. Rev. Biophys.* 7:131–177.

EMERSON, R. L., R. V. CHALMERS, and C. CEDERSTRAND. 1957. Some factors influencing the long-wave limit of photosynthesis. *Proc. Natl. Acad. Sci.* (US) 43:133–143.

FORK, D. C., and J. AMESZ. 1969. Action spectra and energy transfer in photosynthesis. *Ann. Rev. Plant Physiol.* 20:305–328.

GAFFRON, H. 1960. Energy storage: photosynthesis. In F. C. Steward (ed.), *Plant Physiology: A Treatise,* Vol. 1B. Academic Press, New York.

GOVINDGEE (ed.). 1975. *Bioenergetics of photosynthesis.* Academic Press, New York.

GOVINDGEE, and R. GOVINDGEE. 1974. The primary events of photosynthesis. *Sci. Amer.* 231:64–82.

GREGORY, R. P. F. 1977. *Biochemistry of Photosynthesis.* 2d ed. Wiley, London.

INADA, K. 1976. Action spectra for photosynthesis in higher plants. *Plant and Cell Physiol.* 17:355–365.

JAGENDORF, A. T., and E. URIBE. 1966. Photophosphorylation and the chemi-osmotic hypothesis. *U.S. Brookhaven Symp. Biol.* 19:215–245.

KAMEN, M. D. 1963. *Primary Processes in Photosynthesis.* Academic Press, New York.

KIRK, J. T. O. 1971. Chloroplast structure and function. *Ann. Rev. Biochem.* 40:161–196.

KIRK, J. T. O., and R. A. E. TILNEY. 1978. *The Plastids. Their Chemistry, Structure, Growth and Inheritance.* Elsevier, Amsterdam.

KOK, B. 1976. Photosynthesis: the path of energy. In J. Bonner and J. E. Varner (eds.), *Plant Biochemistry,* 3d ed. Academic Press, New York.

LEVINE, R. P. 1969. Analysis of photosynthesis using mutant strains of algae and higher plants. *Ann. Rev. Plant Physiol.* 20:523–540.

MYERS, J. 1974. Conceptual developments in photosynthesis, 1924–1974. *Plant Physiol.* 54:420–426.

RABINOWICH, E. 1948–1954. *Photosynthesis,* Vols. I, IIA, IIB. Academic Press, New York.

RADMER, R., and B. KOK. 1975. Energy capture in photosynthesis: photosystem II. *Ann. Rev. Biochem.* 44:409–433.

SIMONIS, W., and W. URBACH. 1973. Photophosphorylation in vivo. *Ann. Rev. Plant Physiol.* 24:89–114.

THORNBER, J. P. 1975. Chlorophyll proteins: light harvesting and reaction centers. *Ann. Rev. Plant Physiol.* 26:127–158.

TREBST, A. 1974. Energy conservation in photosynthetic electron transport of chloroplasts. *Ann. Rev. Plant Physiol.* 25:423–458.

TREBST, A., and M. AVRON (eds.). 1977. Photosynthesis I. *Encyclopedia of Plant Physiology,* New Series Vol. 5. Springer-Verlag, New York.

Carbon Metabolism

Among the most important chemical reactions on earth are the carbon metabolism reactions of green plants in which CO_2 from the atmosphere is used as the carbon source for synthesis of sugars and ultimately all organic compounds of living organisms. As explained in the previous chapter, light energy is used to drive photochemical reactions in which ATP and NADPH are formed. These compounds are subsequently used to provide the energy necessary to form sugar from CO_2. This chapter discusses the historical basis for our present understanding of carbon metabolism in plants and details the present state of the art.

16.1 History

As indicated in the previous chapter, photosynthesis can be divided into two components: the light reactions or primary processes, and the dark or carbon metabolism reactions. The important aspect of the latter is the fixation of CO_2 into stable organic compounds. The two components can be represented in simple equational form.

(1) $Light + NADP + ADP + P_i \longrightarrow$

$$NADPH + ATP$$

(2) $CO_2 + H_2O + NADPH + ATP \longrightarrow$

$$CH_2O + NADP + ADP + P_i$$

Chapter 15 describes in some detail how NADPH and ATP are formed using the energy of light. In this chapter, it will be shown how such chemical energy (in the form of NADPH and ATP) is used to reduce CO_2 to form carbohydrates and other cellular organic compounds. Basically, this will be a detailed discussion of equation (2) shown above, the dark reactions originally identified by Blackman in 1905.

Since the first experiments of de Saussure in 1804, which demonstrated CO_2 uptake by plants in the light, the question of how CO_2 is incorporated into the organic matter of plants has been of interest. Perhaps the first suggestion was by Liebig, who in 1843 proposed that organic acids were the first products of photosynthesis. He assumed that there was carboxylation of an organic compound to form an organic acid.

$$CO_2 + RH \longrightarrow RCO_2H$$

Subsequent transformation of the organic acid (by reduction) would ultimately yield carbohydrate.

In 1870 Baeyer proposed that formaldehyde was the first product of photosynthesis and that sugars would be produced by polymerization of formaldehyde units. A simplified form of this hypothesis follows.

$$CO_2 + H_2 \longrightarrow HCHO + \tfrac{1}{2}O_2$$

$$6HCHO \longrightarrow C_6H_{12}O_6$$

Much effort was expended to locate formaldehyde in plants but to no avail. Formaldehyde is a rather rare plant constituent.

Finally, in 1943 Gardner proposed that glyceral-

dehyde was the first product of photosynthesis, and with the use of $^{14}CO_2$ as a tracer it was then shown rather conclusively by Calvin, Benson, and co-workers in the 1950s that phosphoglyceraldehyde was, in fact, the first sugar product of photosynthesis. Furthermore, it was quickly shown that both the α and β carbons of phosphoglyceraldehyde became rapidly labeled, suggesting that the overall process was a cyclic one. Such a hypothesis was readily accepted since many felt that the carbon metabolism of photosynthesis would be somewhat comparable to the reversal of aerobic respiration, a cyclic process.

Initially, much effort was expended to locate a hypothetical two-carbon fragment as an acceptor molecule. An initial proposal by Calvin's group was as follows.

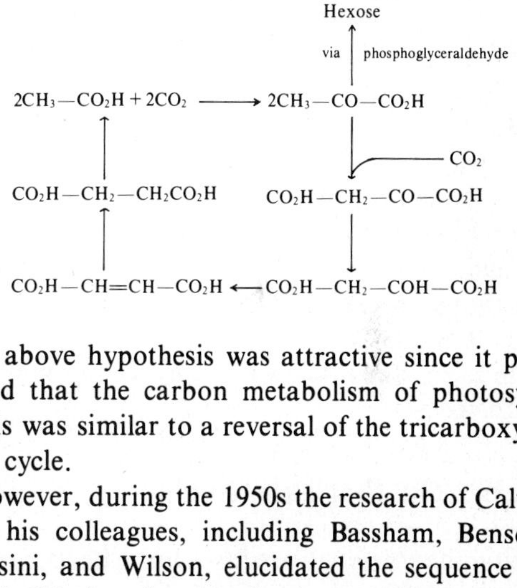

The above hypothesis was attractive since it proposed that the carbon metabolism of photosynthesis was similar to a reversal of the tricarboxylic acid cycle.

However, during the 1950s the research of Calvin and his colleagues, including Bassham, Benson, Massini, and Wilson, elucidated the sequence by which CO_2 is reduced to form sugar. They showed that ribulose bisphosphate was the acceptor molecule for CO_2 that produced phosphoglyceraldehyde, and through a series of sugar-transformation reactions the ribulose bisphosphate was then regenerated and net phosphoglyceraldehyde was produced.

The experimental evidence and the nature of the photosynthetic cycle are presented in the next section.

16.2 The Path of Carbon in Photosynthesis

There were perhaps two important events of the 1940s that allowed Calvin and his colleagues to elucidate the path of carbon from CO_2 to sugar. In 1945 after World War II, ^{14}C, the radioactive isotope of carbon, became available from nuclear reactor experiments at Berkeley. It was thus possible to label carbon dioxide with ^{14}C and actually follow the carbon of CO_2 as it entered the carbon compounds produced during photosynthesis. Prior to this time most research designed to ascertain the path of carbon was done by first chemically identifying compounds within the plant leaf and then attempting to predict pathways on the basis of known chemical reactions. Isolation of enzymes that would catalyze the proposed reactions yielded confirmatory evidence. Being able to actually trace the path of carbon greatly simplified the research procedures.

In addition, the technique of paper chromatography, introduced in the 1940s, allowed the rapid separation and qualitative analysis of a large number of compounds from a complex mixture.

The important questions asked were: (1) what was the initial stable product of photosynthesis; (2) what was the acceptor molecule for CO_2; (3) what was the path of carbon from the initial product to the ultimate sugar; and (4) how was the acceptor molecule synthesized?

16.2.1 The first experiments

Paper chromatography—identification of products

Chromatography, first described in analytical form by Mikhail Tswett in 1903 during experiments separating chlorophyll pigments from leaves, brings about separation of similar compounds based on their partition coefficients between a stationary phase and a mobile phase. The stationary phase is adsorbed to a supporting matrix. Phases can be liquid or gas, but perhaps the simplest is liquid paper chromatography. Here, cellulosic paper functions as the matrix and is frequently prepared from laboratory filter-paper sheets. Simple soluble compounds such as those produced by photosynthesis can be separated by allowing an aqueous alcoholic solution to migrate across the paper on which the sample has been applied. The filter-paper sheet can be of any size, but for the usual photosynthetic experiments to be described below it is on the order of 46 to 56 cm. A small 80% ethanol sample of the components to be separated and identified is applied to one corner about 2 cm in from each side. First, a developing solvent of butanol, propionic acid, and water (or other appropriate mixture) is allowed to migrate across the paper through the sample. The water-hydrated cellulose acts as the stationary phase while the butanol is the mobile phase. The acid maintains a low pH, keeping all ionizable compounds in one form. As the solvent (butanol) migrates across the paper, different compounds in the sample will partition between the stationary hydrated cellulose and the mobile butanol phase according to their individual partition coefficients. Continuous partitioning occurs between butanol and new areas of paper, resulting in different rates of migration and separation of individual compounds from each other.

After drying, the paper is turned 90° and developed in a second direction with phenol and water, improving the separation and creating a two-dimensional chromatogram capable of separating virtually hundreds of compounds.

Compounds can be located and identified on the chromatogram with specific indicator solutions. Thus organic acids can be detected on the paper chromatogram by spraying on pH indicators such as bromophenol blue. Wherever an acid occurs, there will appear a yellow spot on a blue background. Amino acids can similarly be detected by spraying with ninhydrin solution that is specific for amino acids. Neutral carbohydrates can be detected with naphthoresorcinol, as can phosphorylated compounds with acid ammonium molybdate solution. There are, of course, virtually hundreds of indicators that can be sprayed on paper chromatograms for purposes of identification.

When tissue is supplied with $^{14}CO_2$, those compounds synthesized by carboxylations or their products will contain ^{14}C and can be located by scanning the paper with a suitable radioactive detection device, such as a thin-window Geiger counter. Frequently the paper chromatogram with radioactive compounds is placed against x-ray film. Wherever there is radioactivity, the film will be exposed (see Fig. 16.1). The amount of radioactivity associated with each compound can be counted directly as described above, or individual radioactive compounds can be eluted from the paper with suitable solvents and counted by liquid scintillation or other means. Compounds can be identified (after elution) by chemical means.

A rapid method of verifying the identity of compounds on the paper chromatogram is by the technique of cochromatography. Unknown compounds are chromatographed with known substances. Coincidence of migration is evidence for identity.

The success of the research to ascertain the flow of carbon in photosynthesis was based on the ingenuity of the investigators and the availability of the analytical technique of paper chromatography and the ^{14}C isotope of carbon used as a tracer.

When green tissue is allowed to take up $^{14}CO_2$ for a short time, many compounds become labeled with ^{14}C (Fig. 16.1). Among the first compounds of photosynthesis are sugar phosphates, followed by organic acids, amino acids, and carbohydrates.

Experimental design

Much of the research on photosynthetic carbon metabolism was conducted with two genera of green algae, *Chlorella* and *Scenedesmus*. In the basic experiment a flattened flask, called a "lollipop" by the researchers, contained algal cells in a buffer and salt solution. $^{14}CO_2$ was bubbled through the solution containing the algal cells. The design was such that pH, temperature, CO_2 (both $^{12}CO_2$ and $^{14}CO_2$), and light were controlled. Furthermore, the stream containing CO_2 was passed through an ion chamber to continuously monitor $^{14}CO_2$ by detection of radioactivity and then through an infrared gas analyzer to monitor total CO_2. A diagram of the apparatus is shown in Fig. 16.2.

The spout at the base of the lollipop was fitted with a stopcock and solenoid valve so that samples of cells could be collected automatically at intervals of one or two seconds. Samples as large as one milliliter or more were dropped directly into hot ethanol. The hot alcoholic solution killed the cells, stopping further metabolism of carbon; it also extracted the soluble carbon compounds. After centrifugation to remove insoluble cellular debris, the entire sample was applied to the filter-paper sheet for chromatography.

A basic assumption

The experiments to determine the primary and secondary products of photosynthesis were based on one important assumption. When the products of photosynthetic carboxylation were graphed as a percentage of ^{14}C in each compound at each time interval, the initial product would have a decreasing (negative) slope, and all other compounds would appear with increasing (positive) slopes. This is true because the initial product must be present before secondary products can be formed. At first, all ^{14}C from the carboxylation of $^{14}CO_2$ would be in the first product, and secondary products would not receive the label until later. In equational form the reaction is

$$^{14}CO_2 + X \longrightarrow {}^{14}A \longrightarrow {}^{14}B \longrightarrow {}^{14}C,$$

where X is the acceptor, A is the first product, and B and C are secondary products. The assumption, of course, would not be valid if there were more than one carboxylation reaction.

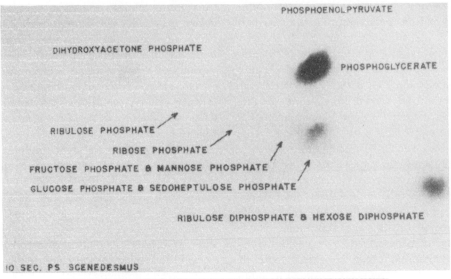

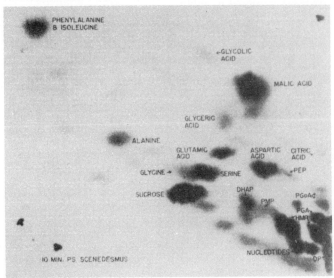

FIGURE 16.1 Autoradiograms illustrating the products of photosynthesis by the green alga *Scenedesmus*. Top: products produced by photosynthesis in the presence of $^{14}CO_2$ after 10 sec in the light. Phosphoglyceric acid is the primary product, consistent with being the first product of photosynthesis. Bottom: products of photosynthesis after 10 min in the light. Many compounds are now present, among the most important of which are sucrose and malic acid. PEP = phosphoenolpyruvate. DHAP = dihydroxyacetone phosphate. PMP = pentose monophosphates. PGoA = phosphoglycolic acid. PGA = phosphoglyceric acid. HMP = hexose monophosphates. DP = pentose and hexose biphosphates. From M. Calvin and P. Massini. 1952. The path of carbon in photosynthesis. *Experientia* 8:445–457.

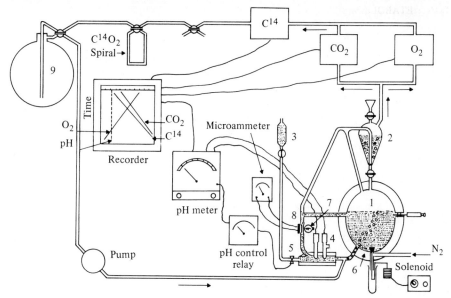

FIGURE 16.2 The experimental setup used in the laboratory of Melvin Calvin during the elucidation of the reductive pentose cycle for CO_2 assimilation. 1 = reaction chamber containing alga cells. 2 = nutrient solution reservoir. 3 = acid and base reservoir used to regulate pH during experiments. 4 = pH electrodes to measure pH during experiments. 5 = solenoid valve to allow acid or base to enter nutrient solution in order to maintain pH during experiment. 6 = solenoid valve for automatic sampling of alga cells after set time intervals of photosynthesis. 7 = light source. 8 = photocell to measure light intensity. 9 = gas reservoir. The boxes labeled C^{14}, CO_2, and O_2 are sensors to monitor radioactive CO_2 (^{14}C), total CO_2, and O_2. The data are recorded on the chart recorder. The apparatus allows for the measurement of photosynthesis by monitoring levels of $^{14}CO_2$, O_2, and total CO_2 in the airstream that is bubbled through the nutrient solution and allows for sampling alga cells at intervals as short as a few seconds to determine the radioactive products after photosynthesis with $^{14}CO_2$. From J. Bassham and M. Kirk. 1960. Dynamics of the photosynthesis of carbon compounds. I. Carboxylation reactions. *Biochim. Biophys. Acta* 43: 447–464.

16.2.2 The first product of photosynthesis

When algal cells are allowed to fix $^{14}CO_2$ in the light, there is a linear accumulation of ^{14}C in the cells for many seconds. When ^{14}C-radioactivity in each labeled compound is graphed as a percentage of the label fixed for each time, only phosphoglycerate has a decreasing slope. All other compounds with a ^{14}C label, including the sugar phosphates, show positive slopes and are assumed to be secondary products. Such an experiment is shown in Fig. 16.3. For this reason, it was concluded that phosphoglycerate, an organic acid, was the first product of photosynthesis.

When CO_2 was maintained at a high level and light was kept low in some experiments, malate appeared to be a primary product of photosynthetic carboxylation. However, experiments using the inhibitor malonate argued against malate as a primary product. Malonate will inhibit malate synthesis without significantly reducing the total amount of $^{14}CO_2$ fixed. Hence malate must be a secondary product of photosynthesis and not a primary product. It is assumed that malate is synthesized subsequent to phosphoglycerate. However, when CO_2 is high and light is low, much malate is synthesized by the dark carboxylation reactions discussed in Chapter 10. Malate synthesis in green

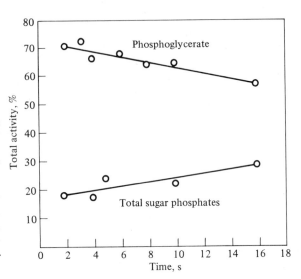

FIGURE 16.3 Graphic illustration of the first products of photosynthesis. Phosphoglycerate graphs as the initial product since it has a decreasing amount of radioactivity with time, whereas the remainder of the phosphate compounds have an increasing amount of ^{14}C, indicating that they are produced after the first product. Reprinted with permission from J. A. Bassham, A. A. Benson, D. K. Lorel, A. Z. Harris, A. T. Wilson, and M. Calvin. 1954. The path of carbon in photosynthesis. XXI. The cyclic regeneration of carbon dioxide acceptor. *J. Amer. Chem. Soc.* 76:1760-1770. Copyright 1954 American Chemical Society.

tissue is further discussed in Sections 16.2.9, 16.4, and 16.5.

16.2.3 The CO_2 acceptor molecule

Once it was established that phosphoglycerate was the first product of photosynthesis, the next question of importance was the nature of the carboxylation reaction. What was the acceptor molecule for CO_2? Much effort was expended in an attempt to locate a two-carbon compound that upon carboxylation would yield phosphoglycerate. Subsequent research revealed that a two-carbon compound was not the acceptor but rather a five-carbon compound, ribulose bisphosphate.

$$
\begin{array}{l}
CH_2OPO_3^= \\
| \\
C=O \\
| \\
CHOH \\
| \\
CHOH \\
| \\
CH_2OPO_3^=
\end{array}
$$

Perhaps the first critical experiment was conducted in 1952 by Massini in Calvin's laboratory. He obtained evidence that ribulose bisphosphate was probably the acceptor molecule.

The Massini experiment

Algal cells were incubated with $^{14}CO_2$ until all components of photosynthesis were in steady state and their specific activities (^{14}C per mole) were equal to the $^{14}CO_2$ supplied. At this point it was assumed that all components were in steady state and that the flow of carbon through the pathway was constant.

After steady state was reached, the lights were turned off and the amount of ^{14}C label in each component was determined for the next several minutes. It was observed that when the cells were transferred from light to dark there was a rapid transient increase in ^{14}C label in phosphoglycerate and a decrease in sugar bisphosphates (see Fig. 16.4), particularly ribulose bisphosphate. Other compounds did not change significantly or did not appear to be of interest with respect to the immediate question. Thus it was concluded that sugar bisphosphates were supplying the carboxylation precursor, and that light was necessary for their synthesis. The experiment also showed that light was necessary for the subsequent utilization of phosphoglycerate. Of course, we now know that light was necessary for the synthesis of the sugar bisphosphate precursor

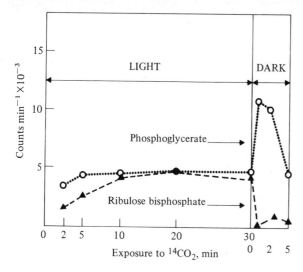

FIGURE 16.4 The Massini experiment, in which light is removed in the presence of CO_2. Ribulose bisphosphate decreases and phosphoglycerate increases. This experiment suggested that ribulose bisphosphate was the CO_2 acceptor molecule and that light was necessary both for its synthesis and the utilization of the first product, phosphoglycerate. From M. Calvin and P. Massini. 1952. The path of carbon in photosynthesis. *Experientia* 8:445–457.

because the ATP for phosphorylation comes from photophosphorylation, and the light requirement for phosphoglycerate utilization is for the NADPH and ATP used in its reduction to sugar.

The sugar bisphosphate that appeared to be the most likely precursor for carboxylation was ribulose bisphosphate. The next important experiment by Wilson and Calvin in 1954 confirmed this latter suspicion.

The Wilson experiment

The experiment that confirmed ribulose bisphosphate as the CO_2 acceptor was similar to the former experiment by Massini in that photosynthesizing cells were allowed to come to steady state in the light. After steady state was reached, the $^{14}CO_2$ was removed but the lights were kept on. In this experiment (shown in Fig. 16.5), when transferred to CO_2 free air, phosphoglycerate decreased and ribulose bisphosphate increased consistent with the notion that ribulose bisphosphate is the acceptor of CO_2. When CO_2 is removed, the acceptor accumulates since it is not used. Its ATP-dependent synthesis can continue in the light until there is depletion of the remaining intermediates. Phosphoglycerate similarly decreases since it is no longer synthesized from CO_2 and ribulose bisphosphate,

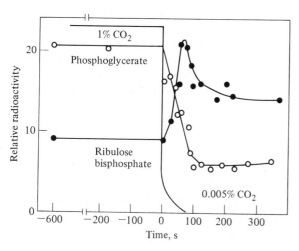

FIGURE 16.5 The Wilson experiment, indicating that when CO_2 is removed in the presence of light, ribulose bisphosphate is no longer used but is continuously produced. Even though phosphoglycerate is no longer produced, it is used in the light. This experiment with the Massini experiment shown in Fig. 16.4, showed that ribulose bisphosphate was the CO_2 acceptor molecule and that light was necessary for its synthesis. Light is also required for the use of phosphoglycerate. Reprinted with permission from J. A. Bassham, A. A. Benson, D. K. Lorel, A. Z. Harris, A. T. Wilson, and M. Calvin. 1954. The path of carbon in photosynthesis. XXI. The cyclic regeneration of carbon dioxide acceptor. *J. Amer. Chem. Soc.* 76:1760–1770. Copyright 1954 American Chemical Society.

but it is still metabolized through the light-dependent reactions. Since the cells are still in the light, ATP and NADPH are still produced and the phosphoglycerate is removed as it is reduced to sugar.

16.2.4 Carboxylation

Subsequent to the Wilson experiment, a protein was isolated from green plant tissue that could catalyze the carboxylation of ribulose-1,5-bisphosphate to yield two moles of 3-phosphoglycerate.

Ribulose bisphosphate $+ CO_2 \longrightarrow$

2 Phosphoglycerate

The enzyme was named carboxydismutase but is now referred to as a carboxylase. The proper name is ribulose-1,5-bisphosphate carboxylase.

The ribulose bisphosphate carboxylase is apparently identical to Fraction I protein, the predominant protein of chloroplasts. In fact, as much as 50% of the total protein of spinach leaves is Fraction I protein.

Ribulose bisphosphate carboxylase is composed of about 16 pairs of two kinds of subunits, a 20,000 and a 10,000 subunit. The total molecular weight of the active protein is about 480,000.* The intact protein has about 80 sulfhydryl groups.

There is evidence that the DNA code for the synthesis of the smaller 10,000 subunit is of nuclear origin and that the larger 20,000 subunit DNA is chloroplast in origin. Thus the information in the nucleus and in the chloroplast cooperate in the assembly of Fraction I protein.

The carboxylation of ribulose bisphosphate takes place by nucleophilic attack of CO_2 on the carbonyl group of the ribulose bisphosphate, yielding two moles of phosphoglycerate. The reaction uses CO_2

as the active species (not bicarbonate or carbonate) and has a pH optimum of about 7.8 when in the presence of $10 mM$ magnesium, which is required for activity. The affinity of the carboxylase for ribulose bisphosphate is rather high, with a K_m of about $0.1 mM$ to $0.24 mM$. The K_m for CO_2 is about $0.05 mM$ CO_2. It should be apparent that in an experiment with $^{14}CO_2$, only one phosphoglycerate molecule will be labeled with ^{14}C.

$$
\begin{array}{c}
CH_2OPO_3^= \\
| \\
C{=}O \\
| \\
CHOH \\
| \\
CHOH \\
| \\
CH_2OPO_3^=
\end{array}
\;+\; {}^{14}CO_2 \longrightarrow
\begin{array}{c}
{}^{14}CO_2^- \\
| \\
CHOH \\
| \\
CH_2OPO_3^=
\end{array}
\;+\;
\begin{array}{c}
CO_2^- \\
| \\
CHOH \\
| \\
CH_2OPO_3^=
\end{array}
$$

16.2.5 The reduction of phosphoglycerate

NADP-specific glyceraldehyde-3-phosphate dehydrogenase is the enzyme that catalyzes the reduction of the first product, phosphoglycerate, to the first sugar of photosynthesis, glyceraldehyde phosphate. The basic reaction was discussed in Section 14.3.1 under glycolysis. Recall that during glycolysis, glyceraldehyde phosphate is oxidized to phosphoglycerate, catalyzed by an NAD-specific glyceraldehyde-3-phosphate dehydrogenase. First the phosphoglycerate is phosphorylated by phosphoglycerate kinase, using ATP as the phosphate donor. The reaction produces 1,3-diphosphoglycerate and ADP. Of course, the ATP comes from photophosphorylation, and thus the reaction is considered to be one of the photosynthetic reactions. The kinase reaction follows.

$$
\begin{array}{c}
CO_2^- \\
| \\
CHOH \\
| \\
CH_2OPO_3^=
\end{array}
\;+\; ATP \longrightarrow
\begin{array}{c}
\overset{O}{\overset{\|}{C}}{-}O{-}PO_3^= \\
| \\
CHOH \\
| \\
CH_2OPO_3^=
\end{array}
\;+\; ADP
$$

The final reaction produces glyceraldehyde-3-phosphate, catalyzed by NADP-glyceraldehyde-3-

---

*As with molecular weights of any proteins, the numbers should only be taken as averages or approximations. Kawashima and Wildman (1970), for example, give the molecular weight of Fraction I protein determined from a variety of references as $5.15–5.75 \times 10^5$.

phosphate dehydrogenase.

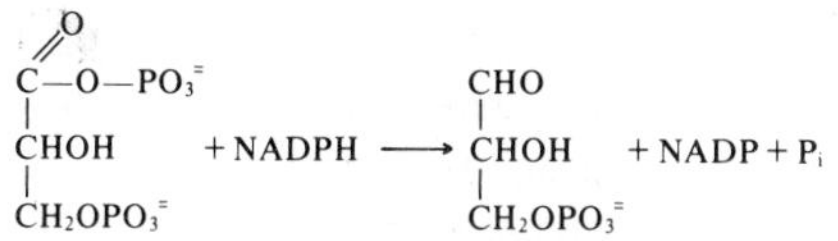

$$
\begin{array}{c}
\overset{O}{\underset{|}{C}}-O-PO_3^= \\
\underset{|}{CHOH} \\
CH_2OPO_3^=
\end{array}
+ NADPH \longrightarrow
\begin{array}{c}
CHO \\
\underset{|}{CHOH} \\
CH_2OPO_3^=
\end{array}
+ NADP + P_i
$$

The NADPH comes directly from photosynthesis.

16.2.6 The regeneration of ribulose bisphosphate

The enzyme ribulose-5-phosphate kinase catalyzes the biosynthesis of ribulose bisphosphate from ribulose-5-phosphate and ATP, regenerating the 5-carbon acceptor for photosynthetic carboxylation.

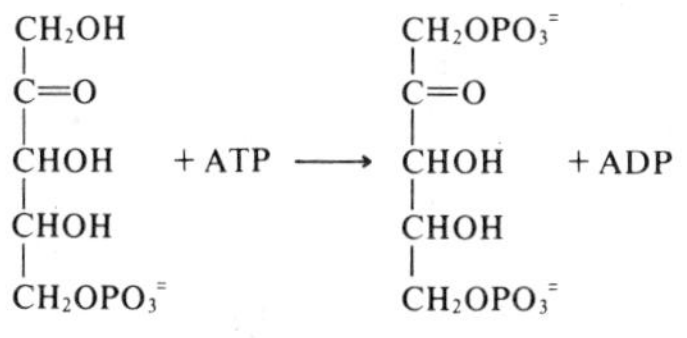

$$
\begin{array}{c}
CH_2OH \\
\underset{|}{C}=O \\
\underset{|}{CHOH} \\
\underset{|}{CHOH} \\
CH_2OPO_3^=
\end{array}
+ ATP \longrightarrow
\begin{array}{c}
CH_2OPO_3^= \\
\underset{|}{C}=O \\
\underset{|}{CHOH} \\
\underset{|}{CHOH} \\
CH_2OPO_3^=
\end{array}
+ ADP
$$

This step is the second phosphorylation of the carbon assimilation reactions.

Overall, the primary reactions of photosynthetic carbon metabolism are: (1) the generation (actually, regeneration) of the ribulose bisphosphate CO_2 acceptor, (2) the carboxylation reactions, and (3) the reduction of phosphoglycerate to glyceraldehyde phosphate. These reactions are summarized as follows.

(1) Ribulose-5-P + ATP →

Ribulose Bis-P + ADP

(2) Ribulose Bis-P + CO_2 → 2 Phosphoglycerate

(3) 2 Phosphoglycerate + 2ATP + 2NADPH→

2 Glyceraldehyde phosphate + 2ADP

+ 2NADP + 2P_i

For the uptake of one mole of CO_2 there is a requirement for three moles of ATP and two moles of NADPH. The production of one net glyceralde- hyde phosphate requires three moles of CO_2, 9 moles of ATP, and 6 moles of NADPH.

For each three molecules of CO_2 fixed that produce one net 3-carbon compound, three molecules of ribulose are consumed. The 15 carbons from the three ribulose appear as 5-glyceraldehyde phosphate, each with three carbons. The remainder of the photosynthetic cycle involves the recycling of the glyceraldehyde phosphates back to ribulose through a series of reactions known as the Calvin–Benson cycle, or the reductive pentose cycle. These reactions are largely equilibrium reactions that do not consume energy, and thus no NADPH or ATP are required. The reductive pentose cycle is discussed below.

16.2.7 The reductive pentose cycle

The reductive pentose cycle, or Calvin–Benson cycle, is outlined in Fig. 16.6 and shows the net assimilation of three moles of CO_2 to produce one net mole of glyceraldehyde-3-phosphate. During the process, three moles of ribulose-1,5-bisphosphate are consumed along with nine moles of ATP and six moles of NADPH. The cycle is completed with the regeneration of the three moles of acceptor ribulose bisphosphate.

In the diagram of Fig. 16.6, the six glyceraldehyde phosphates produced are numbered so that they can be followed through the sugar rearrangement reactions of the cycle. Many of these reactions are common to the pentose phosphate pathway of respiration and were discussed in some detail in Chapter 14. That discussion should be reviewed before proceeding with the study of the reductive pentose cycle. As stated above, the reactions are largely equilibrium steps and no energy is consumed or generated during the cycling. The entire cycle occurs in the stroma of the chloroplast.

Beginning with glyceraldehyde phosphate (GAP #1 of Fig. 16.6), the enzyme triose phosphate isomerase catalyzes the conversion to dihydroxy-acetone phosphate (DHAP). A second GAP (#2)

REDUCTIVE PENTOSE CYCLE

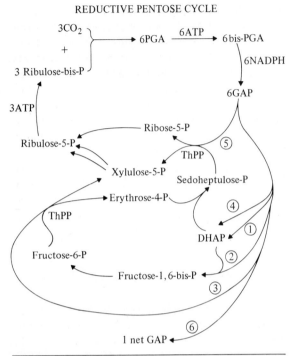

$$3CO_2 + 9ATP + 6NADPH \rightarrow 1GAP + 9ADP + 9NADP$$

FIGURE 16.6 The reductive pentose cycle, as outlined by Melvin Calvin and his associates. Refer to the text for details.

then condenses with the DHAP to form fructose-1,6-bisphosphate, catalyzed by the enzyme aldolase. The fructose-1,6-bisphosphate is hydrolyzed to fructose-6-P by a specific phosphatase. A third GAP (#3) reacts with the fructose-6-P to form the four-carbon sugar erythrose-4-P and the five-carbon ketose sugar xylulose-5-P through the thiamine pyrophosphate-dependent transketolase reaction. To this point three GAP have been consumed (a total of nine carbons), and a five-carbon and a four-carbon sugar have been produced.

A fourth GAP (#4) is converted to another DHAP through triose phosphate isomerase, which condenses with the erythrose-4-P of the transketolase reaction to form the seven-carbon sedoheptulose

phosphate. The reaction is catalyzed by a trans-aldolase enzyme. The sedoheptulose reacts with another GAP (#5) through a thiamine pyrophosphate transketolase reaction to form a second xylulose-5-P and a ribose-5-P. The five GAP with a total of 15 carbons have produced three 5-carbon sugars—two xylulose-5-P and one ribose-5-P. The xylulose-5-P are epimerized through a specific epimerase to form ribulose-5-P, and the ribose-5-P is isomerized by an isomerase to form a third ribulose-5-P. Ribulose-5-P kinase then catalyzes the phosphorylation of the three ribulose-5-P to form the original CO_2 acceptor molecules, ribulose-1,5-bisphosphate, completing the cycle.

Provided that each reaction occurs at least once, the reductive pentose cycle accounts for the assimilation of three CO_2 to give a net production of a three-carbon sugar, glyceraldehyde-3-P; it also regenerates the initial acceptor molecule, ribulose-1,5-bisphosphate.

16.2.8 Photosynthetic bacteria

Some species of bacteria are capable of reducing CO_2 to carbohydrate, using light energy and reduced sulfur compounds such as H_2S as the electron donors. Because water is not the electron donor, O_2 is not evolved. Many will deposit elemental sulfur or even accumulate sulfuric acid in vacuoles. Some of the photosynthetic bacteria, such as the purple sulfur and green sulfur bacteria, use reduced organic compounds and thus are not strictly photoautotrophs but rather photohetero-trophs. They are "photo" because light energy is used and heterotrophic because organic compounds are necessary as electron donors. The carboxylation reaction of the green sulfur bacteria follows.

$$CO_2 + 2H_2S \longrightarrow CH_2O + 2S + H_2O$$

One of the most interesting of the photosynthetic bacteria is the group that uses light to drive the reduction of CO_2 to form carbohydrate, using molecular hydrogen (H_2) as the reductant. Like the hydrogen bacteria, some green algae will photo-reduce CO_2 to carbohydrate, using H_2 when under anaerobic conditions.

$$CO_2 + 2H_2 \longrightarrow CH_2O + H_2O$$

Observations such as the above led van Neil in 1930 to generalize on photosynthesis as follows:

$$CO_2 + 2H_2A \xrightarrow{\text{light}} CH_2O + 2A + H_2O,$$

where H_2A is the electron donor. It was van Neil who proposed that the oxygen came from water, here symbolized as H_2A, and not from CO_2 as previously supposed. It has been confirmed from a variety of kinds of experiments that the oxygen comes from water and not CO_2.

For this reason it is more common and, of course, more correct to write the overall equation for photosynthesis with water on both sides.

$$6CO_2 + 12H_2O \longrightarrow C_6H_{12}O_6 + 6O_2 + 6H_2O$$

16.2.9 Secondary photosynthetic products

Although the initial products of photosynthesis are sugar phosphates, within a very few seconds there are produced neutral sugars, organic acids, amino acids, lipoidal compounds, and storage carbohydrates. The biosynthesis of some of these is discussed below.

Organic and amino acids

The first product of photosynthesis is itself an organic acid, phosphoglycerate. Once phosphoglycerate molecules are formed, some may be converted to pyruvate (see Chapter 14), and entry into the tricarboxylic acid cycle will ultimately produce many organic acids. The predominant ones are oxalacetate and α-ketoglutarate. The latter are readily aminated to form the corresponding amino acids aspartate and glutamate. The photosynthetic production of glycine and serine are discussed in Section 16.3.1.

As stated in Section 16.2.2, malate appears as an early product of photosynthesis, but the experiments to elucidate the carbon reactions of photosynthesis indicated that malate was not a component of the cycle. The important observation was that malonate would inhibit malate synthesis without significantly reducing overall CO_2 uptake. We now know that the malate produced during photosynthesis comes from a secondary carboxylation. The exception, of course, is in C_4-photosynthetic plants discussed in Section 16.4.

3-phosphoglycerate, the first product, is readily converted to 2-phosphoglycerate by catalysis through phosphoglyceryl mutase, and the enzyme enolase catalyzes the production of phosphoenolpyruvate (refer to Section 16.3.1).

$$\text{3-PGA} \longrightarrow \text{2-PGA} \longrightarrow \text{PEP}$$

The ubiquitous enzyme phosphoenolpyruvate carboxylase catalyzes the carboxylation of phosphoenolpyruvate to form oxalacetate, which is rapidly reduced to malate in plant tissues by malate dehydrogenase.

$$\text{PEP} + CO_2 \longrightarrow \text{OAA} \longrightarrow \text{MAL}$$

Thus malate appears as a secondary product that is synthesized through a secondary carboxylation.

Evidence for the secondary carboxylation comes from studies of ^{14}C distribution within the malate molecule after carboxylation with $^{14}CO_2$. Phosphoglycerate formed in photosynthesis will initially be labeled in the carbon-3 position.

$$\begin{array}{l} CH_2OPO_3^= \\ | \\ CHOH \\ | \\ {}^*CO_2^- \end{array}$$

If the above compound is converted to 2-phosphoglycerate and then to phosphoenolpyruvate, the latter will also be carboxyl-labeled at carbon-3.

$$\begin{array}{l} CH_2 \\ \| \\ C-O-PO_3^= \\ | \\ {}^*CO_2^- \end{array}$$

Subsequent carboxylation with phosphoenolpyruvate carboxylase results in a label at the carbon-4 position of oxalacetate (or malate).

$$
\begin{array}{c}
\overset{\displaystyle *CO_2^-}{\underset{\displaystyle *CO_2^-}{\overset{\displaystyle |}{\underset{\displaystyle |}{\underset{\displaystyle C-OPO_3^=}{\overset{\displaystyle CH_2}{\|}}}}}} + CO_2 \longrightarrow
\overset{\displaystyle *CO_2}{\underset{\displaystyle *CO_2^-}{\overset{\displaystyle |}{\underset{\displaystyle |}{\underset{\displaystyle C=O}{\overset{\displaystyle CH_2}{|}}}}}}
\end{array}
$$

Both carbon-1 and carbon-4 will be labeled. Furthermore, since only one of the phosphoglycerate molecules will be labeled during the photosynthetic carboxylation of ribulose bisphosphate, the distribution of ^{14}C in the malate molecule will be two-thirds in carbon-4 and one-third in carbon-1. This ratio has been observed and is taken as evidence for the secondary carboxylation hypothesis for the synthesis of malate during photosynthesis.

In Section 16.4 it will be shown that the C_4-photosynthetic plants do in fact synthesize malate as a first product and that ^{14}C from $^{14}CO_2$ appears initially at carbon-4 of malate.

An important initial organic acid product of photosynthesis is glycolate. Glycolate is the primary substrate for photorespiration and will be discussed in detail in Section 16.3.

Polymeric compounds

The biosynthesis of the polymeric compounds—nucleic acids, protein, carbohydrate, and lipid—was discussed in Chapters 8, 9, 10, and 11. These chapters should be reviewed at this time.

Hydrolysis of hexose phosphates by phosphatases produces neutral hexoses, which are used for the biosynthesis of storage carbohydrate. Neutral carbohydrate molecules or even hexose phosphates are readily converted to pyruvate and acetate, which are substrates for lipid and nucleic acid biosynthesis. Once amino acids are produced, protein will be formed.

Generally, when green plants are growing under conditions that favor carbohydrate biosynthesis (strong light), starch or glucan concentration will be high and lipid and protein synthesis will be low. Competition for substrates will govern ratios of products. For example, in strong light and low nitrogen, starch will accumulate. In the presence of ample nitrogen sources, protein will accumulate at the expense of storage carbohydrate.

16.3 Photorespiration

Much research has shown that certain plants such as soybean, wheat, and oats would produce CO_2 in the light when in the presence of oxygen. The oxygen effect was attributed in part to a depression of photosynthesis and in part to a stimulation of respiration. Because oxygen consumption and CO_2 evolution in the light differ in many respects from respiration in the dark, they are collectively referred to as photorespiration. It was clear that photorespiration contributed to a decrease in CO_2 assimilation and thus a decrease in primary productivity.

Furthermore, it was shown that other plants—in particular, some of the tropical grasses such as maize and sugarcane (C_4 species)—did not evolve CO_2 in the light and either apparently lacked

photorespiration or had the capacity to reassimilate the photorespired CO_2 prior to its loss.

An experiment illustrating photorespiration by tobacco and the lack of apparent photorespiration by maize is shown in Fig. 16.7. Leaf discs were supplied with $^{14}CO_2$ in the light and placed in a chamber lacking CO_2. CO_2 evolution in the light is a measure of photorespiration. In the dark, CO_2 evolution is approximately the same in both species, indicating comparable rates of respiration.

16.3.1 The biochemistry of photorespiration

Many plants, and in particular those referred to as C_3 species, show a marked oxygen inhibition of

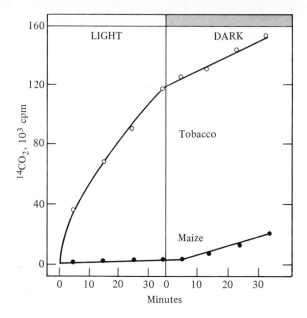

FIGURE 16.7 Production of CO_2 in the light by tobacco and maize as a measure of photorespiration. The experiment indicates that tobacco shows a measurable photorespiration but maize does not. Both species produce CO_2 in the dark by respiration. The experiment was conducted by allowing the plants to photosynthesize $^{14}CO_2$ in the light and then placing them in CO_2-free air to measure $^{14}CO_2$ release by photorespiration. From I. Zelitch. 1968. Investigations on photorespiration with a sensitive ^{14}C-assay. *Plant Physiol.* 43:1829–1837.

CO_2 uptake. Oxygen inhibition of photosynthesis was first observed by Warburg in 1920 and is called the Warburg effect.

When an experiment is conducted in which CO_2 uptake is measured as a function of light, most plants, and in particular C_3 species, show light saturation at about 20% of full sunlight (refer to Fig. 16.19). However, when the same experiment is conducted at 1% or less O_2, little light saturation is observed. This kind of experiment indicates that oxygen inhibition of photosynthesis (the Warburg effect) can be reversed.

A further observation important for an understanding of photorespiration is that at low CO_2 and high O_2 there is an increased glycolate biosynthesis during photosynthesis. Thus there is both a direct inhibition of CO_2 uptake by oxygen and an increased production of glycolate. A hypothesis to explain photorespiration must take into account both of these observations. The important questions are: (1) how is glycolate synthesized; (2) what is the pathway for dissimilation of glycolate; and (3) what is the direct effect of oxygen on photosynthesis?

Because glycolate becomes uniformly labeled very quickly in experiments with $^{14}CO_2$, it was assumed that there was a primary carboxylation to form glycolate. A possible sequence of reactions follows.

$$2CO_2 \longrightarrow CO_2^- {-} CO_2^- \longrightarrow$$
$$CHO{-}CO_2^- \longrightarrow CH_2OH{-}CO_2^-$$

Carboxylation forms oxalate ($CO_2^- {-} CO_2^-$), and subsequent reduction forms glyoxylate ($CHO{-}CO_2^-$) and then glycolate ($CH_2OH{-}CO_2^-$).

Because little oxalate is produced in most species and because it appears that glycolate is formed prior to glyoxylate, such a hypothesis has remained unproven.

Another possible source of photosynthetic glycolate is from one of the ketose sugars of the reductive pentose cycle as a result of oxidation during the transketolase reactions.

$$
\begin{array}{ccc}
CH_2OH & & CH_2OH \\
| & & | \\
C{=}O & \longrightarrow & HOC{=}O \quad +R \\
| & & \\
R & &
\end{array}
$$

However, it is known that when the glycolate is in the presence of $^{18}O_2$, oxygen-18 is incorporated into the carboxyl group of glycolate. Such labeling would not be possible with the ketose sugar hypothesis. Some glycolate could be formed through such a mechanism, but most of it must be produced by an alternative route.

It is now known (see Chollet and Ogren, 1975) that the carboxylating enzyme, ribulose bisphosphate carboxylase, has an oxygenase function as well as a carboxylase function. This oxygenase function, named ribulose bisphosphate oxygenase, forms phosphoglycolate and phosphoglycerate from ribulose bisphosphate.

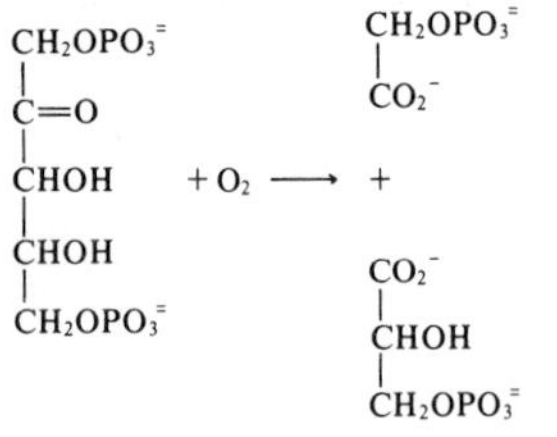

$$
\begin{array}{l}
CH_2OPO_3^= \\
| \\
C{=}O \\
| \\
CHOH \quad + O_2 \longrightarrow \\
| \\
CHOH \\
| \\
CH_2OPO_3^=
\end{array}
\qquad
\begin{array}{l}
CH_2OPO_3^= \\
| \\
CO_2^- \\
\\
+ \\
\\
CO_2^- \\
| \\
CHOH \\
| \\
CH_2OPO_3^=
\end{array}
$$

The phosphoglycolate is then hydrolyzed to glycolate by a specific chloroplast phosphatase.

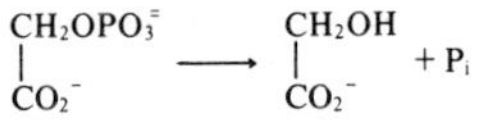

$$
\begin{array}{l}
CH_2OPO_3^= \\
| \\
CO_2^-
\end{array}
\longrightarrow
\begin{array}{l}
CH_2OH \\
| \quad + P_i \\
CO_2^-
\end{array}
$$

The discovery of the oxygenase function of ribulose bisphosphate carboxylase clarified many confusing features of carbon metabolism. First, it was known that the stoichiometry of the carboxylase was not always correct in that there were not two moles of phosphoglycerate produced for each mole of ribulose bisphosphate consumed. Second, the oxygenase function accounts for early glycolate synthesis, and after a few cycles of the reductive pentose cycle, glycolate will be uniformly labeled when in the presence of $^{14}CO_2$. Finally, it accounts for the direct oxygen inhibition of CO_2 fixation by the carboxylase.

Furthermore, inspection of the reaction shown above indicates that oxygen-18 would be incorporated into the carboxyl group of glycolate when the glycolate is in the presence of $^{18}O_2$, consistent with the known facts.

16.3.2 The glycolate pathway

The metabolic pathway for glycolate metabolism in green-leaf tissue is shown in Fig. 16.8. Phosphoglycolate is produced from ribulose bisphosphate in the chloroplasts. In addition, within the chloro-

plast the phosphoglycolate is hydrolyzed to glycolate. The latter is transported to the peroxisomes, the single-membrane-bounded organelles of green leaves, where it is oxidized to glyoxylate. The protein catalyst is glycolate oxidase, a specific marker enzyme for peroxisomes.

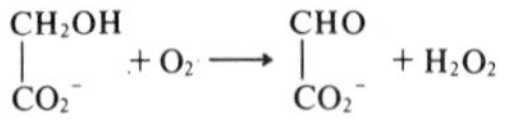

$$
\begin{array}{l}
CH_2OH \\
| \quad + O_2 \longrightarrow \\
CO_2^-
\end{array}
\begin{array}{l}
CHO \\
| \quad + H_2O_2 \\
CO_2^-
\end{array}
$$

The glycolate oxidase reaction consumes oxygen and produces hydrogen peroxide. The peroxide is converted to water and O_2 by the enzyme catalase.

Glycine is formed from glyoxylate through a coupled reaction with glutamate-glyoxylate aminotransferase. The latter enzyme is also localized in peroxisomes. The amino acid glycine, which results from the above transaminase reaction, is transported to the mitochondria where it enters into a reaction requiring tetrahydrofolate (THFA) to produce serine. Two glycines react to produce the three-carbon amino acid serine, CO_2, and ammonia. Because the reaction is readily reversible, continued catalysis results in the conversion of all the glycine carbon to CO_2, representing a loss in photosynthetic CO_2 uptake. Furthermore, there is little or no energy conservation during these oxidations. For this reason, the photorespiration-glycolate pathway is considered to be wasteful.

16.3.3 The significance of photorespiration

There is substantial evidence that a large proportion of the CO_2 assimilated by photosynthesis is lost through the glycolate pathway of photorespiration. Photorespiration contributes directly to a reduction in dry-matter accumulation and ultimately in yield.

Those plants that have reduced photorespiration, such as maize and sugarcane, frequently give very high yields. It would be to our economic advantage from an agricultural standpoint to develop plants with reduced photorespiration. However, approaches to such a goal have been limited, but efforts have been made to select plants that lack the

THE GLYCOLATE AND GLYCERATE PATHWAYS

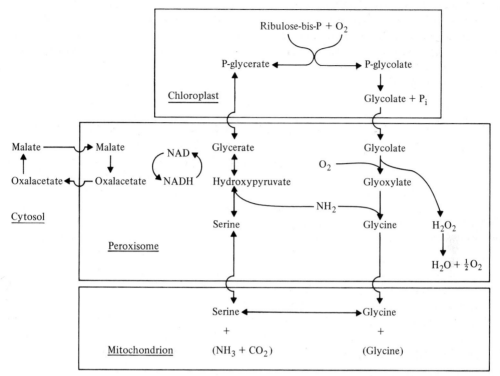

FIGURE 16.8 The metabolic pathway of photo-respiration in green plants. Glycolate and glycerate are produced in the chloroplasts. The glycolate is transported to peroxisomes where it is oxidized, ultimately forming glycine. The glycine is further metabolized in the mitochondria to form serine. Serine can go back to the peroxisomes where it can be converted to glycerate that can enter into the photosynthetic cycle in the chloroplasts. The reactions are such that serine and glycine could be synthesized from glycerate.

ribulose bisphosphate oxygenase function or that have a high capacity for reassimilation of photorespiratory CO_2.

The oxygenase function of the ribulose bisphosphate carboxylase seems to be rather general throughout the plant kingdom, and most efforts are directed toward selecting plants with high CO_2 uptake rates and low photorespiration rates through classical selection processes.

Although photorespiration does bring about a loss of carbon with no conservation of energy (in the form of ATP), the amino acids glycine and serine are synthesized. For this reason, it may be wrong to assume that photorespiration is an en-tirely wasteful process.

$$2 \begin{array}{c} \text{CH}_2\text{—NH}_2 \\ | \\ \text{CO}_2^- \end{array} \xrightleftharpoons{\text{THFA}} \begin{array}{c} \text{CH}_2\text{OH} \\ | \\ \text{CHNH}_2 + \text{CO}_2 + \text{NH}_3 \\ | \\ \text{CO}_2^- \end{array}$$

To complete the sequence, serine is transported back to the peroxisomes where it enters into a reaction catalyzed by serine-glyoxylate aminotransferase, producing hydroxypyruvate. The latter is reduced to glycerate via hydroxypyruvate reductase, an NADH-requiring reaction.

$$\begin{array}{c} \text{CH}_2\text{OH} \\ | \\ \text{C}=\text{O} \\ | \\ \text{CO}_2^- \end{array} + \text{NADH} \rightleftharpoons \begin{array}{c} \text{CH}_2\text{OH} \\ | \\ \text{CHOH} \\ | \\ \text{CO}_2^- \end{array} + \text{NAD}$$

The glycerate is then available for the photosynthetic glycerate pool (phosphoglycerate).

It should be noted that many of these enzyme reactions are reversible and that glycerate from photosynthesis can be converted to serine and glycine by reversal of the appropriate reactions. And indeed, glycine and serine are important early products of photosynthesis. Glycine and serine can be formed through the glycolate pathway as outlined in Fig. 16.8 or by reversal of the reactions from glycerate to serine. This latter process is sometimes referred to as the glycerate pathway and takes place without loss of CO_2 (up to serine) or oxygen

consumption. It is not considered to be photorespiration.

The glycolate pathway of photorespiration consumes oxygen at two different steps. The initial reaction, which produces phosphoglycolate and phosphoglycerate, consumes one mole of O_2. A second mole is consumed in the glycolate oxidase step; however, half is regenerated by catalase. The CO_2 evolution is largely from the glycine–serine conversion reaction. Again it should be pointed out that little or no energy is conserved during photorespiration; thus it is largely wasteful in terms of carbon and energy metabolism.

16.4 Photosynthesis in C_4 Species

In 1965 Kortschak, Hartt, and Burr observed that the photosynthetic products of sugarcane behaved as if malate were the first product and phosphoglycerate a secondary product. Figure 16.9 shows their experiment with sugarcane, clearly indicating

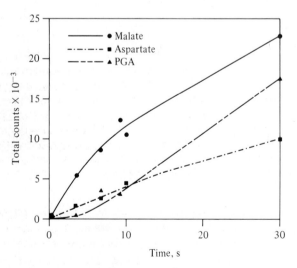

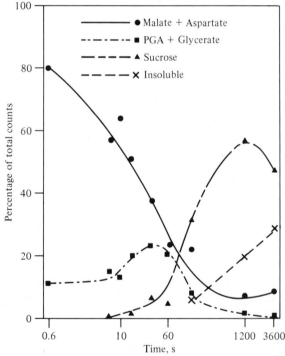

FIGURE 16.9 Left: Time course of malate, aspartate, and phosphoglycerate synthesis by sugarcane in light. Malate and aspartate are predominant. Right: Photosynthesis data of sugarcane, graphed as a percentage of radioactivity in each product at each measurement time (see Fig. 16.3). The data indicate that malate (plus aspartate) is the first stable product. From H. P. Kortshak, C. E. Hartt, and G. O. Burr. 1965. Carbon dioxide fixation in sugarcane leaves. *Plant Physiol.* 40: 209–213.

that malate is the first product. In addition, when malonate was added to a suspension of sugarcane leaf slices, both malate and carbon assimilation were reduced, unlike the process in most plants. These two pieces of evidence indicated that malate was indeed the first product of photosynthesis in sugarcane.

Shortly after the publication of the above experiments, Hatch and Slack (1970) reconfirmed that malate was the first product of photosynthesis in sugarcane. They also were able to draw this conclusion for a large group of plants, including many tropical grasses and some dicotyledonous plants falling within the Centrospermae. These plants were subsequently referred to as C_4 species because the first product was a compound with four carbons, malate. Other plants were called C_3 species because phosphoglycerate, the first product, is a compound with three carbons. Plant families with C_4 species are listed in Table 16.1.

TABLE 16.1 Plant families known to have C_4 species and CAM species

C_4	CAM
Aizoaceae	Agavaceae
Amaranthaceae	Aizoaceae
Asteraceae	Asclepiadaceae
Chenopodiaceae	Asteraceae
Cyperaceae	Bromeliaceae
Euphorbiaceae	Cactaceae
Gramineae	Crassulaceae
Eragrostoideae	Cucurbitaceae
Arundinoideae	Didiereaceae
Panicoideae	Euphorbiaceae
Nyctaginaceae	Geraniaceae
Portulacaceae	Labiatae
Zygophyllaceae	Liliaceae
	Orchidaceae
	Oxalidaceae
	Piperaceae
	Polypodiaceae
	Portulacaceae
	Vitaceae
	Welwitschiaceae

16.4.1 Characteristics of C_4 species

Unlike the mesophyll cells of dicotyledonous C_3 species, which are differentiated into a spongy and a palisade parenchyma, C_4 species are characterized by having Kranz (kranz = wreath or border) leaf anatomy. In Kranz anatomy the vascular bundles of the leaf are bounded by large, chloroplast-containing bundle sheath cells, and radiating out from the bundle sheath cells are the more typical cells with chloroplasts. The latter are simply called mesophyll cells to differentiate them from the bundle sheath cells that also have chloroplasts.

Kranz anatomy is depicted in Fig. 16.10. The chloroplasts in the mesophyll cells of C_4 species have typical granal stacking and in all respects appear similar to the usual chloroplasts of higher green plants. However, they lack the enzyme ribulose bisphosphate carboxylase, the main carboxylating enzyme of CO_2 assimilation.

The large bundle sheath chloroplasts, on the other hand, usually lack grana when mature, having instead an unstacked lamellar membrane system, and they frequently accumulate much starch (see Fig. 16.10). These chloroplasts do have ribulose bisphosphate carboxylase. These anatomical and enzymatic differences between the two cell types result in the most important feature of C_4 species, which is the partitioning of photosynthetic carbon metabolism functions between mesophyll and bundle sheath cells. The concept is outlined in Fig. 16.11 and shows an initial carboxylation by phosphoenolpyruvate carboxylase in the cytosol of the mesophyll cells to form a C_4-acid, malate or aspartate. The complete reaction is as follows.

$$PEP + CO_2 \longrightarrow OAA \begin{cases} \nearrow ASP \\ \searrow MAL \end{cases}$$

The first product of the carboxylation is oxalacetate (OAA), which is rapidly reduced to malate (MAL) or aminated to form aspartate (ASP). The C_4-acid, malate or aspartate, is transported to the bundle sheath cells where it is decarboxylated to yield a C_3 compound and free CO_2. This CO_2 is

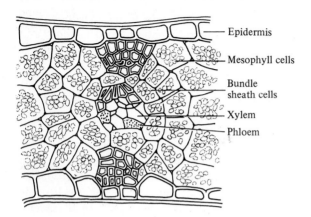

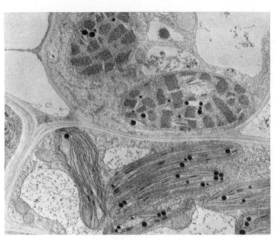

FIGURE 16.10 Kranz anatomy indicative of C_4 species. The drawing shows the typical arrangement of chloroplast-containing bundle sheath cells and relatively undifferentiated mesophyll cells. The inset is an electron micrograph (about ×15,000) showing a section through a bundle sheath cell and adjacent mesophyll cell. The chloroplasts of the bundle sheath cell have a lamellar arrangement. The chloroplasts of the mesophyll cell have typical grana. Photomicrograph of maize courtesy of R. L. Heath, University of California, Riverside.

assimilated photosynthetically through carboxylation with ribulose bisphosphate carboxylase to form phosphoglycerate. The subsequent reactions are identical to those of the reductive pentose cycle, and it is in fact the reductive pentose cycle portion of the process that results in the photosynthetic assimilation of CO_2.

The 3-carbon fragment is transported back to the mesophyll cytosol as pyruvate, phosphoenolpyruvate, or alanine, depending on the species. Both pyruvate and alanine are converted to phosphoenolpyruvate, which is then ready to undergo another cycle of carboxylation. Alanine is first converted to pyruvate through a transaminase reaction. Pyruvate is converted to phosphoenolpyruvate by a unique plant enzyme, pyruvate inorganic phosphate dikinase. The dikinase reaction follows.

$$\text{PYR} + P_i + \text{ATP} \longrightarrow \text{PEP} + \text{AMP} + P_i P_i$$

The reaction, catalyzed by pyruvate inorganic phosphate dikinase, appears improbable because of the unfavorable energetics, but it is driven by the hydrolysis of the pyrophosphate ($P_i P_i$), eliminating one of the products and preventing reversal. The hydrolysis of pyrophosphate, catalyzed by pyrophosphatase, is quite common in plant tissues.

$$P_i P_i \longrightarrow 2 P_i$$

The dikinase enzyme is light-activated and requires

THE C_4 PHOTOSYNTHESIS PATHWAY

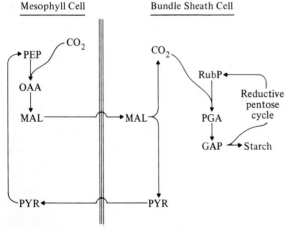

FIGURE 16.11 The metabolic pathway for C_4-photosynthesis.

TABLE 16.2 Enzyme distribution between mesophyll cells and bundle sheath cells of a typical C$_4$ species

Mesophyll Cells	Bundle Sheath Cells
Phosphoenolpyruvate carboxylase	Ribulose bisphosphate carboxylase
Pyruvate P$_i$ dikinase	Malate enzyme
NADP-malate dehydrogenase	Ribose-P isomerase
Pyrophosphatase	Ribulose-P kinase
Phosphoglycerate kinase	Phosphoglycerate kinase
NADP-triosephosphate dehydrogenase	NADP-triosephosphate dehydrogenase

Data from Hatch and Slack (1970).

ATP from photophosphorylation. Formation of phosphoenolpyruvate from pyruvate is a photosynthetically dependent reaction. The entire process, including two carboxylations, is a light-dependent process and will not occur to any appreciable extent in the dark.

The enzyme distribution between mesophyll and bundle sheath cells is shown in Table 16.2. Perhaps the most important to note is the phosphoenolpyruvate carboxylase and pyruvate inorganic phosphate dikinase of the mesophyll cells, and the ribulose bisphosphate carboxylase, the decarboxylating malate enzyme, and many of the reductive pentose cycle enzymes of the bundle sheath cells. Of interest is that triose phosphate dehydrogenase (requiring NADP) and phosphoglycerate kinase occur in comparable quantities in both cell types.

16.4.2 Transport in C$_4$ species

The transport of metabolites from mesophyll cells to bundle sheath cells and the return during C$_4$-photosynthesis presents several problems. However, electron microscopy indicates that there are many plasmodesmata between mesophyll and bundle sheath cells, and indeed calculations suggest that simple diffusion would be sufficient to support continued photosynthesis. A second problem of transport is that pyruvate and phospho-

enolpyruvate are very reactive biochemicals and are not as likely to be transported as are alanine and malate, which are more stable. Nevertheless, one proposal is that transport in some species is by a malate/pyruvate shuttle system between mesophyll cells and bundle sheath cells. The malate synthesized in the mesophyll cells is rapidly transported to the bundle sheath cells where it is decarboxylated by malate enzyme to form CO$_2$ and pyruvate. Pyruvate is then transported back to the mesophyll cells.

There are some species of C$_4$ plants that produce more aspartate than malate and lack significant amounts of NADP-malate enzyme in their bundle sheath cells. These plants are aspartate transporters, and the C$_3$-fragment that returns is alanine rather than pyruvate. There is another group that transports aspartate from the mesophyll cells to the bundle sheath cells but returns phosphoenolpyruvate rather than pyruvate or alanine. On the basis of the metabolite transported and the kinds of enzymes within the cells, C$_4$ species can be grouped into three classes.

NADP-malate enzyme type

Plants that transport malate/pyruvate and decarboxylate with NADP-malate enzyme are referred to as NADP-malate enzyme types. Such species include maize and sugarcane. These plants, as well as being high in NADP-malate enzyme activity in bundle sheath cells, also have high NADP-malate dehydrogenase in chloroplasts of the mesophyll cells. It is assumed that after oxalacetate formation the malate is synthesized by catalysis with NADP-malate dehydrogenase.

$$OAA + NADPH \longrightarrow MAL + NADP$$

Phosphoenolpyruvate carboxykinase type

C_4 plants that transport aspartate/alanine may have phosphoenolpyruvate carboxykinase in place of NADP-malate enzyme. The phosphoenolpyruvate carboxykinase reaction is as follows.

$$OAA + ATP \longrightarrow PEP + CO_2 + ADP$$

In these species the C_4-acid transported from mesophyll to bundle sheath cells is aspartate and the C_3-fragment returned is alanine. Thus the phosphoenolpyruvate is converted to alanine, probably through pyruvate.

Oxalacetate and pyruvate are readily shuttled as amino acids by reactions including aminotransferase enzymes. The general scheme is as follows.

$$OAA \diagdown \diagup GLU \diagdown \diagup PYR$$
$$ASP \diagup \diagdown AKG \diagdown ALA$$

Since these reactions are readily reversible, oxalacetate and aspartate are interconvertible, as are pyruvate and alanine. Because of the reciprocal nature of the aminotransferases, ammonia will not be evolved in the tissue.

Transport of the keto acids as amino acids begins with oxalacetate, synthesized by phosphoenolpyruvate carboxylase. Oxalacetate is converted to aspartate by aspartate aminotransferase. The amino donor is glutamate, which is converted to α-ketoglutarate in the process. The aspartate is transported to the bundle sheath cells and converted back to oxalacetate by catalysis through another aminotransferase. Research has shown that although these transferases catalyze the identical reaction they are different isozymes. Oxalacetate is then either converted back to malate and decarboxylated or, in the case of the phosphoenolpyruvate carboxykinase type species, is decarboxylated directly to form phosphoenolpyruvate. The phosphoenolpyruvate is then converted to pyruvate and transported, or the pyruvate is converted to alanine as described above and transported. Once the alanine is back to the mesophyll cells, it is converted to pyruvate and then back to phosphoenolpyruvate

by pyruvate P_i dikinase to regenerate to the original acceptor molecule.

NAD-malate enzyme type

There is still another group of C_4 species that lack significant activities of both NADP-malate enzyme and phosphoenolpyruvate carboxykinase but have high levels of an NAD-malate enzyme.

$$MAL + NAD \longrightarrow PYR + CO_2 + NADH$$

The NAD-malate enzyme is localized in large mitochondria of the bundle sheath cells.

The C_4-acid and the C_3-fragment transported in these species are aspartate and alanine, like the phosphoenolpyruvate carboxykinase type.

As with all attempts to classify groups of living organisms, there are always many exceptions that make the groups less well defined. Many C_4 species do not fall clearly in one of the three groups above, and some apparently transport more than one kind of molecule.

16.4.3 Photosystems of C_4 species

In most plants the ratio of chlorophyll a to chlorophyll b is about 3.5. In the agranal chloroplasts of the bundle sheath cells of C_4 species and, in particular, in the NADP-malate enzyme types, the a to b ratio is as high as 6. Furthermore, they have a reduced Hill activity and thus appear to be deficient in photosystem II. This would imply a reduction in noncyclic electron flow and a low production of NADPH. Cyclic electron flow and ATP production would be possible, and indeed the evidence suggests this.

It is perhaps significant that these same species that are photosystem II deficient and thus necessarily have a limited capacity for NADP reduction produce NADPH during the decarboxylation of malate in the NADP-malate enzyme reaction.

It should be clear that the overall energy requirement for CO_2 fixation by C_4 species is greater than for C_3 species. In C_3 species there is a theoretical requirement for 3 ATP and 2 NADPH per CO_2

assimilated. In addition to any energy requirement for transport, in C₄ species ATP is required to regenerate the phosphoenolpyruvate that is the substrate for the first carboxylation. Additionally, at least 2 ATP are required for the pyruvate P_i dikinase reaction, and in the case of the phosphoenolpyruvate carboxykinase types an additional ATP is required for the decarboxylation of oxalacetate to form phosphoenolpyruvate. Thus two or three more ATP equivalents are required in C₄ species.

Because the actual reductive steps of photosynthesis are the same in both C₃ species and C₄ species (that is, in the reductive pentose cycle), the NADPH requirements are comparable.

16.4.4 Carbon isotope discrimination

In unpolluted air the percentages of ^{12}C, ^{13}C, and ^{14}C in CO_2 are about 98.9%, 1.1%, and 10^{-10}%, respectively. During photosynthesis, plants do not assimilate the three isotopes equally by carboxylation but rather discriminate against the heavier forms. Therefore, photosynthesis results in carbon isotope fractionation.

The composition of ^{13}C in plant tissues is expressed as a ratio of ^{13}C to ^{12}C relative to a standard obtained from a carbonate fossil skeleton of *Belemnitella*. The latter was obtained from the Peedee formation found in South Carolina. The index ratio is as follows.

$$\delta^{13}C \; ^o/_{oo} = \left(\frac{^{13}C/^{12}C \text{ sample}}{^{13}C/^{12}C \text{ standard}} - 1 \right) \cdot 1000$$

When the ^{13}C isotope composition of clean, unpolluted air is expressed on the basis of the above equation, the ^{13}C to ^{12}C relative ratio is $-7\,^o/_{oo}$, that is, the $\delta^{13}C$ is $-7\,^o/_{oo}$. This figure indicates that unpolluted air free from contamination has about 0.7% less ^{13}C than the carbonate standard.

C₃-photosynthetic plants when completely combusted show a $\delta^{13}C$ mode of about $-27\,^o/_{oo}$, or $20\,^o/_{oo}$ less than clean air. Thus there is approximately a 2% discrimination of $^{13}CO_2$ over $^{12}CO_2$ during photosynthesis by C₃ species. C₄ species, however, show a $\delta^{13}C$ mode of about $-11\,^o/_{oo}$, 1.6% less discrimination than C₃ species. Figure 16.12 shows the carbon isotope composition of C₃ and C₄ species.

Research has shown that the carboxylating enzymes are responsible for much of the isotope fractionation. Phosphoenolpyruvate carboxylase discriminates less against $^{13}CO_2$ than does ribulose

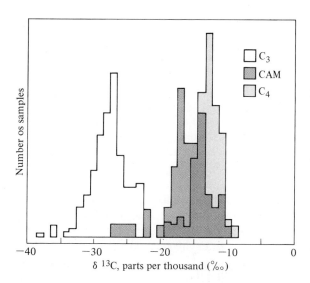

FIGURE 16.12 Bar graph showing the ^{13}C composition of C₃, C₄, and CAM species expressed as $\delta^{13}C$ in $^o/_{oo}$. Unpolluted air has a ^{13}C of $-7\,^o/_{oo}$, indicating that air has less $\delta^{13}C$ than the standard prepared from a fossil carbonate. The mode for C₃ plants is $-27\,^o/_{oo}$, and the mode for C₄ plants is $-11\,^o/_{oo}$. Hence C₄ plants have a higher $\delta^{13}C$ composition than C₃ plants. CAM plants, because of the nature of their carbon-metabolism pathway, show a variable isotope composition. Adapted from J. C. Lerman. 1975. How to interpret variations in the carbon isotope ratio of plants: Biologic and environmental effects. In R. Marcelle (ed.), *Environmental and Biological Control of Photosynthesis*. W. Junk Publ., The Hague, Netherlands.

Figure legend: C₃, CAM, C₄

Number os samples

$\delta^{13}C$, parts per thousand ($^o/_{oo}$)

bisphosphate carboxylase. Plants that fix atmospheric CO_2 by the phosphoenolpyruvate carboxylase reaction have a carbon isotope composition closer to air than plants that assimilate via the ribulose bisphosphate carboxylase enzyme exclusively. It may not be immediately apparent why there should be fractionation when the assimilation of CO_2 to produce sugar is catalyzed by ribulose bisphosphate carboxylase in both C_4 and C_3 species. In C_4 species the fractionation occurs because of the first carboxylation, catalyzed by phosphoenolpyruvate carboxylase. Provided that all the CO_2 generated by decarboxylation of the 4-carbon product is fixed by ribulose bisphosphate catalysis, the isotope ratio will reflect the carboxylation by phosphoenolpyruvate carboxylase.

The isotope composition within a species is so uniform and can be determined with such great precision with the use of a mass spectrometer that it is possible to quickly ascertain if a sample of common table sugar (sucrose) came from sugar beets (a C_3 species) or from sugarcane (a C_4 species). The $\delta^{13}C$ of the sugar from sugar beets will be about -27 $^o/_{oo}$ and the $\delta^{13}C$ of the sugar from sugarcane will be about -11 $^o/_{oo}$.

It is known that there is fractionation of other isotopes by plants during metabolism. Both photosynthesis and respiration result in fractionation of $^{18}O_2$* and $^{16}O_2$. Exactly why such fractionation occurs is not entirely clear, but compounds with isotopes do vary in mass and it is reasonable to assume that isotope effects result because of the slightly different enzyme responses.

16.4.5 The biological significance of C_4-photosynthesis

C_4 species appear to have greater rates of CO_2 assimilation than do C_3 species. Furthermore, they show little or no light saturation of photosynthesis when compared with C_3 species. Unlike C_3 species, oxygen has little effect on CO_2 uptake. Photosynthetic rates are similar in 21% and 1% O_2. This observation has led to the conclusion that C_4 species have reduced or absent photorespiration. Further evidence for this is that the CO_2 compensation point of C_4 species is close to zero whereas it is usually on the order of one third of ambient (i.e., 100 ppm) for C_3 species.

Because it is known that the ribulose bisphosphate carboxylase of C_4 species has oxygenase activity similar to C_3 species, it is assumed that the low photorespiration is due to a reduced oxygenase activity in C_4 species resulting from the high CO_2-to-O_2 ratio in the bundle sheath cells near the carboxylase. It is believed that the transport of the C_4-acids and decarboxylation in the bundle sheath cells keep CO_2 at a high level. The decarboxylation, coupled with the reduced photosystem II activity in some C_4 species, keeps the CO_2/O_2 ratio high, which favors the carboxylase activity over the oxygenase activity. Thus the significance of C_4-photosynthesis appears to be as a mechanism to reduce photorespiration and maximize CO_2 assimilation.

There is also substantial evidence that any CO_2 leakage from the bundle sheath cells, perhaps by glycolate metabolism, is reassimilated by the phosphoenolpyruvate carboxylase activity of the mesophyll cells, further reducing photorespiration.

Because most C_4 species are of tropical origin and occur naturally in arid or physiologically dry environments and because their temperature optima for photosynthesis and growth are higher than for most C_3 species, it is generally believed that C_4 species are adapted to dry and hot climates. Indeed, many of the most aggressive and abundant herbs of the hot tropics and arid lands are C_4 species. In the southwestern United States many of the noxious weeds, including crabgrass, Dallas grass and Russian thistle, among others, are C_4-photosynthetic plants. In many of the deserts in the same region, the winter annuals are C_3 species and the summer annuals are C_4 species. It seems reasonable to conclude that C_4-photosynthesis imparts adaptability to environments with water and heat stress.

*Of course, $^{18}O_2$ is rather improbable—the usual isotope of $^{16}O_2$ would be ^{18}O—^{16}O.

Furthermore, some of the highest-yielding agricultural crops fall within the C$_4$ group. Cereals such as sorghum, maize, and millet are C$_4$ species. Winter-grown or cool-weather cereals such as rice and wheat are C$_3$ species. The high-yielding sugarcane grown in the tropics is a C$_4$ species whereas sugar beets, also grown for high production of sugar, is a more mesic C$_3$ species. Some of the more favored arid-zone turf grasses are C$_4$ species, including Bermuda grass and St. Augustine grass. The more mesic turf grasses requiring more care and water such as the bents, fescues, and bluegrasses are C$_3$ species. Many of the arid prairie grasses such as *Aristida, Andropogon, Bouteloua,*

Buchloe, Hilaria, Paspalum, and *Sporobolus* are C$_4$ species.

Overall, it can be said that C$_4$ species do better in hot climates than in cold climates. For example, not one is found in the Arctic. However, it should be apparent that many C$_3$ species do equally well in hot climates. Just why C$_4$ species should be more adapted to hot, dry climates than C$_3$ species is not entirely clear. A consequence of the higher rates of photosynthesis by C$_4$ species is that transpiration ratios (the ratio of water loss to carbon assimilated) is lower in C$_4$ species than in C$_3$ species. Thus water efficiency in terms of transpiration ratio is greater in C$_4$ species than in C$_3$ species.

16.5 Crassulacean Acid Metabolism (CAM)

Many succulent plants have a carbon metabolism characterized by accumulation of massive quantities of organic acids while in the dark. Acidity increases and the pH of the cell sap decreases. Such plants have stomata open at night and closed during the day while in the light. As a consequence of the nocturnal stomatal opening, much of the transpiration occurs at night, as does the exogenous CO$_2$ uptake. During the night, while CO$_2$ is being fixed, starch or other storage glucan decreases.

The dark uptake of CO$_2$ by succulents was observed as early as 1810 by De Saussure while studying a cactus. In 1815 Heyne noted that the succulent *Bryophyllum* accumulated organic acids at night. In the years that followed, and in particular after the 1930s, when modern physiological and biochemical techniques became available, much effort was devoted by plant physiologists to understand the acid metabolism of succulent plants. After the development of the concept of C$_4$-photosynthesis in the 1960s there was much renewed interest in crassulacean acid metabolism because of the similarity of carbon metabolism between C$_4$ species and CAM species and the realization that CAM was an adaptive modification of basic photosynthetic metabolism.

The term crassulacean acid metabolism derives

from the large succulent plant family, the Crassulaceae, since many of the favorite test plants such as *Crassula, Sedum, Kalanchoe,* and *Bryophyllum* are in the Crassulaceae. In addition, the acronym CAM is easily pronounced and remembered. The term does imply, however, that CAM is unique to the Crassulaceae. Such is not the case, there being about 18 families of flowering plants, one fern family (Polypodiaceae), and *Welwitschia* of the Welwitschiaceae that have species with CAM. The most noteworthy flowering plant families with CAM species are the Crassulaceae, the Cactaceae, the Aizoaceae, and the Euphorbiaceae. Epiphytic species of the Orchidaceae and the Bromeliaceae frequently have CAM. Many of the nonepiphytic bromeliads are CAM species, the most important being pineapple.

Of course, not all species of any one family of succulent plants have CAM with perhaps the exception of the Cactaceae and the Crassulaceae. For example, only succulent euphorbias have CAM, and some families have C$_3$, C$_4$, and CAM species. Families that have all three types are the Aizoaceae, Asteraceae, Euphorbiaceae, and the Portulacaceae. Most families that have CAM species also have C$_3$ species. And, as will be explained below, some species have the capacity of shifting from C$_3$

to CAM in response to environmental perturbation. Not all families classified as succulent in the horticultural sense have CAM species. Refer to Table 16.1 for the known plant families that have species with crassulacean acid metabolism.

The following are features of the nighttime CO_2 metabolism of succulent plants with CAM.

1. Nocturnal stomatal opening
2. Nighttime transpiration
3. Nighttime CO_2 uptake
4. Acidity increase during the night
5. Starch depletion during the night
6. Acidity decrease during the day
7. Starch increase during the day
8. Daytime stomatal closure and little or no gas exchange

Studies with $^{14}CO_2$ have shown that the bulk of the acid fluctuations in succulent plants with CAM can be accounted for by malic acid. As much as 100 to 200 milliequivalents per kilogram fresh weight of malic acid may accumulate in the dark (Fig. 16.13).

During the subsequent light period, stomata close, the acidity decreases, and starch once again accumulates. It is known that malic acid decarboxylation to form CO_2 accounts for the acidity decrease and that the CO_2 is conserved by photosynthesis.

The metabolic pathway of CAM is similar in many respects to C_4-photosynthesis. Figure 16.14 shows carbon flow and should be compared with Fig. 16.10, showing carbon flow in C_4 species. The initial carboxylation at night is catalyzed by phosphoenolpyruvate carboxylase to yield oxalacetate. Oxalacetate is reduced to malate, and malic acid accumulates in the large vacuoles of the green mesophyll cells (see Fig. 16.15). Starch decreases concomitantly when glycolysis generates phosphoenolpyruvate, the substrate for carboxylation.

Light facilitates the release of malic acid from the vacuole, and decarboxylation of malate releases CO_2 into the tissues. Because stomata are now closed, CO_2 release is generally not possible; most CO_2 is refixed photosynthetically through the reductive pentose cycle, as in C_3 species.

It should be apparent that in CAM, malic acid acts as a storage compound for CO_2, whereas in C_4 species malate functions as a direct CO_2 donor and is an active intermediate. The carboxylations in CAM are separated temporally (night–day); the

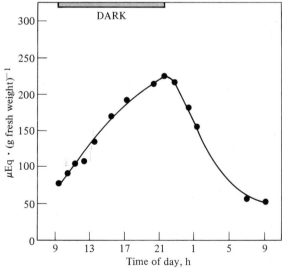

FIGURE 16.13 The diurnal fluctuation of acidity (microequivalents per gram fresh weight) during a 24-hour period. Organic acids build up at night and decrease during the daylight period. The data are from *Kalanchoe tubiflora*, a typical CAM plant. From B. G. Sutton. 1974. Regulation of carbohydrate metabolism in succulent plants. Ph.D. thesis, Australian National University, Canberra.

THE CAM PATHWAY

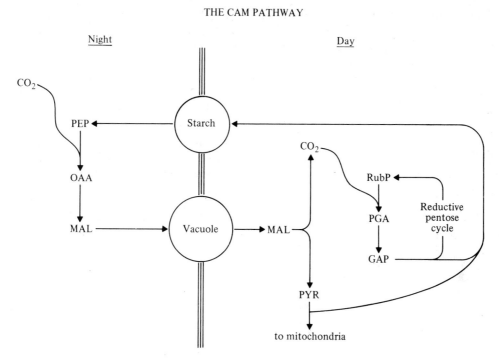

FIGURE 16.14 The metabolic pathway of crassulacean acid metabolism.

phosphoenolpyruvate carboxylase reaction takes place at night and the ribulose bisphosphate carboxylase reaction takes place during the day in the light. In C_4-photosynthesis the carboxylases are separated spatially in different cells, but both reactions take place in the light.

The biological importance of CAM appears to be a mechanism that reduces water loss during gas exchange. These plants are typically found in arid or physiologically dry environments. Many are true desert plants, some grow in shallow soils with little available water, others occur in salt-stressed environments, and still others are epiphytic, such as some bromeliads and certain orchids.

Stomata are open at night when the evaporative demand is low. CO_2 is thus taken up with a minimum of water loss. During the subsequent daylight period, plants are sealed tightly, minimizing water loss, and photosynthesis proceeds because of the internal generation of CO_2 from malic acid decarboxylation.

Evidence for the above generalization is shown in the reduced transpiration ratio of CAM succulents, which is on the order of 50 to 150 grams water loss to grams of CO_2 gained. C_3 species may have transpiration ratios as high as 1000, and C_4 species are around 500 but greater than 150.

The typical pattern of CO_2 exchange in succulent CAM plants is illustrated in Fig. 16.16. CO_2 uptake continues throughout the dark period until near the end, when malic acid is at a high level. Immediately after illumination there is a large burst of CO_2 uptake. Finally, the CO_2 uptake rate falls to zero during the day but may rise again toward the end of the light (day) period.

To a large extent the CO_2-uptake pattern of succulents with CAM is a function of nighttime stomatal opening and daytime closing. Stomata open

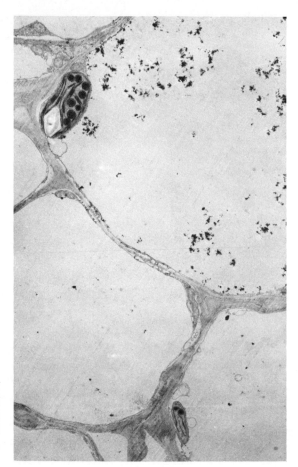

FIGURE 16.15 Electron micrograph of a portion of mesophyll of *Bryophyllum pinnatum*, a typical CAM plant. There is a very large vacuole that stores water and malic acid, and peripheral cytoplasm with a prominent chloroplast. Micrograph courtest of Dr. W. W. Thomson, University of California, Riverside.

FIGURE 16.16 Typical pattern of gas exchange (CO_2) in a CAM plant. CO_2 uptake occurs largely at night. The dashed line shows the expected CO_2 exchange curve for a non-CAM plant. From M. Kluge and I. P. Ting. 1979. *Crassulacean Acid Metabolism: An Analysis of an Ecological Adaptation*. Springer-Verlag, New York.

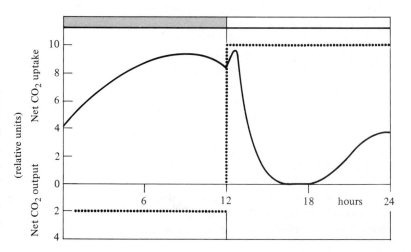

rapidly in the dark with only a slight tendency for closing toward the end of the dark period. They are closed throughout the day, exhibiting some opening toward the late afternoon and evening.

When succulent plants are performing the CAM pathway of photosynthesis with stomata open at night and closed during the day, the carbon isotope composition ($\delta^{13}C$) will be high: in the vicinity of $-14 °/oo$. This is because the first carboxylation at night is catalyzed by phosphoenolpyruvate carboxylase, which discriminates less against ^{13}C than does ribulose bisphosphate carboxylase. Provided that all the CO_2 comes from malic acid decarboxylation and all is fixed, the isotope composition of the sugars will be essentially the same as the

malic acid. However, if stomata are open during the day as well as at night, there may be some dilution of the CO_2 from malic acid with ambient CO_2 such that the fractionation will reflect ribulose bisphosphate carboxylase activity. Thus CAM plants tend to have a rather variable $\delta^{13}C$, unlike C_3 and C_4 plants, which have predictable values. This is clearly illustrated in Fig. 16.12. The mode for C_3 species is about $-27 °/oo$ and for C_4 species about $-11 °/oo$. The succulent species investigated that are capable of CAM have a $\delta^{13}C$ ranging from $-9 °/oo$ to $-27 °/oo$, depending on the source of CO_2 and the extent of carboxylation by phosphoenolpyruvate carboxylase and ribulose bisphosphate carboxylase.

16.6 Environmental Effects on Photosynthesis

Photosynthesis, being partly a physical process and partly a chemical process, is affected by a variety of environmental factors. Perhaps the easiest way to visualize these effects is to observe the response of gas exchange to light, temperature, oxygen, water, and CO_2.

16.6.1 Light

Provided that CO_2, temperature, and other environmental factors are not limiting, photosynthesis will increase when light is increased. An idealized curve is shown in Fig. 16.17. Many plants will show light saturation at about 20 to 30% of full sunlight; others, such as the C_4 species, will respond with an increase in CO_2 uptake far above full sunlight.

In the curve shown in Fig. 16.17, the point indicated as I_s is approximately the value of light saturation. An average I_s for leaves grown in full sunlight is about 0.56 kJ m^{-2} s^{-1}. Leaves that are shade-adapted will usually have an I_s point somewhat lower, in the vicinity of 0.18 or less.

The point on the graph where the curve intersects the abscissa is the point at which CO_2 uptake just balances CO_2 release. This is the point where light

is so low that photosynthesis just balances respiration; it is referred to as the light compensation (I_c). The I_c will vary among plants, but for C_3 species it is usually on the order of 0.02 kJ m^{-2} s^{-1} for sun-adapted leaves and about 0.007 for shade-adapted leaves.

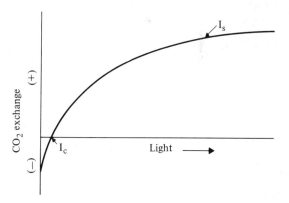

FIGURE 16.17 Typical CO_2 exchange curve for a C_3 species as a function of light. $(+)$ = CO_2 uptake. $(-)$ = CO_2 release. I_s = light value at which further increases are not accompanied by an increase in CO_2 uptake. I_s is called the light saturation point. I_c = light value at which CO_2 uptake equals CO_2 release. I_c is the light compensation point.

C_4 species tend to have an I_c close to zero, largely because of the lack of photorespiration and the high capacity for refixation of any CO_2 generated through respiratory processes.

When a light saturation experiment is conducted at low O_2, neither C_3 nor C_4 species show much light saturation. Such an experiment is illustrated in Fig. 16.18. Two species are compared: *Atriplex rosea*, a C_4 plant, and *A. patula*, a C_3 plant. At 1.5% O_2, both species show an increase in net photosynthesis up to high irradiance, but at 21% O_2 *A. patula* saturates just above 0.1 kJ m^{-2} s^{-1}, and *A. rosea* does not. There is little difference in the response of *A. rosea*, the C_4 species, at low and at normal O_2. As mentioned above, these data are explained on the basis of low photorespiration in C_4 species. At low O_2 concentrations, the C_3 *A. patula* has a reduced photorespiration and hence does not show light saturation to the same extent that it does at normal O_2 concentrations. Thus the light saturation point may be the point where photorespiration and CO_2 uptake by photosynthesis are in equilibrium.

The I_c, or light compensation point, is more nearly related to dark respiration since photorespiration would be less at low light intensities.

16.6.2 Temperature

Temperature affects photosynthesis and gas exchange in many ways. Stomata are temperature-sensitive, and at both low and high temperatures closure may occur. The low-temperature closing response probably occurs because of direct effects on enzymes and transport phenomena. High-temperature closing is almost certainly the result of water stress, which causes guard cells to become flaccid and stomata to close.

The direct effect of temperature on photosynthetic and respiratory gas exchange is shown in Fig. 7.10. As temperature increases, CO_2 uptake increases to a maximum and then begins to decrease. Respiration increases with temperature and reaches a maximum far above the photosynthetic optimum. Apparently the photosynthetic optimum

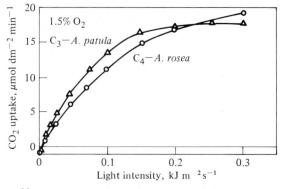

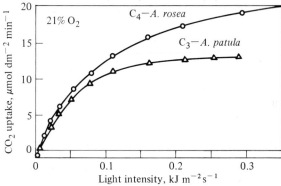

FIGURE 16.18 Net CO_2 uptake in the light by C_3 species *Atriplex patula* and C_4 species *Atriplex rosea* in normal air with 21% oxygen and in air with reduced oxygen. In air the C_3 species shows light saturation, indicative of photorespiration, whereas the C_4 species does not show much evidence of photorespiration. Adapted from O. Bjorkman, R. W. Pearcy and M. A. Nobs. 1970. Physiological ecology investigations. Photosynthetic characteristics. *Carnegie Institution Year Book* 69. Stanford, California.

occurs in part because of the large respiratory response to temperature. Photosynthetic gas exchange reaches a maximum at the optimum temperature, and then positive net gas exchange decreases because of the respiratory losses. A graph of gross photosynthesis (P_g), net photosynthesis (P_n), and respiration (R) is shown in Fig. 16.19. Net photosynthesis is, of course, the difference between gross photosynthesis (total CO_2 uptake before any losses) and CO_2 losses due to respiratory processes,

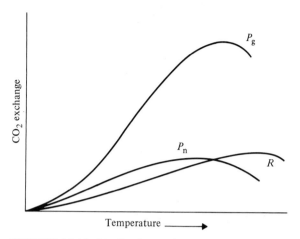

FIGURE 16.19 Idealized graph showing the effect of temperature on gross photosynthesis (P_g = total CO_2 fixed), net photosynthesis (P_n = CO_2 remaining in tissue after photorespiratory and respiratory losses), and respiration (R).

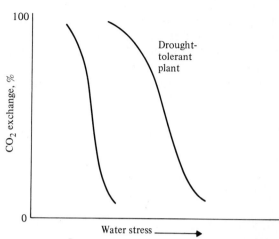

FIGURE 16.20 Idealized curve to show the expected effect of water stress on net photosynthesis. As water stress increases, net CO_2 exchange decreases. A drought-tolerant plant shows much the same curve, but CO_2 uptake is maintained at greater stresses.

either by photorespiration or dark respiration.

$$P_n = P_g - R$$

The optimum temperature for photosynthesis in most plants is about 25° to 30°. Arid zone-adapted plants may have much higher temperature optima.

16.6.3 Water

Photosynthesis decreases when plants are water-stressed. Because water is a substrate for photosynthesis in higher green plants, one might expect that water limitations would restrict the Hill reaction, electron transport, and ultimately NADP reduction. This may be so to a limited extent, but most water-stress effects are secondary and are largely due to poor plant-tissue hydration. Stomatal closure restricts CO_2 uptake and reduces photosynthesis. It is not uncommon for plants to go into incipient wilting during the heat of the day, causing stomatal closure and a temporary cessation of CO_2 uptake—the so-called midday slump.

Water stress will also alter the hydration of enzymatic proteins, affecting their activity. Such

effects will result in altered metabolism and, in the case of photosynthesis, a reduction.

Figure 16.20 shows an idealized graph of the effect of water stress on CO_2 uptake by mesic plants and by drought-tolerant plants (adapted to water stress). The shapes of the two curves are quite similar, but it requires more water stress to bring about the same response in drought-tolerant plants.

16.6.4 Carbon dioxide

Since CO_2 is a primary substrate for photosynthesis, a deficiency will result in a reduction in photosynthesis. Most plants respond to increases in CO_2 levels several times greater than ambient with an increase in CO_2 assimilation. However, high CO_2 concentrations cause stomatal closure, resulting in reduced CO_2 uptake.

Figure 16.21 illustrates typical CO_2 response curves for C_3 species and C_4 species. Both respond to CO_2 increases with increased CO_2 uptake, the rate for C_4 species usually being somewhat greater than for C_3 species. The most interesting feature of the graph is the intercept on the abscissa. This point, called the CO_2 compensation point, γ, is the point where CO_2 uptake equals CO_2 release. It is similar to the I_c except that the limitation here is CO_2 and not light. The CO_2 compensation point is defined by the following.

$$P_n = P_g - R = 0$$

or

$$P_g = R$$

C_3 species have an γ on the order of 50 to 100 parts per million (ppm) CO_2 or as high as one third of ambient CO_2 concentrations, whereas C_4 species have a γ of 0 to 5 ppm.

The observation that ambient CO_2 limits photosynthesis has led to experiments with CO_2 fertilization. Furthermore, since the continued burning of fossil fuels is increasing the atmospheric concentration of CO_2 by about 4% per year, there has been much interest in ascertaining the long-range effects of two to three times ambient CO_2 on plant growth.

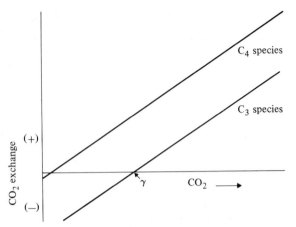

FIGURE 16.21 Net CO_2 exchange as a function of CO_2 concentration. C_4 species show a nearly linear response to CO_2 from near zero to several times ambient. C_3 species show a similar response except at about 100 ppm CO_2 and below, where there is CO_2 evolution. The point where CO_2 uptake matches CO_2 loss is called the CO_2 compensation point (γ) and is taken as an estimate of photorespiration.

16.7 Productivity

A study of plant productivity is beyond the scope of this text, but a few comments are useful for the understanding of plant photosynthesis.

Productivity can be defined as the production of dry matter per area of ground per growing season. Light interaction and the CO_2 exchange surface are functions of the total leaf surface, but of course, shading effects need to be considered. A convenient expression of effective photosynthetic surface relative to plant cover is the leaf-area index (L_i), defined as the total leaf surface per unit of ground area. The units are area of leaves per area of ground surface.

Productivity (P) can be defined as

$$P = P_n \cdot L_i \cdot t,$$

where P_n is net photosynthesis expressed in units of tons hectare^{-1} day^{-1}, t is time in days, and L_i is the leaf-area index in units of hectares of leaf area per hectare of ground.

Thus productivity (P) will be expressed in units of tons ha^{-1} over the growing season and can be used to compare various natural stands and agricultural regions. Table 16.3 summarizes some primary productivity data for plant formations according to Lieth (1972).

Perhaps more important for our understanding of the relationship between plant physiology and plant productivity is the efficiency of radiant-energy conversion into plant dry matter. The efficiency can be defined as follows.

$$E = \frac{\text{Stored energy}}{\text{Absorbed energy}}$$

In the above expression the absorbed radiation can be estimated by assuming that about 40% of the incoming radiation between 0.3 and 3 μm is absorbed, as discussed by Larcher (1975). This assumption takes into account the photosynthetically active radiation. The energy stored

TABLE 16.3 **Examples of some average productivities on the earth**

Formation	Average, tons ha^{-1} yr^{-1}
Earth's average	3
Land mass average	6.7
Ocean's average	1.6
Forest average	13
Woodland average	6
Grassland average	6
Desert average	0.01
Freshwater average	12.5
Cultivated land average	6.5

After Lieth (1972).

within the plant body can be ascertained by combustion and determination of the total energy content of the dry matter. On the average, plant material has a total energy content of about 18.8 kilojoules per gram dry weight.

Calculations have shown that the efficiency of conversion of photosynthetically active radiation is on the order of about 2–3% and up to 6% for efficient agriculture.

It is almost always tacitly assumed that productivity is directly related to photosynthesis. In fact, productivity was defined above as the product of net photosynthesis, leaf-area index, and time. Physiologists believe that this must be true, because only photosynthesis can add carbon to the plant body. But when such a notion is translated into yield of agricultural crops, the relationship may hold true but will not necessarily be linear. Partitioning of carbon among the various sinks of the plant must be taken into consideration. Excess photosynthate may go into root growth rather than into fruit growth. An understanding of growth correlation is necessary to fully sort out the various parameters associated with the relationship between photosynthesis and productivity in an agricultural sense. Growth correlation as a function of carbon partitioning and allocation is discussed in Chapter 19.

Figure 16.22 shows the results of an experiment in which yield of strawberry fruit was related to photosynthesis. In this particular experiment there was a linear increase in the size of fruit and in the number of fruit as the photosynthesis of the plant was increased. Since the rate of change of yield with photosynthesis (that is, the slope of the line) was just less than one (slope = 0.838), there was almost a one-to-one correspondence between leaf photosynthesis and yield. This suggests that much of the photosynthate of strawberry plants goes into fruit production. Furthermore, this experiment gives much support to the goal of developing plants with high photosynthetic rates that will give high yields as well.

Improving photosynthetic efficiency is one way in which physiologists and other agricultural scientists have attempted to increase crop yield. The discovery of C_4-photosynthesis that is associated with a reduced photorespiration has led to much

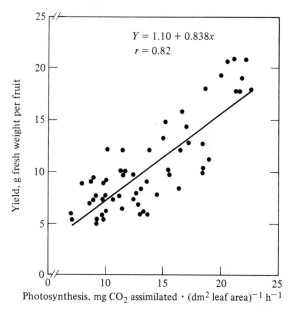

FIGURE 16.22 Relationship between yield, expressed as size of strawberries, and photosynthesis. Photosynthesis was measured at the time of flowering. The graph shows a linear relationship between the size of strawberries and the rate of photosynthesis. From F. Sances. 1980. Spider mite–strawberry interactions and implications to integrated pest management. Doctoral dissertation. University of California, Riverside.

research having the goal of developing C_3 species that have the desirable photosynthetic characteristics of C_4 species. Both selection and genetic manipulation have been used. But, of course, improving photosynthetic efficiency is not the only means by which plants can be manipulated to improve yield and agricultural value.

There is an optimum density of planting that will result in maximum yield for any one crop under specified conditions. Low crop density results in inefficient light gathering as well as poor water and nutrient use. Increasing density will ordinarily increase yield per area until interference between adjacent plants becomes limiting. The limitation either may be the result of competition for light, water, nutrients, and other factors or may be the result of allelopathic phenomena (see Chapter 12).

Much effort has gone into plant canopy studies in an attempt to develop agricultural plant stands that have optimum orientation of leaves for maximum interception of light. Study and manipulation of canopy architecture with the use of computer models is an effective means of improving plant productivity without any change in the inherent photosynthetic capacity of the crop.

Finally, the deliberate manipulation of the microclimate around plant stands can be mentioned as a means to increase productivity. Of course, irrigation and fertilization are common agricultural practices, but more subtle manipulation may be possible. The plant stand and canopy to some extent regulate the microclimate. Interception of radiant energy and reradiation of the absorbed radiation influence the plants' surroundings. The amount of evapotranspiration may influence the humidity, and the canopy itself greatly modifies the mass transfer of air masses. Thus the density of planting, the architecture of the canopy, the kind of plant, and the availability of water and nutrients will have an influence on the microclimate around the stand. The microclimate will in turn affect plant growth and productivity.

Because carbon dioxide limits plant photosynthesis under optimum light conditions, much research has gone into means to fertilize with CO_2. Carbon dioxide fertilization is relatively easy in a closed environment such as a glasshouse but obviously becomes extremely difficult under the more usual field conditions of agriculture.

The cooperation between plant physiologists and agricultural scientists is one way of utilizing basic science to solve practical problems. Knowledge of plant function can be used to manipulate both the plant and the environment to increase crop yield.

Review Exercises

16.1 It has been suggested that C_3 species, C_4 species, and CAM species represent three different groups of photosynthetic plants with three different photosynthetic options. Do you agree with this concept? Do you think it possible that other photosynthetic groups of plants exist in the plant world? How would you design experiments to find them?

16.2 Because as much as 40% of the total CO_2 assimilated in photosynthesis may be lost through photorespiration, it seems reasonable to conclude that photorespiration is wasteful and detrimental to plant productivity. Do you think that such a concept is valid? Do you think it reasonable to look for plants with low photorespiration, hoping that they will produce high yields? How would you design experiments to find such plants?

16.3 Do you see any phylogenetic relationships among the C_3, C_4, and CAM species? Is it possible for such complex metabolic pathways to have originated independently more than once?

16.4 Determine the exact ATP requirement for CO_2 fixation in the several types of C_4 species. Are these ATP-to-CO_2 ratios consistent with our knowledge of the photochemistry of photosynthesis? If additional ATP is required for CO_2 assimilation, where would it come from?

16.5 Discuss in some detail how temperature and water influence photosynthesis.

References

BASSHAM, J. A. 1964. Kinetic studies of the photosynthetic carbon reduction cycle. *Ann. Rev. Plant Physiol.* 15:101–120.

BJORKMAN, O. 1973. Comparative studies on photosynthesis in higher plants. In A. C. Giese (ed.), *Photophysiology.* Academic Press, New York.

BJORKMAN, O., and J. A. BERRY. 1973. High efficiency photosynthesis. *Sci. Amer.* 229:80–93.

BLACK, C. C. 1971. Ecological implications of dividing plants in groups with distinct photosynthetic production capacities. *Adv. Ecol. Res.* 7:87–114.

BLACK, C. C. 1973. Photosynthetic carbon fixation in relation to net CO_2 uptake. *Ann. Rev. Plant Physiol.* 24:253–286.

BLACK, C. C., and R. H. BURRIS (eds.). 1976. *CO_2 Metabolism and Productivity.* University Park Press, Baltimore.

CALVIN, M., and J. A. BASSHAM. 1962. *The Photosynthetic Carbon Compounds.* Benjamin Press, New York.

CHOLLET, R., and W. L. OGREN. 1975. Regulation of photorespiration in C_3 and C_4 species. *Bot. Rev.* 41: 233–258.

DOWNTON, W. J. S. 1975. The occurrence of C_4-photosynthesis among plants. *Photosynthetica* 9:96–105.

GIBBS, M. (ed.). 1971. *Structure and Function of Chloroplasts.* Springer-Verlag, Berlin.

GIVAN, C. V., and J. L. HARWOOD. 1976. Biosynthesis of small molecules in chloroplasts of higher plants. *Biol. Rev.* 51:365–406.

HATCH, M. D., C. B. OSMOND, and R. O. SLATYER (eds.). 1971. *Photosynthesis and Photorespiration.* Wiley-Interscience, New York.

HATCH, M. D., and C. R. SLACK. 1970. Photosynthetic CO_2-fixation pathways. *Ann. Rev. Plant Physiol.* 21:141–162.

HATCH, M. D., C. R. SLACK, and H. S. JOHNSON. 1967. Further studies on a new pathway of photosynthetic carbon dioxide fixation in sugarcane and its occurrence in other plant species. *Biochem. J.* 102:417–422.

JACKSON, W. A., and R. J. VOLK. 1970. Photorespiration. *Ann. Rev. Plant Physiol.* 21:385–432.

KAWASHIMA, N., and S. G. WILDMAN. 1970. Fraction I protein. *Ann. Rev. Plant Physiol.* 21:325–358.

KELLY, G. J., E. LATZKO, and M. GIBBS. 1976. Regulatory aspects of photosynthetic carbon metabolism. *Ann. Rev. Plant Physiol.* 27:181–205.

KLUGE, M., and I. P. TING. 1978. *Crassulacean Acid Metabolism: Analysis of an Ecological Adaptation.* Springer-Verlag, New York.

KORTSCHAK, H. P., C. E. HARTT, and G. O. BURR. 1965. Carbon dioxide fixation in sugarcane leaves. *Plant Physiol.* 40:209–213.

LAETSCH, W. M. 1974. The C_4 syndrome: A structural analysis. *Ann. Rev. Plant Physiol.* 25:27–52.

LARCHER, W. 1975. *Physiological Plant Ecology.* Springer-Verlag, Berlin.

LIETH, H. 1972. Über die Primärproduktion der Pflanzendecke der Erde. *Angew. Botan.* 46:1–37.

LOOMIS, R. S., W. A. WILLIAMS, and A. E. HALL. 1971. Agricultural productivity. *Ann. Rev. Plant Physiol.* 22:431–468.

MARCELLE, R. 1975. *Environmental and Biological Control of Photosynthesis.* W. Junk Publ., The Hague, Netherlands.

MARZOLA, D. L., and D. P. BARTHOLOMEW. 1979. Photosynthetic pathway and biomass energy production. *Science* 205:555–559.

MOONEY, H. A. 1972. The carbon balance of plants. *Ann. Rev. Ecol.* 33:72–86.

SAN PIETRO, A., F. A. GREER, and A. T. ARMY (eds.). 1967. *Harvesting the Sun.* Academic Press, New York.

SETLIK, I. (ed.). 1970. *Prediction and Measurement of Photosynthetic Productivity.* Pudoc, Wageningen, Netherlands.

SZAREK, S., and I. P. TING. 1977. The occurrence of crassulacean acid metabolism in plants. *Photosynthetica* 11:330–342.

iv

Metabolic Processes of Plants

PROSPECTUS Firm hypotheses have been proposed to account for the mechanism of ion uptake by plant tissues. Further research will emphasize the nature of the transport proteins and will attempt to gain evidence that specific ATPase proteins are directly involved. Much nutrition research will be devoted to improve nitrogen availability. Both the improvement of the nitrogen-fixing capacity of legumes and the development of nonlegumes with nitrogen-fixing capacity will be investigated. Much of the research will be done through the cooperation of physiologists and geneticists.

At present, plant respiration is receiving only a small amount of research attention. Nevertheless, there are many important questions that still deserve research attention. The mechanism by which phosphorylation actually occurs is still largely unknown. The alternate oxidase that functions in cyanide-insensitive respiration must still be elucidated. And there is certainly much to be learned about the flow of carbon from storage glucans to the oxidative reactions of the mitochondria. Is pyruvate or malate the major substrate for respiration? These questions and more will be active areas of research in plant respiration for many years to come.

Our overall knowledge of photosynthesis is quite good. Existing theories for electron transport during photophosphorylation and NADP reduction are consistent overall with the available data. However, much needs to be learned about cyclic and pseudocyclic electron transport. We do not even know for certain if these processes that appear to be so important for generating ATP actually occur *in vivo*.

Much effort will be expended to ascertain the electron-transport intermediates of the Z scheme for electron transport in chloroplasts; in particular, the nature of the oxidants and reductants produced by the photosystems will be investigated. More information will be gained about the mechanism of oxygen evolution during water splitting, and effort will be expended to learn about NADP reduction in photosystem I.

The metabolic pathways for C_3, C_4, and CAM species photosynthesis are well documented by experimentation. But there will still be much research designed to further clarify the carbon-assimilation pathways of photosynthesis, with the ultimate goal of manipulation through genetic programs. Attempts will be made to incorporate beneficial C_4-photosynthetic traits into agriculturally important C_3 plants. Much of this research will be done using the techniques developed using microbial systems. In particular, cell-fusion experiments will be done to bring together desirable traits for agricultural use.

Finally, regulation of photosynthesis will become even more important in the future. More will be learned about the enzymes involved and how they are regulated. Research will be undertaken to clarify how the products of photosynthesis regulate photosynthesis and to find out how photosynthate is allocated among the various sinks in the plant.

Allocation of carbon becomes very important with respect to agricultural yields, because it is the allocation of carbon that determines the final yield of crops.

Growth and Development: Integrative Processes

V

PROLOGUE Questions of growth and development of plants are among the most interesting in plant physiology. However, such questions about growth coordination, morphogenesis, and differentiation are by far the most complex that will be encountered in all of biology. Studies of short- and long-distance transport in plants, plant water relations, gas exchange, ion uptake, and metabolism are comparatively simple. For these studies, simple hypotheses can be formulated and frequently tested by the usual scientific methods. Thus reasonably good advances have been made in our understanding of plant biophysics, biochemistry, and metabolism. By contrast, we know very little about plant growth and development. Questions and concepts are so complex that few researchers have been able to formulate testable hypotheses. This is not to say that we do not have a wealth of information about growth and development processes. We do, but most of it is fragmentary and rather specific. Our information is so incomplete that even a trivial understanding of growth and development is not available.

There have been many studies describing the kinetics of plant growth. Such studies are readily treated by mathematics and are extremely adaptable to computer analysis. There are many equations available to describe growth, and under ideal conditions the equations fit experimental observation quite well. For the most part, however, such equations are purely empirical and give little insight into the real questions of growth and development.

Many of the plant hormones have been isolated and characterized. Because they are readily available in pure form, there has been much experimentation; there is a wealth of information about their specific effects, both singly and in combination. Although we have theories to explain plant hormone action, not one is entirely satisfactory.

Cell division, elongation and differentiation, and growth correlation and morphogenesis are aspects of growth and development that have been studied in detail. These are discussed in Chapter 19.

The discovery that certain growth phenomena respond to red light and that they can be reversed by far-red light led to the prediction of the existence of the pigment phytochrome and then its ultimate isolation and characterization. It is now known that as well as induction–reversion phenomena such as seed germination, many other growth processes are regulated by phytochrome. Photomorphogenic phenomena such as stem elongation, leaf expansion, and hypocotyl hook opening are under the regulation of phytochrome. In addition, biological rhythms such as the sleep movements of leaves are phytochrome-controlled. The phytochrome-dependent growth responses are discussed in Chapter 20.

After a period of vegetative growth the plant apex will change from a vegetative structure to a reproductive structure. Such a change results in major alterations in the physiology of the plant that ultimately result in flowering, fruit set, fruit development and maturation. Usually after a dormant period, the seeds will be ready to germinate and begin the life cycle over again. The reproductive growth of the plant is discussed in Chapter 21.

CHAPTER SEVENTEEN

Growth, Growth Kinetics, and Growth Movements

The growth and development of a flowering plant is a complex phenomenon. It can be considered to begin with germination, followed by a large complex series of morphological and physiological events that are called growth and development. A simplified scheme for the life cycle of an annual plant is shown in Fig. 17.1. After germination and vegetative growth, the vegetative apices will differentiate to form reproductive apices. The reproductive apices produce flowers or fruiting structures and ultimately fruit and seeds. The seed that is produced ordinarily goes through a dormant stage prior to germination that will complete the life cycle. In the case of an annual plant, the vegetative portion will die, and only the seed carries on to the following growing season. Perennial plants do not die but frequently go into a dormant stage prior to resuming vegetative growth.

Throughout the life cycle there are many events that govern growth and development. The plant is in part a consequence of its genetic composition and in part a consequence of the environment in

459

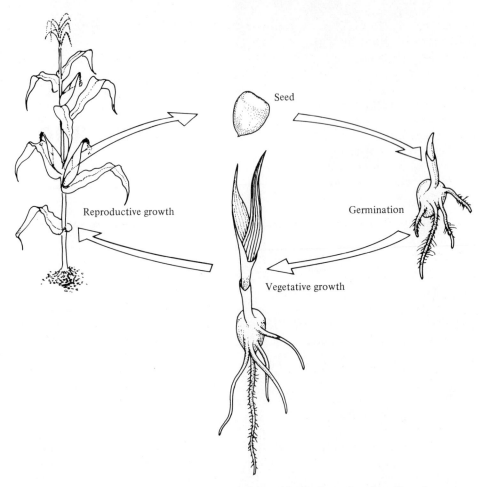

FIGURE 17.1 Diagrammatic representation of the life cycle of a flowering plant from dormant seed through germination, vegetative growth, and flowering.

which it grows and develops. To some extent the genome appears to be preprogrammed, but the program is highly modified by environmental factors. But the development of the plant is regulated by internal growth substances called hormones as well as by environmental factors. Thus the genome has the information for growth and development. The development of the plant is regulated internally by hormones, and the *entire* sequence of events is modified by the environment.

The major classes of plant hormones are auxin, gibberellins, cytokinins, abscisic acid, and ethylene. Specifically, the best-known plant responses to these various hormones are cell enlargement for auxin, stem elongation for gibberellins, cell division for cytokinins, dormancy or arrested growth for abscisic acid, and maturation for ethylene. Of course, for each hormone type there are a variety of growth responses that have been described, and there is much evidence that they function more in concert than alone. Their specific responses are discussed in detail in Chapter 18. They are further discussed in the context of growth in this chapter and in Chapters 19, 20, and 21.

Environmental factors are also important in the growth and development of plants. Not only does environment affect the rate of growth, but also many factors such as light, temperature, humidity, and nutritional factors affect the final morphology of the plant. Light is involved in curvature of stems and in morphogenesis. Temperature, humidity, and nutrition have more subtle effects, but they do alter morphology. These effects and others are discussed in Chapter 19.

In this chapter growth and development are defined, followed by a discussion of the kinetics of growth and the factors that determine the kinetics of growth. In addition, the various plant growth movements are discussed with the exception of the photomorphogenic growth responses, which are discussed in Chapter 20.

17.1 Growth

17.1.1 The problem

As discussed above, the growth and development of a plant takes place by a series of complex, coordinated events that ultimately results in the plant as we know it. In the previous parts of this book the primary concern has been the functioning of the plant, including both the biophysical and biochemical aspects. The objective here is to develop concepts about the actual growth of the plant—how it occurs and how it is regulated and controlled by internal and external factors. In this chapter growth is defined to set a framework for conceptual development. Second, the actual kinetics of growth, that is, how growth and development occur as a function of time, is discussed using simple mathematical formulas. Finally, there is a discussion of a class of growth phenomena called growth movements. Although plants (except for male sex cells) do not move in the same sense as animals, directional growth does occur during the growth and development of the plant. To some extent, such growth appears to be entirely endogenously regulated, such as the nutational growth movements. In others—the tropic and nastic growth movements—growth is in part a response to environment. In the tropic growth responses, growth responds to directional stimuli and the growth is unilateral, toward light, for example. In the nastic growth movements, growth is also in response to environmental signals but is independent of the direction of the signal and is not directional. An example is the opening of a flower bud.

There are other growth phenomena occurring in response to environmental stimuli that do not fall into the tropic or nastic groups. These are the photomorphogenic growth phenomena that are in large part responsive to red light. Leaf expansion and the inhibition of stem elongation (prevention of etiolation) are examples. These growth responses along with the rhythmic sleep movements of leaves are the result of light absorption by the pigment phytochrome.

Other chapters in this part include discussions of the plant growth hormones (Chapter 18), the regulation of vegetative growth (Chapter 19), and reproductive growth (Chapter 21).

17.1.2 A definition of growth and development

Growth is an extremely complex natural phenomenon and as a consequence is quite difficult to define precisely. Perhaps the simplest definition is that growth is an irreversible increase in size or volume accompanied by the biosynthesis of new protoplasmic constituents. Development is a combination of both growth and cellular differentiation. When defined in such a manner, growth is the quantitative aspect of development, representing an increase in the number and size of cells. Differentiation is the qualitative aspect of development.

Qualitative changes in cells and cell constituents is the measure of differentiation.

Growth can be divided into two phases or processes: (1) cell division, and (2) cell enlargement. Cell division, discussed in Chapter 19, results in two daughter cells arising from a mother cell, increasing the number of cells in an individual. It is considered complete when the daughter cells have reached the size of the original mother cell. Cell enlargement or cell elongation results in an increase in the size of newly formed cells beyond that of the mother cell.

Differentiation, the qualitative aspect of development, is the formation of a specialized form or structure and/or the production of specialized substances. In morphological differentiation, there may be formation of subcellular structures such as organelles, formation of specialized cells such as conducting, protective, or storage cells, or, at the organ level, production of leaves, flowers, and other complex structures. With regard to substances, there may be the accumulation of specialized products such as alkaloids, terpenes, latex, cuticle, cellulose, or pigments.

17.2 Growth Kinetics

The kinetics of growth is very complex, depending on many factors, both internal and external. For this reason, a quantitative description of growth has been rather difficult to achieve. But for an idealized situation for a plant growing under constant conditions, a growth curve can be obtained that will adequately describe the growth process.

To obtain a growth curve for a part of a plant (leaf, fruit, epidermal hair, etc.), a seedling, or an entire plant, it is necessary first to decide what index of growth will be used as the measurement and then determine the change in the index as a function of time. Means to measure growth and the analysis of growth curves are discussed next.

17.2.1 The measurement of growth

On first thought it would seem that the measurement of growth by a plant would be a rather simple task. And in some instances this is quite true. Frequently, an estimate of growth can be obtained with a simple ruler or balance. However, our definition of growth is an irreversible increase in volume or size. Growth determinations must, therefore, be indications of irreversible increases and not simple turgor changes that are the result of varying water content. Despite such problems growth is commonly estimated from measurements of length, area, or weight. The measurement depends primarily on the kind of growth analysis being conducted.

Actual volume of tissue can be measured from estimates of dimensions or by displacement in water. Or the fresh weight of a plant or plant part can be estimated by weighing on a balance. However, as indicated above, for most studies measurements other than volume are more common and sometimes more useful.

Linear measurements of stem growth are ordinarily a good indication of the growth of the entire plant since growth is highly coordinated. Growth of one part usually reflects the growth of all other parts. Linear measurements are useful for an estimation of the growth of individual cells. Linear measurements are also useful if nondestructive sampling measurements are necessary. A valuable technique is to obtain a correlation between a volume or weight measurement and a linear measurement, allowing prediction of volume or weight from subsequent linear measurements. The following function will meet most experimental needs:

$$\log y = k \log x + \log b,$$

where

y = volume or weight,

x = linear measurement,

k = correlation coefficient, and

b = constant.

Other functions may be necessary, but this one approximates most growth conditions.

In many kinds of growth-analysis experiments, leaf area is a useful index. Leaf area can be determined by specially designed instruments that use light absorption as a measurement of area or by photographic techniques in which a leaf is placed against light-sensitive paper (for example, in a copying machine) and an image obtained. The area of the image can then be determined. Nondestructive estimates of leaf area can be obtained by ascertaining a correlation between leaf area and the linear dimensions of the leaf, width and length. The function that most frequently fits experimental data is:

$$A = k \cdot L \cdot W,$$

where

A = area of leaf,

L = length of leaf,

W = width of leaf, and

k = correlation coefficient.

In growth studies in which the objective is to determine total increase in dry matter (that is, net productivity), a rather rigorous definition has been established by the International Biological Programme and agreed on by most investigators (see also Chapter 16). The definition of primary productivity is:

$$P_n = \triangle B + L,$$

where

P_n = net dry-matter production during a specified time increment (between two measurements),

$\triangle B$ = total dry-weight increase during a specified time increment, and

L = losses in dry matter during the time increment because of death or consumption by herbivores.

Net dry-matter production (P_n) does not include losses by respiration. If losses by respiration are included, the estimate is gross productivity (P_g). Net productivity is a useful index of growth in physiological studies. In the construction of a growth curve almost any index of growth can be used provided that care is taken to understand the implications of the index. A discussion of growth curves follows.

17.2.2 Growth curves

A typical growth curve is shown in Fig. 17.2. An increase in size is graphed against time. Characteristic for virtually all growth is a lag phase in the beginning in which growth with time is rather slow but still occurs with an increasing rate. A logarithmic phase in which the growth rate is proportional to time follows the lag phase and as early as 1887 was called by J. von Sachs the "Grand Period of Growth." After the Grand Period of Growth a decreasing growth phase occurs, and finally a stationary phase occurs in which there is no longer any increase in size with time. This sigmoid (s-shaped) growth curve is characteristic for virtually all growth in living organisms.

The growth curve shown in Fig. 17.2 was obtained with pinto bean leaves expanding under the controlled growth conditions of an environmental growth chamber. Because of the controlled conditions, expansion was quite uniform, and the curve obtained approximates an ideal curve. Since growth is a combination of cell division and cell expansion, if the beans were grown under more natural conditions with variation in day to day environmental factors the curve would probably not be so symmetrical.

Figure 17.3 is a graph of the growth rate curve obtained from the data shown in Fig. 17.2. The growth per day is graphed as a function of age of the pinto bean leaf. The peak reflects the maximum growth rate obtained during the expansion of the leaf.

Similar growth curves are expected for the growth-by-elongation of single cells, roots, and stems or the increase in size of fruits. In addition, populations or groups of cells, single-celled organisms, or whole plants grow according to the sigmoid

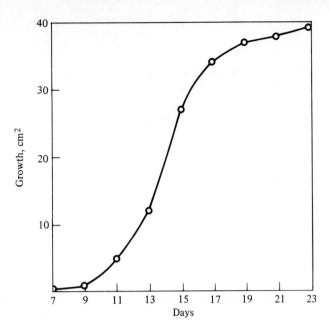

FIGURE 17.2 A sigmoid growth curve. The data are for the expansion of a pinto bean leaf grown in a growth chamber. From a student experiment, University of California, Riverside.

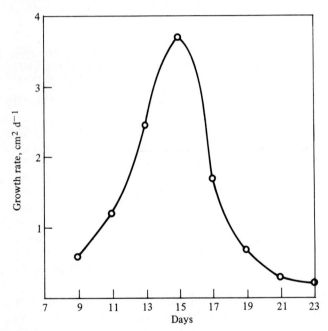

FIGURE 17.3 A growth-rate curve. The data are those shown in Fig. 17.2. In this graph the data are plotted as growth per day as a function of age in days. The maximum growth rate marked by the peak corresponds to the inflection point on the cumulative growth curve shown in Fig. 17.2.

growth curve. Of course, since each is somewhat different there will be differences in the shape of the growth curves. Some growth curves will have little lag and some will have extended lag periods, all modified by genetics and environment.

Frequently, one encounters the term "relative growth rate." Relative growth rate is the amount of growth per time expressed as a function of the amount of existing tissue. Relative growth rates are commonly expressed as percentages. A plant growing 10% per day would be increasing in size at the rate of 10% of its total mass each day. Since the

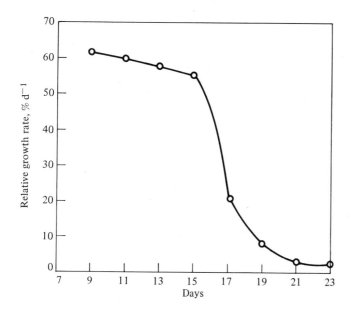

FIGURE 17.4 A relative-growth-rate curve in which the percentage growth per day is graphed against age in days. The data are the same as shown in Fig. 17.2 and 17.3.

overall growth of plant or plant part will generally follow a sigmoid curve, the relative growth rate will clearly change during the lag, log, and stationary phases of growth. Relative growth rates are probably only useful during the Grand Period of Growth. Figure 17.4 shows the relative growth rate curve for the pinto bean leaf expansion. The relative growth rate is nearly constant during the log phase of growth but then falls rapidly when the growth rate slows.

Perhaps the simplest case for growth analysis is the growth accompanying the increase in the number of cells as a result of cell division. A consideration of the curve in Fig. 17.2 suggests that the lag phase is the period in which cell division is slow and erratic with little or no cell enlargement. During the end of the lag phase and during the log phase, cells are all dividing at a constant rate, and there is a geometric increase in the number of cells. Growth throughout this entire period (that is, through the lag and log phases) is exponential. As cell division decreases, the growth rate decreases until no further cell division occurs and steady state is reached.

Attempts to obtain a mathematical model for growth are thwarted because of the immense com-

plexity of growth and growth processes. However, it is possible to describe the ideal growth curve in Fig. 17.2 with simple, familiar mathematics.

During the initial growth period (the lag and log phases), the growth rate (dN/dt) by cell division is proportional to the number of cells present at any time, t. Hence a first-order rate expression is adequate to describe the increasing function:

$$\frac{dN}{dt} = kN,$$

where

N = the number of cells,

t = time, and

k = first-order rate constant.

Integration over time gives

$$\ln N = kt + \ln N_o,$$

where N_o is the number of cells at time = 0. This expression can be recognized as an exponential function:

$$N = N_o e^{kt}.$$

Growth by cell division should obey such a function

provided that no factors are limiting, all cells are dividing at the same rate, and there is no subsequent elongation.

But, of course, such is not the case, and growth begins to cease after a certain period of time as necessary factors become limiting or toxic materials that limit growth begin to accumulate.

If we assume that there is an upper limit to the number of cells in any one dividing population (N_{max}), we can rewrite the rate expression.

$$\frac{dN}{dt} = k(N_{max} - N)$$

Here the growth rate, dN/dt, will approach zero as N approaches the maximum, N_{max}.

However, this function does not adequately describe the lag and log phases of growth but rather just the decreasing growth-rate portion of the entire growth curve.

Combination of the two expressions will result in a new expression adequate to describe the entire sigmoid growth curve.

$$\frac{dN}{dt} = kN(N_{max} - N)$$

By assuming that N_o is small, integration gives an equation in the form

$$\ln \frac{N}{N_{max} - N} = kt.$$

The above solution expresses the symmetrical sigmoid growth curve that is found for the growth of most plants. A graph of $\ln N/(N_{max} - N)$ against time will be a straight line with a slope equal to the growth rate constant, k.

The maximum growth rate is the central inflection point, where

$$N = \frac{N_{max}}{2}.$$

The constants N_{max} (marking the maximum number of cells) and k (the growth-rate constant) are functions of the environment and the plant's genetic composition. In themselves, they are far too complex to completely analyze mathematically.

Although the assumption is rather presumptuous for the growth of higher plants, we can assume that the increase in weight (W) at any one time will be proportional to the weight at that time. Thus the same expression used for the increase in the number of cells with time can be used to describe the growth of almost any plant or plant part, if growth is expressed as an increase in weight or volume.

$$\frac{dW}{dt} = kW(W_{max} - W)$$

Examples of other nearly symmetrical sigmoid growth curves are shown in Fig. 17.5. Figure 17.5a shows the actual growth of a population of single-celled algal cells (*Euglena*) as a function of time. The "growth" is made up of cell division by mitosis and subsequent cell enlargement of the newly formed cells to the size of the parent cell. Although the growth curve is not exactly symmetrical, the sigmoid shape is readily apparent. Cessation of growth results because of nutrient limitation and toxin buildup within the culture solution.

The growth curve illustrated in Fig. 17.5b is for the elongation of cotton fibers when grown *in vitro* on an artificial growth medium. Cotton fibers are single cells, and growth here is represented by cell elongation; virtually all epidermal cells capable of forming fibers were present on the day of anthesis.

The expansion and growth of modified stems (pads) of a beavertail cactus (*Opuntia basilaris*) are illustrated in Fig. 17.5c. Three different pads are shown. Pads A and B were on the same plant, which was growing under cultivated conditions. Both curves are somewhat sigmoidal, which is typical of the expected growth curve, but pad A has a period of cessation after about 10 days. Pad B had a very long and slow lag period prior to the log Grand Period of Growth. Pad C was on the same plant, but expansion was during the summer (in July) rather than during the spring as it was for pads A and B. Despite the small expansion of Pad C in July, a sigmoid curve is still evident.

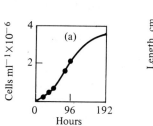

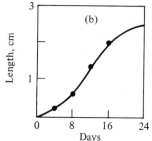

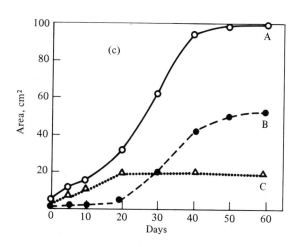

FIGURE 17.5 Some representative growth curves. (a) Cumulative growth of a population of alga cells (*Euglena*). A typical sigmoid curve is shown. From M. Peak. 1972. Intermediary metabolism in *Euglena gracilis* Z undergoing changes between autotrophic and heterotropic metabolism: malate dehydrogenase and malic enzyme. Doctoral dissertation, University of California, Riverside. (b) Growth curve for the elongation of cotton fiber. The fiber is a single cell and elongates, following a typical sigmoid curve. From an experiment by R. Dhindsa, University of California, Riverside. (c) Growth of stems of *Opuntia basilaris*. A and B are two stems on the same plant that were expanding during the spring. Both show sigmoid growth kinetics but the rates are different. C is for a stem that was expanding in the middle of the summer. From a student experiment. University of California, Riverside.

The growth curves illustrated in Fig. 17.5 illustrate that growth of plants and plant parts follows a sigmoidal curve, which has a lag phase, a log phase, and culminates with a stationary phase.

As one might expect, inspection of the sigmoid growth curves illustrated in Fig. 17.5 indicates that plant growth data may not graph to form a symmetrical sigmoid curve; asymmetry about the central inflection point commonly occurs. Equations that are modifications of those above have been developed to describe asymmetric sigmoid growth curves. Such curves can be used for fitting data; they allow predictions of more accuracy than if the above symmetrical functions were to be used.

It should be apparent from the above discussion that various parts of any particular plant are apt to grow at different rates. Stems will grow at rates different from roots, and both will be different from

leaves. Different leaves on any one plant may grow at different rates. The expansion of two stem sections of the cactus shown in Fig. 17.5c illustrates this point.

17.2.3 The plastochron index

Although many indices of growth have been developed, most give an estimate of chronological development. However, in many kinds of studies, analysis according to chronological age is not of sufficient value because different plants will reach the same development stage at very different chronological ages. For this reason Erickson (1976) developed the "plastochron index," which is a measure of developmental age and can be used to compare developmental stages of different plants. The term plastochron was introduced in 1880 by

Askenasy to mean the time between formation of two successive internodes. As used by Erickson, it can refer to the time between the initiation of two successive leaves or the time between any two comparable developmental points.

The plastochron index, or plastochron age, for a plant is the number of leaves that have developed. If exactly five leaves have developed, the plastochron age would be 5; if six leaves have developed, the plastochron age would be 6, and so on. Arbitrarily, Erickson assumed for *Xanthium* that a plant would have a plastochron index of n when the nth leaf to develop was exactly 10 mm long. The plastochron index would be $n + 1$ when leaf $n + 1$ was exactly 10 mm long. Because any leaf would rarely be exactly 10 mm long at the time of measurement, he developed an expression to calculate fractional indices (ages). Assuming that the time period during development of successive leaves is constant, the plastochron index (PI) is:

$$PI = n + \frac{\log L_n - \log 10}{\log L_n - \log L_{n+1}},$$

where

n = serial number of the leaf that comes closest to exceeding 10 mm,

L_n = length of leaf n, and

L_{n+1} = length of the leaf that is just shorter than 10 mm (i.e., the leaf that is one plastochron younger than leaf n).

If the youngest leaf is exactly 10 mm long and is the fifth leaf, the plastochron index would be 5. However, if leaf 5 was just longer than 10 mm and a younger leaf was just shorter than 10 mm, the plastochron index would be somewhere between 5 and 6, depending on the lengths of leaf 5 and leaf 6.

The exact plastochron index for any one leaf can be obtained by subtracting the leaf's serial number from the plastochron age of the plant. This is so because we have assumed a constant period between the development of successive leaves. It should be apparent that a leaf exactly 10 mm long will have a plastochron age of zero. A leaf shorter than 10 mm long will have a negative plastochron age. Leaves longer than 10 mm, of course, will have positive plastochron ages.

The plastochron index has been used in a variety of time-dependent studies that would not have been possible if only chronological age was known. It has been shown, with studies of the development of poplar leaves with different sensitivities to pollutants, studies of anatomical development, and studies of the hormonal regulation of growth, that much variability was eliminated if developmental age was considered (that is, the use of the plastochron index) rather than simply chronological age.

17.2.4 Growth of the entire plant

The estimation of the growth of an entire plant would be a rather formidable task. It would entail the measurement of growth using parts both above and below the surface of the ground. Despite this, we would expect that during the active growing season the growth function would approximate a sigmoid growth curve. However, we would not anticipate that the growth of roots would necessarily be parallel to the growth of shoots or leaves. Partitioning of photosynthate among the various growing parts would govern the growth of the parts, all, of course, modified by environmental parameters. Roots and shoots do not grow at the same rates nor do they necessarily grow at the same time. It is common, for example, that roots begin growth and development prior to shoot and leaf development. Furthermore, periods of intermittent growth of both shoots and roots are known for woody perennials in temperate zones.

In some tree species there is a single growth period after dormancy at the beginning of the growing season. This pattern is common in oaks, ashes, and many of the conifers of temperate zones. In other trees including many conifers there are recurrent periods of growth throughout the growing season. This intermittent growth pattern is also quite common in tropical and subtropical zones.

Complicating these growth patterns is diurnal growth. Much growth occurs at night when temperatures are more favorable and water balance is better.

Roots as well as shoots show a periodicity of growth patterns. As stated above, roots may grow at different times than shoots, but they also have intermittent growth periods similar to shoots. Furthermore, there is evidence that not all root tips of any one plant necessarily grow at the same time. Complicating these observations even further is that roots even in temperate zones may not have dormancy periods in the same sense as shoots do.

During favorable periods, even in winter, roots may grow.

Because of these problems complicating whole-plant growth, many studies of entire-plant growth and plant-population growth use indices other than direct estimates of size. One of the most common is an estimation of net assimilation rate, or primary productivity (P_n), discussed earlier in this chapter (also see Chapter 16). The assumption is that growth can be estimated from measurements of dry weight of aboveground parts, taking into consideration losses because of respiration and translocation to parts below ground.

17.3 Growth Movements

Although vascular plants are not considered to move in the sense that animals are mobile, there are a variety of movements associated with growing plants and with plant parts during normal daily cycles. The plant movements can be divided into three major categories: (1) growth movements, (2) turgor movements, and (3) hydration movements.

Growth movements are movements that occur because of the growth and elongation of cells. Growth movements are not reversible and may occur because of equal or unequal cell elongation. Growth movements are of three basic types. Tropisms are unequal growth movements resulting from environmental stimuli received directionally. The two most common examples are phototropic and geotropic growth movements. Phototropisms are growth or movement toward or away from light. The light is received on one side of the plant, and, in the case of a positive phototropic response, growth is toward the light. For a negative phototropic response, growth would be away from the light.

Nastic growth movements are those growth responses or movements in which the direction of the response is independent of the direction of stimulus. Flower buds open in response to light. Opening

results because of a greater rate of growth on the upper surface than on the lower. The result is curling of the bracts and flower opening.

Nutations are growth movements resulting from unequal growth and are apparently independent of the environment. The best-known example is the circumnutational growth of stems. Stems grow in a pattern approximating an expanding spiral.

Turgor movements occur because of differential turgidity of cells. They are fully reversible. Sleep movements of leaves, the touch movement of the sensitive plant *Mimosa,* and the opening and closing of stomata are turgor movements.

Hydration movements are less common but include (1) opening of certain dried fruits such as in legumes through pressures caused by hydration of dried cells and (2) the rapid opening of fern sporangia resulting in the forced expulsion of spores. Hydration movements are usually associated with nonliving tissue.

17.3.1 Tropic growth responses

Tropisms are basic growth movements that occur in response to directional stimuli. They occur because of unequal growth and cause bending toward (positive) or away from (negative) the direction of

FIGURE 17.6 Mosaic growth pattern of ivy leaves growing on the side of a building. The orientation results in maximum exposure to light.

the response. The unequal growth appears to be in part the result of redistribution of auxin (or other hormone) as a consequence of the stimulus reception.

Phototropism

The most apparent and most easily studied phototropic response is the positive phototropism shown by coleoptiles exposed to unilateral light. When exposed to light on one side, coleoptiles will bend toward the light because of cell elongation on the nonilluminated side.

Another common phototropic response in addition to bending toward light is leaf orientation. Leaves grow and orient such that a mosaic pattern develops (Fig. 17.6). Observation of a leaf canopy or a vine growing on a surface shows that the leaves are positioned such that maximum surface is exposed to the light.

So-called compass plants of the genus *Lactuca* and *Silphium* have leaves oriented at right angles to the sun (lamina face east and west). In other plants there are photoresponses to light that are actually turgor movements in which leaves or flowers follow the sun's angle throughout the day. Perhaps the best example is the sunflower, which gets its name from the fact that the floral head follows the path of sunlight. Actually, since these movements are reversible they are not tropisms.

Sun-tracking in lupine was studied (Fig. 17.7) and shown to be a direct response to light. Although it would persist under natural conditions, it evidently is not a biological rhythm (see Chapter 20) because it will not occur in the dark. In continuous, unilateral light the leaves oriented toward the light source.

Many herbaceous plants are negatively phototropic in full sun and positively phototropic in shade. In sunlight the position of some grasses is prostrate, but when shaded they grow upright and become elongated. If grown in light, some roots will show a negative phototropic response and grow away from the direction of illumination.

WEST　　　　　　　**EAST**

FIGURE 17.7 Sun-tracking of lupine leaves during the day. In the morning (panel 2) the leaves are oriented toward the sun in the east. As the sun passes from east to west (panels 3 and 4), the leaves follow. From C. M. Wainwright. 1977. Sun-tracking and related leaf movements in a desert lupine (*Lupinus arizonicus*). *Amer. J. Bot.* 64:1032–1041.

The phototropic response is maximally sensitive to blue light (Fig. 17.8), unlike photomorphogenic responses that are red light-sensitive (phytochrome responses—see Chapter 20). Phototropism is not a photomorphogenic response since the latter is not dependent on directional or unilateral illumination. At low energies (up to about $0.1 \, J \, m^{-2}$) there is a positive response to light. Above this the phototropic response reverses until the light energy increases and there is a second response. The first, low-intensity response shows reciprocity (intensity $\times$ time = constant), but the high intensity response does not. For this reason it has not been possible to obtain a proper action spectrum for the high-intensity response.

The low-intensity response is most sensitive to blue light below 500 nm, with a maximum at about 450 nm (Fig. 17.8b). Because the action spectrum is similar to the absorption spectra of both carotenoids and flavins it has not been possible to determine unequivocally what pigment absorbs the light energy. There is some evidence that the pigment is a complex consisting of a flavin and a cytochrome.

As will be discussed further in Chapter 18, it is the apex of the coleoptile that perceives the light in the phototropic response, even though the cells in the region of elongation are the ones with the differential growth. Light stimulates lateral transport of auxin from the light side to the dark side, causing the unequal growth and bending toward the light. This is the Cholodny and Went hypothesis proposed independently in the 1920s.

A second hypothesis, which is largely inconsistent with the available data, is that light causes the photodestruction of auxin. This hypothesis was strengthened with the observation that blue-light activation of riboflavin and violaxanthin would bring about auxin destruction. However, *in vivo* light sufficient to cause curvature does not reduce the total auxin content of the tip. A third hypothesis is that light reduces the rate of synthesis of auxin on the lighted side. Experiments showed that the same amount of auxin was produced by coleoptile

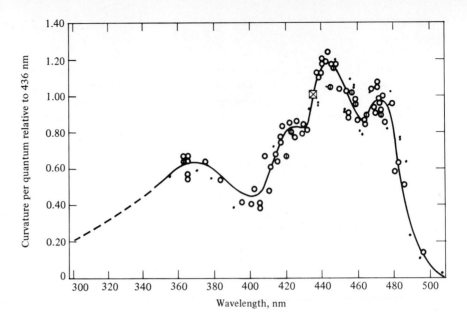

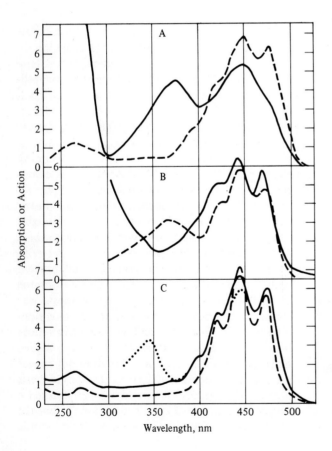

FIGURE 17.8 Top: Action spectrum for phototropic curvature of *Avena* coleoptile. The curve is based on several experiments. Bottom: Absorption spectra: A. Solid line = riboflavin. Broken line = β-carotene. B. Solid line = hexane extract from coleoptile. Broken line = action sepctrum from 17.8 (top). C. Solid line = α-carotene. Broken line = lutein, a carotenoid. Dotted line = 9,9'-mono-*cis*-β-carotene. The action spectrum for curvature compares favorably with the carotenoids and riboflavin, the possible pigments involved in the growth response. From K. V. Thimann and G. M. Curry. 1960. Phototropism and photoaxis. In I. M. Florkin and H. S. Mason (eds.), *Comparative Biochemistry*, Vol. 1. Academic Press, New York.

tips in light and dark, which was evidently at variance with the light-destruction and inhibition-synthesis hypotheses. Another hypothesis is that there are light reactions independent of auxin that bring about the unequal growth of phototropic responses.

It has been proposed that light causes growth inhibition on the lighted side rather than growth stimulation on the opposite side. The interpretation is that phototropism is differential growth inhibition rather than differential growth stimulation. However, the lateral-transport hypothesis proposed by Went and Cholodny is the one most favored.

Geotropism

Ordinarily, roots of plants grow downward and shoots grow upward. Such growth is a normal response to gravitational forces. Shoot growth away from the center of the earth is negative geotropism, and root growth toward the center is positive geotropism. Most plant parts show some geotropic responses. Growth of stems such as rhizomes and stolons at right angles to the gravitational field is called diageotropic growth. Lateral branches frequently become oriented at angles around 90°, a growth movement in response to gravity called plagiogeotropism. Few plant organs are truly ageotropic (no response to gravity).

Although geotropic growth of roots may be partially regulated by auxin, experiments with root caps have implicated a growth inhibitor as being more important. If the root cap is removed, elongation will still occur but geotropism will not. If the lateral half of the root cap is removed, root growth occurs toward the side with the remaining portion. This observation has led investigators to propose that the root cap produces a growth inhibitor (assumed to be abscisic acid) that causes the asymmetric growth of the root that is responding to gravity.

Exactly how gravitational fields are sensed by plant tissues is now known; however, some in-

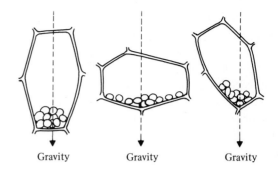

FIGURE 17.9 Drawing to illustrate how statocytes are believed to sense gravity. Geotropic stimulation causes sedimentation of starch grains bringing about the geotropic response. Drawn after L. Hawker. 1932. A quantitative study of the geotropism of seedlings with special reference to the nature and development of their statolith apparatus. *Ann. Bot.* 46:121–157.

formation is available, allowing testable hypotheses to be formulated. It has been proposed that there are gravity-sensing organelles, called statoliths (probably starch grains), within specialized root-cap cells, called statocytes. Gravitational forces cause sedimenting of the heavy starch-grain statoliths against specially oriented endoplasmic reticulum (see Fig. 17.9). Consistent with the concept of statoliths, such cells with large starch grains can be observed in root-cap cells and evidently will reorient in response to changing gravitational fields.

Some roots thus appear to have a gravitational sensing device. Support for such a hypothesis comes from observations that the time required for sedimentation of starch-grain statoliths is approximately the same as that required for gravitational forces to bring about geotropic responses.

Because many plant organs that show geotropism do not have starch grains, alternative hypotheses have been proposed. One is that horizontal organs sense geoelectrical potentials. Potential differences as great as 20 millivolts can be measured across horizontal organs. Such a potential difference would be accompanied by ion redistribution and would perhaps reorient growth direction.

Thigmotropism

Many plants show growth movements in response to mechanical stimulation (for example, touch). Perhaps the best example is the twining of the tendrils of pea plants and other climbing vines (Fig. 17.10). Several minutes after mechanical stimulation of a pea tendril there is contraction on the lower surface and elongation on the upper surface such that the tendril will twine around a small object.

FIGURE 17.10 Thigmotropic growth of tendrils on vine. The tendrils grow around the stake in response to their touching of the stake.

Little is known about the twining response except that auxin treatment will cause the response and internal ATP levels decrease, suggesting an energy requirement.

It is suspected that the differential growth that causes twining in thigmotropic responses is the result of lateral auxin transport similar to that occurring in phototropic responses.

Hydrotropism

A true, positive hydrotropism would be growth toward water, but the water would have to act as the environmental stimulus causing the response. Roots usually grow geotropically and have little or no tendency to grow toward water. An extensive study by Loomis and his students (1936) showed that only a few species of legumes and cucurbits actually showed any tendency to curve toward moist soil and away from dry soil (Fig. 17.11).

The normal growth pattern of roots is that they either grow downward (according to the geotropic principles discussed above) or grow more or less randomly. When they enter dry soil, growth ordinarily ceases. The potential for lateral root development is present throughout the course of root development, but lateral roots rarely develop unless the root is in moist soil above the permanent wilting point (soil moisture level at which plants will wilt).

Conceptually, hydrotropic root growth could occur along water-vapor gradients from wet to dry soils, but evidence for such growth is lacking except for the case of the few legumes and cucurbits studied by Loomis. Most of the plants they investigated showed no tendency for hydrotropism.

If hydrotropic growth movements do occur in plants, it would be assumed that auxin transport and differential growth (causing bending) would be the explantation.

Other tropisms

There are many other types of tropic movements of plants that occur in response to environmental

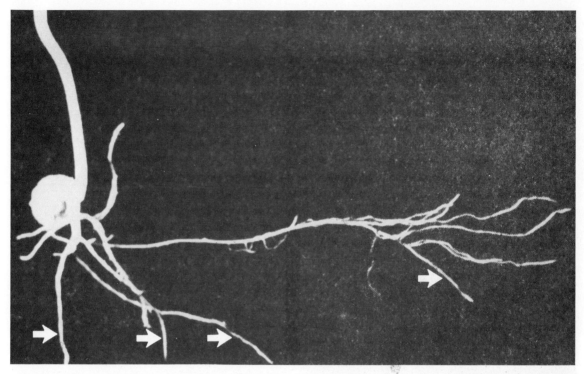

FIGURE 17.11 Photograph of pea roots growing in soil that was wet on the right but dry on the left and below. The arrows mark points where the roots grew toward the dry soil and then stopped. There is no evidence for a hydrotropic growth response. Reprinted from W. E. Loomis and L. M. Ewan. 1936. Hydrotropic responses of roots in soil. *Botanical Gazette* 97:728–743. By permission of The University of Chicago Press.

stimuli. Growth in response to chemicals (chemotropism), electric fields (electrotropism), and temperature (thermotropism) are all possible, but little is actually known about them.

17.3.2 Nutational growth

Nutational growth movements are growth movements that do not appear to be in response to environmental stimuli. A stem growing upward (negative geotropism) does not follow a straight line but elongates along an expanding elliptical pattern. Such growth by stems, perhaps best exemplified by twining plants such as climbing vines, is called circumnutational growth (Fig. 17.12). As the stem twines, there is more cell elongation on the outer, lower portion of the stem than on the inner, upper surface (with respect to the ellipse) such that the twining growth pattern emerges.

The greater cell elongation on the lower surface, which accounts for the circumnutation, may be the result of auxin accumulation on the lower surface of the stem; however, there is no direct evidence. There is evidence that when the plant growth hormone gibberellin is low, circumnutational growth is reduced or absent.

Studies of the circumnutational growth of pea tendrils showed that the average speed of the rotary, helical movement of the tip of the tendril was 1.57 millimeters per minute. For each circumnutation approximately 80 minutes elapsed (Fig. 17.13).

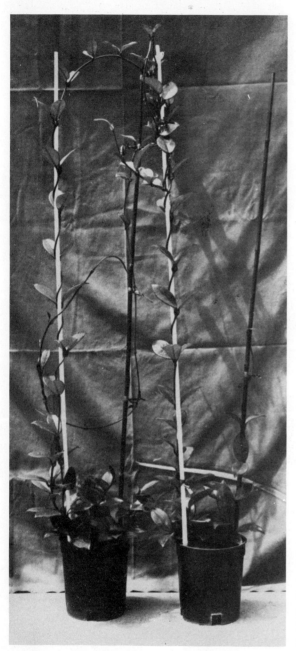

FIGURE 17.12 Circumnutational growth of *Hoya carnosa*, a climbing vine.

FIGURE 17.13 Diagram of the circumnutational growth of a pea tendril. The pattern at the top of the drawing shows the helical growth pattern from the top. From M. J. Jaffe. 1972. Physiological studies on pea tendrils. *Physiol. Plantarum* 26:73–80.

17.3.3 Nastic growth movements

Nastic growth movements* occur independently of the direction of the environmental stimulus. They occur because of unequal growth at the sides or surfaces of the growing organs. A good example is epinasty, the opening of a flower bud by the curling back of bracts and perianth parts. There is a greater rate of cell elongation on the upper surface than on

*Nastic movements are of two types: (1) those that are actual growth movements, and (2) those that are in response to turgor changes occurring in pulvini, e.g., the sleep movements of leaves.

the lower surface, which causes the curl. These growth movements may be in response to light (photonastic), temperature (thermonastic), or other environmental factors.

A nastic response in which bending is downward is called epinasty. Epinasty is common in leaves. The petiole bends downward and the leaf points toward the ground rather than upward. The opening of the flowers is a good example of an epinastic growth response. As explained elsewhere (see Fig. 18.13), ethylene causes epinasty in many plants. The opposite of epinasty is hyponasty, in which plant parts bend upward. Hyponasty can be induced by application of gibberellin. Perhaps the best-known hyponastic growth movement is the unbending of plumular hooks during germination of legumes.

Little is known about the physiology of nastic growth movements, but along with the other known growth movements it is assumed that auxin transport and hormone balance are important factors.

17.3.4 Nongrowth turgor movements

Many of the more obvious plant movements are not true growth movements. They are reversible, occur-ring because of turgor changes in specialized cells or regions of organs. The very rapid closing of the leaves of the Venus flytrap in response to insect touch is a good example. Another example of turgor response is the opening and closing of stomata, discussed in Chapter 5.

The diurnal curling of leaves of grasses is a turgor movement that occurs because of the loss of water from specialized epidermal cells on the upper surface called bulliform cells (Fig. 17.14). When the bulliform cells are fully turgid the leaves are open and relatively flat. When water is lost from the bulliform cells into adjacent tissues, leaf curling occurs. Such curling is a rhythmic phenomenon that occurs daily but also occurs when the plant wilts.

Another good example of a turgor movement in plants is the sun-tracking phenomena of certain flowers (sunflowers) and leaves throughout the day.

Many legumes and certain other plants show characteristic leaf-collapse turgor movements that may be rhythmic, such as the sleep movements (see Chapter 20), or a consequence of mechanical, chemical, or thermal stimulation. The "sensitive plant," *Mimosa pudica,* responds to touch by a rapid folding of its leaves (Fig. 17.15). The response

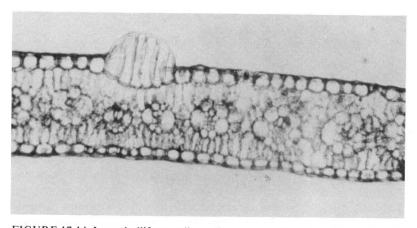
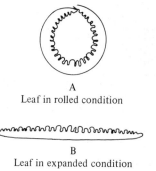

A
Leaf in rolled condition

B
Leaf in expanded condition

FIGURE 17.14 Large bulliform cells on the upper surface of a maize leaf. Collapse of the bulliform cells because of loss of turgor causes leaf curling.

FIGURE 17.15 *Mimosa pudica*, the "sensitive plant," responds to touch by leaf and leaflet folding. Arrow on the left shows leaflet before touch. Arrow on the right shows leaflet after being touched.

begins in less than 0.1 second and is complete in about 1 second. Recovery after stimulation requires 10 to 20 minutes or more. There is long-distance transmission of the stimulus. Leaves other than those receiving the stimulus will respond but more slowly.

Mimosa will respond to mechanical stimulation such as touch and to chemicals, electrical shock, and of course, wilting. Interestingly, after a dark period the unfolding responds to blue light.

At the base of the petioles and sometimes at the base of leaflet rachises there are specialized organs called pulvini (singular—pulvinus). A pulvinus is composed of thin-walled parenchyma cells surrounding vascular tissue and can be observed as a swollen base of the petiole or rachis (Fig. 17.16). When the cells of the pulvinus are fully turgid, the petiole and rachises are erect and the leaf and leaflets are fully unfolded. If water is transferred from the parenchyma cells into the vascular tissue or adjacent tissue, there will be collapse of the petiole and rachises, causing leaf folding. Folding downward results because of turgor loss of the pulvini cells on the lower surface relative to the turgor of the pulvini cells on the upper surface. Folding inward would show the reverse differential turgor.

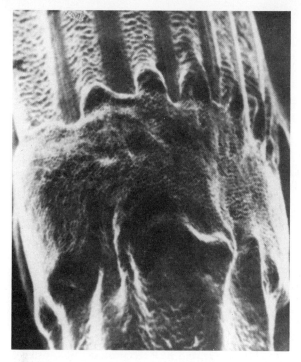

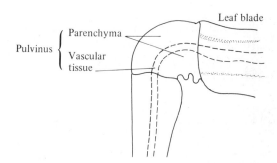

FIGURE 17.16 Diagram is of a pulvinus, a specialized structure composed of vacular tissue surrounded by parenchyma. When water is transferred from the surrounding parenchyma to the vascular tissue the pulvinus becomes flaccid, causing the leaf or leaflet to fold. Photograph is of a pulvinus at the base of a grass blade. Photograph from P. Dayanandan, F. V. Hebard, V. D. Baldwin, and P. B. Kaufman. 1977. Structure of gravity-sensitive sheath and internodal pulvini in grass shoots. *Amer. J. Bot.* 64:1189–1199.

There is evidence that the mechanism of turgor change in the pulvini cells is comparable to stomatal guard-cell turgor changes. Potassium and malate concentration changes are associated with osmotic-potential changes, bringing about water shifts and turgor changes. Chapter 5 discusses these changes that occur in guard cells in detail.

The rhythmic behavior of sleep movements in leaves is discussed in Chapter 20.

Review Exercises

17.1 Plot the following data obtained from the expansion of a plant leaf. Estimate the maximum growth and the maximum growth rate. Also graph the growth-rate curve as estimated from the data (that is, plot the growth rate as a function of time).

Age (weeks)	Leaf area (cm²)
0	12
2	34
4	96
8	214
10	270
12	293
14	300

By graphic analysis, estimate the growth rate constant, k.

17.2 Calculate the plastochron index for a plant with 8 leaves, the youngest of which is 6 mm long and the next to youngest 15 mm long. What would be the plastochron age of the fourth leaf to develop? Of the sixth leaf? What assumptions were necessary for the latter determinations?

17.3 How would you ascertain if a particular plant movement, such as a sunflower tracking the path of the sun, is a growth movement or a turgor movement? In general, what criteria would you use to differentiate between growth movements and turgor movements?

17.4 Design an experiment to test the hypothesis that the plant growth hormone auxin is transported from the illuminated side to the dark side during a positive phototropic growth response. Design an experiment to test the hypothesis that plant roots show hydrotropism.

17.5 In a greenhouse in southern California, the vine *Hoya carnosa* shows a right-handed circumnutational growth. How would you ascertain if the growth pattern is genetic or environmental? In general, would it be possible to determine if nutations are responses that are completely independent of environmental stimuli?

References

AUDUS, L. J. 1973. *Plant Growth Substances,* Vol. 1. 3d ed. L. Hill, London.

BRIGGS, W. R. 1963. The phototropic responses of higher plants. *Ann. Rev. Plant Physiol.* 14:311–352.

DIGBY, J., and R. D. FIRN. 1976. A critical assessment of the Cholodny–Went theory of shoot geotropism. *Current Adv. Plant Sci.* 8:953–960.

ERICKSON, R. O. 1976. Modeling of plant growth. *Ann. Rev. Plant Physiol.* 27:407–434.

EVANS, G. C. 1972. *The Quantitative Analysis of Plant Growth.* University of California Press, Berkeley.

HAUPT, W., and M. E. FEINLIEB (eds.). 1979. *Physiology of Movements. Encyclopedia of Plant Physiology.* New Series Vol. 7. Springer-Verlag, Berlin.

JAFFE, M. J. 1973. Thigmorphogenesis: The response of plant growth and development to mechanical stimulation. *Planta* 114:143–157.

JUNIPER, B. E. 1976. Geotropism. *Ann. Rev. Plant Physiol.* 27:385–406.

LEOPOLD, A. C., and P. E. KRIEDEMANN. 1964. *Plant Growth and Development.* 2d ed. McGraw-Hill, New York.

LOCKHART, J. A. 1976. Plant growth, assimilation, and development: A conceptual framework. *BioScience* 26:332–338.

SATTER, R. L., and A. W. GALSTON. 1973. Leaf movement: Rosetta stone of plant behavior. *BioScience* 23:407–416.

SHEN-MILLER, J., and R. R. HINCHMAN. 1974. Gravity sensing in plants: A critique of the statolith theory. *BioScience* 23:21–27.

STEWARD, F. C. (ed.). 1969. *Analysis of Growth. Plant Physiology,* Vol. VA. Academic Press, New York.

CHAPTER EIGHTEEN

The Plant Hormones

Julius Sachs in 1880 postulated that substances in plants regulate the growth of the different plant organs. But the first complete study of growth substances was that of Charles and Francis Darwin contained in Charles Darwin's book *The Power of Movement in Plants,* also published in 1880. A critical observation of the Darwins was that the tip of a growing seedling, actually a coleoptile of a grass, perceived the light, the result of which was bending toward the light. Yet the curvature was well below the tip, the sensitive point of perception. For this reason it was postulated that there was some kind of communication between the tip and the stalk below.

After the work of the Darwins, Boysen Jensen in 1910 showed that if the tip was removed, no curvature occurred. However, bending would again occur if the tip was returned to its original position but with a layer of gelatin placed between the severed stalk and the tip. This experiment seemed to confirm that a material substance was actually transported from the tip through the gelatin to the region below.

481

Paal in 1919 showed that if the tip were removed and replaced unequally or in an offset position (refer ahead to Fig. 18.2), bending would occur away from the offset position of the tip. The active substance diffused from the offset tip down one side of the coleoptile, causing differential growth. The side receiving the substance grew faster than the opposite side, causing the unequal growth and curvature. When the tip is in the proper place, equal and straight growth occurs. Light evidently causes lateral transport or destroys the substance. Less growth occurs on the lighted side with the result that there is bending toward the light because of the unequal growth.

Further proof of a material substance was demonstrated by Went in 1928 while working in Utrecht. He removed tips, placed them on agar to allow the growth substance to diffuse into the agar, and then placed the agar with growth substance on the cut portion of the coleoptile tip. Curvature would result if the agar block was placed unequally, just as it did with the cut tip. The coleoptiles were still sensitive to light if the agar blocks were in place.

The substance that promotes cell enlargement in plants was named auxin. Kogl in 1934 isolated auxin from human urine and identified it as indole-3-acetic acid.   **[See A in col. 2]**

Subsequently, the research of Haagen-Smit and others in the 1940s showed that indole-3-acetic

Indole-3-acetic acid (IAA)

A

acid (IAA) was a naturally occurring component of higher plants.

Further research by plant physiologists has revealed the existence of several other classes of naturally occurring plant growth substances in addition to auxins. They include the gibberellins, the cytokinins, abscisic acid, ethylene, and some phenolics with growth-regulating activity. The remainder of this chapter will discuss our knowledge of the naturally occurring plant growth substances, including their properties, structures, and individual modes of action. They are treated as if they occur and function alone as specific promoters or inhibitors of growth and development. Of course, this is not true. The plant hormones do not occur in plants alone and they do not function alone. They function primarily in relation to each other such that the balance is more important than the presence or concentration of any one. Chapter 19 will summarize the correlative function of the plant hormones and outline our knowledge of how the hormones act together to bring about coordinated growth and development.

18.1 A Definition

Terminology of the plant growth substances has become somewhat confused because of the similarity to the hormonal concept for animal growth and development. The hormones regulating animal growth and development are endogenously produced substances that are synthesized in one place and have their targets at other places within the body. Compounds such as vitamins and other kinds of animal growth factors are thus not considered to be hormones..In plants, the plant growth substances referred to above, i.e., auxins, gibberellins, cytokinins, abscisic acid, phenolic growth substances, and ethylene, do appear to fit the classical definition.

Throughout this text, the term "plant growth substance" will be used for those compounds that regulate aspects of growth and development and will include the naturally occurring hormones (auxins, gibberellins, cytokinins, abscisic acid, ethylene, growth-regulating phenolics, perhaps florigin (the flowering hormone), the artificial growth-regulating substances such as the herbicide 2,4-D, and growth inhibitors such as maleic hydrazide. Thus "plant growth substance" is an all-inclusive term. The term "hormone" will be used to mean only the naturally occurring plant growth substances.

18.2 Auxin

The early experiments of Charles Darwin (referred to above) are perhaps the best beginning for an understanding of the complex nature of plant growth substances and their mode of action. When grass coleoptiles (usually those of *Avena*) are grown in the dark, they will elongate. If irradiated with unilateral light, they will bend toward the light (Fig. 18.1). The light causes unequal distribution of the plant hormone auxin, with more on the dark side. Because of the greater auxin concentration, there is greater cell elongation on the dark side than on the lighted side, causing differential growth and bending toward the light. However, if the tip is covered so that it remains dark, the light has no effect. Thus the light stimulus is received and auxin is redistributed at the tip, and the plant growth substance is then transported toward the base, where the response is cell elongation.

The experiments of Fritz Went (mentioned above) were instrumental in defining the nature of the plant growth substance that promotes the elongation of coleoptiles. He severed the tips of the coleoptiles and placed them on gelatin, allowing auxin to diffuse into the agar blocks. The blocks could then be placed back on the severed coleoptiles and their function was identical to the actual physiological tips (causing elongation). As mentioned above, it was these experiments that pointed to auxin as an actual material substance. The definition of auxin is the plant growth substance that promotes elongation in coleoptiles. The Went experiments are diagrammed in Fig. 18.2.

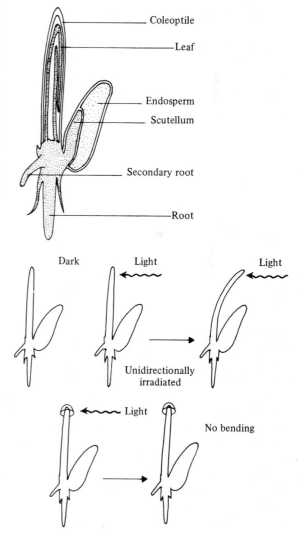

FIGURE 18.1 The light experiment that demonstrates the phototropism of cereal coleoptiles. The coleoptile is illuminated with unilateral light, and subsequent growth is toward the direction of the light. If the coleoptile tip is either severed (and removed) or shaded from the unilateral light, no photocurvature occurs. The experiment demonstrates that the tip is the photosensitive portion of the coleoptile. Also shown is a detailed drawing of a cereal coleoptile.

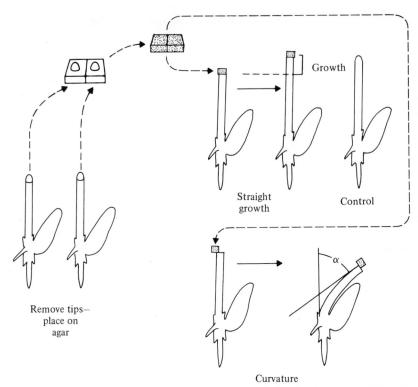

Growth

Straight
growth

Control

Remove tips—
place on
agar

Curvature

α

FIGURE 18.2 The Went experiment, showing the collection in agar of diffusible auxin (unbound) from severed coleoptile tips. If the agar is placed directly on the decapitated coleoptile, normal growth occurs. If it is placed to the side, growth will occur away from the auxin. The angle of curvature can be used as a bioassay for auxin. See Fig. 18.4. (Refer to F. W. Went. 1942. Growth, auxin and tropisms in decapitated *Avena* coleoptiles. *Plant Physiol.* 17:236–249.)

18.2.1 Structure and assay

The first successful attempt to isolate and identify auxin was by F. Kogl working in Holland in 1934. He and his associates isolated some compounds from human urine that had auxin activity, defined by the stimulation of coleoptile elongation. Initially, two compounds, called auxin A and auxin B, were isolated and identified as nonindole compounds. They then identified indoleacetic acid as auxin, and today there is still much evidence that it is the main, if not the only, active naturally occurring auxin.

We know that auxin does not occur only as free auxin in plant tissues. The diffusible auxin isolated on agar from severed coleoptile tips seems to be free auxin and the physiologically active form. There is much evidence that auxin also occurs in bound form, i.e., not physiologically active. Among the bound forms of auxin identified in plants are O-glycosides, including IAA-glucose, IAA-myoinositol, IAA-myoinositol arabinosides and galactosides, and IAA-glucans. In addition, IAA-aspartate has been identified in plants.

Other indole derivatives showing auxin activity are indoleacetonitrile, indoleacetaldehyde, indolepyruvate, indole ethanol, and tryptamine. However, much of the evidence suggests that these compounds are converted to IAA *in vivo* and that they do not in themselves have auxin activity.

The structures of these important derivatives follow.

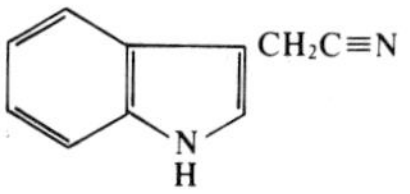

Indoleacetonitrile

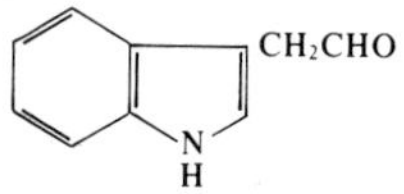

Indoleacetaldehyde

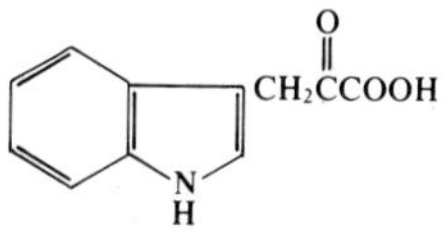

Indolepyruvic acid

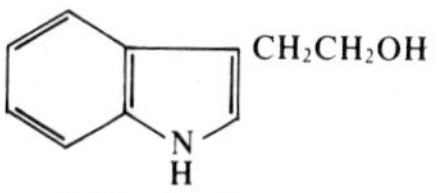

Indole ethanol

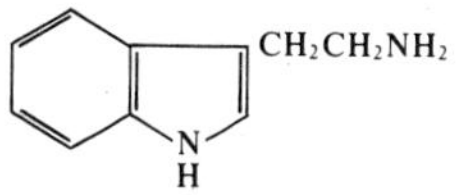

Tryptamine

18.2.2 The metabolism of auxin

Synthesis

The immediate precursor of indoleacetic acid in plants seems to be tryptophan. When ^{14}C-labeled tryptophan is supplied to actively growing plant tissues capable of forming IAA, ^{14}C-IAA is produced.

Wightman and his colleagues (1968) studied the biosynthesis of IAA in plants, showing that tryptophan can be converted to IAA, either by deamination to form indole-3-pyruvic acid followed by decarboxylation and oxidation of the terminal group to a carboxyl, or by decarboxylation to form tryptamine followed by deamination and oxidation to form IAA. The probable biosynthetic sequence is illustrated in Fig. 18.3 (from Wightman's work).

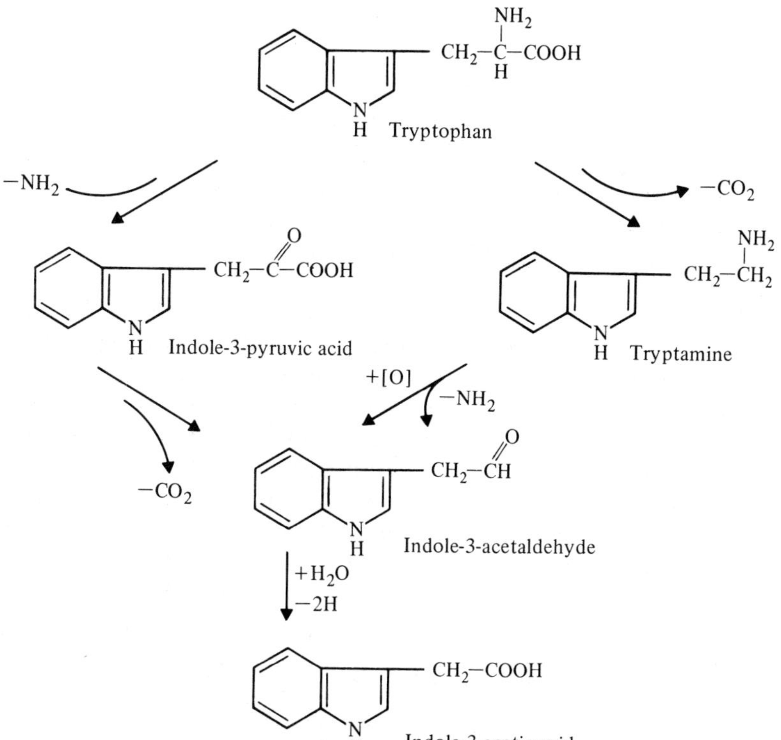

FIGURE 18.3 Possible biosynthetic pathways for the synthesis of indole-3-acetic acid (IAA) from tryptophan in plants.

Degradation

Intimately linked to growth and development in plants is the metabolism of IAA. Exactly how the auxin content is reduced at the termination of a growth and developmental process is not entirely known. Possible means to reduce auxin content are by translocation away from the target region, oxidation by peroxidases such as IAA oxidase to inactive compounds, or condensation with other chemicals. It is known that within a few hours after auxin is supplied to a tissue, much of it is converted to glycosides such as IAA-glucose or IAA-inositol or to the peptide IAA-aspartate.

IAA glucoside

IAA inositol

IAA aspartate

These IAA derivatives are known to be completely without auxin activity and thus are believed to function in the cessation of auxin-dependent growth responses. Of course, they may also function as auxin precursors.

Alternatively, auxin may be inactivated by oxidation. Primarily, oxidation is by one or more peroxidase enzymes, ultimately forming either indolealdehyde or methyleneoxindole. The peroxidases uses H_2O_2 as an electron acceptor, and the products frequently polymerize.

[See **B** in col. 2]

In addition, indoleacetic acid oxidase activity has been described in plants that metabolize IAA in a nonperoxidative reaction.

IAA can be inactivated *in vitro* by light. Photo-

B

oxidation also causes production of indolealdehyde and methyleneoxindole. The importance of photo-oxidation in the metabolism of IAA in the plant is not known.

18.2.3 The assay of auxin

Because the plant growth substances occur in such low concentrations within the plant tissues in which they are active, the assay and preparation require special techniques. The usual gravimetric analytical methods are useful in the milligram range, and colorimetric techniques can be used to assay microgram quantities. Chromatographic techniques are also useful over the microgram range. However, the plant growth substances occur in plant extracts in nanogram quantities (10^{-9} g) or less. Perhaps the most interesting analytical techniques that have been developed are the bioassay methods in which a preparation suspected of having a growth substance is tested in live tissue for the known response. Went in 1937 developed a quantitative bioassay for auxin based on the known curvature of *Avena* coleoptiles. The amount of curvature is a direct function of the amount of IAA, and hence a graph of curvature against IAA concentration can be used as a standard to assess the quantity of auxin in a sample.

Straight growth of the *Avena* coleoptile can be used as a bioassay in place of the curvature test. The disadvantage of the straight-growth test is that the response is logarithmic, whereas the curvature test shows a linear response over the low, sensitive concentrations of auxin. In the straight-growth test the coleoptile section can be immersed in the test solution, and thus transport of auxin is not a bothersome problem, as it is in the curvature test. With the latter there must be transport from the agar block down into the coleoptile section.

In the *Avena* coleoptile curvature test, the coleoptile tip is removed and a section is then removed to expose the true leaf below. An agar block with the test sample is placed on the severed tip, and after a period of incubation the curvature in degrees is measured. The test is sensitive enough to measure 0.08 nanograms of IAA in an agar block 4 cubic millimeters in volume. The *Avena* coleoptile curvature test is shown pictorially in Fig. 18.4, accompanied by a graph.

Because of the low quantities in plant tissues, the actual preparation of growth substances prior to assay is difficult. One method that has proven successful is a combination of paper or thin-layer chromatography and the *Avena* coleoptile growth test. A sample containing the unknown mixture of growth regulators is chromatographed in the appropriate solvent, and then segments of the chromatogram are tested for auxin (or other growth regulator) activity. Such an experiment conducted by Crosby and Vlitos (1961) with tobacco is shown in Fig. 18.5. Two areas of activity were demonstrated with the *Avena* coleoptile growth test, one corresponding to the migration distance of IAA and one to the migration distance of indoleacetonitrile. In addition, the region just "beyond" IAA and IAN had auxin activity, and there are regions of inhibition. It can be concluded from this kind of experiment that this variety of tobacco had IAA and IAN, perhaps other auxinlike substances, and growth-inhibiting substances.

Gas chromatographic techniques coupled with mass spectrographic assay methods have been de-

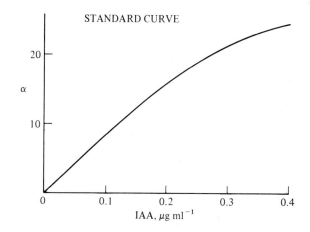

STANDARD CURVE

IAA, μg ml^{-1}

FIGURE 18.4 The *Avena* coleoptile curvature test for the bioassay of auxin. The extent of bending as measured by the angle, α, is a function of the amount of IAA or auxin present in the agar block. First, a concentration response is determined with known IAA solutions. The degree of curvature from an unknown sample contained in an agar block can then be determined and the concentration read from the standard curve.

Remove tip   Expose first leaf   Attach agar block with sample and read curvature after growth

FIGURE 18.5 Chromatographic-bioassay method of analyzing plant tissues. First, a sample is chromatographed to separate auxinlike hormones. The abscissa shows the R_f (relative distance from origin, point 0, to front, point 1.0). Substances are eluted from the chromatogram according to R_f and tested against either the *Avena* coleoptile curvature test or the straight-growth test. Here the data are graphed as growth against R_f. Two areas on the chromatogram showed auxin activity, one corresponding to the R_f of known IAA and the other near the known R_f of indoleacetonitrile. (From an experiment by D. G. Crosby and A. J. Vlitos. 1961. New auxins from "Maryland Mammoth" tobacco. In Anonymous. *Plant Growth Regulation*. Iowa State University Press, Ames, Iowa.)

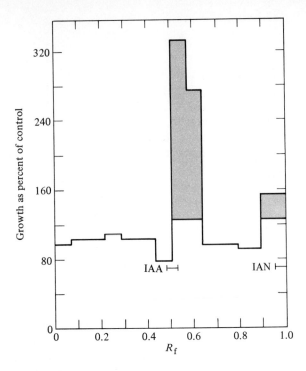

veloped to detect and measure plant growth hormones in plants. Although such methods require complex and expensive equipment, they are very sensitive and quantitative. Smith and Shindy in 1975 developed such a method, allowing them to identify auxins, gibberellins, cytokinins and abscisic acid in cotton bolls (Fig. 18.6).

18.2.4 The transport of auxin

Polar transport

Perhaps one of the most interesting physiological features of auxin is energy-dependent polar transport. In coleoptiles, for example, auxin is transported basipetally (toward the base) at speeds as high as 1 to 1.5 centimeters per hour. Since this speed of transport is greater than can be accounted for by diffusion and since it is energy dependent and occurs against a concentration gradient, it seems quite clear that it is active transport. Transport is apparently through the living parenchyma cells of the coleoptile.

The transport system will discriminate against analogues of auxin. Naphthaleneacetic acid (NAA) and 2,4-dichlorophenoxyacetic acid (2,4-D), two synthetic auxins, are both transported through the polar transport system, but many substituted auxins and inactive isomers are poorly transported or not transported at all.

The transport is mostly toward enlarging or growing cells, although in the case of the root tip transport is mostly toward the tip, i.e., acropetally. Auxin is transported out of the leaf toward the petiole base, and the continual flow prevents leaf abscission.

There is evidence that the polar-transport properties of auxin account in part for its great capacity to regulate growth. This notion is discussed in more detail in Chapter 19.

Lateral transport

When coleoptiles are irradiated unilaterally, there appears to be lateral transport of auxin away from the lighted side. Less auxin can be measured on the lighted side than on the darkened side, and if ^{14}C-IAA is supplied to the tissue and irradiated on one side the total amount of radioactivity remains the same within the coleoptile, but there is a move-

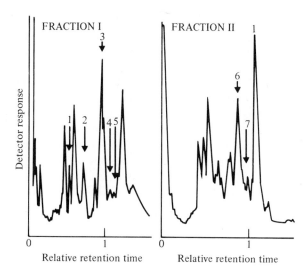

FRACTION I

FRACTION II

Detector response

Relative retention time

Relative retention time

FIGURE 18.6 The plant growth hormones can be assayed by gas chromatographic–mass spectroscopic methods. Shown in the graph is the identification of several plant hormones from an extract of cotton bolls. 1 = indoleacetic acid. 2 = abscisic acid (ABA). 3 = gibberellin (GA_{13}). 4 = gibberellins (GA_4 and GA_7). 5 = gibberellins (GA_1 and GA_3 — GA_3 = gibberellic acid). 6 = dihydrozeatin. 7 = zeatin. (From an experiment by W. Shindy and O. E. Smith. 1975. Identification of plant hormones from cotton ovules. *Plant Physiol.* 55:550–554.)

ment toward the darkened side. Thus the evidence favors transport rather than light destruction. This is the Cholodny–Went hypothesis discussed in Chapter 17.

18.2.5 Function

The regulatory role of indoleacetic acid has been studied intensively but remains poorly understood. In the early work of Went, he was led to conclude that there would be virtually no growth in the absence of IAA. To date, there are no data that contradict this broad generalization for plant growth. IAA is synthesized primarily in growing regions, and there is an excellent correlation between the presence of IAA in a tissue and growth.

The list of growth and developmental processes that are influenced by auxin is long. The following is a list of processes that are in some way influenced by auxin.

Dormancy	Fruit ripening
Flower initiation	Rooting
Fruit growth	Growth rates
Abscission	Fruit set
Juvenility	Tuberization
Sex determination	Senescence

Of course, since the plant hormones largely influence growth through a balance of concentrations, growth substances other than auxin such as gibberellin, cytokinin, abscisic acid, and ethylene also influence these processes.

The three basic types of growth movements that were discussed in Chapter 17—tropisms, nastic movements, and nutations—are probably regulated by auxin. Tropisms are growth movements in response to unilateral stimuli and include phototropic growth (toward or away from light) and geotropic growth (toward or away from the earth).* Tropisms are actual growth movements since they are irreversible. They do not include those movements that are the result of turgor changes, such as leaf folding, stomatal closing, and leaf and flower orientation of "compass" plants. Nastic growth movements are general, nondirectional movements that respond to environmental stimuli. A good example is the light-stimulated opening of flowers. The curling back of floral bracts, sepals, or petals is in response to light, either general or unilateral. The response is not directional. Nutational growth is the spiral upward growth of plants because of unequal growth around the elongating stem. This

*There is some evidence that auxin is only slightly involved in root geotropism.

growth pattern is an expanding ellipse and appears to be independent of the environment.

Phototropic growth movements were discussed in Chapter 17. Phototropism occurs in response to blue light and evidently is the result of light-stimulated transport of auxin away from the light, causing unequal growth on the darkened side.

Geotropic growth responses illustrate the complexity of plant growth. Roots grow toward the earth and stems grow away from the earth. Roots show positive geotropic growth and stems have negative geotropic growth. This polarity of growth remains regardless of how the plant axis is oriented. For example, if a young seedling is placed on its side, the root tip will bend toward the center of gravity and the shoot tip will grow away from the center. Geotropism was discussed in Chapter 17.

The fact that roots and shoots respond to different concentrations of auxin illustrates an important feature of plant growth substances. There is a sub-optimum, optimum, and above-optimum concentration for the hormone-dependent growth responses. Below the threshold level there is no response. There is a suboptimum concentration region where the response is proportional to concentration, an optimum where the response is maximum, and a high concentration in which the growth response is inhibited. Studies indicate (Fig. 18.7) that the threshold, optimum, and inhibitory levels for auxin-dependent growth of roots and of stems differ. Thus the same concentration of auxin can inhibit root growth and stimulate shoot growth.

The curve of growth response to auxin illustrates how the synthetic auxin 2,4-D can be such an effective herbicide. 2,4-D has most of the properties of IAA except that there is little curvature of coleoptiles. 2,4-D is an effective auxin at very low concentrations; high concentrations cause unnatural and inhibitory growth and ultimately death.

As explained above, auxin is involved in the regulation of a variety of growth processes. Any scheme to explain the mechanism by which auxin brings about regulatory responses must take into account the large amplification that accompanies its action. Such amplification is characteristic of many biological reactions that are dependent on synthesis and are regulated or mediated by enzyme catalysis.

There is much evidence that the auxin response and RNA synthesis and RNA-dependent protein synthesis are highly correlated in a cause-and-effect relationship. Auxin enhances protein synthesis and the incorporation experimentally of radioactive precursors into RNA. Inhibitors of growth also frequently inhibit RNA and protein synthesis. If auxin controls processes at the nucleic acid or protein-synthesis level, the amplification can be explained on the basis of enzyme synthesis. Nevertheless, auxin responses and growth are correlated with a myriad of other physiological and metabolic events, and a cause-and-effect relationship between auxin and RNA and protein metabolism is not unequivocal. It would be hard to

FIGURE 18.7 Graph to show the variable response of roots and stems to different concentrations of auxin. Note that the same concentration of auxin that inhibits roots stimulates shoots (about $10^{-7}M$). Redrawn from F. C. Steward. 1966. *About Plants: Topics in Plant Biology*. Addison-Wesley, Reading, Mass.

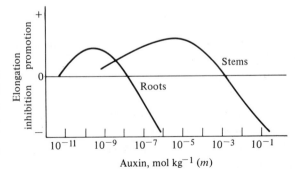

visualize any growth without protein synthesis.

The primary and most rapid response to auxin treatment is an increase in the rate of elongation of cells that occurs within a few minutes after application. Furthermore, it can be shown that there is an increase in the rate of incorporation of cell-wall precursors into cell-wall polymers within 10 to 15 minutes after auxin application to cells capable of expansion. All of this will occur in the presence of inhibitors of protein synthesis. These data indicate that the primary mode of action acts directly on cell-wall expansion and is relatively independent of protein synthesis.

In Chapter 19 the mechanism of cell-wall extension growth will be discussed; it is indicated that low pH causes cell-wall loosening, which marks the beginning of the cell growth process. It is proposed that auxin causes a lowering of pH by stimulating proton secretion. The low pH activates cell-wall loosening enzymes and cell-wall synthetic enzymes, beginning the process of cell expansion. Sustained growth would probably require protein synthesis, but the primary effect of auxin is directly on cell-wall metabolism.

As will be explained in more detail in Chapter 19, there is much evidence that there are specific hormone receptors in cells that play an important role in growth regulation. Because of this idea of specific receptors, it seems reasonable to assume that active auxin (and other hormones) have rigid structural requirements for maximum activity. It is believed that the auxin structural activity requirements are:

1. a ring with at least one double bond,

2. a side chain with a carboxyl group or a group that can be converted to a carboxyl,

3. at least one carbon between the ring and the carboxyl,

4. the proper spatial arrangement between the ring and the carboxyl, and

5. the capacity to form a covalent bond at the ortho position of the ring.

Reference to the structure of indoleacetic acid shows that it is the model for the structural requirement. It is assumed that compounds with this basic structure can bind to the proper receptor and will show auxin activity.

Studies to confirm a rigid structural activity requirement have not been unequivocal. For example, some benzoic acid derivatives do not meet the structural requirement yet have some auxin activity. The hypothesis that there is a structural requirement for auxin activity, however, has aided in the selection of synthetic compounds with auxin activity.

18.2.6 Synthetic auxins

Synthetic auxins that can be rapidly and inexpensively prepared are powerful tools for the regulation of plant growth and development. The desire to obtain such synthetic plant growth regulators for agricultural, horticultural, and even military use led to the research discussed above, which was designed to determine structural requirements for chemicals with auxin activity. Some examples of active synthetic auxins that do meet the structural-activity requirement hypothesis follow.

Indole-3-propionic acid

Indole-3-butyric acid

Naphthaleneacetic acid

2,4-dichlorophenoxyacetic acid

There are several known compounds that in small concentrations will directly interfere with auxin action. Many of these antiauxins are structurally similar to auxins but lack the proper groups or spatial orientation for activity. Compounds such as γ-phenylbutyric acid, *trans*-cinnamic acid, tri-iodobenzoic acid, and 2,4,5-trichlorophenoxy-acetic acid are some of these known antiauxins.

Compounds that are structural analogues of auxin such as indoleisobutyric acid do not have anti-auxin properties.

18.3 Gibberellins

An erratic growth disease of rice, characterized by elongated stem growth, led to the discovery of the naturally occurring plant growth substances called gibberellins. The disease, called "foolish seedling" disease, or "bakanae" disease, was shown by Eiichi Kurosawa in 1926 to be caused by the fungus *Gibberella fujikuroi.* This fungus, when cultured under aseptic conditions, secreted a heat-stable substance that would cause elongation of rice stems. The discovery of auxin occurred at about the same time, in 1928.

Subsequently, in 1934 Yabuta and others were able to isolate and crystallize the active compound.

Their studies were first complicated because of the presence of fusaric acid (5-n-butylpicolinic acid), which acted as a growth inhibitor. Eventually, they isolated and characterized the growth-promoting substance and named it gibberellic acid.

In 1952 gibberellin was crystallized in the United States and then became generally available for experimental work. It was shown at this time that the application of gibberellin to dwarf pea plants would "break" the dwarfness; evidently the dwarfness was due to a deficiency of gibberellin (Fig. 18.8).

There have now been more than 50 different gib-

FIGURE 18.8 Experiment showing the effect of gibberellin on the growth of dwarf peas. The plant in the center was treated with GA_3 72 hours before the photograph was taken. The dwarf $+ GA_3$ elongated to the size of the normal pea. The only apparent difference between the untreated dwarf and the treated dwarf is the elongation of the internodes (note that all three plants have the same number of internodes). These plants were 10 days old at the time of the photograph.

Normal pea   Dwarf pea + GA_3 Dwarf pea

berellins isolated from plants, all causing the very marked growth response of stem elongation. The stem-elongation response to gibberellin is far more dramatic than it is to auxin.

18.3.1 Structure

The gibberellins are acidic diterpenoids found in all higher plants. They are designated in a series: GA_1, GA_2, GA_3, etc. The first to be characterized in the United States was called gibberellic acid and is now known as GA_3.

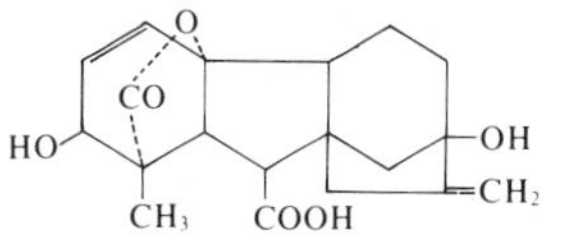

Physiologically they promote stem growth, although many and varied responses are known, as will be discussed.

The parent (base) compound from which all the gibberellins are derived is gibberellane.

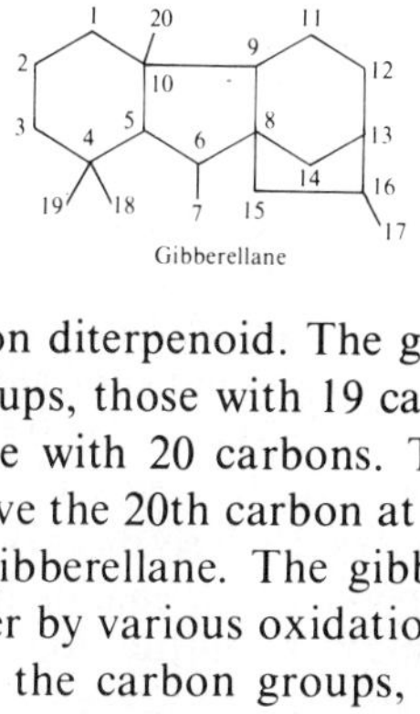

It is a 20-carbon diterpenoid. The gibberellins fall within two groups, those with 19 carbons, such as GA_3, and those with 20 carbons. The 20-carbon gibberellins have the 20th carbon at position 20 as indicated on gibberellane. The gibberellins differ from each other by various oxidation states of the ring structure, the carbon groups, and hydroxyl groups. Gibberellins also occur naturally as glycosides and as bound, inactive forms similar to auxin.

The biosynthesis of the gibberellins is not totally understood but probably occurs through the mevalonate pathway for terpenoid biosynthesis as discussed in Chapter 12. An abbreviated scheme for the possible synthesis of the gibberellins is shown in Fig. 18.9.

Unlike auxin, the gibberellins seem to be rather stable in plants. They are readily interconverted

ABBREVIATED PATHWAY FOR POSSIBLE SYNTHESIS OF GIBBERELLINS

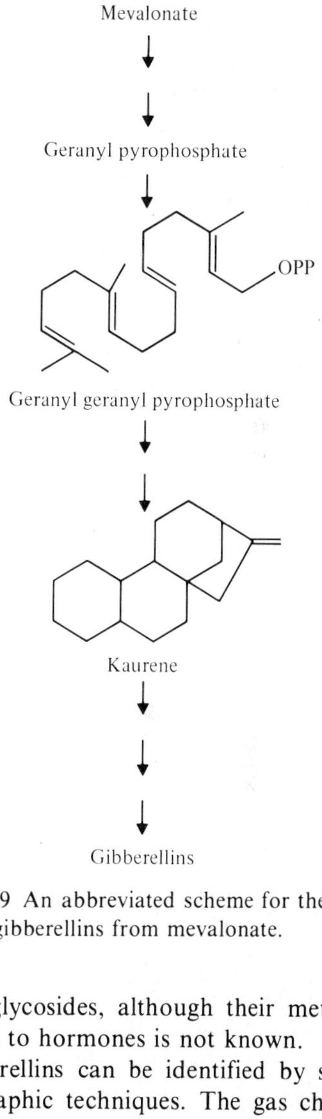

FIGURE 18.9 An abbreviated scheme for the possible synthesis of gibberellins from mevalonate.

and form glycosides, although their metabolism with respect to hormones is not known.

The gibberellins can be identified by standard chromatographic techniques. The gas chromatographic–mass spectrographic technique of Shindy and Smith (1975) is especially sensitive; it is sufficient to detect a few nanograms per gram fresh weight of tissue (see Fig. 18.6). The highly sensitive bioassay techniques using growth responses are more common.

Bioassay of gibberellin has been based on a variety of responses in plants that are deficient in gibberellins. Assays can be based on the breaking of genetic dwarfness in beans, maize, and peas. Other bioassays are based on the growth of lettuce hypocotyls, the prevention of senescence in leaves as indicated by chlorophyll content, and the production of α-amylase by barley endosperm. Any biological response to gibberellin that can be described quantitatively could conceivably be used as a bioassay.

18.3.2 Function

The classical definition of a gibberellin is a compound that stimulates stem elongation. Furthermore, it is known that the stem-elongation response comes about mostly by cell enlargement and not by cell division. Roots and leaves show only a weak response to added gibberellins. Gibberellins, however, like auxin, influence nearly all growth and development processes. And, in fact (as will be discussed in Chapter 19), a balance of plant growth hormones is an important feature of proper correlative growth and development.

Unlike auxins, gibberellins are not polarly transported, and, as a consequence, they bring about a more general type of response, which frequently occurs throughout the plant. Gibberellins are transported both basipetally and acropetally with equal facility, probably to a large extent through the xylem and phloem.

Perhaps the most dramatic and easily demonstrated gibberellin response is the breaking of genetic dwarfness in plants discussed above (see Fig. 18.8). If GA₃ is applied to genetic dwarf pea or corn plants, the dwarfs will show internode elongation comparable to the normal nondwarfs. The normal plants show no response to added gibberellins. Thus the major response appears to be normal internode elongation, and such dwarf plants apparently lack sufficient endogenous gibberellin.

Gibberellin will break dormancy in certain seeds, even in those which ordinarily require a light treatment for germination. In addition, gibberellin can induce flowering in cold-requiring and photo-

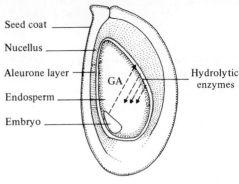

CEREAL SEED ANATOMY

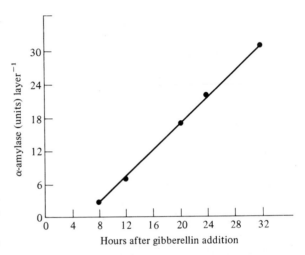

FIGURE 18.10 The production of α-amylase by barley aleurone cells after gibberellic acid treatment. Also shown is a diagram of the anatomy of the cereal seed, showing the hypothesis for gibberellin production by the embryo. The GA is transported to the aleurone layer, stimulating the production of hydrolytic enzymes. Hydrolytic enzymes are secreted into the endosperm tissue, causing hydrolysis. From J. E. Varner and P. Tuan-Hua Ho. 1976. Hormones. In J. Bonner and J. E. Varner (eds.), *Plant Biochemistry*. 3rd ed. Academic Press, New York.

period-dependent plants. Gibberellins also stimulate fruit set and development and under some circumstances prevent leaf senescence.

Gibberellin is a growth- and development-promoting substance. However, one of the most interesting and thoroughly studied responses to exogenously added gibberellin is the stimulation of the production of hydrolytic enzymes during germination in cereals. In barley seeds, germination is accompanied by the hydrolysis of the endosperm, which mobilizes carbohydrate and amino acids used in the growth processes during and subsequent to germination.

Gibberellin produced by the embryo stimulates the triploid aleurone cells of the cereal seed to produce α-amylase, proteases, and other hydrolytic enzymes (Fig. 18.10). Varner and his colleagues in

1967 showed that the production of α-amylase is through *de novo* synthesis. RNA synthesis inhibitors such as actinomycin-D will prevent α-amylase production in the aleurone tissue.

Kinetic studies indicate that the α-amylase activity begins to increase prior to the synthesis of RNA. It is not clear if the effect of gibberellin is directly on RNA synthesis. Nevertheless, there is a correlation between gibberellin, RNA synthesis, and protein synthesis.

18.4 Cytokinins

Compounds that stimulate cell division in plants are called cytokinins. For the most part they are derivatives of purine (adenine) bases.

The plant growth substances, auxin and gibberellin, have cell elongation as a primary response, although virtually all growth and developmental processes are influenced to some degree. Early in modern plant research, investigators such as Wiesner in 1892 proposed the existence of cell-division factors, and Haberlandt in 1913 found that extracts of phloem could induce cell division in parenchyma tissue. Later, research by van Overbeek and his colleagues in the 1940s showed that the fluid from the inside of the coconut in which the embryo is bathed would promote cell division.

In 1955 Carlos Miller prepared a factor by heat treatment of DNA, 6-furfurylamino purine, which would promote cell division.

Kinetin (6-furfurylamino purine)

This factor was named kinetin but is now known not to be a naturally occurring cytokinin. Nevertheless, kinetin is a classical example of a cytokinin, the growth substance that promotes cell division. The cell-division promoters in coconut milk are cytokinins.

In 1964 Letham and his co-workers isolated a

cytokinin from corn seeds, naming it zeatin.

Zeatin

Zeatin was the first naturally occurring cytokinin isolated from plant tissues. It is now known that all plants contain cytokinins. The greatest concentrations seem to be in embryos and developing fruits.

Quite interestingly, most of the naturally occurring cytokinins have been obtained from the hydrolysis of RNA, although there is no firm evidence that cytokinins are incorporated into the RNA molecule during synthesis. Most evidence indicates that the cytokinins are formed by modification of existing bases. The cytokinin isopentenyl-adenosine (IPA) or another similar modified base may be found adjacent to the anticodon in tRNA (see Fig. 8.8). The IPA is not incorporated into the tRNA; rather, the isopentenyl group of adenine is added from mevalonate after the tRNA is produced by polymerization.

Because of the presence of the IPA adjacent to the anticodons that recognize codon triplets beginning with U (uracil), there is the strong suggestion that IPA plays a role in mRNA–tRNA–ribosome recognition. Other than this one possibility, the role of cytokinins in tRNA is not understood.

There is some evidence that when cytokinins are

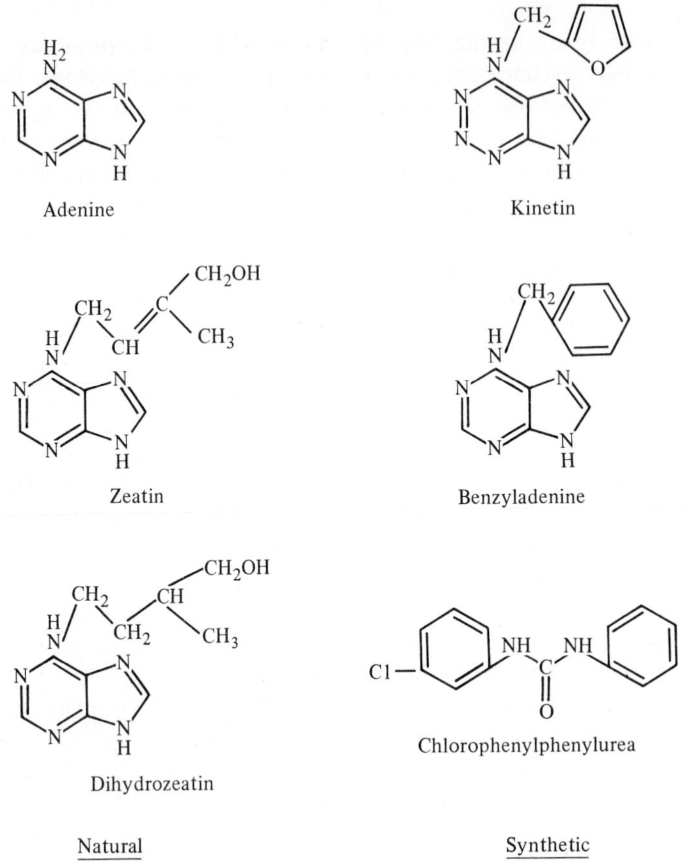

FIGURE 18.11 Diagrams of some natural and synthetic compounds that have cytokinin activity.

prepared from natural sources, they lose activity as the preparations become more pure. This has led to the suggestion that there are factors that enhance cytokinin activity. One such factor that has been identified is myoinositol.

18.4.1 Structure and assay

Most cytokinins, either naturally occurring or synthetic, are adenine derivatives.

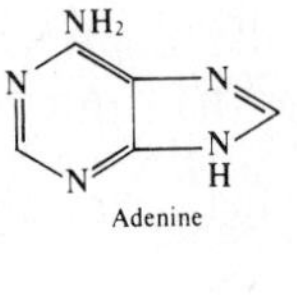

Adenine shows cytokinin activity when added to plant tissues. A few nonadenine compounds are known that have cytokinin activity (Fig. 18.11).

The biosynthesis of cytokinins probably starts with adenine produced by the process explained in Chapter 9 and from the mevalonate pathway for isoprenoid biosynthesis. The side chain of the cytokinins is an isoprenoid unit, and ^{14}C-mevalonate is incorporated into cytokinins.

The inactivation of cytokinins may proceed through glycoside formation since inactive glycosides are known to occur in plants. Alternatively, cleavage of the side chain at the 6-amino bond followed by oxidation or metabolism of the adenine residue would cause inactivation.

There is no firm evidence that structure is related to the activity of the cytokinins despite the fact that most are adenine derivatives similar to kinetin. Substitutions on the adenine ring do alter biological activity, but there are some urea derivatives (chlorophenylphenylurea, for example) that have strong cytokinin activity. There are two hundred or more compounds similar to chlorophenylphenyl-urea with cytokinin activity.

Chlorophenylphenylurea

Cytokinins can be assayed by chromatographic and bioassay techniques. In cotton bolls, Shindy and Smith (1975) identified zeatin, dihydrozeatin, and a zeatin riboside by gas chromatography and mass spectrometry (Fig. 18.6).

Two common bioassay methods that can be used to detect low levels of cytokinin involve (1) the retention of chlorophyll by aging leaves after application of cytokinins and (2) the extent of enlargement of cotyledons. At low concentrations of cytokinin, the loss of chlorophyll by senescing leaves is quantitatively prevented by cytokinin. Standard curves can be prepared to allow measurement of cytokinin in unknown solutions. Similarly, the enlargement of radish cotyledons can be used as a bioassay for cytokinins.

18.4.2 Function

As with both auxin and gibberellins, cytokinin-mediated plant responses are accompanied by enhanced nucleic acid and protein metabolism. However, so far there is no unequivocal evidence that any of these growth-promoting plant hormones

act directly on the protein-synthesizing system of the plant. The mode of operation of the cytokinins is not understood.

The cytokinins regulate growth in a variety of ways. The regulation of cell division is perhaps the most obvious and important role. DNA synthesis and cell division are both stimulated by cytokinins. There is some evidence that cytokinesis stimulation is a unique property of the cytokinins. An interesting effect of cytokinins is the delay of senescence in leaves. If leaves are treated with cytokinins they will remain green beyond the time they normally would become chlorotic and senescent. Evidently storage reserves are mobilized in response to cytokinin from the roots to the shoots decrease during basis for the bioassay mentioned above.

The levels of cytokinin and the supply of cytokinin from the roots to the shoots decreases during senescence of shoots, supporting the notion that cytokinins are antisenescence factors.

In addition to the above, cytokinins will break seed dormancy, stimulate seed germination, enhance flowering, and affect fruit growth. Cytokinins will induce bud formation in a variety of tissues, including callus growing in culture. As explained later, this may be important in apical dominance.

The cytokinins are particularly abundant in embryos, young fruits, and in roots, either in free active form or as a component of tRNA. As explained above, cytokinins do not appear to be incorporated into tRNA but rather occur because of base modification in the tRNA molecule.

In some respects cytokinins are not readily translocated, although movement through the xylem from roots to shoots occurs readily. Cytokinins are rather immobile in leaves. Some polar transport through leaf petioles has been reported.

18.5 Abscisic Acid

Research in 1953 by Osborne led to the discovery of a substance that could accelerate abscission of leaves. This substance was called abscisin for its abscission-stimulating activity. A substance called

dormin was subsequently isolated from tree leaves that were in the process of going dormant; it could induce dormancy in actively growing woody plants. It was quickly realized that abscisin and dormin were the same substance, and Addicott in 1968 proposed the name abscisic acid (ABA) for this growth-inhibiting compound.

Abscisic acid

Now abscisic acid has been isolated from a variety of higher plants and from ferns, horsetails, and mosses.

18.5.1 Structure and assay

Abscisic acid is a terpenoid and unlike most other hormones has an asymmetric carbon. The optically active carbon center has the hydroxyl and the side chain attached to it. The naturally occurring isomer is dextrorotatory ABA, (+)ABA, although levorotatory ABA, (—)ABA, is equally active. The biologically active form of (+)ABA has the *cis*-configuration; the *trans* form is virtually inactive. The free carboxyl group is necessary for activity, but glucosides of ABA that occur naturally can stimulate responses when added to sensitive tissues. Evidently there is hydrolysis of the glucoside prior to activity.

Biosynthesis of ABA probably takes place through the mevalonic acid pathway for iosprenoid biosynthesis. ^{14}C-mevalonate is incorporated into abscisic acid by plant tissues. Most biosynthesis occurs in leaves and fruit.

A few synthetic active analogues of ABA are known. Xanthoxin, an analogue of ABA, may be derived from xanthophylls and has ABA-like activity.

Xanthoxin

Xanthoxin may be either an ABA precursor or an ABA-like hormone in itself.

Inactivation of ABA has not been studied in detail, although the glucoside of ABA is inactive prior to hydrolysis. (+)ABA can be inactivated by conversion to phaseic acid and a few other compounds.

Phaseic acid

The fact that abscisic acid is optically active can be used to advantage in its assay. The Cotton effect (interaction between light absorption and optical rotation) has been used to assay (+)ABA quantitatively. Such assays based on optical activity have revealed ABA in plant tissues in concentrations up to 4 micrograms per gram fresh weight. Abscisic acid can also be detected in plant tissues by using gas chromatography–mass spectroscopic methods (Fig. 18.6).

Bioassays for abscisic acid depend largely on its inhibitory properties. Both the inhibition of germination and the interference with growth responses to the promoting hormones, auxin and gibberellin, have been used to measure abscisic acid in plant extracts. In addition, its property of causing stomatal closure has been proposed as a bioassay, since stomatal closure responds linearly to the logarithm of ABA concentration applied to leaves.

18.5.2 Function

Abscisic acid is a general growth inhibitor and linked to the abscission of leaves and fruits. It will induce dormancy and is believed to be the dormancy factor in plants. It is readily translocated and shows few polar-transport properties.

Abscisic acid inhibits seed germination, counteracting gibberellin in this respect. It has shown that actinomycin-D, a transcription inhibitor preventing DNA-dependent RNA synthesis, will prevent ABA inhibition of seed germination. During development of the cotton boll, young seeds will germinate precociously if ABA is removed by leaching. ABA may also regulate the use of preformed mRNA rather than regulate the synthesis of mRNA. In any case, it appears that the control point for ABA function is at the protein-synthesis level of metabolism.

There is some evidence that ABA affects protein synthesis through activation of ribonuclease activity. Such activation would degrade RNA and consequently reduce RNA-dependent protein synthesis.

An interesting plant response is an increase in the foliar concentration of abscisic acid during water stress. When plants are water-stressed by drought, leaf ABA levels increase with the result that stomata close. There is some reason to believe that ABA is a regulator of stomatal opening, working in opposition to cytokinin, which stimulates stomatal opening. In fact, there is a mutant tomato known that is deficient in ABA and tends to be chronically wilted. Treatment with ABA causes stomatal closure and relieves the wilting, which leads to the conclusion that ABA regulates stomatal opening.

Recent research has somewhat obscured the role of abscisic acid in dormancy and leaf abscission. The original investigations by Wareing and his colleagues in 1965 with sycamore showed that ABA increases when photoperiods become short and led to the conclusion that ABA is involved in dormancy in this species. Furthermore, the treatment of sycamore with ABA causes dormancy, fully supporting the conclusion. However, research with other species has not been so conclusive, and in fact contrary results have been obtained in which dormancy occurs with little or no change in endogenous ABA levels.

As shown by Addicott and his co-workers during the early 1960s, ABA seems to be the factor responsible for abscission and opening of the cotton boll. And, of course, it was this response that led to the naming of abscisic acid. Nevertheless, ABA will not cause abscission in all plants, and there are some experiments suggesting that ABA's effect in promoting abscission in those plants that respond is because ABA stimulates ethylene production. Perhaps the main abscission role of ABA is in fruit drop rather than leaf drop.

Even more confusing are observations that under some circumstances ABA will promote growth rather than inhibit it. However, most responses are ones that seem to counteract the growth-promoting properties of other hormones, especially the gibberellins.

Although many of the growth responses, inhibitory or otherwise, seem to result because of regulation at the nucleic acid–protein synthesis level, there is much evidence that some ABA responses are much quicker than those that could result from such a complex process. One mentioned above was stomatal opening. The response appears to be immediate and may be a direct effect on membrane properties.

18.6 Ethylene

The historical development of ethylene as a plant hormone has been interesting because of its unique gaseous properties, which are unlike any of the other known naturally occurring plant growth substances. Its effects on plant growth, including tropic responses, breaking dormancy, and enhancing fruit ripening, were known since the beginning of the 20th century. In 1934 Gane showed that ripening fruit produced ethylene gas. It was then proposed that ethylene was a hormone, but it was not until the 1950s and 1960s that ethylene was accepted as a naturally occurring growth substance. It is quite unique in that transport is probably exclusively by gaseous diffusion.

Because of the transport and the apparent lack of any directional control, ethylene may not be a hormone in the most stringent classical sense. Nevertheless, it shows growth-regulatory properties and, therefore, can logically be included with the hormones.

18.6.1 Structure and assay

Ethyelene is a simple molecule with a molecular weight of 28.

$$CH_2=CH_2$$

Its biosynthesis is poorly understood, but with the use of radioactive methionine it has been shown that carbons 3 and 4 of methionine are readily converted into ethylene.

$$CH_3-S-\underbrace{CH_2-CH_2}-CHNH_2-CO_2$$
$$\downarrow$$
$$CH_2=CH_2$$

Like methionine, other natural substances such as ethanol, acetate, and even glutamate may be precursors of ethylene.

Being a readily diffusible gas, the endogenous level of ethylene falls rapidly once production ceases. Thus there may be little in the way of inactivation (halting hormonal action) other than simple diffusion away from the target sites.

Indoleacetic acid will promote ethylene production (Fig. 18.12), and thus the interaction of IAA and ethylene in growth regulation is coordinated. As mentioned above in the discussion of ABA metabolism, ABA may stimulate the production of ethylene, especially during abscission.

Ethylene is readily assayed by enclosing tissues in gas-tight containers and sampling the surrounding atmosphere for ethylene using gas chromatography. Bioassay can also be used to test for the presence of ethylene.

18.6.2 Function

Ethylene production and presence is correlated with a variety of growth responses, including fruit ripening, senescence, abscission, epinasty, dormancy, leaf expansion, and flower induction. The wide array of responses to ethylene are not too unlike those caused by auxin, gibberellin, and cytokinin. Fruit ripening in response to ethylene has been studied in detail. Not only will ethylene treatment enhance the ripening of fruits such as banana and avocado, but ethylene is also rapidly and continually produced during ripening (Fig. 18.13).

One of the most interesting ethylene effects is the stimulation of cellulase like activity that brings about leaf abscission. It has been shown that ethylene induces a specific isozyme in bean leaf abscission zones that ultimately results in abscission of the leaf.

It has been known for several decades that ethylene would cause leaf epinasty. As early as 1931, Zimmerman and his associates showed that one ppm ethylene would cause epinasty of rose leaves (Fig. 18.14).

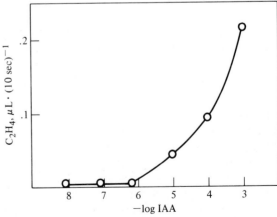

FIGURE 18.12 The production of ethylene gas by etiolated pea stems (over an 18-hour period) as a function of different concentrations of indoleacetic acid. From S. P. Burg and E. A. Burg. 1966. The interaction between auxin and ethylene and its role in plant growth. *Proc. Nat. Acad. Sci.* (U.S.) 55:262–269.

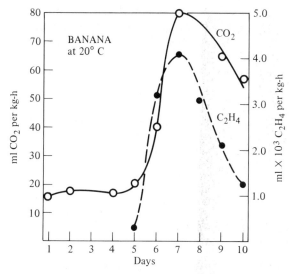

FIGURE 18.13 The production of ethylene gas by ripening banana fruit. From J. B. Biale, R. E. Young, and A. J. Olmstead. 1954. Fruit respiration and ethylene production. *Plant Physiol.* 29:168–174.

In addition to promoting plant growth, ethylene inhibits many important growth phenomena. Studies have shown that ethylene will inhibit stem elongation in pea sections and bud development of etiolated pea cuttings. Thus ethylene is apparently involved in both stimulation and inhibition of growth.

Most plants and tissues are sensitive to ethylene over the range of 0.01 to 10 ppm, with the half-maximal response at about 0.1 ppm ethylene.

Analogues of ethylene, including propylene, acetylene, butene, and other simple, unsaturated organic compounds, have some ethylenelike activity. As an example, propylene has about 1/100 the activity of ethylene. Other analogues are less active than propylene.

Once ethylene is removed, the response seems to stop immediately. Carbon dioxide is a competitive inhibitor for all known ethylene growth responses. For this reason a single ethylene receptor site has been proposed, although its nature is poorly understood. There is some evidence that ethylene may bind to a metal-containing site.

FIGURE 18.14 Experiment demonstrating that the component in illuminating gas (mostly methane) that causes leaf epinasty is ethylene. Panel A shows a control rose (left) and a rose treated with illuminating gas (100 ppm). There is marked epinasty. Panel B shows treatment with 1 ppm ethylene (plant on the right) and similar epinasty. (From P. W. Zimmerman, A. E. Hitchcock, and W. Crocker. 1931. The effect of ethylene and illuminating gas on roses. *Contr. Boyce Thompson Inst.* 3:459–481.)

The exact mode of action of ethylene is as obscure as the mode of action of the other hormones. It is clear that many of the ethylene growth responses are accompanied by increased nucleic acid metabolism and protein synthesis. Thus there is perhaps some effect at the nucleic acid level. Because ethylene alters responses such as tropisms that depend on the polar transport of auxin, it has been proposed that ethylene interferes with auxin transport and ultimately affects polar growth. Just how such alteration of the polar transport of auxin occurs is not known.

The production of ethylene by plants is evidently very sensitive to outside factors. A variety of kinds of stress will cause ethylene production, perhaps the most interesting of which is wounding. The actual production of ethylene by plant tissue requires oxygen and proper temperatures. As stated above, it is stimulated by both IAA and ABA, but ethylene also is produced in response to a variety of treatments by chemicals including cycloheximide and iodoacetate.

Ethephon (2-chloroethylphosphonic acid), a compound that will generate ethylene, has received much commercial attention. The possibility of using such an ethylene generator to regulate fruit ripening, cause fruit abscission during harvesting, and control other growth processes is of interest to commercial growers.

18.7 Phenolics and Other Growth Inhibitors

Many naturally occurring phenolic compounds, including cinnamic acid, coumarin, salicylic acid, caffeic acid, gallic acid, and scopoletin, show some growth-regulation activity but frequently at rather high concentrations. Examples of some of the inhibitory phenolics are shown in Fig. 18.15. Phenolics, when added to growing plant tissue, will inhibit growth in general (Fig. 18.16).

Many if not most of the phenolics occur as glycosides. Their chemistry and biosynthesis were discussed in Chapter 12.

The mode of action of phenolic inhibitors is poorly understood, but phenolics, particularly when polymerized, will bind to proteins and inhibit enzyme activity.

Several important and interesting growth-retarding synthetic substances have been identified. Growth-retarding chemicals have much economic importance in the regulation of undesirable plant growth and for the maintenance of horticultural crops and plants. Some of the useful and well-known growth-retarding chemicals are illustrated in Fig. 18.17. Their mode of action is not known, but they retard stem elongation in many cases. Gibberellin may overcome the growth retardation in some instances.

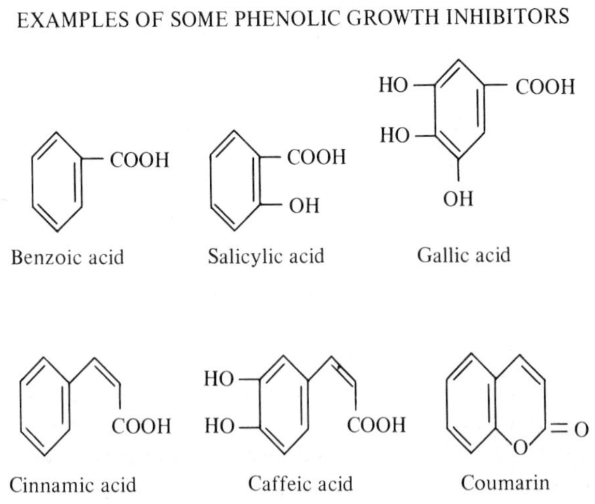

EXAMPLES OF SOME PHENOLIC GROWTH INHIBITORS

FIGURE 18.15 The chemical structures of some known phenolic growth inhibitors.

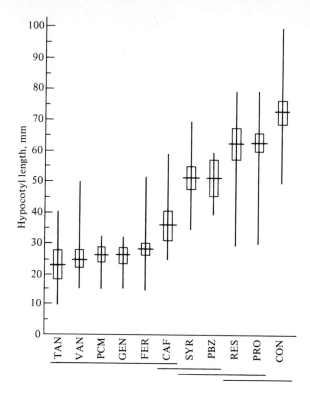

FIGURE 18.16 Experiment showing the inhibition of growth of mung bean hypocotyls by 1mM phenolic solutions. The vertical bars are the standard errors of the means and the vertical lines are the ranges. TAN = tannic acid. VAN = vanillic acid. PCM = p-coumaric acid. GEN = gentisic acid. FER = ferulic acid. CAF = caffeic acid. SYR = syringic acid. PBZ = p-hydroxybenzoic acid. RES = a-3,5-resorcylic acid. PRO = protocatechuic acid. CON = control solution with no added phenolic compounds. (From E. K. Demos, M. Woolwine, R. H. Wilson, and C. McMillan. 1975. The effects of ten phenolic compounds on hypocotyl growth and mitochondrial metabolism of mung beans. *Amer. J. Bot.* 62:97–102.)

EXAMPLES OF SOME SYNTHETIC GROWTH RETARDANTS

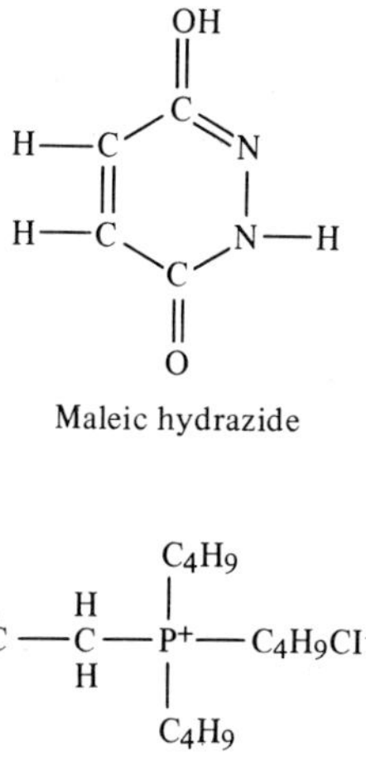

Maleic hydrazide

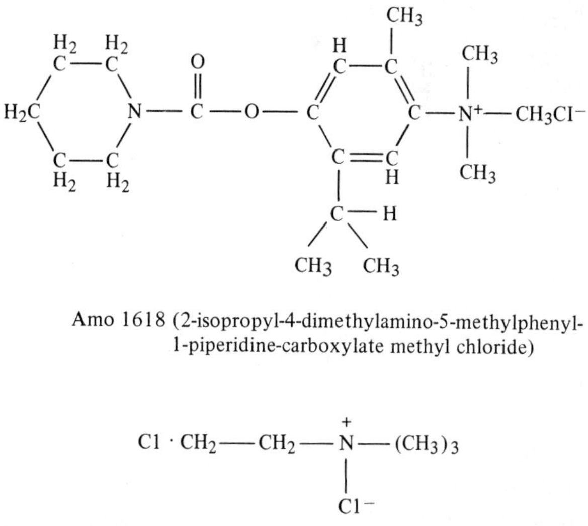

Amo 1618 (2-isopropyl-4-dimethylamino-5-methylphenyl-1-piperidine-carboxylate methyl chloride)

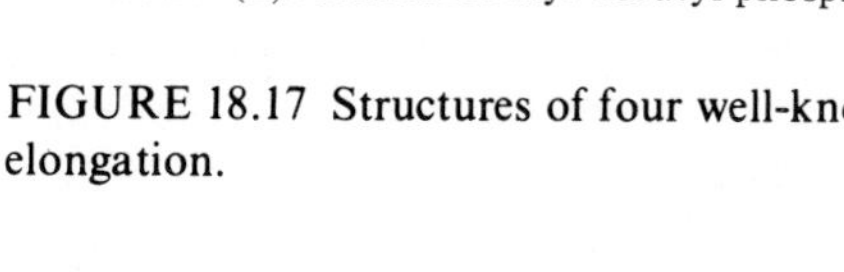

Phosfon-D (2,4-dichlorobenzyl-tributyl phosphonium chloride)

CCC (2-chloroethyl-trimethyl ammonium chloride)

FIGURE 18.17 Structures of four well-known synthetic growth retardants. They function largely by inhibiting stem elongation.

Maleic hydrazide was one of the first useful synthetic growth regulators for plants. It is used commercially to control sprout formation on stored potatoes and onions and to control unwanted excessive plant growth (e.g., along rights-of-way). Chlormequat (CCC) is now being widely used as a growth retardant in cereal agriculture and ornamental plants and topiary plantings.

18.8 Cyclic AMP, Cyclitols, and Vitamins

In animal tissues, many hormones function by regulating the levels of cyclic AMP.

Adenosine-3′,5′-phosphate
(cyclic AMP)

There has been an intensive search for cyclic AMP and enzymes that can metabolize cyclic AMP in plants. To date most of these studies have proven fruitless, and there is little evidence that c-AMP is important in plant growth regulation.

The cyclitols, including myoinositol, are known to stimulate growth of callus tissue culture when in the presence of other growth substances such as cytokinin. Whether inositol is an actual plant hormone is not known with certainty. However, it does have some properties consistent with the hormonal concept.

Myoinositol

Vitamins* are, of course, absolutely necessary for plant growth and development. They function in a variety of ways as components of enzyme reactions by being cofactors or portions of cofactors, by being in electron-transport phenomena, and involvement in other reactions. Many tissues cultured *in vitro* will not grow unless the medium is supplemented with vitamins. Root tissue requires some B vitamins, as do young embryos. Although the vitamins have some properties similar to the plant growth substances (i.e., the hormones), their mode of action is different from the hormones, and vitamins are usually not considered as plant growth substances in the same way as IAA, GA, ABA, and the others.

In brief, the functions of some of the common vitamins follow.

Thiamine (B₁)

As thiamine pyrophosphate (ThPP), thiamine functions as a cofactor for reactions of decarboxylation of α-keto acids and in group transfers, such as in the transketolase reactions of pentose metabolism.

Nicotinic acid

Nicotinic acid (niacin) is a B vitamin that is a component of NAD and NADP and is therefore a necessary factor for some oxidation–reduction reactions.

Riboflavin

Riboflavin (B₂), a component of many flavoproteins, is a contributing factor in electron-transport processes.

*Note that the term "vitamin" may not be appropriate here!

Pantothenic acid

Pantothenic acid is a component of coenzyme A, the factor that forms active derivatives of metabolites such as acetate (acetyl CoA) during metabolism.

Pyridoxine

Pyridoxine (B_6) is a cofactor in transamination reactions and reactions of decarboxylation of amino acids.

Biotin

Biotin, a B vitamin, is a cofactor for some carboxylation reactions such as when acetyl CoA carboxylase forms malonyl CoA during fatty acid biosynthesis. It is not involved in the reactions of the photosynthetic enzymes phosphoenolpyruvate and ribulose bisphosphate carboxylases.

Folic acid

Folic acid is a carrier for one-carbon fragments during one-carbon transfer reactions.

Ascorbic acid

Ascorbic acid (vitamin C) functions as a terminal electron acceptor in some plant tissues in the ascorbic acid oxidase reaction.

Quinones

Quinones, such as vitamin K, function in electron transport.

Tocopherols

Tocopherols, such as vitamin E, are membrane components that may function as antioxidants.

18.9 The Flowering Hormone (Florigen)

There has been an exhaustive search in plants for a hormone (already named florigen) that will trigger the flowering process. The availability of a substance that would allow regulation of flowering in plants would revolutionize agricultural and horticultural practices.

After a sufficient period of vegetative growth, environmental signals may cause floral induction. The process of induction that occurs in leaves is believed to be followed by hormone production. Although there is a report that a substance extracted from the leaves of a plant that was induced to flower promoted flowering in noninduced plants, most evidence for a flowering hormone comes from grafting experiments. The flowering stimulus can be transmitted from induced to noninduced plants through a graft. There are contrary experiments in which a noninduced plant grafted to an induced plant prevented flowering, suggesting the transmission of an inhibitor of flowering. However, most evidence points to a specific flowering stimulus produced in the leaves that eventually promotes flowering. Despite such evidence, efforts to isolate a "florigen" have not succeeded.

Of course, most of the growth substances (e.g., auxin, gibberellin, and ethylene) will promote flowering of some plants under certain circumstances, and the isolation of substances from plants that promote flowering or transmission of a stimulus across a graft does not necessarily indicate the presence of florigen.

There is a possibility that the elusive florigen may be a gibberellin, since flowering in some plants is accompanied by an increase in gibberellins. In addition, gibberellin will overcome both the long-day requirement for flowering in rosette-forming perennials and the cold requirement (vernalization) in other plants. Both of these responses to gibberellin are evidence of the role of gibberellins in the flowering process.

Perhaps at this time it would be better to consider florigen as a concept and not as a specific substance.

But because the leaves are the organs that perceive the environmental stimulus that brings about flowering, it seems reasonable to conclude that there is hormonal transmission from the leaves to the apices, which ultimately form the floral organs.

18.10 The Mechanism of Hormone Action

Until the mechanism by which hormones function is known, much confusion will exist over the bewildering array of their actions. Unlike the animal hormones, which have very definite targets and physiological or developmental roles, the plant hormones bring about different responses depending on the tissues being treated. For example, auxin causes cell division in the vascular cambium, formation of roots on stems, and elongation of coleoptiles. Furthermore, much of the evidence suggests that the individual hormones do not act alone but in concert with other hormones such that the relative balance is more important than individual amounts.

The more information one obtains about the various effects brought about by the different plant hormones, the more it seems that each plant hormone has some effect on every growth process occurring during the development of the plant. This observation of such wide and varied responses to the hormones lends much credence to the concept of a balance of hormones as the important parameter rather than the presence or absolute concentration of any one particular hormone. Undoubtedly certain hormones must be present for a particular function, but it is evidently not the presence of any one hormone but rather its presence in relation to the others. As Went stated: no auxin, no growth. The dictum of Went can be extended to include the statement that without the proper balance of hormones there will not be the proper growth correlation.

Because the hormones are functional at very low concentrations, there must be great amplification in their action. Known biological phenomena that result in amplification include the protein-synthesis mechanism, the activity of enzymes, and membrane-permeability properties. A small change in one of these phenomena can result in a very large effect. For example, the derepression of an enzyme will result in the synthesis of a large number of enzyme molecules. Much amplification will result from a seemingly very small change. Thus most of the effort to discover the mode of action of plant hormones has been concerned with protein synthesis, enzyme activation, and changes in membrane permeability.

Because many of the plant hormones seem to have structural requirements for activity, there is a general impression that molecular configuration is important for recognition. Proper hormone recognition undoubtedly occurs through the presence of specific hormone receptors. For this reason much effort has been expended to identify and isolate hormone receptors in plants.

Studies have shown that many hormones bind to nucleic acids, to proteins, and to membrane fractions of the cells. The binding to nucleic acids, however, appears to be rather nonspecific, and most evidence indicates that nucleic acid–hormone interaction is not the specific mode of action for the plant hormones.

Proteins bind small molecules in a highly specific manner. For example, enzymes are usually very specific for their substrates, and slight changes in the structure of a substrate molecule usually render it completely inactive. For this reason it seems reasonable to assume that the hormone receptor would be a protein. Studies with hormone–protein binding have not been too fruitful, but studies of auxin binding to membrane fractions of cells have shown much specificity, suggesting that a membrane-bound protein may be the hormone receptor.

Studies with radioactive gibberellins, cytokinins,

abscisic acid, and ethylene have not been too successful in elucidating receptors. The gibberellins accumulate at sites where they are the most active, which suggests the presence of specific receptors. Cytokinins occur in tRNA. However, research has shown that the cytokinins are not incorporated into the RNA molecule; rather there is alteration of adenine bases that are already a part of the nucleic acid polymer. Thus most evidence suggests that tRNA is not the receptor for cytokinins.

There is some reason to believe that the ethylene binding site is metallic; however, the evidence is rather circumstantial, being based primarily on model studies of olefin binding to silver. In addition, since ethylene is hydrophobic the receptor site could be lipoidal (i.e., a membrane). There is virtually no information about the receptor site for abscisic acid.

Despite all the uncertainties about receptors for the various plant hormones, it seems reasonable to assume that there are specific receptor sites. Once these are identified and characterized, our knowledge of hormone action in plants will increase greatly. Such information will aid our basic understanding of plant physiology and will guide us in the practical application of growth and development procedures.

For the most part, plant responses to hormones are manifested over relatively long time periods. In the usual experiment, hormone is applied to a plant or plant tissue and some growth response is measured, such as stem elongation, germination, root formation, or other morphological change. These growth responses require hours or even days before being evident. However, there are many responses to hormones that occur rapidly, in minutes or even seconds. Therefore any unifying theory about hormonal action will have to take into account the known rapid responses to plant hormones.

Although auxin will bring about some rapid changes in metabolic activity, such as an increase in respiration in isolated mitochondria and a stimulation of CO_2 evolution by *Avena* coleoptiles, perhaps the most interesting rapid responses are alterations in membrane permeability. Both auxin and abscisic acid will cause a rapid increase in membrane permeability of plant cells. In bean roots auxin causes a fast alteration in membrane potential. There is an increase in water permeability in pea stems and a more rapid uptake of ions. Abscisic acid causes membrane permeability changes that can be measured by an increase in the rate of uptake of potassium. Such observations have led to the suggestion that perhaps one of the primary modes of action of the plant hormones is an alteration of membrane properties. Further research will be necessary to properly incorporate membrane effects into the hypotheses for hormone action in plants.

A unifying theory for plant hormone function has yet to be proposed. The rapid effects that appear to result because of membrane-permeability changes may not be related to effects that clearly involve nucleic acid and protein-synthesis stimulation. But, as indicated above, alterations in membrane permeability bring about a large amplification and could be the beginning of processes that are ultimately manifested at the protein-synthesis level. Future research will greatly enhance our understanding and should eventually lead to a proper theory that can be tested by experimentation.

Review Exercises

18.1 Diagram the probable pathways for the biosynthesis of the following common plant growth regulators: indoleacetic acid, gibberellic acid, zeatin, abscisic acid, ethylene, and cinnamic acid. Some steps, of course, are not known, but you should be able to make reasonable guesses.

18.2 In an assay using the *Avena* coleoptile curvature test, 5 ml of plant extract from 1 g of leaves suspected of having auxin resulted in a curvature of 11°. A series of standard solutions of IAA gave the following curvature data.

IAA (mg/liter)	Curvature (degrees)
0.02	7
0.03	12
0.05	20
0.07	25
0.10	27

From these data, compute the amount of unbound auxin in the plant extract. Give your answer in mg and in mol per g of tissue.

18.3 Do you think that ethylene should properly be considered a plant hormone? Should substances such as vitamin B_1, which promote growth *in vitro*, be considered as hormones?

18.4 When plant tendrils touch a thin object such as a twig, cord, or stake, they tend to coil around the object. This is a growth response called thigmotropism. How would you visualize auxin being involved in such a growth movement?

18.5 Compare and contrast auxin with some of the important animal growth hormones such as the gonadotropins, the estrogens, and insulin.

References

ANONYMOUS. 1961. *Plant Growth Regulation.* Iowa State University Press, Ames, Iowa.

ABELES, F. B. 1972. Biosynthesis and mechanism of action of ethylene. *Ann. Rev. Plant Physiol.* 23:259–292.

AUDUS, L. J. 1972. *Plant Growth Substances.* Leondar Hill [Books] Ltd., London.

DAVIES, P. J. 1973. Current theories on the mode of action of auxin. *Bot. Rev.* 39:139–171.

GOLDSMITH, M. H. M. 1977. The polar transport of auxin. *Ann. Rev. Plant Physiol.* 28:439–478.

GROSS, D. 1975. Growth regulating substances of plant origin (a review). *Phytochemistry* 14:2105–2112.

JONES, R. L. 1973. Gibberellins: their physiological role. *Ann. Rev. Plant Physiol.* 24:571–598.

KEFELI, V. I., and C. S. KAKYROV. 1971. Natural growth inhibitors, their chemical and physiological properties. *Ann. Rev. Plant Physiol.* 22:185–196.

KEY, J. L. 1969. Hormones and nucleic acid metabolism. *Ann. Rev. Plant Physiol.* 20:449–474.

KHAN, A. A. 1975. Primary, preventive, and permissive roles of hormones in plant systems. *Bot. Rev.* 41: 391–420.

KRISHNAMOORTHY, H. N. (ed.). 1975. *Gibberellins and Plant Growth.* Wiley Eastern, New Delhi.

LESHEM, Y. 1973. *The Molecular and Hormonal Basis of Plant Growth and Regulation.* Pergamon Press, Oxford.

LETHAM, D. S., P. B. GOODWIN, and T. J. V. HIGGINS. 1978. *Phytohormones and Related Compounds: A Comprehensive Treatise,* Vols. I, II. Elsevier-North Holland Biomedical Press, New York.

MILBORROW, B. V. 1974. The chemistry and physiology of abscisic acid. *Ann. Rev. Plant Physiol.* 25:259–307.

SCHNEIDER, E. A., and F. WIGHTMAN. 1974. Metabolism of auxin in higher plants. *Ann. Rev. Plant Physiol.* 25:487–513.

SHELDRAKE, A. R. 1973. The production of hormones in higher plants. *Biol. Rev.* 48:509–559.

STEWARD, F. C. (ed.). 1971. *Plant Physiology,* Vol. VIA. *Physiology of Development: The Hormones.* Academic Press, New York.

WIGHTMAN, F., and G. SETTERFIELD (eds.). 1968. *Biochemistry and Physiology of Plant Growth Substances.* The Runge Press, Ottawa.

CHAPTER NINETEEN

Vegetative Growth

The vegetative growth of plants is an extremely complicated but well-coordinated series of events that begins with the growth of the embryo and culminates in a mature plant capable of fruiting and setting seed. Even before the embryo begins to grow, the zygote formed by cell fusion of male and female sex cells has the beginnings of polarity that sets the stage for the form of the mature plant. The initial stages of growth begin by cell divisions and subsequent cell elongation followed by differentiation. As cell division, elongation, and differentiation proceed, morphogenesis of the principal parts of the plant occurs. All of this is coordinated in part by the information preprogrammed in the DNA and in part by the plant growth regulators. All of this development and coordination is modified by the environment in which the plant is developing. The genome has the information for the potential of growth and development, but a good portion of the final plant is the result of environmental modification.

In this chapter the process of vegetative growth and development is discussed, beginning at the

509

cellular level. Aspects of cell division, cell elongation, and cell differentiation are outlined. The correlation of growth is then discussed, primarily from the viewpoint of balance of hormones with modification by nutrition and environment. Shoot, leaf, and root development are discussed as if they were individual, separate events. Finally, environmental influences on final development are covered.

The coordinated growth of plants is the most complex and least understood aspect of plant physiology. Many facts and details are known but there are no integrating theories adequate to explain the entire process. Thus the student will be left with many unanswered questions, some of which will undoubtedly be explored and answered by future generations.

A useful way to begin the study of plant development is to consider the concept of totipotency. Totipotency implies that all cells of the organism have the genetic information for its entire life cycle; that is, all cells have a complete complement of DNA. Experimental evidence for totipotency was shown by F. C. Steward and colleagues, who in 1964 were able to culture an entire carrot plant from isolated single cells obtained from callus-tissue culture of carrot-root tissue.

Despite evidence for totipotency there is much direct and implied evidence that all genes are not functional during all periods of growth and development. Once again, F. C. Steward (Barber and Steward, 1968) showed that the proteins isolated from pea roots along their horizontal axes had different properties (Fig. 19.1). Thus different proteins are present during different developmental stages.

Because most of the proteins are enzymatic, the data of Steward clearly indicate that different enzymes and different metabolic pathways are operative at different developmental stages. One of the important questions is, therefore, what regulates the presence and/or absence of any one enzyme at any one stage of development?

As an answer to the above question, it is worth recalling that the mature plant is partially a consequence of its genetic composition (DNA) and partially a consequence of the environment in which it

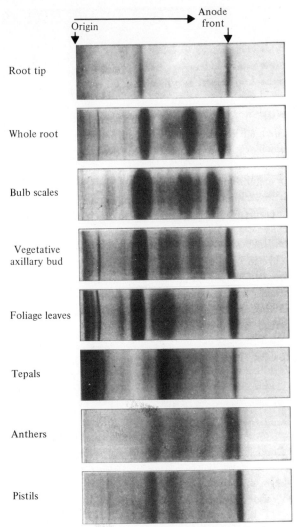

Root tip

Whole root

Bulb scales

Vegetative axillary bud

Foliage leaves

Tepals

Anthers

Pistils

FIGURE 19.1 Electrophoretic separation of proteins from different tissues of *Tulipa*. The experiment shows that different tissues have different proteins despite the fact that the genome is the same. (From J. T. Barber and F. C. Steward. 1968. The proteins of *Tulipa* and their relation to morphogenesis. *Develop. Biol.* 17:326–349.)

developed. Although we have a rather incomplete appreciation of the relative importance of heredity and environment, we can say that the genetic composition governs the potential for development and that the environment will modify the potential. Development occurs within a strict time frame, and hence time is an important feature of development.

Furthermore, within the plant developmental changes depend on an integrated system of hormones, inhibitors, and pigments, all functioning in concert for regulation. The general pattern for growth and development can be summarized as follows.

DNA → RNA → Protein → Enzymes →

Metabolic sequences → Growth and development

The genetic composition specifies the synthesis of particular proteins. The proteins are largely enzymes, and enzymes are the catalysts that regulate metabolism. Metabolism in turn determines growth and development. All of this is modified directly and indirectly by environment. Enzymes are of two basic types, adaptive and constitutive. The constitutive enzymes are present in nearly all cells through most stages of development. Although environmental signals may alter the quantities and the activities of constitutive enzymes, they are always present; they function in the basic metabolism of cells. Examples would be the respiratory enzymes of aerobic metabolism.

The adaptive or induced enzymes are those synthesized only under specific conditions and thus function only during certain stages of development. They may either be controlled by endogenous regulators or be induced in response to environmental signals perceived by the plant.

Jacob and Monod in the 1960s developed an enzyme induction and repression concept that partially explains gene regulation by environmental signals. The classical example, which was the basis of much of the interpretation, involves studies of procaryote cells. The bacterium *Escherichia coli* has very low levels of β-galactosidase, an enzyme that catalyzes the hydrolysis of lactose to galactose and glucose. However, if *E. coli* is grown in the presence of lactose as the primary carbon source, there is a manyfold increase in the amount of β-galactosidase. The increase has been shown to be due to *de novo* synthesis and represents the induction of an enzyme by its own substrate. If the cells are transferred back to a medium lacking lactose, the β-galactosidase enzyme is no longer synthesized

and decreases to the base level. Important to the understanding of enzyme induction of this type is that more than one enzyme is induced by the presence of a single compound, lactose. In *E. coli* as well as β-galactosidase there is also induced a β-galactoside permease and a β-thiogalactoside acetyltransferase; all are enzymes required for the metabolism of lactose. The induction of more than one enzyme by a single substrate or metabolite is called coordinate induction.

Enzyme repression functions somewhat in the opposite manner to enzyme induction. Like enzyme induction it is best known in procaryote cells. In enzyme repression the synthesis of an enzyme or group of enzymes is prevented (repressed) by the presence of one of the products of the enzyme reaction. An excellent example is the repression of the enzymes responsible for the synthesis of histidine if *E. coli* is grown in the presence of histidine. As with enzyme induction, if a group of enzymes that function in a related role are repressed the phenomenon is called coordinate repression. Derepression is, of course, the activation of a repressed gene.

Although it appears that a large part of the gene regulation in procaryotes takes place at the transcriptional level, in eucaryote organisms such as plants the evidence for significant gene regulation by such methods is much less documented. For one reason, the genomes (including the physical arrangement of genes on the chromosomes) are much more complex than those of procaryotes, and furthermore, there is far more differentiation within the higher plant. In the mature tissues (not necessarily meristematic) much of the genome is evidently permanently repressed. This must be true since all evidence indicates that each cell has a complete set of genes; yet most cells, once differentiated, will not dedifferentiate. Thus much of the genome must be inactive. Nevertheless, there is some evidence for gene regulation by the methods outlined by Jacob and Monod in higher plants.

Many enzymes are induced by light. One such enzyme is glycolate oxidase, an enzyme of photorespiration. Very low levels of activity are present in dark-grown etiolated leaves, but upon illumina-

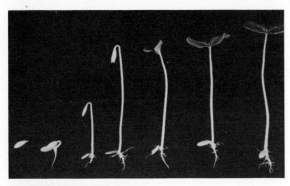

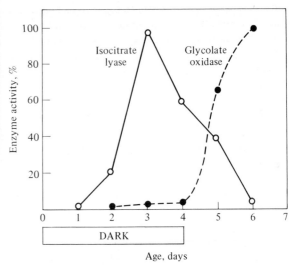

FIGURE 19.2 Experiment to show the appearance of isocitrate lyase (a marker of the glyoxylate cycle) and glycolate oxidase (a marker of photorespiration in peroxisomes) during the development of cucumber cotyledons (above). The seedlings were grown in the dark for four days and were then transferred to the light. The enzyme isocitrate lyase was maximum at day 3. Glycolate oxidase did not increase significantly until the seedlings were exposed to light. (From I. M. Wainwright. 1975. Malate dehydrogenase isozymes in developing cucumber cotyledons. Ph.D. dissertation, University of California, Riverside.)

tion activity increases markedly (see Fig. 19.2). Although we do not know the exact mechanism of light induction, it probably occurs by a process not too different from that proposed by Jacob and Monod.

Care must be taken, however, to differentiate between induction by *de novo* synthesis and activation of existing enzymes. There has been a significant amount of research to study the light activation of enzymes of photosynthesis. Data suggest that chloroplast enzymes such as ribulose bisphosphate carboxylase, NADP-glyceraldehyde-3-P dehydrogenase, and ribulose-5-P kinase are activated by light. The mechanism is rather complex and seems to be mediated through small proteins that function as light-effect mediators (LEMs). The LEMs are evidently reduced by light by electron transfer from the electron-transport system of the chloroplasts. The LEMs then cause a thiol–disulfide exchange on the light-activated enzymes.

Some enzymes appear during specific stages of the life cycle of plants. Such an example is the time-dependent induction of the glyoxylate cycle, which functions in fatty acid mobilization during germination of lipid-storing seeds. During germination, fat is mobilized and converted to carbohydrate. In the cotyledons of cucumber the glyoxylate cycle enzymes appear for a few days prior to the cotyledon becoming green and photosynthetic. It is assumed that the appearance of the glyoxylate cycle enzymes represents an actual *de novo* synthesis of protein. In Fig. 19.2 the transient appearance of isocitrate lyase is shown along with the light-dependent induction of glycolate oxidase as the cotyledons become green.

Plant hormones are known to be involved in the regulation of enzyme activity. An excellent example occurs when seeds break dormancy and begin to germinate. Varner (see Varner and Ho, 1976) studied the induction of amylase when isolated barley aleurone layers are treated with the plant hormone gibberellin. The implication of this experiment is that gibberellin induces amylase, which hydrolyzes starch as carbohydrates are mobilized during germination. Experiments have suggested that gibberellin regulates DNA transcription.

Exactly how such activation and inactivation of specific portions of the plant genome is accomplished is not known. However, Bonner (see Bonner, 1976) has produced evidence that much of the DNA of the cell exists as a DNA–histone complex. When the histone is removed from the DNA, the DNA can again act as a template and enzyme synthesis occurs. When histone was removed from pea embryo chromatin there was a fivefold increase in DNA-dependent RNA synthesis. Furthermore, globulin synthesis was induced by the removal of histones from the chromatin. Thus these experiments suggest that histone represses the synthesis of protein by inhibiting DNA activity.

Because there are only a few different kinds of histones and many different genes, if the histone hypothesis is valid it must be rather nonspecific. The regulation of DNA metabolism by histones should be considered as a tentative hypothesis until further experimentation is completed.

To summarize this brief introduction to growth and development, we can say that growth and development is a coordinated system of processes leading to the development of roots, stems, leaves, and other structures, all of which function as an integrated unit. The basic facts known about growth and development are the following.

1. Cells, at least when young, have a total complement of genetic information; that is, all cells at some time have an entire DNA complement. This is the concept of totipotency.

2. All genes are not functional during all stages of growth and development. Thus there are mechanisms that alter the functioning of the plant genome during the life cycle.

3. Protein synthesis and thus enzymes and metabolism are regulated both endogenously and exogenously by environmental signals.

4. The plant growth regulators (hormones) alter growth and developmental processes and are probably the main chemicals that function in growth integration.

5. The entire growth and development of the plant is highly modified by the environment in which it develops.

Through the efforts of many researchers, we now have much information about the biochemical changes that accompany growth and developmental processes but little knowledge about the complex processes that bring about actual morphogenesis.

19.1 Vegetative Growth Stages

Vegetative growth is a coordinated system of processes that lead to the development of roots, stems, and leaves. Growth occurs in specialized regions of the plant body called meristems. Meristems occur at the growing points (apices) of both roots and shoots and in certain lateral regions called cambia. Primary growth of plants occurs through cell divisions of the apical meristems, and secondary growth results from divisions of cells in the cambia. Primary growth largely gives rise to length and some girth whereas secondary growth gives rise to lateral expansion.

Apical meristems are indeterminate in that they will continue to produce new cells by cell division throughout the life of the plant. Other meristems, such as those of leaves, flowers, and some kinds of thorns, are determinate in that the appendage has a finite size.

In addition to the apical and lateral meristems, some plants (monocots, for example) have special, nonapical meristems called intercalary meristems.

Both the apical and intercalary meristems give rise to primary growth of plants whereas the lateral meristems give rise to secondary growth and only

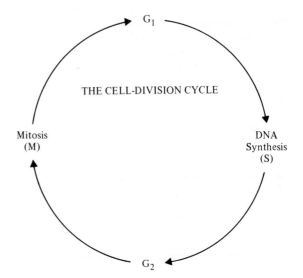

FIGURE 19.3 The cell-division cycle. G_1 = postmitotic stage, presynthesis of DNA; S = DNA synthesis stage; G_2 = postsynthesis of DNA, premitotic stage; and M = mitotic stage.

occur in perennial and woody dicots. Some large, "woody" monocots such as the palms have a kind of a secondary growth resulting from specialized secondary thickening meristems, but these are different from the cambium-type secondary meristems of dicots.

19.1.1 The cell-division cycle

The primary event of growth is the division of cells. It should be apparent that the actual plane of cell division will govern to a large extent the final morphology of the plant. Thus those factors that regulate cell division will be primary components in the establishment of the overall morphology of the plant.

The actual rates of cell division have been estimated by determining the number of new cells produced by growing points during a specified time span. It appears that, at the growing point of a shoot, each cell will divide every 3 to 5 days. Root tips, however, seem to have greater rates of meristematic activity, each cell dividing at intervals of one day or less.

In efforts to understand the cell-division cycle, it

is convenient to divide the cycle into four discrete stages (Fig. 19.3). These stages are as follows.

G_1 = postmitotic, presynthesis period (of DNA)

S = DNA-synthesis period

G_2 = postsynthetic, premitotic period

M = mitotic period

It is known that both hormonal balance and nutrition, particularly carbohydrate nutrition, will regulate the cycle and hence control the overall rate of cell division. The meristematic cell can arrest in either G_1 or G_2, although most species studied seem to rest primarily in G_1. Thus the principal control point for most cells would be the period after mitosis and before DNA synthesis.

Evidence suggests that the plant growth regulator auxin either directly or indirectly affects the S (DNA synthesis) period. While other plant hormones such as the cytokinins are involved, their role is less well understood.

Anaerobiosis will arrest the G_1 period, but if mitosis has begun it will continue in the absence of oxygen. Meristems in a no-oxygen atmosphere will all have resting or interphase nuclei. Light inhibits mitosis, the sensitivity being largely during G_1.

19.1.2 Cell division

The mechanics of cell division are reasonably well understood in plants and are briefly outlined here. A cytology textbook should be consulted for further details.

Cell division is a process whereby cells replicate to form identical daughter cells. Within the plant body the process of cell division increases cell number during growth. After cell division the individual cells will differentiate into the types of cells they will be in the mature plant.

Perhaps the most important feature of cell division is the exact replication of the chromosomes with their genetic information, ensuring continuity among cells. The exact structure of the chromosome is somewhat uncertain, but it appears to be fibrillar and composed of an intimate association of basic proteins (histones) with nucleic acids, largely deoxyribonucleic acid (DNA). Duplication of chromosomes occurs prior to mitosis during the stage called interphase. The process of duplication is not entirely understood, but much information is available about DNA replication. This information is discussed in Chapter 8.

Mitosis is a continuously occurring process of chromosome separation (at the beginning) up to the production of two daughter nuclei (at the end) but is conveniently divided into several stages for the purpose of study. These stages of mitosis, interphase, prophase, metaphase, anaphase, and telophase are discussed below.

Overall, cell division is a complicated two-phase process involving the division of the nucleus and genetic material (mitosis) followed by the division of the rest of the cell. Nuclear division is called karyokinesis, and the cell divides in two by a process called cytokinesis. The stages of mitosis are illustrated in Fig. 19.4.

INTERPHASE

Interphase is the period between mitotic divisions of cells having the capability of cell division. The cell is loosely said to be resting, but resting only with respect to division. It is during interphase that the DNA is replicated and the chromosomes become double-stranded. Each chromosome divides into two identical halves called chromatids that remain attached to each other at a specific point, the centromere. During this phase the chromosomes are extremely thin and barely visible within the nucleus.

PROPHASE

The first actual stage of karyokinesis, the prophase, is marked by the visible appearance of the chromosomes within the nucleus. The nuclear envelope begins to disappear, and the threadlike chromosomes shorten by tight coiling and become distinctly visible with the light microscope. The double-stranded nature of the replicated chromosomes, however, is not yet visible because in addition to being attached to the centromere, the chromatids remain tightly apposed to each other along their entire length. During this stage the spindle fibers (composed of microtubules) are formed. Spindle fibers aid in chromosome migration during anaphase.

METAPHASE

The next stage of mitosis, the metaphase, is noted by the presence of the chromosomes at the equatorial plate of the cell. The individual chromosomes now appear double, clearly showing that the replication process has occurred. This stage of cell division is preparatory to the migration of the chromatids (newly formed chromosomes) to opposite poles of the cell. It is during metaphase that chromosome number and morphology can be determined. Such determination is called karyotyping.

ANAPHASE

The actual migration of newly found chromosomes to the opposite poles is called anaphase.

PHASES OF MITOSIS

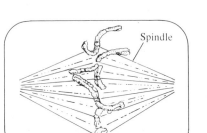

Interphase

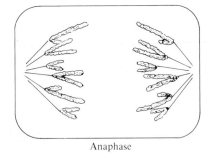

Prophase

Metaphase

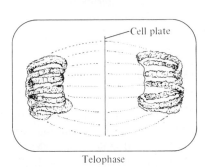

Anaphase

Telophase

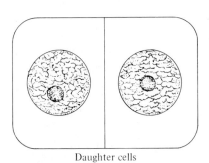

Daughter cells

FIGURE 19.4 The stages of mitosis in plant cells.

Each double-stranded chromosome will separate into daughters that are now no longer called chromatids but rather daughter chromosomes. A member from each pair migrates to an opposite pole. The actual process of migration is not very well understood but is apparently aided by the microtubules extending from the cell equator to each pole. These microtubules form the spindle, which runs from pole to pole through the cell equator. The daughter chromatids appear to be attached to the microtubule spindle fibers at the centromere and perhaps move by microtubule sliding or depolymerization. The microtubules do not become thickened during the process. ATP energy is probably involved in the migration, which takes place at about 1 μm per minute.

The essence of mitosis is clear from the process of anaphase. The complement of chromosomes within the nucleus has doubled to form two identical sets. After anaphase there will be two sets of identical chromosomes that are characteristic for the species.

TELOPHASE

The telophase is marked by the presence of the daughter chromosomes on either side of the equatorial plate of the cell. In principle, telophase appears to be the reverse of prophase. The individual chromosomes become uncoiled and lose their visibility in the light microscope. Nuclear envelopes are synthesized, and eventually the cell will have two distinct nuclei, each with an identical set of chromosomes. The process of karyokinesis is completed.

Cytokinesis begins after telophase. Along the equator of the cell in plants a cell plate forms, evidently by fusion of vesicles containing polysaccharide that was derived from dictyosomes. The cell plate grows from the center of the cell toward the periphery until there are two apparent daughter cells formed. Each daughter cell then deposits primary cell-wall material (largely cellulose) against the cell plate. The cell plate becomes the middle lamella between cells, being composed largely of pectinaceous material, and functions as the "cement" holding the cells together. After primary-wall deposition the process of cytokinesis is completed; the two daughter cells are fully formed.

After daughter-cell formation, one may remain meristematic for subsequent divisions and the other will enlarge by elongation and eventually differentiate into a functional plant cell.

Certain drugs, such as colchicine and caffeine, have been used to study cell division. Colchicine prevents spindle formation, keeping the dividing cells in metaphase since anaphase cannot occur. When the nuclear envelope reforms the number of chromosomes will be doubled, resulting in a ploidy condition. Caffeine suppresses cytokinesis but

allows cell division. Cells treated with caffeine and certain derivatives of caffeine will have multiple nuclei.

The relationship between cell division and growth is not fully resolved. Whether cell division is a prerequisite for subsequent growth is difficult to determine with certainty. Nevertheless, cell division will result in growth.

Little is known about the division of cellular organelles other than nuclei. It seems reasonable to conclude that cell division would be accompanied by division of other organelles. Chloroplasts appear to divide when they are young or in the proplastid stage. Division of mitochondria in plants is uncertain, but there is some evidence for replication and segregation of other subcellular organelles during cytokinesis.

19.1.3 Cell elongation

After cell division, cells normally elongate beyond the size of the mother cell. In growing root tips, this occurs largely in the section from 0.2 to 1.0 mm behind the apex. Although we know little about the actual factors that regulate and control the initiation and termination of extension or elongation of cells, it seems reasonable to surmise that there is an upper limit to cell size. Upper limits are probably determined to a large extent by restrictions of transport and diffusion of materials within and between cells.

Granted, there are some rather large plant cells, but those large cells within the higher plant body are usually elongated and have a relatively small volume. The large plant cells of green algae such as *Nitella* and *Valonia* are actually coenocytes.

Much is known about the physiology of the actual cell-extension process. We know, for example, that a favorable water balance is necessary for any substantial elongation, because the force for cellular expansion is the turgor within the cell itself. The plant cell wall is both elastic and plastic. The elastic properties of the cell allow some expansion when under positive turgor pressure. But when turgor is relaxed the cell will return to its original volume. This expansion is, of course, not growth; growth is defined as an irreversible increase in volume.

Cell elongation and enlargement in plants is accompanied by vacuolation. During elongation the central vacuole becomes quite large and the cytoplasm forms a thin layer closely appressed to the cell wall.

Experiments have shown a positive correlation between cell-wall plasticity and growth of the cell. Thus it seems that during cell expansion there is an increase in cell-wall plasticity. It is visualized that the plasticity comes about by an alteration in ionic and covalent bonds within the cell wall proper. Pectin compounds of cell walls bind calcium ions. Removal of the Ca^{+2} breaks the ionic attraction of the pectins. The ionic bond is through the carboxyls of pectin.

$$\vdash COO^- \quad Ca^{++} \quad {}^-OOC \dashv$$

An enzyme, pectin methylesterase, will methylate the carboxyls.

$$\vdash COOCH_3 \quad CH_3OOC \dashv$$

This enzyme prevents the formation of the ionic bridges with calcium. Although the evidence for the pectin methylesterase hypothesis of increasing cell-wall plasticity is not unequivocal, such a mechanism or a similar one could account for cell-wall loosening during expansion.

An interesting aspect of cell elongation is that low pH will cause extension. Furthermore, low pH (i.e., high hydrogen-ion concentration) brings about cell-wall extension even when the cell is no longer living. This observation has led to the hypothesis that there are acid-labile bonds within the cell-wall matrix that are broken enzymatically during wall extension and that can be broken by acid treatment. Such a notion is consistent with the above hypothesis that breaking the ionic bonds between carboxyls and calcium effectively loosens the cell wall. Protonation of the carboxyls would accomplish the same end.

The plant growth regulator auxin is implicated in the process of cell-wall loosening, and auxin-induced pH changes have been observed, but the precise mechanism is not understood. Auxin could effect a proton pump, altering proton concentrations, and may also act directly at the enzyme or protein level.

Once the cell wall has been made plastic, excess turgor pressure results in cell expansion. Experimentation has shown that if cotton fibers are grown *in vitro* in high osmotic media they will not elongate. In fact, when the water potential of the growth medium reaches the initial turgor of the cell, growth ceases (Fig. 19.5). Further data suggest that potassium and malate act as the *in vivo* osmotic agents that lower the osmotic potential of the cell, resulting in water uptake and increased turgor.

During and after cell expansion, new cell-wall material must be synthesized if there is to be growth. Growth could be by apical extension, that is, by new material deposited just at the growing point, or (as is mostly the case during cell expansion) by material deposited uniformly within the interstices of the existing cell-wall matrix. This is the process of intussusception and probably occurs when cells are increasing in area.

There has been debate about the process of cell-wall deposition during cell elongation. The process of intussusception referred to above occurs by new cell-wall material being deposited within the interstices of the existing cell-wall fibrils. Alternatively, there can be deposition of new cell-wall material directly on existing cell walls. There is some evidence that this direct deposition process (called apposition) is probably more prevalent during cell-wall synthesis than is the process of intussusception.

The actual synthesis of new material is greatly aided by the Golgi apparatus. As can be observed in the micrograph of the elongating cotton-fiber cell shown in Fig. 1.1, there are many dictyosomes of the Golgi apparatus adjacent to the cell wall.

Cleland (1971) has shown that cell growth by expansion is a two-step process consistent with the above hypothesis. First, there is a biochemical step

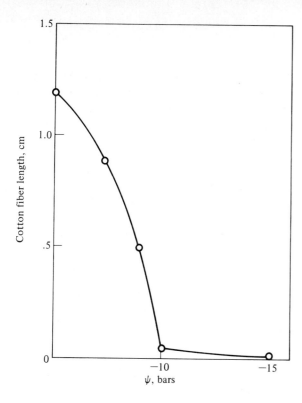

FIGURE 19.5 The elongation of cotton fibers that were cultured in solutions of varying water potential. Once the medium reached about −10 bars the elongation of the single-celled fiber ceased, giving evidence that positive turgor in the cell is a requirement for cell elongation. (From R. S. Dhindsa, C. A. Beasley, and I. P. Ting. 1975. Osmoregulation in cotton fiber. *Plant Physiol*. 56:394–398.)

with a Q_{10} of 2–3 that is inhibited by cyanide. This is the cell-wall loosening stage (increase in plasticity) that is influenced by auxin. Second, there is the physical mechanism of turgor-driven expansion that has a Q_{10} of 1 to 1.2 (in the range of physical processes) and is not inhibited by cyanide. The model for cell expansion follows.

$$\text{Rigid wall} \xrightarrow[\text{Auxin}]{\substack{\text{(biochemical} \\ \text{step)}}}$$

$$\text{Loosened wall} \xrightarrow[\text{Turgor}]{\substack{\text{(physical} \\ \text{step)}}} \text{Cell expansion}$$

Research has shown that RNA synthesis and protein synthesis were necessary for cell expansion of soybean hypocotyls. The inhibitor, actinomycin-A, inhibited cell expansion, particularly auxin-stimulated cell expansion. Actinomycin-A inhibits DNA-dependent RNA synthesis and concomitant protein synthesis. Auxin apparently regulates RNA metabolism by activating a specific RNA polymerase. Additional studies indicate that the extension of a cotton fiber, a single cell, requires active metabolic processes.

An important question with respect to cell elongation is why cells elongate to form cylinders rather than expanding isodiametrically to form spheres, as would be expected. With the use of polarized light and microscopic techniques it has been shown that the cellulose microfibrils within the cell wall are arranged transversely in the form of hoops around the girth of the elongating cell (Fig. 19.6). Such an arrangement prevents spherical growth and, because of the restriction, causes cylindrical growth.

The transverse arrangement of the cellulose microfibrils correlates with the transverse arrangement of microtubules within the cytoplasm of the elongating cell. Thus it is presumed that the arrangement of the microtubules governs the arrangement of the microfibrils of the cell wall. The polarity of the cell is evidently a function of the arrangement of the microtubules.

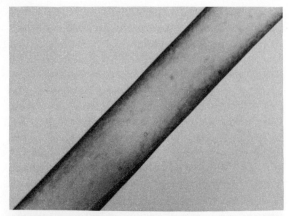

FIGURE 19.6 Photographs of a portion of internode from *Nitella*. Upper: photographed in bright field. Lower: photographed in polarized light. The photograph obtained with polarized light shows a distinctive pattern because of the transverse arrangement of cellulose microfibrils. There is a central bright stripe flanked by dark lateral bands. The curved wall edges are bright. If all the microfibrils were perfectly transverse the cell would appear bright from edge to edge. The band formation is the result of the curved surface and some random organization of the microfibrils in the region of the dark bands. (From an experiment by Dr. Paul A. Richmond. Photographs courtesy of Paul A. Richmond, University of the Pacific, Stockton, Calif.)

To extend this notion further, it is assumed that the orderly arrangement of cells in files that can be observed during shoot and root differentiation comes about because of the cylindrical develop-

ment of the individual cells. Thus the arrangement of the microtubules governs the arrangement of the cellulose microfibrils of the cell wall. The microfibrils govern the shape of the individual cells, and the cells ultimately determine the polarity of the root and shoot.

There is evidence that changes in growth directions such as occur at leaf primordia result because of rearrangement of the cellulose microfibrils. It is assumed that the latter arrangement is a function of the internal arrangement of the microtubules.

19.1.4 Cell differentiation

Considering that all cells of the living plant have the same genome, i.e., the same genetic composition, we may say that a cell is differentiated with respect to another cell when it differs from that cell. Perhaps differentiation is the most enigmatic process of all plant biology since it is one of the most obvious phenomena and yet one of the least understood. From the early days of such observers as Hooke, Malpighi, and Grew, the question of differentiation has been foremost in the minds of plant physiologists. One of the first clear notions of differentiation was that of Vöchting, who stated in 1878 that the fate of a cell is in large part a function of its relative position within the plant body.

The process of differentiation in plants is very orderly. Beginning with the zygote and young embryo, shoots and roots first become differentiated. Following this major change in organization of the young embryo, organ primordia (for leaf, branch, etc.) are formed, followed by differentiation of tissues. Individual cell types such as epidermal cells, cortical cells, and others apparently differentiate last, after tissue types become apparent. In the root, for example, the stele is evident before xylem and phloem are fully developed. Thus differentiation appears to proceed from the higher orders of organization to the lower orders, that is, from organs to tissues and then to cells.

At the cellular level the very first step is the establishment of polarity within the cell. The previous discussion of microtubule arrangement, which evidently governs the polarity of the expanding cell, is important to remember. It seems reasonable to conclude (at least in part) that polarity is determined before cell elongation by the arrangement of microtubules in the cell.

It is clear that polarity begins during the first few divisions. Perhaps polarity is created initially by unequal cell divisions such that the first cells will have a different complement of cellular or cytoplasmic inclusions. Of course, the genome will be identical, but epigenetic influences could alter the development of the cells.

Polarity, once established, is locked into the development of the plant. If a cut stem is rooted, roots will form at the end closest to the physiological root end, and leaves will form at the distal end. This is true even if the stem is inverted. Roots will still form at the true basal end and leaves at the opposite end.

One theory for cytodifferentiation is that the developing organism is preprogrammed through the genetic material. The genetic material (the DNA) is complexed with histones and largely inactive. As histones are removed, certain portions of the genome become active, and hence differentiation proceeds as an orderly progression of events dependent upon the program within the genome. The program is modified by epigenetic factors as well as by environmental signals. We are certain that there can be modification because development is somewhat flexible and can be altered by the environment

With respect to the epigenetic influence of differentiation, the operon theory of Jacob and Monod* may be mentioned. Working with bacteria, they visualized a sequence of enzyme genes and an operator gene making up a unit called an operon. Another gene, a regulator gene, is activated by small metabolites, substrate molecules, or perhaps hormones. The regulator gene, once activated by the environmental stimulus, signals the operator

gene to activate the enzyme genes. Thus through groups of operator genes and operon units there can be epigenetic regulation of differentiation that is preprogrammed within the genome.

The influence of the nucleus on development was demonstrated by elegant experiments with the very large single-celled green alga *Acetabularia*. *Acetabularia* has a rhizoid-like base, a stalk, and a cap (Fig. 19.7). Caps differ between species. Although it is a coenocyte, there is one large nucleus. When the cap of an individual is removed a new one will form. If the cap and base are removed and the base (with a nucleus) from another species is grafted to the severed plant, the new cap will be like the second species. Thus the nucleus and not the cytoplasm of the stalk governs the kind of cap produced.

Because messenger RNA is relatively short-lived, protein synthesis requires new mRNA that is synthesized using the nuclear DNA as a template. Thus differentiation is dependent upon the nucleus as is evident in the *Acetabularia* experiment.

Much of our knowledge about cytodifferentiation in plants comes from studies of xylary elements, xylem and phloem. Callus tissue cultured *in vitro* forms xylem and phloem and offers an opportunity to study differentiation. In the presence of low sugar, xylem forms, whereas in the presence of high sugar, phloem differentiates. Intermediate sugar concentrations yielded cambium tissue. In all cases auxin had to be present. Furthermore, wound tissue, a type of callus, readily formed xylem and phloem.

Roberts (1976) developed a cytodifferentiation sequence for xylem differentiation. The sequence is: cell origination, cell enlargement, secondary wall deposition and lignification, and end-wall lysis (to form the functional xylem element). It is not entirely clear what the relationship between the cell cycle (shown in Fig. 19.3) and the cytodifferentiation sequence is, but cytodifferentiation may begin from the G_1 stage after mitosis.

Furthermore, from Roberts's work, DNA synthesis seems necessary for cytodifferentiation. However, DNA synthesis may occur in the absence of mitosis. Frequently before cytodifferentiation

---

*The Jacob and Monod hypothesis for enzyme induction and repression was mentioned briefly at the beginning of this chapter.

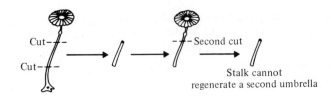

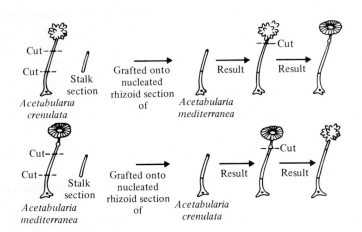

FIGURE 19.7 Diagrammatic representation of the experiment with *Acetabularia* illustrating the importance of the nucleus in differentiation. See J. Haemmerling. 1963. The role of the nucleus in differentiation especially in *Acetabularia. Symposia Soc. Exp. Biol.* 17:127–137. Also J. Haemmerling. 1963. Nucleo-cytoplasmic interactions in *Acetabularia* and other cells. *Ann. Rev. Plant Physiol.* 14:65–92. The plate appears in P. Adams, J. W. Baker, and G. E. Allen. 1970. *The Study of Botany.* Addison-Wesley, Reading, Mass.

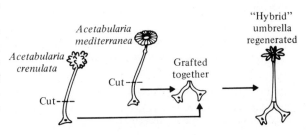

there is replication of DNA and other chromosome components, which leads to polyploidy. Roberts assumed that additional DNA is a requisite for the active protein synthesis accompanying cytodifferentiation.

Despite the complexities associated with cell division, elongation, and differentiation, there is much coordination for the development of the plant. Such coordination is within the concept of growth correlation, the topic of the next section.

19.2 Growth Correlation and the Balance of Hormones

It is quite clear from the discussions of the various classes of hormones (Chapter 18) that their reactions are rather nonspecific. Auxin and gibberellins are growth-promoting substances, the cytokinins are cell-division factors, abscisic acid is mostly a growth-inhibiting factor, and ethylene is a maturation factor. A reasonably sound hypothesis for growth and development is that for proper correlative growth there must be proper balance of hormones. And until that time when we have a comprehensive knowledge of the mode of action of the various plant growth substances we must be satisfied to say that it is the balance of hormones that regulates proper growth and development.

Growth correlation can be defined as the relationship of the growth of one part of a plant to the growth of another part. Thus there is a correlation between root and shoot growth, vegetative and reproductive growth, and even a correlation between growth and differentiation per se. Growth correlation can be simple and can be the result of the competition between parts (or functions) for nutrients, vitamins, minerals, or, probably more commonly, the result of the balance of the various plant-growth substances.

Kefford and Goldacre in 1961 were perhaps the first to challenge the concept of auxin as a cell-enlargement factor; they defined it as a predisposing substance—predisposing cells to change. The nature of the change is dependent upon other plant-growth substances. Auxin interacts with gibberellin to bring about cell enlargement and with cytokinins to stimulate cell division. They proposed that auxin was a correlative regulator of plant growth. Although these ideas may not withstand the test of further scientific inquiry, they do clearly introduce the concept of balance of hormones for proper growth and development.

Some investigators have interpreted hormone-regulated growth as interactions between growth-promoting and growth-inhibiting hormones. The growth and development of the plant is regulated by the balance of growth promoters and inhibitors. As an example, germination of seeds is prevented by inhibitors such as abscisic acid, and after environmental signals such as proper temperature or light are received, gibberellins are produced that overcome the inhibitory power of the inhibitors.

Experiments by Skoog and Miller in 1957 with tobacco callus-tissue cultures clearly illustrate the complex interaction of hormonal balance during growth and development. In their experiments there was no growth of any consequence in the absence of added auxin. As the auxin concentration was increased in the growing medium the callus culture began to differentiate such that roots formed. At still higher concentrations of auxin root formation was inhibited. This shows the usual type of hormone response in which there is a threshold level, an optimum level, and an inhibitory level of hormone.

As cytokinin (in these experiments, kinetin) is increased in the growth medium at intermediate levels of auxin, bud and shoot formation begins. Thus increasing kinetin in the presence of the proper amount of auxin is accompanied by shoot differentiation. At high levels of auxin the optimum amount of kinetin does not result in shoot differentiation but only callus growth. These experiments are illustrated in Fig. 19.8.

The Skoog and Miller experiments show that the balance of hormones is important for growth and

IAA, mg L^{-1}

0   0 0.005   0.03   0.18   1.08   3.0

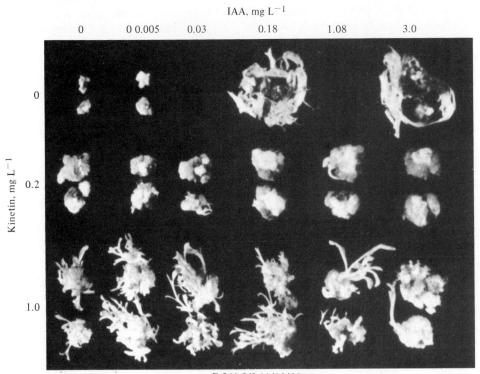

Kinetin, mg L^{-1}

0

0.2

1.0

E 561 9/9-11/10/55
With casein hydrolysate 3g L^{-1}

FIGURE 19.8 An experiment showing how the balance of hormones influences plant growth and development. The plate shows callus tissue grown in casein bydroly-sate with different amounts of IAA and kinetin. With no IAA there is no growth. As the concentration of IAA increases, roots develop. With increasing kinetin, shoots develop. At high IAA and high kinetin, only callus forms, (From F. Skoog and C. Miller. 1957. Chemical regulation of growth and organ formation in plant tissues cultured *in vitro. Symposia Soc. Exp. Biol.* 11:118–131.)

FIGURE 19.9 Experiment performed by C. A. Beasley showing plant hormone interaction during the development of cotton fiber grown in culture (see facing page). The photoghraph shows two flasks with ovules that are producing fiber. Gibberellic acid causes increased length, but indoleacetic acid has no effect on elongation (for the culture of unfertilized ovules. IAA is necessary). Both kinetin and abscisic acid inhibit fiber elongation. Gibberellic acid, however, will overcome the inhibition of fiber elongation caused by both kinetin and abscisic acid. (From C. A. Beasley and I. P. Ting. 1973. The effects of plant growth substances on *in vitro* fiber development from fertilized cotton ovules. *Amer. J. Bot.* 60:130–139.)

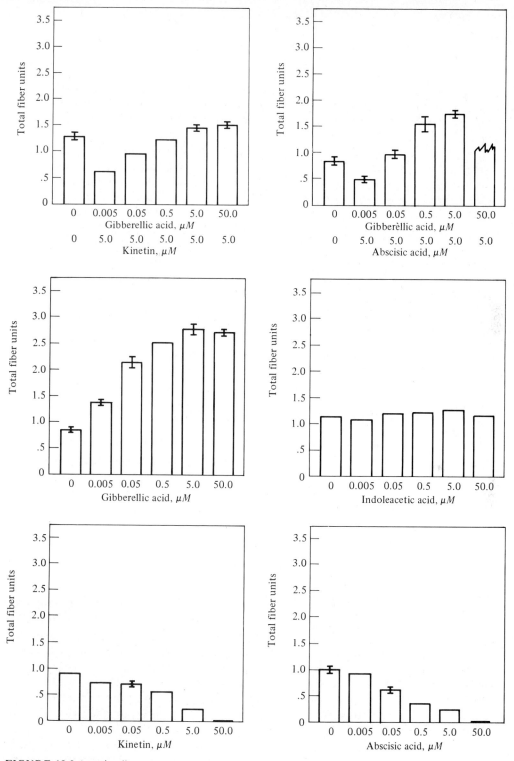

FIGURE 19.9 (contiued)

development and, in addition, seem to indicate that there are not specific substances associated with root and shoot formation, dispelling an older concept. There is not a specific root hormone or shoot hormone; differentiation depends on more than one growth substance.

The development of the cotton fiber, a single epidermal cell of the seed, also illustrates the complex interaction among hormones during growth and development. In the experiments of Beasley and his colleagues (1973), both unfertilized and fertilized cotton ovules were cultured under aseptic conditions to study the development of the cotton fiber. Figure 19.9 shows the young ovule with fibers beginning to develop from the ovule surface. Each fiber is a single cell that elongates for about three weeks. After elongation ceases, secondary wall formation begins and the cell matures and dies, forming the fiber of commerce.

Even when fibers are cultured at the proper temperature and in a nutrient medium containing a carbohydrate energy source such as glucose and necessary inorganic constituents, plant-growth substances must also be added for proper growth to occur. Fertilized ovules do not respond to IAA, only to gibberellin. Evidently the fertilized ovules produce sufficient auxin. Without gibberellin, however, there will be no elongation. Unfertilized ovules do require auxin for elongation, and thus both IAA and GA_3 must be added to the culture for normal cell elongation to occur. Figure 19.9 shows a culture of the fertilized cotton ovules with fiber grown in the presence of GA_3. Also shown in the figure is that the presence of cytokinin (kinetin) and abscisic acid inhibit growth of the cotton fiber.

At $5.0 \mu M$ kinetin, the cotton fibers barely elongate. Increasing concentrations of gibberellin will overcome this inhibition. In addition, gibberellin will overcome the inhibition due to ABA. Thus there is interaction among hormones resulting in normal growth and development.

19.3 Growth Form

19.3.1 Apical dominance

Growth correlation is perhaps one of the most complex of all the growth phenomena. One of the most obvious examples of growth correlation of intact plants is apical dominance.

Apical dominance is the regulatory control of growth that the plant apex exerts on the other potentially meristematic regions. Apical dominance controls the form of the plant body. In most plants there is strict apical dominance such that there is a single "leader" developed from the shoot. A good illustration would be a fir or spruce tree. In others there may be branching, with more than one aerial apex showing some degree of apical dominance.

In addition to playing the role of suppressing the actual growth of other potential apices, apical dominance regulates both the extent and the angle of branching. Apical dominance controls the overall growth form of the aboveground portions of plants.

When the apex of a plant is severed or injured, lateral buds will begin to develop as apical dominance is lost. Eventually one of the new growing apices will assume dominance. Such an observation clearly indicates that apical dominance is the result of apex activity (Fig. 19.10).

The first theory to explain apical dominance was based on the knowledge that actively growing tips produce auxin that is basipetally transported. The high concentration of auxin emanating from the apex suppresses the growth of lateral buds. Support for this hypothesis comes from experiments in which the apex is severed and replaced with an agar block containing auxin. In many plants the "auxin block" will act similarly to the true physiological tip.

It was largely through the studies of Thimann and his colleagues (see Thimann, 1977) that the concepts

FIGURE 19.10 Apical dominance in *Clarkia*. The apex of the plant at the left was pinched off when the plant was about three weeks old. The photograph was taken several weeks later. Note that the major difference between the two plants is that the lateral branches of the plant on the right have not grown but rather have been suppressed. Experiment by Nancy Smith, University of California, Riverside.

roots) or cytokinin synthesis. In any case, when auxin is high as the result of production at the apex, cytokinin is low. When auxin production is reduced, cytokinin increases, with the result that lateral shoots develop.

Roots also show apical dominance. The growing tip will suppress the growth of lateral roots, preventing additional main root tips. Unlike apical dominance in shoots, however, auxin blocks will not replace the physiological tips of roots and maintain apical dominance. For this reason there is some doubt about the role of auxin in apical dominance.

Apical dominance may be the result at least in part of the tip being a sink for reserves such that competition for nutrients prevents other possible apices from becoming dominant. However, this nutritive correlation hypothesis has now largely been replaced by the hormone correlation hypothesis just discussed.

19.3.2 The morphogenesis of plant organs

Morphogenesis, i.e., the origin and development of form, is genetically determined and environmentally modified. Thus morphology is a question of development, and factors affecting and influencing development ultimately determine final morphology. In this section, morphogenesis of shoots, leaves, and roots will be discussed as if they act alone. However, it should clearly be kept in mind that such development is highly coordinated in the intact plant. Growth is most certainly coordinated through hormone balance.

Shoot development

A diagram of a shoot apex is shown in Fig. 19.11. Such a shoot apex can be divided into three active regions: the protoderm, the procambium, and the ground meristem. The protoderm, through a series of anticlinal (right angles to the surface) and periclinal (parallel to the surface) cell divisions, forms the epidermal tissue. The provascular tissue is the meristematic tissue that gives rise to the conducting

of apical dominance and hormone balance developed. In some of their experiments they showed that cytokinins would overcome the auxin inhibition of lateral bud growth, strongly suggesting that a balance between auxin and cytokinin was responsible for apical dominance. It seems reasonable to conclude that apical dominance is the result of the interaction of hormones, largely auxin and cytokinin. The auxin is the correlative signal that either influences cytokinin transport (probably from

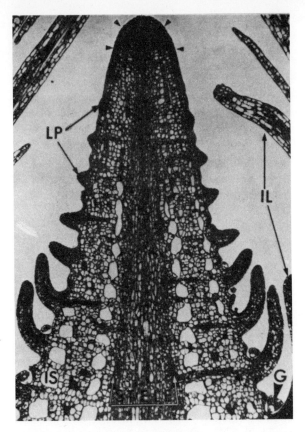

FIGURE 19.11 Photomicrograph of the apex of *Hippuris*, showing leaf primordia (LP), young or immature leaves (IL), intercellular spaces (IS), the apical dome (arrows), and the stele. Reproduced with permission from *Plant Structure and Development* (T. P. O'Brien and M. E. McCully) Collier McMillan 1969.

tissue, the xylem and phloem. Ground meristem cell division results in parenchymatous tissue, such as pith and cortex.

It should be evident from the shoot apex diagram of Fig. 19.11 that lateral appendages (leaves and branches) arise from superficial meristems (the protoderm), unlike the lateral appendages of the root, which develop from deep tissues.

As explained above, the development of any one bud at the shoot apex is a function of apical dominance. Auxin evidently stimulates the actual growth, but the initiation is promoted by cytokinin. The actual elongation is largely a function of auxin that is transported basipetally. It occurs in high concentrations in elongating internodes. Furthermore, the elongation is proportional to the logarithm of the concentration of auxin. Despite this, auxin added to elongating stems usually does not stimulate additional growth, contrary to the addition of gibberellin, which frequently promotes

elongation greater than normal. Internodes attain a greater length, but there are the same number. Because ethylene reduces internode elongation, it is assumed that normal shoot growth results through a proper balance of hormones. Although it is clear that hormone balance regulates growth, what controls the ultimate form of the shoot is not known. A brief discussion of shoot morphogenesis follows.

The shoot apex develops as a spiral cone with a defined arrangement of lateral appendages for each individual shoot. Such arrangement is called phyllotaxy. Although we know little about the factors that govern and control the arrangement of leaves around the shoot, the external phenomenon of phyllotaxy has been studied in much detail. To visualize the concept of phyllotaxy, the leaves may be traced in a spiral that follows the leaves in order of their formation. The spiral can be traced acropetally through successive leaves around the stem, choosing an older leaf as a point for reference. Counting both the number of leaves and the number of turns around the stem to reach a leaf in exact vertical alignment with the starting leaf, the result is expressed as the ratio number of turns: number of leaves. The vertical alignment is called an orthostichy.

The numbers that appear most frequently in expressions of phyllotaxy are numbers in the Fibonacci sequence. The Fibonacci sequence 1, 1, 2, 3, 5, 8, 13, 21, . . . is one in which each number is the sum of the two numbers preceding it, with the exception of the first two. Numbers of this sequence are found in both the numerator and denominator of the phyllotaxy ratios.

Such phyllotaxy systems appear in varying degrees of complexity, with 1/2 being the simplest and the others generally following the Fibonacci sequence 1/3, 2/5, 3/8, 5/13, ... The decimal equivalent of these ratios converges on the value 0.382.

Phyllotaxy may also be expressed in terms of the spirals formed by successive leaves. Two prominent spirals, a left-handed one and a right-handed one, are formed along shoots by successive leaves. When the leaves or other components (such as primordia) are numbered in order of their initiation, the numbers will differ by a constant. The constant will commonly be a Fibonacci number. A spiral that links leaves or primordia differing by some constant in their developmental sequence is called a parastichy. Phyllotaxy systems defined in this manner are expressed by the constants by which the successive leaves of the two opposing spirals differ: 1 + 2, 2 + 3, 3 + 5, 5 + 8, 8 + 13, ...

One of the most common phyllotaxy systems is the 2 + 3 arrangement shown for *Rhododendron* in Fig. 19.12. The photograph shows a scanning electron micrograph of the growing point of *Rhododendron*. On the accompanying diagram the five possible parastichy spirals are drawn. Note that the clockwise spirals (there are three of them, traced with solid lines) trace leaves that differ in position number by three, and the opposing counterclockwise spirals (there are two of them, traced

with dashed lines) trace leaves that differ in position number by two. Thus a phyllotaxis of 2 + 3 is shown.

An additional aspect to be considered in the relationship between lateral appendages and overall shoot morphology is the angle formed between a leaf and the shoot axis. As explained in the last section, these angles are related to apical dominance and are governed by hormonal balance, with auxin being perhaps the most important contributing factor.

On first thought it might appear that the plant shoot is rather symmetrical, but in fact it is quite asymmetrical. Because of the spiral growth and spirally arranged leaves, shoots show either a right-hand or left-hand symmetry. For the most part the asymmetry is genetic, but some plants show a change in spiral symmetry between juvenile and mature forms, and others change by chance or as a function of the environment.

Overall, our best interpretation of the gross morphology of a shoot is in terms of the phyllotaxis symmetry. It is assumed that the angle of the appendages does not change during the growth process. Growth is the result of length and surface-area increases.

The remainder of the gross morphology of the stem will depend on the ratio of anticlinal to periclinal cell divisions and the extent of cell enlargement. Thus some shoots will be extremely long and thin (vines) and others may be short and thick (the succulent cacti).

Furthermore, the shoot will vary in morphology depending on the extent (or presence or absence) of secondary growth. Primary growth largely gives rise to the vertical axis and some girth whereas secondary growth, the result of cambial activity, gives rise to girth. Lateral appendages—numbers, shapes, and size—are the result of primary growth. Except for the lower plants and monocotyledons, most plants have both primary and secondary growth.

Some plants have intercalary rather than apical meristems. If the tip of a common herbaceous plant is removed the tip will no longer elongate by cellular divisions, although it is not uncommon for a new apical shoot to originate from the axil of a lateral branch. In the case of grasses and certain other monocots with intercalary meristems, grazing or removal of the apex does not prevent growth. Turf grasses can be continually mowed because of intercalary meristems at the base of the leaf blade.

Lateral meristems give rise to lateral secondary growth of stems largely through anticlinal cell divisions. Not all plants have secondary growth. Herbaceous dicotyledonous plants and the "non-woody" monocotyledonous plants have only primary growth, and the plant body arises from cell divisions of the apical meristems. Certain large "woody" monocots such as yuccas and palms have secondary thickening meristems that give rise to lateral growth and *appear* to be somewhat woody. The latter is not strictly true; the term "wood" should be restricted to xylem arising from the vascular cambium, the main lateral meristem of plants. The student should refer to a plant anatomy textbook for anatomical details on secondary growth of monocots.

The two main lateral meristems of plants are the vascular cambium and the cork cambium (phellogen). The vascular cambium is a single sheet of

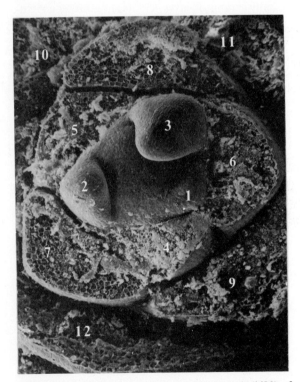 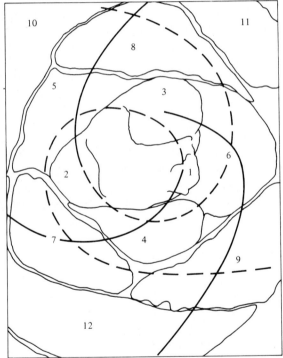

FIGURE 19.12 Scanning electron micrograph (X80) of a dormant shoot apex of *Rhododendron* (left) and a drawing of the apex (right). The leaf primordia and immature leaves are numbered from the youngest to the oldest on both the micrograph and the drawing. The plant has a 2 + 3 phyllotaxy. The five possible parastichy spirals are illustrated on the drawing. There are two counterclockwise spirals (dashed lines), each passing through leaves that are separated by a position number of 2, and three counterclockwise spirals, each passing through leaves that are separated by a position number of 3. Electron micrograph prepared by Richard Mueller and Don Pardoe, University of California, Berkeley.

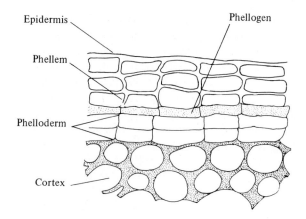

Epidermis

Phellogen

Phellem

Phelloderm

Cortex

FIGURE 19.13 Drawing to show how the cork cambium (phellogen) forms cork (phellem cells) toward the outside of the stem and phelloderm cells toward the inside. The cork cambium produces secondary growth.

meristematic cells that by anticlinal divisions produce phloem tissue toward the surface of the plant body and xylem tissue toward the center.

The outer phloem, which becomes part of the bark, is continuously sloughed off. However, xylem remains since it is located toward the center of the plant body. That xylem tissue most recently produced is referred to as the sapwood and is the functional water-conducting tissue. Older xylem, which is largely occluded and perhaps filled with resins and tannins, is nonfunctional and is referred to as heartwood. The heartwood may decay, leaving the woody stem in the form of a cylinder.

It is not uncommon for xylem to appear in the form of discrete rings of different cell sizes. In temperate environments, xylem cells produced in early spring are large, thin-walled cells. As the growing season progresses into summer, water becomes limiting and day temperatures are more unfavorable for growth. Xylem cells become smaller and have thicker walls such that there is a progression from large, thin-walled cells as the growing season progresses from the favorable springtime to a generally less favorable summertime. Successive seasons produce definite growth rings, and if they are formed annually they are called annual rings. Annual rings can be used to date woody plants.

The science of dendrochronology attempts to use such growth rings to determine tree age and interpret past climatic patterns on the basis of ring appearance. The appearance of the cells in the rings will give some indications of the nature of the growing conditions during the period in which the rings were formed. Even fires can be detected if rings appear charred.

The phellogen (cork cambium) arises from the phloem or adjacent tissues and produces cork tissue by anticlinal divisions. Cells produced to the outside are the cork cells proper and are called phellem, and those produced to the inside are called phelloderm cells. This tissue functions in protection and along with the phloem makes up the bark of woody plants. Since bark is continuously sloughed off, phellogen is not a permanent meristematic tissue but must be periodically renewed. Phellogen-producing cork is illustrated in Fig. 19.13.

Leaf development

As stated above, the leaf arises from superficial meristematic tissue of the conical apex and thus represents primary growth. The arrangement of the leaves around the apex is called phyllotaxis and was discussed under stem morphogenesis. Unlike the shoot, the leaf shows determinate growth, reaching a maximum size.

Overall, the growth and development of the leaf is a complex function of hormonal balance, genetic predetermination, and environment. The latter can

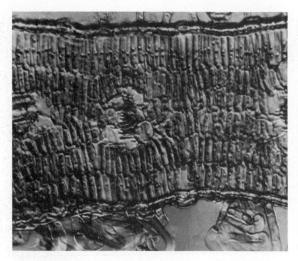 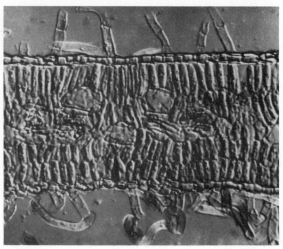

FIGURE 19.14 Transverse sections of leaves of *Encelia farinosa*. Left: section through a leaf that developed in the summer under high-intensity light, showing typical "sun leaf" anatomy. Right: section through a leaf that developed in the winter, showing characteristics of "shade leaf" anatomy (compare with the sun leaf). The sun leaf is thicker (0.26 mm) than the shade leaf (0.16 mm). The sun leaf has a lower surface area-to-volume ratio than the shade leaf. Sun leaves that develop under high-intensity light usually have a higher light-saturation point during photosynthesis. Note that these leaves have mostly palisade parenchyma. The sun leaf has six layers and the shade leaf has four layers. Photographs are X250. Photographs by Robin L. Kobaly. (From Robin L. Kobaly. 1977. Effects of mesophyll surface area on gas exchange and leaf resistances in a desert shrub, *Encelia farinosa* Gray. Master's thesis, University of California, Riverside.)

profoundly alter leaf morphology, as can be seen with some submerged plants with leaves above and below the waterline and the different morphology of sun and shade leaves (Fig. 19.14).

For the most part the ultimate shape of the leaf depends on differential cell division of surface and internal meristematic cells, the positioning of the major venation, and differential enlargement of cells. The leaf primordium begins development by periclinal divisions and very soon marginal and submarginal initial cells develop that will give rise to the leaf lamina. The main axis of the leaf primordium gives rise to the main veins and overall venation. The marginal initials divide anticlinally, giving rise to the epidermis that forms the surface of the leaf. The controlled divisions of the marginal and submarginal initials account for the number of cell layers within the dorsoventral leaf lamina.

The marginal initials cease division first, and subsequent growth is by cell enlargement. Marginal cell division may stop when the cell is only 10% to 20% of its full size. The submarginal initials give rise to the spongy parenchyma of the leaf mesophyll palisade parenchyma. As a result of this unequal cell division of the mesophyll, as the lamina continues to expand (largely by cell enlargement) the spongy mesophyll cells are pulled apart, creating a greater proportion of intercellular space.

If the central primordium and marginal initials divide more or less at the same rate, the leaf shape will be simple and typical. If the meristems associated with venation development divide and enlarge faster than the marginals producing the lamina, a lobed and dissected leaf will develop. Thus the relative rates of cell division and enlargement appear to determine the ultimate shape of the leaf. As previously stated, this will be in part a function of the genetic composition of the species and in part a function of the environment in which the leaf develops.

Although little is known about the hormonal regulation of leaf development, experimentation has shown that auxin is necessary for vein growth but has little effect on the growth and development of the mesophyll. Both the cytokinins and gibberellins promote growth of the leaf mesophyll tissue.

The growth and development of the leaf shows sigmoid growth kinetics. The initial portion of the growth is the result of cell division, whereas the later portions are mostly by cell enlargement.

In the pinto bean leaf, as the leaf expands there is a decrease in both soluble protein and enzymatic activity. Throughout development the leaf has photosynthetic capacity, but the rate of photosynthesis per unit area reaches a peak at about the time of full expansion and then either continues until senescence or in some cases decreases immediately (Fig. 19.15). Such a peak, occurring at full expansion or just after, corresponds to the greatest concentration of chlorophyll and starch in the leaf. Soluble sugars and free amino acids are at a minimum concentration perhaps because of the strong metabolic activity at this time.

Initially the young, developing leaf is a sink for carbohydrates of photosynthesis. Translocation is into rather than out of the leaf. Photosynthate produced by rapidly expanding leaves is retained. Leaves that have just fully expanded and that are high in starch will translocate photosynthate to younger leaves and to roots. Older leaves in the senescent stage translocate photosynthate to roots (Fig. 19.16).

At some time after full size is reached the leaf will begin a normal process of deterioration (senescence). It is marked by the decline in physiological activities, such as a reduction of photosynthesis, a decrease in RNA and protein synthesis, and the mobilization and translocation of carbohydrate and nitrogenous reserves out of the leaf. The application of cytokinin will prevent this mobilization that accompanies senescence, instead maintaining chlorophyll content and physiological activity for a longer period. It appears that proper hormone balance will delay senescence or that imbalance will hasten senescence. Ultimately, the deciduous dicotyledonous leaf will abscise (Fig. 19.17).

Abscission is a physiological process occurring in a specialized region of the leaf petiole base. It is marked by an increase in physiological activity including protein synthesis and increased respiration. The leaf produces auxin, which is transported from the lamina to the petiole. Continuous production of auxin by the leaf during development and maturation apparently prevents abscission. When auxin production slows during aging, the abscission

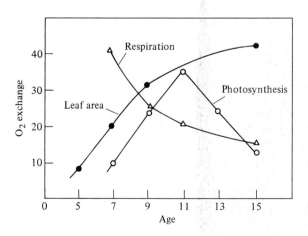

FIGURE 19.15 Change in photosynthesis and respiration of a bean leaf during development (expansion). Respiration (expressed on an area basis) decreases as the leaf expands, whereas photosynthesis reaches a peak at about the time of full expansion. Photosynthesis and respiration were estimated by measuring oxygen evolution and oxygen consumption, respectively. From a student experiment using pinto bean leaves at the University of California, Riverside.

FIGURE 19.16 Autoradiographs showing the translocation of ^{14}C-photosynthate out of older leaves (right) and the retention of photosynthate by young leaves (left). Both plants were the same age. The plant at the left had $^{14}CO_2$ supplied to the youngest leaf in the light. Very little translocation from the young leaf to the older leaves or roots occurred. The plant at the right had $^{14}CO_2$ supplied to the older leaf. Translocation of ^{14}C-photosynthate was to the young leaf and to the roots. The senescent cotyledons can just be seen in the center of the right panel; they did not act as a sink for photosynthate. From a student experiment on cotton plants at the University of California, Riverside.

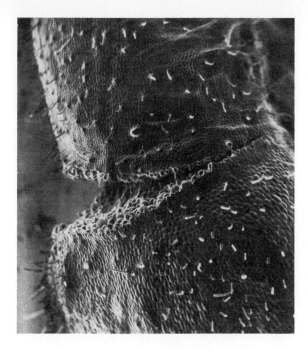

FIGURE 19.17 Electron micrograph of a bean leaf petiole during abscission. Hydrolytic enzymes break down the middle lamella but leave the cells of the petiole relatively intact. During the process, the distal portion (upper section) decreases in size evidently because of water loss, causing a physical force for actual separation. Photograph by Dr. Roy Sexton, Stirling University, Stirling, Scotland.

zone becomes sensitive to ethylene and perhaps abscisic acid. The hormone balance change is accompanied by protein synthesis, which produces pectinases and cellulases. These hydrolytic enzymes dissolve away the cell walls of the petiole in the abscission zone, ultimately bringing about abscission. It has been shown that the cellulase produced in response to ethylene during the abscission process is a different isozyme than is the cellulase associated with other physiological processes.

Environment has a profound influence on the development of leaves. Leaves that develop under stress conditions such as high irradiation or low moisture tend to be more compact, with a high volume to surface ratio. Such leaves are said to be xeromorphic.

Xeromorphic features develop in leaves that are produced in full sun, which are called sun leaves as opposed to those that develop in shade. Sun leaves have a high amount of palisade parenchyma relative to spongy parenchyma. The morphology of sun leaves is compared with shade leaves in Fig. 19.14. The increased palisade parenchyma is the result of additional cell layers. Furthermore, the cells tend to be smaller with thick walls. The epidermis may develop a thick cuticle, and there frequently are more stomata per unit area. Intercellular spaces are smaller.

The CO_2 compensation point and light-saturation point for shade leaves are lower than for sun leaves. Thus it takes relatively more light for the same rate of photosynthesis in sun leaves in comparison with shade leaves. Not only is the light-saturation point lower in shade leaves, but also the maximum rate of photosynthesis reached in shade leaves is usually lower than in sun leaves.

In some species, photoperiod influences leaf development. *Kalanchoe blossfeldiana,* a short-day plant, has more compact, succulent leaves when grown on short days than when grown on long days.

In many species, salt will cause the development of more succulent leaves which may have more palisade tissue and consequently will be thicker.

Root development

In general, the root meristem is more defined than the shoot meristem. Roots have a nondividing, quiescent center plus well-defined regions of meristematic activity. Shoots are less well defined; they have more of a group of meristems such that there are definite meristematic regions that give rise to leaves and floral organs. The primary root developing from the radicle of the embryo usually matures at a rate greater than that of the shoot.

At the very apex there are a few initial cells that give rise to the cells of the root and root cap. Just behind this region of cell division is the region of cell elongation. The region of cell division extends back about 0.2 mm to no more than about 0.5 mm; the region of cell elongation extends back about another 0.5 mm in most roots. Cells in the region of elongation are less dense cytoplasmically than are the dividing cells and have visible vacuoles. After elongation the root cells mature, and it is here in the region of maturation that certain of the root epidermal cells form root hairs. Cells now appear large and highly vacuolated.

Unlike the shoot, which has a rather complicated morphology, the root is rather simple. There is no branching and there are no nodes. Just behind the region of maturation, lateral branches may form from the pericycle.

The food-conducting cells begin differentiation and are functional before the tracheids and vessel elements. Both tracheids and vessel elements begin differentiation in the region of elongation.

The region of greatest metabolic activity is just behind the root tip. There is a maximum uptake of materials, and the greatest respiration rate occurs here. The growing root becomes a sink for carbohydrate translocation, and in some instances roots may even develop into storage organs.

The plant hormones regulate and control many of the growth processes in roots such as lateral root formation, elongation, cambial activity, and geotropism. The root, in fact, is a good experimental organ since it can be grown readily in culture, allowing extensive study of the effect of hormones on physiological processes. An important question with respect to hormonal physiology of roots is the source of hormones. There is some evidence that roots synthesize auxin. However, auxin transport in roots, unlike that in shoots, is largely acropetal;

that is, transport is toward the tip rather than away from the tip. The transport is polar as it is in shoots. There is evidence for some basipetal transport at the very tip.

Cytokinins are synthesized in the roots and transported to the shoots. The evidence for gibberellin synthesis in the roots is more equivocal, but evidence from excised root studies shows that roots can make gibberellin. Whether this is as significant for the economy of the plant as the production of cytokinin is not known.

Perhaps the most important evidence for the synthesis of cytokinins in roots and its transport to shoots comes from studies of hormone concentrations in the xylem exudate of decapitated plants. When stressed by drought or heat, there is a reduction of cytokinin as evidenced by a lower concentration in the xylem exudate. These same kinds of experiments show a reduction of gibberellin in the xylem exudate, suggesting that gibberellins are also synthesized in the roots.

Ethylene is produced by roots, especially in response to auxin treatment. Thus roots are not that much different from other tissues in that they do have an auxin-inducible ethylene production system. Abscisic acid occurs in roots, and there is circumstantial evidence that roots can produce it; it occurs in xylem exudates.

Like cytokinins, abscisic acid levels change in response to drought. Unlike cytokinin, however, it seems that the increase in abscisic acid that occurs in leaves is the result of more production in the leaves. The reduction in cytokinin of leaves in response to drought is evidently the result of reduced root biosynthesis or a reduction in transport from roots to shoots. The interaction of cytokinin and abscisic acid as the result of water stress has created much interest, because cytokinin opens stomata and abscisic acid will cause stomatal closure. Water stress of roots results in a reduction of leaf cytokinin and an increase in leaf abscisic acid, both of which cause stomata to close.

Overall, the evidence is quite conclusive that the roots are a major site of biosynthesis of the plant hormones including auxin, gibberellin, cytokinin, abscisic acid, and ethylene. In the case of cytokinin, at least, the root may be a primary source of supply for the shoot.

The exact role of the plant hormones in root growth is not entirely clear, but it seems reasonable to conclude that as well as specific individual effects such as the stimulation of cell division by cytokinin, there are probably many growth responses that are the result of hormone balance. Actual root elongation may be the result of the acropetal transport of auxin to the elongating region. However, gibberellin may also play an important role. How abscisic acid functions in axial root growth is not known, but it is undoubtedly important in the geotropic response to roots.

Hormone interaction is implicated in the formation of lateral roots. Lateral root formation depends on the presence of auxin and perhaps cytokinin, the cell-division factor. The acropetal polar transport of auxin from root base to tip will stimulate the production of lateral roots provided that a root initiation inhibitor is not present. The assumed inhibitor is abscisic acid because in culture abscisic acid will prevent lateral root initiation. Thus the balance of auxin, cytokinin, and abscisic acid evidently regulates lateral root formation.

In addition to lateral root formation, the hormones auxin and cytokinin play an interactive role in cambial development and activity in roots. Roots are a major site of hormone synthesis, producing hormones and transporting them to shoots and correlating root and shoot growth; in addition, the correlation of axial root growth, directional root growth, and lateral thickening responds to a balance and interaction of plant hormones.

The previous discussions of vegetative growth were for the most part concerned only with intrinsic factors that govern morphology. But perhaps almost as important as the genetically determined phenomena is environment, which plays a major role in the determination of the overall morphology of the plant. Next is a brief discussion of the environment and how it influences plant morphogenesis.

19.4 Environmental Influences on Vegetative Growth

The plant, being tightly coupled to the environment, will be influenced and affected by almost any environmental factor. Light, temperature, soil factors (including gases, water, and nutrients), and atmospheric factors (humidity, gases, pollutants, wind, and even atmospheric pressure and gravity) are apt to alter or govern plant growth. It is within the scope of this book to cover the influences of such factors on growth and development.

Perhaps an easy way to consider the influence of environmental factors on plant growth and development is to recall Liebig's "Law of the Minimum," proposed by Liebig in 1843 while investigating the effect of fertilizers on crop yield. Liebig proposed that only the addition of that factor which is most limiting will influence growth. However, we now know experimentally that an addition of almost any factor which is present at a less than optimum level will result in a positive response. It was Mitscherlich who in 1909 first modified Liebig's Law of the Minimum to a concept of a *limiting factor* in which the response to a deficient factor will be proportional to the extent that the factor is limiting. The idea is that any factor that is present in an amount different from optimum will be accompanied by a response if brought closer to the optimum. This, of course, implies that a reduction in a factor which is in excess, if the excess limits growth, will result in a positive response. Varying those factors that depart from optimum by only small amounts will result in only small responses, but addition of those factors that greatly deviate from the optimum will be accompanied by large responses.

19.4.1 Light

Light influences the vegetative growth of plants in a complex manner. Light affects photosynthesis and hence carbon nutrition, leaf temperature, and water balance, and acts directly in many growth responses, such as stem elongation, leaf expansion, and stem curvature. These phenomena, along with the effect of light on germination and flowering, are discussed in detail in Chapters 20 and 21. An overview of light and vegetative growth is presented below.

Growth of crowded or shaded plants is reduced because of a decrease in light. Light is an extremely important environmental factor in transpiration. Stomata open in response to light, facilitating gas exchange. Light and temperature are frequently correlated, and since temperature of leaves is the most important environmental variable in transpriation, its influence is paramount. Excess transpiration, resulting in extreme water loss, results in plant water stress and decreased growth.

High light intensities cause plants to develop short, stocky stems. Low light intensities cause etiolation, which results in tall, spindly plants. Reduced growth at high light intensities is mostly the result of water stress, but light does affect cell elongation directly by its effect on auxin.

The duration of light influences reproductive growth independent of intensity. Some plants only flower when exposed to short day lengths, and others will only flower on long days. A variety of other growth phenomena respond to day length as well. These growth responses and others are discussed in Chapters 20 and 21.

19.4.2 Temperature

Plant growth shows a minimum, optimum, and maximum temperature response. Most plants grow best at 30° to 35°, but alpine and arctic plants do well at 5° to 10°, and desert and tropical plants may grow well at high temperatures. Thus the growth response to temperature will vary widely with species, variety, and ecotype. Plants can be conditioned to grow within a reasonably wide range of temperatures, a phenomenon called plasticity.

Much of our knowledge about plant growth and development under variable but controlled temperature conditions comes from the work of Went (1957), when he was at the Earhart Growth Laboratory in Pasadena, California. The Earhart facility was a large, controlled-plant-growth building called a phytotron. A phytotron has temperature and light control such that plants can be grown and studied under a variety of experimental conditions.

Much research has been done with temperature effects since it is one of the easiest of all environmental factors to control. Early research showed that many plants grow primarily at night, evidently because water balance and hence plant turgor are more favorable. Night (nycto-) temperature is as important or perhaps even more important than day temperature, as are fluctuating diurnal temperatures.

Figure 19.18 shows the growth of several portions

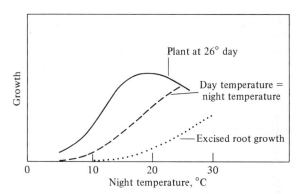

FIGURE 19.18 Growth of tomato plants as a function of temperature. Plants kept under conditions of 26° days and 18° nights showed greater growth. When day temperatures were the same as night temperatures there was less growth. Excised roots showed an increase in growth with increasing temperature. From F. W. Went. 1957. *The Experimental Control of Plant Growth.* Chronica Botanica, Waltham, Mass.

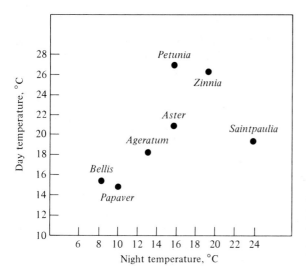

FIGURE 19.19 The optimum day and night temperatures for growth of several plants. Note that *Saintpaulia* has a higher optimum night temperature than day temperature. From F. W. Went. 1957. *The Experimental Control of Plant Growth*. Chronica Botanica, Waltham, Mass.

of a tomato plant as a function of temperature as determined in the Pasadena phytotron. The graph shows growth of the plants when kept at constant day temperature of 26° and variable night temperature, growth when the plants were kept at constant day and night temperature, and growth of excised roots as a function of temperature. Roots grow and respond to temperature in a way very different from shoots. Most plants do better when exposed to a variable day–night temperature regime.

Figure 19.19 shows a graph of the optimum day temperature for growth against the optimum night temperature. A wide range of temperature optima are evident, and except in the case of *Saintpaulia*, the optimum day temperature is higher than the optimum night temperature.

19.4.3 Water

Water is the environmental variable that most often limits plant growth the world over. Water availability is a function of the amount of precipitation and evaporation. The soil type governs the extent to which plants can extract water. Water acts as a substrate for many chemical reactions occurring within the plant, but the importance of water as an environmental factor is in plant hydration and in atmospheric humidity, which influences transpiration (refer to Chapter 4).

Figure 19.20 shows the complex interaction of light, temperature, and atmospheric humidity on the expansion of castor bean leaves. It seems from these data that expansion is more closely correlated with temperature than with light or humidity. During the heat of the day when the humidity decreases because of increased air temperature, leaf expansion decreases. Most expansion occurs during the day and follows the temperature curve.

The lesson from the experiment in Fig. 19.20 is that growth is a complex phenomenon that is influenced in a complex manner by environmental variables.

19.4.4 Mineral nutrition

Nutrition as influenced by soil fertility is one of the most important environmental growth factors. Soil type, soil pH, salinity, texture, atmosphere, and available minerals are among the important factors.

Frequently, the better the aeration of soils the better the plant growth. In poorly aerated soils,

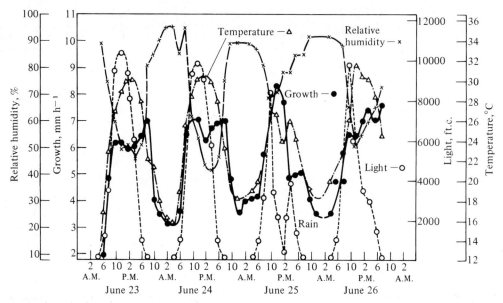

FIGURE 19.20 Graph showing the complex interaction of environmental variables for growth of castor bean leaves. Leaf expansion is closely related to temperature and less closely related to light or humidity. From W. E. Loomis (ed.). 1953. *Growth and Differentiation in Plants*. Iowa State College Press, Ames, Iowa.

oxygen is depleted when CO_2 accumulates. Low O_2 and high CO_2 limit the uptake of nutrients, especially potassium. Soils saturated with water also show poor aeration; roots exhibit limited mineral uptake.

Acidity of soils is an important factor in mineral availability. Acid soils are poor in exchangeable bases, but perhaps the most important problem in acid soils is low calcium and the unavailability of phosphorus.

Iron, manganese, and aluminum are made mobile and may be at toxic levels in acid soils. Many acid-soil plants occur on acid soils precisely because of their low capacity for iron uptake. The increased available iron in acid soils creates a better environment for such plants. Soils with high pH frequently have excess calcium, high levels of sodium, and poor nutrient availability.

Table 19.1 shows the interaction of soil moisture and nitrogen in plant growth. The shoots of tomato do best when grown in high nitrogen and ample water. As water and nitrogen become less available shoot growth decreases, but root growth is relatively unaffected over these concentration ranges. The result is an increase in the root-to-shoot ratio as nitrogen and water become suboptimal for growth of the plant. Once again the complex interaction between environmental variables is evident.

19.4.5 Other environmental factors

Many other environmental variables influence the growth and development of plants. The tropic growth movements in response to directional stimuli such as gravity and physical irritation (also, of course, light and temperature) are discussed in Chapter 17. Electromagnetic radiation in general affects plant growth and development. Ultraviolet and visible light effects are discussed in Chapters 3 and 7. High-energy radiation having frequencies

TABLE 19.1 Growth in grams of shoots and roots of tomato plants with varying water and nitrogen levels

Water level	Nitrogen	Shoots, g	Roots, g	Ratio, shoot:root
Optimum	Optimum	224	43	5.2
Low	Optimum	137	42	3.3
Optimum	Low	93	46	2.0
Low	Low	73	40	1.8

Adapted from Loomis (1953).

greater than ultraviolet light is ionizing and disrupts cellular processes. Radiation of wavelengths longer than that of visible light affects plants primarily as heat. However, there is some evidence that radiation with even longer wavelengths influences plant development.

Audible sound frequencies have been shown to increase growth in some cereals. Energies used are below those sufficient to disrupt chemicals but are in the range of vibrational effects. The mechanism of how sound influences growth processes is not known.

Review Exercises

19.1 If the protein content of a cell is about 10^{-10} mg and the average molar mass of a protein is about 65,000 g mol^{-1}, what would be the average number of protein molecules per cell?

19.2 A callus-tissue culture with about 10,000 cells weighed 2 mg. In two weeks it weighed 130 mg. Compute from these data the average number of days per cell division.

19.3 How do you visualize the functioning of the nutri-

tive regulation concept of growth correlation? Do you think it equally plausible as the hormonal concept of growth correlation?

19.4 Of what practical importance is the concept of apical dominance to agriculture and horticulture? Do you know of any specific examples?

19.5 Select a few simple annual plants and attempt to determine their phyllotaxy.

References

BONNER, J. 1976. The nucleus. In J. Bonner and J. E. Varner (eds.), *Plant Biochemistry*. Academic Press, New York.

CLELAND, R. 1971. Cell wall extension. *Ann. Rev. Plant Physiol.* 22:197–222.

CLOWES, F. A. L. 1961. *Apical Meristems*. Blackwell Scientific Publ., Oxford.

DOWNS, R. J., and H. HELMERS. 1975. *Environment and the Experimental Control of Plant Growth*. Academic Press, New York.

HUMPHRIES, E. C., and A. W. WHEELER. 1963. The physiology of leaf growth. *Ann. Rev. Plant Physiol.* 14:385–410.

KAHL, G. 1973. Genetics and metabolic regulation in differentiating plant storage tissue cells. *Bot. Rev.* 39:274–299.

KALDEWEY, H., and Y. VARDAR. 1972. *Hormonal Regulation in Plant Growth and Development*. Verlag Chemie, Weinheim, W. Germany.

LOOMIS, W. E. (ed.). 1953. *Growth and Differentiation in Plants*. Iowa State College Press, Ames, Iowa.

MERNS, F., and A. N. BINUS. 1979. Cell determination in plant development. *BioScience* 29:221–225.

MILLER, M. W., and C. C. KUEHNERT. 1972. *The Dynamics of Meristematic Cell Populations*. Plenum Press, New York.

O'BRIEN, T. P., and M. E. McCULLY. 1969. *Plant Structure and Development*. Macmillan, New York.

PILET, P. E. (ed.). 1976. *Plant Growth Regulation*. Springer-Verlag, Berlin.

RAY, P. M., P. B. GREEN, and R. CLELAND. 1972. Role of turgor in plant cell growth. *Nature* 239:163–164.

ROBERTS, L. W. 1976. *Cytodifferentiation in Plants*. Cambridge University Press, Cambridge.

STERN, H. 1966. The regulation of cell division. *Ann. Rev. Plant Physiol.* 17:345–378.

STEWARD, F. C., M. O. MAPES, A. E. KENT, and R. D. HOLSTEN. 1964. Growth and development of cultured plant cells. *Science* 143:20–27.

STREET, H. E. (ed.). 1974. *Tissue Culture and Plant Science*. Academic Press, New York.

THIMANN, K. V. 1977. *Hormone Action in the Whole Life of Plants*. University of Massachusetts Press, Amherst, Mass.

TORREY, J. G., and D. T. CLARKSON (eds.). 1975. *The Development and Function of Roots*. Academic Press, New York.

VARNER, J. E., and D. TUAN-HUA HO. 1976. Hormones. In J. Bonner and J. E. Varner (eds.), *Plant Biochemistry*. Academic Press, New York.

WARDLAW, C. W. 1968. *Morphogenesis in Plants*. Methuen & Co., London.

WAREING, P. F., and I. D. J. PHILLIPS. 1978. *The Control of Plant Growth and Differentiation in Plants*. Pergamon Press, Oxford.

YEOMAN, M. M. (ed.). 1976. *Cell Division in Higher Plants*. Academic Press, New York.

CHAPTER TWENTY

Phytochrome Growth Responses

Light plays a major role in the growth and development of plants totally independent of photosynthesis. There are three important plant growth and development phenomena other than photosynthesis that are directly linked to light absorption. First, there are phototropism and nastic movements that respond largely to blue light. These are growth movements and were discussed in Chapter 17. Second, there is photoperiodism, i.e., responses to the seasonal variation in daylight. Photoperiodism, as will be shown below, is the result of light absorption by the ubiquitous plant pigment phytochrome. For the most part, photoperiodism is a response to absorption of red and far-red light. Finally, photomorphogenesis, i.e., growth and development directly controlled by light, is largely a response to high-intensity blue light, but it is also linked directly to phytochrome.

In this chapter the pigment phytochrome will be discussed in relation to photoperiodism and photomorphogenesis. In addition, biological rhythms will be introduced as important plant growth and development phenomena. They are intimately tied to the phytochrome system.

543

20.1 Photoperiodism

As defined above, photoperiodism is a response to the seasonal variation of day length. During this century it was discovered that day length was an important environmental parameter for the timing of flowering. It is known that some plants will flower only if the day length is shorter than some specified time (short-day plants), some will flower only if the day length is longer than some specific time (long-day plants), and others are completely independent of day length (day-neutral plants). The length of the day (the light period) is referred to as the photoperiod.

In the early 1900s G. Klebs of Germany noted that *Sempervivum* would not flower if kept on short days but would flower if kept on long days or in continuous light. He supposed that the long-day requirement was related to growth of the plant. At about the same time in France, Tournois discovered that hop and hemp plants would flower only if grown on short days or on long nights. Thus the early beginning of photoperiodism in plants goes back to Klebs and Tournois, but it was left to two American USDA researchers, W. W. Garner and H. A. Allard to demonstrate and develop in 1920 the concept of photoperiodism as we now understand it. Photoperiodism was a difficult concept to grasp because it requires the plant to measure time.

The first experiments of Garner and Allard were prompted by the observation that a particular variety of tobacco, Maryland Mammoth, would remain vegetative all during the growing season when grown outside at Beltsville, Maryland despite the fact that it would flower if grown farther south. When Maryland Mammoth tobacco was grown in the greenhouse during the winter, flowering was prodigious.

Thorough experiments eliminated temperature and nutrition as the cause of winter flowering in the tobacco. Finally, Garner and Allard designed light-tight boxes to shorten the day length of the plants growing in summer and discovered that tobacco would flower if the day length were shorter than 14 hours. If grown on photoperiods of greater than 14 hours or in continuous light, no flowering occurred. Garner and Allard called the plants that would flower only if grown on photoperiods of less than some specific time "short-day plants" (SDP). Subsequent research showed that other plants would flower only if the photoperiod was greater than some specific time—"long-day plants" (LDP). Still other plants are indeterminate and flower independently of the photoperiod. These are the "day-neutral plants."

Figure 20.1 shows one of the first experiments performed by Garner and Allard with Maryland Mammoth tobacco. The plants in the upper panel were grown indoors with light from 9 A.M. to 4 P.M., a photoperiod of seven hours. The plants in the lower panel were grown outside during the long days of the summer. The plants grown indoors on the short photoperiods flowered, whereas those grown outside on long photoperiods did not flower.

Figure 20.2 shows some other experiments of Garner and Allard in 1920. In the upper panel of Fig. 20.2, the Biloxi soybean plants on the left were grown on day lengths of five hours. The plants on the right were grown outside during the summer. The plants grown indoors on the short photoperiods flowered, whereas those outside did not. As shown in the lower panel of Fig. 20.2, the excessive growth is evidently due to longer photoperiods. Both flowered, but there is more growth by the plants kept on the longer days.

These experiments clearly demonstrated that plants had a mechanism by which time could be measured. Subsequent studies, many of them done in the USDA laboratories at Beltsville, partially answered the question of timing by plants.

There are two considerations for an understanding of photoperiodic events. First, there is the transducer that receives the signal (in this case light) and brings about the response. This transducer is evidently phytochrome. Second, there is a consideration of how time is measured by the plant. This is a more complex question related to the rhythmic phenomena related to phytochrome that occur in plants.

FIGURE 20.1 Experiment that demonstrated photoperiod in the flowering response of Maryland Mammoth tobacco. Upper: plants grown inside on 7-hour photoperiods. Lower: plants grown outside on photoperiods exceeding 14 hours (the critical photoperiod for Maryland Mammoth tobacco is 14 hours). The plants grown on the short photoperiods flowered, whereas those on the long photoperiods did not. This experiment led to our modern concepts of photoperiodism. (From W. W. Garner and H. A. Allard. 1920. Effect of the relative length of day and night and other factors of the environment on growth and reproduction in plants. *J. Agr. Res.* 18:553–606, U.S. Department of Agriculture.)

FIGURE 20.2 Upper: Biloxi soybeans kept indoors on 5-hour photoperiods (left) or outside on long photoperiods (right). Plants on the left flowered and set seed. Plants on the right grew more but did not flower. Lower: plants on the left were grown indoors on 7-hour photoperiods and those on the right were grown on 12-hour photoperiods. The day lengths were short enough for flowering (both trials flowered and set pods), but plants kept on longer photoperiods grew more. Growth per se is not as important for flowering as is the day length. W. W. Garner and H. A. Allard. 1920. Effect of the relative length of day and night and other factors of the environment on growth and reproduction in plants. *J. Agr. Res.* 18:553–606, U.S. Department of Agriculture.)

20.2 The Measurement of Time

Once the principles of photoperiodism were developed and fully appreciated in plants, the crucial question to be answered was the nature of the timing device. How could plants differentiate between short and long days?* How can plants tell time? An initial observation by Hamner and Bonner in 1938 was that the short-day plant *Xanthium* (cocklebur) would not flower if the long night was interrupted with a brief light period. Thus cocklebur grown on inductive photoperiods of less than 15.6 hours will not flower when the dark period is interrupted with a brief period of light. This interruption can be as short as three minutes and still be effective.

The interruption of night by light prevented flowering by short-day plants and was followed by experiments in which it was shown that long-day plants would flower if grown on short-day, noninductive cycles if the long night was interrupted with light. These experiments are illustrated in diagrammatic form in Fig. 20.3.

With the knowledge that light interruption of a long night period would promote the flowering of long-day plants and inhibit the flowering of short-day plants, two USDA research workers, Borthwick and Hendricks, set out in the 1950s to ascertain the light quality that most strongly promoted the response. They designed a spectrograph that could illuminate whole, growing plants with light of varying quality and were ultimately able to obtain an action spectrum for the light response. They found that the action spectrum of mid-night light causing interruption of flowering in short-day plants and promotion of the flowering of long-day plants when grown on short days was the same.

*It must be kept in mind that the designations short-day and long-day are physiological. A short-day plant is a plant that responds to a day length less than some critical length. A long-day plant responds to a day length greater than some critical length. The absolute time is not important. Thus a SDP and a LDP could both have a critical day length of 14 hours. The short-day plant will flower if the photoperiod is less than 14 hours, whereas the long-day plant will flower if the photoperiod is greater than 14 hours.

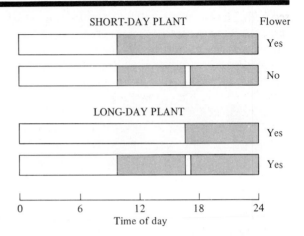

FIGURE 20.3 Diagram illustrating how the light interruption of a long night (short photoperiod) prevents flowering in a short-day plant and promotes flowering in a long-day plant.

Namely, the response in both cases was most sensitive to red light with a wavelength maximum of 660 nm.

Also during the 1950s at Beltsville, researchers were studying the light requirement for germination in lettuce seeds and observed that red light promoted germination but far-red light inhibited it. Sequential light treatments of red:far-red:red: far-red and so on could be given and the germination would respond to the last light exposure (Table 20.1). If the last light exposure was red,

TABLE 20.1 Effect of red and far-red light on the germination of lettuce seeds

LIGHT TREATMENT	GERMINATION, %
Dark control	8.8
Red (1 min)	98
Red : Far-red (4 min)	54
R : F-R : R	100
R : F-R : R : F-R	43
R : F-R : R : F-R : R	99

From Borthwick *et al*. (1952).

germination would occur. If the last light exposure was far-red, there was reduced germination. Subsequently, it was shown that the red-light flowering response was also reversed by far-red light.

It was shown by Borthwick and Hendricks that the flowering of short-day cocklebur and long-day barley was influenced by light with exactly the same action spectrum. Flowering was promoted in barley and suppressed in cocklebur by radiation in the range of 540 to 695 nm with a maximum of 660 nm, and the reverse of this response was caused by light in the range of 695 to 800 nm with a maximum of 730 nm.

In 1952 it seemed clear that there must be a pig-ment in plants that was sensitive to light with wavelengths of 660 nm and 730 nm. The actual discovery of the pigment was made with a dual wavelength spectrophotometer that had the capability of measuring the difference in absorption at 660 nm and 730 nm while irradiating at one or the other. It was shown that this unknown pigment, named phytochrome by Borthwick and Hendricks, would absorb maximally at 730 nm if irradiated with 660 nm light. If it was irradiated with 730 nm, the maximum absorption was at 660 nm. Thus red light (660 nm) would convert (phototransform) the pigment to a far-red absorbing form (730 nm), and the conversion was fully reversible.

20.3 Phytochrome

20.3.1 The pigment

Through the efforts of the workers at Beltsville, a blue-green biliprotein with an open tetrapyrrole chromophore that had the light properties of phytochrome was isolated and purified. Figure 20.4 shows the absorption characteristics of the two forms of phytochrome. This absorption spectrum is comparable to the action spectrum of the red/far-red flowering and germination response. The important feature of the biliprotein is that red light causes phototransformation to a far-red light-

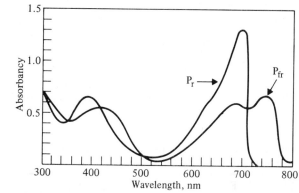

FIGURE 20.4 Absorption spectra of the two forms of phytochrome. Left: the P_r form absorbs maximally at about 660 nm and the P_{fr} form has an absorption maximum at about 730 nm. Right: the rearrangement of the phytochrome chromophore after irradiation. Absorption spectra redrawn with permission from W. L. Butler, S. B. Hendricks, and H. W. Siegelman. 1964. Action spectra of phytochrome *in vitro. Photochem.*

Photobiol. 3:521–528. Copyright 1964 Pergamon Press, Ltd.

absorbing form, and the far-red light-absorbing form can be reverted back to the red light-absorbing form by irradiation with far-red light.

Phytochrome is believed to have a mass of about 120,000 daltons. It has a high proportion of polar amino acids and is water-soluble.

A hypothesis for the red/far-red phototransformation of phytochrome is illustrated in Fig. 20.5. The P_r, or red-absorbing form (wavelength maximum = 660 nm), is the form that is first synthesized. Upon illumination with red light, the P_r form is rapidly converted to the P_{fr} form, which absorbs far-red light with a wavelength maximum of 730 nm. Whereas the P_r form of phytochrome is relatively stable, the P_{fr} form is unstable. In plants such as cauliflower and other dicotyledons, there is dark reversion of the P_{fr} form to the P_r form, but such dark reversion has not been observed in monocotyledons.

When the P_{fr} is illuminated with far-red light, it is rapidly converted back to the P_r form. After a period of hours depending on the species of plant, red light will not cause phototransformation back to the P_{fr} form.

In vivo there is an equilibrium between the two forms of phytochrome, P_r and P_{fr}. In light this equilibrium is a function of the ratio of different wavelengths of light (that is, the light quality) and is not a function of the total energy of the incident light. Red light will form about 80% P_{fr} and 20% P_r, whereas far-red light with wavelengths longer than 737 nm will result in 97% P_r. The photoconversions are not complete and equilibria are established because there are no wavelengths that are absorbed by P_r and not absorbed by P_{fr} (their absorption spectra overlap). The phototransformation from P_r to P_{fr} in red light takes several seconds, but the photoreversion of P_{fr} to P_r in far-red light only requires 20 to 30 milliseconds.

The P_{fr} form, which is evidently the active form, may bring about a physiological response. Or, alternatively, it may be metabolized and destroyed. Destruction is defined as the disappearance of P_{fr} with no P_r being formed. As the P_{fr} form of phyto-

MODEL FOR PHYTOCHROME PHOTOTRANSFORMATION AND PHYSIOLOGICAL FUNCTION

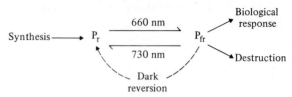

FIGURE 20.5 Simplified model to show the phototransformation of phytochrome and physiological function. The P_r form is the form that is synthesized. Red light (660 nm) converts the P_r form to the P_{fr} form, and far-red light converts the P_{fr} form to the P_r form. In dicots the P_{fr} form can revert back to the P_r form in the dark. The P_{fr} form is considered to be the active form and either brings about the physiological response or is inactivated by metabolic destruction.

chrome disappears, the total concentration of phytochrome decreases. After a period of some hours, the phytochrome concentration increases through renewed biosynthesis.

Physiologically, it is best to consider P_{fr} as the active form and P_r as the inactive form. However, it is also reasonable to consider P_r as an inhibitor that is destroyed by conversion to P_{fr}. The P_r form is the thermodynamically stable form and P_{fr} is the metastable form. P_r can only be converted to P_{fr} through red-light illumination, but in some plant species P_{fr} can revert back to P_r by a dark process as well as by far-red illumination. Since phytochrome is synthesized in the P_r form it is almost always present, but P_{fr} will only be present in active concentrations during or immediately after red-light irradiation.* The dark conversion of P_{fr} to P_r that occurs in some plants is temperature-dependent but independent of most metabolic inhibitors. However, the destruction of P_{fr} is inhibited by azide, cyanide, carbon monoxide, and anaerobiosis. It is not inhibited by the uncoupler dinitrophenol. The

*White light is sufficiently high in red and low in far-red to cause phototransformation. Blue light of sufficient intensity will shift P_r to P_{fr}.

dark reversion is probably a chemical-equilibrium process, but the destruction is dependent on the cellular metabolism of the tissue. Perhaps destruction occurs through proteolysis.

20.3.2 Induction–reversion phenomena

Many of the known induction–reversion phytochrome-mediated responses are listed in Table 20.2. Some of the most notable are the promotion of flowering in short-day plants, the inhibition of flowering in long-day plants, the promotion of germination in certain seeds, the unfolding of the plumule hook in seedlings, and the promotion of leaf expansion. Whether the response is promotive or inhibitory, it should be remembered that it is the P_{fr} form that is the physiologically active form of phytochrome. In short-day plants P_{fr} inhibits flowering, whereas in long-day plants P_{fr} promotes the flowering process.

There are two main hypotheses for how phytochrome regulates such diverse plant growth processes. One hypothesis is that phytochrome functions by alterations in membrane permeability and perhaps other properties. The second hypothesis is that phytochrome operates on the plant genome. Many of the phytochrome responses occur with little time lag, which is suggestive of membrane phenomena. Phytochrome–membrane interactions could result in ion flux alterations, changes in metabolite compartmentation, and differences in enzyme distribution. There is evidence for potassium-flux changes and alterations in ATP concentrations. There are immediate membrane potential changes, and phenomena such as coleoptile elongation occur within one minute after red-light stimulation. Exactly how membrane effects bring about phytochrome-related processes is not known.

Many of the phytochrome-related processes have much longer lag times. The induction of phenylalanine ammonia lyase (PAL) occurs within 60 minutes after stimulation in *Brassica*. Other, more complicated phenomena such as flowering occur with a several-day lag time. Even *Pharbitis nil,* which only requires a single inductive cycle, has a

TABLE 20.2   Examples of some phytochrome-controlled responses in higher plants

GYMNOSPERMS

Seed germination	Internode extension
Hypocotyl hook opening	Bud dormancy

ANGIOSPERMS

Seed germination	Photoperiodism
Hypocotyl hook opening	Flower induction
Internode extension	Expansion of cotyledons
Root primordia initiation	Succulency
Leaf initiation and expansion	Epinasty
Leaflet movement	Leaf abscission
Electrical potentials	Tuberization
Membrane permeability	Bud dormancy
Phototropic sensitivity	Sex expression
Geotropic sensitivity	Unfolding of monocot leaves
Anthocyanin synthesis	Rhythmic phenomena
Plastid formation	

From Wareing and Phillips (1978).

lag of about six days before flowering.* The increase in PAL and the observation that polyribosomes increase is evidence that protein synthesis is involved in the phytochrome mechanism. However, it is rather difficult to ascertain if protein synthesis is a cause or an effect of the phytochrome mechanism. As stated before, it is rather difficult to conceive of any growth phenomena occurring without protein synthesis. Thus it is possible that the initial site of the phytochrome reaction is on membranes, resulting in great amplification of subsequent protein metabolism.

Phytochrome is widespread in the plant kingdom, being found in virtually all higher plants and in algae. In etiolated seedlings it appears to be concentrated in meristems, both apical and cambial. In general, the concentrations are too low to detect by spectrophotometric means in intact leaves, but it has been identified in green leaf extracts.

Immunological techniques indicate that P_r is present throughout the cell except for vacuoles. After red-light illumination, the P_{fr} that is photo-transformed from P_r seems to become more localized.

Still other kinds of studies suggest that phytochrome is localized in the cell between the cytoplasm and the cell wall, evidently as a component of the plasma membrane. It is hypothesized that during the transformation of P_r to P_{fr} the P_{fr} changes orientation within the membrane. Such a hypothesis is quite supportive of the notion that phytochrome acts by altering membrane permeability.

An interesting and little-understood response to light that is believed to be phytochrome-linked is the high irradiation response (HIR) largely associated with photomorphogenetic phenomena. The red/far-red light responses of plants mediated through phytochrome will occur in very low light intensities. Some of these same responses that are

inhibited by red light are activated by high light intensities, with a maximum in the blue region of the spectrum (see Fig. 20.7 later). For example, high-intensity blue light and low-intensity far-red light both inhibit lettuce hypocotyl elongation and promote enlargement of mustard cotyledons. Interestingly, red light will not reverse the HIR response; that is, it is not photoreversible, a condition similar to most phytochrome responses.

20.3.3 The HIR (high-intensity response) system

H. Mohr, a longtime student of photomorphogenesis in plants, assumes (see Mohr, 1972) that the HIR system functions in photomorphogenesis by phytochrome absorbing light in the blue, which causes phototransformation to the effective P_{fr} form. Thus in both the low-intensity, induction–reversion phenomena and the high-intensity, blue-light photomorphogenic phenomena the active form of phytochrome is the far-red light-absorbing form, P_{fr}.

Thus phytochrome is associated with more than one kind of plant response. There are the induction–reversion phenomena such as germination, flowering, and plumular hook opening. There are also the high-intensity responses (HIR) such as the photomorphogenic phenomena, e.g., the inhibition of lettuce hypocotyl lengthening and leaf expansion. Perhaps a basic difference between the two is that the induction–reversion phenomena do not show irradiance dependence after phytochrome equilibrium is established. After P_{fr} is formed, more light has no additional effect. With the HIR, however, even after phytochrome equilibrium is reached further irradiance will bring about a greater response. The more the irradiance the more the inhibition of internode extension.

The third type of plant growth phenomenon intimately associated with phytochrome is represented by the biological endogenous rhythms, discussed at the end of this chapter.

---

*Of course, with most of these responses the photoreversion capability of the system is lost long before the appearance of the main response.

20.4 Photomorphogenesis

Photomorphogenesis is defined as growth and development directly dependent on light but not related to photosynthesis. The photomorphogenic phenomena are high-intensity responses (HIR). Unlike the induction–reversion phenomena (flowering or germination), the HIR show irradiance dependence. Whereas the induction–reversion phenomena respond only to wavelengths of light, the photomorphogenic phenomena respond to both wavelengths and total irradiance.

To understand the difference between the induction–reversion phenomena and the HIR phenomena, note that the induction–reversion phenomena respond to a few moments of light, which are sufficient to bring about phytochrome equilibrium (equilibrium of the two forms, P_r and P_{fr}). It is this equilibrium that is important in the response. The HIR may be initiated by phytochrome pigment equilibrium, but the greater the irradiance the greater the magnitude of the response. The HIR-dependent phenomena do not show reciprocity.

Several HIR phenomena have been studied in detail. These include the inhibition of internode growth, the expansion of cotyledons of mustard, and anthocyanin biosynthesis. Perhaps the most striking example of the phytochrome-dependent high-intensity response is etiolation (Fig. 20.6). The two plants shown in Fig. 20.6 have the same genome and were treated identically except that the etiolated plant was grown in the dark. The excessive growth of the etiolated plant is largely due to an increase in cell expansion. There are similar numbers of cells and the same number of internodes. The HIR inhibits elongation of internodes.

An action spectrum for this type of HIR is shown in Fig. 20.7. There are two types of experiments depicted in the figure. In one, the dark-grown plants were exposed to a short interval of light at various wavelengths. Red light with a wavelength maximum of 660 nm was most effective in inhibiting internode elongation. This shows the phytochrome response to be similar to the induction–reversion

FIGURE 20.6 The effect of light on plant development. The two plants from the same seed lot were grown under identical conditions except that the plant at the left received normal light and dark periods whereas the plant at the right was kept in total darkness.

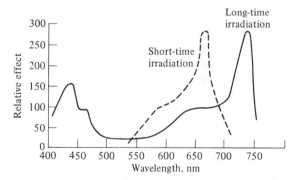

FIGURE 20.7 Action spectra for the high-intensity photomorphogenic response. On a short time basis (for induction) the peak is at about 660 nm. For prolonged irradiation (6 to 12 hours), the action spectrum shows peaks in the blue and in the far red. (From E. Wagner and H. Mohr. 1966. Kinetic studies to interpret "high-energy phenomena" of photomorphogenesis on the basis of phytochrome. *Photochem. Photobiol.* 5:397–406. Copyright 1966 Pergamon Press Ltd.)

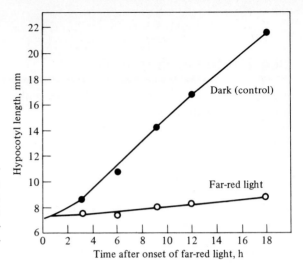

FIGURE 20.8 Experiment by Mohr showing how short-term far-red light will inhibit hypocotyl elongation. In the dark the hypocotyls elongate. The inhibition occurs with a very short lag time. (From H. Mohr. 1966. Differential gene activation as a mode of action of phytochrome 730. *Photochem. Photobiol.* 5:469–483. Copyright 1966 Pergamon Press, Ltd.)

phenomena in which P_r is phototransformed to the active P_{fr}. However, if the dark-grown, etiolated plants are exposed to light for longer periods of 6 to 12 hours, the action spectrum is quite different. Maximum responses are present in the blue at about 550 nm and in the far-red at about 730 nm. The response in the long-duration experiment is the same as in the short-duration treatment—inhibition of internode elongation. It was proposed by Mohr and his colleagues in 1966 that the blue light in this HIR produces P_{fr} that brings about the photomorphogenic response of internode elongation. Just why there should be an efficiency peak in the far-red is not entirely clear since far-red light should bring about phototransformation of the P_{fr} to P_r, preventing the response. The plants should act as if they were in the dark with maximum etiolation.

It was assumed by Mohr on the basis of kinetic experiments that long exposure to far-red irradiation will keep sufficient phytochrome in the P_{fr} form to cause the photomorphogenesis. The action spectrum shown in Fig. 20.7 is typical of the high-energy responses of photomorphogenesis.

A frequently studied photomorphogenic response is the inhibition of elongation of mustard seedling hypocotyls. The inhibition shows the typical HIR. As shown in Fig. 20.8, hypocotyls will elongate linearly up to at least 18 hours if grown in the dark.

If they are given continuous far-red light, there is inhibition of the elongation. The inhibition of lettuce hypocotyl elongation has also been studied. The action spectrum has an efficiency peak in the blue and in the far-red, typical of the HIR.

Thus the elongation of hypocotyls and seedling internodes is a good example of photomorphogenesis controlled by phytochrome. The negative photomorphogenic responses take place with little or no lag. The positive photomorphogenic responses, such as expansion of cotyledons and biosynthesis of anthocyanins, have a lag period of several hours (Fig. 20.9).

In 1966 Mohr studied the phytochrome-related biosynthesis of anthocyanins in an attempt to decipher the mechanism of phytochrome action. Mustard seedlings grown in the dark do not produce any significant levels of anthocyanins. However, if they are treated with continuous far-red light, anthocyanin biosynthesis starts with a lag of several hours (Fig. 20.9). The biosynthesis shows the typical induction–reversion phenomena when in association with short durations of red and far-red light. Furthermore, the amount of anthocyanin produced is dependent on light intensity typical of HIR. Thus anthocyanin biosynthesis is typical of HIR.

In studies to differentiate between membrane effects and effects on the genome it was shown that

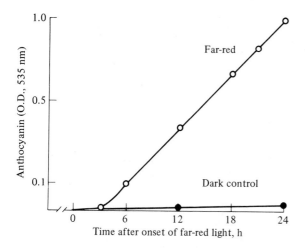

FIGURE 20.9 Far-red induction of anthocyanin biosynthesis in mustard. The response occurs with a marked lag period. No anthocyanin forms in the dark. From H. Mohr. 1966. Differential gene activation as a mode of action of phytochrome 730. *Photochem. Photobiol.* 5:469–483. Copyright 1966 Pergamon Press, Ltd.

the protein-synthesis inhibitor actinomycin D would prevent anthocyanin biosynthesis. It was suggested that the HIR first brings about a photochemical reaction that is followed by a direct effect on the protein-synthesizing mechanism.

In support of such a hypothesis, it is also known that the enzyme phenylalanine ammonia lyase is induced in a HIR in a manner similar to anthocyanin biosynthesis. Quite interesting is the finding that PAL catalyzes the synthesis of *trans*-cinnamic acid from phenylalanine. *Trans*-cinnamic acid is a precursor of anthocyanin.

20.5 Biological Rhythms

20.5.1 Basic principles of endogenous rhythms

It is interesting that one of the first plant physiological functions to be noted and studied was the sleep movements of leaves, a phenomenon known to be a rhythm (Fig. 20.10). As early as 1727, Mairan observed sleep movements in leaves and recorded that they would still occur even if the plant was kept in continuous darkness. However, it was Carl Linnaeus who in 1751 named the nightly folding of leaves "sleep movements." Sleep movements were studied largely in legumes by Darwin, De Candolle, Sachs, and Pfeffer and were well known by the 20th century. Figure 20.11 shows a drawing of an original experiment by Pfeffer conducted in 1875 that illustrates the sleep movements of a young *Acacia* leaf.

Researchers of the 19th century determined that periodic phenomena such as leaf movements occurred in cycles of about 24 hours and that the cycle would continue for some time, even in continuous light or darkness. Furthermore, it was known that the cycle length could be altered by changing the length of the photoperiod, but once the plants were returned to either continuous light or dark, the 24-hour cycle was reestablished. It was apparent to these investigators that the rhythms were governed by endogenous factors.

Interest was renewed in 1930 when Bünning began to study rhythms associated with leaf movements from the point of view of time measurement. During this period the concept of plant photoperiodism was being developed. Today, endogenous plant rhythms have been tightly linked to photoperiod and to phytochrome.

FIGURE 20.10 Photograph of pinto bean plants showing sleep movement of leaves. The plant at the left was taken from a dark chamber several hours after the beginning of the normal dark period. Both the true leaf pair and the cotyledons show the movement. The plant at the right was taken from a light chamber several hours after the beginning of the normal light period.

A few terms and concepts are necessary in order to fully understand the meaning of biological rhythms. First, the concept of a biological rhythm implies endogenously controlled, cyclic events that continue independent of the environment. It does not include repeating events that occur because of daily environmental cycles of light or temperature. The daily variation in leaf temperature is very cyclic but only occurs under normal daily cycles. There will be no daily cycle of leaf temperature under constant thermal conditions. True biological rhythms continue under constant environmental conditions, although they will damp out with time. Rhythms are largely temperature-independent although the extent of a response, however, may be sensitive to temperature.

Figure 20.12 illustrates the basic features of the cyclical biological rhythm. It has no beginning or end, but of course there may be points that are identifiable, such as a leaf completely folded or

FIGURE 20.11 A reproduction of Pfeffer's original experiment in 1915 illustrating the endogenous rhythmic nature of the sleep movements of *Acacia* leaves. After entrainment, the plants were kept in continuous darkness, and the rhythm persisted for several days as it slowly damped out.

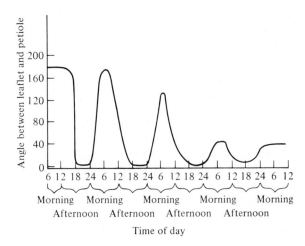

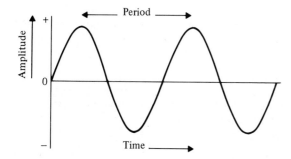

FIGURE 20.12 Diagram to illustrate the features of a rhythm, showing its period and amplitude. Note that the amplitude is the magnitude of the response, measured as the maximum departure from the mean.

completely open. Each point of the cycle repeats at constant intervals. It is in fact a true oscillation. The time between exact repeating points of a single cycle is called its period. For most biological rhythms the period is about 24 hours. These are daily rhythms and will continue independent of environmental influences, although biological rhythmic phenomena change period slightly from 24 hours when placed under constant temperatures. Apparently, normal light and (perhaps) temperature cycles are necessary to renew the cycle.

Biological rhythms such as the sleep movements of leaves will ordinarily have periods of 24 hours, usually measured from noon to noon. Such rhythms are called daily, or circadian (*circa dies* = about a day). Periods, of course, may be of any length. Tidal rhythms are about 12.8 hours, lunar rhythms are 28 days, and annual rhythms are yearly.

The rhythm can be set with light provided that the system is actually oscillating. This setting is called entrainment. The light bringing about the entrainment is called zeitgeber (time-giver).

Two other terms used to describe the rhythm or oscillation are its amplitude and phase. The amplitude is the intensity of the oscillation and is measured as the magnitude of departure from the mean, or midpoint, value. The phase refers to the position of any one point of the cycle in relation to the other points of the cycle. The phase can be expressed in degrees when the oscillation is viewed as a cycle. Thus a change from the highest point to the lowest point would be a phase change of 180°.

True biological rhythms, which are regulated by endogenous factors, can be interpreted as sine waves. The time between any two points of the sine wave having the same phase is called the period, and the intensity or height of the wave is the amplitude. Any stage or point of the wave can be defined as being at a particular phase. In actual practice, biological rhythms may not fit a sine wave exactly. Nevertheless, these same terms are used to describe the rhythm.

20.5.2 Sleep movements

Most rhythms that have been studied have a daily period and thus are circadian rhythms. Common circadian rhythms of plants include the well-known sleep movements of leaves, the daily opening and closing of flowers, and some lesser-known metabolic phenomena such as rhythmic CO_2 exchange and rhythmic variation of enzymatic activity. The best known of the latter is the acid metabolism of succulent plants. Although annual rhythms apparently occur in plants (for example, flowering, seed germination, leaf abscission, and spring growth), it is rather difficult to prove that they are actually endogenous rhythms. Experiments to prove annual rhythms would have to be long-term, with precise environmental controls. Maintenance of plants under constant environmental conditions for periods of years would probably so upset growth processes that it would be difficult to follow endogenous rhythms.

Perhaps the most extensively studied plant

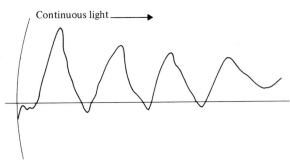

Continuous light ⟶

FIGURE 20.13 The experiment with *Xanthium* leaves showing endogenous rhythm of sleep movements in continuous light. Plants were pretreated with light–dark cycles and then placed on a continuous light regime. The dark curved line marks the beginning of continuous light and sets the phase of the rhythm. The oscillation slowly damps out. The period is about 24 hours. The curves were obtained with a kymograph attached to the leaves. Downward leaf movements are illustrated as an upward slope of the graph. (From T. Hoshizaki, D. E. Brest, and K. C. Hamner. 1969. *Xanthium* leaf movements in light and dark. *Plant Physiol.* 44:151–152.)

rhythm is the sleep movement of leaves that occurs in legumes (see Fig. 20.10). It was shown that bean leaves grown in continuous light do not show a leaf-folding rhythm. If given a single dark exposure during a day period, the rhythm will begin and continue for a week or more even if the plants are kept in total darkness or in complete light. If the plants are kept under constant light conditions, the period will be somewhat greater than 24 hours.

An experiment by Hoshizaki and his colleagues in 1969 illustrates the complexity of rhythmic sleep movements (Fig. 20.13). *Xanthium* leaves were treated with a variety of photoperiods and then placed on continuous light. Regardless of the pretreatment the phase was set by the beginning of the continuous light. The periods in continuous light were about 24 hours, and the amplitude damped out after five or more days.

It is phytochrome that is the factor responsible for perceiving the light signal causing entrainment. Evidence for phytochrome involvement is that red

light will cause entrainment, and in some species far-red light will nullify the red-light effect. Thus P_{fr} is the active form for rhythm entrainment.

20.5.3 Metabolic rhythms

There is much evidence for circadian rhythms of metabolism in plants. Studies with green algae and *Euglena* have demonstrated that the chloroplast activity and associated photosynthetic activity display properties of a true biological rhythm. There are cyclical, periodic changes in chloroplast structure and carbohydrate contents indicative of rhythms. In green algae it was shown that maximum photosynthetic activity occurred near the middle of the light period. Evidence that the periodic peak of activity is a true rhythm is that it can be reset by changes in the light–dark schedule and that it persists in continuous light for several days even if kept at constant temperature.

Net photosynthesis of peanut leaves shows an endogenous rhythm. Pallas, Samish, and Willmer in 1974 pretreated plants either with periods of 14 hours light/10 hours dark or with periods of 4 hours light/4 hours dark and then placed them in continuous light (Fig. 20.14). The rhythm of photosynthesis that followed had a period of 26 hours and slowly damped out after several days. Respiration and transpiration show a rhythm similar to photosynthesis. The rhythm of photosynthesis and transpiration under constant temperature may be the result of the rhythmic activity of stomatal opening, which parallels the gas exchange (refer to Chapter 5).

There is some evidence that the synthesis of NADP glyceraldehyde phosphate dehydrogenase and perhaps even ribulose bisphosphate carboxylase are regulated by phytochrome in higher plants.

In addition to the above it has been shown that ATP levels are rhythmic, being highest near the beginning of the light period, and that RNA synthesis and polysaccharide content reach maxima somewhat after the middle of the light period. Some of the rhythmic phenomena are synchronized and

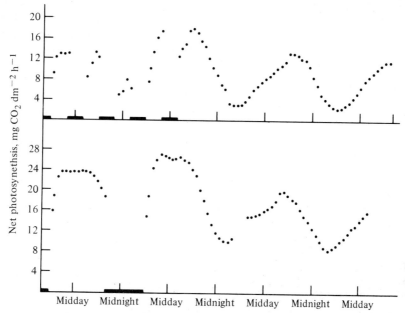

FIGURE 20.14 Biological rhythm of photosynthesis in peanut leaves. The leaves were pretreated with either 4 hour light/4 hours dark (upper curve) or 14 hours light/10 hours dark (lower curve) and then placed on continuous light. Photosynthesis showed a definite endogenous rhythm, with periods of about 26 hours. The maximum is at about midday and the minimum is near midnight. (From J. E. Pallas, Y. B. Samish, and C. M. Willmer. 1974. Endogenous rhythmic activity of photosynthesis, transpiration, dark respiration, and carbon dioxide compensation point of peanut leaves. *Plant Physiol.* 53:907–911.)

others are out of phase. The exact meaning of such rhythms is not known, but there are implications for metabolic control.

The output of CO_2 into CO_2-free air by excised *Bryophyllum* leaves is another example of a biological rhythm. Experiments by Wilkins in 1962 have shown that leaves can be entrained by periods ranging from 24 hours light/24 hours dark to 3 hours light/3 hours dark. If the periods are greater than 24 hours the entrainment is relatively independent of light intensity, but if they are less than 24 hours, entrainment only occurs if light intensity is high (Fig. 20.15). After entrainment, if the plants are returned to continuous dark the period returns to circadian (i.e., about 24 hours) and begins to damp out. During the damping process, the period remains the same but the amplitude decreases.

The upper curve of Fig. 20.15 shows the CO_2 output into CO_2-free air by *Bryophyllum* leaves kept at 26°. Entrainment was with 6 hours light/6 hours dark at 500 lux light intensity. When the leaves were returned to continuous dark, the period of the rhythm returned to 24 hours and slowly damped out after a few periods. The middle panel shows the rhythm when plants were treated with photoperiods of 3 hours light/3 hours dark at 500 lux. Entrainment does not occur. Nor does it occur if the photoperiods are 6 hours light/6 hours dark at low light intensities, as shown in the lower panel. In each of the three experiments shown in the figure, the beginning of the continuous dark period evidently sets the phase.

Following the work of Wilkens, Queiroz (see Queiroz and Morel, 1974) studied the rhythmic,

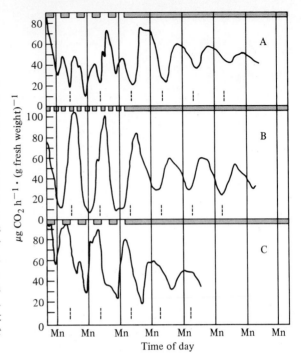

FIGURE 20.15 Experiments illustrating the biological rhythm of CO_2 output by the CAM plant *Bryophyllum*. Excised leaves were placed in CO_2-free air, and the output of CO_2 was measured. Upper curve: entrainment of rhythm when leaves were treated with photoperiods of 6 hours light/6 hours dark at 500 lux and $26°$. The rhythm was entrained. When returned to continuous dark the period returned to circadian. Although the period remained constant in the dark, the rhythm damped, as can be seen from the decrease in amplitude. The middle panel shows 3 hours light/3 hours dark at 500 lux and the lower panel shows 6 hours light/6 hours dark at 100 lux. In neither case did entrainment occur. Mn = midnight. (From M. B. Wilkins. 1962. An endogenous rhythm in the rate of carbon dioxide output of *Bryophyllum*. IV. Effect of intensity of illumination on entrainment of the rhythm by cycles of light and darkness. *Plant Physiol*. 37:735–741.)

diurnal cycle of crassulacean acid metabolism. The rhythm, as explained in Chapter 16, is characterized by a diurnal fluctuation of malic acid. In addition, the two important enzymes of CAM, P-enolpyruvate carboxylase and malate enzyme, also fluctuate diurnally. The graph in Fig. 20.16 shows that malic acid accumulates in the dark and disappears in the light. The activity of the carboxy-lating enzyme, P-enolpyruvate carboxylase, begins to increase at the end of the light period and reaches a maximum at the beginning of the dark period. Malate enzyme activity is low at the beginning of the dark period and reaches a maximum at the end of the dark period. The activity decreases during the light period.

There is a definite circadian rhythm of CO_2 fixa-

FIGURE 20.16 Endogenous rhythms of the activities of the carboxylating enzyme P-enolpyruvate carboxylase, the decarboxylating enzyme malate enzyme, and the concentration of malic acid in the CAM plant *Kalanchoe*. The two enzymes are out of phase and evidently account for the rhythmic change in concentration of malic acid in the tissue. It is postulated that the enzyme rhythm is controlled by an endogenous oscillator. PEPC = P-enolpyruvate carboxylase. ME = malate enzyme. MA = malic acid. Redrawn from O. Queiroz and C. Morel. 1974. Photoperiodism and enzyme activity. *Plant Physiol*. 53: 596–602.

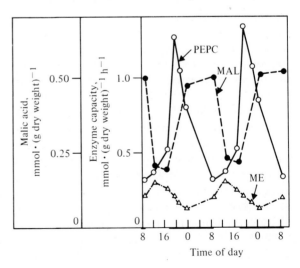

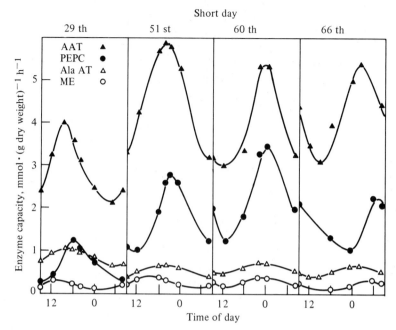

FIGURE 20.17 The induction of enzyme activity by short photoperiods in the CAM plant *Kalanchoe*. Aspartate aminotransferase (AAT) and P-enolpyruvate carboxylase (PEPC) increase in their activity with an increasing number of short days. Note also the very distinct rhythmic activity that changes phase with the increasing number of days. Ala AT = alanine amino transferase. ME = malate enzyme. From O. Queiroz and C. Morel. 1974. Photoperiodism and enzyme activity. *Plant Physiol.* 53:596–602.

tion that is the result of the circadian rhythm of P-enolpyruvate carboxylase activity. Malic acid is the primary product of the CO_2 fixation and hence tends to follow CO_2 fixation and P-enolpyruvate carboxylase activity. Moreover, there is a diurnal rhythm of CO_2 output resulting from the activity of malate enzyme, which decarboxylates malate. The decarboxylation accounts for the decrease of malic acid and partially for the circadian oscillation of acid.

Experiments have implicated phytochrome as the control for the circadian rhythm of P-enolpyruvate carboxylase activity. Evidence for phytochrome being involved in the CAM pathway is that CAM is induced by short days (Fig. 20.17). If plants from long days are placed on a short-day regime, CAM increases after about seven days and continues to increase up to 50 days. The CAM activity is manifested by the circadian oscillation of enzymatic activity and malic acid levels.

The mechanism by which the circadian rhythm of CAM is controlled is unknown. A possibility is that oscillations begin by malate inhibition of

P-enolpyruvate carboxylase, an assumed allosteric reaction (see Chapters 9 and 16). But the rhythm seems far too complex for such a simple explanation, and it is assumed that the primary control is a yet-undefined basic oscillator, the biological clock. The nature of such an oscillator is completely unknown.

20.5.4 The endogenous timing mechanism

The mechanism by which plants measure time has intrigued physiologists for several decades. The search probably began in earnest with the research of Bünning during the 1950s (see Bünning, 1973) in Germany. At present there are two different hypotheses for time measurement. One is an hourglass-type timing mechanism in which some endogenous event proceeds through a sequence to completion. Such a continuous sequence is believed to measure the duration of light or dark. It appears that the duration of the dark period is most important.

Perhaps we can best visualize the hourglass-type

hypothesis by assuming that the dark reversion of P_{fr} is the hourglass that measures the duration of light. This hypothesis, although quite attractive, is largely discounted because the dark reversion of P_{fr} to P_r does not seem to be a universal event. It is known only in some dicotyledonous plants.

The alternative hypothesis is a clock-type theory in which the external stimulus, e.g., light or dark, is measured against an internal oscillator, or clock. The internal oscillator has a period of 24 hours. It was proposed by Bünning (see Bünning, 1973) that there are two phases within the plant, a light phase and a dark phase. Each of the phases is differentially sensitive to light, which would account for the fact that red light could inhibit or stimulate a response. P_{fr} acts differently depending on the phase. Thus the internal oscillator shifts from light-sensitive to dark-sensitive. The internal oscillator could be visualized as a series of cyclic metabolic events. It is important that the internal oscillator be endogenously regulated.

It is not clear how phytochrome is related to the internal oscillator. Since there is a general opinion that all timing devices are the same, and because phytochrome is not present in all tissues that show rhythmic behavior, phytochrome itself cannot be the internal oscillator.

It should be apparent that light has two roles in the theory. First, light acts as the zeitgeber that entrains a particular response. Second, light acts as the external stimulus that induces or inhibits responses.

Although we now know much about phytochrome, timing mechanisms in plants, and rhythms, our basic knowledge about the nature of the biological clock (the endogenous oscillator) is very scanty. Physiologists, using models based on analog computer simulations, have supposed that there is a basic oscillator that could be an enzymatic reaction or an entire pathway regulated by feedback control. This is somewhat consistent with the discussion above concerning the regulation of CAM. The basic oscillating system is regulated internally by feedback control and externally by environmental parameters. Light is the most important, being the best-known factor for entrainment of biological rhythms. However, much more research will be necessary before the exact timing mechanism of plants is understood.

Review Exercises

20.1 In terms of the mechanism of the response, differentiate between phototropisms and photomorphogenic growth responses.

20.2 Do you think there is any adaptive significance in the fact that phytochrome-dependent plant growth phenomena are red/far-red-dependent reactions rather than, for example, blue- or green-dependent?

20.3 Explain how the high-intensity radiation (HIR) growth responses can be sensitive to blue light and yet be linked to phytochrome.

20.4 In some plants the enzyme aspartate aminotransferase shows rhythmic activity, being present at high levels in the dark and low levels in the light. How could you demonstrate whether this diurnal fluctuation in enzymatic activity is an endogenously controlled rhythm or is the result of diurnal environmental changes? What are some possible environmental diurnal changes that could influence enzymatic activity? How would they cause the activity of the enzyme to fluctuate? If neither environment nor endogenous rhythms are functioning, what other possibilities are there?

20.5 In a succulent plant, the diurnal variation in the activity of P-enolpyruvate carboxylase was 1.2 nmol g^{-1} (dry weight) h^{-1}. It was shown that this fluctuation in activity was the result of a biological rhythm. What is the maximum intensity or amplitude of the rhythm?

References

BORTHWICK, H. A., S. B. HENDRICKS, M. W. PARKER, E. H. TOOLE, and V. K. TOOLE. 1952. A reversible photoreaction controlling seed germination. *Proc. Natl. Acad. Sci.* (U.S.) 38:662–666.

BRIGGS, W. R., and H. V. RICE. 1972. Phytochrome: Chemical and physical properties and mechanism of action. *Ann. Rev. Plant Physiol.* 23:293–334.

BROWN, F. A. 1972. The "clocks" timing biological rhythms. *Amer. Sci.* 60:756–766.

BÜNNING, E. 1973. *The Physiological Clock.* 3d ed. Academic Press, New York.

CUMMING, B. G., and E. WAGNER. 1968. Rhythmic processes in plants. *Ann. Rev. Plant Physiol.* 19:381–416.

EVANS, L. T. 1975. *Daylength and the Flowering of Plants.* W. A. Benjamin, Menlo Park, Calif.

HAMNER, K. C., and T. HOSHIZAKI. 1974. Photoperiodism and circadian rhythms: An hypothesis. *BioScience* 24:407–414.

HILLMAN, W. S. 1976. Biological rhythms and physiological timing. *Ann. Rev. Plant Physiol.* 27:159–179.

KENDRICK, R. E., and B. FRANKLAND. 1976. *Phytochrome and Plant Growth.* Edward Arnold, London.

MITRAKOS, K., and W. SHOPSHIRE (eds.). 1972. *Phytochrome.* Academic Press, New York.

MOHR, H. 1972. *Lectures on Photomorphogenesis.* Springer-Verlag, Berlin.

SMITH, H. 1975. *Phytochrome and Photomorphogenesis.* McGraw-Hill, New York.

SMITH, H. 1976. *Light and Plant Development.* Butterworth Pub., Woburn, Mass.

SWEENEY, B. M. 1969. *Rhythmic Phenomena in Plants.* Academic Press, London.

VINCE-PRUE, D. 1975. *Photoperiodism in Plants.* McGraw-Hill, London.

WAREING, P. F., and I. D. J. PHILLIPS. 1978. *The Control of Plant Growth and Differentiation in Plants.* Pergamon Press, Oxford.

CHAPTER TWENTY-ONE

Reproductive Growth

21.1 Life Cycle of Flowering Plant

For the purposes of discussion the life cycle of a flowering plant can be considered to start with germination. The seed, which represents a dormant stage, germinates and grows into the mature plant, or sporophyte generation. The sporophyte generation has the diploid number of chromosomes (2N), whereas the gametophyte or sexual generation is haploid (1N). Sporophytes or tissues of the sporophyte may, of course, have multiples of the haploid set of chromosomes, such as triploid (3N), tetraploid (4N), and even larger multiples.

To complete its life cycle, the sporophyte will eventually shift from purely vegetative to reproductive growth. Vegetative apices will change, frequently in response to environmental stimuli, and will form reproductive apices. First, the floral organs are formed. Second, mother cells will form 1N reproductive spores through reductional (meiotic) division. The female spore is called the micromegaspore and the male spore is called the microspore. Subsequent mitotic divisions produce the megagametophyte, or embryo sac, with the egg cell and the microgametophyte, or pollen grain, which will produce sperm. After pollination, union of sperm and egg cell (fertilization) forms the zygote. The zygote, which is 2N, will develop into the

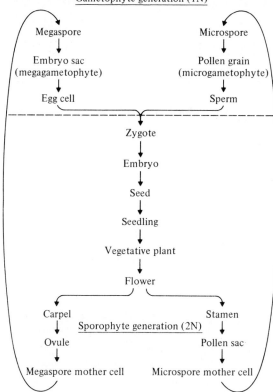

Gametophyte generation (1N)

Megaspore → Microspore

Embryo sac (megagametophyte) → Pollen grain (microgametophyte)

Egg cell → Sperm

Zygote

Embryo

Seed

Seedling

Vegetative plant

Flower

Carpel → Stamen

Sporophyte generation (2N)

Ovule → Pollen sac

Megaspore mother cell → Microspore mother cell

FIGURE 21.1 Generalized life cycle of a flowering plant

embryo and ultimately into the mature sporophyte. A generalized life cycle of a flowering plant is illustrated in Fig. 21.1.

21.1.1 Floral organs

Flowers, the sexual organs of the plant, are the diagnostic feature of the angiosperms. Gymnosperms have reproductive organs but not flowers.

The basic parts of the flower are illustrated in Fig. 21.2. The complete flower is composed of four definite whorls of parts that are homologous to leaves and arise from the basal receptacle at the apex of the floral stem, the peduncle. The outermost whorl is the calyx, composed of leaflike sepals that function primarily to protect the young, developing bud. The sepals are frequently green but may have other pigments besides chlorophyll. A second similar whorl, situated just to the inside of the calyx, is the corolla, made of individual petals. Petals are usually colored, making up the showy part of the flower. The corolla functions in the attraction of pollinators and may aid in pollination through its shape or position. The calyx and corolla together are called the perianth.

The inner two whorls are the male (outer) and female (inner) reproductive organs. The outer male reproductive whorl is called the androecium and is made of individual stamens. Each stamen is composed of a filament (stalk) and the saclike anthers containing the male gametophytes, or pollen. The male gametophytes produce the sperm by mitotic divisions from their generative nuclei.

The innermost whorl, the female reproductive part of the flower, is called the gynoecium and is composed of individual carpels. Carpels are arranged into a lower, usually swollen, ovary enclosing the ovules (seeds-to-be) and a terminal receptive portion, the stigma, which receives the pollen during pollination. The stalked portion of the carpel between the ovary and stigma is the style. Carpels may be fused to form a single pistil or they may be separate.

A cross section through an ovary shows the arrangement of the enclosed ovules. Chambers within the ovary are called locules. A carpel will have one locule, but the number of ovules varies depending on the species. In the case of fused carpels, the number of locules is an indication of the number of carpels. Ovules are attached to the carpel wall on tissue called placenta. The actual connection between placenta and ovule is the funiculus.

Perianth and androecium may either be attached below the ovary in the more usual superior ovary situation or the attachment may be on the ovary, a condition termed inferior. The terms inferior and superior refer to the ovary position relative to the other flower parts.

Flowers that have both essential parts, i.e., androecium and gynoecium, are called perfect flowers. If either male or female whorl is missing the flower is said to be imperfect.

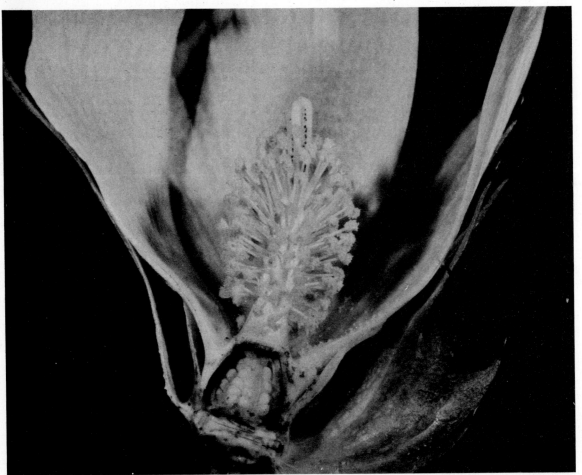

FIGURE 21.2 Typical angiosperm flower and associated parts. The photograph is of a cotton flower on the day of anthesis. From C. A. Beasley. 1975. Developmental morphology of cotton flowers and seed as seen with the scanning electron microscope. *Amer. J. Bot.* 62:584–592.

Plants with imperfect flowers are monoecious if both the pistillate and staminate flowers are on the same plant and are dioecious if they are on separate plants. Perfect flowers are bisexual.

A cross section of an anther is shown in Fig. 21.11. Within the pollen sacs of the anther are the young pollen grains. Each is made of a tetrad of microspores, which arose from meiotic divisions of a microspore mother cell. Pollination occurs when the pollen grain lands on a stigma of the pistil. Germination of the pollen produces a pollen tube, which grows through the style to the embryo sac within the ovule. The pollen has two nuclei: a tube nucleus, which functions as the nucleus of the pollen, and a generative nucleus, which upon mitotic division produces two sperm nuclei.

Division of the megaspore mother cell by meiosis within the ovary also produces four haploid cells (megaspores), but three abort, leaving a single cell. The remaining cell divides mitotically three times to form an embryo sac with eight nuclei. Further development of the embryo sac and surrounding tissues produces the ovule.

The tissue immediately surrounding the egg sac is called the nucellus and, unlike the haploid (1N) egg sac, is diploid (2N). Surrounding the nucellus, the remainder of the 2N tissue will develop into ovule tissue, and ultimately (after fertilization) into the seed. The fruit develops from the remainder of the ovary tissue plus other floral parts.

Of the eight nuclei within the egg sac, the one closest to the micropylar end (opening) is the egg nucleus proper and is situated between two synergid nuclei. At the opposite end of the egg sac are three nuclei called antipodals. The two remaining nuclei near the center of the egg sac are called polar nuclei.

The pollen tube grows to the ovule and enters through the micropyle, where it discharges the two sperm nuclei into the embryo sac. One sperm nucleus fuses with the egg nucleus, the process of fertilization, and the other fuses with the two polar nuclei. The 2N zygote formed through fertilization will ultimately grow into the embryo, and the 3N nucleus develops into endosperm tissue, which is the storage tissue of the young seed. The double-fertilization process in plants thus produces a 2N zygote and a 3N nucleus that forms storage tissue. Therefore, seeds are composed of 1N, 2N, and 3N tissue.

The seed is the mature ovule and has three basic parts, the seed coat or testa, the endosperm or storage tissue, and the embryo or young seedling. The testa functions to protect the enclosed embryo. It may be completely impervious to gases and water, thus preventing germination and ensuring longevity. The endosperm of monocots such as maize contains the storage carbohydrate or lipid that supplies the growing embryo during development before autotrophic life begins. In seeds of dicots such as bean, stored material is in the cotyledons, and the cotyledon supplies nutrition directly to the developing embryo.

The embryo is a vertical axis. That portion above the point of attachment of the cotyledons is called the epicotyl and that portion below the attachment of the cotyledons is the hypocotyl. The terminal portion of the epicotyl is frequently called the plumule and the apex of the hypocotyl is the radicle. The plumule, so called because of the featherlike appearance of immature leaves, develops into the shoot axis and the radicle of the root axis. In monocots the fleshy part of the cotyledon is called the scutellum. In addition, in monocots the epicotyl is protected by a leaflike structure called the coleoptile and the hypocotyl tip is covered with a sheath called the coleorhiza.

After germination, cotyledons may remain below ground, such as in most monocots and in peas, or they may be aboveground, as in the common bean, depending on whether the epicotyl or the hypocotyl elongates. Cotyledons may be fleshy and transfer all their storage products to the developing seedling, or, in some aboveground instances, they may become green and autotrophic and function in photosynthesis prior to the appearance of true leaves.

21.1.2 Meiosis

Plants as a rule have a high capability for asexual reproduction. Except perhaps in annuals, vegetative propagation is an extremely effective means of increasing numbers of individuals and is the most frequent form of reproduction in many habitats. Regardless of such effectiveness, virtually all living organisms reproduce sexually at some stage, ensuring sufficient variability within the taxon for continued survival in the changing environment of the earth. Meiosis functions in the production of sex cells (sperm and eggs) during sexual reproduction.

Prior to sexual reproduction, the vegetative shoot apex differentiates into a floral apex, a flower primordium. During floral production, microspore mother cells within the anther and megaspore mother cells within the ovary divide by meiosis to produce four haploid sex cells each.

Meiosis begins with diploid mother cells, and the process continues until four haploid sex cells are formed (Fig. 21.3). Three of the four haploid cells from the megaspore mother cell abort, leaving one

egg cell. In the male the microspore mother cell divides through meiosis to form a tetrad of haploid cells, which differentiate into pollen.

PROPHASE I

During the early portion of prophase I, meiosis is quite similar to mitosis. The chromosomes become threadlike and distinctly visible with the aid of the light microscope. The chromosomes are not visibly double-stranded, but experiments with radioisotopes indicate that they have replicated. By late prophase I the chromosomes are distinctly double-stranded, paired, and seemingly intertwined. This intertwining is a process known as crossing over and results in exchange of genetic material between chromosome pairs. An actual pairing is called a synapsis. The crossing-over phenomenon at a chiasma (actual point of contact between homolo-

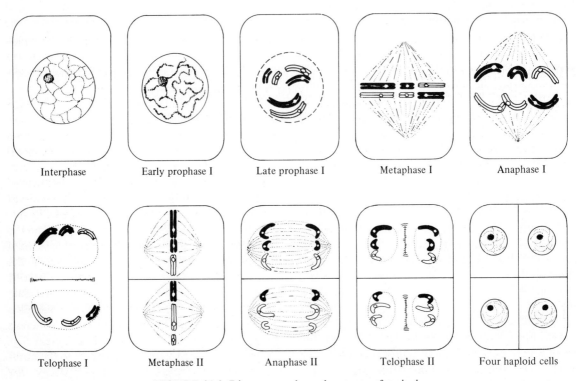

Interphase Early prophase I Late prophase I Metaphase I Anaphase I

Telophase I Metaphase II Anaphase II Telophase II Four haploid cells

FIGURE 21.3 Diagram to show the stages of meiosis.

gous chromosomes) and consequent exchange of genetic material produces variability among offspring by preventing genes from being inherited in blocks. Of course, the closer genes are on chromosomes, the greater the probability that they will be inherited together. In addition, during late prophase I the nuclear envelope disappears and the spindle begins to organize.

METAPHASE I

During metaphase I of meiosis the pairs of chromosomes orient together along the cell equator. This process is unlike mitosis, in which the chromosome pairs are more independent.

ANAPHASE I

During anaphase I, the members of each pair will migrate to opposite poles. Each daughter cell will eventually have an entire set of chromosomes (one member from each pair) and will be haploid, unlike the spore mother cell, which had the 2N diploid number. A most important feature is that the direction of migration of any one member of a pair is entirely independent of any other pair member; the migration process is completely randomized. Such randomized migration is the second way in which variability is produced, the first being the crossing-over process discussed under prophase I. Other variability, of course, arises through mutation.

TELOPHASE I

During telophase I the process appears similar to the mitotic telophase in that the double-stranded chromosomes lengthen by uncoiling and become indistinct under the light microscope.

INTERPHASE II

The interphase between the two division sequences of meiosis is similar to the interphase between mitotic divisions except that it is short and no replication of genetic material occurs, since replication occurred prior to prophase I. The chromosomes are already double-stranded.

PROPHASE II

The second divisional sequence of meiosis is similar to the sequence in mitosis. In prophase II the double-stranded chromosomes become visible, the nuclear envelope disappears, and the spindle forms.

METAPHASE II

During metaphase II the single set of double-stranded chromosomes align along the equator of the cell in preparation for migration.

ANAPHASE II

In anaphase II as in mitotic anaphase, the double-stranded chromosomes divide, and each daughter chromosome migrates to a pole.

TELOPHASE II

During telophase II the single-stranded chromosomes lose their thick appearance and slowly lengthen to the point of microscopic invisibility while the nuclear envelopes re-form.

The meiotic process produces four cells, each having half the number of chromosomes of the spore mother cells. It is not an entirely random set, however, since the set is composed of one member of each of the different kinds of chromosomes in the cell. It is important to recall that because of the crossing-over phenomenon and random migration to opposite poles there are two genetically different kinds of cells. The second division of the meiotic sequence formed a second set of cells identical to the first set. As explained above, in the stamen a tetrad is produced such that the four cells remain viable and functional, whereas in the female three will abort, leaving a single egg cell. The pollen cells produce sperm by mitotic divisions, whereas the egg cell is the functional female gamete.

Species are quite specific with respect to the number of chromosomes, although polyploidy, the condition of multiple sets of chromosomes, is common in plants. As examples, onion has 16 chromosomes, pea has 14, and maize has 20.* Certain hybrids, either natural or otherwise, may have additional pairs or even entire extra sets. The point to consider here is that chromosomes exist in pairs. Thus in maize the entire complement of chromosomes is 20, made up of 10 pairs. Genes, the genetic entities having the genetic information for traits, are located on the chromosomes; thus each pair will have two genes for each trait, one on each member of the pair. As is well known, the alternative genes may be identical, coding for the exact same trait, or they may be different (alleles), coding for variations on a trait. During reductional division (meiosis), the newly formed daughter cells (sex cells) will each have one member of each pair and will thus have an entire complement of alternative genes. It becomes clear that the primary function of meiosis is to ensure variability among sex cells and hence among offspring. Reductional division followed by reproduction by means of fertilization results in the mix-up and recombination of genetic information.

An important feature for physiological considerations is that plant material obtained from seed is apt to be extremely variable. Such variation must be taken into account during experimentation. Moreover, all plants and plant cells have an entire complement of genetic information, i.e., the entire genome. Thus developmental considerations are concerned with the regulation of genetic activity. Reproductive growth in particular represents a dramatic alteration in growth form that must involve many complex genetic controls.

21.2 Flowering

The phenomenon of flowering in plants marks the transition from vegetative growth to reproductive growth. The floral organs are the sex organs of the plant and function exclusively in sexual reproduction. Sexual reproduction, of course, ensures variability within a species through the exchange of genetic material between compatible plants.

After vegetative growth (reaching "ripeness to flower"), plants can be induced to flower. The term induction has a rather specific meaning in this context and refers to those changes taking place in leaves that ultimately lead to flowering. It is known that the leaves receive the environmental stimulus for the proper timing of flowering. Changes occur in the receptive leaf, presumably leading to the production of a hormone or hormones loosely identified as "florigen"† (see Chapter 18), which is transmitted to the vegetative apex. The transition of the vegetative apex to a reproductive apex, i.e., floral induction, has been referred to as evocation by Evans (1971). The process of flowering can be visualized as follows.

Induction (leaves) ⟶ Hormone transmission ⟶
Evocation (apex)

According to Evans the three most important questions related to the flowering phenomenon are the following.

1. In the context of photoperiodic induction, how is time measured by plants in the dark and in the light? (See Chapter 20.)
2. What is the nature of the hormone (growth stimulus) involved in the flowering process?
3. How do the hormones initiate the transformation from vegetative growth to reproductive growth at the apex?

These features of the flowering process are discussed below.

---

*See C. D. Darlington and A. P. Wylie. 1956. *Chromosome Atlas of Flowering Plants.* Macmillan, New York. See the journal *Taxon* for recent determinations.

†The use of the term "florigen" should be considered as conceptual only and not necessarily a specific plant growth substance. Florigen could be a specific hormone or it could be a gibberellin or even a balance of hormones.

21.2.1 Floral bud formation

In most flowering plants, vegetative growth is indeterminate, continuing more or less indefinitely. The transition from a vegetative apex to a reproductive apex is accompanied by a shift from indeterminate to limited, determinate-type growth. In most plants, determinate growth occurs just at the particular apices that form reproductive buds; however, some plants become determinate and die after flowering. These plants are called monocarpic.

Classically, the floral organs of plants are considered to be homologous to vegetative shoots in that the sepals, petals, stamens, and carpels of flowers appear to be modified leaves. Thus the transition from vegetative growth to reproductive growth occurs by a change in the nature of the vegetative apex. Alternatively, the flower is interpreted as a unique plant organ and not derived directly from vegetative organs.

The transition from vegetative to reproductive growth occurs in response to environmental signals; it is the induction phase of flowering. Provided that the plant has matured to the extent that it is "ripe" to flower and depending on the plant species, either day length, temperature, or water regime can act as an environmental stimulus, signaling the shift from vegetative to reproductive growth.

The induction phase is followed by hormonal changes. This stage must be accompanied by transmission of a substance or substances (presumably a hormone) from the leaves to the apices.

There is some experimental evidence that sex expression in flowers is related to hormone balance. In monoecious cucumbers, during ordinary development the male flowers are produced before the female flowers. If the plant is treated with auxin, the female flowers appear sooner. Furthermore, gibberellin treatment causes an increase in the number of male flowers.

In dioecious hemp plants having male flowers, auxin causes the production of female flowers. Thus there is much evidence that hormones are related to sex expression and that a balance of hormones is probably necessary for normal development.

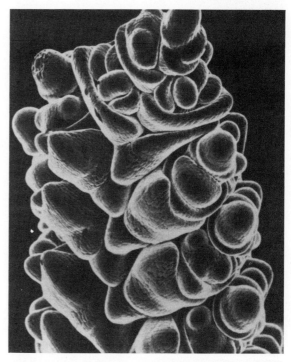

FIGURE 21.4 The floral apex of wheat. Scanning electron micrograph from J. Troughton and L. A. Donaldson. 1973. *Probing Plant Structures*. A. H. and A. W. Reed, Wellington, New Zealand. Photograph reproduced courtesy of The New Zealand Department of Scientific and Industrial Research.

During evocation, the apex frequently becomes enlarged and may become broad or elongated and conical in shape (see Fig. 21.4). The number of cells going through mitotic divisions increases markedly. The inactive, quiescent center known as the "meristeme d'attente" begins cell division. There is an increase in the number of ribosomes, an increase in the amount of RNA, and increased protein synthesis. Thus floral production is a very active, metabolic process.

21.2.2 The timing of flowering

With most plants, flowering is a sequential event of the life cycle occurring at a definite, set time of the year. All individuals of a species in the same area tend to flower at the same time. The proper timing results in outcrossing and ensures that pistillate plants are receptive at the time pollen is produced by staminate plants in dioecious species. Such timing may be set by environmental signals, including light or day length, temperature, and water availability.

"Ripeness to flower"

Because it was known that plants will not flower in response to environmental stimuli until they have reached a certain degree of development, the concept of "ripeness to flower" was introduced. However, the concept is rather vague and difficult to define with any degree of precision. There are, for example, plants such as bamboo and *Agave* (century plant) that will not flower until after many years of vegetative growth. Such plants are monocarpic, dying after they flower. Other species may flower only a few days after germination.

Woody plants in general must grow through several seasons before they are ripe to flower; however, such "ripening" can be accelerated with appropriate treatments. Girdling, the process of bark and phloem removal, tends to force flowering in young stems of apple and citrus. The girdling prevents outward translocation of carbohydrates; the excess carbohydrate concentration apparently stimulates flowering in stems that ordinarily would not flower because of their age.

Biennials and perennials normally do not flower until the second year of growth. Some annuals, such as the Japanese morning glory, *Pharbitis nil*, are ripe to flower within a few days after germination. *Pharbitis nil* requires short days to flower, is receptive to short days after about four days of growth, and can be induced to flower with a single long night (or short day) (Fig. 21.5).

It can be concluded that most plants require a certain amount of vegetative growth before they are "ripe" to flower. Only after this period of vegetative growth can they be induced to flower. The processes occurring during this period are largely

FIGURE 21.5 Photograph of *Pharbitis nil*, a short-day plant that will flower after a single inductive cycle. From S-I. Imamura. 1967. *Physiology of Flowering in Pharbitis nil.* Japanese Society of Plant Physiologists, Tokyo.

unknown and may perhaps differ markedly among species.

Vernalization

Vernalization is the promotion of flowering (or other process) in plants by exposure to low temperatures (Fig. 21.6). Summer varieties of wheat planted in the spring will flower in early summer. However, if winter varieties are planted in the spring rather than in the fall, they will not flower until late summer, well after the summer varieties have flowered. If the winter varieties are planted at the usual time in the fall, they will grow through the winter and flower in early summer like the summer varieties.

It was discovered in Russia that if the moistened

FIGURE 21.6 Photograph showing vernalization in carrot. The plant at the left is the control. The plant at the right was treated with 8 weeks of cold, resulting in bolting and flowering. The middle plant was treated like the control with no cold exposure but was sprayed with gibberellin. The gibberellin replaced the vernalization requirement. From A. Lang. 1957. The effect of gibberellins upon flower formation. *Proc. Nat. Acad. Sci.* 43:709–717.

seed of a winter wheat is given a cold treatment prior to planting in the spring, flowering would occur in the early summer. This process of a cold treatment to induce flowering early, or "making springlike," was named vernalization by the famous Russian botanist Lysenko.

There are two basic types of plants that respond to vernalization. Winter annuals such as winter wheat require a cold treatment after the seed has imbibed water or is in the seedling stage before they will flower. These plants will normally germinate in late fall or winter and be subjected to the low temperatures of the winter. They will flower in the spring or early summer.

In addition, biennials and some perennials re-quire a cold treatment between seasonal growth prior to flowering. Biennials, for example, form rosettes the first year. During the winter the rosette leaves may die, but the root remains alive. Following the cold of the winter, the biennial will flower. The cold winter, i.e., the vernalization process, is a requirement for flowering.

Many perennials also require a cold winter prior to flowering in the spring or early summer. In some plants, cold stimulates flowering directly. However, those plants that form floral buds and flower during cold periods are not considered to be vernalized. Vernalization is an inductive process in which the response follows the treatment.

Little is known about the process of vernalization, but some experiments have indicated that the growing apex is the target tissue for the cold treatment. Furthermore, the apex has to be growing with cell divisions to be receptive. If apices are transplanted from vernalized plants to the tips of nonvernalized plants, flowering will occur as if they were vernalized. These and other types of grafting experiments have led to the suggestion that a specific vernalization hormone (vernalin) exists. But, like florigen, vernalin may not be a specific substance. Since gibberellin will cause flowering in some biennials in the absence of cold treatments, gibberellin may be the hormone with "vernalin" activity (Fig. 21.6).

It can be visualized that there is a substance that is converted to a precursor of "vernalin" in the cold. In the cold the precursor can then be converted to "vernalin," but at warm temperatures it reverts back to the original substance.

$$X \underset{\text{warm}}{\overset{\text{cold}}{\rightleftharpoons}} X' \longrightarrow \text{"Vernalin"}$$

Experiments have shown that after vernalization by cold treatment, grain could be devernalized by high-temperature treatment during only one day at 35°. It is visualized that X' is converted back to X. The nature of these substances is completely unknown.

Photoperiodism and flowering

PHOTOPERIOD-SENSITIVE PLANTS

The concept of photoperiodism and phytochrome was discussed in Chapter 20. The early work of Garner and Allard at Beltsville and others established that there were plants that would not flower unless the photoperiod was less than some critical period (Fig. 21.7). These plants were called short-day (SD) plants. In addition, there are plants called long-day (LD) plants that will not flower unless the photoperiod exceeds a critical length (Table 21.1). Because of this definition, it is possible that the photoperiod for a short-day species and for a long-day species could be identical, e.g., 12 hours. If the critical photoperiod is 12 hours, the short-day plant will flower if the photoperiod is less than 12 hours and the long-day plant will flower if the photoperiod is greater than 12 hours. There is

nothing in the definition saying that a short-day plant will necessarily flower if the photoperiod is short although this is frequently the case.

For example, winter rice, a short-day plant, and orchid grass, a long-day plant, both have a critical photoperiod of 12 hours. As another example, both short-day and long-day plants may flower on exactly the same photoperiod. Common bread wheat, a long-day plant with a critical photoperiod of 12 hours, will flower if the photoperiod is 13 hours. Similarly, the short-day Maryland Mammoth tobacco will flower on 13-hour photoperiods since the critical photoperiod is 14 hours.

Many photoperiod-sensitive plants can measure photoperiods to within 15 minutes. For example, the critical day length for the short-day plant *Xanthium* is 15¾ hours; it will not flower unless the photoperiod is less than 15¾ hours.

Photoperiodism can be illustrated with the graph in Fig. 21.8. The diagram illustrates how the flower-

Short day   Long day

Approx. 60 days

Zinnia

Short day   Short day → Long day

Approx. 110 days

Henbane

FIGURE 21.7 Photographs to illustrate photoperiodism of flowering in plants. Left: zinnia, a short-day plant. Right: henbane, a long-day plant. Photographs from F. B. Salisbury. 1971. *The Biology of Flowering.* Natural History Press, Garden City, New York.

TABLE 21.1   Examples of short-day, long-day, and day-neutral flowering plants

SHORT-DAY PLANTS	CRITICAL TIME, HOURS
Bryophyllum pinnatum	12
Chrysanthemum	15
Xanthium (cocklebur)	15.6
Kalanchoe blossfeldiana	12
Euphorbia (poinsettia)	12.5
Oryza sativa (winter rice)	12
Fragaria (strawberry)	10
Nicotiana tabacum (Maryland Mammoth tobacco)	14

QUANTITATIVE SHORT-DAY PLANTS

Cannabis sativa
Gossypium hirsutum (cotton, one variety)
Saccharum officinarum (sugarcane)
Solanum tuberosum (potato, one variety)

LONG-DAY PLANTS	
Anethum graveolens (dill)	10
Hibiscus syriacus	12
Avena sativa (oats)	9
Dactylis glomeratus (orchid grass)	12
Lolium perenne (ryegrass)	9
Sedum spectabile	13
Triticum aestivum (wheat)	12

QUANTITATIVE LONG-DAY PLANTS

Brassica rapa (turnip)
Hordeum vulgare (spring barley)
Nicotiana tabacum (Havana tobacco)
Secale cereale (spring rye)
Sorghum vulgare
Triticum aestivum (spring wheat)

DAY-NEUTRAL PLANTS

Cucumis sativus (cucumber)
Gomphrena globosa (globe amaranth)
Ilex aquifolium (English holly)
Poa annua
Zea mays (maize)

From Salisbury (1963).

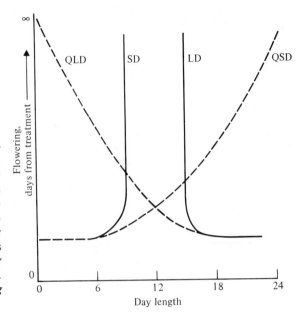

FIGURE 21.8 Graph to illustrate the nature of photoperiodic flowering response in short-day (SD) plants, long-day (LD) plants, and in quantitative long-day (QLD) and short-day (QSD) plants. The graph shows days from beginning of treatment on the ordinate graphed against day length. The graph is idealized to illustrate the phenomenon. Both short-day and long-day plants show an "all or none" type of response, whereas the quantitative LD and SD plants flower progressively earlier as the day length approaches a critical time. Adapted from F. B. Salisbury. 1963. *The Flowering Process.* The Macmillan Co., New York.

ing of short-day and long-day plants responds to photoperiod. The diagram shows the number of days until flowering after photoperiod treatment begins, plotted against the length of the photoperiod. The days are adjusted to arbitrary days so that comparison is easier. Day-neutral plants, those that flower independently of the photoperiod, begin flowering at a specific time of year regardless of the length of the photoperiod. Short-day plants of a species all flower at the same time when grown on short photoperiods up to the point at which the photoperiod exceeds the critical day length. Beyond the critical day length, flowering ceases. In contrast, long-day plants will not flower until the day length exceeds the critical time. Then all plants of a species begin to flower.

As might be expected, not all plants that are sensitive to photoperiod show such clear flowering responses. There is a group of species called quantitative short-day plants that will flower more quickly when the photoperiods become shorter. As shown in Fig. 21.8, the quantitative short-day (QSD) plants will not flower when grown on very long days. When the photoperiod is decreased, fewer days are required to reach flowering. Finally,

a critical photoperiod is reached (on the graph of Fig. 21.8 at about six hours) at which plants will flower in a minimum length of time. Similarly, there are quantitative long-day plants (QLD). As the photoperiod is increased there are fewer days required prior to flowering. When the critical photoperiod is reached, a minimum number of days is then required before flowering.

In addition to the above there are some plants that require a specific photoperiodic treatment (exposure) prior to the short or long day. Thus plants such as *Aloe* and *Kalanchoe* require a long-day exposure before the short days. These plants are called long-short-day plants. In contrast, there are short-long-day plants such as *Poa pratensis* (Kentucky bluegrass) and *Dactylis glomerata* (orchardgrass). For these plants the short-day exposure is the inductive treatment, but the inflorescence will not develop unless the plants are on long days. In some short-long-day plants such as *Campanula medium,* low temperature will substitute for the short-day requirement.

Still other plants do not appear to flower if the days are too long or too short. The common lambs-quarters (*Chenopodium album*) and *Coleus hybrida* will flower only when exposed to intermediate day lengths.

An obvious question is: how do plants differentiate between the short days of late winter or early spring and the short days of the fall? For some plants a long day is required before they will respond to the short photoperiod. These are the long-short-day plants mentioned above. Thus these plants would not be sensitive to the short days of spring when they have just germinated. The long days of the summer predispose them to the short days of the fall. Similarly, there are plants that require a short photoperiod before being receptive to long photoperiods, the short-long-day plants.

In addition, there is frequently a correlation with environmental temperature. Some short-day plants will not flower in the spring, because by the time the temperature is high enough for significant growth the days are too long.

A question frequently asked is whether it is the length of the daylight period or the length of the night period that is important. Experiments with soybeans have shown that it is not the total duration of light each day that is important. Soybean is a short-day plant with a critical photoperiod of 14 hours. If soybeans are given 3 cycles of 12 hours of light and 12 hours of dark, they will be induced to flower because the photoperiod is less than 14 hours. If they are given 6 hours of light and 6 hours of dark they will not flower even though the photoperiod is sufficiently short. Evidently the dark period is not long enough. If they are given 36 hours of light and 36 hours of dark they will not be induced to flower. Evidently the light period is too long.

Further experiments have shown that if soybean was grown on light cycles of 4 to 16 hours, the night period had to be 10 hours or longer for flower induction. Thus the critical dark period is 10 hours or longer. Similar experiments with *Xanthium* showed a critical dark period of 8¼ hours.

If the dark period was kept at 16 hours for soybean and the light was varied, flowering increased up to 12 hours of light. Light periods greater than 12 hours induced fewer flowers, and there were none at 20 hours of light.

Thus it was concluded that the dark period for soybean must be longer than 10 hours and that the light period could not be excessive. It should be evident from these experiments why the critical photoperiod for soybean is 14 hours under natural conditions of a 24-hour day. Furthermore, short-day plants such as soybean could just as well be called long-night plants.

For the short-day plant *Xanthium* with a critical photoperiod of 15.6 hours, it is possible to visually determine that flower induction has occurred after three days. Figure 21.9 illustrates several stages of floral induction in *Xanthium* as outlined by Salisbury (1963). The figure shows the transition from a vegetative apex to a floral apex. In *Xanthium,* the longer the dark period (provided that the plants are exposed to the inductive short days, less than 15.6 hours), the greater the rate of floral induction. Three days after induction the floral apex can be observed. Plants exposed to 16 hours of dark (8 hours of light) develop flowers much more rapidly than those exposed to 10 hours of dark (14 hours of light).

Although temperature will profoundly influence the flowering process, the interaction of photoperiod and temperature is not important for the most part. Photoperiodic induction of flowering is a photochemical reaction and is thus largely temperature-independent. Induction occurs over a wide range of temperatures. However, there are some plants that respond to changing temperature with a change in the length of the critical photoperiod. As an example, *Poinsettia* is a short-day plant at high temperature and a long-day plant at low temperature. In his book on the flowering process Salisbury (1963) lists 48 categories of flowering plants, many of which show extremely complex photoperiod and temperature interactions. His book should be consulted for additional information on this interesting topic.

The inductive cycle

Much of our knowledge of photoperiodism comes from plants that will respond to a single inductive cycle, i.e., a single treatment that will induce flower-

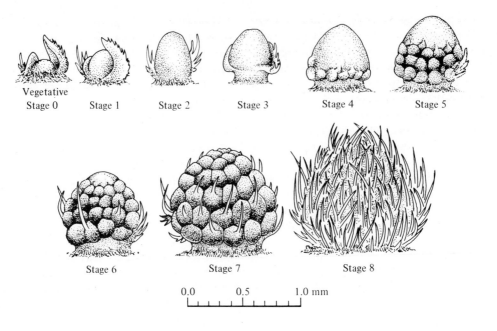

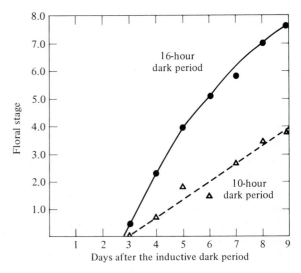

FIGURE 21.9 Drawings to show the change from a vegetative apex to a floral apex and flowering in *Xanthium*. *Xanthium*, a short-day plant, will flower if exposed to day lengths of less than 15.6 hours. However, the longer the dark period (the shorter the photoperiod), the more rapidly apices shift from vegetative to floral and the greater the rate of flowering. Note that floral apices can be detected in both treatments 3 days after the inductive cycles began. From F.B.Salisbury. 1955. The dual role of auxin in flowering. *Plant Physiol.* 30:327–334.

ing. Cocklebur (*Xanthium strumarium* (*X. pennsylvanicum*)). Japanese morning glory (*Pharbitis nil*), *Chenopodium album*, and some duckweeds will flower in response to a single short day. As mentioned previously, *Pharbitis nil* is particularly useful because it is sensitive to photoperiod after about four days of growth. *Chenopodium rubrum* will respond to a single photoperiod while still in the cotyledon stage.

A few long-day plants are known that will respond to a single inductive cycle. *Lolium temulentum* will flower while on short days if given a single long-day induction after five weeks of short-day growth. Pimpernel (*Anagallis arvensis*) will respond to a single inductive long-day treatment while in the seedling stage. Others that will respond to single long-day inductive cycles are dill (*Anethum graveolens*), white mustard (*Sinapis alba*), rape (*Brassica campestris*), and barley (*Hordeum vulgare*). However, all of these will eventually flower if kept on short days for extended periods.

There is some evidence suggesting that those plants that respond to one or only a few inductive cycles are those found in the more northern climates of the northern hemisphere. For example, it is known that northern races or strains of *Polygonum thunbergii* respond to fewer short-day inductive cycles than those occurring more to the south.

PHOTOPERIOD AND HORMONE INTERACTION

Many biennial plants require a cold period between their initial vegetative growth and the long-day inductive cycles. Research with the rosette-forming biennials—henbane, carrot, parsley, beet, and turnip—showed that gibberellin treatment would bypass the cold requirement when these plants were grown on long days (see Fig. 21.6). Gibberellin will also bypass vernalization in oats. The long-day plants *Samolus parviflorus, Crepis tectorum, Silene armeria,* and *Hyoscyamus niger* will flower if grown on short days and treated with gibberellin. In *Bryophyllum,* a genus that has species requiring long-day treatments before the short-day inductive cycles, gibberellin will bypass the long-day requirement but not the short-day requirement.

In the experiments mentioned above, the short-day plants cocklebur, rye, and soybean, when grown on long days would not flower with gibberellin treatment. These experiments indicate that gibberellin may overcome the cold requirement and long-day requirements for flowering in some plants.

Gibberellin is apparently not effective in promoting flowering in short-day plants.

Auxin treatment will not induce flowering except in pineapple. If it is treated with synthetic auxins (but not IAA), pineapple will flower. Ethylene will also promote flowering in pineapple. Perhaps the auxin stimulates ethylene production, which then causes the flowering.

In some plants abscisic acid treatment will induce flowering, and in others it inhibits flowering. The interpretation is that the promotion or inhibition of flowering accompanying abscisic acid is indirect.

21.2.3 Florigen, the flowering hormone

Although it was suggested nearly 40 years ago by the Russian scientist Chailakhyan, the presence of a flowering hormone distinct from the other plant hormones has not been unequivocally demonstrated. Nevertheless, it is tacitly assumed that there is a hormone that transmits information from the vegetative portion of the plant to the apex, ultimately bringing about the flowering response. Using the example of a photoperiod-sensitive plant, it is the leaf that detects the photoperiod, and thus it appears quite evident that something is transmitted from the leaf to the apex. Clearly this unknown entity fits the classical definition of a hormone.

An interesting experiment that is well known to plant physiologists was conducted by R. G. Lincoln and his colleagues in 1964 with *Xanthium*. They exposed *Xanthium* to inductive photoperiods and after a period of time extracted the leaves with organic solvents. An acidified fraction of this extract then caused flowering of noninduced plants when applied to the leaves. Although this experiment has been very difficult to repeat, it remains as evidence for a chemical substance functioning as the flowering hormone.

Still another attractive demonstration that gives much evidence for a material substance functioning as florigen comes from grafting experiments. Hamner and Bonner (1938) placed *Xanthium* on inductive photoperiods and then grafted induced

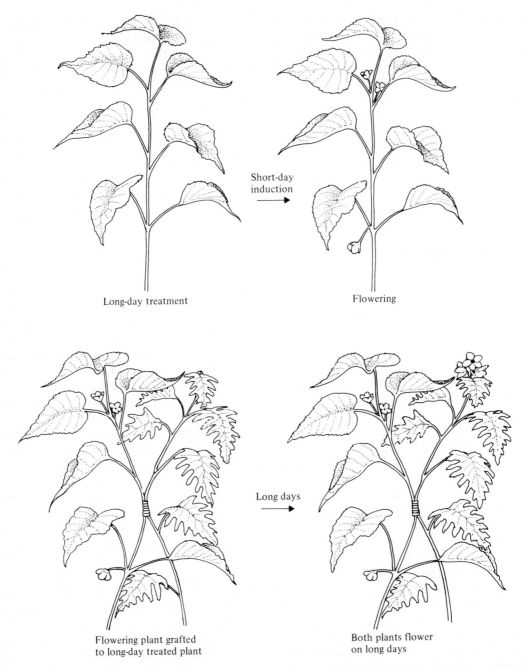

Long-day treatment

Short-day
induction

Flowering

Flowering plant grafted
to long-day treated plant

Long days

Both plants flower
on long days

FIGURE 21.10 Diagram to show experiment in which *Xanthium* plants induced to flower are grafted onto non-induced plants. Evidently the flowering hormone "florigen" is transmitted from the induced flowering plant to the noninduced plant, causing flowering in the latter. (From K. Hamner and J. Bonner. 1938. Photoperiodism in relation to hormones as factors in floral initiation and development. *Botanical Gazette* 100: 388–431.)

plants to those kept on long noninductive photoperiods. When the grafted pair was kept on noninductive photoperiods both flowered, indicating that there was transmission through the graft union of a chemical substance (Fig. 21.10).

Despite such strong circumstantial evidence for the presence of florigen, it has not been isolated, meaning that its existence has not been proven. As stated in Chapter 18, florigen could be a gibberellin or it could be a balance of hormones. There is some evidence that rather than being a promotive substance, the flowering hormone is actually an inhibitor and flowering comes about when it is inactivated or destroyed. Future research will no doubt clarify the concept of florigen and explain the hormonal basis for the flowering process.

21.2.4 Pollination

The pollen grain is the male gametophyte of the vascular plant. Once the tetrad of four microspores is formed, the tetrad becomes relatively isolated by the development of an impervious callose wall. The individual microspores will subsequently develop into the pollen grains. Initially there is a mitotic division of each microspore to form two nuclei. The pollen grains of many of the more primitive orders of flowering plants are binucleate (having two nuclei). One is the vegetative nucleus, which functions as the nucleus for the germinating pollen grain, and the other is the generative nucleus, which through a subsequent mitotic division forms the sperm.

In the more advanced flowering plant families the pollen grains may be trinucleate. The generative nucleus divides mitotically prior to pollen germination to form two generative nuclei. Each will develop into sperm such that the pollen tube, ready for the fertilization of the egg and the polar nuclei of the female gametophyte, will have two sperm.

The mature pollen grain is a highly resistant structure with an impervious wall. It may have a variety of surface features (Fig. 21.11).

Pollination, which is the act of the pollen landing on the stigma of the pistil, occurs in a variety of ways. Usually, plants are pollinated by means of the wind, which carries the pollen from mature anthers to the receptive stigma, or by insects visiting flowers. There are several ways in which outcrossing, as opposed to self-pollination, is ensured. Self-incompatibility is probably the primary mechanism. In some cases there is differential ripening of the pollen and the pistil such that pollen will not pollinate the plant of its origin. Some other examples are: (1) the pistil may prevent germination of the pollen from the same plant but not from a different plant; (2) inhibitors in the pistil may prevent growth of the pollen tube if germination of pollen from the same plant occurs; and (3) even if the pollen tube grows, fertilization may be prevented.

After pollination, the pollen grain will germinate if the stigma is receptive. Enzymes secreted by the pollen enhance the growth of the pollen tube through the style and (eventually) in through the micropyle (Fig. 21.12). Substances in the pistil cause positive chemotropic growth of the pollen tube.

In the case of binucleate pollen, the generative nucleus divides mitotically after germination to form the sperm. Trinucleate pollen already has two generative nuclei that differentiate into functional sperm. Sperm are discharged from the pollen tube into the embryo sac, where the double-fertilization process takes place. One sperm nucleus unites with the egg cell to form the zygote, and the other unites in a triple-fusion process with the two polar nuclei of the embryo sac to form the 3N endosperm nucleus.

Germination of pollen grains has been extensively studied. Binucleate pollen readily germinates in sucrose solutions. The sucrose functions in part as an energy source and in part as an osmoregulatory mechanism to maintain nearly isotonic conditions, preventing excess water uptake and subsequent lysis. Boron and calcium are known to aid in the germination and pollen-tube growth processes (Fig. 21.13). There is little evidence that auxin or gibberellin will enhance pollen-tube growth, although certainly there are endogenous growth substances associated with elongation. The stigma probably supplies growth substances that aid germination and pollen-tube growth.

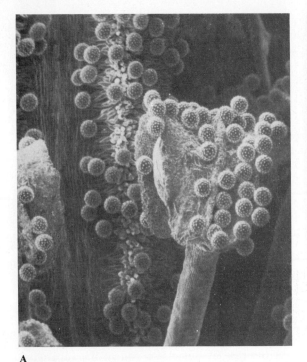

A

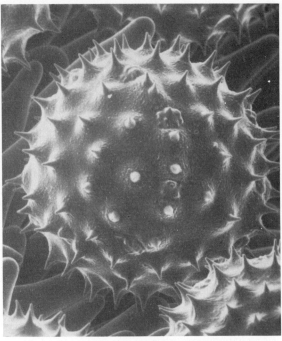

B

FIGURE 21.11 (A) Scanning electron micrograph of anther just after dehiscence (×70), and (B) an enlarged view of pollen grain (×500) of cotton. (C) A drawing of a cross section through an anther and an enlarged pollen grain. Electron micrographs from C. A. Beasley. 1975. Developmental morphology of cotton flowers and seed as seen with the scanning electron microscope. *Amer. J. Bot.* 62:584–592.

Pollen grains

Pollen sac

Cross section, mature anther

Generative cell

Tube cell

Pollen grain

C

21.2.5 Fertilization

The most distinctive feature of fertilization in plants is the double-fertilization process. Fertilization and embryogenesis have been studied extensively in cotton by Jensen and his colleagues (see Jensen, 1973).

Even before pollination, some of the events associated with fertilization begin to occur in the ovule. The two polar nuclei of the ovule begin fusion as the pollen tube grows through the style toward the micropyle, and the nucellar tissue forms a channel through which the tube enters. The small sperm, which are mostly cells with just a nucleus and little or no cytoplasm, enter through one of the synergid cells. One sperm leaves the synergid and enters the egg, where it fuses with the egg nucleus. The other sperm fuses with the two polar nuclei to form the 3N nucleus to complete the double-fertilization process.

integument to the embryo. The very first division is unequal, and polarity of the embryo is quite evident, although even before the first division there is some polarity of the egg cell and zygote. In the two-celled stage a small terminal cell and a larger basal cell are formed. Subsequent cell division by the basal cell forms a three-celled embryo. Eventually there is formed a globular embryo separated from the basal cell by suspensor cells (Fig. 21.14). The suspensor apparently pushes the developing

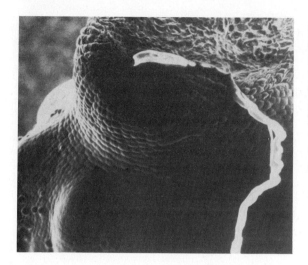

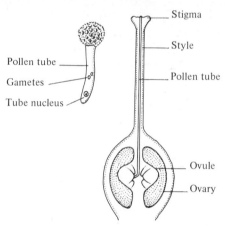

FIGURE 21.12 Electron micrograph of a pollen tube entering a micropyle two days after anthesis (refer to Fig. 21.13). Drawings are of a pollen grain during germination and of a pistil and entering pollen tube. Scanning electron micrograph (X 150) from C. A. Beasley. 1975. Developmental morphology of cotton flowers and seed as seen with the scanning electron microscope. *Amer. J. Bot.* 62:584–592.

The 3N endosperm nucleus through a series of divisions forms the endosperm tissue. Concomitantly, the zygote begins division to form the embryo. The synergids function largely in nutrition of the zygote and the young developing embryo, evidently transferring nutrients from the cells of the

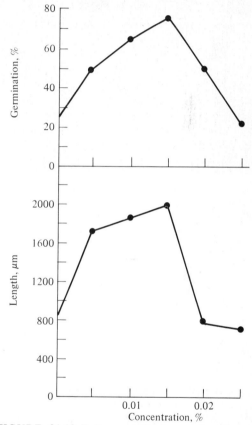

FIGURE 21.13 Pollen germination and pollen tube growth as a function of boron in the medium. Boron appears to be a requirement for optimum germination and growth. Also required is an energy source such as sucrose (the sucrose functions as an osmoticum as well), calcium, oxygen, and perhaps gibberellin. (Redrawn from I. Vasil. 1964. In H. F. Linskens (ed.), *Pollen Physiology and Fertilization*. North-Holland, Amsterdam.)

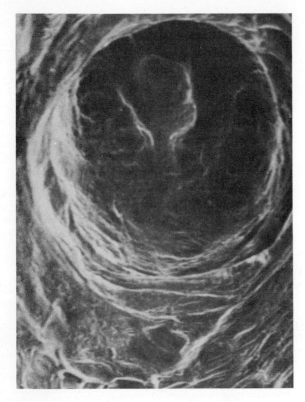

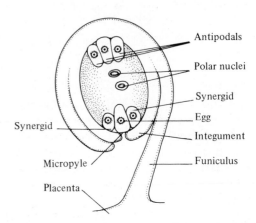

FIGURE 21.14 Scanning electron micrograph of the inside of an embryo sac, viewing the egg apparatus (× 1300). The drawing is a mature ovule showing the egg apparatus (egg plus synergid nuclei). Micrograph from C. A. Beasley. 1975. Developmental morphology of cotton flowers and seed as seen with the scanning electron microscope. *Amer. J. Bot.* 62:584–592.

embryo into the endosperm tissue and functions partially in the absorption of nutrients. Further divisions form a young embryo with visible cotyledons and a radicle.

21.2.6 Fruit set and development

In most flowering plants pollination stimulates the development of the ovary into a mature fruit. It is peculiar that pollination, not fertilization, is the stimulus. Accompanying fruit development is the abscission of other floral parts such as sepals and petals.

Exactly how pollination stimulates initial fruit development is not entirely understood; however, the pollen is an excellent source of auxin, and in all probability auxin and other growth regulators from the pollen initiate the process. In addition, the ovary after pollination begins auxin production, which undoubtedly contributes to development.

In some plants both cytokinins and gibberellins will stimulate fruit development in the absence of pollination. It should be recalled that auxin and gibberellin have little effect on pollen-tube growth.

Fruit development without fertilization is called parthenocarpy, a process quite common in fruit-bearing plants. Parthenocarpy can occur (1) without any pollination, (2) after pollination but without fertilization, or in some cases (3) with both pollination and fertilization but with abortion of the ovules or young seeds. In any case, the fruit that develops is without seeds.

Parthenocarpic fruits are common and economically useful. Edible fruit such as pineapple, tomato, banana, many kinds of melons, and seedless grapes are common commercial parthenocarpic fruits.

21.2.7 Fruit growth

Once fruit is set and the ovary begins to enlarge, the process of fruit maturation occurs. Maturation can be defined as the growth process of the fruit up to full size. After maturation there are qualitative changes that occur that can be referred to as ripening.

Maturation

Fruit maturation (development) is a double process. In one process, the pericarp tissue develops to full size through cellular division and cellular enlargement. In the other, the tissues formed through union of the pollen gametes and the embryo-sac nuclei develop. That there are two processes is evident from the observations of the growth of parthenocarpic fruit. Certainly in fertilized fruit there is coordination and balance brought about by hormonal and metabolite transport.

Many fruits develop according to a typical sigmoid growth curve. Others, in particular the "stone" fruits such as cherries, plums, peaches, and apricots, exhibit a double-sigmoid growth curve resulting because of a quiescent period of growth during maturation. In part this apparent quiescent period corresponds to a period of rapid seed maturation and may be the result of competition between the developing ovary (fruit) and developing ovules (seeds) for nutrients.

As stated above, fruit set and initial development of the fruit is correlated with auxin and may also involve gibberellins, cytokinins, and abscisic acid. In addition, some maturing fruits produce ethylene, which may promote fruit maturation. However, ethylene appears to play a more important role in fruit ripening (refer to Fig. 18.13).

A very important aspect of fruit maturation is the mobilization of energy reserves of leaves and massive transport into the fruit. Developing fruits are major sinks for the translocation of carbohydrate reserves, amino acids, and other materials.

The ultimate size obtained by developing fruit is limited by the genetic composition of the plant species but varies within rather wide limits, depending on environment and certain endogenous factors. In part, the size of the fruit is a function of the number of cells. There is a good relationship as well between the number of seeds and the size of any one fruit of a particular plant species.

An important aspect of fruit development from an agricultural and horticultural viewpoint is fruit drop. Navel oranges, for example, that set fruit in southern California in April and May will have significant fruit abscission in June if hot weather occurs. This fruit drop may be related to excessive respiration because of the hot weather or to impaired translocation into the developing fruit. Auxin apparently decreases, and there may be an increase in abscisic acid accompanying this precocious fruit abscission.

Fruit drop can be significant, especially in tree crops. For example, *Macadamia* may lose up to 90% of the crop because of precocious abscission. Auxin spray prevents fruit drop in some species.

Ripening

At or near the end of fruit maturation, many qualitative changes occur within the fruit. These qualitative changes are collectively called fruit ripening. The general changes that occur are softening of the fruit pulp, hydrolysis of storage compounds and structural cellulose, a decrease in phenolic compounds, an alteration in the composition of pigments, and frequently changes in the rate of respiration.

Softening comes about largely by hydrolysis of pectins through the action of pectin methylesterase. Methylation of the hydroxyls of the polygalacturonic acid residues of pectin prevents crosslinkages. Hydrolysis of the polymers is also associated with softening. Hydrolysis of cellulose further breaks down cell walls during the softening process.

Hydrolysis of storage carbohydrates, lipids, and proteins produces soluble sugars, which results in the fruit becoming sweet in taste. A loss of astringent phenolics further improves quality with respect to flavor.

Many fruits are green because of the presence of chlorophyll. During ripening, chlorophyll disappears and fruits become yellow, orange, or red because of the presence of carotenes, xanthophylls, anthocyanins, and other pigments. There may be synthesis of these pigments; or they may be simply unmasked as chlorophyll is lost. Color changes, of course, may occur during the maturation process as well as during ripening.

Respiratory changes during ripening are perhaps the most thoroughly studied of all the events known to be associated with ripening. Some fruits such as citrus show a steady rate of respiration throughout the entire ripening period. Some, such as pepper and peanut, show a decrease in respiration. Others, such as avocado and banana, have a characteristic change in respiration during the ripening process called the climacteric. Ripening in avocado and banana first shows a decrease in respiration, followed by a rapid increase to a peak (the climacteric) and then a general decline (Fig. 21.15). The climacteric peak may or may not correspond to the most desirable flavor. Apparently the climacteric is associated with the hydrolysis of reserves. In fruits such as avocado the climacteric rise will not begin until the fruit is picked.

It was through studies of the climacteric that ethylene was finally shown to be associated with ripening. In the 1920s kerosene was burned in the presence of lemons to improve their appearance, and later it was shown that ethylene would do the same. Eventually it was demonstrated that ethylene would stimulate the climacteric, and finally ethylene was proposed as the ripening hormone. Ethylene is now generally accepted as the fruit-ripening hormone. During ripening, fruits produce ethylene, which will stimulate the climacteric and accelerate ripening.

Respiration is thought to be very tightly linked to ripening, at least in those fruits showing a definite climacteric. Respiratory inhibitors will prevent ripening and the climacteric. Protein-synthesis inhibitors such as cycloheximide will also prevent ripening. Storage under high CO_2 or low O_2 prevents ripening.

Even though there is substantial evidence that ethylene is the ripening hormone, it is not considered to act alone. Like most growth and development processes, a balance of hormones is implicated in ripening. Low auxin and high ethylene stimulate ripening, but gibberellin and cytokinins will retard ripening in some instances.

21.2.8 Seed formation

The actual formation of the mature seed is a rather complex and poorly understood process. Anatomical and morphological studies have provided information about the development of the embryo sac and the subsequent formation of the embryo. After fertilization, embryo formation (embryogenesis) occurs. However, in most if not all seeds embryogenesis lags behind the development of the other

FIGURE 21.15 Oxygen consumption by avocado fruit after harvest, showing the climacteric, evidently a function of mobilization of reserves during ripening. The climacteric in avocado does not begin until harvest. (From A. Millerd, J. Bonner, and J. B. Biale. 1953. The climacteric rise in fruit respiration as controlled by phosphorylative coupling. *Plant Physiol.* 28:521–531.)

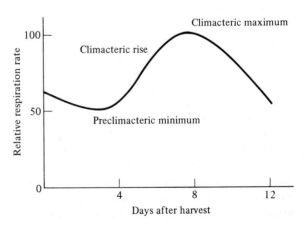

portions of the seed. First the integuments and endosperm and then the embryo will develop at the expense of the endosperm.

During embryogenesis the initial stages undergo much cell division. DNA synthesis is rapid and even continues beyond the period of cell division into the period of cell enlargement. This frequently results in polyploidy and large amounts of RNA. Some of the mRNA carries over into the first stages of germination. This long-lived mRNA was discussed in Chapter 8.

Species differ with respect to the method of embryogenesis and seed development. In the cereals, which have been studied most extensively,

seed development is accompanied by deposition of carbohydrate reserves (mostly starch) in the extensive endosperm, and protein is deposited in the aleurone layer. The single cotyledon, known as the scutellum, is not a primary reserve depot.

During embryogenesis in many dicotyledonous species such as the legumes and the cotton plant, carbohydrate, lipid, and protein reserve material is transferred from the endosperm tissue and stored in the cotyledons of the embryo. Thus during germination in these species, mobilization of reserves to support seedling development occurs directly in the tissues of the embryo and does not come from the endosperm.

21.3 Seed Germination

The mature seed, marking the beginning stage of a new plant, normally will be dormant prior to germination. After maturation, some seeds require an after-ripening stage before they will germinate. Such after-ripening may be a period of further development of the embryo or a time for biochemical changes before germination.

In some seeds the after-ripening period must occur when they are fully imbibed with water, a process called stratification. For the latter, seeds must be stored in a moist, usually cold environment prior to germination. Subsequent to the after-ripening events, some seeds may still not germinate if placed in a favorable environment of temperature and water because of dormancy.

Actual dormancy is the quiescent period between maturation and germination during which the seed will not germinate even if placed under optimum and favorable conditions. Dormancy may be the result of (1) a seed coat impervious to gases or water, (2) internal inhibitors, or (3) a requirement for a proper thermoperiodic or photoperiodic treatment.

The longevity of seeds varies markedly with species. Some tree seeds, such as willows and maples, may remain viable for only a few days or weeks. Elm seeds remain viable for several months,

and most seeds will survive for many years provided that they are stored under conditions of low moisture. Five percent to 10% moisture on a dry-weight basis is the usual level required for optimum storage survival. During storage, respiratory activity is present but barely detectable. Much of the respiration may be anaerobic.

During germination, reserves stored either in the endosperm or in the cotyledons are mobilized. In some cases endosperm reserves are transferred to cotyledons; in seeds such as the orchid seed there are virtually no reserves.

The actual germination process can conveniently be divided into three distinct stages. First, subsequent to after-ripening and breaking dormancy (if the seed is dormant), the seed takes up water by the physical process of imbibition. Second, there is the appearance of metabolic processes directly associated with germination. Finally, emergence and subsequent growth and development of the seedling occur.

Figure 21.16 illustrates the three stages of water uptake during germination. During stage I, water uptake is by imbibition and is a function of the imbibitional component of the water potential (ψ_m). This stage will occur in dead seeds and is purely physical. Stage II water uptake occurs along

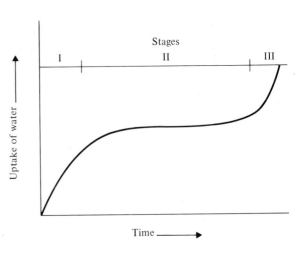

FIGURE 21.16 Stages of water uptake by a seed during germination. Stage I is imbibition, a physical process. Stage II occurs only in living seeds and comes about by a decreased osmotic potential. Stage III marks the beginning of visible germination. (From J. D. Bewley and M. Black. 1978. *Physiology and Biochemistry of Seeds.* Vol. I. Springer-Verlag, Berlin.)

with active metabolism and evidently is a function of osmotic potential (ψ_π). Stage III marks the appearance of visible germination and is accompanied by marked growth. During this stage, reserves are actively mobilized and the seedling emerges. Water uptake is largely a function of ψ_π.

21.3.1 Imbibition

Imbibition as a physical process was discussed in Chapter 4. Any dry material that has the properties of a hydrophilic colloid will adsorb water when wetted. This process of adsorption of water by colloids is termed imbibition and is accompanied by marked swelling and heat liberation. The increased volume after adsorption is not, however, equal to the sum of the volumes of the adsorbed water and the dry colloid because of the orientation of water molecules on the surface. Heat liberation occurs because of the decrease in free energy of the water molecules after adsorption. The forces of imbibition are so great that early masons could quarry rock with dry wood inserted into holes drilled in rock. Pouring water on the dry wood pegs caused swelling and rock fracture.

After colloidal polymers of the dry seed adsorb water during imbibition, enzymatic proteins are synthesized, induced, or activated, and the second phase of germination begins.

21.3.2 The metabolism of germination

During the imbibition process there is a marked increase in weight because of the uptake of water. Throughout the imbibition stage there is an increase in the rate of respiration. There is a good correlation between the percentage of moisture in the seed and the rate of respiration. After a few hours imbibition ceases, and the seed reaches a plateau stage of active metabolic activity prior to the emergence of the radicle. Figure 21.17 illustrates respiration during the imbibition and plateau stages prior to seedling emergence.

Respiration during the plateau stage is characterized by a low respiratory quotient because there is much more CO_2 evolved than O_2 consumed. In many seeds much of the respiration during the plateau stage is cyanide-resistant and thus is accomplished apparently by the alternate respiratory pathway (see Chapter 14).

Little DNA synthesis takes place in the imbibitional and plateau stages of germination since there is not much cell division occurring. However, there is active RNA synthesis and *de novo* synthesis of proteins with enzyme activity. In addition to the *de novo* synthesis of enzymes, there is some activation of latent enzyme activity. For example, the enzyme β-amylase is known to preexist in the seed and is activated during the initial phase. Its production is not prevented by inhibitors of RNA metabolism and of protein synthesis such as actinomycin D and chloramphenicol. Furthermore, proteases increase the activity of β-amylase, apparently because it is stored in the ungerminated seed as a zymogen. Zymogen is a large inactive protein that is activated by proteolytic cleavage of portions of the polypeptide. Unlike β-amylase, however, α-amylase, another common polysaccharide hydrolytic enzyme, is known to be synthesized *de novo* during the germination of certain seeds. The details have been discussed in Chapter 19.

As previously discussed in Chapter 18 (refer to Fig. 18.10), during the germination of barley seeds gibberellins stimulate the *de novo* synthesis of α-amylase, proteases, and other hydrolytic enzymes that function in the mobilization processes accompanying germination. Inhibitors of RNA synthesis such as actinomycin D inhibit the synthesis of α-amylase, and deuterium labeling experiments have shown that the label appears in the α-amylase consistent with the notion of *de novo* synthesis.

The amylases and other hydrolytic enzymes appear during germination of seeds that store carbohydrate reserves. In seeds such as castor bean and cucurbits, which store lipoidal compounds, germination is accompanied by the appearance of the enzymes of the glyoxylate cycle. This metabolic cycle, explained in Chapter 11, couples fatty acid oxidation to carbohydrate biosynthesis during the mobilization of food reserves. Such seeds store reserves in the form of fats and oils. During germination the fats and oils are hydrolyzed to free fatty acids, which are oxidized through the β-oxidation pathway (Chapter 11) to acetate. The glyoxylate cycle synthesizes malate from acetate, catalyzed by the two important enzymes isocitrate lyase and malate synthetase. These two enzymes are synthesized *de novo* during the germination of lipid-storing seeds. The appearance of isocitrate lyase during the germination of cucumber seeds has been illustrated in Fig. 19.2. The malate is converted to phosphoenolpyruvate and ultimately to soluble carbohydrates such as sucrose. Sucrose is the primary energy source during germination.

21.3.3 Seedling emergence

After the plateau stage during which the enzymatic machinery is produced, the young embryo will begin to grow and the seedling will emerge. In most seeds the first part to appear is the radicle, the portion of the hypocotyl that will ultimately develop into the root. The plumule, or terminal portion of

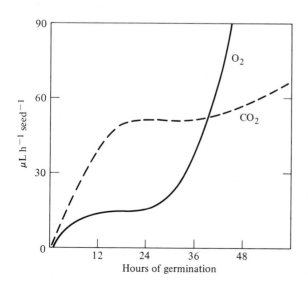

FIGURE 21.17 Respiration of seeds during the first few hours of germination. After wetting during imbibition, respiration, as measured by either oxygen consumption or carbon dioxide evolution, increases markedly. This is followed by a steady level of respiration and finally another dramatic increase, marked by visible germination when the seed coat breaks. This figure should be compared with the water-uptake curve shown in Fig. 21.16. The changing respiratory quotient (RQ = CO_2 evolution/O_2 uptake) is evidently the result of changing respiratory substrates. (Redrawn from S. P. Spragg and E. W. Yemm. 1959. Respiratory mechanisms and the changes of glutathione and ascorbic acid in germinating peas. *J. Exp. Bot.* 10:409–425.)

the epicotyl, usually appears after the radicle. Of course, species vary markedly in the manner of emergence.

After the radicle emerges in the common bean *Phaseolus,* the remainder of the hypocotyl forms an aboveground hook. As this hypocotyl hook unbends, which is a typical phytochrome response, the epicotyl and cotyledons are pulled from the soil. In bean, stored food is in the fleshy cotyledons, having been transferred from the endosperm during embryo development. Although the fleshy cotyledons with stored food become green with chlorophyll, probably very little significant photosynthesis occurs. The plumule produces true leaves rapidly, and the plant matures.

The germination of pea, *Pisum,* is somewhat similar to bean in that the radicle emerges first, the food reserves are stored in the fleshy cotyledons, and a hook is formed; unlike bean, however, the cotyledons remain below ground. Thus bean has epigeal (aboveground) cotyledons and pea has hypogeal (below-ground) cotyledons.

In corn (maize), which is a monocot, the single cotyledon is hypogeal and the radicle appears first during germination. However, there is no hypocotyl hook formed. The epicotyl emerges and grows directly upward. In such monocots the epicotyl tip, or plumule, is protected by a special leaflike sheath, the coleoptile. The radicle tip is covered by a comparable protective sheath called the coleorhiza.

It should be apparent that the portion of the embryo above the cotyledons, the epicotyl, does not develop completely into shoot tissue, nor does the hypocotyl (embryonic tissue below the cotyledons) develop completely into root tissue.

21.3.4 The timing of germination

From the standpoint of seedling survival, perhaps one of the most important aspects is the proper timing of germination. On first thought, it might appear that once the soil warms and there is ample water, germination would be initiated and the young seedling would have a reasonable chance of survival. However, soil could warm for a few days

at the surface during midwinter or in deserts, and an irregular rain could fall with sufficient moisture to permit germination. A change of weather back to normal would result in seedling death and perhaps loss of an entire year's crop of seeds.

Agricultural plants have largely been selected for uniform germination. A farmer or home gardener expects that virtually all seeds planted will germinate and develop into mature plants. Native species, however, frequently produce seeds with very erratic and irregular germination patterns. This irregularity coupled with the longevity of seeds means that seeds from any one year's crop are apt to remain in the soil for many years. A good year for germination and a bad year for establishment will thus not necessarily result in the loss of an entire year's crop of seeds. Mechanisms (physiological devices) that affect the timing of germination include different kinds of seeds produced by a single plant, seed coats that are impervious to water or gases such as O_2, chemicals that when leached measure precipitation ("rain gauges"), temperature-sensitive factors that can measure thermoperiods, and phytochrome, which is sensitive to light.

Differential seed types

Many of the legumes, such as mesquite, acacia, clovers, and lupines, produce seeds with varying degrees of germination potential. In most cases such seeds have seed coats that are impermeable to water. Weathering and actual scratching of the seed coat (a process called scarification) may be necessary before imbibition is possible. The seed is otherwise capable of germination.

Cocklebur (*Xanthium*) and some *Franseria* species produce two different types of seeds. One type has an oxygen-impermeable seed coat and will not germinate until the permeability is increased by weathering or chemical treatment. The other has a seed coat that is permeable to oxygen and will germinate immediately after reaching maturity.

In some mustards, the pod containing the seeds is divided into two sections. One section dehisces, thereby freeing the seeds, which will germinate under favorable conditions. The other section does not dehisce and seeds remain inside until weathering frees them.

The detection of water (hydroperiod)

In *Atriplex,* the fruit remains attached to floral bracts. If the mature seeds remain with the bracts, germination does not occur until sufficient water has passed over them. If the seeds are removed from the bracts they will germinate. It was discovered that the bracts contained a high concentration of chloride, which acts as an osmotic sink. When placed in a moist environment, the bracts take up water preferentially and the seeds will not imbibe. After leaching the chloride from the bracts with the equivalent of about 2.5 cm of water, the seeds will germinate. Such a leaching requirement prior to germination has been termed a "rain gauge."

There is some evidence for other kinds of germination inhibitors in addition to the osmotic type discussed above.

Temperature requirements (thermoperiod)

Provided that the moisture content is low (5% to 10%), most seeds are relatively resistant to both high and low temperatures. After imbibition, germination increases with increasing temperature over the range of 10° to 35° or 40°. In species such as the winter-germinating plants there may be a rather low temperature optimum for germination. For example, the optimum temperature for winter-germinating mustard (*Brassica arvensis*) is about 16°. As would be expected, the optimum temperature for winter-germinating plants is lower than the optimum temperature for summer- or spring-germinating species.

There is an interaction between moisture and temperature. A summer-germinating species will not germinate in the winter even if sufficient moisture is present because the temperature is not right.

Little is known about the physiology of water and temperature interactions, but it can be surmised that there are endogenous germination inhibitors that are metabolized away under the proper hydro- and thermoregimes.

Light requirements

Many seeds require a light exposure after imbibition and before germination. In early studies at Beltsville in the 1950s and elsewhere it was shown that the light requirement had a maximum sensitivity in the red, and it is now known to be the result of the phytochrome system (Chapter 20). A particular variety of lettuce, Grand Rapids, is particularly sensitive to red light over a narrow temperature range close to room temperature. After imbibition, a single red-light exposure for about one minute is sufficient to bring about germination. A single exposure to far-red light for a few minutes will prevent germination (see Table 20.1). The last light exposure, either red or far-red, determines whether the seeds will germinate. If red they will germinate; if far-red the amount of germination is far less than 100%.

In some seeds phytochrome is associated with the seed coat; if the seed coat is removed, germination is independent of light. In others there is evidence that phytochrome is in the embryo.

There is at least one plant—*Citrullus*—that has a light requirement for germination with a maximum wavelength sensitivity in the green and blue. This light requirement is apparently not a phytochrome response.

It has been shown that certain seeds not only have a light requirement for germination but also have a photoperiod requirement. These seeds appear to have a timing mechanism that regulates seasonal germination. The photoperiod-dependent response seems to be an adaptation that permits germination during a particular season. However, the light requirement per se is somewhat more difficult to interpret. It has been proposed that the light requirement ensures that seeds will not germinate unless they are close to the soil surface, even though there is sufficient moisture and adequate heat.

FIGURE 21.18 Experiment to demonstrate the stimulation of protein synthesis in lettuce seeds by gibberellin. Seeds were treated with gibberellin (GA) or water and given ^{3}H-leucine at 4, 8, and 12 hours after imbibition. The amount of ^{3}H-leucine incorporated into protein expressed as radioactive (^{3}H) counts per minute (cpm) per mg of protein is taken as an indication of the amount of protein synthesis. Drawn after J. D. Bewley. 1979. Dormancy breaking by hormones and other chemicals—action at the molecular level. In I. Rubenstein, R. L. Philips, C. E. Green, and B. G. Gengenbach (eds.), *The Plant Seed: Development, Preservation, and Germination.* Academic Press, New York.

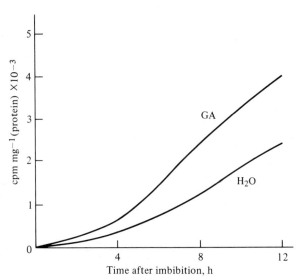

21.3.5 Hormonal and chemical effects on germination

As stated above, there are believed to be endogenous inhibitors of germination probably functioning as dormancy factors. There is strong evidence that abscisic acid is a seed-dormancy factor. Other chemicals known or suspected to be dormancy factors are the phenolics, such as coumarin and salicylic acid.

With respect to promotion of germination, gibberellin is the most prominent of the naturally occurring growth regulators. Gibberellin, for example, will overcome the light requirement for germination in lettuce seeds and in some species will substitute for the low or alternating temperature requirements. It is suspected that gibberellin may be the main regulatory substance produced by the various treatments that promote or permit germination.

The promotion of germination by gibberellin may be the result of the stimulation of RNA and protein synthesis. As shown in Fig. 21.18, treatment of lettuce seeds by gibberellin stimulates the incorporation of ^{3}H-leucine into protein after the seeds have taken up water. The uptake and incorporation of ^{3}H-leucine into protein is interpreted as evidence for active protein synthesis, supporting the hypothesis that the action of gibberellin in promoting germination is at the level of RNA and protein metabolism.

Some species will germinate in response to cytokinin treatment. Ethylene is produced by some germinating seeds and will enhance germination in some seeds.

The most commonly used synthetic organic compound for breaking dormancy and promoting germination is thiourea (Fig. 21.19).

$$\underset{NH_2-C-NH_2}{\overset{\overset{\textstyle S}{\|}}{}}$$

Thiourea will overcome the light requirement for lettuce seed germination. It perhaps stimulates the synthesis of gibberellin. Interestingly, a common inorganic compound, KNO_3, will stimulate germination.

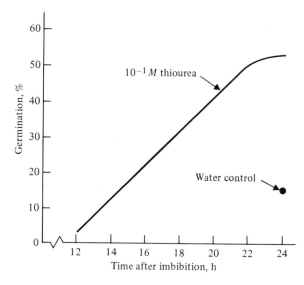

FIGURE 21.19 Experiment to demonstrate the stimulation of germination by thiourea. The black point labeled water control is the percentage of germination in water 24 hours after imbibition. Drawn after J. D. Bewley. 1979. Dormancy breaking by hormones and other chemicals—action at the molecular level. In I. Rubenstein, R. L. Philips, C. E. Green, and B. G. Gengenbach (eds.), *The Plant Seed: Development, Preservation, and Germination.* Academic Press, New York.

21.4 Asexual Reproduction—Tuberization and Bulb Formation

Although sexual reproduction contributes to plant propagation, its primary biological role seems to be in genetic variation. Much plant propagation is through asexual or vegetative processes. Vegetative reproduction may be by runners or stolons produced from stem tissue or by rhizomes of root tissue. Some of the *Opuntia* cacti propagate by abscission of terminal stem sections that will root and grow into new plants. Other plants such as *Bryophyllum* form plantlets at leaf edges. These plantlets fall from the parent and grow as genetically identical plants.

Perhaps one of the most interesting asexual processes is the formation of storage organs that can function in vegetative reproduction. Corms, bulbs, and tubers from stem tissue develop in a manner analogous and perhaps even homologous to fruits. The formation of fleshy storage roots in carrot and radish, for example, is somewhat comparable.

Tubers develop from stems, rhizomes, and stolons, usually at an apex that has ceased to elongate. Perhaps one of the best examples is the potato (*Solanum*) tuber. The meristem developing into a tuber acts as a strong sink for carbohydrate mobilization and grows by both cell division and cell enlargement. Ripening of the tuber is accompanied with massive starch deposition and some protein synthesis. In some plants inulin rather than starch accumulates. That the tuber is actually an enlarged storage stem can be seen from the "eyes," which are nodes showing leaf scars. During sprouting, apical dominance is shown, and only one node ordinarily develops.

The formation of storage and vegetative reproductive organs is regulated in the same manner as reproductive growth. Tuberization responds to photoperiod just as flowering does.

Potato tubers and the tubers of *Dahlia* and *Helianthus tuberosus* (Jerusalem artichoke) are produced in response to short days. Onions usually form bulbs when grown on long days. The photoperiodic response of tuberization appears to be the same as for flowering, even to the extent that it is the leaves that are the receptive organs.

Auxin has been reported to stimulate tuberization in potato, whereas gibberellins appear to inhibit it. Cytokinins enhance tuber growth probably through their action on cell division. Abscisic acid may stimulate tuberization. The synthetic growth regulator, thiourea, causes sprouting of potatoes.

Whereas the ripening process is accompanied by the deposition of the insoluble carbohydrates starch and inulin, the sprouting of tubers is accompanied by marked hydrolytic enzyme activity and a sweetening of the tissues. Hydrolysis of the stored carbohydrates is largely by the action of amylases and phosphorylases.

21.5 Dormancy

Dormancy occurs in most plant species at one period or another during the life cycle. It is a developmental stage induced usually by environmental signals and is accompanied by some desiccation, reduced respiration, low RNA metabolism, and reduced protein synthesis. As previously stated, the remaining respiration is probably mostly anaerobic and is frequently cyanide-resistant (refer to Chapter 14). Dormancy ordinarily functions in carrying the plant through adverse environmental conditions and is most frequent in plants that occur in temperate and cold habitats.

Accompanying dormancy in woody plants is bud formation. During bud formation, apical meristems cease growth and scales and bracts are formed that enclose the bud. Bud formation and the onset of dormancy are related to protein synthesis and thus are active physiological and metabolic events in the life cycle of the plant.

Environmental signals such as water limitation, photoperiod, and either low or high temperature may induce dormancy. Many desert shrubs become dormant when temperatures are high and water is unavailable. Some temperate-zone plants go dormant when exposed to cold temperatures, and many trees become dormant in response to short photoperiods.

In many species dormancy cannot be broken without a cold exposure. Many trees go dormant after experiencing short photoperiods and will not come out of dormancy until they are placed on long photoperiods. After going into dormancy, others must experience cold treatments before breaking dormancy.

Some information is available on the hormonal regulation of dormancy. It is generally assumed that endogenous growth regulators of the phenolic type are involved in the induction of the dormant state. And, of course, abscisic acid is believed to be an important dormancy hormone in higher plants. Abscisic acid levels increase markedly in most dormant tissues. Gibberellins are also known to break dormancy in buds and overcome dormancy of many seeds. Thus it appears that dormancy is probably regulated by the levels of gibberellins and abscisic acid.

Ethylene is also implicated in breaking dormancy. Gibberellin application to dormant tissues frequently stimulates the production of ethylene, eventually stimulating growth.

21.6 Juvenility, Maturation, Senescence, and Death

After breaking dormancy, tissues usually go into a rapid growth period accompanied by high rates of respiration, nucleic acid metabolism, and protein synthesis. This growth stage, marked by the logarithmic and "grand" periods of growth, can be called juvenility. In some active juvenile tissues there are obvious morphological features. In many woody plants, juvenile leaves are distinctly different from leaves developing from more mature regions. In ivy, for example, juvenile leaves are entire and large but mature leaves are distinctly palmate.

After the juvenile stages of growth, which vary in extent and time depending on the species, the tissues will mature, and metabolic and physiological events are slower and more uniform. Photosynthetic rates may decrease as will respiration rates. Rather than being sinks for carbohydrates, leaves tend to accumulate reserves when mature.

Ultimately, the aging process reaches a stage of senescence, in which normal deteriorative processes that lead finally to death take place. During senescence, metabolic activities fall rapidly, reserves are translocated out, and color changes frequently appear. In the case of senescencing leaves of woody, deciduous plants, many metabolites are trans-located out to other regions of the plant body.

Whereas the juvenile and maturation stages are marked by relatively high levels of auxin and gibberellins, the senescent stage shows a decrease in these hormones and frequently an increase in phenolics and abscisic acid. Application of cytokinins to leaves will prevent mobilization of reserves and will delay senescence.

The final stage is death of the leaf or, in the case of annuals and biennials, death of the entire plant. Dead tissues fall to the ground, where the organic and inorganic materials act as substrates for beneficial microorganisms, causing soil improvement. The inorganic elements are then recycled to the subsequent generations.

Review Exercises

21.1 In 1920 Garner and Allard discovered that Maryland Mammoth tobacco would not flower when grown outside at Beltsville, Maryland, during the growing season because it is a short-day plant. Why did it not flower in the fall, when the days became short?

21.2 What do you think would be the probable photoperiodic classification of the following plants: a plant that flowers in the Arctic; a plant that flowers in the winter in the desert; a plant that flowers in the fall in a temperate climate; a plant that flowers in the late spring or early summer in a temperate climate; a plant that flowers in the tropics; and a plant that flowers in the greenhouse in the middle of winter in a temperate climate?

21.3 Propose a hypothesis to account for the observation that the fruit of the Elberta peach shows a double-sigmoid growth curve. Consider what the tissues of the fruit are in your answer.

21.4 Clearly differentiate between pollination and fertilization. Define parthenocarpy. It is usually pollination and not fertilization that stimulates growth of the ovary into the fruit. What do you visualize is the mechanism?

21.5 If you had some seeds that would not germinate when placed in a moist environment, what procedures would you use in an attempt to cause germination?

References

BEWLEY, J. D., and M. BLACK. 1978. *Physiology and Biochemistry of Seeds,* Vol. I. Springer-Verlag, Berlin.

CLELAND, C. F. 1978. The flowering enigma. *BioScience* 28:265–269.

COOMBE, B. G. 1976. The development of fleshy fruits. *Ann. Rev. Plant Physiol.* 27:207–208.

DURE, L. S. 1975. Seed formation. *Ann. Rev. Plant Physiol.* 26:259–278.

EVANS, L. T. 1971. Flower induction and the florigen concept. *Ann. Rev. Plant Physiol.* 22:365–394.

EVANS, L. T. 1975. *Daylength and the Flowering of Plants.* W. A. Benjamin, Menlo Park, Calif.

INAMURA, S-I. 1967. *Physiology of Flowering in* Pharbitis nil. Japanese Society of Plant Physiologists, Tokyo.

JENSEN, W. A. 1973. Fertilization in higher plants. *BioScience* 23:21–27.

KAHN, A. A. (ed.). 1977. *The Physiology and Biochemistry of Seed Dormancy and Germination.* North-Holland, Amsterdam.

KIVILAAN, A., and R. S. BANDURSKI. 1973. The ninety-year period for Dr. Beal's seed viability experiment. *Amer. J. Bot.* 60:140–145.

LINSKENS, H. F. (ed.). 1964. *Pollen Physiology and Fertilization.* North-Holland, Amsterdam.

LINSKENS, H. F. (ed.). 1974. *Fertilization in Higher Plants.* North-Holland, Amsterdam.

MAYER, A. M., and A. POLJAKOFF-MAYBER. 1975. *The Germination of Seeds.* 2d ed. Pergamon Press, Oxford.

MAYER, A. M., and Y. SHAIN. 1974. Control of seed germination. *Ann. Rev. Plant Physiol.* 25:167–193.

PERRY, T. O. 1971. Dormancy of trees in winter. *Science* 171:29–36.

SACHER, J. A. 1973. Senescence and postharvest physiology. *Ann. Rev. Plant Physiol.* 24:197–224.

SALISBURY, F. P. 1963. *The Flowering Process.* Macmillan, New York.

SALISBURY, F. P. 1971. *The Biology of Flowering.* Natural History Press, Garden City, New York.

STEWARD, F. C. (ed.). 1972. *Plant Physiology,* Vol. VIC. *Physiology of Development.* Academic Press, New York.

TAYLORSON, R. B., and S. B. HENDRICKS. 1976. Aspects of dormancy in plants. *BioScience* 26:95–101.

TAYLORSON, R. B., and S. B. HENDRICKS. 1977. Dormancy in seeds. *Ann. Rev. Plant Physiol.* 28:331–354.

WOOLHOUSE, H. W. 1978. Senescence processes in the life cycle of flowering plants. *BioScience* 28:25–31.

ZEEVAART, J. A. D. 1976. Physiology of flower formation. *Ann. Rev. Plant Physiol.* 27:321–348.

V

Growth and Development: Integrative Processes

PROSPECTUS There is no doubt that the growth and development discipline of plant physiology is the most complex. It relies heavily on knowledge gained from physics, chemistry, and metabolism, and it attempts to integrate all of this knowledge into coherent theories to account for the various growth and development phenomena. The ultimate question must be how a simple, single cell such as a zygote can grow and develop into a mature plant with all of its complexities. We know something about cell division and cell elongation. Something is known about cell maturation and differentiation. We know that, even though all cells at some point in their life history have an entire complement of DNA, not all the DNA is functional at any one time. Is the sequential and ordered functioning of the DNA preprogrammed, being modified by the environment, or is growth and development more complicated?

Future studies of growth and development will bring more sophisticated mathematical treatments of growth kinetics that rely more on computer programs. More specific problems that will be approached and very likely answered are: (1) the mode of action of the plant hormones; (2) the mode of action of phytochrome; and (3) the nature of the endogenous clock that regulates biological rhythms.

The whole field of growth and development will almost literally explode when we discover how auxin, gibberellin, or the cytokinins regulate complex phenomena such as stem growth and cell division. But until such time as an acceptable theory is formulated we must be content to spray and observe and try to put a bewildering array of observations into a coherent picture. Very closely allied to the mode of action of a growth hormone is the mode of action of phytochrome. Does it function at the DNA–RNA–protein level or are membrane phenomena involved? A breakthrough here will give much insight into the conceptual development of growth theories. And finally, a most intriguing question is how plants tell and measure time. The nature of the biological clock deserves much attention by research workers in the future.

Much work will be devoted to formulate chemicals that can alter plant growth and development. Such chemicals are invaluable to industry and agriculture. But the most interesting is the flowering hormone, florigen. If florigen or a comparable compound can be discovered, there will be the potential for great advances in agriculture and horticulture. To be able to precisely control the developmental stages of plant growth and development, minimizing some and maximizing others to the benefit of mankind, will be a major step forward by civilization.

Appendix I

The International System of Units

The International System of Units (Système International d'Unités = SI), which was adopted in 1960, is a means to standardize units of physical quantities. The seven basic units along with their accepted symbols follow.

PHYSICAL QUANTITY	UNIT
Length	meter (m)
Mass	kilogram (kg)
Time	second (s)
Quantity of substance	mole (mol)
Current	ampere (A)
Temperature	degree Kelvin (°K)
Luminous intensity	candela (cd)

Each of the above quantities is defined rigorously, either by actual physical dimensions or from a sample kept in custody. Many of the units are inconvenient in size for certain descriptions or uses and thus too complex to use in ordinary procedures. The wavelength of blue light, for example, is 4×10^{-7} m. It is, however, more convenient for most purposes to use 400 nanometers (nm) rather than a value expressed in meters with an exponent. Thus there are a variety of common prefixes in use to indicate multiples or fractions of the basic units. Some of the most useful are the following.

PREFIX

deci (d)	10^{-1}	deka (da)	10^{1}
centi (c)	10^{-2}	hecto (h)	10^{2}
milli (m)	10^{-3}	kilo (k)	10^{3}
micro (μ)	10^{-6}	mega (M)	10^{6}
nano (n)	10^{-9}	giga (G)	10^{9}
pico (p)	10^{-12}	tera (T)	10^{12}

Certain other units that are derived from the fundamental units are frequently used in physiology. Some of these follow.

DERIVED UNITS

joule (J)	$kg\ m^2\ s^{-2}$	(energy)
Newton (N)	$kg\ m\ s^{-2}$	(force)
Watt (W)	$kg\ m^2\ s^{-3}$	(power)
Coulomb (C)	$A\ s$	(electric charge)
Volt (V)	$kg\ m^2\ s^{-3}\ A^{-1}$	(electric potential difference)

Other units derived from the SI nomenclature used in this text are the following.

hectare (ha)	$10^4\ m^2$
liter (L)	$10^{-3}\ m^3$
metric ton (t)	$10^3\ kg$
gram (g)	$10^{-3}\ kg$
dyne (dyn)	$10^{-5}\ N$
bar (bar)	$10^5\ N\ m^{-2}$
erg (erg)	$10^{-7}\ J$
poise (P)	$10^{-1}\ kg\ m^{-1}\ s^{-1}$
calorie (cal)	4.184 J
minute (min)	60 s
hour (h)	3600 s

Appendix II

Bibliography

Here is a list of the most common and readily available plant journals that publish original research in plant physiology. In addition to those listed, plant physiological research can also be found in the many botanical, biological, biochemical, and general science journals.

AMERICAN JOURNAL OF BOTANY. Botanical Society of America, Columbus, Ohio.

ANNALS OF BOTANY. The Annals of Botany Co., Schenectady, New York.

AUSTRALIAN JOURNAL OF BOTANY. CSIRO, Melbourne, Australia.

AUSTRALIAN JOURNAL OF PLANT PHYSIOLOGY. CSIRO, Melbourne, Australia.

BOTANICAL GAZETTE. The University of Chicago Press, Chicago, Illinois.

CANADIAN JOURNAL OF BOTANY. National Research Council of Canada, Ottawa, Canada.

INDIAN BOTANICAL SOCIETY JOURNAL. The Indian Botanical Society, Meerut, India.

JOURNAL OF EXPERIMENTAL BOTANY. Society for Experimental Biology, Oxford, Great Britain.

PHOTOSYNTHESIS RESEARCH. Dr. W. Junk bv Publishers, The Hague, Netherlands.

PHOTOSYNTHETICA. Czechoslovak Academy of Sciences, Praha (Prague).

PHYSIOLOGIA PLANTARUM. Scandinavian Society for Plant Physiology, Munksgaard, Copenhagen.

PHYSIOLOGIE VEGETALE. French Society for Plant Physiology, Paris.

PHYTOCHEMISTRY. Pergamon Press, Ltd., Oxford, Great Britain.

PLANTA. Springer-Verlag, Berlin and New York.

PLANT AND CELL PHYSIOLOGY. Japanese Society of Plant Physiology, Kyoto.

PLANT PHYSIOLOGY. American Society of Plant Physiology, Baltimore, Maryland.

PLANT SCIENCE LETTERS. Elsevier Scientific Publishing Co., Amsterdam.

SOVIET PLANT PHYSIOLOGY (translation). Academy of Science of the USSR, Moscow.

THE BOTANICAL MAGAZINE. The Botanical Society of Japan, Tokyo.

THE NEW PHYTOLOGIST. Blackwell Scientific Publishing Co., Oxford, Great Britain.

ZEITSCHRIFT FUR PFLANZENPHYSIOLOGIE. Gustov Fischer Verlag, Stuttgart, W. Germany.

In addition to the periodicals listed above, there are many review publications that have timely articles. Three of these publications are listed below.

ANNUAL REVIEW OF PLANT PHYSIOLOGY. Annual Reviews, Inc., Palo Alto, Calif.

ENCYCLOPEDIA OF PLANT PHYSIOLOGY. Springer-Verlag, Berlin and New York.

THE BOTANICAL REVIEW. The New York Botanical Garden, Bronx, New York.

Appendix III

Answers and Notes for Chapter Review Exercises

Chapter 1

1.1 See pages 6–25. Note the presence of plastids, cell walls and large vacuoles in plants.

1.2 See pages 8–12. Answer from the viewpoint of structure and selectivity.

1.3 See pages 12–20. Consider the outer limiting membrane and internal organization as well as size.

1.4 The cell has a total volume plus content of 10^6 μm^3. Assume the chloroplast is a sphere $= 1767$ μm^3. Assume the mitochondrion is a box $= 5$ μm^3. Assume the microbody is a sphere $= 1.8$ μm^3. Assume the vacuole is a sphere $= 2.68 \times 10^5$ μm^3. The volumes by percent are: chloroplasts $= 1.8\%$, mitochondria $= 0.015\%$, microbodies $= 0.005\%$, and the vacuole $= 26.8\%$.

1.5 See pages 23–25. Consider a nonliving cell versus a living cell and a perforated conduit versus a cell with a primary wall.

Chapter 2

2.1 The vapor pressure at 30° is 42.0 mbar and the $2m$ solution will be reduced to 40.5 mbar. The estimated freezing point of the $0.2m$ solution is $-0.37°$ and the estimated boiling point of the $0.05m$ solution is 100.026°. The estimated osmotic pressure of the $0.5M$ solution is 10.9 bar.

2.2 Calculate the ionic strength of the solution to be $0.006M$ and the activity of the 3 mM sodium from 1 mM sodium citrate to be 2.73 mM. As the solution becomes more concentrated, the calculation will deviate more from reality.

2.3 The pK_a of glycolic acid is 3.83 and the pH would be 2.92. The $0.01M$ HCl would be 2 and the $0.01M$ NaOH would be 12. The pH of the $0.01M$ sodium acetate, $0.02M$ acetic acid buffer would be 4.45.

2.4 The standard free energy of hydrolysis ($\triangle G^{\circ\prime}$)

of glucose-6-phosphate is -12.7 kJ mol^{-1}. The standard free energy of the oxidation of NADH is -219 kJ mol^{-1}.

2.5 Using ratios, the diffusion coefficient for carbon dioxide is 0.15, for hydrogen gas it is 0.72, and for oxygen it is 0.18 cm^2 s^{-1}.

Chapter 3

3.1 See pages 72–76.

3.2 Converting to centimeters, $50\,\text{nm} = 5 \times 10^{-6}\,\text{cm}$, $640\,\mu\text{m} = 6.4 \times 10^{-2}\,\text{cm}$, $800\,\text{m}\mu = 8 \times 10^{-5}\,\text{cm}$, $1000\,\text{Å} = 1 \times 10^{-5}\,\text{cm}$.

3.3 $\mathscr{E}^{200} = 9.93 \times 10^{-19}\,\text{J} = 2.37 \times 10^{-19}\,\text{cal} = 9.93 \times 10^{-12}\,\text{ergs}$

$\mathscr{E}^{400} = 4.97 \times 10^{-19}\,\text{J} = 1.89 \times 10^{-19}\,\text{cal} = 4.97 \times 10^{-12}\,\text{ergs}$

$\mathscr{E}^{600} = 3.31 \times 10^{-19}\,\text{J} = 7.91 \times 10^{-20}\,\text{cal} = 3.31 \times 10^{-12}\,\text{ergs}$

$\mathscr{E}^{1000} = 1.99 \times 10^{-19}\,\text{J} = 4.75 \times 10^{-20}\,\text{cal} = 1.99 \times 10^{-12}\,\text{ergs}$

Expressed in J einstein^{-1}, $200\,\text{nm} = 5.97 \times 10^5$, $400\,\text{nm} = 2.99 \times 10^5$, $600\,\text{nm} = 1.99 \times 10^5$, and $1000\,\text{nm} = 1.2 \times 10^5$.

3.4 In solution there is 8.9×10^{-3} mg chlorophyll ml^{-1}. There are 0.45 mg g^{-1} of leaf tissue, and 0.009 mg cm^{-2} of leaf tissue.

3.5 Yes. 1.2 eV is equivalent to 110.4 kJ, and red light (600 nm) has 219 kJ einstein^{-1} (see 3.3 above). With an efficiency of 50%, there would be sufficient energy.

Chapter 4

4.1 The potentials are as follows.

	Initial		Equilibrium	
	Beaker	Cell	Beaker	Cell
ψ	0	-8	0	0
ψ_p	0	$+8$	0	$+16$
ψ_π	0	-16	0	-16

Water will flow from the beaker into the cell until the water potentials in both are equal, here 0. See pages 87–88 for assumptions.

4.2 When the cell from Review Exercise 4.1 (before equilibrium) is placed in a beaker with -7 bars, the water will flow from the beaker into the cell. The potentials are as follows.

	Initial		Equilibrium	
	Beaker	Cell	Beaker	Cell
ψ	-7	-8	-7	-7
ψ_p	0	$+8$	0	$+9$
ψ_π	-7	-16	-7	-16

If the cell is placed in the -7 beaker after it reached equilibrium in Review Exercise 4.1, water will flow from cell into beaker. The potentials are as follows.

	Initial		Equilibrium	
	Beaker	Cell	Beaker	Cell
ψ	-7	0	-7	-7
ψ_p	0	$+16$	0	$+9$
ψ_π	$\cdot7$	-16	-7	-16

4.3 Water will flow from Cell B to Cell A. The potentials are as follows.

	Initial		Equilibrium	
	Cell A	Cell B	Cell A	Cell B
ψ	-14	-4	-9	-9
ψ_p	$+6$	$+12$	$+11$	$+7$
ψ_π	-20	-16	-20	-16

4.4 The initial potentials are as follows.

	Soil	Root
ψ	-1	-3
ψ_p	—	$+7$
ψ_π	—	-10
ψ_m	-1	—

The equilibrium potentials in the roots are: $\psi = -1$, $\psi_p = +9$, $\psi_\pi = -10$. After the addition of salt to the soil, the soil ψ will go to -5 and water will be lost by the roots to the soil. Wilting may occur. If salt is then taken up by the roots, osmotic adjustment will occur and new potentials in the root will be: $\psi = -5$, $\psi_p = +10$, $\psi_\pi = -15$.

4.5 Assuming that both glucose and sucrose act as ideal solutions, the two cells will be in equilibrium since osmotic potential is a colligative property and independent of the kind of solute. In actual fact, sucrose will create a lower osmotic potential. See pages 48–49 and 92. Yes, it is possible for water to move from a cell that has a high solute concentration to one that has a lower solute concentration. The important feature is the water potential—refer to Review Exercise 4.2 above.

Chapter 5

5.1 Calculate a $\triangle e$ (using Fig. 5.8) of 29.5×10^{-6} g cm^{-3} for the cold day and a $\triangle e$ of 15.2×10^{-6} g cm^{-3} for the warm day. The ratio of 1.9 (for cold day versus warm day) will be directly proportional to the evaporation rate. Answer: more on the cold day. Note that if you use the graph in Fig. 2.6, your $\triangle e$ values will be in millibars but your final conclusion will be the same.

5.2 First compute the vapor pressure of the atmosphere (e) using data from Fig. 2.6. Assume that this is a constant (note that the relative humidity will change with temperature of the air but the vapor pressure will stay about the same unless the air mass changes). When the leaf temperature rises 5° in sunlight, compute a new $\triangle e$ in the 20° chamber of 21 mbar and a new $\triangle e$ in the 30° chamber of 28 mbar. Provided that stomatal opening (that is, leaf resistance) stays the same in both chambers, transpiration will increase more in the 30° chamber.

5.3 Using Fig. 5.8 to obtain vapor densities, calculate a transpiration rate of 12×10^{-6} g cm^{-2} s^{-1} for the sun plant and 4.8×10^{-6} g cm^{-2} s^{-1} for the shade plant.

5.4 There will not be much change since vapor pressure of the air would not change much unless the air temperature causes a change in leaf temperature, as it usually does.

5.5 For species 1 (a C_3 species), the transpiration ratio in air is 421 and in reduced oxygen it is

262. For species 2 (a C_4 species), the transpiration ratio in air is 205 and in reduced oxygen it is 225. Since transpiration is similar for both species in air and in reduced oxygen, evidently stomatal resistances are similar. In species 1 oxygen inhibits photosynthesis, and since stomatal resistance is unchanged it is mesophyll resistance that is higher in oxygen than in its absence (in C_3 species). The biochemistry of the phenomenon of oxygen inhibition of photosynthesis is discussed in Chapter 16 under photorespiration.

Chapter 6

6.1 See pages 132–135.

6.2 One atmosphere is equivalent to 10.2 meters (33.48 feet of water) of water at 15.5° C. Assuming no resistance, about 10.2 meters. With a maximum root pressure of 3 bars and assuming no resistance, water could be pushed up about 31 meters.

6.3 Ignoring the contact angle (cos θ), about 29.36 centimeters. Refer to Chapter 2.

6.4 See page 165 for explanation of the units 1 mg cm^{-2} s^{-1}. Refer to page 161 for flux versus velocity. Flux is an estimate of amount whereas velocity is speed.

6.5 See pages 151–161.

Chapter 7

7.1 Estimating from Wien's displacement law, about 13.4° C.

7.2 Approximately 0.61° C g^{-1} s^{-1}. Why doesn't the leaf ignite?

7.3 When the leaf is 5° C above air temperature, the predicted energy loss by convection is about 0.15 kJ m^{-2} s^{-1}. When 1° C above the air temperature, the predicted loss is about 0.05 kJ m^{-2} s^{-1}.

7.4 The Q_{10} for respiration would be about 2.1.

7.5 Use $Q_a = R + C + LE$ (the energy budget equation) and compute a loss through $R = 0.46$ kJ m^{-2} s^{-1} and loss through $C = 0.15$ kJ m^{-2} s^{-1}. Then by difference, determine an energy loss by $LE = 0.89$ kJ m^{-2} s^{-1}. Converting to water loss in grams using the latent heat of evaporation, compute a predicted transpiration rate of 0.37 g m^{-2} s^{-1} or 0.13 g cm^{-2} h^{-1}.

Chapter 8

8.1 See pages 195–196.
8.2 Using the known base pairing and Table 8.3, determine the proper base sequence in DNA as AAT, AAC, GAA, GAG, GAT, and GAC for leucine.
8.3 See pages 59–61 in Chapter 2.
8.4 Refer to Chapter 2 and find the standard free energy of hydrolysis for P-enolpyruvate and ATP in the *Handbook of Chemistry and Physics* (References, Chapter 14) or by referring to Chapter 14. For PEP, the $\triangle G^{\circ\prime} = -62$ kJ mol^{-1} and for ATP the $\triangle G^{\circ\prime} = -30$ kJ mol^{-1}. The reaction proceeds from PEP to ATP with a $\triangle G^{\circ\prime} = -32$ kJ mol^{-1}.
8.5 Use the Beer–Lambert law given in Chapter 3 ($A = \epsilon l C$) and compute an expected rate of reaction of 0.016 μmol min^{-1}.

Chapter 9

9.1 See pages 51 and 219.
9.2 The energy charge is 0.65, indicating that ATP-generating metabolism is probably occurring.
9.3 Assuming an average molecular weight of 100 for the amino acid residues, approximately 50,000.
9.4 By graphic analysis, the K_m is about 0.082 mM ($V_{max} = 1.14$ A min^{-1}).
9.5 By graphic analysis, the slope of $A_{750 \text{ nm}}$ versus the concentration of BSA (mg ml^{-1}) is 1.86 A per mg ml^{-1}, giving a protein concentration in the extract of 0.134 mg ml^{-1}. In the cotyledon, there are 3.36 mg of protein.

Chapter 10

10.1 Consider reducing versus nonreducing sugars.
10.2 See pages 258–262 and Fig. 10.1.
10.3 See page 263.
10.4 Consider the linkages between the residues, α $(1 \rightarrow 4)$ and β $(1 \rightarrow 4)$, their sizes, their shapes, and branching. Also consider the stability of cellulose in nature.
10.5 See pages 219 and 277.

Chapter 11

11.1 See pages 283, 284, 288, and 290.
11.2 Yes. Plants from warm climates tend to have fatty acids with more saturation. See J. Lyons. 1973. Chilling injury in plants. *Ann. Rev. Plant Physiol.* 24:445–466.
11.3 Compute a saponification number for tristerin by noting that 3 moles of KOH react with one mole of tristerin. Saponification number = 188.8. Triglycerides smaller than tristerin will have larger numbers since on a molar basis the same amount of KOH is required per mole of triglyceride.
11.4 No, because they are complex mixtures of various-sized fatty acids. In order of decreasing melting point: coconut oil (25.1), cottonseed oil (−1.0), olive oil (−6.0), castor oil (−18). The iodine number of linoleic acid (two double bonds) is 181, computed by noting that 4 moles of iodine react per mole of linoleic acid.
11.5 The formation of the initial steryl CoA requires 1 mole of ATP. For each acetyl CoA generated, a maximum of 5 moles of ATP can be generated (see page 381). Note that there will be 8 moles of acetyl CoA generated and one left at the end. These 8 moles can result in a maximum of 40 ATP. Complete oxidation of the 9 acetyl CoA generates 12 ATP for a total possible of 147 moles of ATP per mole of stearic acid.

Chapter 12

12.1 Two notions are that (1) secondary or natural products function in the defense of the plant against predators, and (2) they play important metabolic roles. Refer to these contrasting papers: D. Seigler and P. W. Price. 1976. Secondary compounds in plants: primary functions. *Amer. Naturalist* 110:101−105; and T. Swain. 1977. Secondary compounds as protective agents. *Ann. Rev. Plant Physiol.* 28:479−501.

12.2 See page 226. The nucleosides are N-glycosides. The aglycone is the base (see pages 195–197). The nucleic acids fit the definition of alkaloids because they have a basic (the presence of nitrogen) ring structure (see pages 196 and 313).

12.3 In the laboratory it is most common to manipulate the tissue under anaerobic conditions or add an agent that will chelate the copper in the polyphenol oxidases, preventing the reaction from occurring. In the kitchen, vegetables prone to tanning can be placed under water after cutting; or, vinegar might be added.

12.4 Coincidentally they may be inhibitory analogues or direct poisons or they may have developed during evolution as protective agents. Products that alter human physiology fall into many categories, but perhaps the most striking are the hallucinogens. See R. E. Schultes. 1969. Hallucinogens of plant origin. *Science* 163:245−254; and W. Emboden. 1979. *Narcotic Plants.* Macmillan, New York.

12.5 You may refer to the Lynn Index as a useful listing of secondary products found in plants. The Lynn Index is composed of three files: (1) a listing of compounds and the plants they are found in, (2) a listing by plant species, and (3) a listing by authors of articles. See N. R. Farnsworth *et al.* (eds.). Several volumes published by The College of Pharmacy, University of Illinois, Chicago, Illinois.

Chapter 13

13.1 $Ca(NO_3)_2 = 0.3 M$. There are 0.6 equivalents of calcium and of nitrate in the solution. The dilution gives 122 parts per million (mg per L) of calcium.

13.2 On the basis of the Nernst equation and assuming that the entire weight of the tissue is water, it is predicted that there would be 2340.3 μEq K^+ per g fresh weight. If the actual is 500 μEq g^{-1}, there must be exclusion.

13.3 By graphic analysis, the K_s for potassium is 0.43mM.

13.4 Assuming $\frac{1}{3}$ N_2 reduced per acetylene because N_2 reduction is a 6-electron process whereas acetylene reduction is a 2-electron process, the estimated rate is 0.028 μmol h^{-1} g^{-1} (because of competing and complicating reactions, the conversion factor is actually closer to $\frac{1}{4}$).

13.5 Prepare a nutrient solution by deleting the iron, being careful not to alter the concentration of other essential ions. You could measure growth or other processes that are influenced by iron, but perhaps the simplest is to measure chlorophyll content since the primary visual symptom is chlorosis.

Chapter 14

14.1 38 ATP are possible. Refer to Table 14.2.

14.2 The complete oxidation of the six-carbon fatty acid, caproic acid, yields 45 ATP, somewhat more than the six-carbon sugar, glucose. See Review Answer 14.1 above and Review Answer 11.5.

14.3 See B. J. D. Meeuse. 1975. Thermogenic respiration in aroids. *Ann. Rev. Plant Physiol.* 26:117−126.

14.4 The reaction seems to be more feasible when the pyrophosphate is hydrolyzed. Predict your answer on the basis of the standard free energies of hydrolysis for the various reactants and products.

14.5 Being a kinase, ATP would be produced. Because of the standard free energy of hydrolysis, the reaction is not feasible. Reversal is ordinarily by a phosphatase hydrolyzing the phosphate at position 1 to give fructose-6-P.

Chapter 15

15.1 Usually, NADP reduction is measured by observing the increase in absorbance at 340 nm, but with cells or membrane fragments, special equipment is needed because of light scattering and extraneous absorption. See the papers by R. P. Levine and R. M. Smillie. 1963. The photosynthetic electron transport chain of *Chlamydomonas reinhardi. J. Biol. Chem.* 238:4052–4057; and D. S. Gorman and R. P. Levine, 1965. Cytochrome *f* and plastocyanin: their sequence in the photosynthetic electron transport chain of *Chlamydomonas reinhardi. Proc. Natl. Acad. Sci.* 54:1665–1669.

15.2 Cyclic electron flow and photophosphorylation are assumed to occur *in vivo* because of the high energy demand for photosynthesis and cellular product formation. *In vivo* cyclic electron flow is difficult to measure, but there are some inhibitors such as antimycin, DSPB, and DBMIB. See the article by H. Gimmler. 1977. Photophosphorylation *in vivo*. In A. Trebst and M. Avron. *Photosynthesis I. Encyclopedia of Plant Physiology.* New Series 5. Springer-Verlag, Berlin.

15.3 Other inhibitors that are useful are antimycin A, which blocks cyclic electron flow, and DBMIB, which blocks plastoquinone. DCPIP (reduced) can be used to donate electrons to photosystem I and DCPIP (oxidized) can be used as an electron acceptor from photosystem II. Far-red light can be used to

drive photosystem I with no effect on photosystem II. Subcellular fractionation has potential for future research. See B. S. K. Sun and K. Sauer. 1971. Pigment systems and electron transport in chloroplasts. *Biochim. Biophys. Acta* 234:399–414.

15.4 First consider the transport properties of the outer envelope and then where the various reactions occur in the chloroplasts. See the paper by R. B. Park and N. G. Pon. 1961. Correlation of structure with function in *Spinacea oleracea* chloroplasts. *J. Molecular Biol.* 3:1–10.

15.5 Using the equation on page 56, −92 mV.

Chapter 16

16.1 Refer to the paper by C. C. Black. 1971. Ecological implications of dividing plants in groups with distinct photosynthetic production capacities. *Adv. Ecol. Res.* 7:87–114.

16.2 See the paper by I. Zelitch. 1975. Improving the efficiency of photosynthesis. *Science* 188:626–633.

16.3 For phylogenetic discussions of C_4 and CAM plants see the following: L. T. Evans. 1970. Evolutionary, adaptive, and environmental aspects of the photosynthetic pathway: Assessment. In M. D. Hatch, C. B. Osmond, and R. O. Slatyer (eds.). *Photosynthesis and Photorespiration.* Wiley-Interscience, New York; and M. Kluge and I. P. Ting. 1978. *Crassulacean acid metabolism. Analysis of an Ecological Adaptation.* Springer-Verlag, New York.

16.4 According to analysis by Edwards, the NADP-malate enzyme type requires 5 ATP and 2 NADPH per mole of CO_2 fixed, the NAD-malate enzyme type requires 5 ATP and 2 NADPH per mole of CO_2, and the P-enol-pyruvate carboxykinase type requires 6 ATP and 2 NADPH per mole of CO_2. See the paper by G. E. Edwards, S. C. Huber, S. B. Ku,

C. K. M. Rathnam, M. Gutierrez, and B. C. Mayne. 1976. In R. H. Burris and C. C. Black (eds.). *CO₂ Metabolism and Plant Productivity.* University Park Press, Baltimore.

16.5 See R. Marcelle (ed.). 1975. *Environmental and Biological Control of Photosynthesis.* W. Junk Publishers, The Hague, Netherlands.

Chapter 17

17.1 By inspection of the cumulative growth curve (cm^2 graphed versus days), estimate the maximum growth to be about 320 cm^2. From the growth rate curve, estimate the maximum growth rate (between days 5 and 7) to be about 60 cm^2 day^{-1}. By graphic analysis, estimate the growth rate constant (k) for the sigmoid curve to be about 0.434.

17.2 The plastochron index for the plant is 7.44. The plastochron age for the fourth leaf is 3.44 and for the sixth leaf it is 1.44.

17.3 A tropism is nonreversible, but care must be taken in the interpretation because some tropisms may appear to reverse by differential growth. See, for example, the article by L. Reinhold. 1978. Phytohormones and the orientation of growth. In D. S. Letham, P. B. Goodwin, and T. J. V. Higgins (eds.). *Phytohormones and Related Compounds. A Comprehensive Treatise,* Vol. II. *Phytohormones and the Development of Plants.* Elsevier-North Holland Biomedical Press, New York. (See the section covering thigmotropism.)

17.4 The usual kind of experiment to study transport of auxin relies on radioactive auxin (see pages 488–489). Care must be taken to ensure that a low activity on one side is the result of transport away and not destruction. Hydrotropism experiments are difficult to design because it must be demonstrated that growth is directional in response to water and not that growth occurs just because of the presence of water. Refer to the article by W. E. Loomis

and L. M. Ewan. 1936. Hydrotropic responses of roots in soil. *Bot. Gazette* 97:728–743.

17.5 Perhaps the simplest would be to look for a mutant and then do standard genetic crosses. Otherwise, one might want to manipulate the environment. Consider having the light come up from the west and go down in the east, or grow the plant in the Southern Hemisphere Other environmental variables could also be manipulated.

Chapter 18

18.1 Indoleacetic acid; see Fig. 18.3. Gibberellic acid; see Figs. 12.9, 12.10, and 18.9. Zeatin; see Fig. 8.13. Abscisic acid; see Fig. 12.10. Ethylene; see page 500. Cinnamic acid; see Figs. 12.2 and 12.3.

18.2 By graphic analysis (ignoring the last point at 0.1 mg IAA L^{-1}), a curvature of 11° would correspond to about 0.029 mg IAA L^{-1}. From the 1 g of tissue, there would be 1.44×10^{-4} mg of IAA or 8×10^{-10} moles (MW of IAA = 175.2).

18.3 Most plant physiologists consider that ethylene is a proper plant hormone, but the vitamins are not. See D. J. Osborne. 1978. Ethylene. In D. S. Letham, P. B. Goodwin, and T. J. V. Higgins (eds.). *Phytohormones and Related Compounds. A Comprehensive Treatise,* Vol. 1. *The Biochemistry of Phytohormones and Related Compounds.* Elsevier-North Holland Biomedical Press, New York.

18.4 The hypothesis of Reinhold is that contact stimulus brings about an increase in auxin, but it does not affect the side that is in contact with the stimulus. Production of ethylene on the contact side inhibits growth there, while the auxin on the opposite side stimulates growth. See L. Reinhold. 1978. Phytohormones and the orientation of growth. In D. S. Letham, P. B. Goodwin, and T. J. V. Higgins (eds.). *Phytohormones and Related Com-*

pounds. A Comprehensive Treatise, Vol. II. *Phytohormones and the Development of Plants.* Elsevier-North Holland Biomedical Press, New York.

18.5 This is a complicated question. Refer to any animal physiology book to understand how the animal hormones function.

Chapter 19

19.1 Computing the average number of moles per cell and using Avogadro's number, the estimate is 927,500 protein molecules per cell (i.e., about one million).

19.2 First assume that there is a constant number of cells per milligram of callus and estimate that 130 mg of callus has about 650,000 cells. Then using the first-order growth equation, compute the growth-rate constant (k) of 0.298 day^{-1}. Then note that the doubling time (T_2) is $0.693/k = 2.33$ or an average of 2.33 days per cell division.

19.3 In the specific case of apical dominance, it refers to preferential transport of nutrients to the apex. The present theory, of course, involves the production of auxin by the apex that inhibits lateral bud growth, but nutrition could be a factor. For a more generalized notion of the concept, see F. M. Eaton and H. G. Joham. 1944. Sugar movement to roots, mineral uptake, and the growth cycle of the cotton plant. *Plant Physiol.* 19:507–518; and F. M. Eaton and N. E. Rigler. 1945. Effect of light intensity, nitrogen supply, and fruiting on carbohydrate utilization by the cotton plant. *Plant Physiol.* 20:380–411.

19.4 Some specific examples are: (1) the suppression of root stock growth after bud grafting of the scion by removal of the apex of the root stock (or by simply bending the leader of the root stock below the apex of the scion); (2) the suppression of lateral bud growth on potatoes by allowing the apical bud to begin growth

after harvest and then storing at cool temperatures; (3) the stimulation of multiple flower heads in *Chrysanthemum* by pinching off the apical buds; (4) the stimulation of full growth by pruning such as is done when shaping a hedge; and (5) the production of high-quality tobacco leaves by removal of the terminal growth of the plant, allowing the more desirable basal leaves to develop.

19.5 If you attempt to determine the phyllotaxy of a plant, try cutting the apex such as shown in Fig. 19.12 and making a drawing. Refer to R. F. Williams. 1975. *The Shoot Apex and Leaf Growth.* Cambridge University Press, Cambridge.

Chapter 20

20.1 Both are sensitive to blue light, but there are different pigments involved. Refer back to Chapter 17 for a discussion of phototropism. In your answer, consider the question of reversibility and direction of the slight stimulus as well as the kind of growth responses.

20.2 There is speculation that phytochrome detects shading in plant canopies, that it measures the depth seeds are in soil, and that it is a general light detector inasmuch as all green plants have phytochrome but fungi do not. See the article by H. Borthwick. 1972. The biological significance of phytochrome. In K. Mitrakos and W. Shropshire, Jr. *Phytochrome.* Academic Press, New York; and H. Smith. 1975. Section 9.3 "Why do plants have phytochrome?" of *Phytochrome and Photomorphogenesis. An Introduction to the Photocontrol of Plant Development.* McGraw-Hill, London.

20.3 Refer to the absorption spectra of the phytochrome forms in Fig. 20.4 and compare with the action spectra for photomorphogenic responses in Fig. 20.7.

20.4 Ordinarily, placing the plant under constant

environmental conditions will allow differentiation between an endogenous rhythm and an environmentally regulated cyclic event. Consider in your answer light-activated enzymes and enzymes that may use substrates that are immediate products of photosynthesis. Temperature could also be an important factor. There is always the possibility of an artifact when *in vitro* assays are done. Refer to the review article by O. Queiroz. 1974. Circadian rhythms and metabolic patterns. *Ann. Rev. Plant Physiol.* 25:115–134.

20.5 Note that the amplitude is the intensity of the oscillation measured as the magnitude of departure from the mean or midpoint; hence the amplitude of the P-enolpyruvate carboxylase in this experiment is 0.6 nmol g^{-1} (dry weight) h^{-1}.

Chapter 21

21.1 By the time the days were short enough it was too cold.

21.2 This obviously is a question filled with problems and ambiguities, but the first approximation would probably be: Arctic = long-day plant, winter desert = short-day plant, fall temperate = short-day plant, late spring or early summer temperate = long-day plant, tropics = indeterminate, greenhouse in winter = short-day plant.

21.3 First the ovary grows and then the embryo grows, resulting in a double-sigmoid curve. See H. B. Tukey. 1933. Embryo abortion in early ripening varieties of *Prunus avium*. *Bot. Gazette* 94:433–468.

21.4 Pollination is the landing of pollen on the stigma, and fertilization is the union of gametes to form a zygote. Parthenocarpy is fruit development without fertilization. It may require pollination, but not necessarily. Pollination may stimulate the production of auxin or other growth hormones, resulting in fruit development.

21.5 Consider quiescent versus dormant seeds and then consider anatomical, genetic, metabolic, and hormonal responses. See A. A. Khan (ed.). 1977. *The Physiology and Biochemistry of Seed Dormancy and Germination*. North-Holland, Amsterdam.

Index